Nanoscale Spectroscopy with Applications

Nanoscale Spectroscopy with Applications

Edited by
Sarhan M. Musa

CRC Press is an imprint of the
Taylor & Francis Group, an **informa** business

CRC Press
Taylor & Francis Group
6000 Broken Sound Parkway NW, Suite 300
Boca Raton, FL 33487-2742

First issued in paperback 2017

ISBN-13: 978-1-4665-6853-2 (hbk)
ISBN-13: 978-1-138-07265-7 (pbk)

Library of Congress Cataloging-in-Publication Data

Nanoscale spectroscopy with applications / editor, Sarhan M. Musa.
pages cm
Summary: "This book introduces the key concepts of nanoscale spectroscopy methods used in nanotechnologies in a manner that is easily digestible for a beginner in the field. It discusses future applications of nanotechnologies in technical industries. It also covers new developments and interdisciplinary research in engineering, science, and medicine. An overview of nanoscale spectroscopy for nanotechnologies, the book describes the technologies with an emphasis on how they work and on their key benefits. It also serves as a reference for veterans in the field"-- Provided by publisher.
Includes bibliographical references and index.
ISBN 978-1-4665-6853-2 (hardback)
1. Nanostructured materials--Analysis. 2. Spectrum analysis. I. Musa, Sarhan M.

TA418.9.N35N34564 2014
620.1'127--dc23 2013036870

Visit the Taylor & Francis Web site at
http://www.taylorandfrancis.com

and the CRC Press Web site at
http://www.crcpress.com

Dedicated to my late father, Mahmoud, my mother, Fatmeh,

and my wife, Lama

Contents

Preface

Nanotechnology and nanoscience have now emerged as the most important growth areas in the new millennium, and spectroscopy has become the most common and powerful method that uses radiation to obtain information on the structure and properties of matter. Therefore, the challenges posed by nanoscale spectroscopy continue to attract researchers from all around the world and have led to many new discoveries concerning the interaction between light and matter at dimensions much smaller than the wavelength of electromagnetic radiation. Nanoscale spectroscopy and its applications shed light on interdisciplinary research, including engineering, physics, chemistry, biology, medicine, and aerospace. The study of the interaction between light and matter at the nanometer scale is facilitated by the rapid progress in nanotechnology and nanoscience. Nanoscale spectroscopy is of great benefit to industry as it requires little labor and low cost. This book reports on recent progress and provides an overview of nanoscale spectroscopy and its applications.

This book consists of 13 chapters and 3 appendices. Chapter 1 presents tip-enhanced spectroscopy at the nanoscale along with its practical issues and solutions. The authors focus on tip-enhanced Raman scattering (TERS) for imaging and spectroscopy at the nanoscale (lateral dimension: XY)-system configuration, nanoanalysis of carbon nanomaterials, nanoanalysis of biomaterials, nanoanalysis of semiconductors, precise distance dependence (longitudinal dimension: Z)-TERS system using tapping mode atomic force microscopy (AFM) with time-gated detection, TERS system using tapping mode AFM with time-gated illumination, nanopositioning stabilization (lateral dimension: XY)-precise tip positioning in laser focus spot, real-time compensation of thermal drift, nanopositioning stabilization (longitudinal dimension: Z)-experimental details, stability for microscopy, and stability for nanoscopy.

Chapter 2 presents a brief review of micro- and nanostructures/systems and their applications in certain directions. The authors focus on the technological importance with an emphasis on photonics, especially on guides for selective photonic transmission, twisted clad fibers, chiral fibers, and nanophotonic communication systems. They also discuss ventures into medicine.

Chapter 3 presents the spectroscopy of polymer-based nanocomposite dielectrics with tailored interfaces and the structured spatial distribution of fillers. It looks into different aspects of the authors' research efforts over the last years aimed at identifying critical structural parameters that can mitigate some of the detrimental effects associated with these interfaces and can expand the operating and application possibilities of polymers. The size of the particles comparable to polymer length scales, the particle functionality, and the highly ordered self-assembled configurations are key parameters in designing multifunctional composites with markedly enhanced capabilities. The phenomenal growth of research and commercialization of hybrid nanostructured materials necessitates the development and utilization of techniques that accurately probe properties at the nanoscale. Dielectric relaxation spectroscopy combined with proper analysis tools is a versatile method for dynamically measuring nanoscale properties, separating synergistic contributions, and predicting the macroscopic behavior of composites. This chapter focuses on such spectroscopic studies for composites with potential dielectrical applications and with promise to store electrostatic energy. Across a broad range of polymer matrices, the controlled distribution of fillers and tailored interfaces according to dielectric measurements has resulted in outstanding achievements in the field of nanodielectrics that provide grounds for further development and optimization.

Chapter 4 presents nanoscale spectroscopy with applications to chemistry, focusing on the characterization of nanostructures. The authors describe spectroscopic techniques used for the characterization of nanomaterials, characterization of nanomaterials using Raman spectroscopy, characterization of nanomaterials using nuclear magnetic resonance (NMR) spectroscopy, characterization of mechanical properties of nanomaterials, characterization of surface properties of nanomaterials for AFM and scanning probe microscopy (SPM), and x-ray nanodiffraction.

Chapter 5 provides a general overview of the major developments in localized surface plasmon resonance (LSPR) technology. The main application areas are outlined, and examples of applications of LSPR sensor technology are presented. The authors first introduce the Mie theory (the analytic theory of LSPR), the physical origin of the LSPR itself, and its dependence on the material properties of noble metals and the surrounding refractive index. They then explain the basics of nanoparticle-based LSPR sensing, which includes a description of single-particle and ensemble measurements as well as a comparison of scattering, absorption, and extinction. The theories, applications, and the main challenges of four kinds of major LSPR spectroscopies including absorption spectroscopy, surface-enhanced Raman scattering, LSPR-based Rayleigh scattering, and LSPR controlled fluorescence spectroscopy are covered. The applications show that a variety of chemically and biologically relevant molecules can be detected—from proteins and oligonucleotides to the direct detection of metal ions and anions. In addition, the results of these studies demonstrate the necessity for fundamental spectroscopic studies in guiding experimental design to achieve the largest overall signal or to observe signals at all. They conclude that as the plasmonic response of nanoparticles continues to grow, these LSPR-based sensing experiments will improve as well, leading to higher sensitivity, faster and more reversible responses, and an ever-broadening scope of applicability.

Chapter 6 evaluates nanostructures of ionic liquid aggregates by spectroscopy techniques. The authors conclude that there is a lack of depth in the study of the formation and characterization of dicationic ionic liquids (ILs) using spectrometric methods, which demonstrates the need for the systematic study of ionic liquids that have high potential as a new soft material. Spectrometric methods analyzed in the chapter complement the investigation of the formation and characterization of IL aggregates, which shows that there is no absolute spectrometric experiment that allows the complete characterization of an IL aggregate. It is evident that NMR and LS are the methods that furnished the most complete and detailed information about aggregated formation and characterization. It is also evident that these same methods have not been exploited to their potential. There has been a deficiency in investigating the influence of temperature and solvents in the formation, characterization, and stability of aggregates using the spectrometric methods covered in this chapter. This is indicative of the fact that a great deal of studies is still needed in this area to determine the properties of IL aggregates at temperatures and in dissolution media, which are strongly linked to its applications. Finally, the chapter presents a clear idea of the use of important spectrometric tools for the investigation of the formation and characterization of IL aggregates. It concludes with an optimistic view for the future expansion of the development of IL aggregates as a new soft material. This positivity is derived from the certainty that the results reported here are the beginning of a great advance in this promising field.

Chapter 7 presents controlling reversible self-assembly paths of amyloid beta peptide over gold colloidal nanoparticles' surfaces. The author identifies two distinctly different self-assembly paths in an initial stage of fibrillogenesis. The potential surface over the 20 nm gold colloid supports an oligomer form with dimer unit. The potential surface for $A\beta_{1\text{-}40}$ dimer formation is considered to possess a relatively high barrier height for the transition of temperature, and it conducts a reversible transformation for a conformation, which is resilient to temperature change. The potential surface over the 30 or 40 nm gold colloid, however, supports an oligomer form with trimer unit. The potential surface for $A\beta_{1\text{-}40}$ trimer formation is regarded to have a minimum for a denatured form of $A\beta_{1\text{-}40}$ with lower barrier, which allows conformation converged into a denatured form. The discovery of two different self-assemblies indicates that an initiating core size and temperature condition determine the type of oligomer to be constructed and govern the path of the entire self-assembly process. The selection of core size and temperature condition enables us to pinpoint a particular oligomer form, which is significantly associated with a mechanism of Alzheimer's disease.

Chapter 8 presents nanoscale spectroscopy in the infrared with applications to biology, focusing on scattering near field optical microscopy (s-SNOM) and nanoscale infrared absorbing spectroscopy with atomic force microscopy based infrared (AFMIR).

Chapter 9 presents spectral interference fluorescence microscopy to study the conformation of biomolecules with nanometer accuracy. It introduces a high-throughput platform combining spectral self-interference fluorescence microscopy (SSFM) and DNA microarray technology as a novel tool to study surface-immobilized DNA conformation and DNA–protein interactions. SSFM maps the spectral oscillations emitted by fluorophores located above a reflecting surface into a precisely determined height relative to the surface. In contrast to earlier fluorescence interference microscopy techniques that relied on intensity variation of total emission, SSFM utilizes spectral information to provide subnanometer accuracy with a single measurement. In the chapter, SSFM is used to estimate the conformation of surface-immobilized single-stranded and double-stranded DNA of different lengths and to quantify the level of hybridization when combined with white light reflectance spectroscopy. The authors demonstrate the application of a novel smart polymeric surface to modulate the orientation of immobilized double-stranded DNA by varying the buffer pH and ionic strength. SSFM is used to precisely quantify both the DNA orientation and the polymer conformation with subnanometer resolution. Recently, SSFM was upgraded to a dual-color modality, which determines two axial positions at the same location on the surface. The chapter ends by drawing the conclusion that the quantification of DNA conformation and conformational changes, when combined with new surface functionalization techniques and label-free quantification of biomass density on surfaces, provides critical information for studying DNA–protein interactions.

Chapter 10 describes functional magnetic resonance imaging (fMRI) and nanotechnology. The authors focus on magnetic resonance imaging, contrast agents for MRI, functional MRI, nanoparticles in imaging, and nanoparticles as contrast agents in fMRI. They show that nanoparticles, due to the unique structure and properties that arise from their size, offer a large variety of applications in medicine, especially in targeted drug delivery and medical imaging. As direct drug delivery and MRI contrast agents, magnetic nanoparticles are the optimal choice as their quality and quantity of biocompatibility and level of toxicity are easily interpreted. Also, these magnetic nanoparticles produce large field gradients and shorten both T1 and T2. By altering the surface chemistry and functionalization of these nanoparticles using various processes such as polymer coating, their behavior can be further enhanced.

Chapter 11 presents a brief overview of nanoscale spectroscopy in medicine, focusing on electronic spectroscopy and vibrational spectroscopy. The authors discuss different techniques of nanoscale spectroscopy of clinical importance. These techniques are applied in many different disciplines in biology and medicine. Spectroscopy in general can be categorized as electronic, vibrational, and rotational. However, only electronic and vibrational spectroscopy are of biological importance. Fluorescence spectroscopy is used for the noninvasive early diagnosis of various types of cancers, atherosclerosis, and arrhythmia and for monitoring many different metabolites in cellular culture such as pH, calcium concentration, and glucose. Due to the very high signal-to-noise ratio, fluorescence spectroscopy enables one to distinguish the spatial distribution of even low concentrations of substances. Fluorescence correlation spectroscopy can be used to investigate a variety of nanoscale biological processes such as protein interactions, binding equilibria for drugs, and clustering of membrane-bound receptors. Noninvasive phosphorescence measurements can be used to determine tumor oxygen concentrations. Atomic absorption spectroscopy is applied for the diagnosis and monitoring of Wilson's disease, leukemia, hypozincemia, acute lead poisoning, and several other diseases.

Chapter 12 presents a comprehensive study of medical nanoscale spectroscopy in terms of concepts, principles, and applications. The author summarizes important reports on the applications of nanoscale spectroscopy, nanoscale spectroscopy for manipulating biomedical work, usefulness of nanoscale spectroscopy, importance of the tools of nanoscale spectroscopy in the biomedical field, and common applications of nanoscale spectroscopy in general nanomedicine practice. He also gives examples of nanomedicine studies based on the application of nanoscale spectroscopy.

Chapter 13 discusses nanoscale spectroscopy in defense and national security. It covers the detection of chemical and biological warfare and spectroscopic techniques for nanomaterials, surface-enhanced Raman spectroscopy for defense purposes, bulk detection techniques for explosives, and trace detection.

Finally, the book concludes with the appendices. Appendix A provides common material and physical constants, with the considerations that the materials constants values varied from one published source to another due to many varieties of most materials and conductivity is sensitive to temperature, impurities, moisture content, as well as the dependence of relative permittivity and permeability on temperature and humidity and the like. Appendix B provides equations for photon energy, frequency, and wavelength and electromagnetic spectrum, including the approximation of common optical wavelength ranges of light. In addition, it provides a figure for the wavelengths of commercially available lasers. Finally, Appendix C provides common symbols and useful mathematical formulas.

Sarhan M. Musa

MATLAB® is a registered trademark of The MathWorks, Inc. For product information, please contact:

The MathWorks, Inc.
3 Apple Hill Drive
Natick, MA, 01760-2098 USA
Tel: 508-647-7000
Fax: 508-647-7001
E-mail: inf@mathworks.com
Web: www.mathworks.com

Acknowledgments

I would like to express my sincere appreciation and gratitude to all the book's contributors. Thanks to Brain Gaskin and James Gaskin for their wonderful hearts and for being great American neighbors. It is my pleasure to acknowledge the outstanding help and support of the team at Taylor & Francis Group/CRC Press in preparing this book, especially from Nora Konopka, Michele Smith, Kari Budyk, and Cynthia Klivecka. Thanks to Christine Selvan at SPi Global for her outstanding suggestions. Thanks also to Dr. Fadi Alameddine for taking good care of my mother's health during the course of this project. I would also like to thank Dr. Kendall Harris, my college dean, for his constant support. Finally, this book would never have seen the light of day if not for the constant support, love, and patience of my family.

Editor

Sarhan M. Musa, PhD, currently serves as an associate professor in the Department of Engineering Technology at Prairie View A&M University, Texas. He has been the director of Prairie View Networking Academy, Texas, since 2004. Dr. Musa has published more than 100 papers in peer-reviewed journals and conferences. He is a frequently invited speaker and has consulted for multiple organizations, both nationally and internationally. Dr. Musa is a featured author and editor of several books including *Computational Nanotechnology*. Dr. Musa is a senior member of the Institute of Electrical and Electronics Engineers (IEEE) and is also an LTD Sprint and a Boeing Welliver Fellow.

Contributors

Mohit Agarwal
Department of Electrical Engineering
The University of Akron
Akron, Ohio

S.M. Ashraf (retired)
Materials Research Laboratory
Department of Chemistry
Jamia Millia Islamia
(A Central University)
New Delhi, India

Saurabh Basu
Department of Physics
Indian Institute of Technology
Guwahati, India

Lilian Buriol
Núcleo de Química de Heterociclos
Department of Chemistry
Federal University of Santa Maria
Santa Maria, Brazil

P.K. Choudhury
Institute of Microengineering and
Nanoelectronics
Universiti Kebangsaan Malaysia
Bangi, Selangor, Malaysia

Aditi Deshpande
Department of Biomedical
Engineering
The University of Akron
Akron, Ohio

Krishna Kanti Dey
Department of Chemistry
The Pennsylvania State University
University Park, Pennsylvania

David S. Freedman
Department of Electrical and
Computer Engineering
Boston University
Boston, Massachusetts

Clarissa P. Frizzo
Núcleo de Química de Heterociclos
Department of Chemistry
Federal University of Santa Maria
Santa Maria, Brazil

George C. Giakos
Department of Biomedical
Engineering
and
Department of Electrical Engineering
The University of Akron
Akron, Ohio

Izabelle M. Gindri
Núcleo de Química de Heterociclos
Department of Chemistry
Federal University of Santa Maria
Santa Maria, Brazil

Chintha C. Handapangoda
Department of Electrical and
Computer Systems Engineering
Monash University
Clayton, Victoria, Australia

Norihiko Hayazawa
Near-Field Nanophotonics Research
Team
RIKEN
Wako, Japan

Eamonn Kennedy
NanoPhotonics & Nanoscopy Research Group
School of Physics
University College Dublin
Dublin, Ireland

E. Manias
Department of Materials Science and Engineering
The Pennsylvania State University
University Park, Pennsylvania

Marcos A.P. Martins
Núcleo de Química de Heterociclos
Department of Chemistry
Federal University of Santa Maria
Santa Maria, Brazil

Saeid Nahavandi
Centre for Intelligent Systems Research
Deakin University
Waurn Ponds, Victoria, Australia

G. Polizos
Measurement Science and Systems Engineering Division
Oak Ridge National Laboratory
Oak Ridge, Tennessee

Malin Premaratne
Department of Electrical and Computer Systems Engineering
Monash University
Clayton, Victoria, Australia

Wen-Gang Qu
Division of Nanomaterials and Chemistry
Hefei National Laboratory for Physical Sciences at Microscale
University of Science and Technology of China
Hefei, Anhui, People's Republic of China

C.A. Randall
Department of Materials Science and Engineering
The Pennsylvania State University
University Park, Pennsylvania

Ufana Riaz
Materials Research Laboratory
Department of Chemistry
Jamia Millia Islamia (A Central University)
New Delhi, India

James Rice
NanoPhotonics & Nanoscopy Research Group
School of Physics
University College Dublin
Dublin, Ireland

I. Sauers
Fusion Energy Division,
Oak Ridge National Laboratory
Oak Ridge, Tennessee

Suman Shrestha
Department of Electrical Engineering
The University of Akron
Akron, Ohio

Philipp S. Spuhler
Department of Biomedical Engineering
Boston University
Boston, Massachusetts

Aniele Z. Tier
Núcleo de Química de Heterociclos
Department of Chemistry
Federal University of Santa Maria
Santa Maria, Brazil

V. Tomer
Department of Materials Science and Engineering
The Pennsylvania State University
University Park, Pennsylvania
and
Dow Chemical Company
Spring House, Pennsylvania

E. Tuncer
Fusion Energy Division,
Oak Ridge National Laboratory
Oak Ridge, Tennessee
and
3M Corporate Research Materials Laboratory
3M Austin Center
Austin, Texas

M. Selim Ünlü
Department of Biomedical Engineering
and
Department of Electrical and Computer Engineering
Boston University
Boston, Massachusetts

Marcos A. Villetti
Department of Physics
Federal University of Santa Maria
Santa Maria, Brazil

Viroj Wiwanitkit
Wiwanitkit House
Bangkok, Thailand
and
Hainan Medical College
Haikou, Hainan, People's Republic of China
and
Joseph Ayobabalola University
Ikeji-Arakeji, Nigeria
and
Faculty of Medicine
University of Nis
Nis, Serbia

An-Wu Xu
Division of Nanomaterials and Chemistry
Hefei National Laboratory for Physical Sciences at Microscale
University of Science and Technology of China
Hefei, Anhui, People's Republic of China

Taka-aki Yano
Department of Electronic Chemistry
Tokyo Institute of Technology
Yokohama, Japan

Kazushige Yokoyama
Department of Chemistry
Geneseo College
The State University of New York
Geneseo, New York

Edward Yoxall
EXSS Group
Imperial College London
London, United Kingdom

Nilo Zanatta
Núcleo de Química de Heterociclos
Department of Chemistry
Federal University of Santa Maria
Santa Maria, Brazil

Xirui Zhang
Department of Biomedical Engineering
Boston University
Boston, Massachusetts

1

Tip-Enhanced Spectroscopy at the Nanoscale: Its Practical Issues and Solutions

Norihiko Hayazawa and Taka-aki Yano

CONTENTS

1.1 Introduction

Variety of tip-enhanced spectroscopies, such as fluorescence (Azoulay et al. 1999; Gerton et al. 2004; Hamann et al. 2000; Hayazawa et al. 1999), two-photon excited fluorescence (Hayazawa et al. 2009a,b; Sanchez et al. 1999), infrared absorption (Hillenbrand et al. 2002; Knoll and Keilmann 1999), Raman (Hayazawa et al. 2000; Pettinger et al. 2002; Stockle et al. 2000), and nonlinear Raman (Furusawa et al. 2012; Hayazawa et al. 2004a; Ichimura et al. 2004a,b) are very powerful techniques for the in situ chemical analysis of organic/inorganic materials in nanometer scale. Raman scattering and infrared absorption spectroscopies allow direct observation of molecular vibrations without necessarily photobleaching and quenching the sample. While conventional Raman spectroscopy is relatively straightforward to carry out with well-established light sources and instruments in the visible region, the Raman scattering cross sections (~10^{-30} cm^2) are much smaller than that of the fluorescence (~10^{-16} cm^2) and infrared absorption (~10^{-20} cm^2). Moreover, in a near-field scanning optical microscope (NSOM) setup, the observed volume of the sample must be confined in a nanometer scale corresponding to a very small number of molecules. Thus, making it difficult to realize local probing by spectroscopic techniques because the number of molecules becomes smaller as the spatial resolution gets higher, resulting to an extremely weak signal. Thus, tip-enhanced spectroscopy is more advantageous than conventional aperture type NSOM (Betzig and Chichester 1993; Xie and Dunn 1994) because it provides not only high spatial resolution but also high sensitivity due to the signal enhancement. The tip-enhancement has been known as surface plasmon polariton (SPP) resonance at the tip apex (Kawata 2001). Among the previously mentioned tip-enhanced spectroscopies, the combination with Raman spectroscopy has turned out to be the most promising technique in terms of analytical power because of its high chemical sensitivity, which has been recently recognized as tip-enhanced Raman scattering (TERS) spectroscopy because of the analogy with surface-enhanced Raman scattering (SERS) (Chang and Furtak 1982). TERS has been so far applied for nano-analysis of various materials such as carbon nano-materials (Hartschuh et al. 2003; Hayazawa et al. 2003; Saito et al. 2009), bio-materials (Bailo and Deckert 2008; Domke et al. 2007; Xie and Dunn 1994; Yeo et al. 2008), and semiconductors (Berweger et al. 2009; Hayazawa et al. 2007; Zhang et al. 2010a,b). However, in practice, the reproducibility of the tip-enhancement is one of the big issues that prevent TERS from being a versatile tool for chemical analysis. Moreover, having a good but instant tip-enhanced spectrum at one position has not been straightforward to tip-enhanced imaging on the same level of the tip-enhancement at each position. This is originated mainly from the two practical issues of "tip preparation" and "positioning stability," both of which affect tip-enhancement. In this chapter, we review a variety of tips reported so far and focus on the tip positioning stability issue.

The stability issue arises from the combination of (i) thermal, (ii) vibrational, and (iii) electrical noises/fluctuations, and becomes a major restriction factor of spatial resolution in the submicron scale. The latter can be resorted by managing the state-of-the-art electronics circuits. On the other hand, the thermal and vibrational factors often become more complex as the source of instability may heavily depend on the mechanical designs of the system and also on the surrounding environment. As to (ii) vibrational one, which is often coupled with sound waves, plenty of vibrational isolation schemes and sound proof enclosures have been proposed and commercialized, and so most of the microscopes are often properly installed in such an isolated environment. While ultra-high vacuum (UHV) and low temperature (LT) are the extreme environmental conditions that realize the lowest noises and are useful for fundamental physical or chemical researches (Hanaguri et al. 2004; Kim et al. 2002; Stipe et al. 1998; Wildoer et al. 1998), the observation capability in an ambient condition is one of the greatest advantages of optical microscopy over an electron-based microscopy/spectroscopy such as scanning electron microscopy (SEM) and inelastic electron tunneling spectroscopy (IETS) using scanning tunneling microscopy (STM), which requires a UHV condition and/or LT. Assuming that it is nearly impossible to eliminate the temperature fluctuation in ambient condition, in which even the operator (human) of the experiment can affect temperature fluctuation, the compensation of (i) thermal drift is of crucial importance for the reliable measurement of optical microscopic measurements. Here in this chapter, we review a technique for real-time and in situ compensation of thermal and mechanical drifts in TERS microscopy, which is robust against long-time measurements.

In Section 1.2, variety of tips so far reported is reviewed, and several issues in terms of reproducibility are raised. In Section 1.3, the power of TERS is shown based on several demonstrations of TERS imaging and spectroscopy, which visualize inhomogeneous distribution of materials in the lateral dimensions. The unique feature of TERS over SERS is not limited to the lateral dimensions but applicable to the longitudinal dimension, which actively controls the distance between the tip and the sample. The precise control of the longitudinal dimension is introduced in Section 1.4. Good tips become useless if the tip is not at the proper position within the light field for tip-enhancement. Nano-positioning stabilizations in lateral and longitudinal dimensions are discussed in Sections 1.5 and 1.6, respectively.

1.2 TERS: Variety of Tips

The metallic tip is the most important component of the TERS system. The tip radius, material, structure, and its SPP resonance have to be properly controlled to perform TERS measurements successfully with high resolution and enhancement in a reproducible manner. The resolution of TERS imaging is basically determined by the sharpness of the tip apex. The diameter of the tip is usually required to be in the order of several tens of nanometers or even less. The TERS enhancement, on the other hand, comes from the SPP resonance at the tip (Kawata 2001). Thus, it is also important to control the spectral response of the metallic tip to match the excitation light frequency. The control of the plasmon resonance at the probe apex having the size of the order of nanometer is still a challenging issue in TERS developments.

As tip materials, noble metals such as silver (Ag) and gold (Au) are mostly used. These metals are good plasmonic materials in visible to near-infrared (NIR) because of the small imaginary part of the dielectric function while realizing a negative value of the real part in this spectral range (Palik 1991). On the other hand, Au and Ag are no longer "metallic" but "dielectric" in ultraviolet (UV) where photon energy exceeds the plasma frequencies present on them. In this spectral range, aluminum (Al) is used as a plasmonically active material (Dorfer et al. 2007; Taguchi et al. 2009a, 2012).

1.2.1 Tips

So far, two types of metallic tips have been widely utilized. One is metal-coated cantilever and the other is etched metallic wire, each of which is used in a suitable combination with the feedback scheme of tip–sample distance such as AFM and STM. The metal-coated cantilever is fabricated conveniently from a commercially available silicon (Si) or silicon nitride (Si_3N_4) cantilever of AFM by depositing thin metallic film onto the probe surface (Figure 1.1a) (Anderson 2000; Hayazawa et al. 2000, 2001). Metal with high purity (>99.999%) is thermally evaporated under vacuum condition and deposited onto the probe surface slowly with a rate of less than 1 Å/s in order not to damage the tip apex. The deposition thickness is typically of several tens of nanometers. Small diameter of several tens of nanometers at the tip is easily obtained.

The second type of the metallic probes, etched Ag and Au wire, respectively, are shown in Figure 1.1b and c. These tips are used either in STM-based TERS system (Pettinger et al. 2000, 2002) or in AFM-based system operated by shear-force feedback mechanism in which an etched wire is glued to one of the prongs of a tuning fork (Hartschuh et al. 2003; Stockle et al. 2000; Zhang et al. 2008). Silver wires are etched using perchloric acid, ethanol/methanol, and water solution (Pettinger et al. 2002; Zhang et al. 2011). Gold wires are etched using hydrochloric acid and ethanol solution (Eligal et al. 2009; Ren et al. 2004).

The reported Raman enhancement factor experimentally obtained using metallic probe is typically 10^3–10^4 (Yeo et al. 2006); however, the enhancement factor also depends on the number of molecules contributed to the net enhancement in the size of tip-enhanced field. This fact suggests that a much higher factor could be achievable (e.g., $\sim 10^{14}$) similar to single-molecule

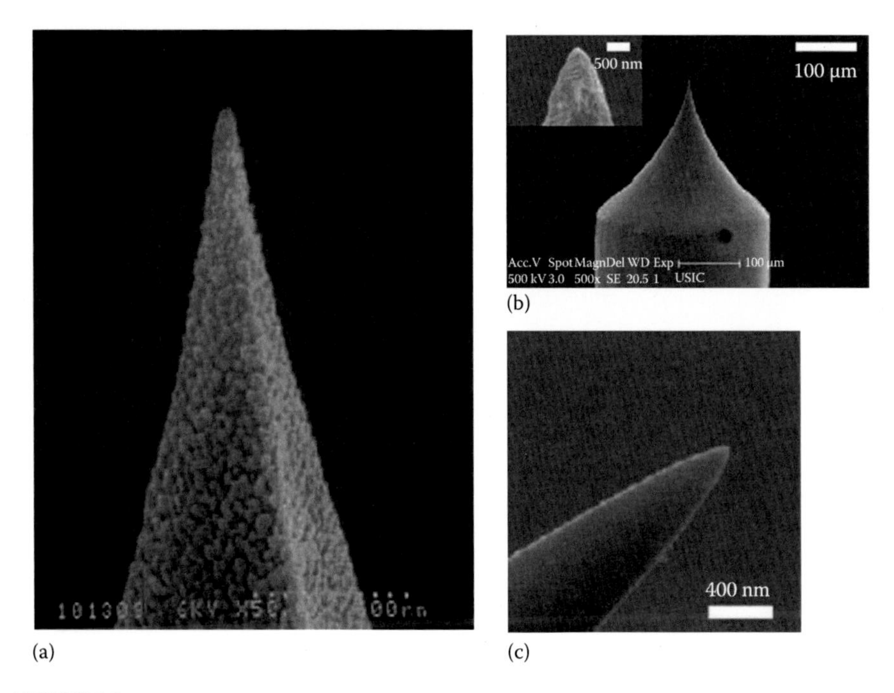

FIGURE 1.1
SEM images of (a) metal-coated Si tip. (From *Chem. Phys. Lett.*, 335(5–6), Hayazawa, N., Inouye, Y., Sekkat, Z., and Kawata, S., Near-field Raman scattering enhanced by a metallized tip, 369–374, Copyright 2001, from Elsevier.) (b) Etched Ag tip. (From Zhang, C., Gao, B., Chen, L.G. et al., Fabrication of silver tips for scanning tunneling microscope induced luminescence, *Rev. Sci. Instrum.*, 82(8), 083101. Copyright 2011, American Institute of Physics.) (c) Etched Au tip. (From Eligal, L., Culfaz, F., McCaughan, V., Cade, N.I., and Richards, D., Etching gold tips suitable for tip-enhanced near-field optical microscopy, *Rev. Sci. Instrum.*, 80(3), 033701. Copyright 2009, American Institute of Physics.)

detection by SERS (Nie and Emery 1997). In any case, the enhancement is related to the SPP of the metallic tip; thus, the spectral investigation of tip plasmon is quite important. One of these works is the spectral measurement of the light scattered by the metallic tip. Typically, the measured spectra shows a single resonance peak with the peak width of approximately 100 nm, and the peak position appears in the visible–NIR between 500 and 800 nm when gold or silver tip is used (Mehtani et al. 2006; Neacsu et al. 2005). Gold tip shows the resonance peak at relatively longer wavelength compared with that of silver because of the difference in the dielectric functions (Johnson and Christy 1972).

1.2.2 Coupling Efficiency: Far-Field Background Issue

The other important aspect in considering TERS probe is the coupling efficiency between the tip plasmon and the far-field light. This is because the tip plasmon is excited initially by the light coming from the far-field, and the Raman signal is consequently measured by a detector placed also in the far-field. In most of the cases including the tips shown in Figure 1.1, the tips are irradiated by a focused laser beam, which subsequently illuminate not only the tip but also a larger sample area determined by the diffraction limit as shown in Figure 1.2a. The illuminated sample area generates unwanted far-field background signal competing with near-field tip-enhanced signal, which may degrade signal-to-noise ratio. In order to reduce such far-field noise generated from the diffraction limit, pure near-field nano-light source has been proposed by either grating-coupled or prism-coupled SPP along the metallic tip as shown in Figure 1.2b. In this case, the laser irradiates only the portion of the tip shaft having grating (Ropers et al. 2007) or prism (Sanchez et al. 2002). Incoming far-field light is efficiently converted to the SPP propagating toward the tip apex through the grating or prism by fulfilling the wavevector mismatch so that SPP propagates along the tip resulting to near-field nano-light source at the tip apex. In addition to SPP-based coupling for pure near-field nano-light source, photonic crystal coupling has been proposed (Carlson and Woehl 2008; De Angelis et al. 2010). Figure 1.3 shows SEM images of the tips based on photonic crystal structure. In Figure 1.3a, the proposed tips consist of a two-dimensional dielectric photonic crystal cavity patterned on a Si_3N_4 AFM cantilever tip, which is fabricated by means of focused ion beam (FIB) and electron-beam induced deposition (De Angelis et al. 2010). The photonic crystal cavity enables an efficient coupling between the external laser source and the tapered waveguide. In this way, generation and localization of tip-enhanced field based on SPPs are realized by means of adiabatic compression through a metallic tapered waveguide to create strongly enhanced Raman excitation in a region just a few nanometers across. Figure 1.3b shows another approach of tip directly fabricated from photonic crystal fiber (PCF) by sealed-tube etching (Carlson and Woehl 2008). These types of tips are expected to be free from far-field background and the Raman signal is generated only at the tip apex. These tips are still young and are expected to improve the reproducibility so as to be one of the promising choices of tips particularly for bulk samples and/or fragile samples, which are not durable against a long-time exposure of light.

1.2.3 Polarization Issue

The tip-enhanced field is generally parallel to the tip axis, which is perpendicular to the sample surface in most of the cases (Bouhelier et al. 2003; Hayazawa et al. 2004b). While this strong longitudinal field can provide the selective excitation of desired vibrational modes (e.g., to phonon modes [Berweger et al. 2009; Tarun et al. 2011] of crystals), this is also regarded as a limitation of sensitivity when targeting the modes parallel to the

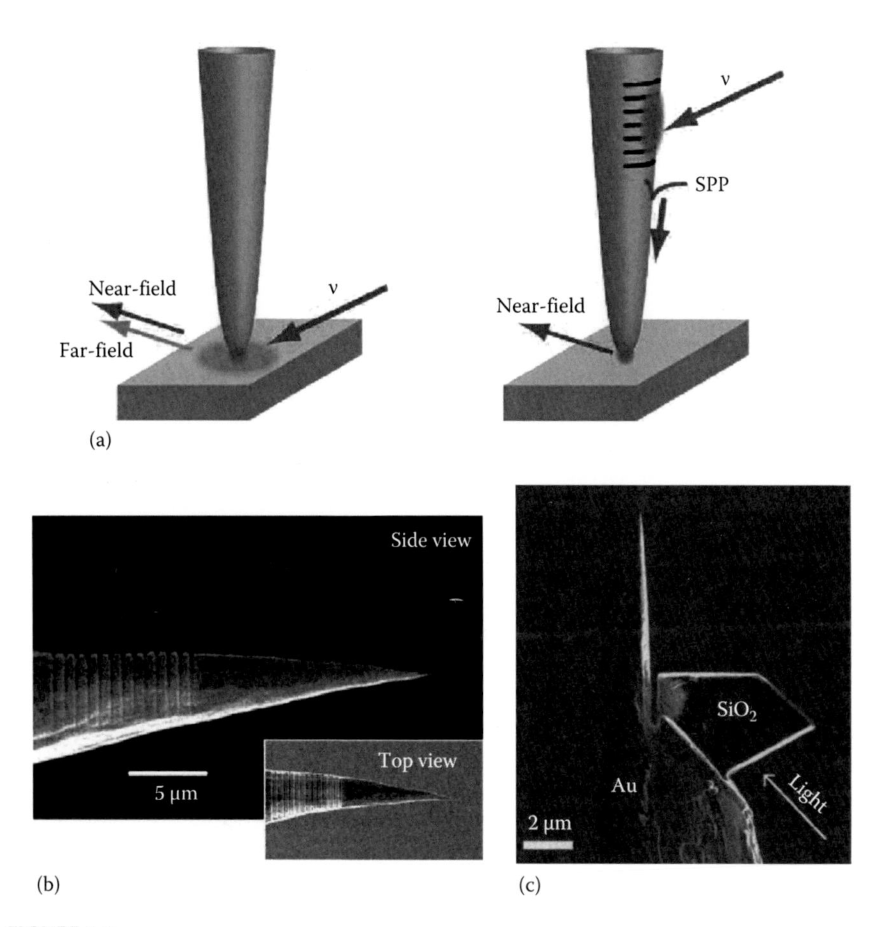

FIGURE 1.2
(a) Schematic of the effective far-field background suppression by grating coupled SPP propagation along the metallic tip. (From Berweger, S. et al., Light on the tip of a needle: Plasmonic nanofocusing for spectroscopy on the nanoscale, *J. Phys. Chem. Lett.*, 3, 2012, 945. Copyright 2012 American Chemical Society.) SEM images of (b) grating coupled and (c) prism coupled SPP tips. (From Ropers, C., Neacsu, C.C., Elsaesser, T., Albrecht, M., Raschke, M.B., and Lienau, C., Grating-coupling of surface plasmons onto metallic tips: A nanoconfined light source, *Nano Lett.*, 7(9), 2007, 2784–2788. Copyright 2007 American Chemical Society; Sanchez, E.J., Krug, J.T., and Xie, X.S., Ion and electron beam assisted growth of nanometric SimOn structures for near-field microscopy, *Rev. Sci. Instrum.*, 73(11), 3901–3907, 2002. Copyright 2002, American Institute of Physics.)

surface. In reality, the tip apex has some finite size and taper angle so that the tip-enhanced field contains certain amount of electric field perpendicular to the tip axis. This idea has been used to visualize single-walled carbon nanotubes (SWNTs) lying on a cover glass (Hartschuh et al. 2003; Hayazawa et al. 2012b). However, the tip-enhanced transverse field is much weaker than the tip-enhanced longitudinal field. In addition, the higher

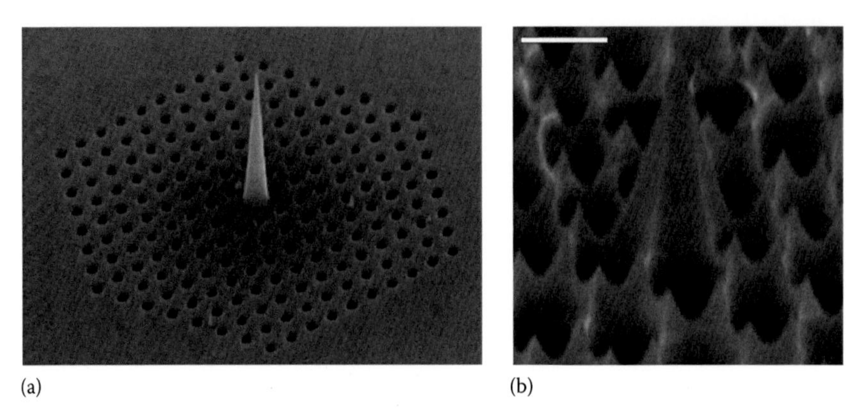

(a) (b)

FIGURE 1.3
SEM images of photonic crystal-based tips. (a) Based on Si3N4 AFM cantilever tip. (From Macmillan Publishers Ltd. *Nat. Nanotechnol.*, De Angelis, F., Das, G., Candeloro, P. et al., Nanoscale chemical mapping using three-dimensional adiabatic compression of surface plasmon polaritons, 5(1), 67–72, 2010. Copyright 2010.) (b) Based on photonic crystal fiber. (From Carlson, C.A. and Woehl, J.C., Fabrication of optical tips from photonic crystal fibers, *Rev. Sci. Instrum.*, 79(10), 103707, 2008. Copyright 2008, American Institute of Physics.)

the spatial resolution is (the smaller the tip end is), the extremely smaller the transverse field becomes. In order to actively control and generate tip-enhanced field parallel to the sample surface, several schemes have been proposed (Farahani et al. 2005; Sanchez et al. 1999). Figure 1.4a shows SEM image of a Au tip with an end diameter of 15 nm fabricated with etching and subsequent FIB milling (Sanchez et al. 1999). Since tip-enhancement arises from a high surface charge density at the tip that is induced by the incident light polarized along the tip axis, the incident light with polarization perpendicular to the tip axis results in no tip-enhancement. The idea here is to bend the tip end parallel to the sample surface so that tip-enhancement can be simply excited by light polarized in the horizontal direction. The other proposed approach is to make a nano-gap parallel to the sample surface so as to efficiently couple with the electric field parallel to the sample surface. Figure 1.4b shows so-called bow-tie antenna fabricated by FIB milling on top of the metal-coated Si_3N_4 cantilever tip (Farahani et al. 2005). Since these tips allow for the efficient detection of modes parallel to the sample surface at the cost of the sensitivity to the modes perpendicular to the surface, the ideas for compatible tips for both polarizations are much awaited in this sense.

1.2.4 Reproducibility Issue

The tip-enhancement has been known as SPP resonance at the tip apex (Kawata 2001). However, in practice, the reproducibility of the tip-enhancement is one of the big issues that prevent tip-enhanced spectroscopy from being

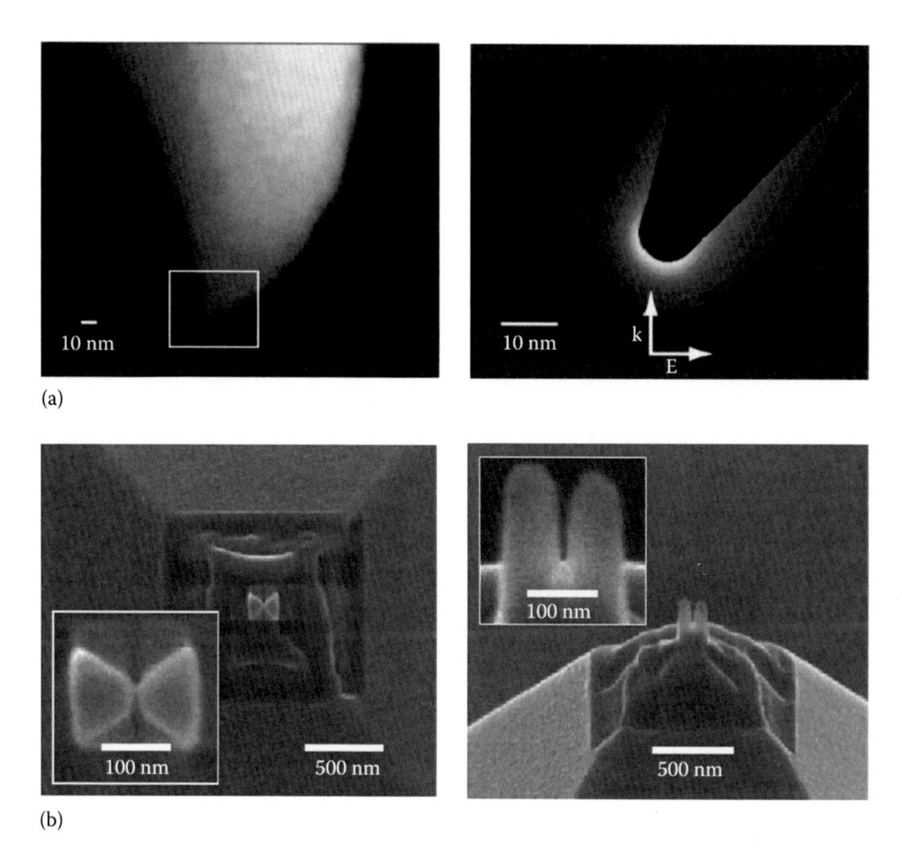

FIGURE 1.4
(a) SEM image and electric field distribution at the bent Au tip. (From Sanchez, E.J., Novotny, L., and Xie, X.S., *Phys. Rev. Lett.*, 82(20), 4014–4017, 1999. Copyright 1999 by the American Physical Society.) (b) SEM images of "bow-tie" antenna tip. (From Farahani, J.N., Pohl, D.W., Eisler, H.J., and Hecht, B., *Phys. Rev. Lett.*, 95(1), 017402, 2005. Copyright 2005 by the American Physical Society.)

a versatile tool for chemical analysis. The maximum tip-enhancement factor is expected when the tip plasmon and the excitation laser are spectrally matched. In practice, the operation wavelength is fixed to the available laser facility used for TERS experiments, while the tip plasmon should be controlled. The other option is to tune the excitation wavelength relative to the SPP resonance, which is practically difficult due to the limitation of available optics such as filters. In any cases, controlling the plasmon resonance is of practical importance to improve the reproducibility of tip-enhancement. Moreover, a robust and simple tip fabrication is favorable for high reproducibility. At this point, there are several reports on the improvement of the reproducibility.

Figure 1.5a shows the SEM image of a highly reproducible tip fabricated by thermal oxidization and subsequent metallization of commercial Si

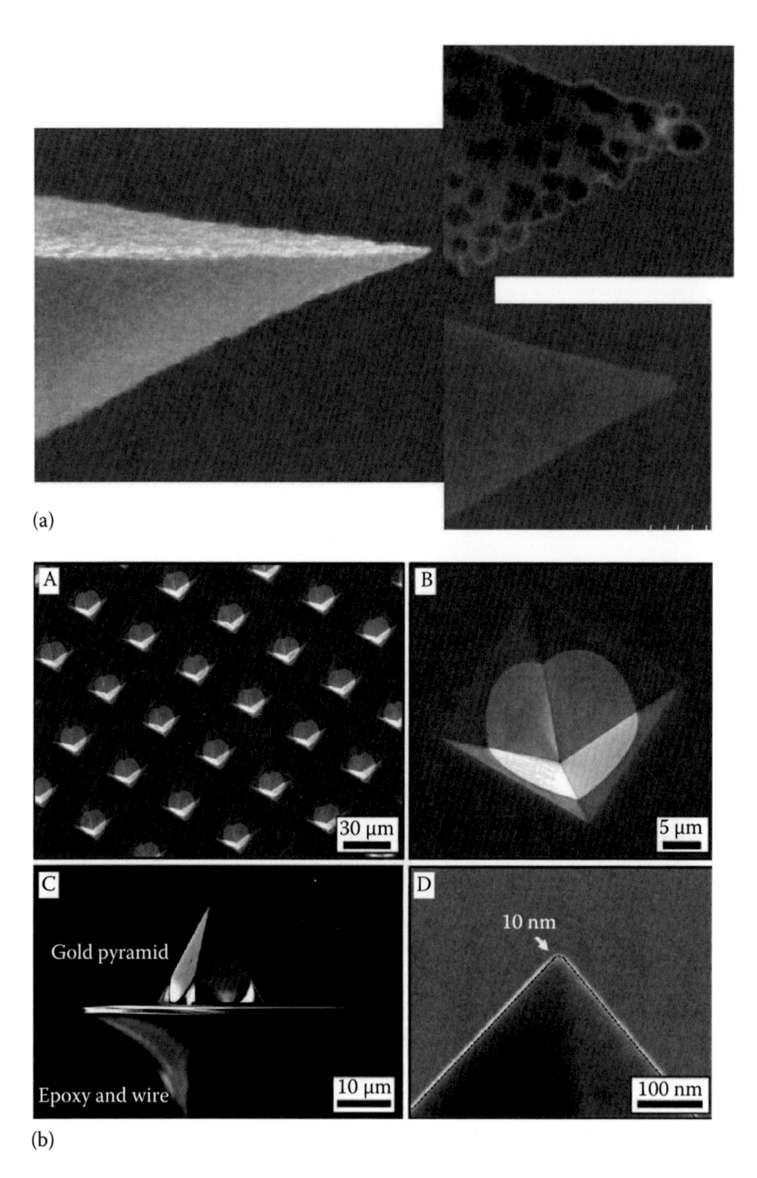

FIGURE 1.5
SEM images of highly reproducible tips based on (a) thermal oxidization and metallization of Si tip, (b) template-stripping technique. (From Hayazawa, N., Yano, T., and Kawata, S.: Highly reproducible tip-enhanced Raman scattering using an oxidized and metallized silicon cantilever tip as a tool for everyone. *J. Raman Spectrosc.* 2012b. 43(9). 1177–1182. Copyright Wiley-VCH Verlag GmbH & Co. KGaA; Johnson, T.W., Lapin, Z.L., Beams, R. et al., Highly reproducible near-field optical imaging with sub-20-nm resolution based on template-stripped gold pyramids, DOI: 10.1021/nn303496g, *ACS Nano*, 6(10), 2012, 9168–9174. Copyright 2012 American Chemical Society; You, Y.M., Purnawirman, N.A., Hu, H.L. et al.: Tip-enhanced Raman spectroscopy using single-crystalline Ag nanowire as tip. *J. Raman Spectrosc.* 2010. 41(10). 1156–1162. Copyright Wiley-VCH Verlag GmbH & Co. KGaA.)

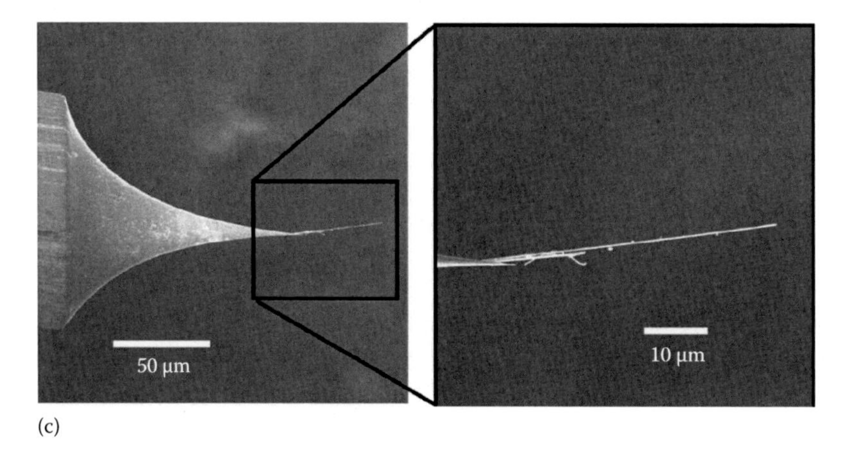

FIGURE 1.5 (continued)
SEM images of highly reproducible tips based on (c) Ag nanowire attached to W tip. (From Hayazawa, N., Yano, T., and Kawata, S.: Highly reproducible tip-enhanced Raman scattering using an oxidized and metallized silicon cantilever tip as a tool for everyone. *J. Raman Spectrosc.* 2012b. 43(9). 1177–1182. Copyright Wiley-VCH Verlag GmbH & Co. KGaA; Johnson, T.W., Lapin, Z.L., Beams, R. et al., Highly reproducible near-field optical imaging with sub-20-nm resolution based on template-stripped gold pyramids, DOI: 10.1021/nn303496g, *ACS Nano*, 6(10), 2012, 9168–9174. Copyright 2012 American Chemical Society; You, Y.M., Purnawirman, N.A., Hu, H.L. et al.: Tip-enhanced Raman spectroscopy using single-crystalline Ag nanowire as tip. *J. Raman Spectrosc.* 2010. 41(10). 1156–1162. Copyright Wiley-VCH Verlag GmbH & Co. KGaA.)

cantilever tips (Hayazawa et al. 2012b). Owing to the change of the refractive index of the tip from Si to silicon dioxide (SiO_2), the plasmon resonance of the Ag coated tip is blue-shifted, showing an enormous enhancement at 532 nm excitation (Taguchi et al. 2009b). Highly reproducible tips exhibit an enhancement factor of >1000 with a 100% yield. Since the tips are fabricated from commercially available Si cantilever tips in a simple and robust way, this approach provides an important step of "TERS for everyone."

The other approach for high reproducibility is to separately prepare a bunch of reproducible metallic nano-structures and attach the nano-structure to a scanning probe as a TERS tip. Figure 1.5b shows a bunch of metallic tip templates fabricated by template-stripping technique (Johnson et al. 2012). With this technique, a massively parallel fabrication up to 1.5 million identical tips over a wafer and 95% of reproducibility is reported. Another candidate of such a template is nanowires (NWs). Since the synthesis of metal NWs has been already well developed, and different diameters of NWs can be easily achieved, the control of the tip size can also be ensured (Chen et al. 2007; Feng et al. 2009; Graff et al. 2005; Huo et al. 2008; Nabais et al. 2009; Sun and Xia 2002; Tang and Kotov 2005; Wang et al. 2007; Wiley et al. 2005a,b; Zhang et al. 2007). By combining the achievement in NW fabrication and the high surface plasmon efficiency of single crystalline noble-metal nanostructures, the reproducibility of TERS can be greatly improved

(Brodard et al. 2012; You et al. 2010). Figure 1.5c shows a SEM image of Ag NW attached to tungsten (W) tip by alternating current dielectrophoresis (You et al. 2010).

We will focus on the most widely used tips described in Section 1.2.1 for the following sections.

1.3 TERS: Imaging and Spectroscopy in the Nanoscale (Lateral Dimension: XY)

TERS has been applied for nanoscale surface analysis of various materials such as nano-carbon materials (Anderson et al. 2007; Hartschuh et al. 2003; Hayazawa et al. 2003; Saito et al. 2006b, 2009; Stadler et al. 2011; Stockle et al. 2000; Verma et al. 2006; Yano et al. 2006), semiconductors (Berweger et al. 2009; Hayazawa et al. 2007; Lee et al. 2007; Matsui et al. 2007; Ogawa et al. 2011; Pan et al. 2006; Saito et al. 2006a; Tarun et al. 2009; Zhang et al. 2009; Zhu et al. 2007) and even biomolecules (Bailo and Deckert 2008; Bohme et al. 2009; Budich et al. 2008; Cialla et al. 2009; Domke et al. 2007; Ichimura et al. 2007; Neugebauer et al. 2007; Taguchi et al. 2009a) due to super-resolution capability in lateral dimension. Several types of system configurations for TERS spectroscopy and imaging are employed depending on sample condition.

1.3.1 System Configuration

TERS systems for imaging and spectroscopy in the nanoscale are commonly configured with a combination of inverted optical microscopy and scanning probe microscopy (SPM) such as atomic force microscopy (AFM) and scanning tunneling microscopy (STM). This configuration allows for the use of high numerical aperture (NA) objective lens in order to tightly focus incident laser light onto a metallic tip through sample as shown in Figure 1.6a. The tightly focused laser spot benefits high collection efficiency of TERS as well as suppression of far-field Raman background coming from the sample in the focus spot. Since the longitudinal field of the incident light efficiently excites SPP at the tip apex, radial polarization is preferably utilized as incident polarization because of much stronger longitudinal field component than linear polarization (Hayazawa et al. 2004b). The radially polarized light is provided by passing the linearly polarized laser light through a radial-waveplate consisting of a segmented half-waveplate, with each segment having a different orientation of the optical axis. TERS signal excited by the radially polarized incident light is efficiently collected by the same high NA objective lens, and directed to a spectrophotometer. The dispersed Raman signal through the spectrometer is detected by a liquid nitrogen cooled CCD

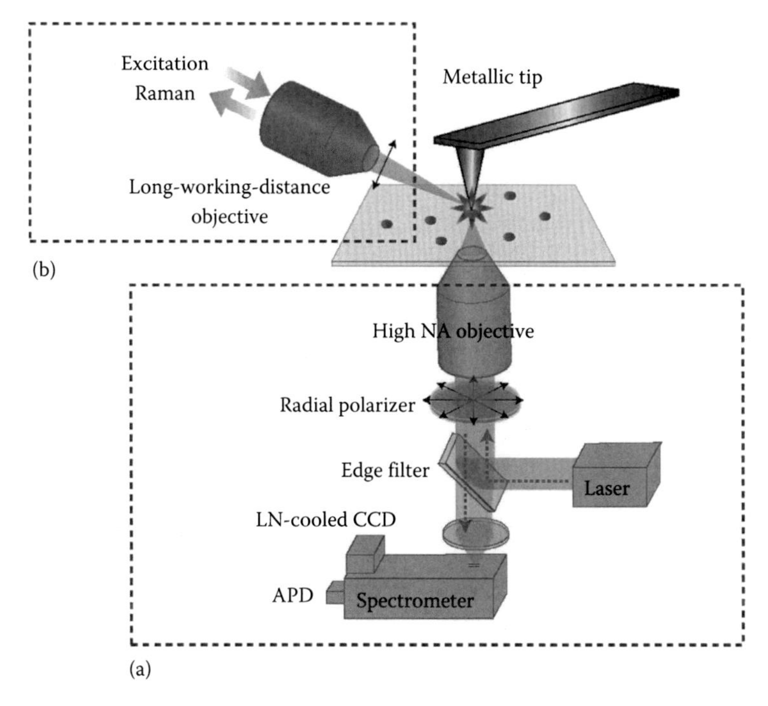

FIGURE 1.6
Schematics of TERS system configurations: (a) transmission mode and (b) reflection mode.

camera for acquiring a TERS spectrum or by a single photon counting module such as avalanche photodiode (APD) and a photo multiplier (PMT) for acquiring a TERS image at a certain Raman frequency.

Although the transmission illumination-collection mode in Figure 1.6a is commonly utilized due to the high detection sensitivity, this mode is limited to transparent or thin samples. For opaque samples (bulk, crystals, etc.), reflection mode is employed (Hayazawa et al. 2002, 2007; Saito et al. 2006a; Sun and Shen 2003), in which the tip is illuminated from the side with the linear polarization parallel to the tip axis as shown in Figure 1.6b. The reflection mode provides poorer collection efficiency than the transmission mode because only a low NA objective lens with long working distance can be utilized due to space restrictions around the SPM head. For highly crystalline samples, depolarization effect enables us to reduce strong far-field Raman background caused by the large focus spot, which results in getting a high-contrast TERS signal (Lee et al. 2007; Motohashi et al. 2008; Poborchii et al. 2005).

1.3.2 Nano-Analysis of Carbon Nano-Materials

An SWNT is one of the best samples for TERS imaging and spectroscopy due to the unique and ideal one-dimensional nature in which its electronic and

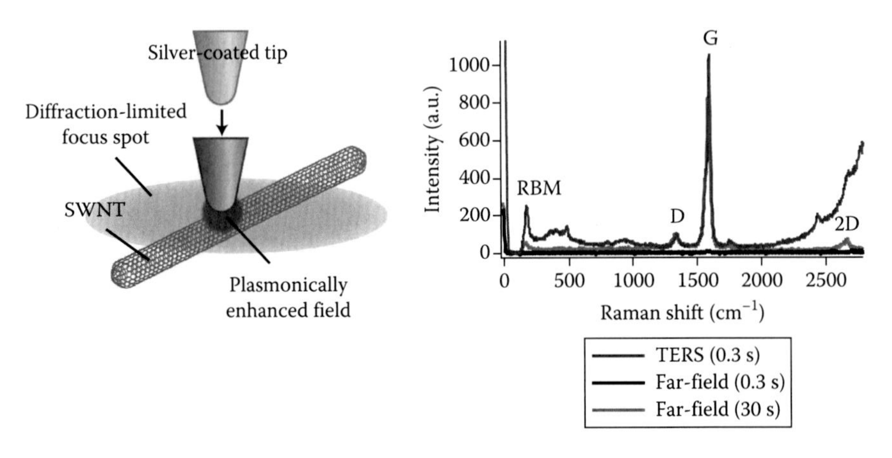

FIGURE 1.7
(See color insert.) TERS spectrum of SWNT measured by positioning a silver-coated tip onto SWNT lying in the focus spot. Far-field Raman spectrum, measured with the tip away from the nanotube, is shown as well for comparison. (From Hayazawa, N., Yano, T., and Kawata, S.: Highly reproducible tip-enhanced Raman scattering using an oxidized and metallized silicon cantilever tip as a tool for everyone. *J. Raman Spectrosc.* 2012b. 43(9). 1177–1182. Copyright Wiley-VCH Verlag GmbH & Co. KGaA.)

structural properties drastically change depending on chirality and diameter of nanotubes. Figure 1.7 shows one of the typical TERS spectra of SWNTs in comparison with far-field micro-Raman spectrum (without the tip). Almost no peak can be observed in the far-field spectrum except for a very small peak at G-band. In the TERS spectrum, all the active peaks, such as RBM, D-band, G-band, and 2D-band, are strongly enhanced by a silver-coated SiO_2 tip with excitation wavelength of 532 nm. Assuming that only one nanotube is in the diffraction-limited focus spot (ϕ: 400 nm) and the size of the tip-enhanced field is 40 nm in diameter corresponding to the silver grain size at the tip end, the estimated enhancement factors for RBM, D-band, G-band, and 2D-band are 250, 1000, 700, and 1000, respectively. The difference of the enhancement factor could be partly due to the different Raman tensors relative to the electric field distribution of the tip-enhanced field, e.g., G-band is not efficiently coupled with the tip-enhanced longitudinal field perpendicular to the tube axis, resulting in the underestimation of the enhancement factor. Thus, the obtained enhancement factor varies for each nanotube having a different chirality.

Figure 1.8 shows a dataset of TERS imaging of SWNTs measured by raster scanning on a cover glass while detecting G-band intensity by APD. The TERS image is shown in comparison with micro Raman and topographic images obtained simultaneously with the TERS image at the same area. Owing to the high and stable tip-enhancement effect, far-field contribution in TERS contrast is negligible so that relatively high concentration area of SWNTs

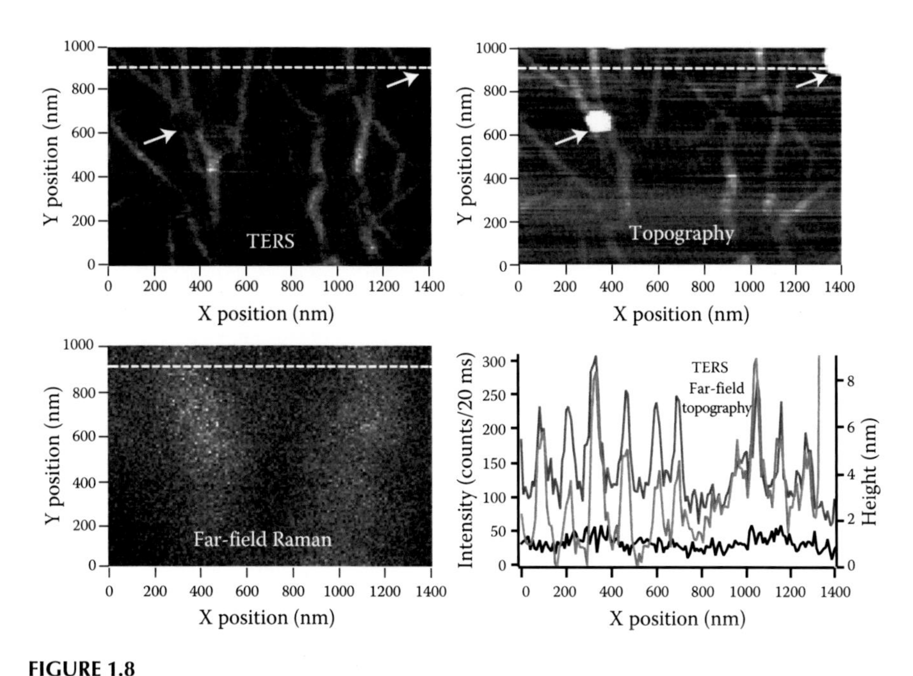

FIGURE 1.8
(See color insert.) TERS image of SWNTs on a cover glass as well as simultaneously obtained topographic image. Far-field Raman image at the same area is also shown for comparison. The cross section at the dashed line clearly exhibits a good agreement between TERS and topography. The arrows in the figure indicate the positions of catalysts which are blind in the TERS image. (From Hayazawa, N., Yano, T., and Kawata, S.: Highly reproducible tip-enhanced Raman scattering using an oxidized and metallized silicon cantilever tip as a tool for everyone. *J. Raman Spectrosc.* 2012b. 43(9). 1177–1182. Copyright Wiley-VCH Verlag GmbH & Co. KGaA.)

is spatially resolved in Figure 1.8. Moreover, all the nanotubes observed in topography that are randomly oriented are all visualized by TERS. In addition, two catalyst Ni/Y particles for arc-discharge method are also seen in the topographic image as an impurity among the nanotubes, providing no TERS signal at the corresponding area. The cross-sectional profile of TERS image exhibits clearly that each SWNT is in good agreement with the topographic profile while micro Raman image is diffraction-limited showing as if only two SWNTs were present in the scanned area.

Another important application of TERS to nano-analysis of SWNTs is to probe local features along an isolated individual SWNT (Anderson et al. 2007; Hartschuh et al. 2003; Hayazawa et al. 2003). It has been demonstrated (Anderson et al. 2007) that TERS measurements of different Raman modes along the length of an isolated SWNT can clearly show a variation of Raman frequency for both RBM and G-band, revealing the variation of chirality along the length of an isolated SWNT. The analysis of spectral shapes of

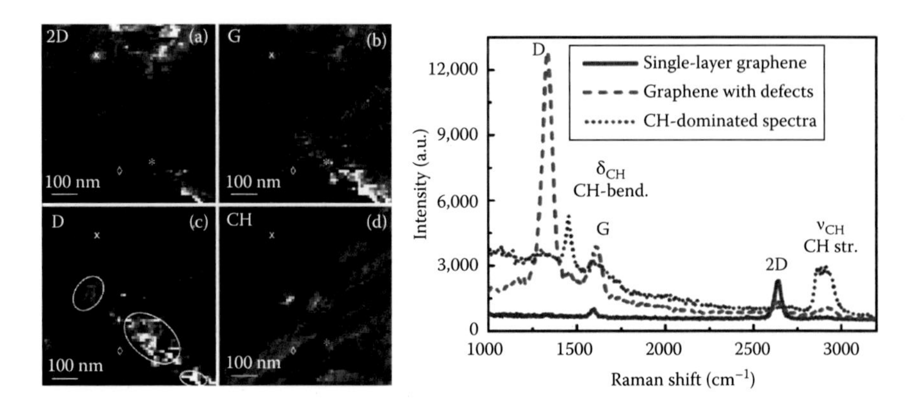

FIGURE 1.9
TERS images of graphene measured with Raman intensity of the 2D-band at 2634 cm^{-1}, the G-band at 1580 cm^{-1}, the D-band at 1350 cm^{-1}, and the CH stretching modes at 2800–3000 cm^{-1}. TERS spectra, measured at the positions (x, *, ◇ indicated in the TERS images), are also shown. (From Stadler, J., Schmid, T., and Zenobi, R., Nanoscale chemical imaging of single-layer graphene, *ACS Nano*, DOI: 10.1021/nn2035523, 5(10), 2011, 8442–8448. Copyright 2011 American Chemical Society.)

G-mode along the nanotube in the same experiment showed that the electronic properties of nanotube changed from semiconducting to metallic along the nanotube. The chirality variation was projected with a spatial resolution of a few tens of nanometers. In another study, the ratio between D-band and G-mode intensities was utilized to quantify the local defect density in an isolated SWNT (Georgi and Hartschuh 2010).

Graphene, which has been attracting a lot of attention over the past few years due to its unique electronic properties, has been also successfully analyzed by TERS measurements (Saito et al. 2009; Stadler et al. 2011). As shown in Figure 1.9, nanometric defects created on single-layer graphene films are clearly imaged in the nanometer scale, which is impossible to be detected by far-field Raman measurement (Stadler et al. 2011). In addition to the localized defects, edges, hydrogen-terminated areas, and contaminated areas are selectively and efficiently detected by measuring tip-enhanced Raman spectral changes of graphene.

1.3.3 Nano-Analysis of Biomaterials

Biomaterials have been also well studied in the nanometer scale by TERS. Especially, TERS investigation of DNA nucleobases has been performed by many groups so far (Domke et al. 2007; Hayazawa et al. 2003; Ichimura et al. 2004a,b; 2007; Taguchi et al. 2009a; Watanabe et al. 2004; Zhang et al. 2010a). Huge tip-enhancement enables to detect DNA molecules with high sensitivity down to the single-molecule level (Ichimura et al. 2007). TERS measurements using deep ultraviolet (DUV) excitation wavelength (Taguchi et al. 2009a) further improve the sensitivity due to the additional resonant Raman effect of DNA bases in the DUV region. High-resolution nonlinear TERS imaging of DNA bases was also performed with a spatial resolution of 15 nm (Ichimura et al. 2004a,b). In addition to the DNA bases, TERS was also utilized to detect a single RNA strand where spectral differences were observed along the strand at nanoscale steps (Bailo and Deckert 2008). These results open up a promising possibility of a label-free DNA and RNA sequencings with base-to-base resolution.

More complex biological samples such as immobilized cells, cell organelles, single viruses, and bacteria are also measured by TERS (Bohme et al. 2009, 2010, 2011; Budich et al. 2008; Hermann et al. 2011; Opilik et al. 2011; Richter et al. 2011; Yeo et al. 2008). Although these investigations clearly observed spectral differences on a nanometer scale, a careful analysis of complicated TERS spectra of complex mixtures is always an important issue. Further improvement of spatial resolution would help to simplify TERS spectra of bio-samples due to further reduction of detection volume.

1.3.4 Nano-Analysis of Semiconductors

Semiconducting materials have been effectively analyzed by TERS for future application on nano-electronic devices (Anderson et al. 2007; Hayazawa et al. 2007; Lee et al. 2007; Saito et al. 2006a; Tarun et al. 2009; Zhu et al. 2007). One of the most successful applications of TERS is the nanoscale analysis of strained silicon, which enhances carrier mobility compared to unstrained silicon. TERS is utilized to measure the local distribution of strain by selectively enhancing the LO phonon vibrations originating from a thin silicon layer, which is under strain due to lattice mismatch with underlying Ge-doped silicon buffer layer. The solid line spectrum in Figure 1.10a shows the TERS spectra of 30-nm-thick strained Si layer grown on a buffer SiGe layer measured with the silver tip in contact with the ε-Si layer. The dotted line spectrum represents the spectrum measured without the tip (i.e., far-field spectrum). The higher-frequency peak represents the one that was obtained from the ε-Si layer while the lower Raman shift frequency peak represents the signal obtained from the underlying SiGe layer. The subtracted spectrum, illustrated by the gray patterned area, exhibits a peak shift from an unstrained Si peak. From the value of frequency shift, the amount of stain can be quantitatively evaluated. Figure 1.10b shows a two-dimensional TERS image of the lattice stress of ε-Si surfaces constructed by the strain-sensitive frequency shift. The TERS frequency image clearly addresses the local distribution of strain with a spatial resolution down to 25 nm. On the other hand, far-field Raman frequency image in Figure 1.10c does not show a distinct frequency shift because the measured far-field Raman shift values show only the average over the diffraction-limited focusing spot.

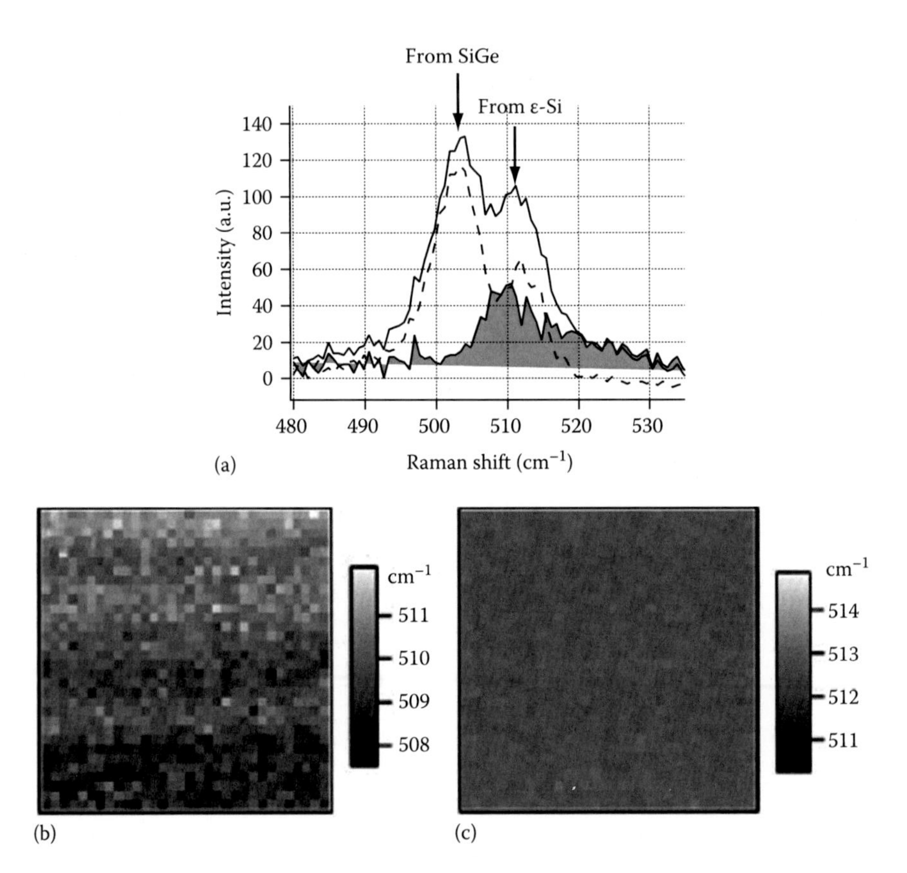

FIGURE 1.10
(a) TERS spectrum (solid line) of 30-nm-thick strained Si layer grown on a buffer SiGe layer measured with the silver tip in contact with the ε-Si layer in comparison with far-field Raman spectrum (dotted line). TERS image (b) exhibits local variation of ε-Si Raman frequency comparing with the far-field image (c). (From Saito, Y. et al.: Stress imaging of semiconductor surface by tip-enhanced Raman spectroscopy. *J. Microscop*. 2008. 229. 217. Copyright Wiley-VCH Verlag GmbH & Co. KGaA.)

1.4 TERS: Precise Distance Dependence (Longitudinal Dimension: Z)

In TERS microscopy and spectroscopy, the tip-enhancement resulting from SPP resonance plays the most essential role both for signal sensitivity and spatial resolution. However, the tip-enhancement effect (electromagnetic interaction) is not the only one affecting Raman spectra. Between the metallic tip and the sample molecules, there are other interactions such as chemical (Kambhampati et al. 1998; Lombardi and Birke 2008; Otto et al. 1992) and mechanical (Yano et al. 2006, 2009) interactions which co-exist with the electromagnetic interaction. The former two interactions appear only when sample molecules are in a close vicinity within the tip. In the TERS system using a contact mode AFM, an experimentally observed TERS spectrum is a complex combination of the contributions of these three interactions making it difficult to interpret experimental TERS spectra. Therefore, elucidation and discrimination of the tip–sample interactions are of scientific and practical importance. This can be realized by measuring a tip–sample distance dependence of TERS, since those three interaction mechanisms have different dependencies on the tip–sample distance. The active control of the distance between the tip and sample is a unique feature only possible in TERS not in SERS. Two system configurations, time-gated detection, and time-gated illumination are discussed and illustrated later.

1.4.1 TERS System Using Tapping Mode AFM with Time-Gated Detection

In the first scheme using the time-gated detection, as shown in Figure 1.11a, TERS signals from molecules are detected by an avalanche photodiode through a monochromator. A dual-gated photon counter captures Raman scattering signal of a certain vibrational mode, while the gates are triggered synchronously with the tapping oscillation of the AFM cantilever (Yano et al. 2007). The concept was first demonstrated in apertureless near-field fluorescence microscope (Yang et al. 2000), in which spectral information is not a big deal. The modification to the TERS system in Figure 1.11a is a substitution of a tapping mode AFM and a photon-counting APD for the contact mode AFM and the CCD detector, respectively. A metal-coated tip is longitudinally vibrated at a certain frequency over sample molecules. This technique is specialized for probing distance-dependent change of TERS intensity, as a single channel detector is used. Figure 1.11b shows a schematic of dual-gated photon counting process utilized for the measurement of the dependence of TERS on longitudinal distance (d) between a tip and a sample. The sinusoidal electric signal (upper curve in the figure) from the AFM controller corresponds to the longitudinal vibration of the tip at a certain frequency. The two gates (gate A and gate B) are independently opened within the tapping period. The time delay of the gate A (τ_A) is swept relative to the other (τ_B) fixed at the time when the tip is farthest from the sample surface. The Raman intensity measured through the gate A (I_A) includes both near-field and far-field Raman components from the sample, while the intensity measured through the gate B (I_B) includes only the far-field Raman scattering component. Therefore, we can extract pure near-field contribution by subtracting I_B from I_A. Sweeping τ_A over the tapping period, one could obtain the tip–sample distance dependence of the TERS intensity. Figure 1.12 shows the typical tip–sample distance dependence of TERS intensity of SWNTs.

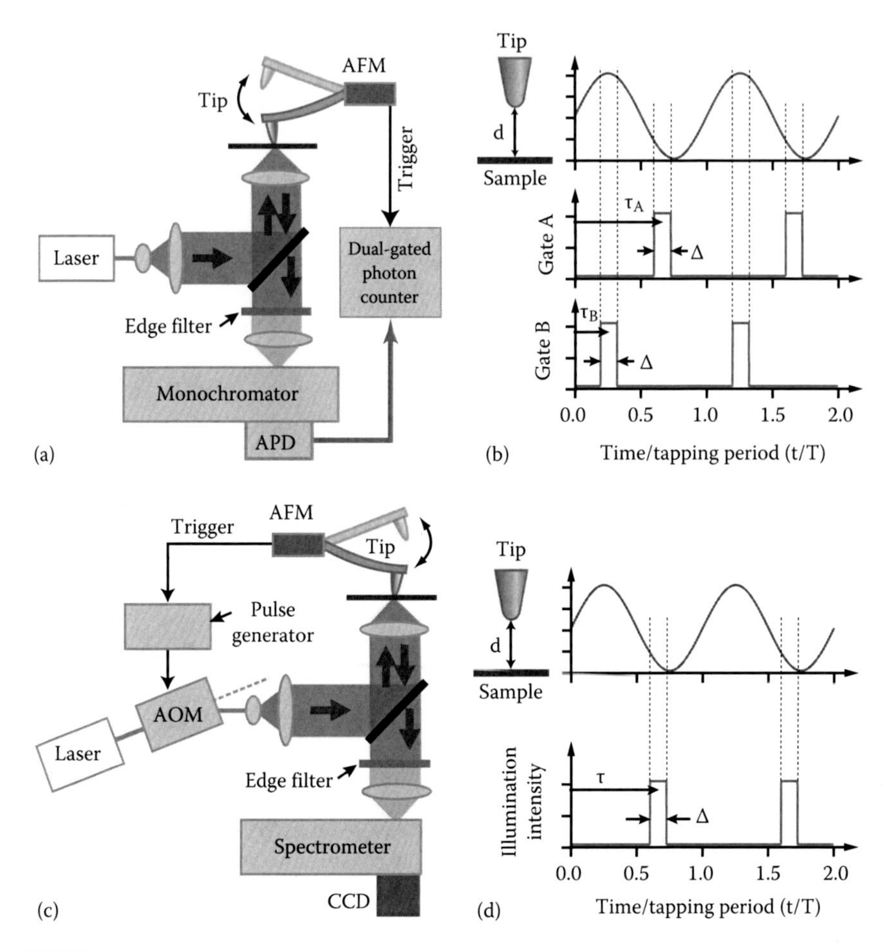

FIGURE 1.11
TERS systems using tapping-mode AFM with time-gated detection (a and b) and time-gated illumination (c and d).

1.4.2 TERS System Using Tapping Mode AFM with Time-Gated Illumination

The first scheme sufficiently meets one's demand when the intensity change is of major interest. However, this scheme uses a single channel detector detecting only Raman signal at a specific wavenumber, enabling us to probe only the intensity changes. Therefore, it is not easy to investigate changes in spectral shape, such as peak frequency shift, emergence of new peaks, and relative intensity change, which are induced by chemical and mechanical interactions between a metal tip and the sample molecules. For this particular purpose, another scheme using a time-gated illumination has been developed (Ichimura et al. 2009). The excitation laser intensity is time-gated in this

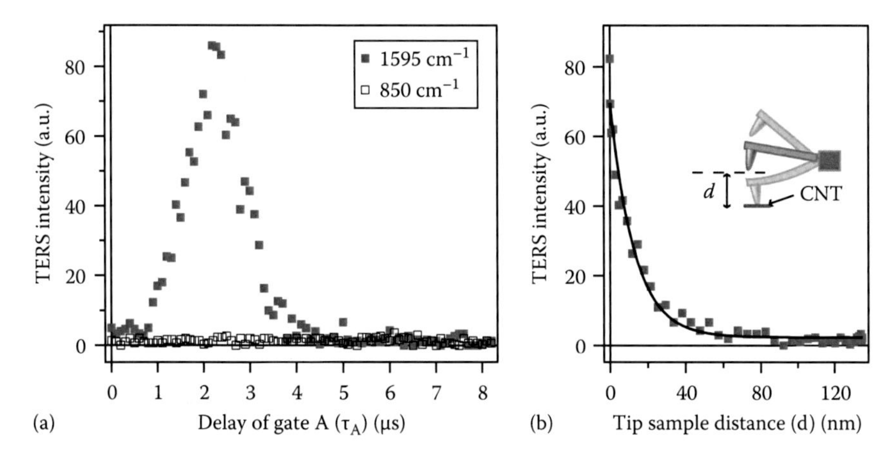

FIGURE 1.12
(a) TERS intensity of SWNTs measured while scanning one of two gates through the entire tapping period (spherical dots: a vibrational frequency of the G band at 1595 cm^{-1}, squared dots: none of vibrational frequency at 850 cm^{-1}). (b) Dependence of tip-SWNTs distance on TERS intensity of the G-band represented from (a). (From Yano, T.A., Ichimura, T., Taguchi, A. et al., Confinement of enhanced field investigated by tip-sample gap regulation in tapping-mode tip-enhanced Raman microscopy, *Appl. Phys. Lett.*, 91(12), 121101. Copyright 2007, American Institute of Physics.)

scheme so that one can record an entire Raman spectrum by a multichannel detector (CCD camera). Figure 1.11c schematically shows the configuration of this scheme. An acousto-optic modulator (AOM) is employed to switch excitation laser intensity so that the tip–sample arrangement is selectively illuminated only for a particular tip–sample distance. Figure 1.11d illustrates the time course of the sinusoidal oscillation of the tip and the synchronized opening of the time-gate. By selecting a particular value of the time delay (τ) of the time gate, one can preselect a desired tip–sample distance. TERS spectra corresponding to a desired tip–sample distance can be recorded with high accuracy. By sweeping the value of τ, one can acquire a distance-dependent dataset of TERS spectra. Figure 1.13 shows the typical tip–sample distance dependence of TERS spectra of adenine nanocrystals. Spectral shift is clearly observed as the distance changes (Hayazawa et al. 2006; Watanabe et al. 2004).

1.5 TERS: Nano-Positioning Stabilization (Lateral Dimension: XY)

Precise adjustment of tip-position relative to the focus spot is crucial for stable TERS measurement. Since tip-enhancement is affected by the tip position inside of the tightly focused laser spot, nanoscale tip-positioning is required

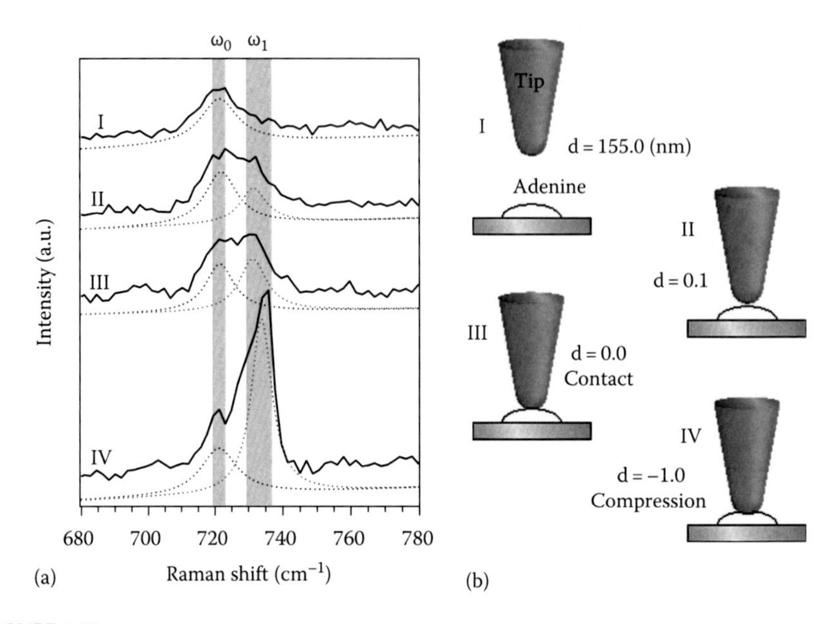

FIGURE 1.13
(a) Distance-dependent TERS spectra of an adenine nano-crystal measured by the time-gated illumination system. (From Ichimura, T., Fujii, S., Verma, P., Yano, T., Inouye, Y., and Kawata, S., *Phys. Rev. Lett.*, 102(18), 186101, 2009. Copyright 2009 by the American Physical Society.)

to gain efficient tip-enhancement. In this chapter, we describe an efficient and practical method to precisely set and stabilize a metallic tip in the focus spot on a nanometer scale.

1.5.1 Precise Tip-Positioning in Laser Focus Spot

In order to realize precise tip-positioning with respect to the tight focus spot, additional piezo scanners are usually installed to raster-scan the tip laterally, which are controlled independently by conventional piezo scanners used for lateral movement of a sample stage in TERS imaging. The precise tip-positioning is done by scanning the metallized tip over the focused spot while detecting the scattered light from the tip as schematically drawn in Figure 1.14a. Figure 1.14b displays the tip-scattering image over a diffraction-limited focus spot of radially polarized light obtained by a silver-coated tip. The intensity profile with a single spot represents the longitudinal electric field distribution, but displays a negative contrast. The positive contrast can be expected when a sufficiently high tip-enhancement is induced (Bouhelier et al. 2003; Hayazawa et al. 2004b). However, most of the metallized tips show the negative contrast except for few. Mapping of the field distribution allows precise positioning of the metallized tip to obtain optimum tip-enhancement effect not only for Raman spectroscopy but also for any other tip-enhanced spectroscopic techniques.

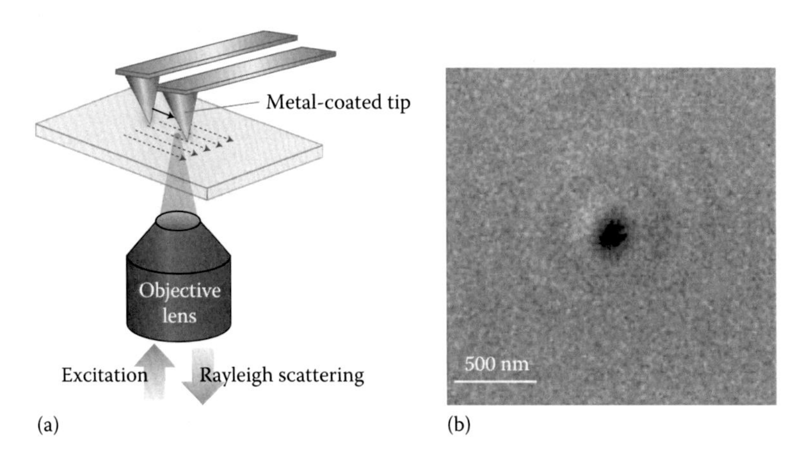

FIGURE 1.14
(a) Schematic of tip-positioning and (b) Rayleigh scattering from the tip measured while raster-scanning a tip over the diffraction-limited spot focused by radially polarized incident light.

1.5.2 Real-Time Compensation of Thermal Drift

Same as many other SPMs, TERS microscopy also suffers from thermal drift due to the shrinkage or expansion of the system under the ambient condition. There have been a number of techniques developed for compensating the thermal drift in SPMs, which have shown excellent results for topographic imaging (Abe et al. 2005; Mantooth et al. 2002; Mokaberi and Requicha 2006; van Noort et al. 1999; Pohl and Moller 1988). These techniques are effective for compensating the thermal drift of relative position between tip and sample. However, in the case of TERS microscopy, the relative position of tip and focus spot is more crucial than that of the tip and sample since even a slight change in the tip position with respect to the focal spot would affect tip-enhancement. Therefore, it is required to in situ compensate the drift, so that the tip stays stationary with respect to the focus spot during the entire TERS measurement period.

Figure 1.15a shows a schematic display of the measurement to optically sense the time-dependent thermal drift of a metallic tip at subnanometer scale and also to compensate the drift in real time (Yano et al. 2012). This system is equipped with a quadrant photodiode (QPD) for sensing position of the tip. Laser light is tightly focused using a high NA objective onto the metal-coated cantilever tip placed on the glass substrate. Tip-enhanced Rayleigh and Raman scattering from the tip-apex are collected by the same objective, and are divided into two optical paths by the edge filter. Tip-enhanced Rayleigh scattering signal is initially focused onto the center of the QPD detector by the tip-positioning technique described in Section 1.5.1. This is to make sure that the tip is positioned in the center of the focus spot. Displacement of the tip with respect to the center of the focus spot is sensed by the difference signals of the QPD, i.e., the normalized difference

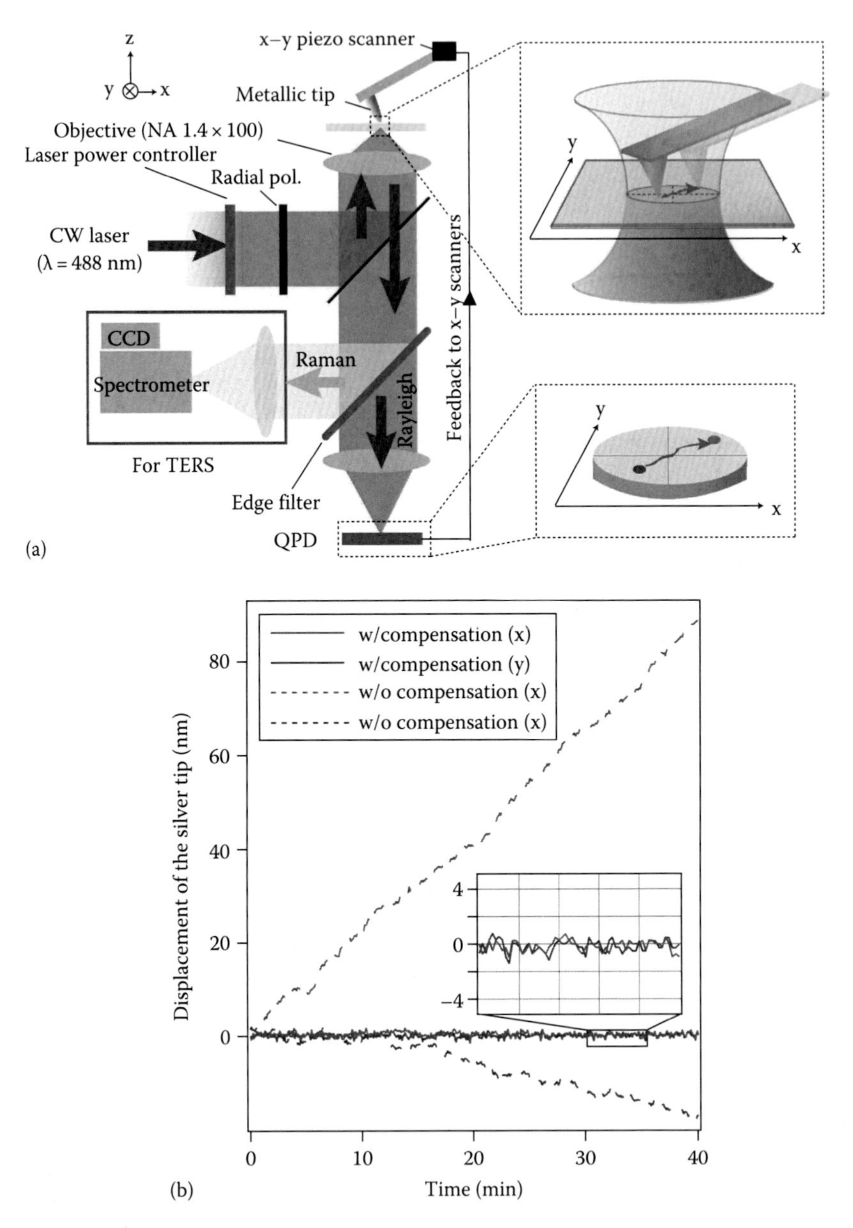

FIGURE 1.15
(a) Schematic of a TERS system for the real-time drift sensing and compensation, (b) time-dependent lateral displacement of the tip with (solid lines) and without (dotted lines) the scheme of drift compensation. (From Yano, T., Ichimura, T., Kuwahara, S., Verma, P., and Kawata, S., Subnanometric stabilization of plasmon-enhanced optical microscopy, *Nanotechnology*, 23(20), 205503, 2012. Copyright 2012, The Institute of Physics.)

signals from the left and right halves of the QPD for sensing displacement in x-direction, and from the top and bottom halves of the QPD for sensing displacement in y-direction. The lateral displacements in x- and y-directions are fed to piezo scanners of the tip in x- and y-directions, respectively, so that the tip is stabilized stationarily at the center of the focus spot.

Figure 1.15b shows a dataset of sensing and compensating time-dependent thermal drift in real time (Yano et al. 2012). The dotted lines in Figure 1.15b show the time-dependent lateral displacement (x- and y-directions) of the silver-coated tip measured through time-dependent change of the tip-enhanced Rayleigh scattering. The tip time-dependently drifted in both x- and y-directions, and exhibited the lateral drift of ~0.5 and ~2.5 nm/min, respectively. This tendency degrades long-time TERS imaging since it usually takes more than 1 h to complete it. However, as shown in the solid lines in Figure 1.15b, when the drift was being compensated in real time with the use of feedback control, the displacement was drastically suppressed. The standard deviation of the displacement during drift-compensation is estimated to be 0.7 nm in both x- and y-directions (see the inset of Figure 1.15b), which enabled us to have subnanometric control over the drift of the tip in the center of the focus spot. This technique enables us to perform robust TERS imaging without any degradation of optical contrast.

1.6 TERS: Nano-Positioning Stabilization (Longitudinal Dimension: Z)

The focus stabilization is crucial for microscopic/nanoscopic measurement at an interface, particularly when scanning a large surface area because there is always a certain amount of the mechanical tilt of the sample substrate, which degrades the contrast of the image. When imaging nanoscopic materials such as carbon nanotubes or silicon nanowires, more stringent nanometric stabilization of the focus position relative to such samples is required; otherwise, it is often difficult to interpret the results from the observation. Moreover, the smaller the sample volume is, the smaller the signal becomes, resulting in a long exposure time at each position. In this sense, long-term stability of the tight focus is essential for both microscopic large area scan and nanosized sample scan (high-resolution/large-area imaging). Moreover, the tip-enhanced spectroscopy requires long-term stability of the relative position of the tip, sample, and the focus position. Based on this fact, a number of autofocusing techniques have been reported (Bravo-Zanoguera et al. 1998; Fein et al. 1990; Mckeogh et al. 1995). However, the achieved position accuracy is 50–80 nm and all of these are based on optical sensing system, in which the light used for sensing itself may become the background source for spectroscopic measurements. In this section, a robust non-optical feedback scheme with ±1 nm position accuracy for virtually unlimited time duration is presented (Hayazawa et al. 2012a). Time-dependent thermal drift of the tight focus and the mechanical tilt of the sample surface were simultaneously sensed by non-optical means and were compensated for in real time. The proposed system is a simple add-on of any kind of conventional optical microscopes using objective lenses. This non-optical and highly precise focus positioning control is of critical importance for microscopic/nanoscopic measurement at an interface and surface-sensitive nonlinear spectroscopy (e.g., second harmonic generation) (Shen 1984) due to its nonlinear dependence on the photon density at the focus. We claim the importance of precise focusing for tip-enhanced spectroscopy (Kawata and Shalaev 2007), in which the strict positioning of the tip relative to the tight focus is an essential requirement for reliable tip-enhancement.

1.6.1 Experimental Details

Focus locking schematic design is depicted in Figure 1.16. The system is based on the tip-enhanced microscopy (Hayazawa et al. 2002, 2004b) consisting of an inverted optical microscope, atomic force microscope (AFM) head, and a capacitive sensor. The capacitive sensor is rigidly attached to the high numerical aperture (NA) objective lens (NA1.49), which is mounted

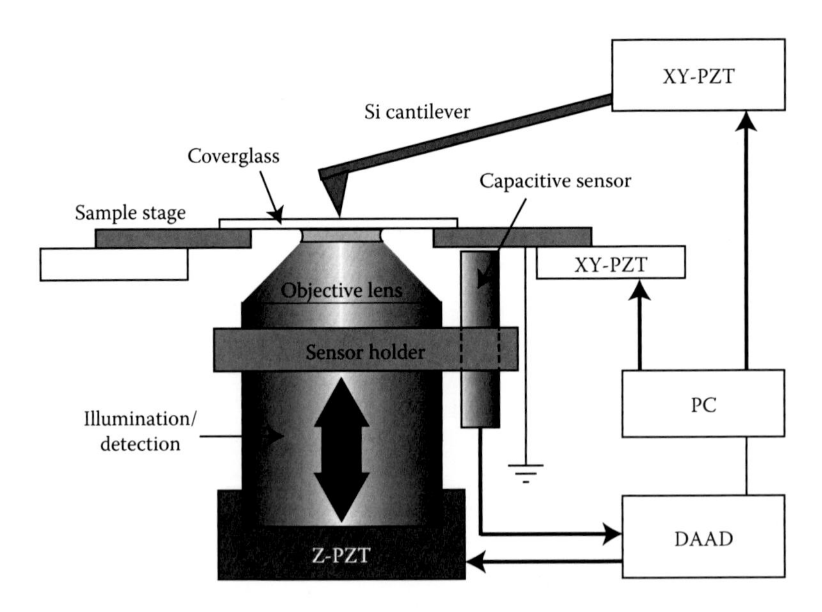

FIGURE 1.16
Schematics of the focus locking system consisting of an objective lens, a sample stage, and a capacitive sensor. The capacitive sensor senses the gap between the sensor electrode and the counter electrode (sample stage).

on a Z-PZT stage. Note that this is a simple add-on to any kind of objective lens; however, it is more effective in high NA objective lens due to its tight focusing power. The gap is sensed by the capacitive sensor and converted to a corresponding DC voltage (low-pass-filter: 1 kHz), which is monitored by an analog-digital (AD) convertor (16 bit resolution). Once the focus position of the objective lens is adjusted, the corresponding DC voltage of the sensor is locked. The displacement of the focus spot along the optical axis due to thermal drift or the tilt of the sample is compensated by applying a feedback voltage to the Z-PZT (DC ~ 1 kHz) via simple *p*-gain for the analog feedback scheme. It should be noted that the surrounding environment of the system is in an ordinary air-conditioned room showing the temperature fluctuation ±0.1°C for long-term; however, the local temporal fluctuation of temperature could generally be even worse because of the flow of air. Figure 1.17 shows the time-dependent gap displacement and the corresponding Z-PZT movement for 24 h. The feedback is switched off after 12 h. We can clearly see the well-stabilized focus position for the first 12 h while showing the PZT movement in order to compensate the thermal drift. The inset shows the close up profile of the dashed area, which clearly shows the stability within ±1 nm by the simple analog feedback. When the feedback is off at 12 h, the PZT movement becomes zero whereas the gap displacement suddenly shows a continuous thermal drift. Based on the observation, the gap displacements of ±300 nm due to the thermal drift are always present in our current experimental condition, which should be compensated. It should be noted that the achieved stability, ±1 nm, is sufficiently smaller than the focal depth of the high NA objective lens so that the signal changes due to defocusing are

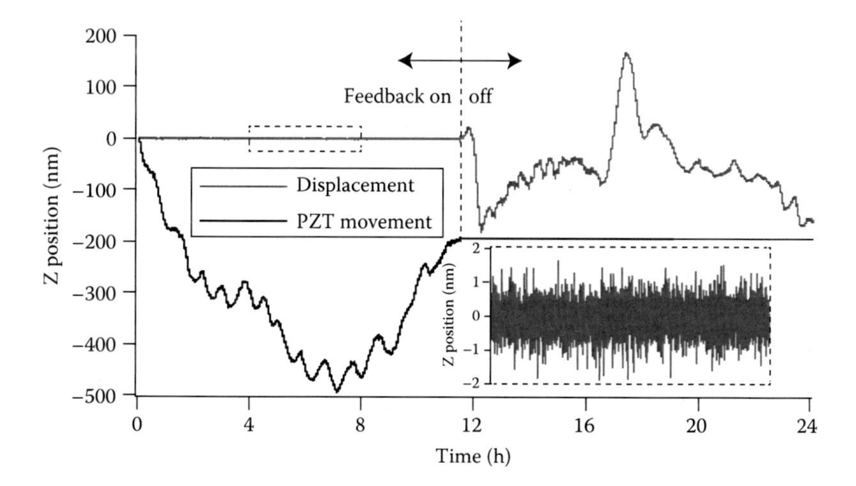

FIGURE 1.17
Time-dependent gap displacement and the corresponding Z-PZT movement for 24 h. The feedback is switched off after 12 h. The inset shows the close up profile of the dashed area with a feedback.

negligible for microscopic observations. It should be noted that scan speed can be further improved by a common PID feedback scheme when there is a sudden change in the z-position during the scan, which is promising for fast microscopic scanning at larger area.

1.6.2 Stability for Microscopy

In order to show the power of nanometric locking of the tight focus for a large-area microscopic imaging, confocal Raman imagings on dispersed single-walled carbon nanotubes (SWNTs) are demonstrated. Figure 1.18 shows the comparison of Raman images (a) with and (b) without feedback at the same scan area. Scanning is done via raster scanning from left bottom along

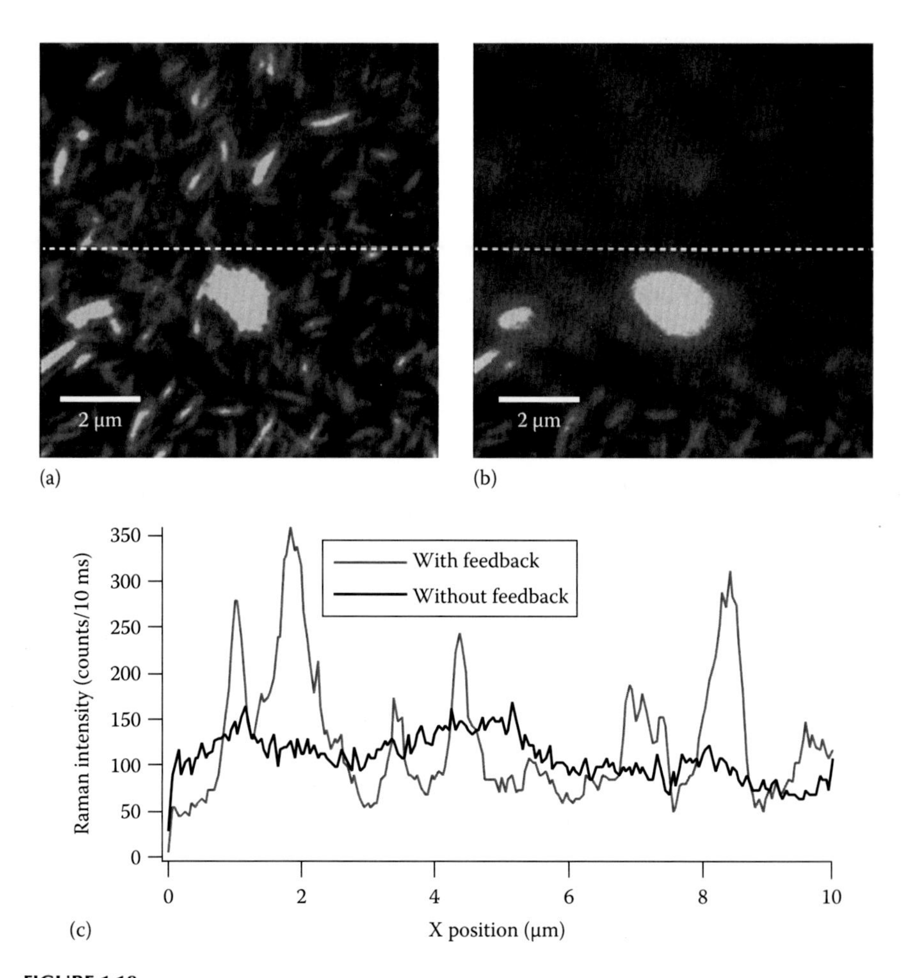

FIGURE 1.18
Confocal micro-Raman images of semiconductive SWNTs (a) with and (b) without a feedback. The cross-sectional profile at the dashed lines are plotted in (c).

the x-axis. Figure 1.18a shows a constantly high contrast throughout the scanning while in Figure 1.18b the contrast becomes obviously degraded during the scan. Figure 1.18c shows the cross-sectional profile of both the images at the dashed line. This is partially due to the thermal drift but mostly to the tilt of the sample substrate. When the scanning area is sufficiently small (such as <2 μm in this sample case), the tilt can be almost negligible; however, this becomes critical for a large-area scanning since it is practically difficult to exclude the mechanical slight tilt of the sample. Indeed, the other way to circumvent the degradation of the image contrast due to the tilt is to calibrate the tilt prior to the imaging. However, when it is combined with the thermal drift, the real-time feedback scheme is required. In most of confocal microscopic study, imaging is often applied on three-dimensional and biological samples. In this case, such tilt is not at all the issue; however, when applied to the surface/interface study, the locking the focus at the exact surface is essential for a reliable data acquisition.

1.6.3 Stability for Tip-Enhanced Nanoscopy

In addition to the microscopic study of the surface/interface, nanoscopic observation in an optical microscope imposes even more stringent requirement for positioning. Particularly in the case of tip-enhanced nanoscopy, the tip has to be precisely positioned onto the tight focus (Bouhelier et al. 2003; Hayazawa et al. 2004b) in order to induce an efficient tip-enhancement due to SPP excitations at the metallic tip-apex (Kawata 2001). The lateral position of the tip relative to the focused spot/sample is another important factor for tip-enhancement, for which we can employ a number of effective techniques for compensating the tip drift in SPMs under ambient conditions (Abe et al. 2005; King et al. 2009) including the discussion in Section 1.5, as well. However, when defocused due to tilt or thermal drift similar to the microscopic case, the tip-enhancement can be drastically decreased because it requires not only the tip to be positioned within the focused spot area but also at a position where there is a strong longitudinal component of the electric field within the focused spot (Bouhelier et al. 2003; Hayazawa et al. 2004b). Because of this requirement, radially polarized beam focused by a high NA objective lens is becoming a promising approach to induce the strong longitudinal field (Anderson et al. 2006; Hayazawa et al. 2004b). In this case, the focused spot is even becoming tighter/smaller than the case of linearly polarized beam (Quabis et al. 2000) so that the positioning stability becomes further strict. Moreover, the slight defocusing may induce strong reduction of the longitudinal field at the center of the focused spot since the strong longitudinal field at the center is the result of constructive interference of all the *p*-polarized components. Accordingly, nanometric locking of the tight focus at the sample surface is of critical importance not only for a high spatial resolution but also for a high tip-enhancement effect. The Rayleigh scattering from the tip and the surface can monitor the focused spot profile, particularly for longitudinal

FIGURE 1.19
Schematic of a silicon cantilever tip on a tight focused spot. Calculated longitudinal electric field profile at the diffraction limit is also plotted. The tip is adjusted at the peak position of the longitudinal field for the best tip-enhancement effect.

electric field since the image profile reflects the longitudinal field profile as it is utilized for feedback of lateral displacement in Section 1.5. When the tip is within the focused spot, the longitudinal field polarized parallel to the tip axis is efficiently scattered to the free space so that the monitored Rayleigh scattering intensity will be decreased (Bouhelier et al. 2003; Hayazawa et al. 2004a, 2009b). This configuration is schematically shown in Figure 1.19. The calculated longitudinal field profile of the diffraction-limited spot is also in the figure. The tip should be adjusted onto the center of the focused spot showing the maximum of the longitudinal field. Thus, it is of essential importance to keep the correct focus with a nanometric precision for a long term because nanoscopic imaging/spectroscopy generally takes much longer acquisition time than microscopic imaging/spectroscopy due to the smaller number of molecules in the nanoscale volume.

Figure 1.20a shows the imaging of Rayleigh scattering intensity when the tip is scanned over the focused spot with several z-positions. Figure 1.20b is the calculated longitudinal field profile. As you can see, even z-position is varied by ±300 nm, each image contrast is quite different and the best contrast is indeed obtained at the right focus. Figure 1.20c is the cross-sectional profile at the center, which clearly reflects the inverted contrast between Rayleigh scattering intensity and longitudinal field. The higher the longitudinal field is, the deeper the dip becomes at the center. This dip contrast at the right focus can be kept for more than 24 h with the feedback system. This is the proof that the same level of tip-enhancement can be expected for virtually unlimited time duration and is important not only for single-point tip-enhanced spectroscopy but also for tip-enhanced spectroscopic imaging (Hayazawa et al. 2012b).

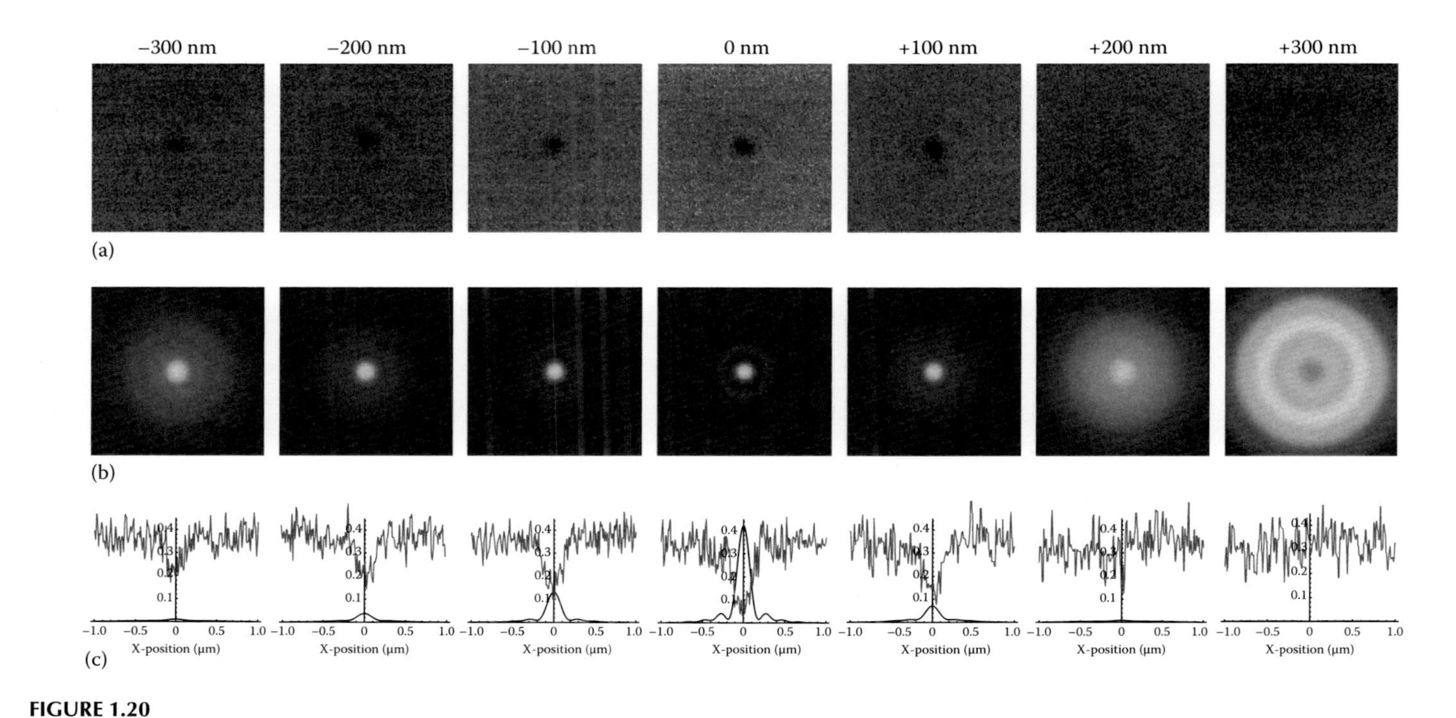

FIGURE 1.20
(a) Intensity of Rayleigh scattering from a silicon cantilever tip scanned over a focused spot as a function of axial focus positions varied from −300 to +300 nm. 0 nm corresponds to the exact focus position for a diffraction limit. "−" and "+" correspond to smaller and larger gap between the glass surface and the lens relative to the exact focus position. (b) Calculated longitudinal field at the tight focus just above the cover glass surface. The contrast is normalized for each image. (c) Cross-sectional profile of (a) gray lines and (b) black lines.

References

Abe, M., Y. Sugimoto, O. Custance, and S. Morita. 2005. Atom tracking for reproducible force spectroscopy at room temperature with non-contact atomic force microscopy. *Nanotechnology* 16(12):3029–3034.

Anderson, M. S. 2000. Locally enhanced Raman spectroscopy with an atomic force microscope. *Applied Physics Letters* 76(21):3130–3132.

Anderson, N., A. Bouhelier, and L. Novotny. 2006. Near-field photonics: Tip-enhanced microscopy and spectroscopy on the nanoscale. *Journal of Optics A—Pure and Applied Optics* 8(4):S227–S233.

Anderson, N., A. Hartschuh, and L. Novotny. 2007. Chirality changes in carbon nanotubes studied with near-field Raman spectroscopy. *Nano Letters* 7(3):577–582.

Azoulay, J., A. Debarre, A. Richard, and P. Tchenio. 1999. Field enhancement and apertureless near-field optical spectroscopy of single molecules. *Journal of Microscopy-Oxford* 194:486–490.

Bailo, E. and V. Deckert. 2008. Tip-enhanced Raman spectroscopy of single RNA strands: Towards a novel direct-sequencing method. *Angewandte Chemie-International Edition* 47(9):1658–1661.

Berweger, S., C. C. Neacsu, Y. B. Mao, H. J. Zhou, S. S. Wong, and M. B. Raschke. 2009. Optical nanocrystallography with tip-enhanced phonon Raman spectroscopy. *Nature Nanotechnology* 4(8):496–499.

Berweger, S. et al. 2012. Light on the tip of a needle: Plasmonic nanofocusing for spectroscopy on the nanoscale. *The Journal of Physical Chemistry Letters* 3, 945.

Betzig, E. and R. J. Chichester. 1993. Single molecules observed by near-field scanning optical microscopy. *Science* 262(5138):1422–1425.

Bohme, R., D. Cialla, M. Richter, P. Rosch, J. Popp, and V. Deckert. 2010. Biochemical imaging below the diffraction limit—Probing cellular membrane related structures by tip-enhanced Raman spectroscopy (TERS). *Journal of Biophotonics* 3(7):455–461.

Bohme, R., M. Mkandawire, U. Krause-Buchholz et al. 2011. Characterizing cytochrome c states—TERS studies of whole mitochondria. *Chemical Communications* 47(41):11453–11455.

Bohme, R., M. Richter, D. Cialla, P. Rosch, V. Deckert, and J. Popp. 2009. Towards a specific characterisation of components on a cell surface—Combined TERS-investigations of lipids and human cells. *Journal of Raman Spectroscopy* 40(10):1452–1457.

Bouhelier, A., M. R. Beversluis, and L. Novotny. 2003. Near-field scattering of longitudinal fields. *Applied Physics Letters* 82(25):4596–4598.

Bravo-Zanoguera, M., B. Von Massenbach, A. L. Kellner, and J. H. Price. 1998. High-performance autofocus circuit for biological microscopy. *Review of Scientific Instruments* 69(11):3966–3977.

Brodard, P., M. Bechelany, L. Philippe, and J. Michler. 2012. Synthesis and attachment of silver nanowires on atomic force microscopy cantilevers for tip-enhanced Raman spectroscopy. *Journal of Raman Spectroscopy* 43(6):745–749.

Budich, C., U. Neugebauer, J. Popp, and V. Deckert. 2008. Cell wall investigations utilizing tip-enhanced Raman scattering. *Journal of Microscopy-Oxford* 229(3):533–539.

Carlson, C. A. and J. C. Woehl. 2008. Fabrication of optical tips from photonic crystal fibers. *Review of Scientific Instruments* 79(10):103707.

Chang, R. K. and T. E. Furtak. 1982. *Surface Enhanced Raman Scattering*. New York: Plenum Press.

Chen, J. Y., B. J. Wiley, and Y. N. Xia. 2007. One-dimensional nanostructures of metals: Large-scale synthesis and some potential applications. *Langmuir* 23(8):4120–4129.

Cialla, D., T. Deckert-Gaudig, C. Budich et al. 2009. Raman to the limit: Tip-enhanced Raman spectroscopic investigations of a single tobacco mosaic virus. *Journal of Raman Spectroscopy* 40(3):240–243.

De Angelis, F., G. Das, P. Candeloro et al. 2010. Nanoscale chemical mapping using three-dimensional adiabatic compression of surface plasmon polaritons. *Nature Nanotechnology* 5(1):67–72.

Domke, K. F., D. Zhang, and B. Pettinger. 2007. Tip-enhanced Raman spectra of picomole quantities of DNA nucleobases at Au(111). *Journal of the American Chemical Society* 129(21):6708–6709.

Dorfer, T., M. Schmitt, and J. Popp. 2007. Deep-UV surface-enhanced Raman scattering. *Journal of Raman Spectroscopy* 38(11):1379–1382.

Eligal, L., F. Culfaz, V. McCaughan, N. I. Cade, and D. Richards. 2009. Etching gold tips suitable for tip-enhanced near-field optical microscopy. *Review of Scientific Instruments* 80(3):033701.

Farahani, J. N., D. W. Pohl, H. J. Eisler, and B. Hecht. 2005. Single quantum dot coupled to a scanning optical antenna: A tunable superemitter. *Physical Review Letters* 95(1):017402.

Fein, M. E., H. F. Kelderman, A. P. Neukermans, A. E. Loh, and D. Wolze. 1990. A new precise optical autofocus system. *Journal of Vacuum Science & Technology B* 8(6):2017–2022.

Feng, H. J., Y. M. Yang, Y. M. You et al. 2009. Simple and rapid synthesis of ultrathin gold nanowires, their self-assembly and application in surface-enhanced Raman scattering. *Chemical Communications* (15):1984–1986.

Furusawa, K., N. Hayazawa, F. C. Catalan, T. Okamoto, and S. Kawata. 2012. Tip-enhanced broadband CARS spectroscopy and imaging using a photonic crystal fiber based broadband light source. *Journal of Raman Spectroscopy* 43(5):656–661.

Georgi, C. and A. Hartschuh. 2010. Tip-enhanced Raman spectroscopic imaging of localized defects in carbon nanotubes. *Applied Physics Letters* 97(14):143117.

Gerton, J. M., L. A. Wade, G. A. Lessard, Z. Ma, and S. R. Quake. 2004. Tip-enhanced fluorescence microscopy at 10 nanometer resolution. *Physical Review Letters* 93(18):180801.

Graff, A., D. Wagner, H. Ditlbacher, and U. Kreibig. 2005. Silver nanowires. *European Physical Journal D* 34(1–3):263–269.

Hamann, H. F., A. Gallagher, and D. J. Nesbitt. 2000. Near-field fluorescence imaging by localized field enhancement near a sharp probe tip. *Applied Physics Letters* 76(14):1953–1955.

Hanaguri, T., C. Lupien, Y. Kohsaka et al. 2004. A "checkerboard" electronic crystal state in lightly hole-doped $Ca_{2-x}Na_xCuO_2Cl_2$. *Nature* 430(7003):1001–1005.

Hartschuh, A., E. J. Sanchez, X. S. Xie, and L. Novotny. 2003. High-resolution near-field Raman microscopy of single-walled carbon nanotubes. *Physical Review Letters* 90(9):095503.

Hayazawa, N., K. Furusawa, and S. Kawata. 2012a. Nanometric locking of the tight focus for optical microscopy and tip-enhanced microscopy. *Nanotechnology* 23(465203):5203.

Hayazawa, N., K. Furusawa, A. Taguchi, and S. Kawata. 2009a. One-photon and two-photon excited fluorescence microscopies based on polarization-control: Applications to tip-enhanced microscopy. *Journal of Applied Physics* 106(11):113103.

Hayazawa, N., K. Furusawa, A. Taguchi, S. Kawata, and H. Abe. 2009b. Tip-enhanced two-photon excited fluorescence microscopy with a silicon tip. *Applied Physics Letters* 94(19):193112.

Hayazawa, N., T. Ichimura, M. Hashimoto, Y. Inouye, and S. Kawata. 2004a. Amplification of coherent anti-Stokes Raman scattering by a metallic nanostructure for a high resolution vibration microscopy. *Journal of Applied Physics* 95(5):2676–2681.

Hayazawa, N., Y. Inouye, and S. Kawata. 1999. Evanescent field excitation and measurement of dye fluorescence in a metallic probe near-field scanning optical microscope. *Journal of Microscopy-Oxford* 194:472–476.

Hayazawa, N., Y. Inouye, Z. Sekkat, and S. Kawata. 2000. Metallized tip amplification of near-field Raman scattering. *Optics Communications* 183(1–4):333–336.

Hayazawa, N., Y. Inouye, Z. Sekkat, and S. Kawata. 2001. Near-field Raman scattering enhanced by a metallized tip. *Chemical Physics Letters* 335(5–6):369–374.

Hayazawa, N., M. Motohashi, Y. Saito et al. 2007. Visualization of localized strain of a crystalline thin layer at the nanoscale by tip-enhanced Raman spectroscopy and microscopy. *Journal of Raman Spectroscopy* 38(6):684–696.

Hayazawa, N., Y. Saito, and S. Kawata. 2004b. Detection and characterization of longitudinal field for tip-enhanced Raman spectroscopy. *Applied Physics Letters* 85(25):6239–6241.

Hayazawa, N., A. Tarun, Y. Inouye, and S. Kawata. 2002. Near-field enhanced Raman spectroscopy using side illumination optics. *Journal of Applied Physics* 92(12):6983–6986.

Hayazawa, N., H. Watanabe, Y. Saito, and S. Kawata. 2006. Towards atomic site-selective sensitivity in tip-enhanced Raman spectroscopy. *Journal of Chemical Physics* 125(24):244705.

Hayazawa, N., T. Yano, and S. Kawata. 2012b. Highly reproducible tip-enhanced Raman scattering using an oxidized and metallized silicon cantilever tip as a tool for everyone. *Journal of Raman Spectroscopy* 43(9):1177–1182.

Hayazawa, N., T. Yano, H. Watanabe, Y. Inouye, and S. Kawata. 2003. Detection of an individual single-wall carbon nanotube by tip-enhanced near-field Raman spectroscopy. *Chemical Physics Letters* 376(1–2):174–180.

Hermann, P., A. Hermelink, V. Lausch et al. 2011. Evaluation of tip-enhanced Raman spectroscopy for characterizing different virus strains. *Analyst* 136(6):1148–1152.

Hillenbrand, R., T. Taubner, and F. Keilmann. 2002. Phonon-enhanced light–matter interaction at the nanometre scale. *Nature* 418(6894):159–162.

Huo, Z. Y., C. K. Tsung, W. Y. Huang, X. F. Zhang, and P. D. Yang. 2008. Sub-two nanometer single crystal Au nanowires. *Nano Letters* 8(7):2041–2044.

Ichimura, T., S. Fujii, P. Verma, T. Yano, Y. Inouye, and S. Kawata. 2009. Subnanometric near-field Raman investigation in the vicinity of a metallic nanostructure. *Physical Review Letters* 102(18):186101.

Ichimura, T., N. Hayazawa, M. Hashimoto, Y. Inouye, and S. Kawata. 2004a. Application of tip-enhanced microscopy for nonlinear Raman spectroscopy. *Applied Physics Letters* 84(10):1768–1770.

Ichimura, T., N. Hayazawa, M. Hashimoto, Y. Inouye, and S. Kawata. 2004b. Tip-enhanced coherent anti-Stokes Raman scattering for vibrational nanoimaging. *Physical Review Letters* 92(22):220801.

Ichimura, T., H. Watanabe, Y. Morita, P. Verma, S. Kawata, and Y. Inouye. 2007. Temporal fluctuation of tip-enhanced Raman spectra of adenine molecules. *Journal of Physical Chemistry C* 111(26):9460–9464.

Johnson, P. B. and R. W. Christy. 1972. Optical-constants of noble-metals. *Physical Review B* 6(12):4370–4379.

Johnson, T. W., Z. J. Lapin, R. Beams et al. 2012. Highly reproducible near-field optical imaging with sub-20-nm resolution based on template-stripped gold pyramids. *ACS Nano* 6(10):9168–9174.

Kambhampati, P., C. M. Child, M. C. Foster, and A. Campion. 1998. On the chemical mechanism of surface enhanced Raman scattering: Experiment and theory. *Journal of Chemical Physics* 108(12):5013–5026.

Kawata, S. 2001. *Near-Field Optics and Surface Plasmon Polaritons, Topics in Applied Physics*. New York: Springer.

Kawata, S. and V. M. Shalaev. 2007. *Tip Enhancement, Advances in Nano-Optics and Nano-Photonics*. Amsterdam, the Netherlands: Elsevier.

Kim, Y., T. Komeda, and M. Kawai. 2002. Single-molecule reaction and characterization by vibrational excitation. *Physical Review Letters* 89(12):126104.

King, G. M., A. R. Carter, A. B. Churnside, L. S. Eberle, and T. T. Perkins. 2009. Ultrastable atomic force microscopy: Atomic-scale stability and registration in ambient conditions. *Nano Letters* 9(4):1451–1456.

Knoll, B. and F. Keilmann. 1999. Near-field probing of vibrational absorption for chemical microscopy. *Nature* 399(6732):134–137.

Lee, N., R. D. Hartschuh, D. Mehtani et al. 2007. High contrast scanning nano-Raman spectroscopy of silicon. *Journal of Raman Spectroscopy* 38(6):789–796.

Lombardi, J. R. and R. L. Birke. 2008. A unified approach to surface-enhanced Raman spectroscopy. *Journal of Physical Chemistry C* 112(14):5605–5617.

Mantooth, B. A., Z. J. Donhauser, K. F. Kelly, and P. S. Weiss. 2002. Cross-correlation image tracking for drift correction and adsorbate analysis. *Review of Scientific Instruments* 73(2):313–317.

Matsui, R., P. Verma, T. Ichimura, Y. Inouye, and S. Kawata. 2007. Nanoanalysis of crystalline properties of GaN thin film using tip-enhanced Raman spectroscopy. *Applied Physics Letters* 90(6):061906.

Mckeogh, L. F., J. P. Sharpe, and K. M. Johnson. 1995. A low-cost automatic translation and autofocusing system for a microscope. *Measurement Science & Technology* 6(5):583–587.

Mehtani, D., N. Lee, R. D. Hartschuh et al. 2006. Optical properties and enhancement factors of the tips for apertureless near-field optics. *Journal of Optics A—Pure and Applied Optics* 8(4):S183–S190.

Mokaberi, B. and A. A. G. Requicha. 2006. Drift compensation for automatic nanomanipulation with scanning probe microscopes. *IEEE Transactions on Automation Science and Engineering* 3(3):199–207.

Motohashi, M., N. Hayazawa, A. Tarun, and S. Kawata. 2008. Depolarization effect in reflection-mode tip-enhanced Raman scattering for Raman active crystals. *Journal of Applied Physics* 103(3):034309.

Nabais, C., R. Schneider, C. Bellouard, J. Lambert, P. Willmann, and D. Billaud. 2009. A new method for the size- and shape-controlled synthesis of lead nanostructures. *Materials Chemistry and Physics* 117(1):268–275.

Neacsu, C. C., G. A. Steudle, and M. B. Raschko. 2005. Plasmonic light scattering from nanoscopic metal tips. *Applied Physics B—Lasers and Optics* 80(3):295–300.

Neugebauer, U., U. Schmid, K. Baumann et al. 2007. Towards a detailed understanding of bacterial metabolism—Spectroscopic characterization of *Staphylococcus epidermidis*. *ChemPhysChem* 8(1):124–137.

Nie, S. M. and S. R. Emery. 1997. Probing single molecules and single nanoparticles by surface-enhanced Raman scattering. *Science* 275(5303):1102–1106.

Ogawa, Y., Y. Yuasa, F. Minami, and S. Oda. 2011. Tip-enhanced Raman mapping of a single Ge nanowire. *Applied Physics Letters* 99(5):053112.

Opilik, L., T. Bauer, T. Schmid, J. Stadler, and R. Zenobi. 2011. Nanoscale chemical imaging of segregated lipid domains using tip-enhanced Raman spectroscopy. *Physical Chemistry Chemical Physics* 13(21):9978–9981.

Otto, A., I. Mrozek, H. Grabhorn, and W. Akemann. 1992. Surface-enhanced Raman-scattering. *Journal of Physics—Condensed Matter* 4(5):1143–1212.

Palik, E. D. 1991. *Handbook of Optical Constants of Solids II*. Boston, MA: Academic Press.

Pan, D. H., N. Klymyshyn, D. H. Hu, and H. P. Lu. 2006. Tip-enhanced near-field Raman spectroscopy probing single dye-sensitized TiO_2 nanoparticles. *Applied Physics Letters* 88(9):093121.

Pettinger, B., G. Picardi, R. Schuster, and G. Ertl. 2000. Surface enhanced Raman spectroscopy: Towards single molecular spectroscopy. *Electrochemistry* 68(12):942–949.

Pettinger, B., G. Picardi, R. Schuster, and G. Ertl. 2002. Surface-enhanced and STM-tip-enhanced Raman spectroscopy at metal surfaces. *Single Molecules* 3(5–6):285–294.

Poborchii, V., T. Tada, and T. Kanayama. 2005. Subwavelength-resolution Raman microscopy of si structures using metal-particle-topped AFM probe. *Japanese Journal of Applied Physics Part 2—Letters & Express Letters* 44(1–7):L202–L204.

Pohl, D. W. and R. Moller. 1988. Tracking tunneling microscopy. *Review of Scientific Instruments* 59(6):840–842.

Quabis, S., R. Dorn, M. Eberler, O. Glockl, and G. Leuchs. 2000. Focusing light to a tighter spot. *Optics Communications* 179(1–6):1–7.

Ren, B., G. Picardi, and B. Pettinger. 2004. Preparation of gold tips suitable for tip-enhanced Raman spectroscopy and light emission by electrochemical etching. *Review of Scientific Instruments* 75(4):837–841.

Richter, M., M. Hedegaard, T. Deckert-Gaudig, P. Lampen, and V. Deckert. 2011. Laterally resolved and direct spectroscopic evidence of nanometer-sized lipid and protein domains on a single cell. *Small* 7(2):209–214.

Ropers, C., C. C. Neacsu, T. Elsaesser, M. Albrecht, M. B. Raschke, and C. Lienau. 2007. Grating-coupling of surface plasmons onto metallic tips: A nanoconfined light source. *Nano Letters* 7(9):2784–2788.

Saito, Y. et al. 2008. Stress imaging of semiconductor surface by tip-enhanced Raman spectroscopy. *Journal of Microscopy* 229, 217–222.

Saito, Y., M. Motohashi, N. Hayazawa, M. Iyoki, and S. Kawata. 2006a. Nanoscale characterization of strained silicon by tip-enhanced Raman spectroscope in reflection mode. *Applied Physics Letters* 88(14):143109.

Saito, Y., P. Verma, K. Masui, Y. Inouye, and S. Kawata. 2009. Nano-scale analysis of graphene layers by tip-enhanced near-field Raman spectroscopy. *Journal of Raman Spectroscopy* 40(10):1434–1440.

Saito, Y., K. Yanagi, N. Hayazawa et al. 2006b. Vibrational analysis of organic molecules encapsulated in carbon nanotubes by tip-enhanced Raman spectroscopy. *Japanese Journal of Applied Physics Part 1—Regular Papers Brief Communications & Review Papers* 45(12):9286–9289.

Sanchez, E. J., J. T. Krug, and X. S. Xie. 2002. Ion and electron beam assisted growth of nanometric SimOn structures for near-field microscopy. *Review of Scientific Instruments* 73(11):3901–3907.

Sanchez, E. J., L. Novotny, and X. S. Xie. 1999. Near-field fluorescence microscopy based on two-photon excitation with metal tips. *Physical Review Letters* 82(20):4014–4017.

Shen, Y. R. 1984. *The Principles of Nonlinear Optics*. New York: John Wiley & Sons.

Stadler, J., T. Schmid, and R. Zenobi. 2011. Nanoscale chemical imaging of single-layer graphene. *ACS Nano* 5(10):8442–8448.

Stipe, B. C., M. A. Rezaei, and W. Ho. 1998. Single-molecule vibrational spectroscopy and microscopy. *Science* 280(5370):1732–1735.

Stockle, R. M., Y. D. Suh, V. Deckert, and R. Zenobi. 2000. Nanoscale chemical analysis by tip-enhanced Raman spectroscopy. *Chemical Physics Letters* 318(1–3): 131–136.

Sun, W. X. and Z. X. Shen. 2003. Near-field scanning Raman microscopy using apertureless probes. *Journal of Raman Spectroscopy* 34(9):668–676.

Sun, Y. G. and Y. N. Xia. 2002. Large-scale synthesis of uniform silver nanowires through a soft, self-seeding, polyol process. *Advanced Materials* 14(11):833–837.

Taguchi, A., N. Hayazawa, K. Furusawa, H. Ishitobi, and S. Kawata. 2009a. Deep-UV tip-enhanced Raman scattering. *Journal of Raman Spectroscopy* 40(9): 1324–1330.

Taguchi, A., N. Hayazawa, Y. Saito, H. Ishitobi, A. Tarun, and S. Kawata. 2009b. Controlling the plasmon resonance wavelength in metal-coated probe using refractive index modification. *Optics Express* 17(8):6509–6518.

Taguchi, A., Y. Saito, K. Watanabe, S. Yijian, and S. Kawata. 2012. Tailoring plasmon resonances in the deep-ultraviolet by size-tunable fabrication of aluminum nanostructures. *Applied Physics Letters* 101(8):081110.

Tang, Z. Y. and N. A. Kotov. 2005. One-dimensional assemblies of nanoparticles: Preparation, properties, and promise. *Advanced Materials* 17(8):951–962.

Tarun, A., N. Hayazawa, H. Ishitobi, S. Kawata, M. Reiche, and O. Moutanabbir. 2011. Mapping the "forbidden" transverse-optical phonon in single strained silicon (100) nanowire. *Nano Letters* 11(11):4780–4788.

Tarun, A., N. Hayazawa, and S. Kawata. 2009. Tip-enhanced Raman spectroscopy for nanoscale strain characterization. *Analytical and Bioanalytical Chemistry* 394(7):1775–1785.

van Noort, S. J. T., K. O. van der Werf, B. G. de Grooth, and J. Greve. 1999. High speed atomic force microscopy of biomolecules by image tracking. *Biophysical Journal* 77(4):2295–2303.

Verma, P., K. Yamada, H. Watanabe, Y. Inouye, and S. Kawata. 2006. Near-field Raman scattering investigation of tip effects on C-60 molecules. *Physical Review B* 73(4):045416.

Wang, C. G., Y. Chen, T. T. Wang, Z. F. Ma, and Z. M. Su. 2007. Biorecognition-driven self-assembly of gold nanorods: A rapid and sensitive approach toward antibody sensing. *Chemistry of Materials* 19(24):5809–5811.

Watanabe, H., Y. Ishida, N. Hayazawa, Y. Inouye, and S. Kawata. 2004. Tip-enhanced near-field Raman analysis of tip-pressurized adenine molecule. *Physical Review B* 69(15):155418.

Wildoer, J. W. G., L. C. Venema, A. G. Rinzler, R. E. Smalley, and C. Dekker. 1998. Electronic structure of atomically resolved carbon nanotubes. *Nature* 391(6662):59–62.

Wiley, B., Y. G. Sun, J. Y. Chen et al. 2005a. Shape-controlled synthesis of silver and gold nanostructures. *Mrs Bulletin* 30(5):356–361.

Wiley, B., Y. G. Sun, B. Mayers, and Y. N. Xia. 2005b. Shape-controlled synthesis of metal nanostructures: The case of silver. *Chemistry—A European Journal* 11(2):454–463.

Xie, X. S. and R. C. Dunn. 1994. Probing single-molecule dynamics. *Science* 265(5170):361–364.

Yang, T. J., G. A. Lessard, and S. R. Quake. 2000. An apertureless near-field microscope for fluorescence imaging. *Applied Physics Letters* 76(3):378–380.

Yano, T., T. Ichimura, S. Kuwahara, P. Verma, and S. Kawata. 2012. Subnanometric stabilization of plasmon-enhanced optical microscopy. *Nanotechnology* 23(20):205503.

Yano, T., P. Verma, S. Kawata, and Y. Inouye. 2006b. Diameter-selective near-field Raman analysis and imaging of isolated carbon nanotube bundles. *Applied Physics Letters* 88(9):093125.

Yano, T., P. Verma, Y. Saito, T. Ichimura, and S. Kawata. 2009. Pressure-assisted tip-enhanced Raman imaging at a resolution of a few nanometres. *Nature Photonics* 3(8):473–477.

Yano, T. A., T. Ichimura, A. Taguchi et al. 2007. Confinement of enhanced field investigated by tip–sample gap regulation in tapping-mode tip-enhanced Raman microscopy. *Applied Physics Letters* 91(12):121101.

Yano, T. A., Y. Inouye, and S. Kawata. 2006a. Nanoscale uniaxial pressure effect of a carbon nanotube bundle on tip-enhanced near-field Raman spectra. *Nano Letters* 6(6):1269–1273.

Yeo, B. S., S. Madler, T. Schmid, W. H. Zhang, and R. Zenobi. 2008. Tip-enhanced Raman spectroscopy can see more: The case of cytochrome C. *Journal of Physical Chemistry C* 112(13):4867–4873.

Yeo, B. S., W. H. Zhang, C. Vannier, and R. Zenobi. 2006. Enhancement of Raman signals with silver-coated tips. *Applied Spectroscopy* 60(10):1142–1147.

You, Y. M., N. A. Purnawirman, H. L. Hu et al. 2010. Tip-enhanced Raman spectroscopy using single-crystalline Ag nanowire as tip. *Journal of Raman Spectroscopy* 41(10):1156–1162.

Zhang, C., B. Gao, L. G. Chen et al. 2011. Fabrication of silver tips for scanning tunneling microscope induced luminescence. *Review of Scientific Instruments* 82(8):083101.

Zhang, D., K. F. Domke, and B. Pettinger. 2010a. Tip-enhanced Raman spectroscopic studies of the hydrogen bonding between adenine and thymine adsorbed on Au (111). *ChemPhysChem* 11(8):1662–1665.

Zhang, D., U. Heinemeyer, C. Stanciu et al. 2010b. Nanoscale spectroscopic imaging of organic semiconductor films by plasmon-polariton coupling. *Physical Review Letters* 104(5):056601.

Zhang, D., X. Wang, K. Braun et al. 2009. Parabolic mirror-assisted tip-enhanced spectroscopic imaging for non-transparent materials. *Journal of Raman Spectroscopy* 40(10):1371–1376.

Zhang, J. H., H. Y. Liu, Z. L. Wang, and N. B. Ming. 2007. Shape-selective synthesis of gold nanoparticles with controlled sizes, shapes, and plasmon resonances. *Advanced Functional Materials* 17(16):3295–3303.

Zhang, W. H., T. Schmid, B. S. Yeo, and R. Zenobi. 2008. Near-field heating, annealing, and signal loss in tip-enhanced Raman spectroscopy. *Journal of Physical Chemistry C* 112(6):2104–2108.

Zhu, L., C. Georgi, M. Hecker et al. 2007. Nano-Raman spectroscopy with metallized atomic force microscopy tips on strained silicon structures. *Journal of Applied Physics* 101(10):104305-1.

2

Micro- and Nanoscale Structures/ Systems and Their Applications in Certain Directions: A Brief Review

P.K. Choudhury, Krishna Kanti Dey, and Saurabh Basu

CONTENTS

2.1 Introduction

Modern times witness the usage of silicon and other semiconducting materials for various technological applications. Advancements in micron- and nanoscale-sized semiconductors attract even more to the researchers [1]. Indeed, the world is now passing through the age of *nano-mania* where everything *nano* seems to be greatly exciting and worthwhile. In this stream, *nanotechnology* remains in the forefront, now widely considered to be the latest key technology able to change human lives in many ways, which can be witnessed through the revolutionary technological progresses that have been realized in many areas including medicine, architecture, transportations, communications, memory devices, etc. [2–5].

Nanotechnology is an emerging field of study built upon modern science and advanced technologies, constituting thereby a multidisciplinary area representing the convergence of several technologies, some of which would include physics, chemistry, optics, electronics, materials science, biology, and bio-inspired technology. Apart from the others, the field of photonics too is no longer untouched of the new initiatives [6–8]. Concepts of photonics are, when applied to the suitable elements of the nanoscale size, the resulting phenomena that fall into the area of nanophotonics—a mixed blend of nanotechnology and photonics—the fusion of the two making an exciting new frontier providing significant challenges for fundamental research [9].

Nanophotonics forms the basis of many optics/photonics-based devices. However, the modern nanotechnology-based devices refer to human-designed systems with some essential elements having sizes as 0.1 nm to thousands of nanometers. Thus, a kind of overlap of nanotechnology can be noticed with *microtechnology* at the micron scale.

The geometry of mediums greatly governs the propagation of electromagnetic (EM) waves, and can be exploited to construct artificial materials, as experimented for fabricating a number of metamaterial-based composite mediums [10–12]. Indeed, the spectroscopic views of materials play vital roles in such developments. Within the context, Michel Faraday's experiment [13] on colored glasses (in the 1850s) remains highly exciting, which demonstrates the color of glass to happen due to the dispersion of nanoparticles; the experiment alludes to artificially constructed particulate composite mediums. Faraday devoted his work as Bakerian lecture—*The Experimental Relations of Gold (and Other Metals) to Light*—which throws a glimpse of the optical consequences upon dispersal of electrically small metallic particles in some host medium; this was a breakthrough achievement toward the theory of nanoparticles. By the end of nineteenth century, many artificial materials and composites with microstructured shapes had been conceived and some had even been realized and experimented upon following spectroscopic techniques. During the twentieth century, phenomenal progress in material science continued to highlight the importance of microstructured materials in technological applications [14,15].

In the present chapter, after introductory remarks, the authors initially put efforts to shine spotlight on the discussions related to the importance of nanoscale technology in the current scenario. After a cursory glimpse, descriptions are turned to the investigations of wave propagation features in some complex mediums. Needless to say that the studies related to the EM behavior of micro- and nanostructured mediums remain indispensable to look at prior to the thoughts of their prudent technological applications; and the spectral characteristics of such mediums are greatly important concerns in this regard [16]. Keeping this in mind, a major part of this chapter is devoted to the descriptions of the EM aspects of some micro- as well as nano-sized optical structures/systems, specifically in terms of their propagation characteristics observed upon penetrations by the EM waves in the optical frequency regime. In this stream, some optical microstructures are touched upon, which include guides with periodicity either in the form of refractive index (RI) variations or twists/chiralities in terms of the EM features they exhibit [17]. It has been found that, in spectroscopic views, periodic stratified guides would be greatly useful in filtering out the desired range of frequencies—a much useful component in many photonic integrated systems [18]. Also, twisted guides can be used in sensing technology as the angle of twists in such microstructures possesses the capability to control their spectral characteristics; a change in angle of twist due to the surrounding environment would be used to sense the measurand [19], and/or a deliberately introduced twist can make the system more sensitive as the transmission of light may then be greatly affected. Apart from these, some illuminations are also thrown on the nanophotonic waveguidance that includes qubit transmissions implementing the theory of cavity quantum electrodynamics (QED) [20]. In communication processes, the quantum state transfer from the sending node to the receiving node is achieved by generating a photon with energy and polarization states corresponding to the logic state of the node; simulations exhibit efficacy of the implemented approach, which confirms its technological applications.

Promising potentials of nanoscale technology to deliver benefits have been reported to cover almost every sector of human life [21], among which the medicinal applications have been one of the greatly valued societal impacts [22]. The idea of nanomedicine is not limited to curing diseases. Mankind has the tendency to use tools and technology to enhance the natural ability of humans. Nanomedicine is expected to play well the role in the development of human enhancement as nanotechnology has the ability to enhance many different aspects of human performance, from memory to the physical ability.

Nanomedicine can be used to improve treatment options in respect of surgery and drugs [23]. The later part of this chapter gives highlighting remarks on the medicinal applications of nanotechnology followed by some experimental as well as theoretical observations in the context of the dynamics of nanoparticles, which would possibly find a springboard place in the study of the delivery of drugs after further rectifications of the arriving problems.

In the course of absorption of drugs, the speed of absorption remains greatly important for a successful drug delivery. In this stream, catalytic locomotion of microstructures (or microbots) is known to be largely impeded by the viscosity of the medium [24–26]. Furthermore, the movement of microbots can be controlled by manipulating the pH gradient of the medium. The results surprisingly reveal the ability for the microbot to self-propel through the medium. An amalgamation of the two distinct features of the medium with regard to the velocity of locomotion of objects would be ultimately viewed in the form of efficient drug transport for therapeutic applications linked with moderate to high blood viscosity, owing to hyperviscosity syndrome found in cardiovascular patients. It is medically known that the human body does not permit even a moderate pH imbalance, making thereby the conjecture invalid for a large pH range. Nevertheless, the investigative approach can still make a significant contribution within the tolerance limit of blood alkalinity. Besides the applicability of the stated investigation to drug transport in biological systems, it addresses an important issue of controllability in the targeted delivery of microbots. However, it further remains to be seen how the previous scenario receives a challenge to describe nanoscale objects from the physical principles that are operative at lower length scales.

The present chapter is organized in this way—after the introductory remarks (Section 2.1), the technological importance of the area is highlighted in Section 2.2, which elaborates the essence of the fusion of micro- and nanoscale technologies. Photonics constitutes one of the prime focus areas under the engineering of smart materials and structures. Keeping this in mind, Section 2.3 provides illuminating ideas and the associated principles followed by the discussions of the EM wave propagation through some microscale optical structures followed by the use of cavity quantum electrodynamics (QED) in nanophotonic waveguidance. Interests are then turned toward highlighting the medicinal applications of nanotechnology. In this context, study of the dynamics of nanoparticles remains greatly important as that would find possible applications in drug delivery or related technologies; the relevant discussions are presented in Section 2.4 in terms of some experimental results followed by the theoretical background. Section 2.5 briefs on the concluding remarks based on the topics discussed in the present chapter.

2.2 Technological Importance

Nanotechnology establishes the understanding and the control of matter at dimensions of roughly from 1 to 100 nm, where unique phenomena enable novel applications. It includes the formation and usage of materials, structures, devices, and systems that have unique properties because of their small size. Thus, the technology essentially enables to engineer materials at the nanoscale with great degree of control, allowing thereby the evolution of new *nano-engineered* materials.

To talk about materials in the nanoscale is not at all new. However, scientific investigations with specific eyes on the nanoscale size are certainly new, and nowadays remain on the research forefront. Nanotechnology involves scientists from many different disciplines, including physicists, chemists, and engineers. Apart from these, biologists also proved their key contribution in the advent of this emerging field because the only possibility for a viable complex nanotechnology is that represented by biology [27]. In fact, materials consisting of nanostructures or possessing morphology at the nanoscale are commonly found in nature. For example, the smallest forms of life, bacteria, and cells do have nanometer-scaled sizes. Also, human body contains nanomaterials such as proteins and DNA. Investigators have seen the present understanding of molecular biology as an existence of proof for nanotechnology. Apart from biology, microelectronics is the most advanced present technology, the complex operating units of which are on a scale as small as micrometers.

The basis of nanotechnology is nanoscience or somewhat more specifically nanophysics, which describes the physics of nanoscaled systems—the scale at which the matter exhibits continuum characteristics. However, the molecules and their clusters of small size can still display their individuality. From a pedagogical point of view, the length scale of 1 nm essentially requires the concepts of quantum mechanics to implement for the behavioral analysis. Interestingly, quantum mechanics possesses the key to understand behavior even down to the femtometer scale of the atomic nucleus; classical theories fail to explain the varieties of root and cause at such a tiny length scale.

It is interesting to note that there are often advantages in making devices smaller (or miniaturized), as found in the modern semiconductor electronics. However, there must be some limits to the miniaturization. That is, how small a device can be made? The idea of limiting size of a miniaturized technology remains interesting for many reasons. Much of today's research on nanotechnology is designed to achieve a better understanding of how matters behave on this miniature scale. Quantum mechanical laws of nanophysics dominate the classical ones as sizes approach the atomic scale. The factors that govern the physical behavior of larger systems do not necessarily apply at the nanoscale too as nanomaterials essentially possess large surface areas relative to their volumes. This makes the phenomena like friction and sticking properties of materials more important than they are in larger systems. While the changes from classical physics to nanophysics may mean the failure of some existing devices, the same changes would open up enormous possibilities for exciting new devices.

During the past couple of decades, the ability to fabricate and characterize nanostructures and nanoscale systems has tremendously improved. These essentially involve a great degree of control using a diversity of techniques, which allowed the fabrication of many systems. The evolutionary steps of nanotechnology open up the avenues to change human lives in many ways as it has been increasingly linked with the advancements in biotechnology, information technology, and cognition science, the symbiosis properties of which would fruitfully be implemented in several everyday applications, as follows monitoring systems for structures vulnerable to natural calamities and terrorisms, medical treatments, and energy-efficient production systems with very little waste—a factor of much demand for bio-green technology [3,28].

As stated previously, modern nanotechnology-based devices refer to human-designed systems comprising some essential elements having sizes of the order of 0.1 nm to thousands of nanometers, indicating thereby a kind of overlap with *microtechnology*. Although the literatures on nanotechnology may refer to nanoscale machines or devices, hardly nanoscale machines are presently available. Further, scarcely any micron-scaled machines exist either, and the smallest mechanical machines available in a wide variety of forms are really on the millimeter scale. As such, nanoscale technology is presently more like a concept than a fact, although the name certainly attracts media and funding realities.

2.3 Highlights to Photonics

Nanophotonics deals with the interaction of light with matter on the nanoscale, and opens up challenging opportunities for interactions among many traditional scientific lexicons [9]. Applications of nanophotonics would include sensing, on-chip and chip-to-chip interconnects, and Si photonics, which could be possible after exploring different materials ranging from polymers to semiconductors grown on a variety of substrates for both passive and active photonic devices [7,8,29]. Apart from these, applications in the medicinal area are also quite attracting as there has been growing need for new technologies that enable to rapidly detect and treat diseases at an early stage [22,23,30]. This essentially throws challenging demands for environment-friendly compact, economic, energy-saving, and fast responding technologies, which can possibly be met by implementing photonic-based themes with an imprint of micro and/or nanotechnology. Photonic-based diagnostics would be able to recognize certain diseases (e.g., detection of cancer cells) in their pre-stages with minimal side effects as these are noninvasive and perform on molecular-based recognitions.

Previous discussions, indeed, indicate the technological impact of nanophotonics in the marketplace. Another significantly important application has been in the communication industry as many modern communication technologies rely on light (or photons) as the information carrier. Scientific breakthroughs have brought tremendous momentum to the field of nanophotonics research, which resulted into an explosive growth of the information age during the past couple of decades. Light can be effectively utilized for telecommunication, sensing etc., which makes nanophotonics as indispensable for the current trends of communication systems.

Amalgamation of nano- and microtechnologies can be greatly important in optics (or photonics). For example, the propagation characteristics of micron-sized optical fibers are much affected due to the presence of nanocoatings of foreign materials and/or composites. Their importance in communication systems has also been realized. For example, nanophotonic waveguidance implementing the concept of cavity QED would be the one that illuminates the application of nanophotonics in quantum communication systems [20].

In this context, a few remarks must be made on the EM interactions of matter. Electromagnetics is truly a microscopic science. Matter can be treated as a continuum for most of the technological purposes when the wavelengths considerably exceed molecular dimensions. Linear and isotopic dielectric materials are the simplest types that can be thought about as possible propagation mediums to sustain EM waves. The medium essentially acts on the progress of waves in it by introducing a *delay* with respect to the propagation in matter-free space or vacuum, and also, *absorbing* the EM energy—the two frequency-dependent dispersive phenomena. Complex materials would even impose more effects on the progress of EM waves. Anisotropy, inhomogeneity, nonlinearity, chirality, etc. can be expressed as some forms of complexities of medium that would affect the propagation of waves [6,31].

Enhanced growth of interest in research related to the action of EM fields on matter has been witnessed after the realization of novel fabrication techniques and understanding of the relationship between the macroscopic properties and the microstructural morphology of materials. Indeed, the computational techniques too played vital roles in the rapid development of research. Considering these remarks, the preceding sections present some interesting and useful outcomes that project the importance of the aspects of amalgamation of micro- and nanotechnologies in photonics. In this stream, some micro- as well as nanostructures/systems are considered in respect of their EM characteristics, and possible useful implications are highlighted.

2.3.1 Guides for Selective Photonic Transmission

Metallic and dielectric are the two forms of mirrors that are commonly used in photonic applications, for example, imaging, solar energy collection, and in laser cavities. Metallic mirrors have the capability to reflect light over a broad range of frequencies incident from arbitrary angles. In metallic coaxial waveguides, the EM field remains confined between two coaxial metallic cylinders forming thereby transverse EM (i.e., TEM) modes. However, these guides are not suitable in the infrared (IR) and the optical frequency regimes, owing to the absorption of the incident power causing thereby very high amount of loss. This severely limits the ultimate bandwidth that metallic guides can transmit. Hybrid metal–dielectric waveguides have been developed for IR wavelength transmissions by incorporating a dielectric coating inside the metallic waveguide. Such guides are primarily used for laser power delivery, but proved to be inefficient at optical wavelengths because of heavy absorption losses in metals.

The use of dielectric waveguides, however, remains a better option for the transmission of optical frequencies. As the mechanism of light guidance is essentially due to the microstructural features of the medium, a wide variety of photonic structures have been proposed to meet the rapid increase in telecommunication needs. Within the context, the propagation of EM waves in stratified periodic structures has been a subject of great interest in optics due to the interestingly important role they play in many potentially useful applications. These stratified structures are the special class of layered mediums wherein dielectric layers are stacked in a periodic fashion. Mathematical formulations in respect of symmetric as well as asymmetric periodic stratified waveguides have appeared in the literature that deal with the evolution of explicit dispersion relations to determine guided modes [32–38].

Propagation of photons in dielectric mediums can similarly be treated as that for electrons in crystals [39,40]. This led to the evolution of many new artificial photonic materials and/or optical micro- and nanostructures yielding specific spectroscopic properties. In fact, spectroscopic features of such systems/devices can be tailored by manipulating structural design. The idea of omniguides was first proposed by Yeh et al. in 1978 [41]. At a later stage, several designs have been proposed for such optical microstructures to present photonic bandgaps [17,42–44] for controlled optical transmission. A typical kind of omniguiding fiber would be the one having a high-index core surrounded by dielectric cylindrical Bragg mirrors (Figure 2.1), consisting of alternate layers of high and low RIs. In these structures, the EM energy remains confined to the core by Bragg reflections off the dielectric mirror due to the existence of a complete photonic bandgap regime in the phase space above the light cone of the ambient mediums. More precisely, multilayer dielectric mirrors can be used to reflect a narrow range of frequencies incident from a particular angle (or angular range). Unlike metallic mirrors, dielectric mirrors used in bandgap fibers have extremely low losses at optical and IR frequencies. Bandgap materials are also multilayered structures capable to reflect EM waves, if the frequency of operation falls within the gap. Therefore, such multilayered structures can exhibit the property of spectral filtering [45].

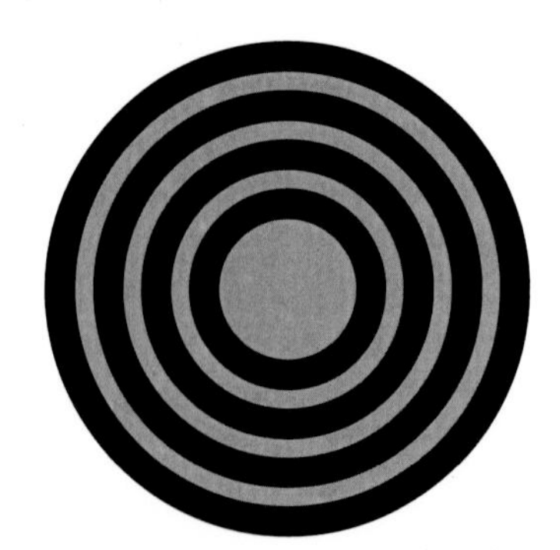

FIGURE 2.1
Cross-sectional view of a typical omniguiding Bragg fiber having low (gray portions) and high (black portion) index layers.

FIGURE 2.2
Transverse geometry of omniguiding Bragg fiber with wider strip of thickness a and RI n_1, and narrower strip of thickness as b and RI n_2.

Different theoretical approaches have been reported to determine the excitation of omniguiding Bragg fibers [18]. The quantum electronic theory is also a concept that can be used to analyze the working principle of omniguiding Bragg fibers (Figure 2.2), and the existence of bandgaps (i.e., the allowed and the forbidden bands) can be demonstrated. It has been observed that the number of allowed (or forbidden) bands essentially depends on the RI values as well as the widths of layers—it increases with the increase in RI difference between different layers and corresponding to larger thickness values of the layer thickness. These bandgap fibers can, therefore, be used in the area of spectral filtering and other integrated optic applications by suitably controlling the values of RI as well as the thickness of mediums.

The estimations of electric fields E in different fiber sections have been reported [46], and it has been observed that fields exhibit a smooth match at the layer interfaces with decaying characteristic with the increase in radial distance (Figure 2.3), justifying thereby the validity of analytical approach of quantum electronic theory. In Figure 2.3, a and b are the

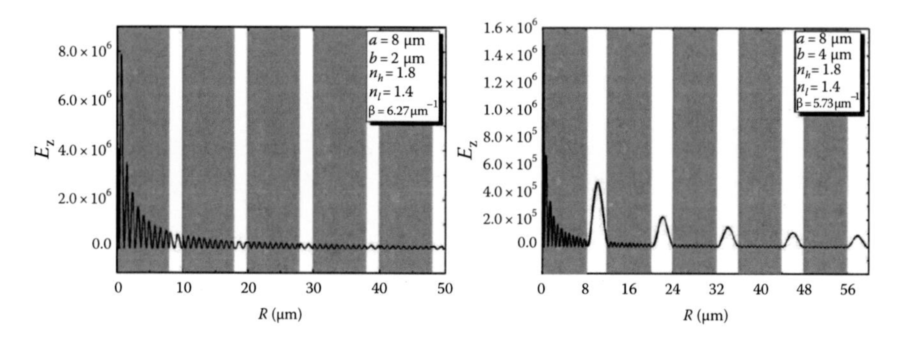

FIGURE 2.3
Plots of the electric field E against the radial distance R.

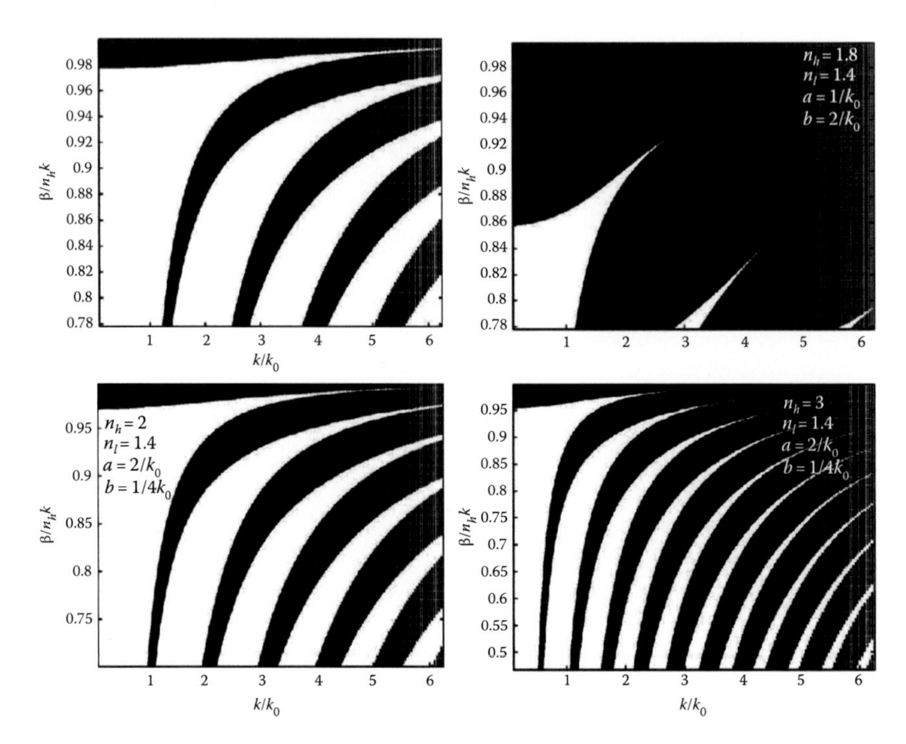

FIGURE 2.4
Plots of the dispersion patterns.

widths of alternating regions, having their respective RI values as n_h and n_l (with $n_h > n_l$).

Apart from electric field distributions, dispersion patterns for omniguiding bandgap fibers have been reported [47], as shown in Figure 2.4, with specific eyes on the dependence of dispersion characteristics on design parameters. In this figure, light and dark regions represent the allowed and the forbidden spectral bands, respectively. Furthermore, k_0 represents the free-space wave number, and the meanings of other symbols are as stated earlier. It has been observed that the widths of the allowed ranges generally decrease with the increase in normalized propagation constant k/k_0, and the number of propagation modes enhances in the guide corresponding to the case when the thickness of the high-index layer is increased.

As stated earlier, several forms of Bragg fibers have been proposed in the literature. In this context, one may also consider a Bragg fiber of the type as shown in Figure 2.5, wherein the cross section consists of concentric dielectric rings arranged in a periodical fashion with a hollow core at the center of the guide. The electric/magnetic field distributions along with the power patterns for such guides have been reported following simulation steps [48].

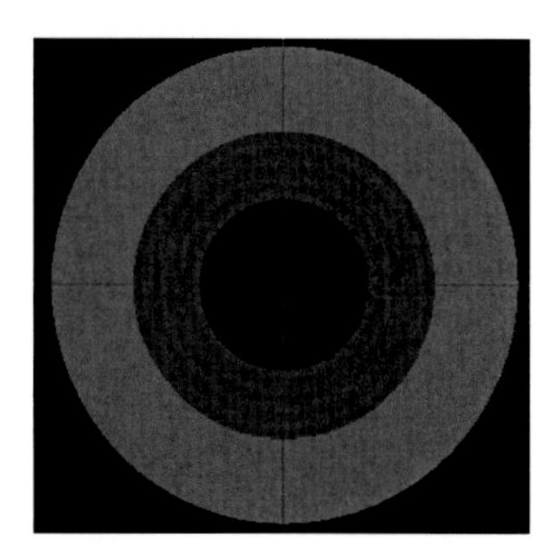

FIGURE 2.5
Cross-sectional view of a typical Bragg fiber.

Simulations of propagation characteristics have been performed by using the software *Mode Solutions* from Lumerical Solutions, Inc. In the simulation process for Bragg fiber of Figure 2.5, materials with RI values 1.2 and 1.4 are assumed to be periodically arranged, with each of the concentric rings having thickness as 0.025 μm. Also, the radius of (hollow) core at the center (of the guide) is taken to be 0.25 μm and the operating frequency is taken as 258 THz (wavelength 1161.99 nm). Under these assumptions, Figures 2.6 and 2.7, respectively, illustrate the electric and the magnetic field patterns corresponding to the *x*-, *y*-, and *z*-components.

Figure 2.6 throws the information that, corresponding to the case of electric field distribution patterns, potential modes exist in the central core region of the guide. However, no such occurrence is seen in the case of magnetic fields (Figure 2.7). Furthermore, in the case of electric field distribution, in each of the *x*- and the *y*-components, instead of a single mode, there are two modes aligned horizontally and vertically. However, a single mode is found corresponding to the *z*-component of electric field. In the case of magnetic field distributions, as shown in Figure 2.7, hollow core contains two modes in each of the *x*- and the *y*-components, whereas the *z*-component does not correspond to any mode in the fiber core.

Figure 2.8 illustrates the propagation of power by the Bragg guide along the *x*-, *y*-, and *z*-directions. Power propagation toward the *z*-direction (the direction of wave propagation) is clearly understood. However, propagation modes exist in the *x*- and *y*-directions. A close inspection of Figure 2.8 reveals that power along the *x*- and the *y*-directions are obvious, owing to the reason that the corresponding modes are confined in small regions well within the hollow core. At the periphery of the core, flow of power toward the radial direction is clearly insignificant, as is evident from the dark shade in the peripheral region of the fiber core.

Omniguiding Bragg fibers have been found to exhibit great potentials for various applications since they drive one further beyond the mechanism of total internal reflection (TIR)—the basic phenomenon for lightwave

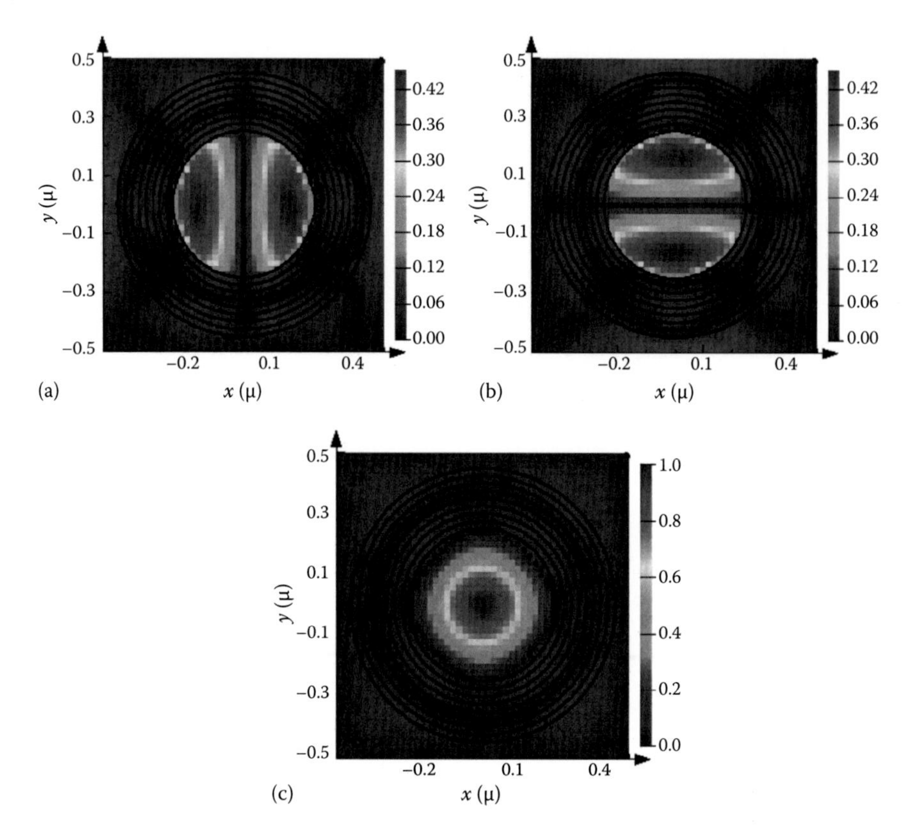

FIGURE 2.6
(See color insert.) Electric field distribution of the (a) *x*-, (b) *y*-, and (c) *z*-components.

transmission through photonic guides. Related simulations yield that such fibers combine some of the best features of metallic coaxial cables and dielectric waveguides. For instance, investigations revealed that omniguiding fibers support a truly single mode in the low-index air core—the feature very similar to the TEM mode of metallic coaxial cables. Furthermore, these fibers can be designed to behave as single mode over a wide range of frequencies, which facilitates a pulse to retain its shape during propagation at the point of intrinsic zero dispersion. The electric field distribution in these fibers is also radially symmetric along the cross section, allowing thereby a maintained polarization state during the propagation of EM waves through the guide.

2.3.2 Twisted Clad Fibers

Complex waveguide structures have been of considerable interest among the R&D community, owing to their possible versatile applications. Among those, twisted clad fibers are new in their class, based on the conceptual

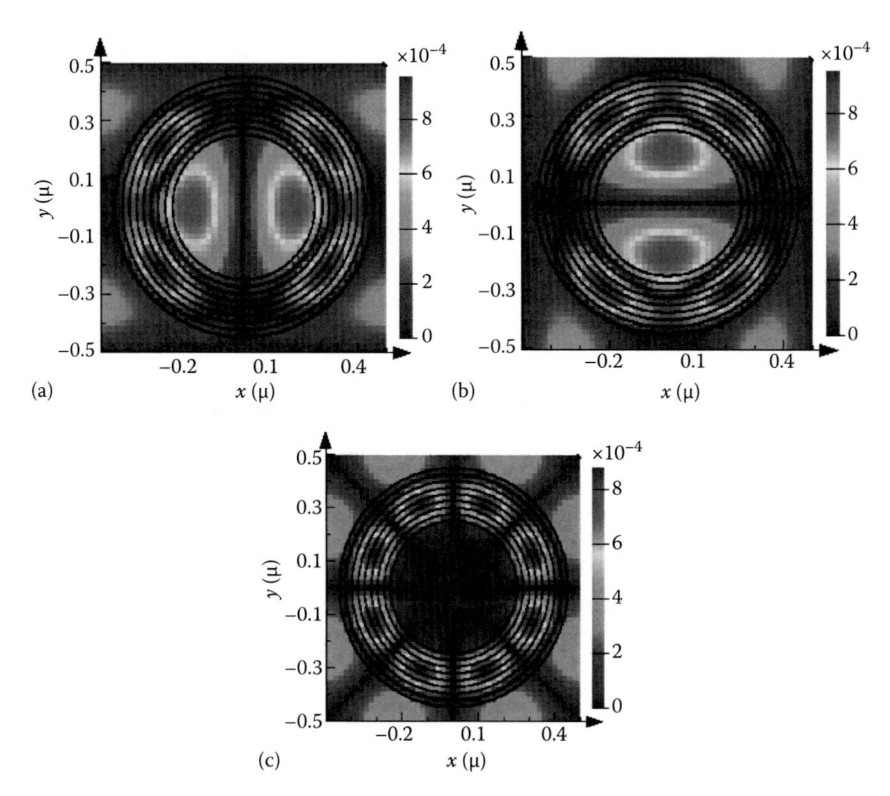

FIGURE 2.7
(See color insert.) Magnetic field distribution of the (a) *x*-, (b) *y*-, and (c) *z*-components.

understanding of the usage of helical structures in low- and medium-power traveling wave tubes [49]. The analysis of helical waveguides generally includes waveguides under slow-wave consideration with conducting sheath and tape helixes—the concept that has been implemented in the case of optical fibers having windings introduced at the core–clad interface [50]. Figure 2.9 illustrates a very simple example of such fibers, wherein a single helical turn makes an angle ψ with the surface normal. However, multiple turns in the guide allow it to acquire the property of periodicity in structure.

The use of helical windings in optical fibers allows control over dispersion characteristics of the guide, which can be manipulated on demand [51,52]. Thus, the spectroscopic properties of the guide can be altered by suitably used helical wrap mounts. It has been found that, under fast-wave consideration, twisted clad elliptical fibers show the existence of band gaps corresponding to a 0° helix pitch, which is attributed to the existence of periodicity in structure [53–55] due to helical turns. In contrast, corresponding to 90° pitch angle, such band gaps are not seen owing to the elimination of periodicity in this case. Furthermore, under the fast-wave consideration, the

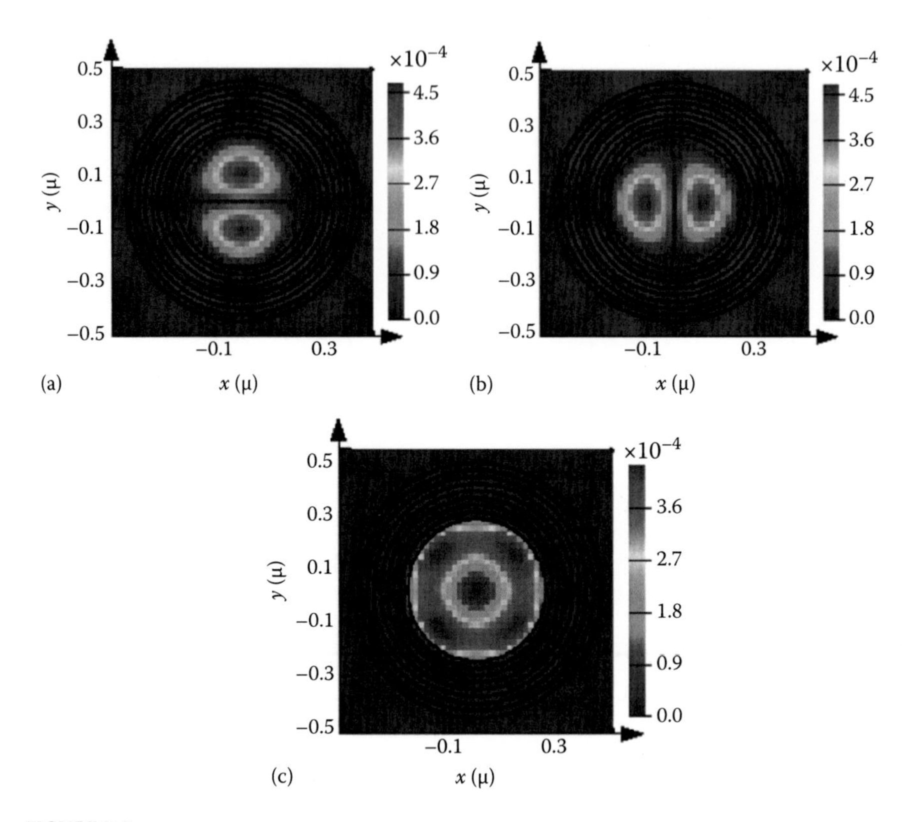

FIGURE 2.8
Power propagation toward the (a) x-direction, (b) y-direction, and (c) z-direction.

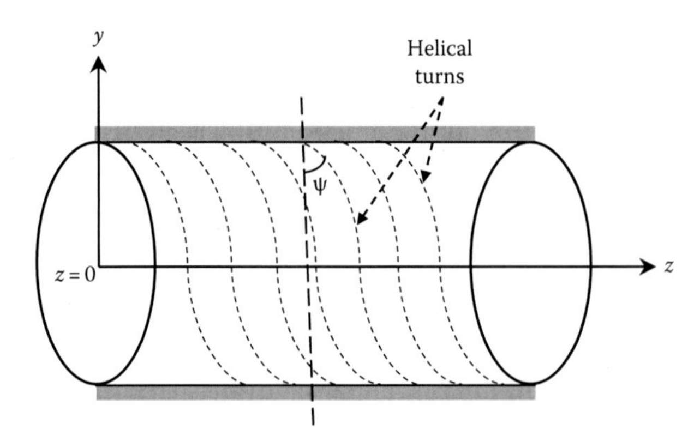

FIGURE 2.9
Longitudinal view of a helical clad fiber.

number of propagating modes depends much on the helix pitch angle, which plays in shifting the modal cutoff to a lower value as compared to the case when the helical turns are parallel to the optical axis.

Efforts have been made to present a comparative study of dispersion relations of dielectric circular fiber loaded with a conducting sheath helix between the core and the clad regions under slow- and fast-wave considerations [56]. Results have been reported in respect of the radial field distributions corresponding to hybrid modes considering variations in helix pitch angles, modal propagation constants, and fiber dimensions. The investigation revealed that a change in either of these parameters introduces prominent effect on field patterns within the fiber structure. As stated earlier, the winding pitch angle plays determining role in the estimation of propagation characteristics, and it is observed that the case of transverse windings essentially controls radial field distributions to reduce their intensities in the different fiber sections.

The propagation constants also have the effect to control field amplitudes—corresponding to lower propagation constants, field amplitudes also become lesser. In this connection, some of the results are illustrated in Figures 2.10 (corresponds to the case of helical turns to be perpendicular to the optical axis) and 2.11 (for helical turns parallel to the optical axis). Considering 10 μm core diameter, it can be observed from these figures that fields are a little higher in the helical clad section of the guide, which essentially indicates their possible applications in evanescent wave optical sensing [57]. Studies have been performed for a rigorous boundary-value problem, and fields are observed to match smoothly at the layer interface, which rather confirms the validity of the analytical approach.

Kumar [58] has reported results corresponding to the case of twisted fibers with elliptical cross sections, considering the core anisotropy to be very small so that the core structure can be thought as approximately cylindrical. It has been found in this case that the effect due to anisotropy plays its determining role in the form of polarization. A kind of modal attenuation has also been observed as the dispersion relations become complex for 0° and 90° pitch angles.

In a later stage, investigations of twisted clad fibers have been carried out by following more exact analytical treatment [59], which led to the development of more accurate dispersion relations. This has been followed by characterizing the structure by considering the effect of core eccentricity on the EM wave propagation under different operating wavelengths as well as the magnitudes of core major axis. It has been reported that, under the fast-wave consideration, number of guided modes in the waveguide structure decreases with increasing eccentricity of the fiber [59]. Apart from core eccentricity, pitch angle also leaves pronounced effect on the dispersion relations for the guide. However, in the case of slow-wave consideration, a rather interesting result has been observed that the characteristic equations are ultimately independent of the controlling parameter ψ. Essentially it may be attributed

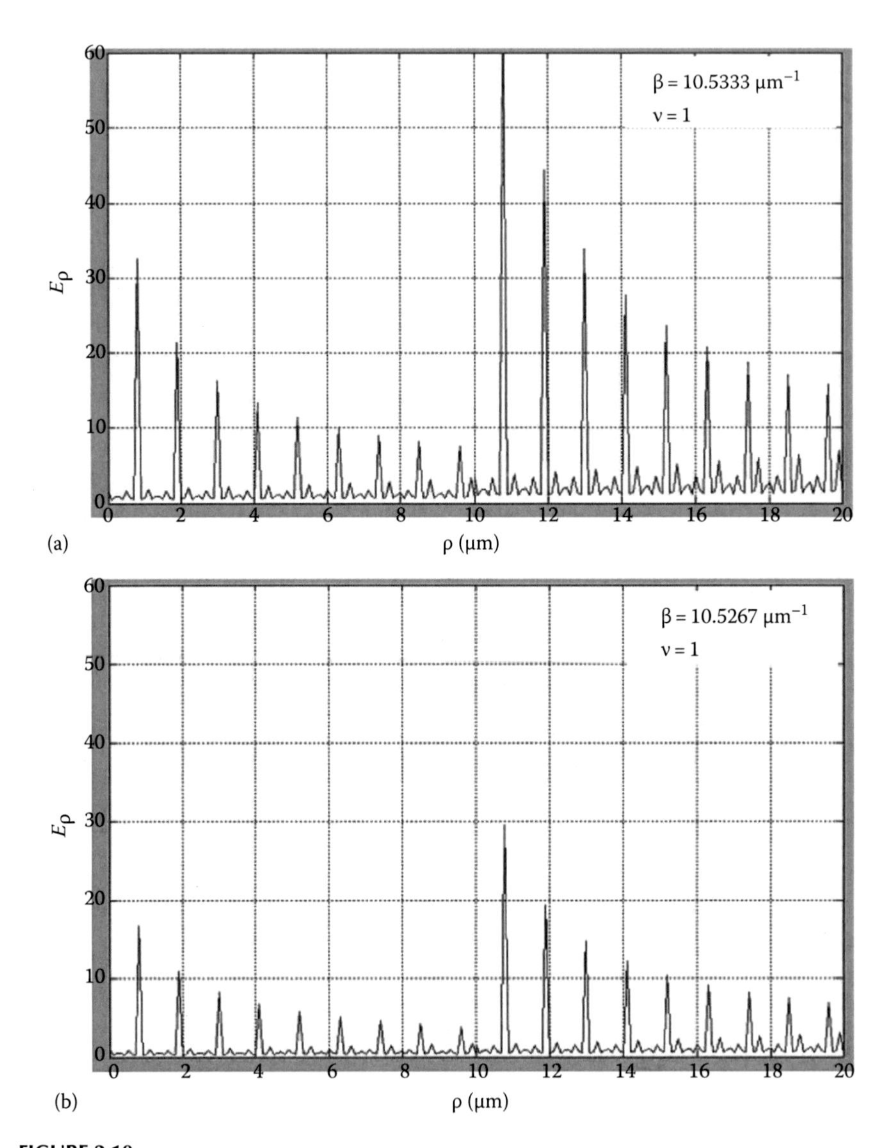

FIGURE 2.10
Plots of the radial component of electric field against fiber radius corresponding to 10 μm core diameter, $\psi = 0°$ and (a) $\beta = 10.5333\ \mu m^{-1}$, (b) $\beta = 10.5267\ \mu m^{-1}$.

to the implementation of slow-wave approximation, which ultimately led to similar dispersion relations for 0° and 90° pitch angles.

Putting emphasis on the dispersion behavior of helical clad fibers under different values of the core ellipse eccentricity e as well as the operating wavelength λ, plots of the normalized propagation constant (k/β) against the ellipse eccentricity e would be of major requirement. Figure 2.12 represents

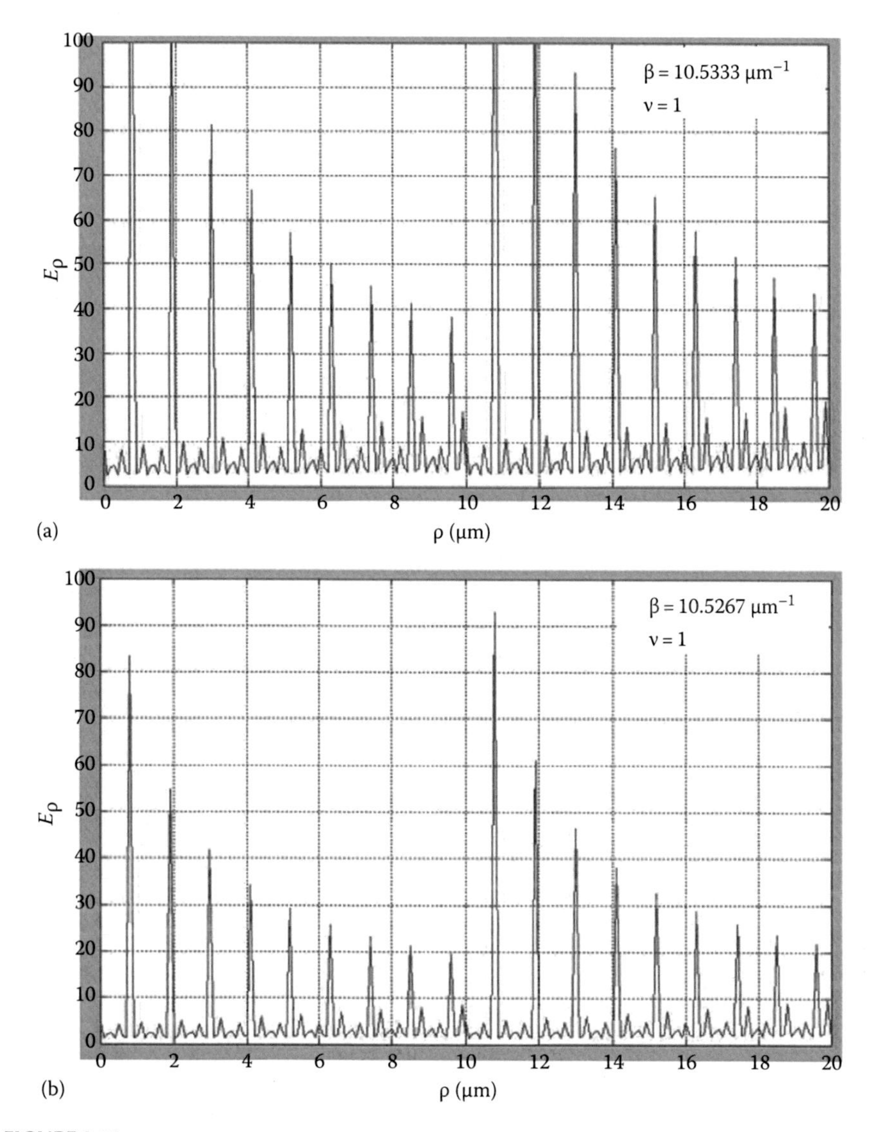

FIGURE 2.11
Plots of the radial component of electric field against fiber radius corresponding to 10 μm core diameter, $\psi = 90°$ and (a) $\beta = 10.5333\ \mu m^{-1}$, (b) $\beta = 10.5267\ \mu m^{-1}$.

the illustrative plots to show this feature corresponding to $\lambda = 1.31$ μm and the core semimajor axis 50 μm and 100 μm. In these plots, F represents the basic dispersion relation for this kind of special fiber. It has been found that the values of k/β generally increase with the increase in core eccentricity.

To assess the variation of k/β with changing core ellipse eccentricity under the slow-wave approximation, illustrative plots are shown in Figure 2.13,

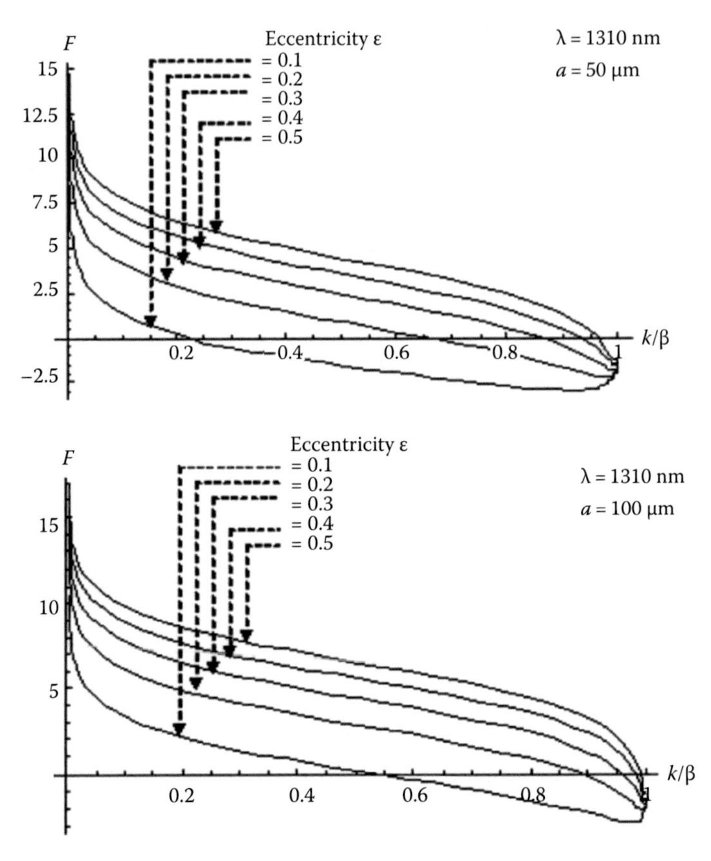

FIGURE 2.12
Variation of F with k/β for $\lambda = 1310$ nm and $a = 50$ and 100 μm.

corresponding to 1310 nm operating wavelength. It has been observed that the effect of change in wavelength on the propagation constant is more corresponding to the larger size of fiber core. However, this effect is less pronounced when the core ellipse eccentricity is fairly increased.

A closer look at the dispersion characteristics of helical clad fibers with tapered core would also be rather interesting as the helix pitch angle essentially plays a determining role. Dispersion relations for such guides have been reported corresponding to helical windings perpendicular and parallel to the propagation axis under the assumption of a small variation of core radius in the longitudinal direction. The study emerged to conclude that the modes attain higher propagation constants under the situation when the helical windings are perpendicular to the fiber axis than the case of parallel windings. However, a comparison of propagation features with those without helical turns revealed that the introduction of helix has the essential characteristic to increase the propagation constants of modes. This reflects the important dominance of pitch angle to control

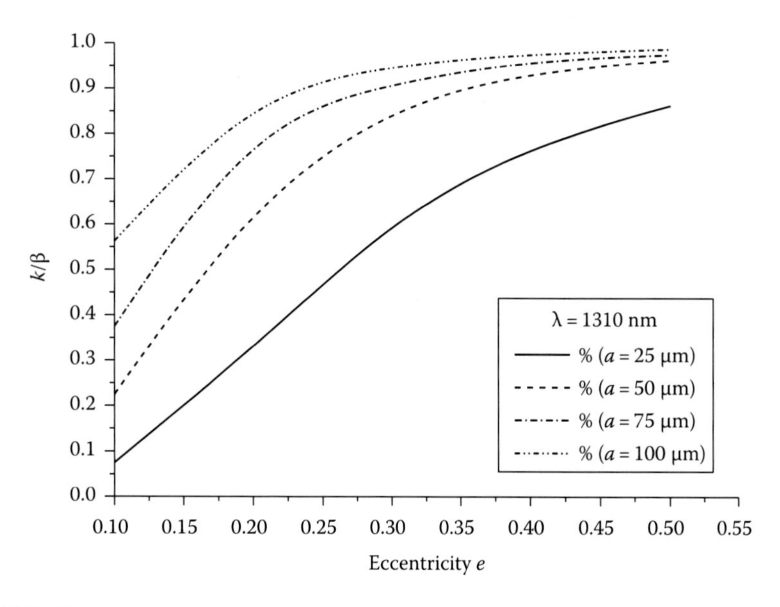

FIGURE 2.13
Plot of (k/β) against the core ellipse eccentricity.

the propagation characteristics of the guide, so far as the fiber design is considered [60,61].

2.3.3 Chiral Fibers

Chirality stands as the technical term for handedness. Objects existing in two distinct states that are identical upon reflection but are not the same through any set of translations or rotations are said to have chirality in their response nature. Typical examples of chiral objects include hands, ears, feet, amino acids, and many salts and sugars. In EM terms, chiral mediums possess the property to discriminate between left- and right-handed EM fields [62], and they are optically active in nature—the phenomenon first discovered by F. Arago while experimenting with quartz. This property ensures that chiral objects interact differently with left- or right-circularly polarized light, depending on their kind of handedness. This produces circular dichroism (i.e., differential absorption of the left- and the right-circularly polarized light) and birefringence (i.e., double or direction-dependent refractive index properties).

Chiral mediums can be classified into the categories of isotropic and structurally ones. The molecules of an isotropic chiral medium can be formed by randomly dispersed and oriented, electrically small, handed inclusions in an isotropic achiral host medium. The molecules of a structurally chiral medium, such as a chiral nematic liquid crystal, are randomly positioned, but they exhibit helicoidal orientation. Structurally chiral mediums can also

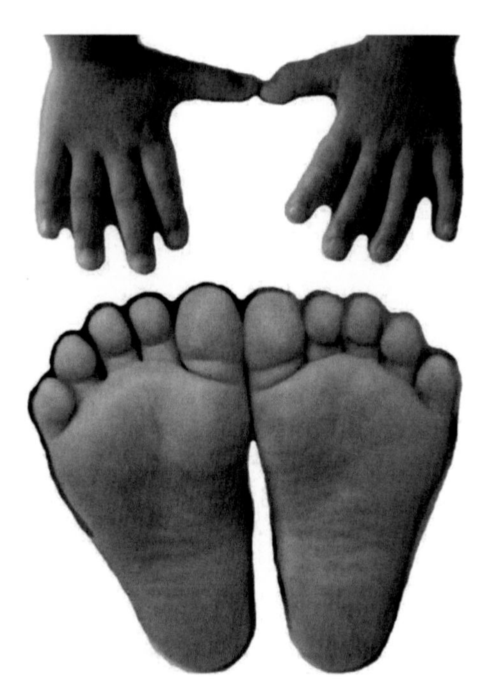

FIGURE 2.14
Chirality in the pairs of hands and feet.

be artificially fabricated either as stacks of uniaxial laminas or using thin film technology.

Different versions of chiral molecules are called as enantiomers; they can have different properties, although they contain identical atoms in identical numbers. Figure 2.14 illustrates the pairs of hands and feet, and Figure 2.15 depicts the enantiomers of limonene. It is interesting to note

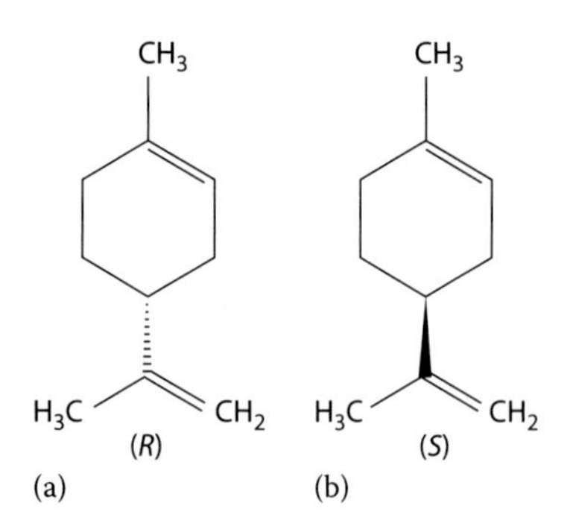

FIGURE 2.15
Enantiomeric limonene molecules: (a) *R*-limonene with fresh citrus or orange-like smell and (b) *S*-limonene with harsh turpentine-like smell.

that a chiral structure and its mirror image may have completely different functional properties. As an example to state drastic difference in behaviors of the two enantiomers of Figure 2.15, the *R*-limonene presents fresh citrus or orange-like smell, whereas that of the *S*-limonene is harsh turpentine-like.

The property of chirality also stems from nanostructured orientation of molecules. These attained a renewed interest because of their possible applications in new fields, and the availability of new types of chiral materials. This drove the R&D community to venture for the investigation of different types of chirowaveguides. The presence of chiral materials in various devices, e.g., printed circuit antennas, multichannel guides, directional couplers, and other integrated optic devices, opens up the possibility of many new applications [63–66].

Analysis of the propagation of EM waves through chiral mediums becomes much formidable owing to the complicated nature of constitutive relations [67,68]. Chirality introduces a coupling between the propagating electric/magnetic fields, which ultimately causes the modes to be hybrid [63,69–72], instead of being simply TE, TM, or TEM modes, as observed in the case of conventional optical fibers [63].

As stated previously, chiral guides have novel integrated optic applications, e.g., directional couplers and photonic switches. Directional couplers, which can also be used as optical switches, are vital for the transfer of energy from one fiber to another adjacent one; the maximum amount of energy transfer takes place in the case when the condition of phase-matching is truly achieved [68]. However, otherwise, only at a specific frequency, the maximum transfer of energy can be performed. The condition of phase-matching is highly sensitive, and depends on many factors including the parameters related to the environment. But, for any frequency of operation, chiral fibers enable easy achievement of the phase-matching condition, leading thereby to attain the situation of easy coupling of light.

Investigators have presented studies related to the different aspects of EM wave propagation in varieties of chiral waveguides; nevertheless, the subject needs a more complete coverage. Keeping this in view, chirowaveguides with some new forms of structures have been introduced [73–77] with the emphasis on the sustained field patterns and the power confinements in the guide corresponding to different values of the chirality admittance parameter.

Figure 2.16 illustrates the electric field patterns in a chiroguide for different combinations of core/clad chirality admittance values [78]. It has been observed that a change in core chirality brings in prominent effect on fields within the fiber structure, whereas such effects are not much observed when the clad chirality is altered. However, a change in core/clad chirality introduces pronounced effects on field intensity within the chirofiber, which is attributed to the effect of chirality on the polarization state of field. The analysis is essentially a rigorous boundary value problem, and the fields are

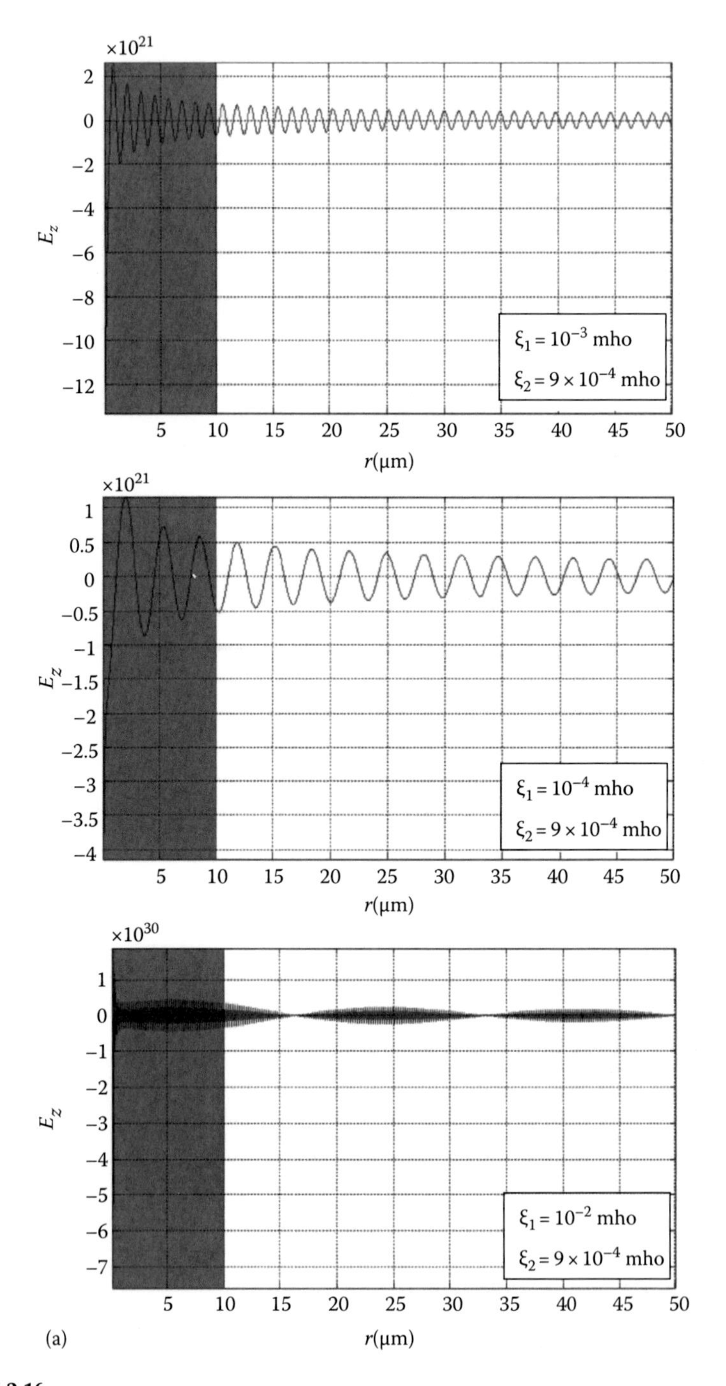

FIGURE 2.16
Variation of electric fields in chirofibers with (a) changing core chirality with clad chirality fixed.

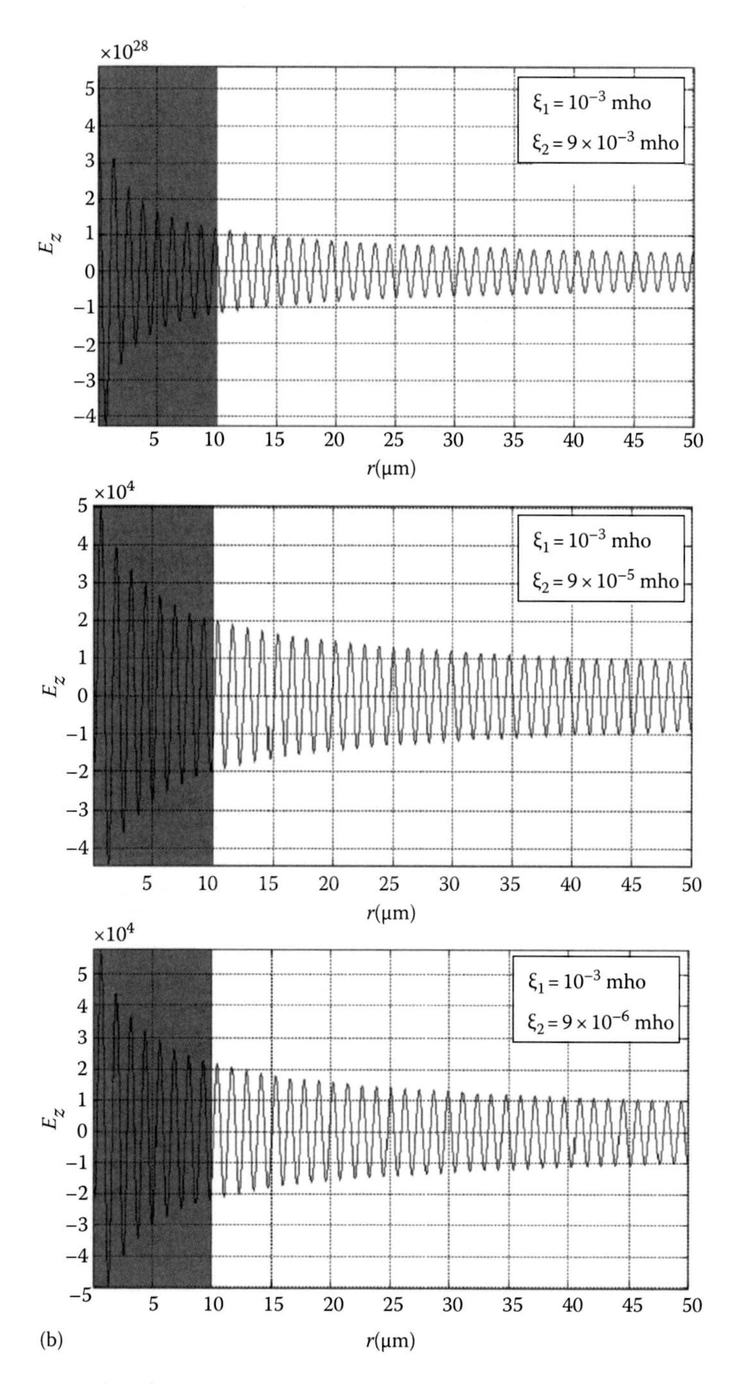

FIGURE 2.16 (continued)
Variation of electric fields in chirofibers with (b) changing clad chirality with core chirality fixed.

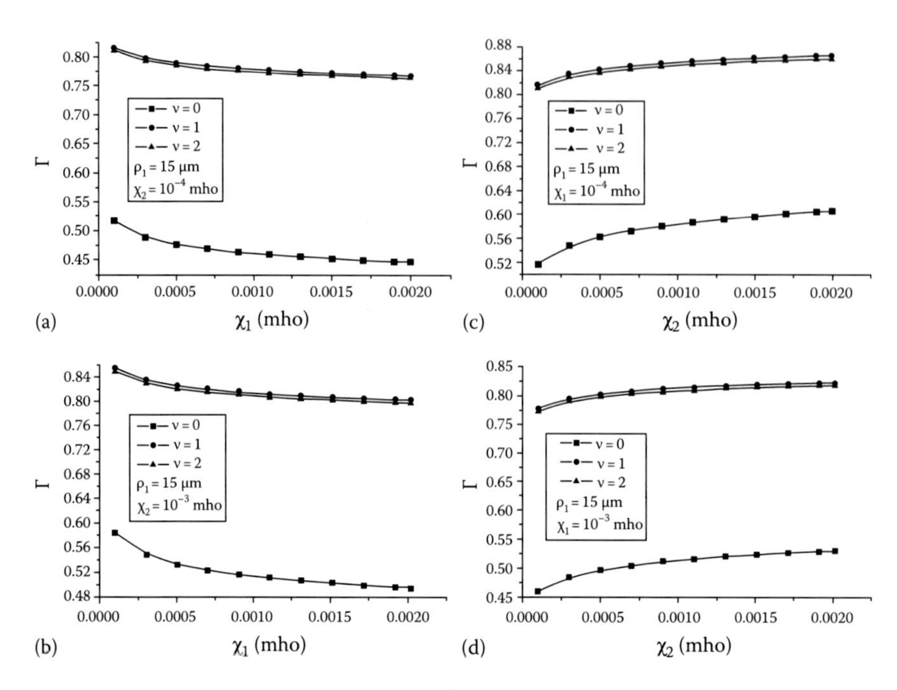

FIGURE 2.17
Variation of Γ with (a, b) the core chirality parameter χ_1 corresponding to the fixed clad chirality parameters χ_2 as 10^{-4} and 10^{-3} mho, and (c, d) clad chirality parameter χ_2 corresponding to the fixed core chirality parameter χ_1 as 10^{-4} and 10^{-3} mho.

observed to match smoothly at the layer interface in the structure, which rather confirms the validity of steps implemented in the analytical treatment.

A few illustrative graphs of the power confinement factor are shown in Figure 2.17; the ones on the left are against the variation in core chirality admittance χ_1, keeping the clad chirality admittance χ_2 fixed, whereas the ones on the right are against the clad chirality admittance χ_2, keeping the core chirality admittance χ_1 fixed. It has been observed that the introduction of core chirality has the influence of reducing the optical power confinement in a chirofiber. Furthermore, with the increase in fiber core size, skew modes present a condition close to degeneracy, and this feature has prominent importance in the areas of integrated optic applications. It is to be remembered that, in Figure 2.17, ρ_1 represents fiber core radius and ν the azimuthal mode index of the guide.

Chiral waveguides are also indispensable in slow-wave structures, which essentially incorporate helical twists in their geometries. The investigation of EM effects in a chirofiber, wherein a chiral cylinder is embedded in the chiral cladding of infinite extent in the transverse direction (both the sections are having different chiralities), have been reported [19]. Further, a helical winding is introduced at the interface of the core–clad sections, as shown in Figure 2.18. With the advent of nanotech-based age, it may be

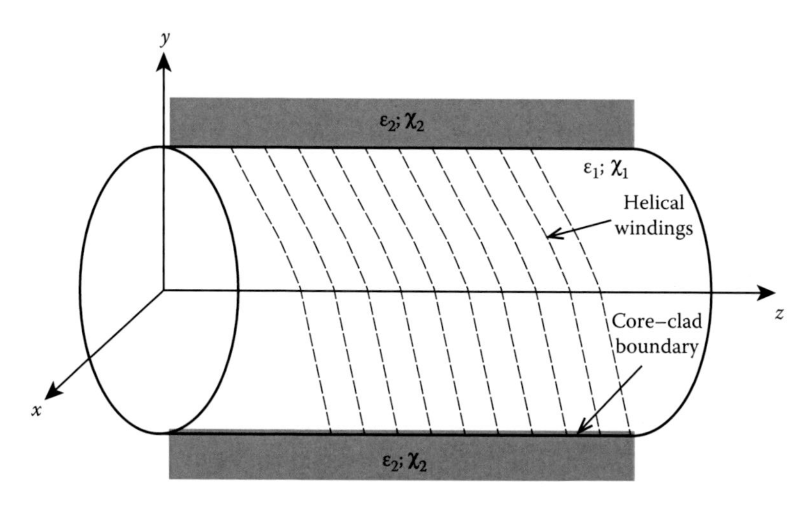

FIGURE 2.18
Section of a chirofiber with helical winding at the core–cladding boundary.

possible in the near future to fabricate fine insulated windings around a chirofiber core.

Studies related to the aforesaid waveguide structure of helical clad chirofibers have been reported for the first time by Lim et al. [19], and it has been found that the introduction of a sheath helix structure in the guide effectively plays a determining role in its propagation characteristics. In this stream, some results related to the confinement of power are illustrated in Figure 2.19. Clearly, there will be two different controlling parameters in this case—chirality admittance and helix pitch of the loaded windings. Figure 2.19 shows the variation of power confinement factor against the allowed values of propagation constant β corresponding to different modal indices and helix pitches 0° and 90°, keeping the core/clad chirality admittance values fixed. Plots were made assuming 15 μm core radius, and for the azimuthal mode indices 0, 1, and 2.

Observations of results revealed that the chirofiber with a helix wound in a direction perpendicular to the optical axis (of the fiber) essentially has the net influence of increasing power confinement in the fiber core. Furthermore, an event very close to degeneracy has also been reported, which becomes more significant corresponding to fibers with larger core dimension—a phenomenon of prominent importance in integrated optic applications. Strong degeneracy has also been observed for fibers with helical windings parallel to the *z*-axis—an inherent property of chirofibers.

Generally, as stated earlier, the analysis of EM wave propagation through chiral mediums becomes much complex particularly due to the complicated nature of constitutive relations in this case. The usefulness of chiroguides certainly exists in many applications related to integrated optics as well as optical sensing.

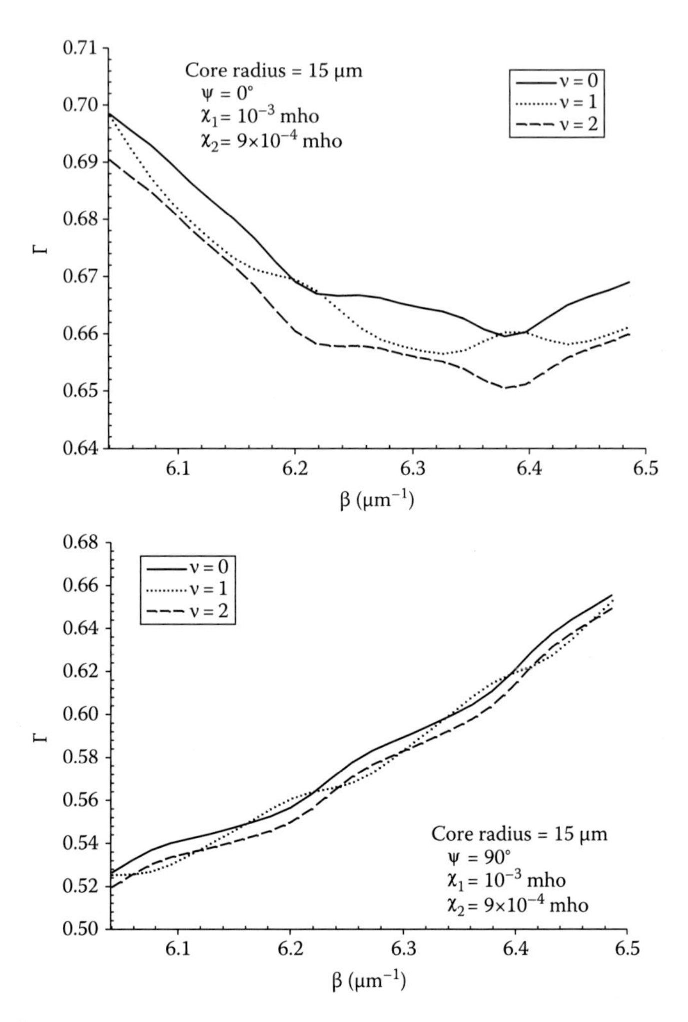

FIGURE 2.19
Variation of the core power confinement for helix pitch angles 0° and 90°.

2.3.4 Nanophotonic Communication Systems

Communication industry has been the major driving force behind the extraordinary progress that photonics has come across. The applications of nanophotonics in communications are progressively becoming an area of major technological challenge. Several exotic experimental results and intriguing theoretical analyses have been growing rapidly in the area of nanophotonics [79–81], which makes the subject to become ubiquitous in the field of communication technologies. This is demonstrated by the revolutionary applications of nanoscale materials and devices in photonic structures and systems. Some of the noted examples would be high-index contrast optical waveguides, photonic crystals, plasmonic devices and circuits, and low-dimensional photonic structures. Investigators have presented the necessary physical understanding along with the state-of-the-art review of nanoscale materials and devices for communication applications. Advancements of enabling photonic components for fiber-optic communication systems began nearly 40 years ago, while the advantages of optical computing systems have been discussed for several decades.

Quantum communication has been one of the important applications of nanophotonics [82–84]. It has drawn considerable attention of the relevant R&D community. The motivation for quantum communication comes from the current exponentially increasing trends of computing and communication devices, which makes it increasingly important to implement quantum devices as replacements of the classical ones. It is very likely that quantum processing and quantum communication will be the dominating aspects of future data processing and networking technologies. In this stream, nanophotonic waveguidance has been an important aspect of the envisioned quantum networks wherein the incorporation of a quantum interface interconverting local and flying qubits remains vital.

Quantum networks are composed of quantum nodes and quantum channels; nodes being the quantum systems to store and process local quantum information (in quantum bits) using quantum gates, and the exchange of information among the network nodes is achieved by the use of quantum channels (or waveguides). Stationary qubits (or photons) have been identified as the fundamental elements to be intelligently manipulated in quantum networks. Thus, photons have been identified as ideal carriers for logic transport from one node to another of the envisioned quantum networks incorporating optical fibers (or waveguides) as these would act as sufficiently lossless quantum channels for flying photons.

In order to transform the concept of quantum network to scalable implementation framework, efforts have been made to find suitable quantum structures and processes for the generation of photons as flying qubits [79–81,85]. Various models have been put forward along with their evolutions, and attempts have been made to propose framework for the construction of nanophotonic waveguidance system using special types of fibers (e.g., photonic crystal fibers) as the channel for flying photons (or qubits). Generally, two-level or multilevel atoms or nanodots are used to represent stationary qubits while photons are used as flying qubits. Photon is emitted as a result of transition from a higher energy level to a lower one of an atom or a nanodot. Right- and left-circular polarization states of photon can effectively represent logic states of the flying qubit, which are binary in nature and can propagate over long distances without significant degradation of the signal.

A nanophotonic waveguidance scheme may incorporate the system having three major components—transmit spin–photon interface, waveguide, and receive photon–spin interface, as shown in Figure 2.20 [86,87]. In this

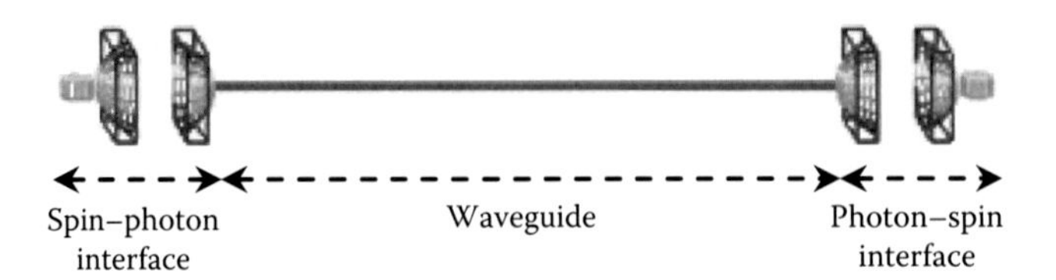

FIGURE 2.20
Components of a nanophotonic waveguide.

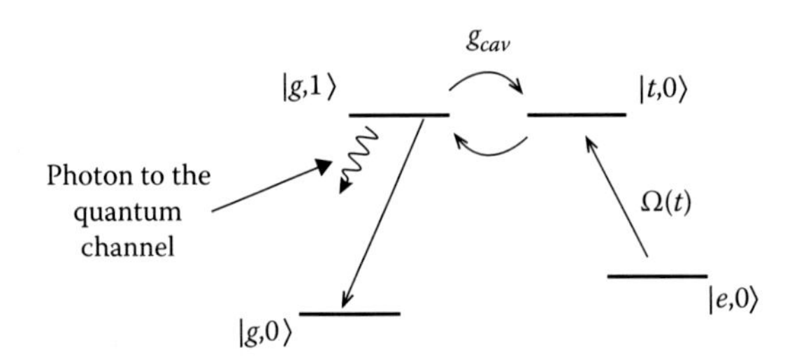

FIGURE 2.21
The spin–photon interface.

figure, the spin–photon interface consists of a nanodot and a coupled optical cavity connected to the waveguide. The qubit is stored in a superposition of the two degenerate spin states, as follows $|g\rangle$ and $|e\rangle$, as illustrated in Figure 2.21, which are the ground states split upon the application of a static magnetic field. The state $|t\rangle$ represents the intermediate state in the Raman process, also called as the trion state.

In a spin–photon interface system, when a laser field of appropriate pulse shape and frequency is made to incident upon the nanodot (i.e., the sending node), a cavity-assisted Raman process raises the nanodot from the ground state $|e,0\rangle$ to a higher energy state $|t,0\rangle$ with a time-dependent complex Rabi frequency $\Omega(t)$. In the process of excitation, the trion state $|t,0\rangle$ is resonantly coupled to the cavity state $|g,1\rangle$ through the coupling constant g_{cav}. While being in the excited state, the system is transformed to the ground spin state $|g,0\rangle$ through the emission of a photon, which leaks out to the connected quantum channel (with the leakage constant k). Figure 2.21 illustrates the transfer of states of photon qubits. The two ground spin states $|e,0\rangle$ and $|g,0\rangle$ are separated by a strong static magnetic field, and also, $\Omega(t)$ is the fundamental controlling input depending on the pulse shape of laser field.

Interestingly, quantum communication is an emerging application of cavity QED, the concepts of which have been implemented for the investigation of an enhanced framework for nanophotonic waveguidance system [48,88–91]. In this connection, a suitable simulation scheme for spin–photon interface has been developed, and the results reveal that spin–photon interface converts the nanodot spin qubit to photon qubit, which is transmitted

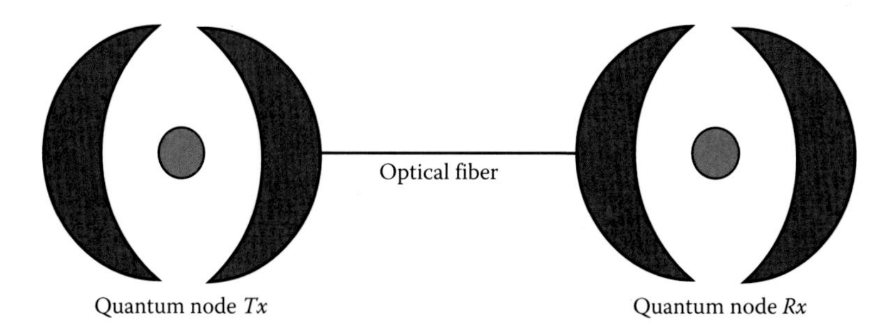

FIGURE 2.22
Transmit and receive nodes as quantum cavities.

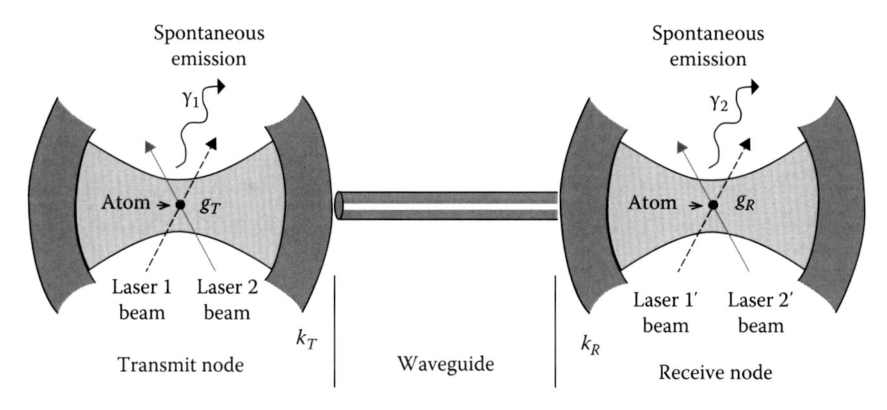

FIGURE 2.23
Nodes under treatments through lasers.

over quantum channel. In this work, each node consists of a Rubidium (^{87}Rb) atom trapped inside a two-mode optical cavity (Figure 2.22). An enhanced approach for transferring quantum state between quantum nodes is proposed based on cavity QED wherein a system of lasers is applied on the atom for generating photons through the process of Raman transition (Figure 2.23).

For simulation purpose, logic states "0" and "1" are represented by two subspaces of the hyperfine energy levels with magnetic sublevels of Rubidium (^{87}Rb) atom. A static magnetic field is applied upon the atoms so that the hyperfine states of ^{87}Rb atom are split into magnetic sublevels. Depending on the logic state of the transmit node, a right- or left-circularly polarized photon with designated frequency is produced through a cavity-assisted Raman process. When the photon is received at the receive node via a quantum channel (i.e., optical fiber), the logic state of the transmit node is restored (through a cavity QED process) into the receive node. A desirable feature of the approach is that, during the transmission of logic state, the transmit node itself should not significantly change its quantum state; this is successfully validated through simulations. This essentially represents the efficacy of the simulation approach used for the demonstration of qubit communication.

2.4 Ventures into Medicine

Applications of nanotechnology in medicinal areas offer tremendous promise, stimulating thereby the development of *nanomedicine* or, more generally, *bionanotechnology* [22,23]. Nanomedicine can be used to improve treatment options in respect of surgery and drugs [92]. By *surgery*, one basically means the operation to be done manually, that is, literally by hand. Many cardiac reasons need manual surgery wherein a surgeon has to intervene physically on the heart of a patient. Some of such cardiac-related techniques can now be remotely performed through automations. However, the basic need remains as the direct intervention on a physical system. Usually, drugs are less direct than surgery as the patient may swallow some pills that have physiological effects in the body, mediated by other processes.

Precision always remains of great importance in surgery, by which one roughly means that the damaged portion of the patient's body must be properly accessed by the surgeon with the surety of not damaging anything else. Surgical applications of nanotechnology float the term nanosurgery, which enables more precise techniques than the traditionally implemented ones. Nanosurgery may be exemplified by a kind of surgery on living cells by the atomic force microscopy (AFM) technique using a nanoneedle of 6–8 μm in length and 200–300 nm in diameter, which is able to penetrate both cellular and nuclear membranes without fatal damage to cells [30]. High-energy femtosecond near-infrared laser pulses can also be used for surgery on nanoscale structures inside living cells and tissues, without much damage; such laser pulses simply vaporize the affected tissue without attempting to destroy them by heat. Gold (Au) nanoparticles are used for various cancerous treatments; these absorb light differently, and the laser ablation process can destroy the attached cancer cells without harming the adjacent ones. In all these examples, the common fact remains that the applications allow for direct intervention at the cellular level. It is to be noted that the applications not only enable to act on individual cells, they can be effective within cells without damaging them.

There are novel nanodevices being developed that have potentials to improve cancer detection, diagnosis, and treatment [93]. However, many challenges are there too specifically to the use of nanostructures within biological systems. This is because nanostructures would be too small to be effective in detection or imaging and/or the body may clear them too rapidly. Larger nanoparticles may accumulate in vital organs, causing thereby problems related to toxicity. For example, Au-nanoparticles can be used for treating cancer cells, but the acquired toxicity from nanoparticles cannot be ignored at the same time. Chemotherapy uses cytotoxic drugs for the treatment of cancer; the technique has the downside that these drugs are toxic to benign as well as malignant cells. Apart from that, there are side effects such as vomiting, nausea, etc. As such, a justification of the risks is to be confirmed before a patient is advised to undergo chemotherapy. These issues remain essential to consider before creating nanodevices suitable for human body.

Coming to the point of drug delivery, there exist myriad advantages that nanomedicine can provide. Much interest has been shown by pharmaceutical companies in terms of incorporating nanotechnology into their product lines, which act as the primary driving force in the rise of nanomedicine. In the course of absorption of drugs, the speed of absorption remains of great importance for a successful drug delivery, and nanotechnology can offer a remarkable difference in this regard. More explicitly, nanotechnology would facilitate the ability to reduce the size of drugs, increasing thereby the surface-to-volume ratio—the key factor to determine the reactivity of nanomaterials. As the body absorbs drugs, it acts on the surfaces of drugs; a large value of the surface-to-volume ratio will essentially make drugs to dissolve faster. This would eventually improve the drug delivery mechanism as the surfaces of materials are most reactive, and the improved surface-to-volume ratio would yield better reactivity.

However, it is not always the case that a fast absorption is ideal. In some cases, drugs would need to be released slowly over time. Time-release vitamins can be good examples of it. Vitamins B and C are good soluble in water, and therefore, they would quickly flush from the body if not administered in a timely released fashion. Nanotechnology can be used for this reason in order to attain better time-release capacities. To attain a clearer view, nanotechnology would facilitate to create smaller apertures through which the pharamacological molecules would pass through. In other words, the technology would enable to form lattice structures with openings through which drug molecules would pass. Apart from the speed of drug acceptance by the body, nanotechnology would also be able to engineer smaller drugs, thereby allowing them to traverse various membranes or other biological barriers. This essentially opens up new possibilities of treatments, which were previously unthinkable.

In the context of drug delivery, it would be interesting to give a glance on some primitive experiments related to the dynamics of nanoparticles that would possibly find great technological usage in pharmaceutical needs. In this stream, the attainment of controlled autonomous transport of micron and nanoscale objects within fluidic media can prove to be enormously beneficial to a variety of emerging fields such as biosensing, self-assembly, fluidics, robotics, and targeted drug transport. Before taking a plunge into applying the ideas and concepts to real systems, preparing simple inorganic prototypes that are capable of morning autonomously in a liquid by exploiting a variety of locally available features may constitute interesting pilot projects that are expected to provide useful inputs. Creating self-propelling prototypes has transcended into an active research field by itself where driving forces out of catalytic chemical reactions—mainly in the form of recoil thrust [94], electrokinetic tension [95], and other interfacial forces [96,97] play pivotal roles in shaping up the research endeavor in this field. However, in the case of systems involving larger catalytic structures, such as micron-sized composite particles functionalized with catalytic nanostructures, which react with liquid medium and generate reaction product(s) in gaseous phase, forces such as buoyancy, interfacial tension, and local distribution of reaction heat, etc. will start playing decisive roles in the overall motion induced in these structures. While exact estimation of all these factors on the overall particle dynamics is rather challenging, one can, however, investigate the effect of buoyancy on the motion of self-powered micron-scale structures, and can subsequently look for the scope of using this *size-dependent* thrust as a possible manipulator of motion.

In this section, discussions are made of the system of nanoparticles under consideration, and some of the relevant experimental details about the self-propulsion in a dilute aqueous peroxide solution are touched upon. Investigation is also made of the effects of viscosity and pH of solution on the motion of micron-scale prototypes. Interestingly, these have contradicting effects on motion in the sense that viscosity dampens the particle velocity, whereas an increase in solution pH (in a certain range) accelerates the motion considerably. As such, the central theme remains to exploit this diverse behavior, and proposes models for targeted transport at the submicrometer regime within systems where both viscosity and pH act as variables.

One can consider a polymer-supported micron-scale structure, functionalized with metal nanoparticles, which can catalytically decompose dilute solution of hydrogen peroxide (H_2O_2). In synthesizing such a composite microstructure, nanoparticle form of metals such as Palladium (Pd), Nickel (Ni), and Gold (Au)—that are catalytically active toward the decomposition of H_2O_2, can be selected. For the polymer support, microbeads of cation exchange resins (polystyrene divinylbenzene copolymer, Amberlite—IR 120, Merck) can be used owing to their ion exchange properties, thermal stability, and immunity toward various solvents, pH, and ionic conditions. Besides, the use of these larger polymer–nanoparticle composite particles adds to the flexibility of observing and analyzing their propulsion within H_2O_2 solution clearly.

In experimental measures, metal nanoparticles (of Pd, Ni, and Au) were deposited on the surface of ion exchange polymer resin beads following known chemical routes reported in the literature [98]. The average diameter and weight of spherical beads used in experiments were about 800 μm and 0.8 mg, respectively. In brief, the deposition of catalytic nanoparticles over these polymer spheres involved the activation of polymer in dilute hydrochloric acid (HCl) solution, followed by the exchange of selected cation on polymer surface using dilute solution of the respective metal salt. The final step of synthesis involved the reduction of deposited metal cations over resin surface using sodium borohydride ($NaBH_4$) solution to form nanoparticles of the corresponding metal [24–26]. Figure 2.24 illustrates various processes involved in the experiment.

Nanoparticle-deposited polymer beads were characterized using various characterization facilities. For example, Pd nanoparticle–deposited polymer microbeads were investigated using a LEO 1430 VP scanning electron microscope (SEM) operating at a maximum voltage of 15 kV. Particles of average diameter of around 80 nm were observed to get formed over the surface of polymer beads. The formation of nanoparticles was finally confirmed by x-ray diffraction measurements using a Bruker AXS, Advance D8 Diffractometer, with a Cu-K_α source of x-ray wavelength 1.54 Å. In the XRD spectra, characteristic reflections from the (111) planes of Pd at a 2θ value of 41° established the presence of Pd in the metallic state over polymer surface.

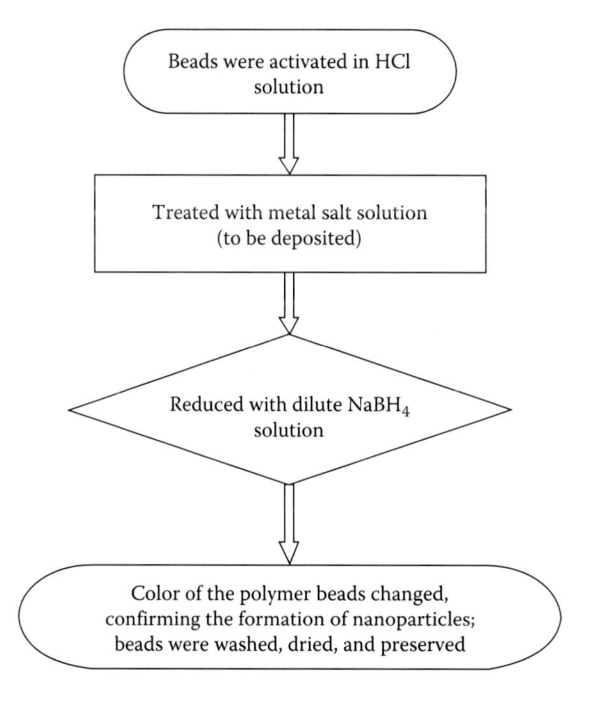

FIGURE 2.24
Experimental steps for depositing catalytic metal nanoparticles over the cation exchange polymer beads.

Further, Au-nanoparticle-deposited polymer beads were, however, studied using a Carl Zeiss Sigma field emission scanning electron microscope (FESEM) operating at a maximum voltage of 30 kV. Owing to very little deposition of Au-nanoparticles over polymer, the oxidation states of deposited structures were confirmed using transmission electron microscopy (TEM) together with the selected area electron diffraction (SAED) measurements. For such needs, micron-sized polymer resins were crushed using a pestle and a mortar, and then dispersed in aqueous HCl for activation, following the previously mentioned procedure. The activated resins were dispersed in 10 mL water and then centrifuged at a speed of 5000 rpm. The supernatant solution was collected and again centrifuged at 12,000 rpm for 5 min. 1.5 g of the precipitate was then collected and redispersed in 10 mL water followed by centrifugation at 12,000 rpm for 5 min. The cycle was pursued for eight times and the finally collected precipitate was redispersed in 10 mL water, followed by the treatment with 100 μL of 0.017 M $HAuCl_4$ solution. The sample was then kept for 48 h in a mechanical shaker operating at 200 rpm and at 25°C. The color of the crushed resins was then found to turn pink, which was subsequently centrifuged, washed, and collected after air-drying for TEM analysis [26]. A dispersion of Au-nanoparticle containing crushed resin beads was drop-cast on carbon-coated Cu grid and then left overnight for air-drying. The grid was then analyzed using a JEOL 2100 TEM, operated at a maximum voltage of 200 kV. The final SEM micrographs of polymer surface coated with Pd and Au-nanoparticles are shown in Figure 2.25.

Before proceeding further, let us take stock of the situation. In experiments, metal nanoparticles, deposited over polymer surface, will act as catalysts for the chemical reaction that will provide energy for propulsion, and the polymer structure provides size and geometry of the motor; the added advantage being clear visibility that aids the experimental process, as mentioned earlier.

The catalytic activity of metal nanoparticles in dilute (usually between 5% and 9%) aqueous H_2O_2 solution is guided by the reaction

$$2H_2O_2(l) \rightarrow O_2(g) + 2H_2O\ (l)$$

where (l) and (g) refer to liquid and gaseous phases, respectively. Thus, the availability of a large number of metal nanoparticles on the polymer surface produces enough oxygen (O_2) bubbles, following the catalytic reaction on it. Some of the formed bubbles get adhered to the surface due to capillary force and the rest get detached. In experiments, enough of them could be seen getting tethered on the surface and growing slowly in size with time. Eventually, a stage arrives when the total buoyancy force inside bubbles supersedes the weight of the composite structure. The motor can then overcome the force of gravity and starts moving up through liquid. The motion could be recorded easily by a simple video recorder in the foreground of a graph sheet where scales are marked (Figure 2.26), and it has been recorded

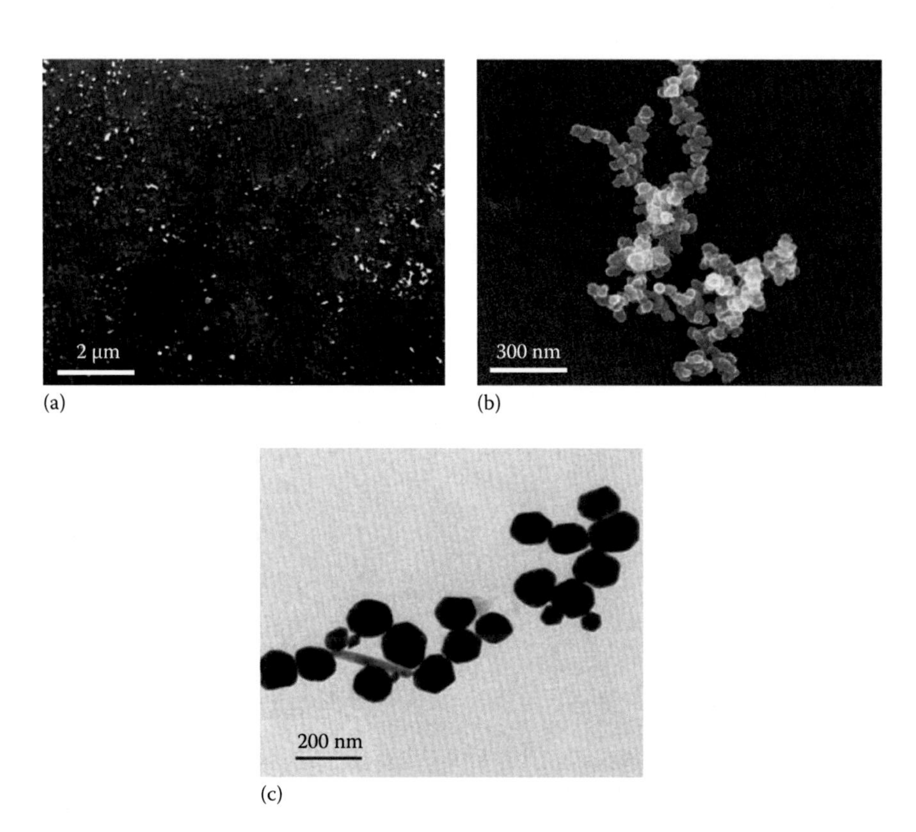

FIGURE 2.25
(a) SEM micrograph of Pd nanoparticle-deposited polymer surface (b) FESEM image of a polymer resin bead containing Au-nanoparticles and (c) TEM image of Au-nanoparticles embedded in polymer matrix.

with a maximum speed of 0.59 cm s^{-1} using the modest, yet accurate, device. The measurements are shown at three instances in Figure 2.26 where the time is measured using a stop watch. The measured order of magnitude of propulsion velocity (of nearly a cm s^{-1}) added excitement to the present scenario since speeds, usually observed in the case of submicroscopic, self-propelled organic structures are of the order of μm s^{-1}. The present system is thus expected to kindle newer horizons toward the development of smart micro/nanobots capable of quickly transporting useful molecules within the liquid phase.

Control over the motion of such a microscopic composite object within a liquid essentially demands to have a great control over the formation of O_2 bubbles on its surface. In order to achieve this, we used the dependence of interfacial tension on the viscosity of a liquid. A careful manipulation of viscosity essentially controls the rate of bubble detachment from the polymer bead surface—subsequently tuning the overall driving force. We manipulate the viscosity of fuel solution by adding calculated amount of

FIGURE 2.26
Time-dependent vertical motion of Pd NP-coated polymer bead placed in 5% aqueous H_2O_2 solution. Starting from rest, in the first 4 s, it covered a distance of $l_2 = 1.42$ cm while in the next 4 s it covered a total distance of $l_3 = 2.25$ cm.

glycerol in it from outside. It is needless to mention that the strength of H_2O_2 during the process of glycerol addition is kept to be the same for different viscosities of solution. To quote some experimental figures, when a volume of 40 μL of glycerol is added to 30 mL of 5% H_2O_2 solution, the velocity of a polymer bead slows down from 0.37 cm s^{-1} (without glycerol) to 0.27 cm s^{-1} (with the concentration mentioned earlier). The polymer bead could be almost held stationary within the solution when the viscosity of solution is increased to about 0.2 Pa s. Quantitatively, the interfacial tension can be related to the viscosity of solution by the empirical formula [99]

$$\ln \gamma = \ln A + \frac{B}{\eta}$$

where
- A and B are constants
- γ is the interfacial tension
- η is the viscosity of the liquid

The reduced interfacial tension actually resulted in a decrease in adhesive force between bubble and polymer surface, detaching thereby bubbles faster from the surface soon after their nucleation. The resultant minimization in bubble coalescence, and hence the net buoyancy, led to a reduced velocity of the bead moving upward, permitting thereby a control over the motion of beads. At higher values of viscosity, owing to the lesser interfacial tension, jets of smaller bubbles were observed to get detaching from beads, thus severely inhibiting the motion.

It is to be reminded at this point that, often in hypertensive patients, blood gets thicker due to cardiovascular diseases. The prototype system presented in the preceding discussion, if properly designed and studied to unearth the dynamics governing the motion accurately, could be of

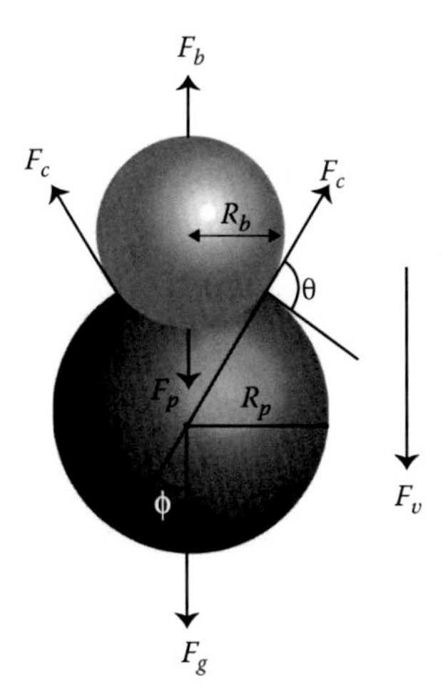

FIGURE 2.27
Schematic of forces acting on a polymer catalytic bead containing a single bubble on top.

immense help in simulating the delivery of useful biomolecules within a viscous serum.

To have quantitative inroads into the dynamics, the prototype can be modeled under the assumptions that the bubble does not change size or shape throughout the course of motion. To keep the things simpler, we further consider that the bubble does not detach from the bead at any point of time during its flight. Finally, instead of taking into account the effect of all the bubbles nucleated around the polymer sphere, we consider a single, large bubble attached on the top of polymer (as shown in Figure 2.27), which produces the equivalent effect of all the smaller bubbles generated around the bead. This would further enable to understand the dynamics accurately, and the scenario of a large number of bubbles taking part in propulsion could possibly be incorporated by a simple scaling.

If R_b represents the radius of bubble, R_p is the radius of nanoparticle-coated beads, F_g is the combined weight of bead with bubble, F_v is the viscous drag, F_c is the capillary force, and F_b is the total buoyancy force, a proper force balance equation would yield

$$F_b = F_v + F_g$$

If the polymer bead is assumed to move upward with a uniform speed, the drag force can be estimated using Stokes' formula $F_v = 6\pi\eta r v$, where we considered

$$r = \frac{R_p + R_b}{2}$$

Further, the viscosity of a binary liquid mixture can be represented as

$$\eta = x_1^2\eta_1 + x_2^2\eta_2 + x_1x_2\eta'$$

where
η is the viscosity of the mixture
x_i $(i=1,2)$ and η_i $(i=1,2)$ are the mole fraction and viscosity of the *i*th component in the mixture, respectively
η' is a *composition* independent cross-coefficient [100]

Using previous relations, it can be easily shown that the composite particle will move uniformly in the upward direction if

$$6\pi\eta rv = F_b - \frac{4}{3}\pi R_p^3\rho_p g$$

where ρ_p is the density of polymer material. The earlier relation can further be simplified to obtain

$$\eta v = \frac{\left(F_b - 4/3\pi R_p^3\rho_p g\right)}{6\pi r} = \text{Constant}$$

Thus, the upward velocity of the composite particle is expected to vary inversely with the viscosity of solution. This is what precisely we observed in experiments where the velocity decreased following a somewhat rectangular hyperbolic profile with the increase in viscosity of fuel solution. The results are shown in Figure 2.28.

Following the concern raised earlier regarding possible application of self-powered micro-/nanobots navigating through fluids for targeted delivery of useful drugs, we take a further step forward by exploring a scenario where Au-nanoparticle functionalized inorganic microparticles are made to self-propel in dilute H_2O_2 solutions. We aim here to study the propulsion of such composite particles and investigate the specificity of such motion on pH values of the fuel solution. Within the context, it is to be mentioned that Au nanoparticles are extensively used in nanoscale science and technology, owing to their stability, catalytic properties, and biocompatibility [101]. Further, protocols for functionalization of Au-nanoparticles with a large number of organic and biological molecules are also available. Thus, the use of Au-nanoparticles could possibly be a natural choice for the construction of model autonomously moving objects. In this stream, we utilize the pH-dependent catalytic decomposition of H_2O_2 by Au-nanoparticles (within alkaline pH regime) to design such a prototype. The decomposition of H_2O_2 produced enough O_2 bubbles, all of which did not get detached immediately from the surface, and rather grew continuously in size. The total buoyant force induced by the attached

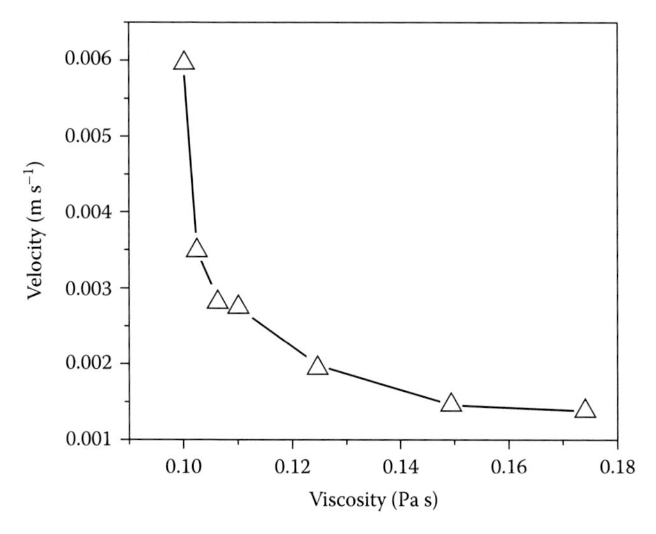

FIGURE 2.28
Variation of the vertical velocity (measured in m s^{-1}) of a polymer resin bead with the viscosity (in Pa s) of H_2O_2 solution, when mixed with various amounts of glycerol.

O_2 bubbles, when sufficiently large, caused beads to move upward, similar to what had been observed previously for Pd-coated particles. The rate of decomposition of H_2O_2 (which, in turn, is proportional to the rate of production of O_2) was observed to be strongly dependent on the pH of solution. The average velocity of a bead, measured at various pH values, followed the same trend as that followed by rate constants corresponding to the catalytic decomposition of H_2O_2 at these pH values [26].

The procedure of depositing Au-nanoparticles on ion exchange polymer resin beads has already been mentioned earlier. pH measurements were performed with an Orion 3 Star pH meter manufactured by Thermo Electron Corporation (Waltham, MA). The velocity of moving bead was calculated from videos captured by a webcam manufactured by Creative Technology Limited (Singapore). At a particular pH, the vertical velocity of polymer resin through liquid was recorded, and the mean of three readings was defined to be the average velocity.

The increase in pH, in essence, increases the catalytic decomposition rate of H_2O_2 over polymer surface. To determine experimentally the rate constants corresponding to H_2O_2 decomposition at various values of pH, several samples of 8.75% (w/v) H_2O_2 solution, each measuring 5 mL, were prepared. The pH of each of these samples was adjusted to a different value using aqueous NaOH solution. These solutions were then titrated with standardized $KMnO_4$ solution at equal intervals of time, both in the presence/absence of catalytic Au-nanoparticle. 0.2 g of Au-nanoparticle deposited/uncoated resin beads was used in each of these solutions, in order to have measurable decomposition of fuel. For every pH, an ensemble of samples was prepared from where one sample at a time was titrated after a definite interval of time. Just before titration, the decomposition of H_2O_2 in these samples (both with and without Au-nanoparticle containing polymer beads) was minimized by adding 0.3 g boric acid (Merck). 1 mL of the resultant solution was then separated out from each of the original samples and was titrated against standardized $KMnO_4$. This was done to avoid possible oxidation of nanoparticles (formed over the polymer surface) by $KMnO_4$. The concentration of $KMnO_4$ solution used for titration was measured using 0.2 N oxalic acid as the primary standard. The strength of H_2O_2 solution thus recorded was plotted in logarithmic scale against the time of titration. The corresponding graphs were found to be straight lines, indicating the decomposition of H_2O_2 following first-order kinetics. The values of rate constants at various values of pH were thus obtained as the slopes of these straight lines.

Figure 2.29 illustrates the pH-dependence of rate constants of H_2O_2 decomposition, measured both in the presence and absence of catalytic nanoparticles. As can be seen, in the pH range of 9.1–10.8, the rate constant for decomposition of H_2O_2 in the presence of Au-nanoparticle deposited beads was higher than the same measured with noncatalytic polymer beads. In addition, it is important here to mention that the pH dependence of rate constant in the presence of Au-nanoparticles followed the typical trend of metal and metal ion–catalyzed decomposition of H_2O_2, reported previously by other investigators [102,103]. The rate constant was measured to be in the range of 0–0.08 min^{-1} within the pH range 9.1–10.8.

In order to find an estimate for the pH-dependent vertical motion of Au-nanoparticle-coated polymer beads in H_2O_2, a model was developed

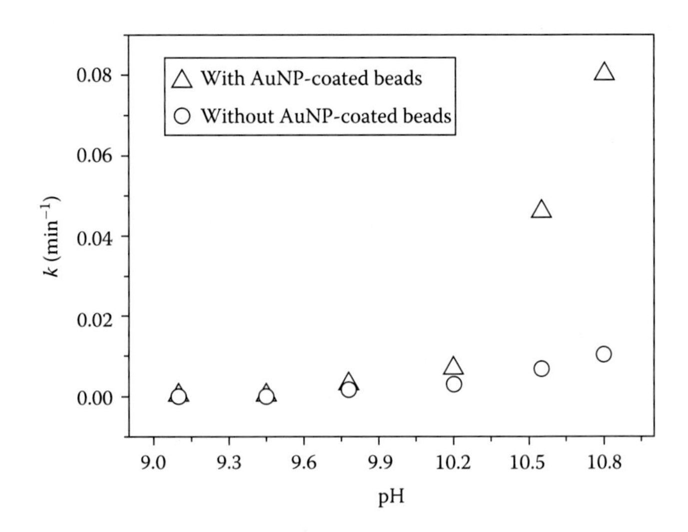

FIGURE 2.29
H_2O_2 decomposition rate versus pH plot in the presence and absence of catalytic Au-nanoparticles.

following Weiss' theory [104] on metal-catalyzed decomposition of alkaline H_2O_2. Essentially, it was assumed that the motion of bead is solely due to the buoyancy of O_2 gas generated on the surface of beads (from the catalytic decomposition of H_2O_2), while O_2 produced from the self-decomposition of H_2O_2 would not contribute to the motion in any way. In addition, it was considered that O_2 was being continuously produced on the bead, leading thereby to its accelerated motion. Finally, in the theoretical analyses, the used rate constants (corresponding to the decomposition of H_2O_2) were obtained from experimentally measured values. The resultant model predicts that the average velocity of a catalytic polymer particle should depend on the pH of solution, obeying the relation of the form

$$V_{av} = \frac{1}{2}\left[2.62\times10^5\left(1-e^{-k_{eff}t}\right)-0.45\right]t_{flight}$$

where, k_{eff} and t_{flight} are, respectively, the *effective catalytic rate constants* (which take into account the effect of only the catalytic nanoparticles on the polymers toward the decomposition of H_2O_2) and the *time of flight* of bead through the solution.

In experiments, velocities of polymer beads were measured at various pH values of solution, and the results are shown in Figure 2.30. Looking at the behavior of rate constant as a function of pH, it is intuitive to understand that, as the solution is made more alkaline, polymer beads are found to move more rapidly. Above a pH value of 9.1, the rate constant starts shifting up, indicating thereby an enhanced nucleation of bubbles on the polymer

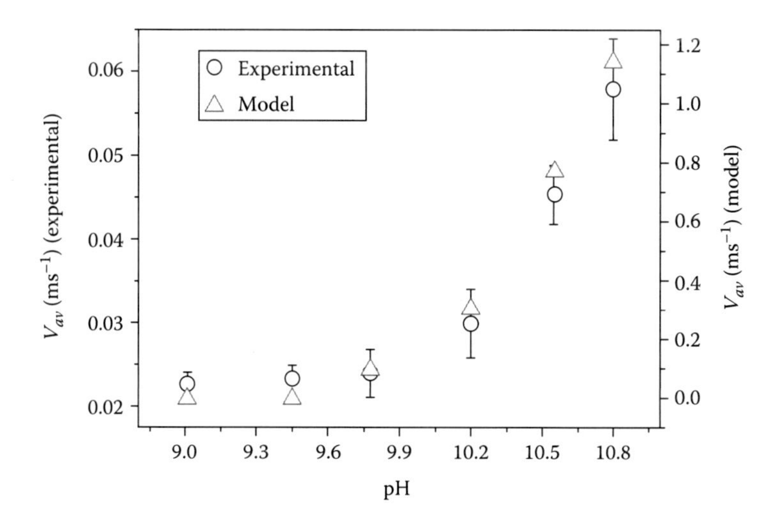

FIGURE 2.30
Comparison of theoretically estimated and experimentally observed velocity profiles of an Au-nanoparticle-coated polymer resin bead as a function of pH of H_2O_2 solution.

surface, resulting in a monotonic increase in bead velocity all the way up to a pH value of 10.16. Beyond this value, there is a sharp jump in O_2 bubble formation, and consequently, we see a sharp increase in the average particle velocity. It is clear from the figure that the trend followed by calculated velocity matched quite well with the experimentally observed one. Thus, the present analyses reasonably account for the pH-specific self-propulsion of Au-nanoparticle-coated polymer microspheres in H_2O_2 fuel—expressing analytically the possible control of catalytic microbots in a liquid simply by tuning the pH of solution.

The two contradictory effects may now be exploited—namely, the decrease and the increase in particle velocity due to viscosity and increased pH of fuel solution, respectively. A couple of simple questions can now be raised—*Is there a way to compensate for the loss in particle velocity through a viscous solution? Would a calculated increase in pH be of any help in this regard?*

The pH of human blood varies within the range 7.30–7.60 [105], which means that it is slightly alkaline. This fact is relevant to us as we have observed that the enhanced delivery of metal nanoparticle-coated polymers occurs in the alkaline regime. However, the severe deterrent is that the tolerance in pH of blood (that the human body can allow) is very narrow. Therefore, the direct applications of the ideas to the proposed field of drug transport may essentially be limited.

Without further trepidation, the academic interest would drive to figure out the extent to which the pH of solution needs to be tuned for a given change in viscosity of the medium, and the exercise may be repeated for a number of viscosity values, in order to keep the velocity of the polymer bead constant. The change in pH (ΔpH) as a function of change in viscosity ($\Delta\eta$) should provide the required intuition in this regard; this is illustrated in Figure 2.31. The plot essentially shows that, within the range of viscosity variation of blood, observed especially for borderline hypertensive patients [106], the equivalent pH changes required to ensure unperturbed delivery of drugs follow a linear profile.

It is widely believed that incisively directed self-propelled autonomous carriers—once fabricated—would definitely set a landmark in the progress of human civilization. In recent years, prototypes of artificial small-scale autonomous carriers have been developed by several active research groups across the globe, using a variety of local/global force fields [107–112]. However, the development of these tiny machines would put forward intimidating challenges when one would be interested to interface them with the macroscopic world. To have an impact in the field of pharmaceutics, these inorganic molecules should meet the crucial feature of biocompatibility. The idea of designing and developing devices that can interact intimately with biological systems at the cellular level is quite exciting, provided they work properly within such environments without affecting the activity of the biological host in any undesirable way. This essentially requires the development of approaches using materials that are able to operate in physiological

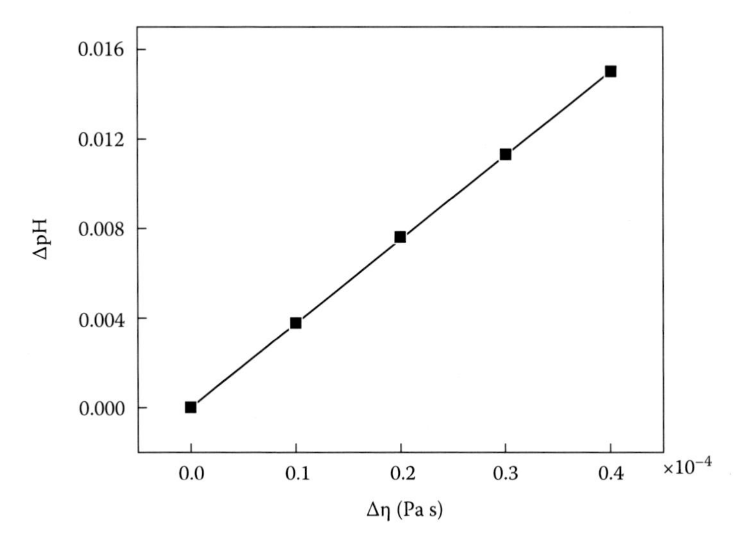

FIGURE 2.31
Variation of estimated change in pH (ΔpH) that counteracts the effect of viscosity change (Δη) of human blood—observed for borderline hypertensive patients, in order to ensure unperturbed delivery of drugs.

conditions. In other words, we need systems that would work following *the laws of small scale physics*, and would harness the energy required for its self-propulsion by *softly* interacting with the physiologically compatible environments around them.

The aim of making artificial motors, which would be able to carry out complex functions, would require a self-driven system of particles, where each of the entities could be independently *programmed*—the particles possessing the ability to communicate with each other in different environmental conditions. Such an ideal system is expected to be featured with ease in fabrication, simplicity in functionality, and also, flexibility, as far as the fabrication of hierarchical structures are concerned. Additionally, the preparation of such ensembles has to be commercially profitable with an impressive degree of operational efficiency. Progress has to be made in the direction of developing systems exhibiting fast and repetitive movements over considerably longer time frames, efficient cargo transportation ability, and the capacity to accomplish an assigned task with the minimum scope of error. To address the issue of optimum dimension of the components of such systems, one might look for the dimensional limit under different conditions of the medium, at which the driving force on the particle arising out of chemical reactions would just be sufficient to overcome the Brownian influence, allowing the movements to be still manipulated as desired. Principles of optical characterization techniques, such as that of dynamic light scattering (DLS), might facilitate such an endeavor. DLS measures Brownian motion and relates it to the size of the particles. Brownian motion is estimated by measuring the rate at which the intensity of the scattered light fluctuates when detected using a suitable optical arrangement. The small particles cause the intensity to fluctuate more rapidly than the large ones. Optical detection of motion of active microparticles of known dimensions dispersed in substrate solution might provide an insight into the size limit up to which artificial motors could be fabricated.

2.5 Concluding Remarks

The makings of miniaturized structures have the ability to alter the physical as well as the electrical and electronic properties of materials—the measurements that can be performed on the micro- and the nanoscales, the exploitation of which will even result into many more products with improved features for novel applications. This lucrative idea is essentially based on efficient transformation of the current nanotech practices in several disciplines, as follows, integrated electronics, optical electronics, medicine, etc. In this stream, the invention of several microtech devices, particularly applied in the area of semiconductor and optical electronics, have been proved to be of immense need.

Among the others, nanophotonics constitutes one of the areas of research in the forefront. The modern research trends in photonics (or nanophotonics) greatly include the spectral behavior of micro- and nanostructured composites, which can be implemented with efficacy in many technological needs. A major part of descriptions made in the present chapter is in-line with the photonics stream, wherein the electromagnetic characteristics of some micro- and nanostructured systems are touched upon. Among the microstructured systems, the propagation characteristics of some special types of guides are taken into account, as follows multilayered bandgap fibers, helical (or twisted) clad fibers, and chiral fibers. Keeping in mind the spectroscopic applications, highlights are given to the bandgap properties of omniguides with dimensional properties on the microscale—the feature that can be effectively exploited for selective transmission of optical signals. Helical clad and chiral fibers can also be effectively used for optical sensing as the effect due to pitch in the case of former one brings in changes in the propagation characteristics of guides, whereas in the case of latter one, the property of chirality makes the guide to be optically active. The importance of nanophotonics in the area of quantum communication is also discussed. In this stream, illustrations are made of the nanophotonic waveguidance for the transfer of qubit states by using the theory of cavity QED.

There are several directions wherein nanotech ventures have turned fantasies into realities for the mankind. Among those, the implementation in medicinal area is one of the prime attractions. The chapter partly gives a cursory understanding of the relevant importance, and emphasizes the dynamics of nanoparticles in the course of drug delivery. In this context, discussions are made on the preparation of metal nanoparticles–coated polymer resin beads followed by the investigation of their locomotion in an aqueous H_2O_2 solution. Clear signatures of autonomous locomotion are observed, where the catalytic activity of metal nanoparticles ultimately drives the motion. Interesting control on the motion is achieved by tuning the viscosity of solution, which impedes the velocity of nanoparticle-coated beads. In an effort to gain a control over the motion of nanoparticles, a very dilute H_2O_2 solution is used with a careful control over the pH of solution. It is particularly seen that the velocity of micron-sized objects rises rapidly in the alkaline regime. These two *apparently* opposing phenomena are put together, and an increased targeted delivery of the intelligent microbots is proposed through a viscous solution where the loss of speed is proposed to be compensated by a pH gradient. Though a direct application of this idea to drug delivery through human blood remains elusive, other competing mechanisms, not considered here, may influence our findings, hopefully in the desired direction.

Acknowledgment

The chapter is dedicated to the highly esteemed professor, the late Prof. Prasad Khastgir, of the Institute of Technology, Banaras Hindu University (Varanasi, India). Though he is no more, his inspirations and mentorship will always keep him alive.

References

1. E.K. Drexler, 1987, *Engines of Creation: The Coming Era of Nanotechnology*, Broadway Books, New York.
2. E.L. Wolf, 2004, *Nanophysics and Nanotechnology: An Introduction to Modern Concepts in Nanoscience*, Wiley-VCH, Weinheim, Germany.
3. A. Lakhtakia, 2005, Whither nanotechnology? *Economic Perspectives* 10(4), 27–28.
4. R.J. Martin Palma and A. Lakhtakia, 2010, *Nanotechnology: A Crash Course*, SPIE Press, Bellingham, WA.
5. D. Faber and A. Lakhtakia, 2009, Scenario planning and nano-technological futures, *European Journal of Physics* 30, S3–S15.

6. B.E.A. Saleh and M.C. Teich, 1991, *Fundamentals of Photonics*, Wiley-Interscience, New York.
7. L. Pavesi, 2003, Will silicon be the photonic material of the next millennium? *Journal of Physics: Condensed Matter* 15(26), 1169–1196.
8. L. Pavesi, 2007, Photonics applications of nanosilicon, in *Frontiers in Optical Technology: Materials and Devices*, P.K. Choudhury and O.N. Singh (Eds.), Nova Science Publishers, Inc., New York.
9. Y. Shen, C.S. Friend, Y. Jiang, D. Jakubczyk, J. Swiatkiewicz, and P.N. Prasad, 2000, Nanophotonics: Interactions, materials, and applications, *The Journal of Physical Chemistry B* 140, 7577–7587.
10. R.M. Walser, 2003, Metamaterials: An introduction, in *Introduction to Complex Mediums for Optics and Electromagnetics*, W.S. Weiglhofer and A. Lakhtakia (Eds.), SPIE Press, Bellingham, WA.
11. J.B. Pendry, 2006, Metamaterials in the sunshine, *Nature Materials* 5(8), 599–600.
12. A. Lakhtakia and T.G. Mackay, 2006, Meet the metamaterials, *OSA Optics & Photonics News* 18(1), 34–39.
13. M. Faraday, 1857, The Bakerian lecture: Experimental relations of gold (and other metals) to light, *Philosophical Transactions of the Royal Society of London* 147, 145–181.
14. R. Vaia and J. Baur, 2008, Adaptive composites, *Science* 319, 420–421.
15. W. Withayachumnankul and D. Abbott, 2009, Metamaterials in the terahertz regime, *IEEE Photonics Journal* 1(2), 99–118.
16. P.K. Choudhury and O.N. Singh, 2005, Electromagnetic materials, in *Encyclopedia of RF and Microwave Engineering*, K. Chang (Ed.), Wiley, New York.
17. P.K. Choudhury and O.N. Singh, 2000, Some multilayered and other unconventional lightguides, in *Electromagnetic Fields in Unconventional Structures and Materials*, O.N. Singh and A. Lakhtakia (Eds.), Wiley, New York.
18. A.B.M.A. Ibrahim and P.K. Choudhury, 2006, Omniguiding Bragg fibers: Recent advances, in *Frontiers in Lasers and Electro-Optics Research*, W.T. Arkin (Ed.), Nova Science Publishers, Inc., New York.
19. K.Y. Lim, P.K. Choudhury, and Z. Yusoff, 2010, Chirofibers with helical windings—An analytical investigation, *Optik* 121, 980–987.
20. M.M. Rahman and P.K. Choudhury, 2011, Cavity quantum electrodynamics for photon mediated transfer of quantum states. *Journal of Applied Physics* 109(11), 113110-1–113110-8.
21. C. Tahan, 2007, Identifying nanotechnology in society, *Advances in Computers* 71, 251–271.
22. F. Allhoff, 2009, The coming era of nanomedicine, *American Journal of Bioethics* 9(10), 3–11.
23. M. Ebbsen and G. Jenen, 2006, Nanomedicine: Techniques, potentials, and ethical implications, *Journal of Biomedicine and Biotechnology* 2006, 1–11.
24. A. Agrawal, K.K. Dey, A. Paul, S. Basu, and A. Chattopadhyay, 2008, Chemical locomotives based on polymer supported catalytic nanoparticles, *The Journal of Physical Chemistry C* 112, 2797–2801.
25. K.K. Dey, D. Sharma, S. Basu, and A. Chattopadhyay, 2008, Veering the motion of a magnetic chemical locomotive in a liquid, *Journal of Chemical Physics* 129, 121101-1–121101-4.
26. K.K. Dey, B.R. Panda, A. Paul, S. Basu, and A. Chattopadhyay, 2010, Catalytic gold nanoparticle driven pH specific chemical locomotion, *Journal of Colloid and Interface Science* 348, 335–341.

27. J.G. Vikram, 2011, Biomimetics: Mimicking nature for technological breakthroughs, *Electronics for You* April Issue, 28–35.
28. P. Murphy, D. Munshi, P.A. Kurian, A. Lakhtakia, and R.V. Bartlett, 2011, Nanotechnology, society, and environment, in *Comprehensive Nanoscience and Technology*, D.L. Andrews, G.D. Scholes, and G.P. Wiederrecht (Eds.), Elsevier, Amsterdam, the Netherlands.
29. U. Hilleringmann, 2007, Integrated optics on silicon, in *Frontiers in Optical Technology: Materials and Devices*, P.K. Choudhury and O.N. Singh (Eds.), Nova Science Publishers, Inc., New York.
30. I. Obataya, C. Nakamura, S.W. Han, N. Nakamura, and J. Miyake, 2005, Nanoscale operation of a living cell using an atomic force microscope with a nanoneedle, *Nano Letters* 5(1), 27–30.
31. A. Ghatak and K. Thyagarajan, 1998, *Introduction to Fibre Optics*, Cambridge University Press, Cambridge, U.K.
32. Y. Xu, R.K. Lee, A. Yariv, J.G. Fleming, and S.-Yu. Lin, 2000, Asymptotic analysis of Bragg fibres, *Optics Letters* 25, 1756–1758.
33. Y. Xu, A. Yariv, J.G. Fleming, and S.-Yu. Lin, 2003, Asymptotic analysis of silicon based Bragg fibers, *Optics Express* 11, 1039–1049.
34. M. Ibanescu, S.G. Johnson, M. Soljacic, J.D. Joannopoulos, Y. Fink, O. Weisberg, T.D. Engeness, S.A. Jacobs, and M. Skorobogatiy, 2003, Analysis of mode structure in hollow dielectric waveguide fibres, *Physical Review E* 67, 1–8.
35. D. Englund, D. Fattal, E. Waks, G. Solomon, B. Zhang, T. Nakaoka, Y. Arakawa, Y. Yamamoto, and J. Vuckovic, 2005, Controlling the spontaneous emission rate of single quantum dots in a two-dimensional photonic crystal, *Physical Review Letters* 95(1), 013904-1–013904-4.
36. H. Shin, P.B. Catrysse, and S. Fan, 2005, Effect of the plasmonic dispersion relation on the transmission properties of subwavelength cylindrical holes, *Physical Review B* 72(8), 085436-1–085436-7.
37. S. Noda, M. Fujita, and T. Asano, 2007, Spontaneous-emission control by photonic crystals and nanocavities, *Nature Photonics* 1(8), 449–458.
38. J.D. Joannopoulos, S.G. Johnson, J.N. Winn, and R.D. Meade, 2008, *Photonic Crystals: Molding the Flow of Light*, Princeton, NJ.
39. E. Yablonovitch, 1994, Photonic crystals, *Journal of Modern Optics* 41, 173–194.
40. P. Russell, 2003, Photonic crystal fibres, *Science* 299, 358–362.
41. P. Yeh, A. Yariv, and E. Marom, 1978, Theory of Bragg fibre, *Journal of the Optical Society of America* 68, 1196–1201.
42. E. Yablonovitch, 1993, Photonic band-gap structures, *Journal of the Optical Society of America B* 10(2), 283–295.
43. J.C. Knight, J. Broeng, T.A. Birks, and P.S.J. Russell, 1998, Photonic bandgap guidance in optical fibres, *Science* 282, 1476–1478.
44. R.F. Cregan, B.J. Mangan, J.C. Knight, T.A. Birks, P.S.J. Russell, P.J. Roberts, and D.C. Allan, 1999, Single-mode photonic bandgap guidance of light in air, *Science* 285, 1537–1539.
45. J.C. Chen, A. Haus, S. Fan, P.R. Villeneuve, and J.D. Joannopouls, 1996, Optical filters from photonic band-gap air bridges, *IEEE Journal of Lightwave Technology* 14, 2575–2580.
46. A.B.M.A. Ibrahim and P.K. Choudhury, 2006, On the analytical investigation of fields and power patterns in coaxial omniguiding Bragg fibres, *Optik* 117, 33–39.

47. A.B.M.A. Ibrahim and P.K. Choudhury, 2005, Analytical design of photonic band-gap fibres and their dispersion characteristics, *Optik* 116, 169–174.
48. M.M. Rahman and P.K. Choudhury, 2011, On the investigation of field and power through photonic crystal fibres—A simulation approach, *Optik* 122, 963–969.
49. J.R. Pierce, 1950, *Travelling Wave Tubes*, D. Van Nostrand, Princeton, NJ, pp. 229–230.
50. P.K. Choudhury, and D. Kumar, 2010, On the slow-wave helical clad elliptical fibres, *Journal of Electromagnetic Waves and Applications* 24, 1931–1942.
51. U.N. Singh, O.N. Singh II, P. Khastgir, and K.K. Dey, 1995, Dispersion characteristics of a helically cladded step-index optical fibre: An analytical study, *Journal of the Optical Society of America B* 12, 1273–1278.
52. D. Kumar, P.K. Choudhury, and F.A. Rahman, 2007, Towards the characteristic dispersion relation for step-index hyperbolic waveguide with conducting helical winding, *Progress in Electromagnetics Research (PIER)* 71, 251–275.
53. D. Kumar and O.N. Singh II, 2001, Some special cases of propagation characteristics of an elliptical step-index fibre with a conducting helical winding on the core-cladding boundary—An analytical treatment, *Optik* 112, 561–566.
54. D. Kumar and O.N. Singh II, 2002, Modal characteristic equation and dispersion curves for an elliptical step-index fibre with a conducting helical winding on the core-cladding boundary—An analytical study, *IEEE Journal of Lightwave Technology* 20, 1416 –1424.
55. D. Kumar and O.N. Singh II, 2002, An analytical study of the modal characteristics of annular step-index waveguide with elliptical cross section with two conducting helical windings on the two boundary surfaces between the guiding and the non-guiding regions, *Optik* 113, 193–196.
56. D. Kumar, P.K. Choudhury, and O.N. Singh II, 2008, Towards the dispersion relations for dielectric optical fibres with helical windings under slow- and fast-wave considerations—A comparative analysis, *Progress in Electromagnetics Research (PIER)* 80, 409–420.
57. A.H.B.M. Safie and P.K. Choudhury, 2009, On the hybrid field patterns of helical clad dielectric optical fibres, *Progress in Electromagnetics Research (PIER)* 91, 69–84.
58. D. Kumar, 2004, A preliminary ground work for the study of the characteristic dispersion equation for a slightly elliptical sheath helix slow wave structure, *Journal of Electromagnetic Waves and Applications* 18, 1033–1044.
59. P.K. Choudhury and D. Kumar, 2010, On the slow-wave helical clad elliptical fibres, *Journal of Electromagnetic Waves and Applications* 24, 1931–1942.
60. C.C. Siong and P.K. Choudhury, 2009, Propagation characteristics of tapered core helical clad dielectric optical fibres, *Journal of Electromagnetic Waves and Applications* 23, 663–674.
61. P.K. Choudhury, 2012, Tapered optical fibers—An investigative approach to the helical and liquid crystal types, in *Fiber Optic Sensors*, Y. Moh, S.W. Harun, and H. Arof (Eds.), In Tech, Rijeka, Croatia, pp. 185–232.
62. A. Lakhtakia, 1990, *Selected Papers on Natural Optical Activity*, SPIE Optical Engineering Press, Bellingham, WA.
63. N. Engheta and P. Pelet, 1989, Modes in chirowaveguides, *Optics Letters* 14, 593–595.
64. N. Engheta and D.L. Jaggard, 1988, Electromagnetic chirality and its applications, *IEEE Transactions on Antennas and Propagation* 30, 6–12.

65. A. Lakhtakia, 2000, A mini-review on isotropic chiral mediums, in *Electromagnetic Fields in Unconventional Materials and Structures*, O.N. Singh and A. Lakhtakia (Eds.), Wiley, New York, pp. 125–149.
66. A. Lakhtakia, 2001, Enhancement of optical activity of chiral sculptured thin films by suitable infiltration of void regions, *Optik* 112, 145–148.
67. S. Bassiri, C.H. Papas, and N. Engheta, 1988, Electromagnetic wave propagation through a dielectric-chiral slab interface and through a chiral slab, *Journal of the Optical Society of America A* 5, 1450–1459.
68. H. Cory and I. Rosenhouse, 1991, Electromagnetic wave propagation along a chiral slab, *IEE Proceedings—H* 138, 51–54.
69. D.L. Jaggard, R.A. Mickelson, and C.H. Papes, 1979, On electromagnetic waves in chiral media, *Journal of Applied Physics* 18, 211–216.
70. C. Eftimiu and L.W. Pearson, 1989, Guided electromagnetic waves in chiral media, *Radio Science* 24, 351–359.
71. P. Pelet and N. Engheta, 1990, The theory of chirowaveguides, *IEEE Transactions on Antennas and Propagation* 38, 90–98.
72. H. Cory and T. Tamir, 1992, Coupling processes in circular open chirowaveguides. *IEE Proceedings—H* 139, 165–170.
73. Kh.S. Singh, P.K. Choudhury, V. Misra, P. Khastgir, and S.P. Ojha, 1993, Field cutoffs of three-layer parabolically deformed planar chirowaveguides, *Journal of the Physical Society of Japan* 62, 3778–3782.
74. P.K. Choudhury and T. Yoshino, 2002, Dependence of optical power confinement on core/cladding chiralities in chirofibres, *Microwave and Optical Technology Letters* 32(5), 359–364.
75. P.K. Choudhury and T. Yoshino, 2002 Characterization of the optical power confinement in a simple chirofibre, *Optik* 113(2), 89–96.
76. P.K. Choudhury, 2002, On the propagation of electromagnetic waves through parabolic cylindrical chiroguides with small flare angles, *Microwave and Optical Technology Letters* 33(6), 414–419.
77. P.K. Choudhury, 2002, Partial electromagnetic wave guidance by a parabolic cylindrical chiroboundary with small flare angle, *Optik* 113(4), 177–180.
78. A. Nair and P.K. Choudhury, 2007, On the analysis of field patterns in chirofibres, *Journal of Electromagnetic Waves and Applications* 21(15), 2277–2286.
79. J.I. Cirac and P. Zoller, 1995, Quantum computations with cold trapped ions, *Physical Review Letters* 74, 4091–4094.
80. J.I. Cirac, P. Zoller, H.J. Kimble, and H. Mabuchi, 1997, Quantum state transfer and entanglement distribution among distant nodes in a quantum network, *Physical Review Letters* 78, 3221–3224.
81. X.D. Fan, P. Palinginis, S. Lacey, H.L. Wang, and M.C. Longeran, 2000, Coupling semiconductor nanocrystals to a fused-silica microsphere—A quantum-dot microcavity with extremely high Q factors, *Optics Letters* 25, 1600–1602.
82. H.J. Carmichael, 1993, Quantum trajectory theory for cascaded open systems, *Physical Review Letters* 70, 2273–2276.
83. C.W. Gardiner, 1993, Driving a quantum system with the output field from another driven quantum system, *Physical Review Letters* 70, 2269–2272.
84. B.B. Blinov, D.L. Moehring, L.-M. Duan, and C. Monroe, 2004, Observation of entanglement between a single trapped atom and a single photon, *Nature* 428, 153–157.

85. P. Chen, C. Piermarocchi, L.J. Sham, D. Gammon, and D.G. Steel, 2004, Theory of quantum optical control of a single spin in a quantum dot, *Physical Review B* 69, 075320–075327.
86. W. Yao, R.B. Lin, and L.J. Sham, 2004, Nanodot-cavity electrodynamics and photon entanglement, *Physical Review Letters* 92, 217402–217405.
87. W. Yao, R.B. Lin, and L.J. Sham, 2005, Theory of control of the spin–photon interface for quantum networks, *Physical Review Letters* 95, 030504–030507.
88. M.M. Rahman and P.K. Choudhury, 2009, Polarized photon generation for the transport of quantum states—A closed-system simulation approach. *Progress in Electromagnetics Research M* 8, 249–261.
89. M.M. Rahman and P.K. Choudhury, 2010, Nanophotonic waveguidance in quantum networks—A simulation approach for quantum state transfer, *Optik* 121, 1649–1653.
90. M.M. Rahman and P.K. Choudhury, 2011, Towards a novel simulation approach for the transport of atomic states through polarized photons, *Optik* 122, 84–88.
91. M.M. Rahman and P.K. Choudhury, 2011, On the quantum link for transport of logic states, *Optik* 122, 660–665.
92. R.A. Freitas, 2007, Personal choice in the coming era of nanomedicine, in *Nanoethics: The Social and Ethical Implications of Nanotechnology*, F. Allhoff et al. (Eds.), Wiley, Hoboken, NJ.
93. J.F. Kukowska-Latallo, K.A. Candido, Z. Cao, S.S. Nigavekar, I.J. Majoros, T.P. Thomas, L.P. Balogh, M.K. Khan, and J.R. Baker Jr., 2005, Nanoparticle targeting of anticancer drug improves therapeutic in animal model of human epithelial cancer, *Cancer Research* 65(12), 5317–5324.
94. R.F. Ismagilov, A. Schwartz, N. Bowden, and G.M. Whitesides, 2002, Autonomous movement and self-assembly, *Angewandte Chemie International Edition* 41, 652–654.
95. Y. Wang, R.M. Hernandez, D.J. Bartlett Jr., J.M. Bingham, T.R. Kline, A. Sen, and T.E. Mallouk, 2006, Bipolar electrochemical mechanism for the propulsion of catalytic nanomotors in hydrogen peroxide solutions, *Langmuir* 22, 10451–10456.
96. R. Golestanian, T.B. Liverpool, and A. Ajdari, 2005, Propulsion of a molecular machine by asymmetric distribution of reaction products, *Physical Review Letters* 94, 220801–220804.
97. R. Golestanian, T.B. Liverpool, and A. Ajdari, 2007, Designing phoretic micro- and nano-swimmers, *New Journal of Physics* 9, 2–8.
98. G. Majumdar, M. Goswami, T.K. Sarma, A. Paul, and A. Chattopadhyay, 2005, Au nanoparticles and polyaniline coated resin beads for simultaneous catalytic oxidation of glucose and colorimetric detection of the product, *Langmuir* 21, 1663–1667.
99. A.H. Pelofsky, 1966, Surface tension–viscosity relation for liquids, *Journal of Chemical & Engineering Data* 11, 394–397.
100. M.P. Saksena and H.S. Kumar, 1975, Viscosity of binary liquid mixtures, *Journal of Physics C: Solid State Physics* 8, 2376–2381.
101. R. Shukla, V. Bansal, M. Chaudhary, S. Basu, R.R. Bhonde, and M. Sastry, 2005, Biocompatibility of gold nanoparticles and their endocytotic fate inside the cellular compartment: A microscopic overview, *Langmuir* 21, 10644–10654.
102. J. de Laat and H. Gallard, 1999, Catalytic decomposition of hydrogen peroxide by Fe(III) in homogeneous aqueous solution: Mechanism and kinetic modeling, *Environmental Science & Technology* 33, 2726–2732.

103. N. Kitajima, S. Fukuzumi, and Y. Ono, 1978, Formation of superoxide ion during the decomposition of hydrogen peroxide on supported metal oxides, *The Journal of Physical Chemistry* 82, 1505–1509.
104. J. Weiss, 1952, The free radical mechanism in the reactions of hydrogen peroxide, *Advances in Catalysis* 4, 343–365.
105. G. Brecher and E.P. Cronkite, 1950, Morphology and enumeration of human blood platelets, *Journal of Applied Physiology* 3, 365–377.
106. R.L. Letcher, S. Chien, T.G. Pickering, and J.H Laragh, 1983, Elevated blood viscosity in patients with borderline essential hypertension, *Hypertension* 5, 757–762.
107. W.F. Paxton, K.C. Kistler, C.C. Olmeda, A. Sen, S.H. St. Angelo, Y. Cao, T.E. Mallouk, P. Lammert, and V.H. Crespi, 2004, Catalytic nanomotors: Autonomous movement of striped nanorods, *Journal of the American Chemical Society* 126, 13424–13431.
108. R. Dreyfus, J. Baudry, M.L. Roper, M. Fermigier, H.A. Stone, and J. Bibette, 2005, Microscopic artificial swimmers, *Nature* 437, 862–865.
109. A. Ghosh and P. Fischer, 2009, Controlled propulsion of artificial magnetic nanostructured propellers, *Nano Letters* 9, 2243–2245.
110. H.S. Muddana, S. Sengupta, T.E. Mallouk, A. Sen, and P.J. Butler, 2010, Substrate catalysis enhances single-enzyme diffusion, *Journal of the American Chemical Society* 132, 2110–2111.
111. S. Sanchez, A.A. Solovev, Y. Mei, and O.G. Schimdt, 2010, Dynamics of biocatalytic microengines mediated by variable friction control, *Journal of the American Chemical Society* 132, 13144–13145.
112. D. Kagan, S. Balasubramanian, and J. Wang, 2011, Chemically triggered swarming of gold microparticles, *Angewandte Chemie International Edition* 50, 503–506.

3

Dielectric Spectroscopy of Polymer-Based Nanocomposite Dielectrics with Tailored Interfaces and Structured Spatial Distribution of Fillers

G. Polizos, E. Tuncer, V. Tomer, I. Sauers, C.A. Randall, and E. Manias

CONTENTS

3.1 Introduction

Emerging technologies in the areas of electronics and energy storage require the design of next-generation dielectric-component materials with well-defined structure and properties with higher performance under voltage and temperature. Polymers filled with inorganic nanoparticles are potential candidates for breakthrough advances, since they enable the integration of constituents with desirable properties, particularly when hierarchically structuring the filler phase. However, implementation of highly innovative material elements necessitates the fundamental understanding of each element's properties and interactions at the nanoscale and, further, how those are manifested in the macroscopic performance of the nanocomposites. Dielectric spectroscopy, coupled with performance measurements, is crucial in quantifying such properties and unveiling the molecular-level interdependencies and their correlations with device performances.

The chapter summarizes recent studies on the nanostructure–property interdependencies in tailored/structured polymer nanocomposites, and underlines the nanoscale principles leading to the design and synthesis of high-performance nanomaterials for energy storage applications. In principle, inorganic materials of high permittivity combined with polymers of high breakdown strength should improve the energy storage capacity, as energy density is directly proportional to permittivity and the square of the highest applied electric field. This is not always the case, and real composites rarely exhibit a behavior that is a straight-forward addition of filler and matrix response. The main drawbacks in such composite approaches most often relate to the polymer–filler interfaces, in particular, when these interfaces are characterized by properties adverse to the desired macroscopic behavior. With increasing filler surface area, and depending on the polymer–filler interactions, the presence of such interfaces can govern the macroscopic properties of the composite and also determine properties such as the dielectric breakdown strength. The research work in this chapter provides clues on the mitigation of such detrimental effects, which are commonly responsible for the poor dielectric performance of the composites. Three key factors are emphasized for synthesizing multifunctional nanocomposites:

1. *Nanoparticle functionality*: The uniform and controlled dispersion of the inorganic fillers is a challenge in the fabrication of nanocomposites. The fundamental thermodynamics of mixing for polymers and nanofillers are described through a balance of entropic and enthalpic factors. Two fabrication routes are presented: (a) functionalization of the particle surface in order to tailor the polymer/particle interfacial interactions and (b) in situ nucleation of nanoparticles in a polymer solution. The polymer dynamics (relaxations) in the vicinity of the nanoscale boundaries are investigated by means of dielectric relaxation spectroscopy (DRS). Of particular interest is when the size of the particles (~5 nm) is comparable to the length-scale of the polymer conformations.
2. *Self-assembly of desired nanostructures*: The mismatch in the permittivity and conductivity between the organic and the inorganic phases generates concentration in the local electric field that may lead to premature electrical failure of the nanodielectric. Self-assembled structures of fillers with different dielectric properties can be employed

to grade the local electric field and, thus, increase the energy density of the composites. We functionalized and covalently bonded two different inorganic constituents to yield a combined filler that has a thermodynamically favorable dispersion and a capability to covalently bond to the polymer matrix. In this section, we summarize our results on the dynamical behavior of the polymer phase, including those polymers adsorbed on filler surfaces or confined in between fillers. The dynamics of the polymer phase relaxing at the inorganic interfaces is correlated with the macroscopic polarization of the composites under nonlinear electric fields.

3. *Hierarchical nanostructures*: Nanocomposites with controlled spatial distribution of fillers are, arguably, the most promising route to optimize macroscopic response. Such nanocomposites, with fillers structured appropriately over the various length scales, can be synthesized using anisotropic dielectrophoretic assembly or flow/stress-gradient induced alignment techniques. The orientation of the fillers is quantitatively analyzed in terms of Hermans orientation factors by employing two-dimensional x-ray diffraction measurements. We emphasize on the important role of conductivity, complex permittivity, and, particularly, local cluster distribution and relative orientation in controlling high-field dielectric behavior. It is shown that filler alignment can be used to markedly improve the high electric-field breakdown strength and, consequently, the recoverable energy density.

A wide range of materials are investigated to gain insights into the generally applicable mechanisms that control the dielectric breakdown strength and the nonlinear conduction, namely, thermoset (epoxy, polydimethylsiloxane) and thermoplastic (polyethylene, polyurethane) polymers are studied, reinforced by model inorganic nanofillers such as barium titanate, titanium dioxide, and montmorillonite. In the following sections, we present a brief introduction to the dielectric spectroscopy and analysis tools. The remainder of the chapter focuses on the fundamental interrelationships between the nanoscale dielectric properties and the macroscopic properties of the nanocomposites with an emphasis on the dielectric breakdown strength.

3.2 Broadband Dielectric Relaxation Spectroscopy

3.2.1 Dielectric Relaxations

DRS is a dynamic technique quantifying the molecular dynamics and conductivity processes in dielectric (insulating or semiconducting) materials due to their interaction with electromagnetic fields. In a linear system, when an external ac electric field is applied to a dielectric medium with permanent dipole moments, the complex dielectric function $\varepsilon^*(\omega) = \varepsilon'(\omega) - i\varepsilon''(\omega)$ is associated with the correlation function $\Phi(t)$ of the polarization fluctuations [1–3]

$$\frac{\varepsilon^*(\omega) - \varepsilon_\infty}{\varepsilon_s - \varepsilon_\infty} = \int_0^\infty \frac{-d\Phi(t)}{dt} \exp(-i\omega t) \tag{3.1}$$

where

$$\Phi(t) = \frac{\langle \Delta P(t) \Delta P(0) \rangle}{\langle \Delta P(0)^2 \rangle} \tag{3.2}$$

The angular brackets denote a statistical ensemble average of the fluctuation of the polarization P around its equilibrium value; $\omega(= 2\pi f)$ stands for the angular frequency and ε_s, ε_∞ are the low- and high-frequency limits of the real part of permittivity. Dielectric measurements can be performed either in the frequency domain (i.e., the sample is excited/polarized by an external sinusoidal electric field and the complex impedance, $Z(\omega)$, is measured) or in the time domain (i.e., the sample is subjected to a step-function voltage and the dielectric functions are derived from the reflection and transmission coefficients). As an example, the correlation function $\Phi(t)$ can describe the decay of electric polarization when an external electric field is removed from a dielectric material. For a single relaxation time τ (i.e., identical non-interacting dipoles), $\Phi(t)$ is an exponential of the form $\exp(-t/\tau)$ and the dielectric dispersion (Debye) is obtained from Equation 3.1

$$\frac{\varepsilon^*(\omega) - \varepsilon_\infty}{\varepsilon_s - \varepsilon_\infty} = \frac{1}{1 + i\omega\tau} \tag{3.3}$$

where the real and imaginary parts are, respectively,

$$\varepsilon'(\omega) = \varepsilon_\infty + \frac{\varepsilon_s - \varepsilon_\infty}{1 + \omega^2\tau^2} \tag{3.4}$$

$$\varepsilon''(\omega) = \frac{\varepsilon_s - \varepsilon_\infty}{1 + \omega^2\tau^2} \omega\tau \tag{3.5}$$

The correlation function for materials with dipole groups of different reorientation polarization will decay in multiple steps giving rise to several relaxation processes. Moreover, heterogeneous systems or systems with strongly interacting dipoles, such as polymers and polymer composites, are described by the Kohlrausch–Williams–Watts (KWW) function (stretched exponential)

$$\Phi(t)=\exp\left[-\left(\frac{t}{\tau_{KWW}}\right)^{\beta_{KWW}}\right],\quad 0<\beta_{KWW}\leq 1 \tag{3.6}$$

that leads to broad and asymmetric distribution of relaxation times. Analysis of the dielectric dispersions can quantify the dynamic parameters associated with the material's structure in the molecular level and specifically (1) the dielectric relaxation strength ($\Delta\varepsilon=\varepsilon_s-\varepsilon_\infty$), which is a measure of the orientation polarization and can be expressed in terms of the dipole number density and mean square dipole moment; (2) the time scale and activation energy of relaxations arising either from the rotational motion of polar chain segments or from cooperative reorientations associated with the glass transition; (3) the distribution of relaxation times and their correlation with intra- and intermolecular interactions; (4) the ac and dc conductivity values, activation energy, and length-scale of the translational diffusion of mobile ions.

DRS supplements other dynamic techniques such as mechanical spectroscopy, nuclear magnetic resonance, neutron and light scattering; however, the main advantage of dielectric spectroscopy is its ability to follow relaxation processes over an extremely wide range of characteristic times (10^5–10^{-12} s). The following sections present the fundamental concepts of the mathematical formalisms and analysis tools used herein for the interpretations of our experimental results.

3.2.2 Dielectric Formalisms

3.2.2.1 Permittivity Formalism

A superposition of Havriliak–Negami (HN) expressions [1,4–7] was employed for the analysis of the complex permittivity function

$$\varepsilon^*(\omega)=\varepsilon_\infty+\sum_{i=1}^{n}\frac{\Delta\varepsilon_i}{\left[1+(i\omega\tau_{0,i})^{1-\alpha_i}\right]^{\beta_i}} \tag{3.7}$$

Least-square fitting to the experimental data was carried out by sharing the common parameters $\Delta\varepsilon_i$, $\tau_{0,i}$, α_i, β_i, in the real

$$\varepsilon'(\omega)=\sum_{i=1}^{n}\Delta\varepsilon_i\frac{\cos(\beta_i\phi_i)}{\left[1+2(\omega\tau_{0,i})^{1-\alpha_i}\sin\left(\frac{1}{2}\alpha_i\pi\right)+(\omega\tau_{0,i})^{2(1-\alpha_i)}\right]^{\beta_i/2}} \tag{3.8}$$

and imaginary part

$$\varepsilon''(\omega)=\sum_{i=1}^{n}\Delta\varepsilon_i\frac{\sin(\beta_i\phi_i)}{\left[1+2(\omega\tau_{0,i})^{1-\alpha_i}\sin\left(\frac{1}{2}\alpha_i\pi\right)+(\omega\tau_{0,i})^{2(1-\alpha_i)}\right]^{\beta_i/2}} \tag{3.9}$$

of the complex permittivity function. A conductivity power law term (σ_0/ε_0) ω^{-s} was added in the imaginary part (Equation 3.9) when contribution from conductivity was present in the datasets. In the earlier equations α, β are the shape parameters defining the symmetrical and asymmetrical broadening of each mode $(0<(1-\alpha)\beta, s\leq 1)$. The argument $(\beta\,\phi)$ denotes the angle from ε_∞ to any value of the $\varepsilon^*(\omega)$ function in a complex locus diagram [5,8] ($\varepsilon''(\omega)$ vs. $\varepsilon'(\omega)$) where

$$\varphi_i=\arctan\left[\frac{(\omega\tau_{o,i})^{1-\alpha_i}\cos(\alpha_i\pi/2)}{1+(\omega\tau_{o,i})^{1-\alpha_i}\sin(\alpha_i\pi/2)}\right] \tag{3.10}$$

and τ_o is a characteristic time that is associated to the relaxation time $\tau_{\max}(=1/2\pi f_{\max})$ of each mode by

$$\tau_{\max,i}=\tau_{o,i}\left[\frac{\sin\left[(1-a_i)\beta_i\pi/(2+2\beta_i)\right]}{\sin\left[(1-a_i)\pi/(2+2\beta_i)\right]}\right]^{1/(1-\alpha_i)} \tag{3.11}$$

3.2.2.2 Electric Modulus Formalism

The electric modulus was introduced by McCrum et al. [2] and is defined as the reciprocal complex permittivity [9–11]

$$M^*(\omega)=\frac{1}{\varepsilon^*(\omega)}=M'(\omega)+iM''(\omega)=\frac{\varepsilon'(\omega)}{\varepsilon'^2(\omega)+\varepsilon''^2(\omega)}+i\frac{\varepsilon''(\omega)}{\varepsilon'^2(\omega)+\varepsilon''^2(\omega)} \tag{3.12}$$

It is widely used to suppress conductivity and polarization phenomena and to reveal the dipolar contributions. Macedo et al. [12–14] applied this formalism to ionic conductors. The authors approached the dielectric response by an equivalent electric circuit of a parallel capacitance and conductance element configuration. The Debye type relaxation time that corresponds to this circuit (or equivalently to the diffusion of the ions in the material) is expressed as the conductivity relaxation time

$$\tau_\sigma=\frac{\varepsilon_0\varepsilon_s}{\sigma_{dc}},\quad \mathbf{E}=\mathbf{E}_o\varphi(t)=\mathbf{E}_o\exp\left(\frac{-t}{\tau_\sigma}\right) \tag{3.13}$$

and describes the decay of the electric field, **E**, when the electric displacement vector, **D**, remains constant. In a more general case, the single relaxation time (Debye) model can be extended to a distribution of relaxation times [12].

The complex electric modulus in the frequency domain can be written as the Fourier transform of the time derivative of the electric field decay function,

$$M^*(\omega) = M_\infty \left[1 - \int_0^\infty -\frac{d\varphi(t)}{dt} \exp(-i\omega t) dt \right] \tag{3.14}$$

In the frequency domain, a corresponding to the HN expression can be obtained in the modulus level. Tsangaris et al. [15,16] resolved the imaginary and real part for different distributions of relaxation times. A HN distribution will reduce to

$$M'(\omega) = M_\infty M_s \frac{\left[M_s A^\beta + (M_\infty - M_s)\cos\beta\varphi \right] A^\beta}{M_s^2 A^{2\beta} + 2A^\beta (M_\infty - M_s) M_s \cos\beta\varphi + (M_\infty - M_s)^2} \tag{3.15}$$

and

$$M''(\omega) = M_\infty M_s \frac{\left[(M_\infty - M_s)\sin\beta\varphi \right] A^\beta}{M_s^2 A^{2\beta} + 2A^\beta (M_\infty - M_s) M_s \cos\beta\varphi + (M_\infty - M_s)^2} \tag{3.16}$$

where

$M_s = 1/\varepsilon_s$

$M_\infty = 1/\varepsilon_\infty$

and

$$A = \left[1 + 2(\omega\tau_o)^{1-\alpha} \sin\left(\frac{\alpha\pi}{2} \right) + (\omega\tau_o)^{2(1-\alpha)} \right]^{1/2} \tag{3.17}$$

The argument φ and the characteristic time τ_0 have similar dependencies as those in Equations 3.10 and 3.11.

3.2.2.3 Impedance Formalism

The conductivity processes in multiphase systems can be identified and separated in terms of the complex impedance formalism, $Z^*(\omega) = Z'(\omega) - iZ''(\omega)$, where the measured real and imaginary parts are connected with the complex permittivity and the geometric capacity of the free space (C_0)

$$Z'_{meas.}(\omega) = \varepsilon''(\omega) \left(\left[\varepsilon'^2(\omega) + \varepsilon''^2(\omega) \right] \omega C_0 \right)^{-1} \tag{3.18}$$

$$Z''_{meas.}(\omega)=\varepsilon'(\omega)\left(\left[\varepsilon'^{2}(\omega)+\varepsilon''^{2}(\omega)\right]\omega C_0\right)^{-1} \quad (3.19)$$

A typical data representation is based on plots of the imaginary versus the real part, also known as Nyquist plots. In such plots, the overall measured complex impedance can be expressed as the superposition of the intrinsic impedance (bulk contribution described by semicircles) and a dispersive capacitance (constant phase element, CPE) contribution describing electrode polarization effects at low frequencies [17].

$$Z^{*}_{meas.}(\omega)=Z^{*}_{intrinsic}(\omega)+A_{CPE}(i\omega)^{-s} \quad (3.20)$$

In the previous expression, the values of the exponent s vary between 0 and 1 for ideal resistive and capacitive behavior, respectively. The real and imaginary components of the intrinsic impedance can be fitted with analytical expressions similar to Equations 3.8 and 3.9

$$Z''_{intrinsic}(\omega)=R_{\infty}+\sum_{i=1}^{n}\frac{\Delta R_i\cos(\beta_i\phi_i)}{\left[1+2(\omega\tau_{0,i})^{1-\alpha_i}\sin\left(\frac{1}{2}\alpha_i\pi\right)+(\omega\tau_{0,i})^{2(1-\alpha_i)}\right]^{\beta_i/2}} \quad (3.21)$$

$$Z''_{intrinsic}(\omega)=\sum_{i=1}^{n}\frac{\Delta R_i\sin(\beta_i\phi_i)}{\left[1+2(\omega\tau_{0,i})^{1-\alpha_i}\sin\left(\frac{1}{2}\alpha_i\pi\right)+(\omega\tau_{0,i})^{2(1-\alpha_i)}\right]^{\beta_i/2}} \quad (3.22)$$

where

R_{∞} is the resistance at the high-frequency limit

$\tau_{0,i}$ and ΔR_i are the characteristic time and the resistance values for each impedance mode, respectively

The ΔR_i values correspond to the intersection of the extrapolated to low-frequencies semicircle and the real Z' axis. Conductivity σ is obtained by normalizing ΔR values to the geometric parameters according to

$$\sigma=(\Delta R)^{-1}\frac{d}{S} \quad (3.23)$$

where

d is the thickness

S is the surface of the sample

Phenomenological models based on equivalent circuits can simulate the polarization mechanisms and the dissipative contributions due to mobile charges.

3.3 Nanoparticle Functionality

3.3.1 Segmental Dynamics and Conductivity Mechanism

Nanoparticle-filled polymers (nanodielectrics) have gained remarkable scientific attention because of their potential use in a variety of applications ranging from polymer electrolytes to electrical insulation [18–22]. In this section, we report on the correlation between the cooperative segmental dynamics and the conductivity mechanism. Presynthesized titanium dioxide (TiO_2) nanoparticles functionalized with polyethylene glycol (PEG) were dispersed in a thermoplastic polyurethane (PU) with polar soft segments [23]. The particles were approximately 10 nm in size and composites were obtained at low-weight fractions (0.42 and 0.10 wt%) to avoid the formation of pronounced aggregates.

The dynamical properties of the unfilled PU were investigated by means of DRS. Three relaxations are shown in Figure 3.1 for temperatures below room temperature. In the order of increasing temperature, the first two subglass modes, γ and β, are associated with local motions of chain segments. Specifically, the γ process is associated with the rotation of polar groups due to the crankshaft motion of $(CH_2)_n$ sequences, whereas the β process is due to the orientation fluctuations of ester complexes

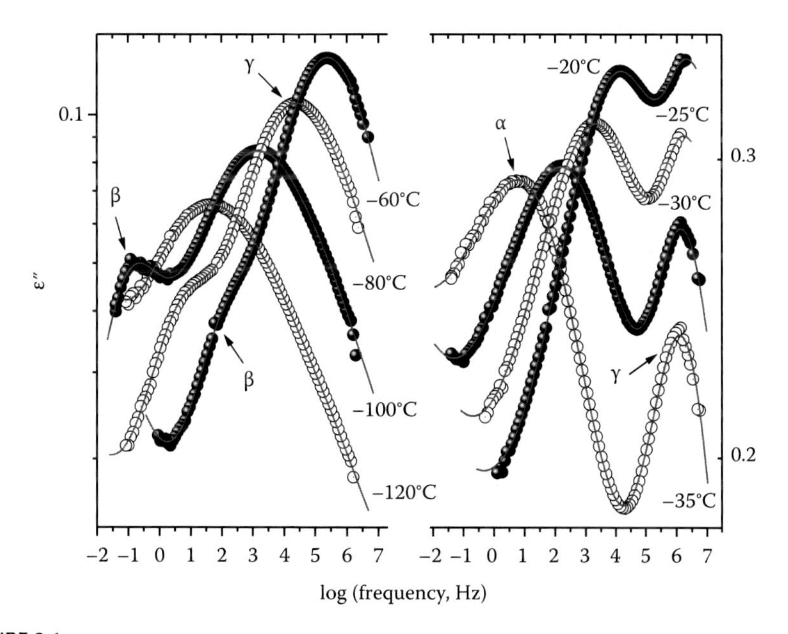

FIGURE 3.1
Imaginary part of permittivity at several temperatures for the unfilled polyurethane matrix. The cooperative α (associated with the glass transition) and the local β, γ relaxations are indicated on the plot. The lines are the best fits of Equation 3.9 to the data.

$-(CH_2)_4-O(C=O)N-$ with participation of attached water molecules [24–29]. The segmental α relaxation at higher temperatures is positioned close to the calorimetric glass transition temperature, T_g, and is attributed to dipole moment cooperative reorientations, during the glass transition of the soft segment microdomains. In the composites, this process is overlapping with a low-frequency conductivity contribution due to variations in the surface charge density of the TiO_2 [23].

Significant changes have been observed in the segmental dynamics of the composites and a comparison plot is shown in Figure 3.2. At high temperatures, the composites exhibit the same dynamical behavior as the pristine PU. However, upon temperature decreasing toward the glass transition temperature, the segmental α mode is found to depend on the particle content;

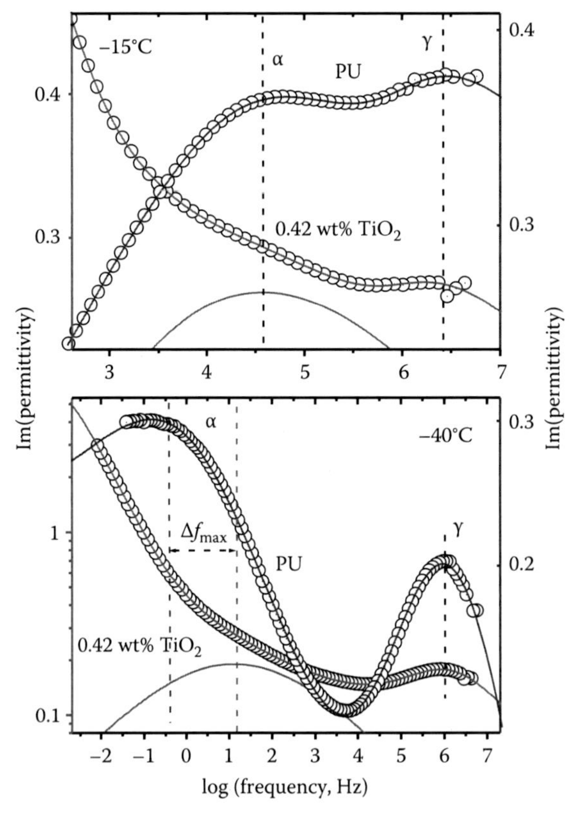

FIGURE 3.2
Comparison plot of the dielectric relaxations for the pristine polyurethane and the 0.42 wt% in TiO_2. No shift is observed in the peak maximum frequencies of the local γ relaxations. However, upon temperature decreasing toward the glass transition temperature, the segmental α mode of the composite shifts 1.5 decades toward higher frequencies and becomes faster compared to the α relaxation in the pristine polyurethane. (From Polizos, G. et al., *Polymer*, 53, 595, 2012.)

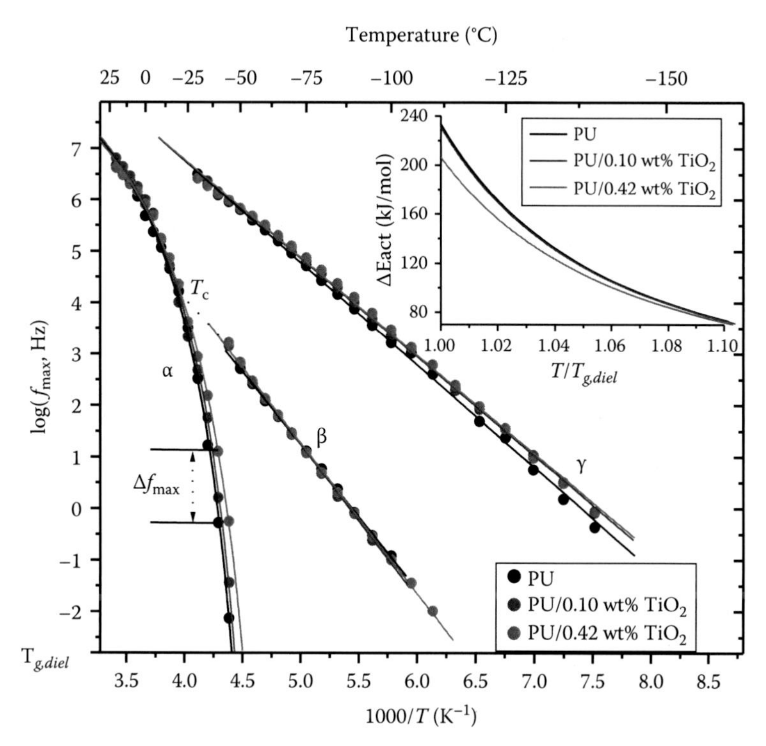

FIGURE 3.3
(See color insert.) Summarized Arrhenius plot of the cooperative (α) and local (β,γ) relaxations at temperatures below room temperature along with the fitting lines according to Equations 3.24 and 3.25. T_c corresponds to the merging temperature of the α and β relaxations. Δf_{max} is the frequency shift of the α relaxation (for the 0.42 wt% composite at −40°C) according to Figure 3.2. The dielectric glass transition $T_{g,diel}$, obtained at 1.6 mHz (τ = 100 s), is also indicated on the plot. The normalized (by T_g) apparent activation energies that correspond to the α relaxations were calculated from Equation 3.27 and are shown in the inset plot. (From Polizos, G. et al., *Polymer*, 53, 595, 2012.)

and specifically for the 0.42 wt% composite, it shifts 1.5 decades toward higher frequencies and becomes faster compared to the α relaxation in the unfilled PU. The time scale dependence for all the relaxations is shown in Figure 3.3. The temperature dependence of f_{max} for the subglass modes, γ and β, is described by the Arrhenius equation

$$f_{max,\beta,\gamma}(T) = f_\infty \exp\left(\frac{-\Delta E_A}{RT}\right) \tag{3.24}$$

where

ΔE_A is the activation energy of the process
f_∞ is the relaxation rate in the high-temperature limit
R is the universal gas constant

TABLE 3.1

Summarized Fitting Analysis Results according to Equations 3.24 and 3.25 for the Frequency Dependence of the Relaxations in Figure 3.3

	α Relaxation					β, γ Relaxations	
Sample	[a]T_g^{DSC}(C)	[b]T_g^{diel}(C)	[c]T_0(C)	[d]m	[e]D	[f]ΔE_β(kJ/mol)	[g]ΔE_γ(kJ/mol)
PU	−43.6	−46.1	−69.8	124	3.5	54.0±0.3	38.1±0.3
0.10 wt%	−45.3	−47.5	−71.2	123	3.5	57.0±0.6	36.8±0.2
0.42 wt%	−48.3	−50.9	−76.7	112	3.9	56.0±0.6	36.5±0.3

Source: Reproduced from Polizos, G. et al., *Polymer*, 53, 595, 2012.
[a]T_g^{DSC} is the calorimetric glass transition temperature according to differential scanning calorimetry thermograms (not shown here).
[b]T_g^{diel} is the dielectric glass transition temperature obtained from the extrapolation of Equation 3.7 to 1.6 mHz ($\tau = 100$ s).
[c]T_0 is the Vogel fitting parameter in Equation 3.25.
[d]m is the fragility index calculated from Equation 3.26.
[e]D is the strength parameter that is proportional to the inverse fragility.
[f]ΔE_β is the activation energy of the β relaxation obtained from Equation 3.24.
[g]ΔE_γ is the activation energy of the γ relaxation obtained from Equation 3.24.

The respective activation energies were calculated from Equation 3.24 and are summarized in Table 3.1. The local dynamics and the activation energies in the pristine PU and in the composites were found to be similar, indicating that the local conformations are independent of the PU–particle interactions.

The time scale of the α relaxation in Figure 3.3 can be described by the Vogel–Tammann–Fulcher (VTF) equation

$$f_{\max,\alpha}(T) = A\exp\left[\frac{-B}{(T - T_0)}\right] \tag{3.25}$$

where A, B, and T_0 (Vogel temperature) are temperature-independent empirical constants listed in Table 3.1. The ideal glass transition temperature, T_0, has been found for glass-forming polymers to be approximately 40–50 K lower than the T_g, and it is defined as the temperature where the configurational entropy of the undercooled liquid is zero. In that respect, T_0 values correlate rather well with the calorimetric and dielectric glass transition temperatures (T_g^{DSC} and T_g^{diel} in Table 3.1). The variation in the T_g values is inherently related to the distinct dynamics of the α relaxation, previously discussed in Figure 3.2. The fragility (m) can be used to quantify this deviation [30,31]

$$m = \left.\frac{d\log\langle\tau\rangle}{d(T_g/T)}\right|_{T=T_g} = BT_g(T_g - T_0)^{-2}\log e \tag{3.26}$$

where m is a dimensionless steepness index and is defined as the slope of the relaxation times ($\tau = 1/2\pi f$) versus the inverse temperature at the glass transition temperature. The fragility is associated with the inter- and intra-molecular interactions, and index m is a measure of the bond interactions during the vitrification of a glass-forming liquid (i.e., the extreme values are $m = 16$ for strong Arrhenius-like glass-forming liquids with directional bonds, and $m = 250$ for fragile polymers with nondirectional bonds [32–34]). The fragility values, derived from the VTF fitting parameters, as well as the strength parameter (D), which is proportional to the inverse fragility ($D = B/T_0$), are summarized in Table 3.1. A significant decrease in the fragility of the polymer matrix is observed for the 0.42 wt% nanocomposite. Typically, the fragility depends on the particle–polymer interactions and is increasing or decreasing for attractive and repulsive interactions, respectively [35–37]. In spite of the hydrogen bond formation between the TiO_2 particles and the PU matrix [23], both the T_g and the fragility were found to decrease. The favorable formation of hydrogen bonds between the functionalized TiO_2 fillers and the hard segments of the PU [23] hinder the cross-linking between the soft and the hard segments and results to soft segment domains with faster segmental dynamics that change more gradually than the dynamics of the corresponding domains in the unfilled PU during the glass formation. It should be noted that the decrease of the fragility index is indirectly related to the TiO_2 nanoparticles. Typically, variations in the fragility of nanocomposites are due to changes in the geometrical frustration and polymer packing density. Herein, the observed changes originate from the weaker interactions between the hard and the soft segments, which also lead to a decrease in the glass transition temperature. This is better illustrated in the inset plot (Figure 3.3), showing the normalized (by T_g) temperature dependence of the apparent activation energy (ΔE^*_{act}) for the α relaxation. It is calculated from the derivative of Equation 3.25 and is expressed as a function of the VTF parameters and the universal gas constant (R) [38]

$$\frac{\Delta E^*_{act}}{R} = \frac{d\log f_{\max}}{d(1/T)} = BT^2(T - T_0)^{-2}\log e \tag{3.27}$$

The apparent activation energy of the 0.42% composite is decreased approximately 11% when the temperature is equal to the glass transition temperature due to the weaker interactions between the hard and the soft segments. On the contrary, when the temperature is higher than the critical α–β merging temperature (T_c), all samples exhibit identical barriers related to the cooperative segmental motion. Since no noticeable changes occur in the local dynamics, this variation in the activation energy is probably associated with fluctuations in the dynamic heterogeneity and cooperativity due to the addition of the nanoparticles [39,40].

The mobility of the polar groups is intrinsically connected to the conduction process, and conductivity was found to dominate the dielectric spectra for temperatures above the critical merging temperature T_c. The complex impedance formalism was employed to analyze the dielectric response of the samples and the respective Nyquist plots are shown in Figure 3.4 for selected temperatures. The pristine PU matrix is characterized by a single mode due to the relaxation of polar groups, whereas a bimodal relaxation is shown in the impedance response of the composites.

Based on the cluster morphology of the composites [23], we may assume a shell model configuration in which the individual TiO_2 particles of a cluster are surrounded by a shell-interfacial polymer phase. In this case, the high-frequency relaxation in the composites can be attributed to the intracluster mobility of the ions and specifically to a short length scale motion involving subdiffusion of ions between neighboring particles within a cluster. This motion takes place through the polar segments of the polymer shell, and is characterized by a resistance $\Delta R_{intracluster}$. In view of the previous discussions, the low frequency-relaxation can be assigned to an intercluster diffusion of the ions. At longer times, the ions will diffuse through the bulk polymer phase and migrate to neighboring clusters. This process is characterized by a resistance $\Delta R_{intercluster}$ and the overall (bulk) resistance of the composite is the sum of the intra- and intercluster resistance as indicated on the plot. At temperatures greater than 50°C, the bulk conductivity in the composites

$$\sigma_{bulk} = (\Delta R_{intracluster} + \Delta R_{intercluster})^{-1} \frac{d}{S} \tag{3.28}$$

is approximately 2.5 orders of magnitude higher than the conductivity of the unfilled matrix [23].

On the contrary, for temperatures below T_g, the nanocomposites exhibit significant lower dielectric losses (therefore and conductivity) compared to the pristine PU. The tan δ values in Figure 3.5 are approximately 1 order of magnitude lower than those in the PU at −263°C. This variant dependence of the conductivity mechanism can be utilized in polymer electrolytes as well as in the design of electronic devices operating at broad temperature windows. Similar behavior has been observed in thermoset-based nanocomposites, which is the subject of the following sections.

3.3.2 Interfacial Dynamics

The formation of interfaces due to the incorporation of nanoparticles into a polymer host may determine the macroscopic properties of the nanodielectric and result to significant enhancements or detrimental effects depending on the size of the interfaces and their interactions with the polymer matrix. Dielectric spectroscopy is a powerful tool for detecting and analyzing the dynamics of polar segments that relax at the vicinity of nanoparticles even when the particles are at small volume fractions. In this section, we present our recent studies on the interfacial dynamics of epoxy resin composites

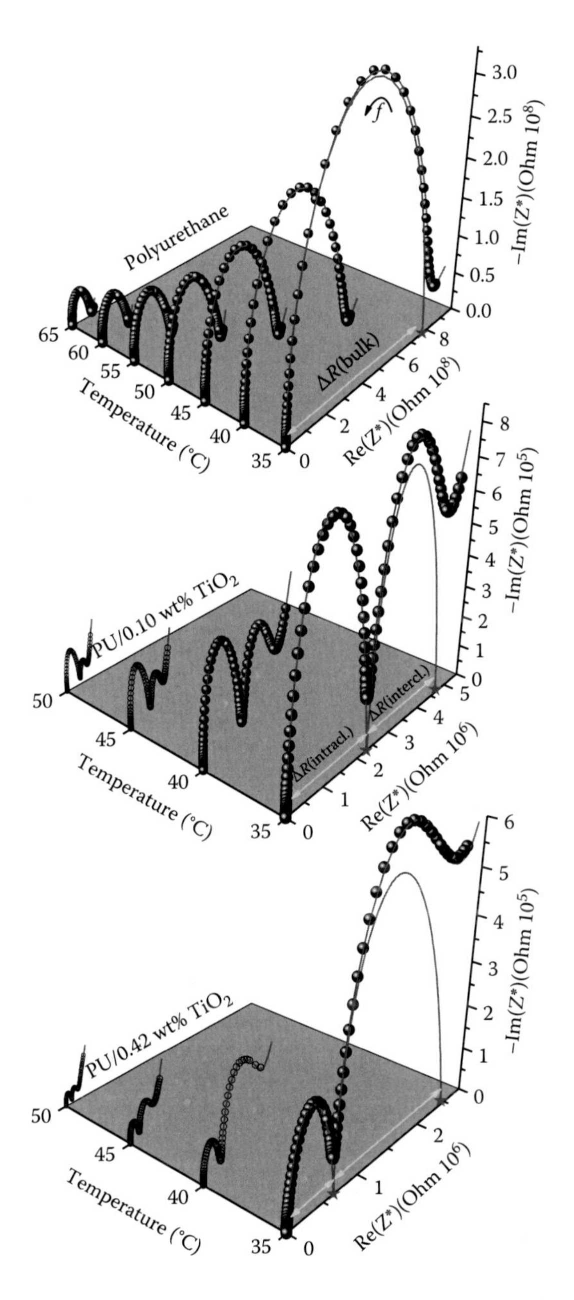

FIGURE 3.4

(See color insert.) Complex impedance Nyquist plots for the unfilled polyurethane and the nanocomposites for several temperatures. The fitting curves to the data are according to Equation 3.20. The conductivity process in the unfilled matrix is described by a single relaxation whereas the nanocomposites are characterized by a bimodal relaxation due to the intra- and intercluster conduction processes. (The top two plots are reproduced from Polizos, G. et al., *Polymer*, 53, 595, 2012.)

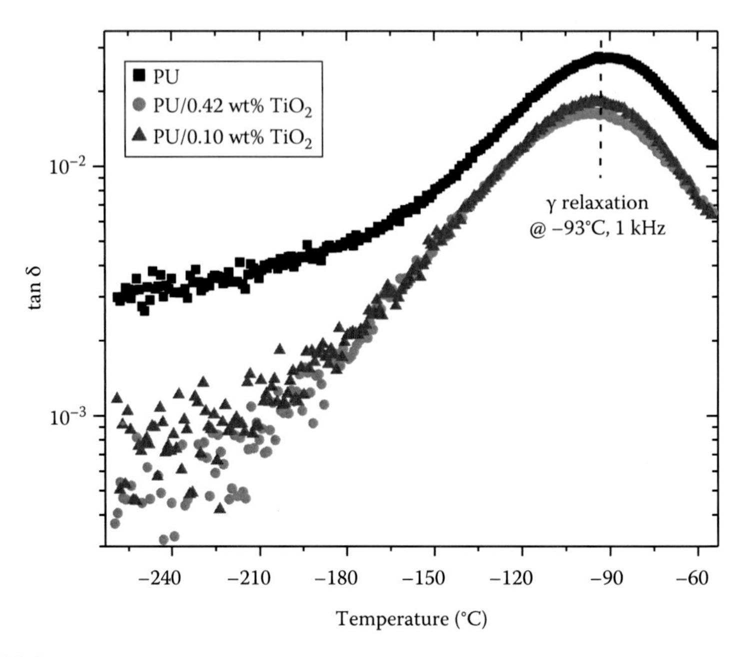

FIGURE 3.5
Isochronal plot for the dielectric tanδ at 1 kHz. The measurements were carried out on a homemade setup, and the γ relaxation is in excellent agreement with the measurement results in Figures 3.1 and 3.3. In the low-temperature limit, the composites exhibit significant lower losses compared to the unfilled polyurethane.

based on organically modified montmorillonite (OMMT) and barium titanate (BT). The nanofillers were selected to exploit the high aspect ratio of the montmorillonite (MMT) and the high permittivity of the BT particles, which improve the mechanical performance and the operating electric field of the composites, respectively. Hybrid organic/inorganic composites of two and three phases were prepared by dispersing BT or/and OMMT particles in the epoxy resin [41]. The interfacial dynamics below the glass transition temperature were found to depend on the particle functionalization and weight fraction.

In the temperature range between T_g and 40°C, a new relaxation mode was evident in the composites with 3 and 6 wt% OMMT, 30 vol% BT as well as in the three-phase system with both BT and OMMT fillers. This mode was absent in the pristine cross-linked epoxy matrix and in the 1 wt% OMMT and 10 vol% BT composites (a comparison plot at 90°C is shown in Figure 3.6). Since the dynamics of this mode are dependent on the particle type and weight fraction, we may associate its origin to the formation of interfaces and specifically to the polar segments relaxing at the vicinity of the inorganic fillers. In good agreement with the IR results [41] that show the formation of an uncross-linked polymer phase that becomes pronounced with weight fraction increasing, we can assign the nature of this relaxation to the interfacial

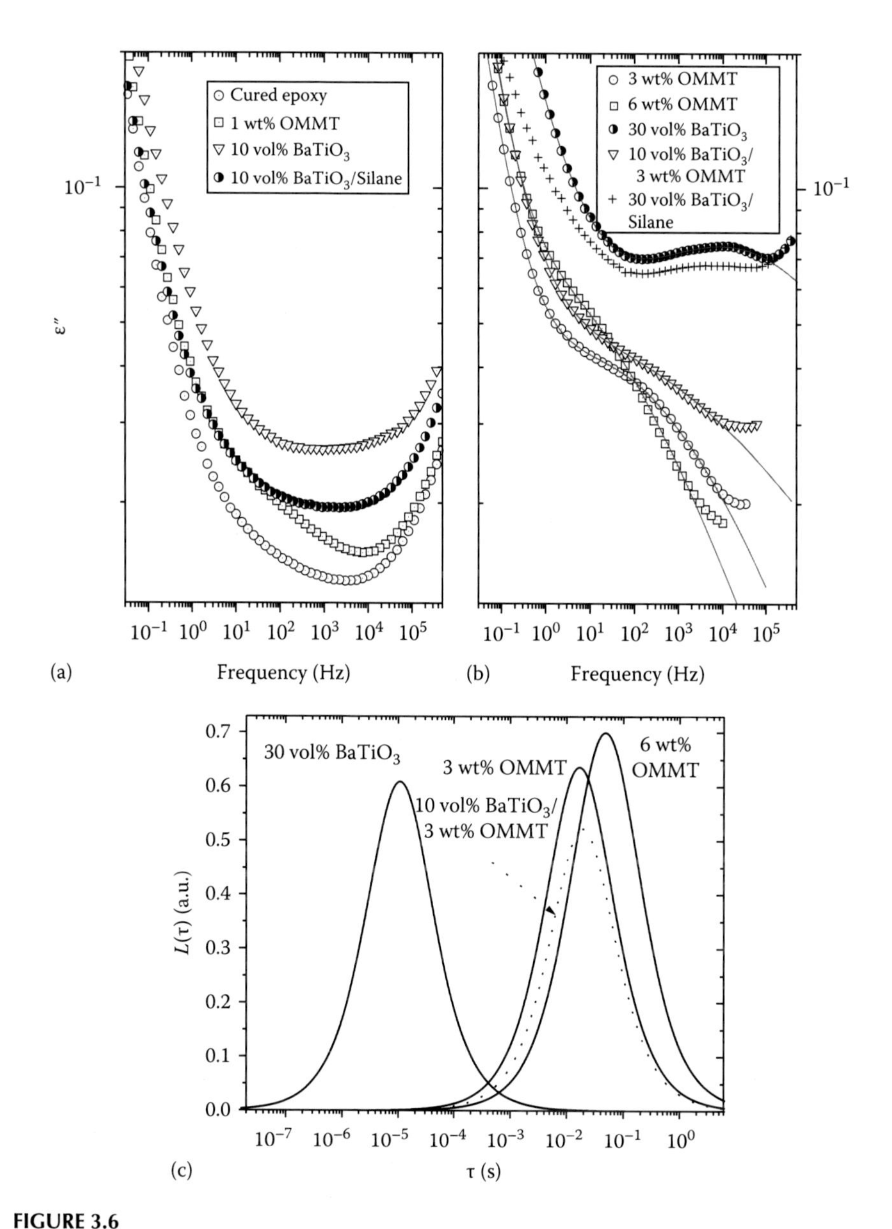

FIGURE 3.6
(a) Summarized dielectric loss plot at 90°C for the pristine epoxy matrix and the two- and three-phase composites. (b) A relaxation process is evident in the composites with high filler contents. The Cole–Cole dielectric function was found to give the best fitting to the data and the corresponding distributions of relaxation times are presented in plot (c). (Reproduced from Tomer, V. et al., *J. Appl. Phys.*, 108, 074116, 2010.)

unreacted (mobile) epoxy monomers. This argument is further supported by the surface modification of the BT, which was found to improve the interfacial properties. Despite the high inorganic content of the 30 vol% BT composite, the dielectric relaxation strength of the interfacial process is suppressed and becomes negligible when the BT surface is modified with organic coupling agents based on silane functional monolayers (Figure 3.6b). The proper functionality of the fillers allows their integration with the polymer matrix and prevents the formation of mobile interfaces.

To separate the dipolar from the conductivity contributions and therefore better illustrate the dielectric response of the composites, the distributions of relaxation times [1,2], $L(\tau)$, are also presented in Figure 3.6c. This interpretation assumes that the broadening of a relaxation is due to the superposition of Debye processes

$$\varepsilon''(\omega) = \int_{-\infty}^{\infty} \frac{\omega\tau L(\tau)}{1+(\omega\tau)^2} d\ln(\tau) \tag{3.29}$$

$L(\tau)$ can be calculated from the fitting parameters of the imaginary and real parts of the permittivity function and is analytically written as

$$L(\tau) = \frac{1}{2\pi} \frac{\sin(\pi - \alpha\pi)}{\cosh\left(\ln\left(\tau/\tau_0\right)\right) + \sin(\pi - \alpha\pi)} \tag{3.30}$$

In the same figure, it is clearly shown that the OMMT-based composites are characterized by approximately 3 orders of magnitude slower dynamics compared to the BT composites (shift of the distribution toward higher relaxation times). This shift is the result of the stronger interactions between the MMT surface and the interfacial polymer layer. The latter is geometrically restricted (physisorbed) and is less mobile than in the BT-based composites. The relative position of the interfacial distributions is a measure of the interfacial strength. Additional information can be derived from the shape of the distribution (i.e., the symmetric Cole–Cole distribution in Figure 3.6c is suggesting polymer segments relaxing in homogeneous environments that possibly result from the good dispersion of the fillers). The particle dispersion and the morphology of the particle clusters may also affect the interfacial dynamics. In Figure 3.6c is shown that the interfacial dynamics in the 6 wt% OMMT composite is slower than in the 3 wt% composite, despite the fact that both composites are based on the same type of fillers. To understand this behavior, we can assume that the measured interfacial relaxation is the average dielectric response. The hydroxyl groups, formed during the epoxide ring conversion, are possible to hydrogen bond to the silicate surface. Therefore, the observed retardation in the dynamics of the 6 wt% may reflect an average increase in the population of the hydroxyl

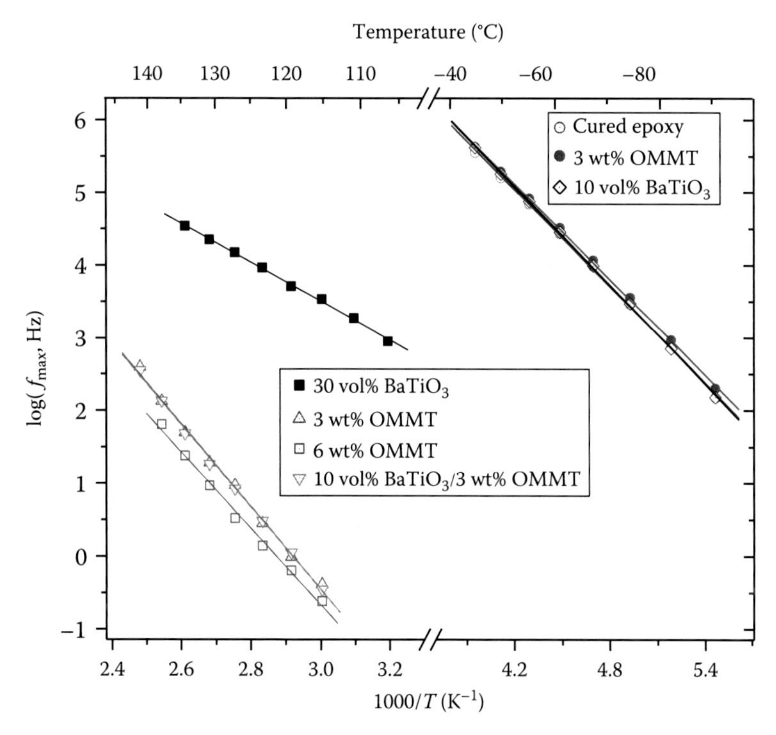

FIGURE 3.7
Summarized Arrhenius plot of the relaxations in selected composites and in the pristine epoxy network. The ultra-fast modes in the subzero temperature range were found to overlap and only three representative compositions are given. A distinct behavior is shown in the interfacial dynamics of the OMMT- and BT-based composites. (Reproduced from Tomer, V. et al., *J. Appl. Phys.*, 108, 074116, 2010.)

groups being attached to the silicate surface, due to a possible increase of the cluster's surface area.

The processes in Figure 3.6b as well as those in the subzero temperature range [41] were analyzed and the summarized Arrhenius plot is presented in Figure 3.7. The subzero modes exhibit dynamics that are several orders of magnitude faster (ultra-fast) compared to those arising from the interfaces. They were found to overlap in all specimens and therefore are independent of the filler type and weight fraction. The respective activation energy for the ultra-fast mode was calculated from Equation 3.24 and was found to be 42 kJ/mol. This value is in good agreement with previous studies that have identified this relaxation to local reorientations of the hydroxyl groups in the cross-linked epoxy network [42].

The interfacial modes were not only characterized by distinct dynamics depending on the filler type but were also found to exhibit different activation energies. The ΔE_A value for the OMMT-based composites is approximately 105 kJ/mol, whereas for the 30 vol% BT composite is 51 kJ/mol.

This noteworthy decrease in the energy barriers and the corresponding speed up of the dynamics (previously discussed) are a clear evidence of the weaker interfacial strength between the BT fillers and the epoxy network.

Heterogeneities in the permittivity and conductivity properties between the different constituents in a multiphase composite restrict the mobility of the ionic species (space charge) at the boundaries of different phases. The resulting space-charge interfacial dynamics [43–45] are inherently related to the bulk conductivity of the composite [41]. An example is shown in Figure 3.8 where a relaxation process is evident in the ac conductivity, $\sigma_{ac}(f)$. In this region, conductivity is frequency-dependent and describes a short-range motion of the space-charge carriers (i.e., a subdiffusive type of transport due to sublinear dependence of the mean square displacement on time [46,47]). For such short time scales, the charges are confined at heterogeneous boundaries. The thermally activated hopping process of the charges bound at the interfaces gives rise to relaxation phenomena that

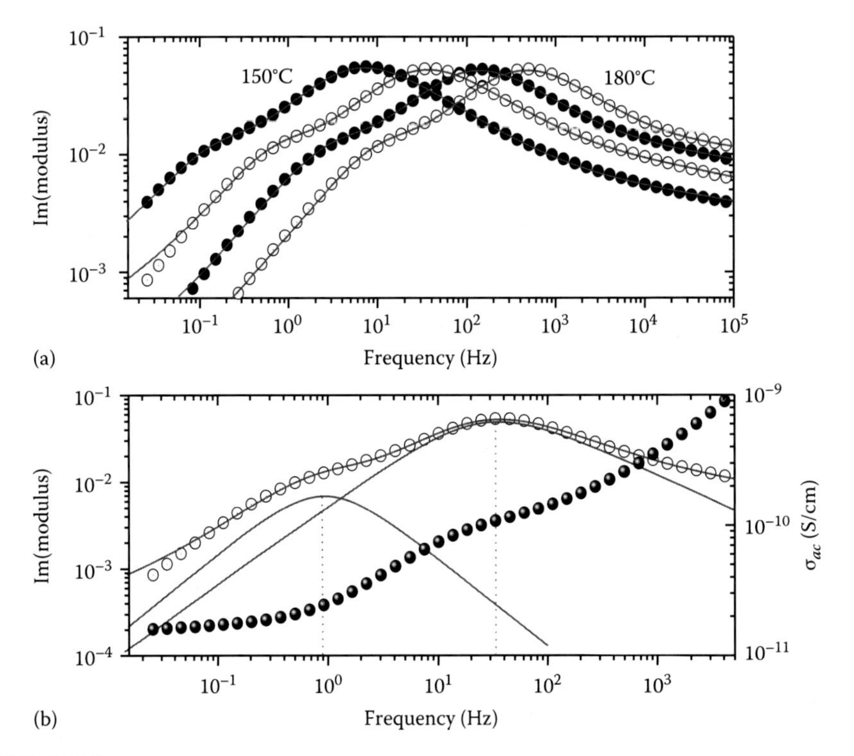

FIGURE 3.8
(a) The imaginary part of electric modulus for the 10 vol% $BaTiO_3$ composite at representative temperatures in steps of 10°C. The lines are the best fits of Equation 3.16 to the experimental data. (b) A comparison plot showing the correspondence between the electric modulus and the ac conductivity at 160°C. The low-frequency process in the electric modulus corresponds to the onset of dc conductivity, whereas the high-frequency process corresponds to the space-charge interfacial relaxation. (Reproduced from Tomer, V. et al., *J. Appl. Phys.*, 108, 074116, 2010.)

at longer time scales (lower frequencies) are followed by a diffusive type motion (i.e., dc conductivity; longer times give the ability to the charges to escape from the interfaces and diffuse in the bulk). This is clearly shown in Figure 3.8 where the low-frequency relaxation is in good agreement with the onset of the dc conductivity. In addition to the impedance formalism analysis (see Section 3.1), the electric modulus formalism can also be utilized to separate the conductivity contributions. The dynamics of the different length scale processes depend on the nanoparticle size and functionality and can be correlated with the recoverable energy density of the nanodielectric [41]. To eliminate or suppress the interfacial polarization effects, we present in the following two sections nanocomposites fabricated by in situ and self-assembly techniques.

3.3.3 In Situ Particle Nucleation

Titanium dioxide (TiO_2) nanoparticles, smaller than 5 nm in diameter, were in situ nucleated in epoxy resin by the hydrolysis and oxidation of $TiCl_3$ [48–53]

$$TiCl_3 + 3H_2O \rightarrow Ti(OH)_3 + 3HCl \tag{3.31}$$

$$2Ti(OH)_3 + \frac{1}{2}O_2 \rightarrow 2TiO_2 + 3H_2O \tag{3.32}$$

The particle size, which is comparable to the length scale of the polymer entanglements, is anticipated to restrict the mobility of the polymer chains [54]. This is clearly shown in the dielectric losses at low temperatures in Figure 3.9. A composite with similar content in TiO_2 nanoparticles that were ex situ synthesized and later dispersed in the epoxy is also included for comparison. The absence of pronounced interfaces and particle agglomerations in the in situ composite [54] resulted in dielectric losses that are approximately 1 order of magnitude lower than those in the unfilled epoxy.

The restricted mobility of the polymer chains was found to improve the dielectric breakdown strength of the in situ composite and potentially increase the energy density that can be stored in the nanodielectric. The failure probability distributions of the samples are presented in Figure 3.10. A standard two parameter Weibull model was employed for the data analysis [55,56]

$$P(\alpha_W, \beta_W, E_{BD}) = 1 - \exp\left[-\left(\frac{E_{BD}}{\alpha_W}\right)^{\beta_W}\right] \tag{3.33}$$

In the previous expression, P is the cumulative probability function; E_{BD} is the dielectric breakdown field; and the parameters α_W and β_W denote the

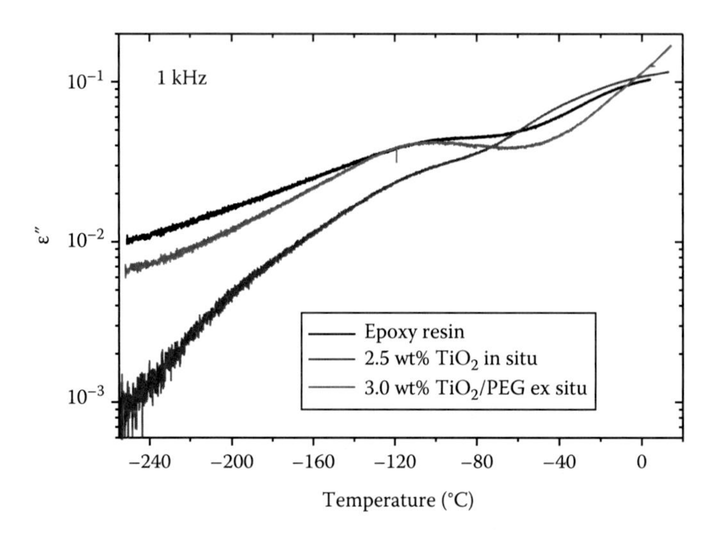

FIGURE 3.9
Dielectric losses at 1 kHz for the pristine epoxy and nanocomposites with similar contents in TiO_2 prepared with in situ and ex situ techniques. A significant decrease of the dielectric losses is shown for the in situ composite.

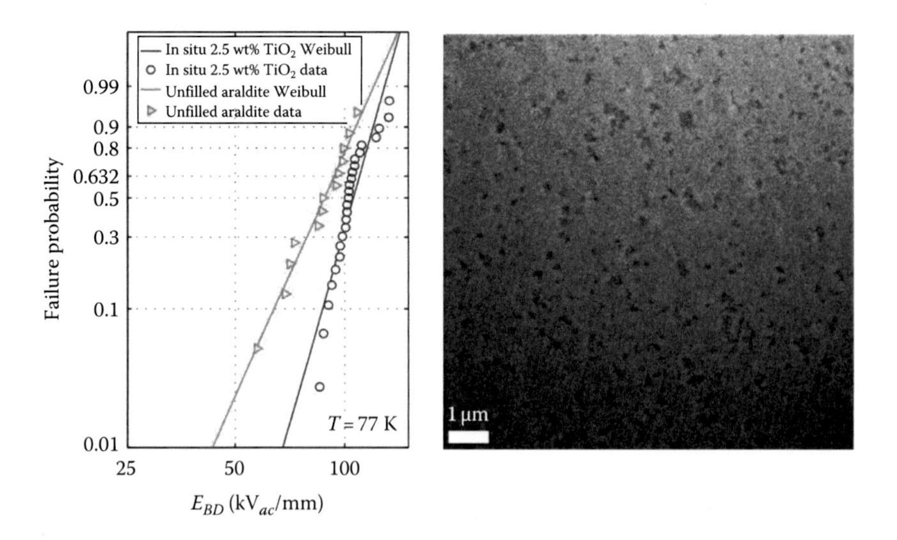

FIGURE 3.10
Dielectric breakdown failure probability analysis based on Weibull statistics. The samples were submerged in liquid nitrogen and the frequency of the applied ac electric field was 60 Hz. The TEM image corresponds to the in situ 2.5 wt% composite. (Reproduced from Polizos, G. et al., *Appl. Phys. Lett.*, 96, 152903, 2010.)

characteristic breakdown strength when the failure probability is 0.6321 and the scattering in the breakdown values, respectively. The improvement in the reliability of the 2.5 wt% in situ composite (β_W values) is approximately 60% whereas the corresponding benefit in the breakdown strength is 20% at the characteristic Weibull probability and 56% at 1% failure probability. Considering the small weight fraction of the TiO_2 phase, the associated improvements are remarkable and extremely significant in the design of high-voltage systems.

The small size and uniform dispersion of the fillers decreased the dielectric losses and narrowed the distribution of relaxation times (narrower peaks in Figure 3.9). However, no shift in the position of the peaks, compared to the unfilled matrix, was observed for the in situ composite. This is also evident in the mechanical relaxations over a broad temperature range, shown in Figure 3.11. The α and the local β, γ, δ relaxations are activated in identical temperature windows for all samples, indicating that the nanoparticles did not change the cross-linking density of the polymer matrix and resulted in composites with thermodynamical properties similar to those in the pristine epoxy. Moreover, a 30% increase in the storage modulus of the 2.5 wt% composite was observed at the lowest measured temperature.

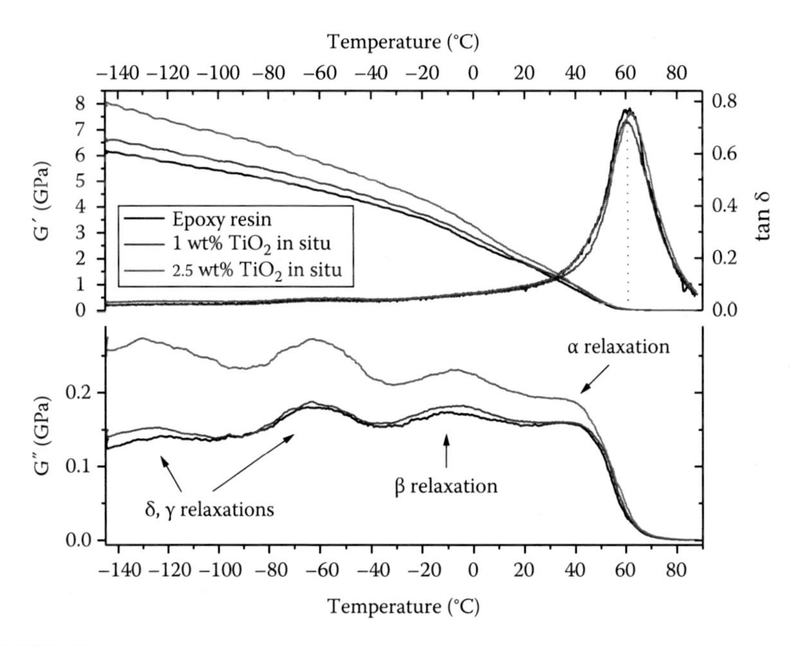

FIGURE 3.11
Storage modulus (G'), loss modulus (G''), and tan δ according to dynamic mechanical analysis at 1 Hz. The nanoparticles were in situ synthesized in the polymer matrix. The α relaxation is associated with the glass transition and the tan δ peak corresponds to T_g.

3.4 Self-Assembly of Nanostructures

Nanoparticle functionality and nanoparticles with controlled size have shown notable benefits against detrimental effects associated with the interfaces; however, certain applications may necessitate the integration of two or more inorganic phases. Particles functionalized with reactive agents to form self-assembled nanostructures can strengthen the interfaces and provide vital enhancements in the dielectric properties of the composites.

Interfacial polarization effects predominantly occur due to the mismatch in the permittivity and conductivity values between the organic and the inorganic phases, and typically generate concentration in the local electric field, which may lead to electrical failure of the nanodielectric. Self-assembly nanostructures of fillers with different dielectric properties are possible to grade the local electric field and increase the operating field and therefore the stored electrical energy in the composite (the stored energy is a function of the permittivity and the square of the operating electric field). In view of the earlier discussion, BT and OMMT nanoparticles were covalently bonded and the integrated two-phase hybrid nanofillers yield a thermodynamically favorable dispersion of the particles in the epoxy resin [57]. The morphology of the resultant structures is shown in Figure 3.12. Assembled layered silicates are attached on the surface of the spherical BT particles. The high organic content (~25 wt%) of the silicates and their ability to cross-link with the epoxy matrix improved the filler dispersion and created robust interfacial environments that resemble those in the unfilled epoxy. This is evident in the polarization curves (Figure 3.12e) where the high field losses are similar for the pristine epoxy and the composite with reactive fillers. The intrinsic high permittivity of the BT particles increased the high field permittivity (higher slope of the polarization loop) and the recoverable energy density of the composite. Additional improvements related to the mechanical performance (68% increase in the storage modulus) and the glass transition temperature were also found [57].

Dielectric spectroscopy can provide insights on the correlation between the properties of the composites and their interfacial dynamics. In Figure 3.13a is shown a comparison plot for temperatures below the glass transition temperature. A relaxation process is present only in the composite with the highest filler content (20 wt%). Since the relaxation strength ($\Delta\varepsilon$) is proportional to the number density of the relaxing units, this process can be ascribed to the formation of a mobile polymer phase that is loosely bound to the inorganic surfaces. At lower particle contents where the particle agglomeration is less pronounced, the interfacial polymer phase is less mobile (more tightly bound on the particle surface [58]) and the corresponding dielectric process is absent or negligible. The strength and dynamics of this process are in good agreement with the thermo-mechanical properties of the composites [57] (i.e., heat capacity, onset and width of the glass transition, cross-linking density, and mechanical modulus) and can be used as a measure of the polymer-filler interfacial strength. The dynamics for the 20 wt% composite in Figure 3.12b were obtained from the best fit analysis of the experimental data with a superposition of Equation 3.9 and a conductivity contribution. The time scale dependence of the relaxation process over the entire measured temperature range is shown in Figure 3.13c and the calculated activation energy was found to be 226 kJ/mol. This process is formed at considerably

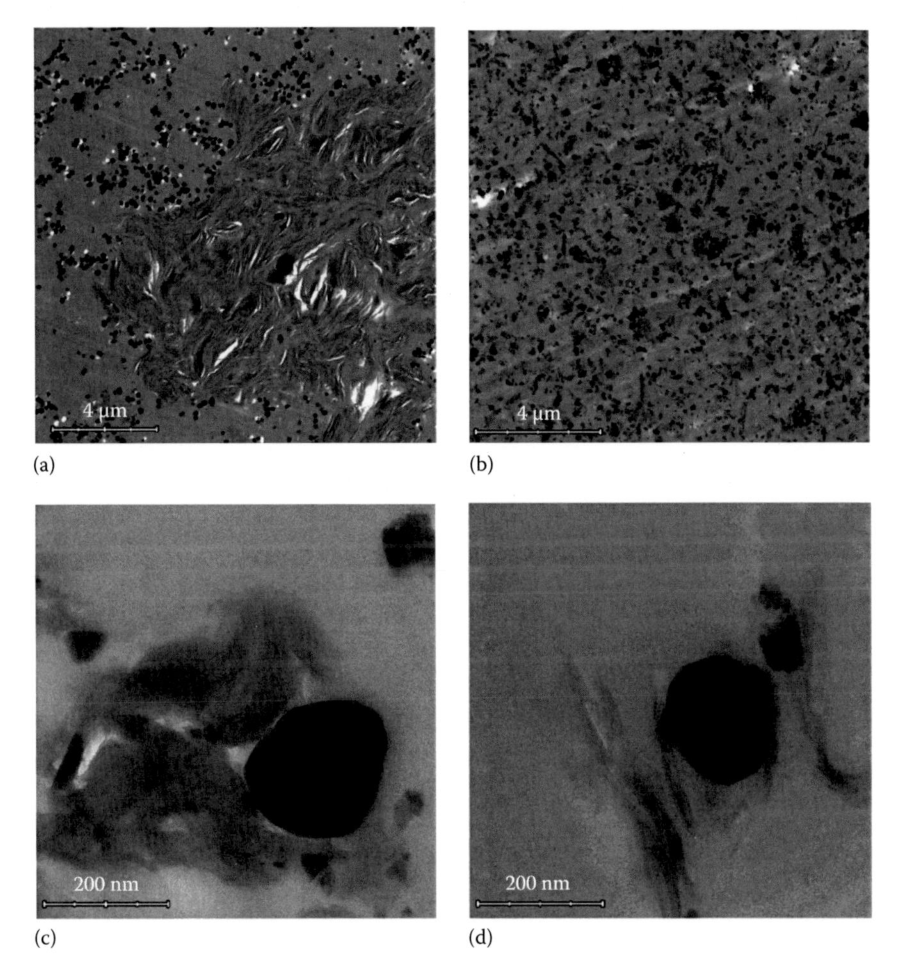

FIGURE 3.12
TEM images of $BaTiO_3$/OMMT-based nanocomposites. (a) A phase separated morphology of noncovalently bonded BT and OMMT; (b) a thermodynamically favorable dispersion of covalently bonded $BaTiO_3$/OMMT fillers; (c and d) higher magnification images of the covalently bonded $BaTiO_3$/OMMT nanofillers in the same nanocomposite.

(continued)

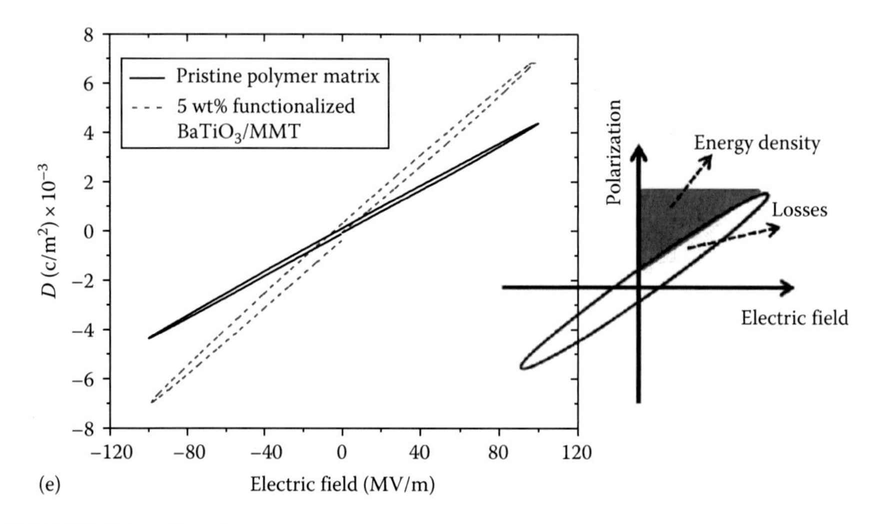

FIGURE 3.12 (continued)
TEM images of $BaTiO_3$/OMMT-based nanocomposites. (e) a comparison plot of dielectric displacement versus electric field and a schematic illustration of the recoverable energy density. (Reproduced from Polizos, G. et al., *J. Appl. Phys.*, 108, 074117, 2010.)

higher weight fractions in hybrid fillers than the corresponding process in the nonreactive single-filler composites (Figure 3.6), and the activation energy value is 150% higher than the activation energy for the single BT/epoxy composites and 20% higher than the value for the single OMMT/epoxy composites. Furthermore, the interfacial dynamics of the covalently bonded hybrid composites are several orders of magnitude slower than in the single-phase BT composites. The slower dynamics as well as the increase in the barriers associated to the interfacial relaxations manifest a significant improvement in the strength of the interfaces.

Similar dependencies were observed in the activation energy for the dc conductivity at temperatures above T_g. A comparison plot between the hybrid filler and the single OMMT composites is shown in Figure 3.14. The hybrid filler composites are characterized by higher activation energy values for all weight fractions; and interestingly, the conduction process in the 5 and 10 wt% composites is associated with energy barriers higher even than those in the pristine epoxy. When the particle content is lower than the percolation threshold, the cross-linked filler configuration reinforces the interfacial strength and results in marked improvements. In good agreement with the thermoplastic-based composites in Section 3.1, the interfacess and the particle agglomeration morphology dominate the conduction process. Tailoring the nanoscale properties is an effective route for the design and fabrication of composites with optimized performance. In this section, the nanoscale synergistic effects of the slower interfacial dynamics and the respective higher activation energy barriers restricted the mobility of

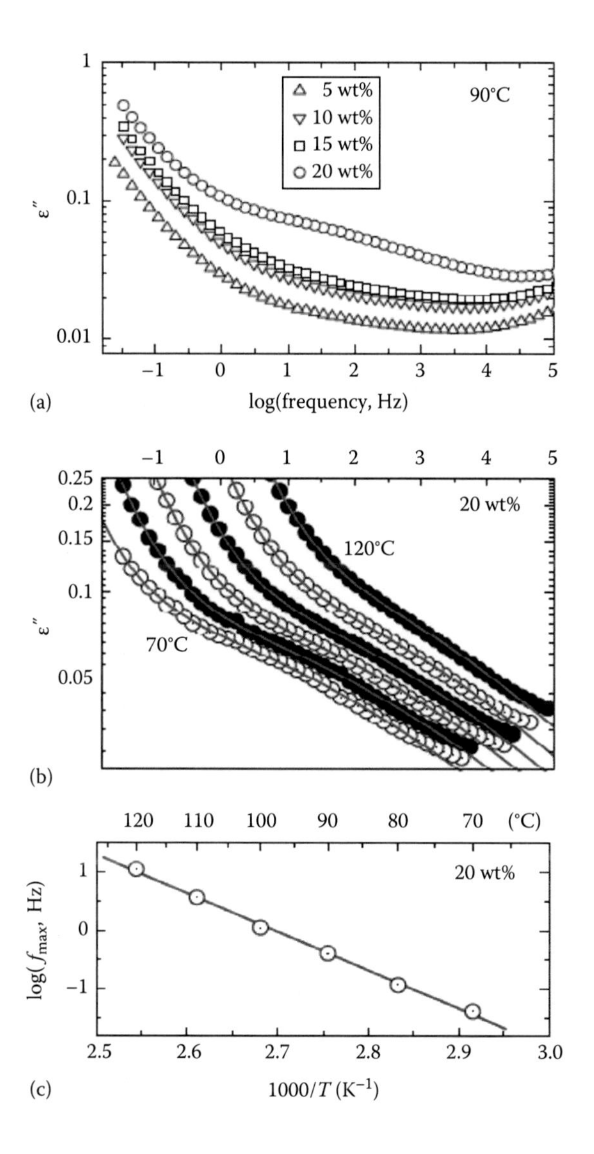

FIGURE 3.13
(a) Dielectric loss curves for the hybrid (covalently bonded) composites at different weight fractions for a representative temperature below T_g. A relaxation process is shaped when the particle content is 20 wt%. (b) This process is due to the formation of a polymer phase loosely bound to the particle surface and is analyzed in the entire temperature range. The lines are the best fits of Equation 3.9 and a conductivity contribution to the experimental data. (c) The corresponding Arrhenius plot obtained from the loss peak maximum frequency. The line is a linear fit of Equation 3.24 to the data. (Reproduced from Polizos, G. et al., *J. Appl. Phys.*, 108, 074117, 2010.)

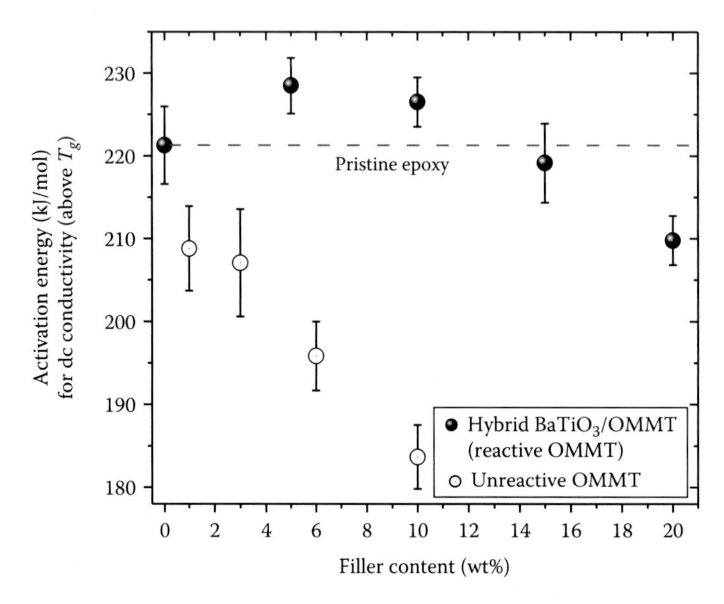

FIGURE 3.14
Activation energies of the conductivity mechanism for the hybrid (covalently bonded) and single filler composites at weight fractions indicated on the plot. (Reproduced from Polizos, G. et al., *J. Appl. Phys.*, 108, 074117, 2010.)

the hybrid composites and resulted to better dielectric breakdown probability distributions [57] that can be utilized in energy storage and electrical insulation applications.

3.5 Hierarchical Structures

3.5.1 Thermoset-Based Nanocomposites

The hierarchical dispersion of nanoparticulates is an effective method to combine nanometer-scale filler dispersion with long-range ordered structures and optimize certain properties in a desired direction. Herein, we present anisotropic composites with controlled particle distribution and correlate their ordered structure with the dielectric properties [59,60]. A thermoset and a thermoplastic matrix were studied. For the thermoset matrix, a cross-linkable silicone elastomer (polydimethylsiloxane) was used and BT nanoparticles were dielectrophoretically assembled [61–65] by applying high-amplitude ac electric field to the polymer/particle suspension during the curing process. This method utilizes induced polarization effects that arise from the contrast between the high permittivity of the particles and the low permittivity of the elastomeric matrix. The time average dielectrophoretic force exerted on the particles is [66,67]

$$\left\langle \vec{F}_{DEP}(t) \right\rangle = 2\pi\varepsilon'_m R^3 \,\mathrm{Re}[K^*(\omega)]\vec{\nabla}E^2_{rms} \tag{3.34}$$

where

- ε'_m denotes the medium's (silicone elastomer) real part of the permittivity function
- R is the particle's radius
- E_{rms} is the rms (root mean square) amplitude of the applied field
- $K^*(\omega)$ is the complex Clausius–Mossotti function that is expressed as

$$K^*(\omega) = \frac{\varepsilon^*_p - \varepsilon^*_m}{\varepsilon^*_p + 2\varepsilon^*_m} \tag{3.35}$$

For both, the particle and the medium ($i = p,m$), the complex permittivity functions are interpreted in the general form, including also the direct current (dc) conductivity, σ_{dc}, contribution: $\varepsilon^*_i(\omega) = \varepsilon'_i(\omega) - \varepsilon''_i(\omega) - i\omega^{-1}\sigma_{i,dc}$. Depending on the dielectric functions $\varepsilon^*_p(\omega)$ and $\varepsilon^*_m(\omega)$, the frequency of the applied field can be tuned to maximize the $K^*(\omega)$ function and therefore the dielectrophoretic force. The resultant column-like aligned structures are shown in Figure 3.15. We follow the sample notation introduced by Newnham [68].

FIGURE 3.15
SEM and TEM (inset) images of aligned $BaTiO_3$ (10 vol%) in silicone elastomer matrix (polydimethylsiloxane). The applied electric field was 1.6 kV/mm at 100 Hz. (Reproduced from Tomer, V. et al., *J. Appl. Phys.*, 103, 034115, 2008.)

Specifically, 0–3 denotes a composite with randomly dispersed fillers and 1–3 a composite with fillers aligned in one dimension. 1–3 composites with fillers aligned in the direction of the measuring field (z-axis) are noted as z-aligned composite while x–y aligned composites refer to fillers aligned perpendicular to the measuring field.

All composites exhibit higher permittivity than the unfilled polymer matrix. Permittivity values were found to depend on the BT concentration as well as on the spatial distribution of the fillers. A representative plot for the 22.5 vol% composites is shown in Figure 3.16a. The z-aligned (parallel) composite has the highest values over the entire frequency range. At small weight fractions, there is no notable change in the values of the composites; however, at higher fractions, the anisotropy difference is evident (Figure 3.16b) and the z-aligned 25 vol% composite is characterized by approximately four times higher permittivity than the unfilled polymer.

Theoretical models can predict the effective permittivity of a dielectric mixture [69,70]. An example is shown in Figure 3.16c for the 0–3 (randomly dispersed) composites. Among several models, the best fit to the experimental data was obtained by the model of Wakino et al. [71]. This model considers the influence of dielectric and/or infringing of the electric flux due to discontinuity at the boundary of constituent phases and is given by the expression

$$\varepsilon_r = \exp\left[\frac{\ln\left\{V_c\varepsilon_c^{(V_c-V_0)} + (1-V_c)\varepsilon_p^{(V_c-V_0)}\right\}}{V_c - V_0}\right] \tag{3.36}$$

where

V_c is the particle volume fraction

ε_c and ε_p are the particle and polymer permittivities, respectively

V_0 is the critical volume fraction at which the dispersed phase becomes continuous

In addition to conventional models, the effective permittivity may also be expressed in terms of the spectral density representation by taking into account depolarization factors related to the topology of the interfaces [72]. The hierarchical structure of inorganic fillers with high permittivity and the integration of their interfaces with the polymer matrix is another route for the design of nanocomposites with predicted dielectric properties.

3.5.2 Thermoplastic-Based Nanocomposites

Layered silicates modified with dimethyl-dioctadecyl-ammonium were dispersed in maleic anhydride functionalized polyethylene (PE), and nanocomposite films were extruded using a commercial blown-film line [60,73].

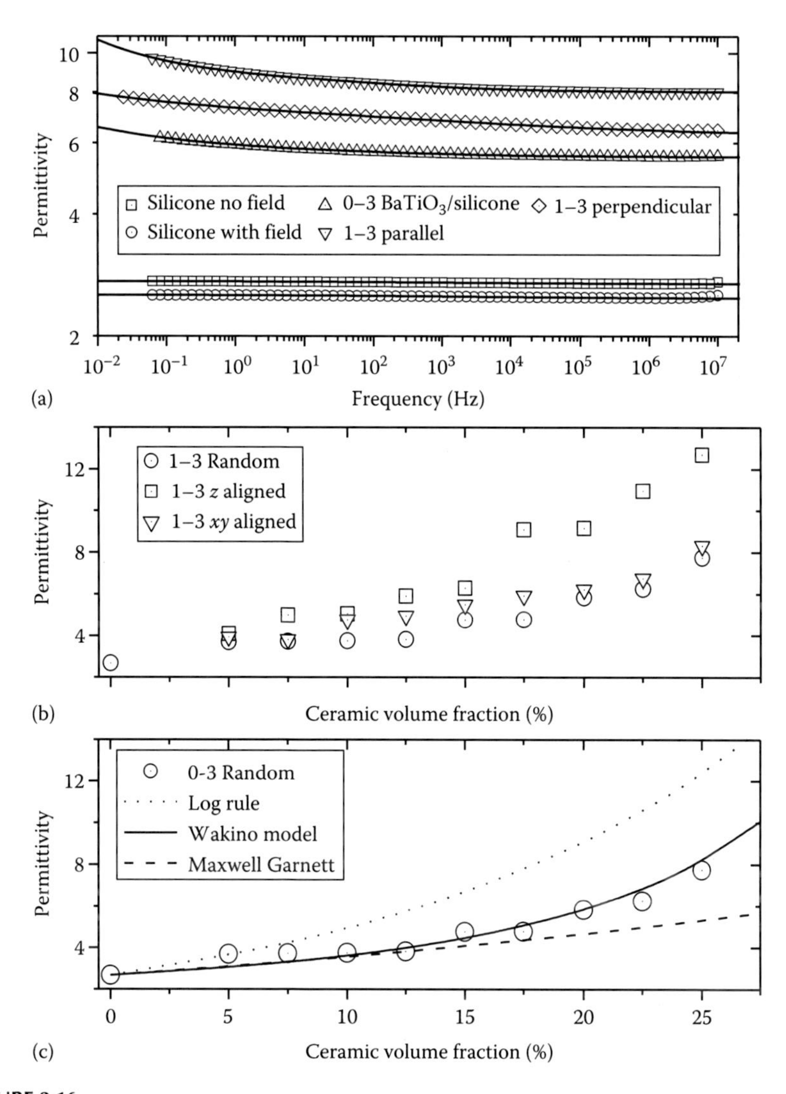

FIGURE 3.16
(a) Real part of permittivity at room temperature for the 22.5 vol% composites and the polydimethylsiloxane matrix (exposed and not exposed to the electric field used for the particle alignment). The lines are the best fits of Equation 3.8 to the data. (b) Permittivity values for the randomly dispersed and structured composites at several weight fractions in $BaTiO_3$. (c) Effective permittivity fitting curves for the randomly dispersed composites according to different mixing rules. (Reproduced from Tomer, V. et al., *J. Appl. Phys.*, 103, 034115, 2008.)

The stress-induced filler orientation was measured by wide-angle x-ray diffraction. The structure of the composites, the diffracted patterns, and the azimuthal profiles are shown in Figure 3.17. The radial intensities $I(\phi)$ were used to determine the order parameter in terms of the Hermans orientation function [74],

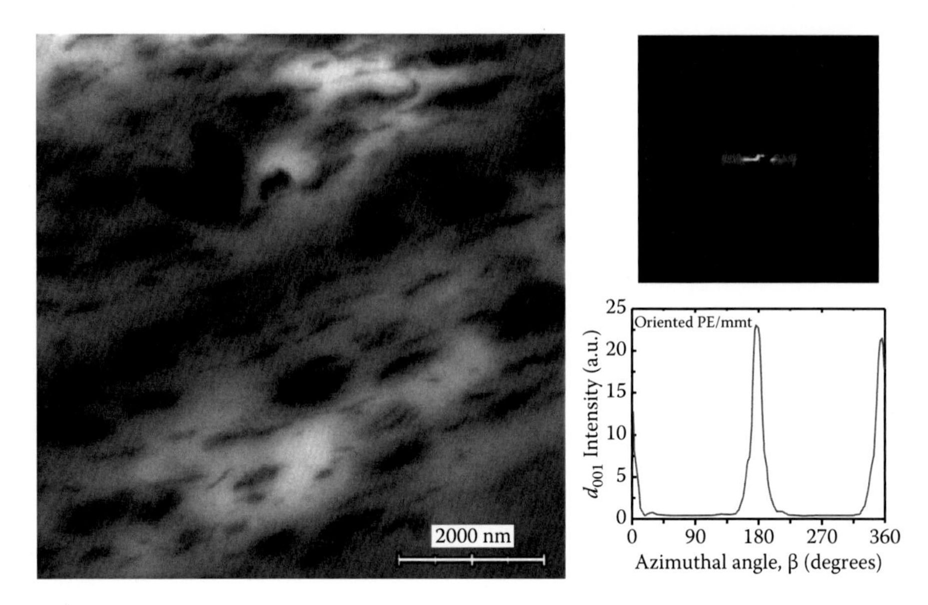

FIGURE 3.17
TEM image of blown-molded polyethylene composites filled with organically modified silicates. The diffracted patterns and the azimuthal profiles are shown on the right plots. The Hermans orientation factor is 0.8. (Reproduced from Tomer, V. et al., *J. Appl. Phys.*, 109, 074113, 2011.)

$$S_d = \frac{3\langle \cos^2 \phi \rangle - 1}{2} \tag{3.37}$$

where

$$\langle \cos^2 \phi \rangle = \frac{\int_0^{\pi/2} I(\phi) \cos^2 \phi \sin \phi d\phi}{\int_0^{\pi/2} I(\phi) \sin \phi d\phi} \tag{3.38}$$

The S_d values for perfectly aligned and randomly dispersed particles are 1 and 0, respectively; and the calculated S_d factor for the oriented fillers in Figure 3.17 was found to be 0.8, manifesting a high-degree orientation at the direction of blow-molding.

In good agreement with the impedance analysis in Section 3.1, dielectric spectroscopy in the permittivity formalism probes two relaxation mechanisms that are present in the composites only and shown in Figure 3.18. They involve interactions between space-charge and polar segments at the

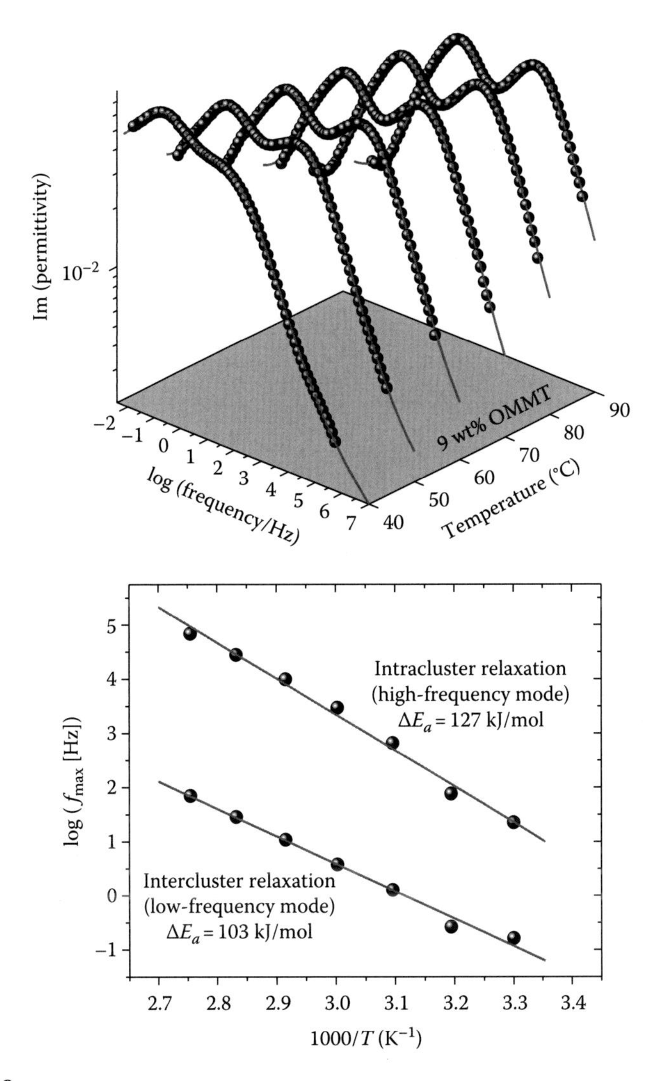

FIGURE 3.18
Dielectric relaxations for the aligned polymer nanocomposite in Figure 3.17. The lines are the best fits of a superposition of two Havriliak–Negami expressions (Equation 3.9) to the data. The corresponding Arrhenius plots and activation energies are shown on the bottom plot. (Reproduced from Tomer, V. et al., *J. Appl. Phys.* 109, 074113, 2011.)

silicate interfaces (intracluster relaxation at high frequencies) as well as interactions between space-charge and polar maleic anhydride segments in the bulk polyethylene phase amid neighboring silicate clusters (intercluster relaxation at low frequencies). The energy barriers associated with these relaxations are quantified by the respective activation energies. Their values, presented in Figure 3.18, are typical for ion conduction mechanisms and the activation energy for the intracluster relaxation was found to be

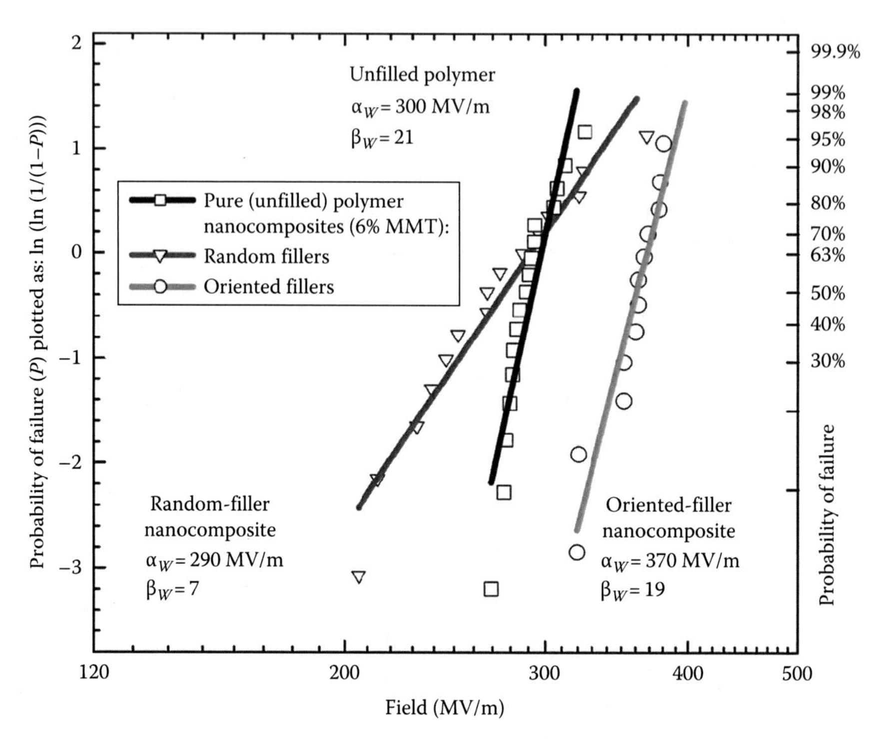

FIGURE 3.19
(See color insert.) Dielectric breakdown failure distributions for the unfilled polymer matrix and the nanocomposites based on randomly and aligned layered silicates. The lines are linear fits to the Weibull cumulative probability and the characteristic parameters are indicated on the plot. (Reproduced from Tomer, V. et al., *J. Appl. Phys.*, 109, 074113, 2011.)

significantly higher due to the strong electrostatic interactions with the silicate surface.

The aligned configuration of the silicate particles has shown outstanding improvement in the dielectric breakdown failure distributions of the nanocomposites, presented in Figure 3.19.

The characteristic breakdown strength was found to qualitatively change behavior with the particle orientation and was markedly increased for the oriented structures, where random filler structures showed the typical onset of failure decreasing to much lower fields than for the matrix. This improvement was the result of synergistic effects related to the ordered structure of high aspect ratio silicates (that force more tortuous paths for the propagation of charges), and to the modification of the interfaces (that create traps for the space-charge, increasing the energy barriers for their diffusion and, therefore, minimizing the bulk charge accumulation).

3.6 Summary

The introduction of inorganic nanoparticles into a polymer matrix unavoidably defines interfaces, which often determine the properties and performance of the composite. In this chapter, we presented aspects of our research efforts over the past years on identifying critical structural parameters that mitigate some of the detrimental effects associated with these interfaces and can expand the operating and application windows of the polymers. The size of the particles comparable to polymer length scales, the particle functionality, and the highly ordered self-assembled configurations are key parameters in designing multifunctional composites with markedly enhanced capabilities. The phenomenal growth of research and commercialization of hybrid nanostructured materials necessitates the development and utilization of techniques that accurately probe properties at the nanoscale. DRS combined with proper analysis tools are a versatile method for dynamically measuring nanoscale properties, separating synergistic contributions, and predicting the macroscopic behavior of composites. This chapter focused on such spectroscopic studies for composites with potential dielectrical applications and with promise to store electrostatic energy. Across a broad range of polymer matrices, the controlled distribution of the fillers and the tailored interfaces according to dielectric measurements resulted to improvements that are outstanding in the field of nanodielectrics and provide the grounds for further development and optimization.

Acknowledgments

The work presented in this chapter was supported by the Office of Naval Research (Grant No. MURI-00014-05-1-0541), the National Science Foundation (Grant No. DMR-0602877), and the U.S. Department of Energy-Office of Electricity Delivery and Energy Reliability, Advanced Cables and Conductors Program for Electric Power Systems (under contract DE-AC05-00OR22725 with Oak Ridge National Laboratory, managed and operated by UT-Battelle, LLC).

Part of this work was performed at Oak Ridge National Laboratory's Center for Nanophase Materials Sciences, sponsored by the Scientific User Facilities Division, Office of Basic Energy Sciences, and U.S. Department of Energy, The National Science foundation I/UCRC Center for Dielectrics Studies.

References

1. F. Kremer, A. Schönhals (eds.), *Broadband Dielectric Spectroscopy*, Springer-Verlag, Berlin, Germany (2002).
2. N.G. McCrum, B.E. Read, G. Williams, *Anelastic and Dielectric Effects in Polymeric Solids*, Wiley, London, U.K. (1967).
3. G. Williams, D.K. Thomas, *Novocontrol Appl. Note Dielectr*. 3, 1–28 (1998).
4. C.J.F. Boettcher, P. Bordewijk, *Theory of Electric Polarization*, 2nd edn., vol. 2, Elsevier, Amsterdam, the Netherlands, p. 72 (1978).
5. S. Havriliak, S. Negami, *J. Polym. Sci., Polym. Symp*. 14, 89 (1966).
6. F. Alvarez, A. Alegria, J. Colmenero, *Phys. Rev. B* 44, 7306 (1991).
7. A. Boersma, J. van Turnhout, M. Wübbenhorst, *Macromolecules* 31, 7453 (1998).
8. K.S. Cole, R.H. Cole, *J. Chem. Phys*. 9, 341 (1941).
9. M. Wübbenhorst, J. van Turnhout, *J. Non-Cryst. Solids* 305, 40 (2002).
10. F.S. Howell, C.T. Moynihan, P.B. Macedo, *Bull. Chem. Soc. Jpn*. 57, 652 (1984).
11. K. Pathmanathan, G.P. Johari, *J. Chem. Phys*. 95, 5990 (1991).
12. P.B. Macedo, C.T. Moynihan, R. Bose, *Phys. Chem. Glasses* 13, 171 (1972).
13. C.T. Moynihan, L.P. Boesch, N.L. Laberge, *Phys. Chem. Glasses* 14, 122 (1973).
14. J.H. Ambrus, C.T. Moynihan, P.B. Macedo, *J. Phys. Chem*. 76, 3287 (1972).
15. G.M. Tsangaris, G.C. Psarras, N. Kouloumbi, *J. Mater. Sci*. 33, 2027 (1998).
16. G.C. Psarras, E. Manolakaki, G.M. Tsangaris, *Compos. Part A: Appl. Sci. Manuf*. 33, 375 (2002).
17. J.D. Jacobs, H. Koerner, H. Heinz, B.L. Farmer, P. Mirau, P.H. Garrett, R.A. Vaia, *J. Phys. Chem. B* 110, 20143–20157 (2006).
18. J.K. Nelson, Y. Hu, *J. Phys. D: Appl. Phys*. 38, 213–222 (2005).
19. Y. Cao, P.C. Irwin, K. Younsi, *IEEE Trans. Dielectr. Electr. Insul*. 11, 797–807 (2004).
20. E. Tuncer, I. Sauers, D.R. James, A.R. Ellis, M.P. Paranthaman, T. Aytug, S. Sathyamurthy, K.L. More, J. Li, A. Goyal, *Nanotechnology* 18, 025703 (2007).
21. M. Volel, M. Armand, W. Gorecki, M.L. Saboungi, *Chem. Mater*. 17, 2028–2033 (2005).
22. J.-D. Jeon, M.-J. Kim, S.-Y. Kwak, *J. Power Sources* 162, 1304–1311 (2006).
23. G. Polizos, E. Tuncer, A.L. Agapov, D. Stevens, A.P. Sokolov, M.K. Kidder, J.D. Jacobs, H. Koerner, R.A. Vaia, K.L. More, I. Sauers, *Polymer* 53, 595–603 (2012).
24. P. Pissis, G. Polizos, Molecular dynamics and ionic conductivity studies in polyurethane thermoplastic elastomers, Chapter 14, in *Handbook of Condensation Thermoplastic Elastomers*, S. Fakirov (ed.), Wiley-VCH, Weinheim, Germany (2005).
25. P. Pissis, L. Apekis, C. Christodoulides, M. Niaounakis, A. Kyritsis, J. Nedbal, *J. Polym. Sci., Polym. Phys*. 34, 1529–1539 (1996).
26. E. Malmstrom, A. Hult, U.W. Gedde, F. Liu, R.H. Boyd, *Polymer* 38, 4873–4879 (1997).
27. P.W. Zhu, S. Zheng, G. Simon, *Macromol. Chem. Phys*. 202, 3008–3017 (2001).
28. M.S. Kim, P.K. Sekhar, S. Bhansali, J.P. Harmon, *J. Nanosci. Nanotechnol*. 9, 5776–5784 (2009).
29. G. Turky, S.S. Shaaban, A, Schönhals, *J. Appl. Polym. Sci*. 113, 2477–2484 (2009).

30. R. Böhmer, K.L. Ngai, C.A. Angell, D.J. Plazek, *J. Chem. Phys.* 99, 4201–4209 (1993).
31. R. Richert, C.A. Angell, *J. Chem. Phys.* 108, 9016–9026 (1998).
32. R. Böhmer, C.A. Angell, *Disorder Effects on Relaxational Processes*, R. Richert and A. Blumen (eds.), Springer, Berlin, Germany (1994).
33. E. Rössler, A.P. Sokolov, *Chem. Geol.* 128, 143–153 (1996).
34. H. Couderc, A. Saiter, J. Grenet, J.M. Saiter, G. Boiteux, E. Nikaj, I. Stevenson, N. D'Souza, *Polym. Eng. Sci.* 49, 836–843 (2009).
35. H.C. Wong, A. Sanz, J.F. Douglas, J.T. Cabral, *J. Mol. Liq.* 153, 79–87 (2010).
36. J. Jancar, J.F. Douglas, F.W. Starr, S.K. Kumar, P. Cassagnau, A.J. Lesser, S.S. Sternstein, M.J. Buehler, *Polymer* 51, 3321–3343 (2010).
37. Y. Ding, S. Pawlus, A.P. Sokolov, J.F. Douglas, A. Karim, C.L. Soles, *Macromolecules* 42, 3201–3206 (2009).
38. G. Polizos, V.V. Shilov, P. Pissis, *J. Non-Cryst. Solids* 305, 212–217 (2002).
39. C.A. Solunov, *Eur. Polym. J.* 35, 1543–1556 (1999).
40. A. Saiter, J.M. Saiter, J. Grenet, *Eur. Polym. J.* 42, 213–219 (2006).
41. V. Tomer, G. Polizos, E. Manias, C.A. Randall, *J. Appl. Phys.* 108, 074116 (2010).
42. J. Mijovic, H. Zhang, *Macromolecules* 36, 1279 (2003).
43. J.C. Maxwell, *A Treatise on Electricity and Magnetism*, Clarendon Press, Oxford, U.K. 1, p. 452 (1892).
44. R.W. Sillars, *J. Inst. Electron. Eng.* 80, 378 (1937).
45. K.W. Wagner, *Electr. Eng. (Arch. Elektrotech.)* 2, 371 (1914).
46. K. Funke, B. Roling, M. Lange, *Solid State Ionics* 105, 195 (1998).
47. B. Roling, C. Martiny, K. Funke, *J. Non-Cryst. Solids* 249, 201 (1999).
48. E. Tuncer, I. Sauers, D.R. James, A.R. Ellis, M.P. Paranthaman, A. Goyal, K.L. More, *Nanotechnology* 18, 325704 (2007).
49. D.H. Lee, J.G. Park, K.J. Choi, H.J. Choi, D.W. Kim, *Eur. J. Inorg. Chem.* 6, 878–882 (2008).
50. C.D. Lokhande, E.H. Lee, K.D. Jung, O.S. Joo, *J. Mater. Sci.* 39, 2915–2918 (2004).
51. S. Cassaignon, M. Koelsch, J.P. Jolivet, *J. Phys. Chem. Solids* 68, 695–700 (2007).
52. F. Pedraza, A. Vazquez, *J. Phys. Chem. Solids* 60, 445–448 (1999).
53. M. Kiyama, T. Akita, Y. Tsutsumi, T. Takada, *Chem. Lett.* 1, 21–24 (1972).
54. G. Polizos, E. Tuncer, I. Sauers, K.L. More, *Appl. Phys. Lett.* 96, 152903 (2010).
55. W. Weibull, *J. Appl. Mech.* 18, 293–297 (1951).
56. S.M. Rowland, R.M. Hill, L.A. Dissado, *J. Phys. C: Solid State Phys.* 19, 6263–6285 (1986).
57. G. Polizos, V. Tomer, E. Manias, C.A. Randall, *J. Appl. Phys.* 108, 074117 (2010).
58. Manias, G. Hadziioannou, G. ten Brinke, *J. Chem. Phys.* 101, 1721–1724 (1994).
59. V. Tomer, C.A. Randall, G. Polizos, J. Kostelnick, E. Manias, *J. Appl. Phys.* 103, 034115 (2008).
60. V. Tomer, G. Polizos, C.A. Randall, E. Manias, *J. Appl. Phys.* 109, 074113 (2011).
61. C.P. Bowen, T.R. Shrout, R.E. Newnham, C.A. Randall, *J. Int. Mater. Syst. Struct.* 6, 159 (1995).
62. C.P. Bowen, R.E. Newnham, C.A. Randall, *J. Mater. Res.* 13, 205 (1998).
63. J.E. Martin, C.P. Tigges, R.A. Anderson, *Phys. Rev. B* 60, 7127 (1999).
64. C.A. Randall, D.V. Miller, J.H. Adair, A.S. Bhalla, *J. Mater. Res.* 8, 899 (1993).
65. S.A. Wilson, G.M. Maistros, R.W. Whatmore *J. Phys. D: Appl. Phys.* 38, 175 (2005).
66. H.A. Pohl, *Dielectrophoresis*, Cambridge University Press, Cambridge, U.K. (1978).

67. T.B. Jones, *Electromechanics of Particles*, Cambridge University Press, Cambridge, U.K. (1995).
68. R.E. Newnham, *Ferroelectrics* 68, 1 (1986).
69. P.S. Neelakanta, *Handbook of Electromagnetic Materials: Monolithic and Composite Versions and Their Applications*, CRC Press, Boca Raton, FL (1995).
70. A. Sihvola, *Electromagnetic Mixing Formulas and Applications*, The Institution of Electrical Engineers, London, U.K. (1999).
71. K. Wakino, T. Okada, N. Yoshida, K. Tomono, *J. Am. Ceram. Soc.* 76, 2588 (1993).
72. E. Tuncer, *Appl. Phys. A* 107, 575–582 (2012).
73. J. Zhang, E. Manias, G. Polizos, J.-Y. Huh, A. Ophir, P. Songtipya, M.M. Jimenez-Gasco, *J. Adhes. Sci. Technol.* 23, 709 (2009).
74. H. Koerner, Y. Luo, X. Li, C. Cohen, R. Hedden, C. Ober, *Macromolecules* 36, 1975 (2003).

4

Nanoscale Spectroscopy with Applications to Chemistry

Ufana Riaz and S.M. Ashraf

CONTENTS

4.1 Spectroscopic Techniques Used for the Characterization of Nanomaterials

The recent trend of transition from micro- to nanomaterial requires the adoption of new techniques for their characterization. Nanotechnology deals with the structures, properties, and processes involving materials having organizational features on the scale of 1–300 nm. At this scale, devices lead to enhanced performance, sensitivity, and reliability with dramatically decreased size. These scales can lead to new phenomena providing opportunities for new levels of sensing, manipulation, and control. The fundamental problem from the experimental point of view, in the nanoscale region, is that the units are too small to see and manipulate. Because it is difficult to visualize the nanoscale, it is essential to develop theoretical and instrumentation approaches that are sufficiently fast and accurate so that the structure and properties of materials can be predicted by varying the conditions of temperature, pressure, and concentration as a function of time.

4.2 Characterization of Nanomaterials Using Raman Spectroscopy

One of the vibrational spectroscopy techniques commonly used for the identification of chemical structure is Raman spectroscopy, which is based on Raman scattering. It is a vital spectroscopic tool that has been conventionally used to analyze the chemical structure and conformation at micron scale. However, its application for nanoscale analysis is limited by the diffraction limit of light and weak signal intensity.

The two main challenges faced in the development of Raman spectroscopy for nanoscale characterization are as follows:

- Overcoming the diffraction limit of light to attain nanoscale resolution
- Achieving good signal statistics from nanomaterials for the inherently weak Raman effect

The resolution of optical setup of Raman system is defined by the lens or objectives used in the illumination/collection optics. One of the most important factors that dictate optical resolution is the diffraction limit of light. This forms the basis of the first challenge facing nanoscale Raman spectroscopy. This resolution limit also known as the phenomenon of limiting the resolution of optical elements was identified by Ernst Abbe, a German physicist, in the 1870s also known as the "Rayleigh criterion." It defines the minimum distance "r" between two objects at which they are resolvable and is given by [1]

$$r \geq \frac{1.22\lambda}{2n\sin\theta} \tag{4.1}$$

where

λ is the wavelength of light
n is the refractive index of the medium
θ is the semiangle of the aperture of the objective

The diffraction limit of optical resolution is the limit for resolving two objects when the first diffraction minimum of the image of one object coincides with the maximum of the other object [2]. Equation 4.1 shows that one of the factors influencing the resolution is the incident wavelength. Hence, for microscopy and spectroscopy techniques operating in the visible range, the best possible resolution is 200–300 nm. Additionally, according to Equation 4.1, the different ways by which resolution can be increased are as follows:

1. Using high refractive index materials
2. Decreasing the wavelength of the incident beam
3. Increasing the aperture angle of the objective

E.H. Synge [3] in 1928 proposed a means of overcoming the diffraction limit of light. He illuminated a sample using a microscopic aperture in an opaque screen as shown in Figure 4.1 [4,5]. By holding the aperture very close to the sample surface, the light emerging from the aperture was forced to interact with the sample before diffracting away. In this arrangement, resolution was defined by the dimensions of the aperture, which were much smaller than the wavelength of light. However, at that time the realization of Synge's approach was hindered due to practical limitations in designing such a small aperture and positioning, it so close to the sample surface. The first near-field microscope, based on this principle, was realized more than 50 years later in the 1980s independently by two groups: Pohl's group [6] at IBM Zurich and Lewis's group [7,8] at Cornell University, United States. Since then, extensive research led to the development of near-field scanning optical microscopy (NSOM) for optical and spectroscopic imaging beyond the diffraction limit. The light beam impinging on an object is converted into propagating components that are able to propagate to the detector and evanescent components confined on the surface. The first beams are associated to the low spatial frequencies of the object, whereas the second ones are connected to their high frequencies. Near-field zone is defined as the extension outside a given material of the field existing inside the material. This is a complex field consisting of both propagating and non-propagating components [1]. The non-propagating component of the near field is defined as the evanescent field, which decays within a few nanometers from the surface and follows closely the structural variations on the surface. The far-field zone exists beyond the near-field zone and is observed using conventional

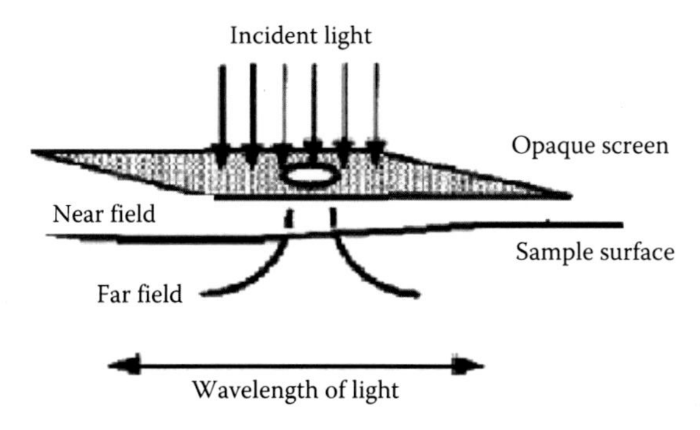

FIGURE 4.1
Synge's principle of diffraction limit.

optical elements. There are two different elements capable of converting the high-frequency local field into propagating waves. The first is based on Synge's approach where the limited object was considered to be a nano-aperture [1]. Generally, the aperture acts as a nano-emitter, where the evanescent fields of the source are used to illuminate the sample and the propagating fields are detected using conventional devices. In the second approach, the limited object is a solid tip that acts as a "nano-antennae" [1] and converts the non-propagating evanescent waves into propagating ones by acting as a scattering locally perturbing the electromagnetic field. The tip can be dielectric, semiconducting, or metallic. In case of metallic tips, additional effect of localized enhancement of the electric field is present. The nano-aperture (or nano-antennae) has to be brought very close to the sample for it to interact with the evanescent field on the surface. During illumination and collection, delineation is required [9]. In illumination probe devices, the tip acts as a local emitter and the sample illumination is locally confined. The transmission NSOM techniques using apertured probes belong to this category. In collection probe devices, far-field sources are employed before illuminating the sample and the tip acts as a local probe. Techniques of this category include the photon scanning tunneling microscope (PSTM) [10], the scanning tunneling optical microscope (STOM), [11] and the NSOMs based on aperture-less probes.

NSOM provides a means of extending the advantages of conventional optical microscopy beyond the limit of diffraction. The probes are generally made from optical fibers using the heating and pulling method [8], chemical etching [12], or a combination of both, Figure 4.2. The purpose of the first step in the fabrication process is to attain a tapered probe with a sharp apex. In the second step, metal (generally aluminum) is deposited on the probe while it is rotated at an angle, Figure 4.3. Opaque aluminum coating covers the probe walls, which reflects light to ensure transmission only from the opening at the tip end [13–14]. The most important aspect of NSOM probes is the possibility of subwavelength resolution. NSOM has been used for analyzing silicon wafers (Figure 4.4a), diamond particles (Figure 4.4b and c), polymeric samples like polyphenylenevinylene (PPV) film, and polydiacetylene crystal. Figure 4.4d shows the near-field Raman scan of resonantly excited polydiacetylene crystal with 633 nm laser with 30 s accumulation time for each spectrum [14–18]. The accumulation time for the near-field Raman spectra, for a reasonable signal-to-noise ratio, is of the order of 30–100 s depending on the Raman cross section of the sample. Hence, a 1 μm × 1 μm scan with a step size of 100 m takes about 1–3 h, which is relatively unreasonable for practical applicability of the technique. The end diameters of the apertured probes used in these experiments range from 100 to 400 nm. Hence, the best resolution attainable is always >100 nm. Other than the weak statistics of the Raman effect, the major factor responsible for these limitations is the low transmission coefficient of the apertured NSOM probes [12,17]. The dependence of Raman signal on the

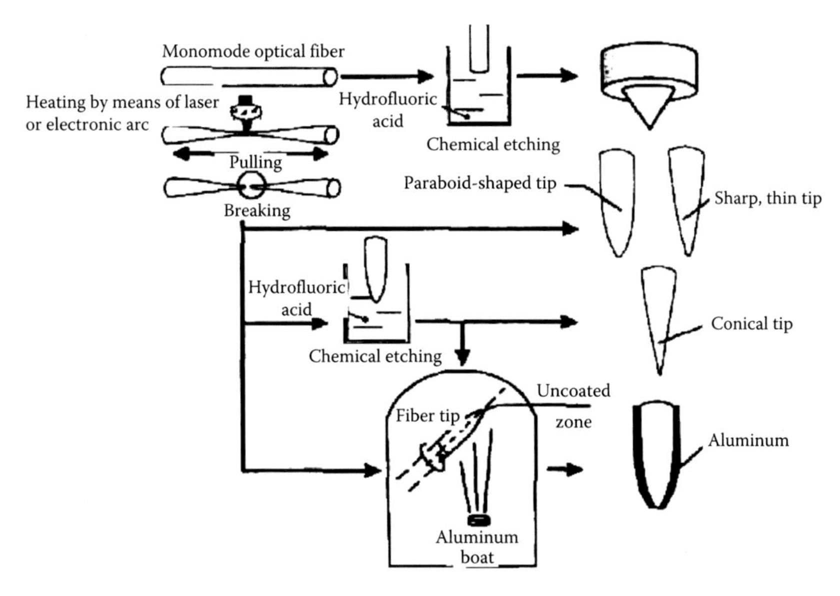

FIGURE 4.2
NSOM setup.

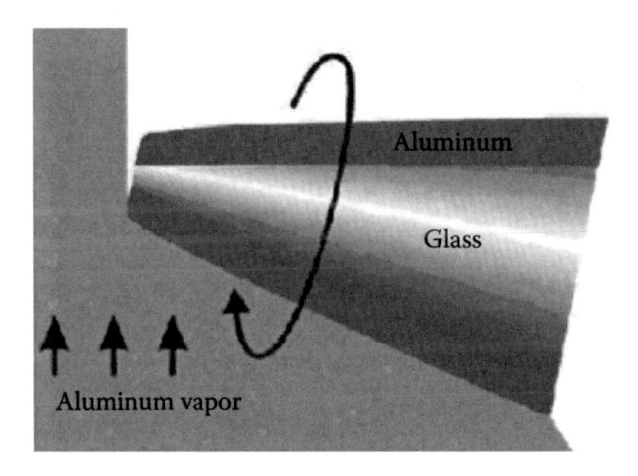

FIGURE 4.3
Aluminum deposition on apertured probes.

incident field intensity results in a low signal-to-noise ratio, hence requiring long accumulation times. The difference in light propagation through the taper and the aperture region determines the transmission coefficient [12]. Based on the theoretical model for transmission through probes, known as "Bethe–Bouwkamp" model [19,20], the transmission coefficient exhibits a fourth power dependence on the aperture diameter for a given wavelength. The cone angle is another geometrical factor influencing light transmission [21]. In general, transmission is higher for larger apertures and larger

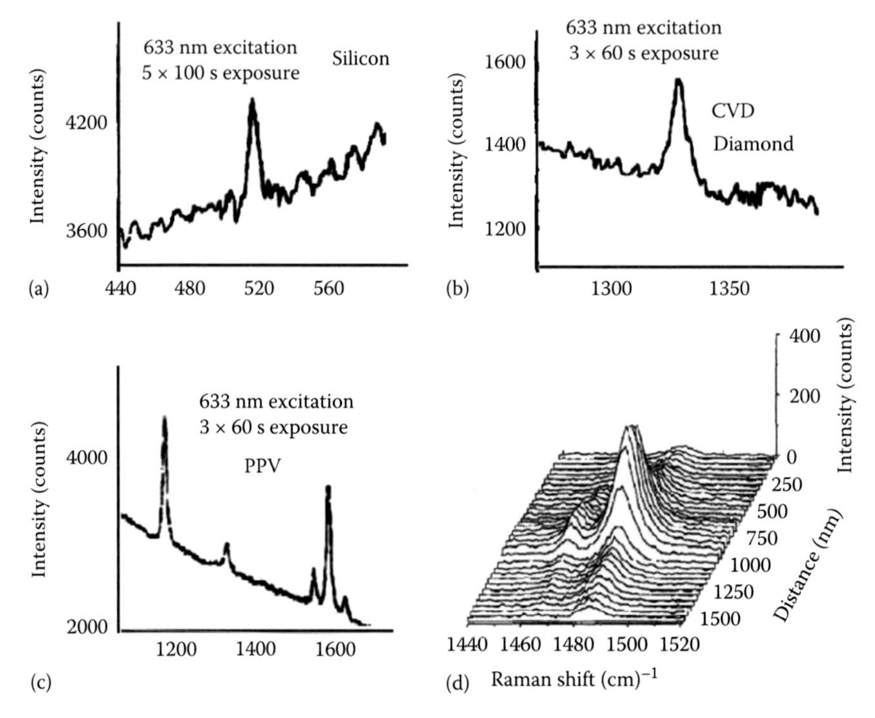

FIGURE 4.4
Raman signal obtained (a) silicon, (b) CVD diamond, (c) PPV, and (d) Raman mapping of polydiacetylene crystal using apertured NSOM probes with 30 s acquisition time for each spectra.

cone angles. Therefore, the aperture size is between the signal statistics and the desired resolution, limiting practical apertures to a size of 80–100 nm. Fleischmann et al. [22] observed an increase in the Raman signal intensity of pyridine deposited on roughened silver electrodes. The reason for this increase was due to an increase in the number of adsorption sites of pyridine molecules with increasing roughness of the electrode. Jeanmaire and Van Duyne [23] and Albrecht and Creighton [24] independently showed that the large increase in the Raman signal intensity was much more than what could be attributed to an increase in the number of adsorption sites and reasoned that it was caused due to Raman scattering itself. This effect was termed as "surface-enhanced Raman scattering" (SERS) [25] and has been observed over the last two decades for a number of molecules adsorbed on metals like silver, gold, copper, lithium, sodium, potassium, aluminum, platinum, and rhodium [26]. SERS is defined as a phenomenon resulting in strongly increased Raman signals from molecules that are attached to nanometer-sized metallic particles [27]. SERS experiments also delineate some significant characteristics of surface-enhanced Raman spectra [28,29] like the following:

- Silver, gold, and copper are the only metals known to give a large increase in signal intensity.
- Surface roughness of 50–200 nm is important for the observation of a large increase.
- The distance over which enhancement effect extends from the surface could be a few tens of nanometers and depends on the substrate morphology.
- The intensity of SERS signal decreases for higher vibrational frequencies.
- Modes forbidden in Raman scattering might appear in the SERS spectra due to different selection rules. Raman spectra tend to be depolarized in SERS.

These mechanisms are shown in Figure 4.5, which compares the factors determining Raman intensity for normal Raman scattering (NRS) and SERS [30]. NRS depends on the Raman cross section (¾*R free*), the incident laser intensity (*I*(°*L*)), and the number of molecules (*N*) in the probed volume. In SERS, only the molecules adsorbed on SERS-active sites on the substrate undergo enhancement (Figure 4.5b). Hence, the intensity also depends on the number of molecules involved in SERS (*N*), the electromagnetic enhancement of the electric fields for both the incident (*A*(°*L*)) and the Raman scattered fields (*A*(°*S*)), and NRS and SERS. The idea for realizing Raman spectroscopy with nanoscale resolution is initiated by J. Wessel [31], inspired by the development of stunning tunnel microscope (STM) at IBM [32]. He proposed the use of a metal particle (MP) tip to combine the advantages of nanoscale resolution with signal enhancement achieved in SERS. He termed the concept as surface-enhanced optical microscopy (SEOM).

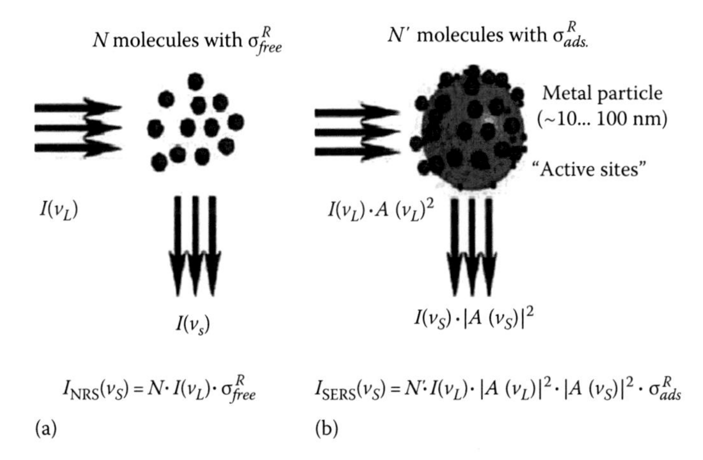

FIGURE 4.5
Schematic comparing the factors influencing signal intensity for (a) NRS and (b) SERS.

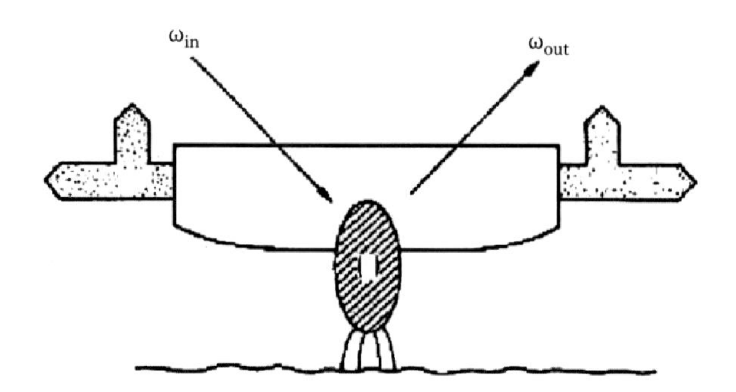

FIGURE 4.6
Wessel's idea of SEOM.

Figure 4.6 shows a schematic of the proposed instrument in which excitation of surface plasmons in the MP with an incident laser beam results in the spatial confinement of the electromagnetic field under the particle. Consequently, there is enhancement of the Raman scattered signal from the sample under the MP tip (i.e., near-field SERS). The tip is mounted in an optically transparent holder that is moved by piezoelectric translators to scan the sample. Hence, Wessel suggested a single solution for both the challenges associated with nanoscale Raman analysis. A metal (silver or gold) probe approaches the sample from above and excitation of the surface plasmons in the metal tip gives localized enhancement of the Raman signal from the sample under the tip. The enhanced signal is collected using the same objective in backscattering geometry. The tip scans the sample either by using one of the atomic force microscopy (AFM) modes or by using a tuning fork arrangement [32,33].

Batchelder and coworkers [34] observed enhancement for C60, Figure 4.7, using 100 nm Au-coated AFM tips. Futamata and coworkers [35] observed enhancements in case of diamond, using 50 nm Au-coated AFM tips, while Novotny and coworkers [36,37] show tip-enhanced spectra of single-walled nanotube (SWNT) using electrochemically etched silver tips, Figure 4.8, by illuminating from below. They reported the best results with an unprecedented resolution of 14 nm observed for SWNT [38]. The enhancement depends on the size of the particle (initial assumption $r \ll \lambda$) [39]. Figure 4.9 shows the variation of enhancement for 5, 50, and 500 nm silver particles. Enhancements decrease for very small particles whose size is smaller than the mean free path of electrons in the metal ≈300°A for silver [40]). As the size of the particle increases, broader resonances are obtained due to the excitation of multiple plasmons and higher dissipation effects that lead to smaller enhancements, Figure 4.9 [39]. The shape of the particle also influences the plasmon resonances associated with it. Spherical particles, due to their symmetry, exhibit a single plasmon resonance, while r elliptical particles exhibit two resonances associated with

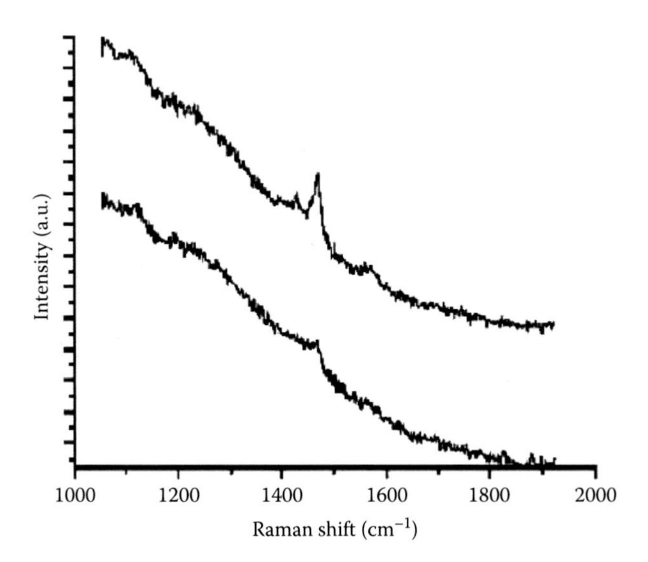

FIGURE 4.7
Raman enhancement for C60 using electrochemically etched 100 nm gold-coated AFM tips.

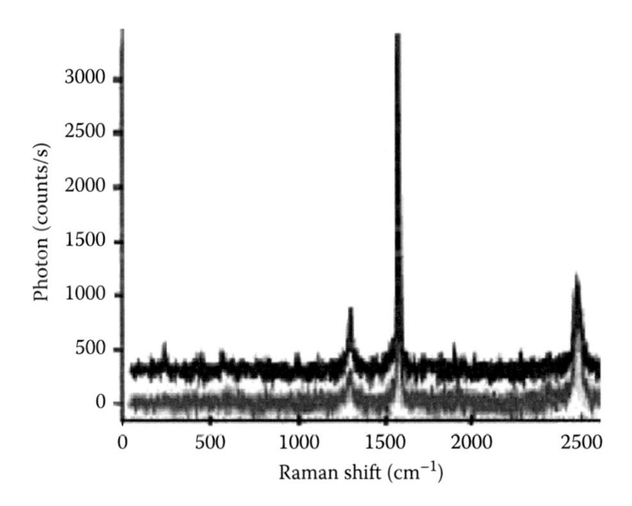

FIGURE 4.8
Raman enhancement for SWNT using electrochemically etched silver tips.

the two axes, depending on the illumination direction, Figure 4.10 [41]. Martin and coworkers [41,42] performed simulations comparing the variation in the resonance spectrum for Ag nanowires [42] with different cross sections corresponding to that of a circle, a hexagon, a pentagon, a square, and a triangle, Figure 4.10. They plotted the scattering cross section (SCS) as a function of wavelength since SCS is an optical property closely related to the plasmon resonances excited in a particle. The resonance spectrum

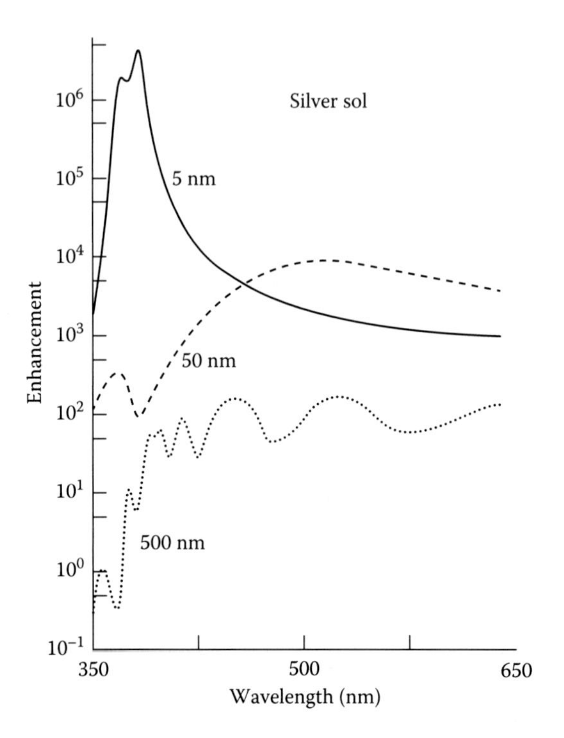

FIGURE 4.9
Dependence of Raman enhancement on particle size illustrated by comparing enhancement for 5, 50, and 500 nm silver particles.

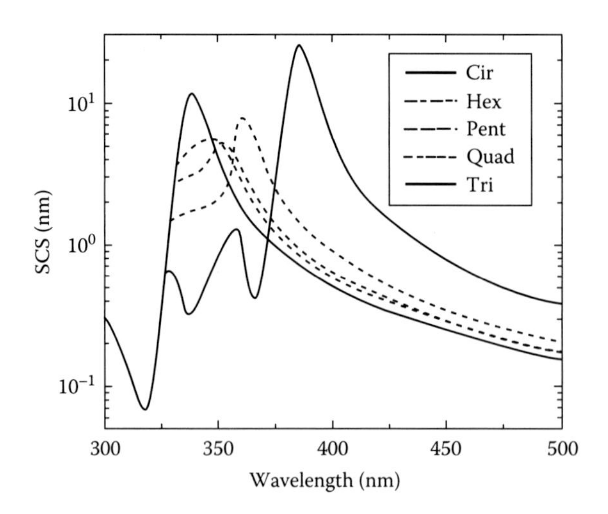

FIGURE 4.10
Resonances excited in MPs as a function of the shape of the particle.

was observed to be more complicated as the symmetry of the particle was reduced, Figure 4.10. The circle showed a single resonance, and as the heterogeneity in the cross section was increased, an additional resonance was observed for the noncircular particles, while a third resonance was also observed for the triangle.

SERS can be used to attain enhancements ranging from 10^4 to 10^{14} by combining different enhancement mechanisms depending on the metal used (resonance effect), the size and shape of the particle (lightning-rod effect), and formation of fractal clusters (interparticle coupling effects). SERS can thus increase the Raman signal statistics from nano-volumes.

Tip-enhanced Raman spectroscopy (TERS) is based on the SEOM proposed by J. Wessel [31]. It is a combination of scanning probe microscopy (SPM) and Raman, which uses metal or metal-coated tips for localized nanoscale enhancement of the Raman signal from a sample placed under the tip. The principle underlying this technique is based on the excitation of plasmons on a metallic tip that gives gigantic localized enhancement of the Raman signal. It is a promising technique for analyzing the chemical composition, structure, and conformational states on the nanoscale. However, its widespread application requires optimization of the technique to achieve high enhancements. In TERS the sample is illuminated using conventional optics (i.e., a microscope objective) from either the side or bottom. This results in diffraction-limited (≈1 μm) focus spot on the sample surface, Figure 4.11. The Raman signal without the aperture-less probe is attained from this illuminated spot and is correspondingly termed as the "far-field signal." When a metallic tip interacts with the electric field of the incident laser beam, it causes an excitation of the surface plasmons in the MPs on the tip. This results in localized enhancement of the electric field within a few nanometers from these particles known as "the near-field signal." Since Raman signal intensity scales as $\gg E^4$, there is a strong increase in the Raman signal intensity from the sample when the tip is brought either in contact with or within a few (≈2–3) nanometers from the sample. This enhanced signal is termed as the "near-field signal," Figure 4.11. From this relation it can be estimated that a 10 times increase in the electric field would correspond to a 10^4 times

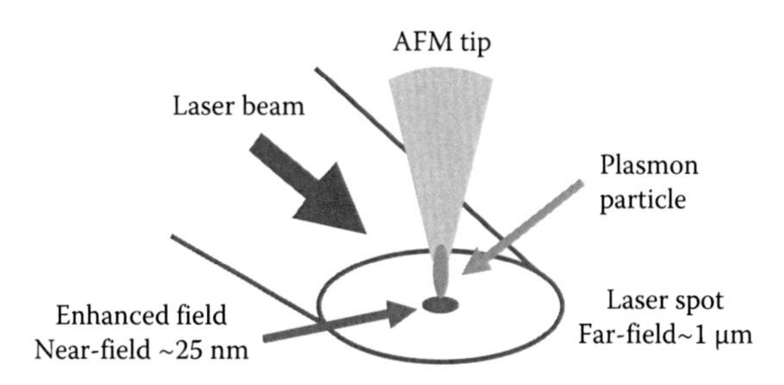

FIGURE 4.11
Schematic representing near-field and far-field in TERS.

localized enhancement of the Raman signal. The localization of this signal is limited by the tip end radius, thus defining the spatial resolution of the system. Raman intensity calculations, based on the difference in the far-field and near-field areas, exhibit a reduction of signal intensity with the tip to 10^{-4} X counts/s. In order to differentiate the near-field component from the far-field signal, it should at least be equal to or higher than X. Hence, the metallic tip should result in an enhancement of the near-field signal by a factor of at least 10^4 to perform nanoscale analysis. An optimum TERS setup requires a versatile system capable of characterizing a number of different samples within reasonable acquisition times and with high spatial resolution. Both these factors are eventually associated with maximizing the enhancement of the electric field with a metallic tip.

4.3 Characterization of Nanomaterials Using NMR Spectroscopy

The spatial resolution of conventional magnetic resonance imaging (MRI) and localized nuclear magnetic resonance (NMR) spectroscopy is limited to 0.1 mm and can be enhanced down to a few micrometers. Further improvement is restricted by the sensitivity. Inductive detection of magnetization using a radio-frequency (rf) pickup coil—as part of every commercial NMR instrument—requires presently at least 10^{12} nuclear spins per volume element to detect a signal, and further improvement seems difficult. In the early nineties, Sidles and coworkers [43] proposed an alternative detection scheme that was based on a modified AFM and could reach a far better sensitivity. The sample was attached to a microscale cantilever and placed in the strong field gradient produced by a small ferromagnet. When the magnetic force acting on the sample was inverted periodically, by rf irradiation, the cantilever began to resonate. The mechanical vibrations could be measured by the help of a laser beam that gets reflected from the cantilever surface. With this technique named magnetic resonance force microscopy (MRFM), Rugar et al. [42] succeeded in measuring the very small ≈aN force from a single electron spin.

The combination of MRFM with high-resolution NMR spectroscopy holds great promises. NMR spectroscopy can benefit from the high sensitivity of mechanical detection to investigate the molecular structure of very small objects. Imaging by MRFM, on the other hand, gains detailed information about the chemical composition of the imaged sample when combined with spectroscopy. Together they allow obtaining NMR spectra for each volume element of the sample, adding superb chemical contrast to imaging at the unprecedented micro- to nanoscale resolution of the MRFM. The field gradient of the MRFM that enables force detection can also be used for MRI.

Due to their high strengths (10^3–10^6 T/m) compared to the ones involved in conventional MRI (<10 T/m), resolutions down to the low nanoscale can be reached [42,43].

Making MRFM applicable for high-resolution NMR spectroscopy offers enormous potentials and is challenging. Exploiting the exquisite sensitivity and the very high imaging resolution of the MRFM enables investigation of the objects in the nanometer scale, which is much smaller than inductive detection. The strong field inhomogeneity associated with the MRFM gradient is detrimental for high-resolution magnetic resonance applications. While superconducting magnets used in standard NMR are presently able to provide magnetic fields that vary less than $\approx 10^9$ over the entire sample volume, resonant slices in MRFM have a width of $\approx 10^4$, that is, spectral resolution is reduced by about five orders of magnitude. There is a possibility to measure the mechanical torque on a sample by transfer of spin angular momentum. This method, which does not require a field gradient, was investigated by Alzetta et al. for ESR [44,45]. Leskowitz et al. and Madsen et al. [46,47] developed an alternative approach that was used with NMR. They measured the field gradient of the sample rather than that of the gradient magnet. Both attempts however employed force detection for sensitivity enhancement and essentially lacked the high imaging capacity of the genuine MRFM invented by Sidles. The key challenge was to overcome the resolution-limiting effect of the gradient magnet, which is the core strategy in molecular structure determination by NMR via the selective suppression or enhancement of specific spin interactions. The presence of dipolar couplings manifests close spatial proximity between two nuclear spins, and by measuring the coupling constants, quantitative distance information can be obtained. Chemical bonds often result in contact couplings (J-couplings), which can be used to investigate the bonding network within the molecule. Chemical shifts are often characteristic for their molecular environment and allow in many cases to assign the atom to a certain functional group. The effect of a certain interaction can be suppressed by canceling the associated Hamiltonian during the observed coherent evolution of the spins. This can be done either by continuously decoupling or by introducing dephasing and rephasing periods while changing the sign of the Hamiltonian in between ("echo" methods). Sophisticated rf-pulse sequences have been developed that allow manipulating the Hamiltonian nearly at will.

Hahn spin echoes [48,49] are widely used to refocus the static interaction between a spin and the external field (or generally, any odd-order interaction). For MRFM experiments, they are particularly useful since the strong field inhomogeneity of the gradient material often leads to the strongest dispersion of the spin signal and therefore overlaps with remaining local fields. A Hahn echo is produced when the coherent evolution of a spin ensemble is refocused by a 180° rf pulse, that is, when a 180° pulse is applied after τ, an echo is formed at time 2τ.

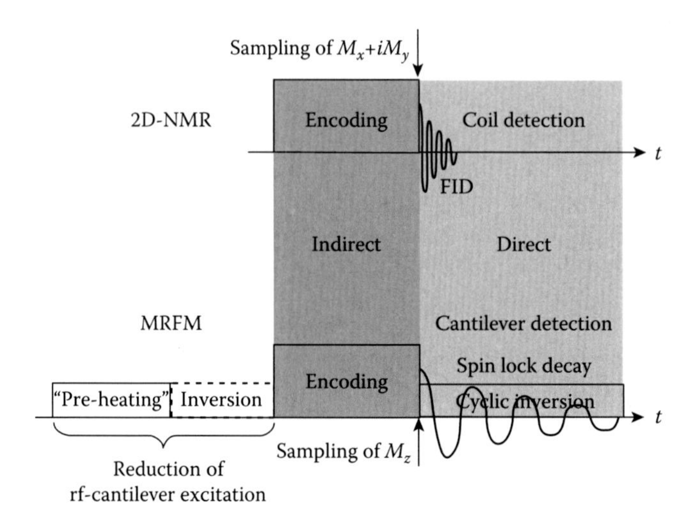

FIGURE 4.12
Spin encoding and signal (coil) detection for inductive 2D-NMR versus MRFM. A "preheating" period and a single inversion can precede the experiment to improve rf immunity.

MRFM detection of NMR spectra is closely related to conventional 2D spectroscopy, Figure 4.12. Both approaches use the same sequential technique. Spectral information is encoded into magnitude, and eventually the phase of the spin magnetization in the first period and in the second period the state is obtained. The concepts known from inductive 2D-NMR spectroscopy are therefore naturally extended to MRFM by simply switching to mechanical detection in the second period. In this way, MRFM can benefit from the highly developed pulsed NMR techniques while having the high sensitivity of mechanical detection, Figure 4.12.

However, there are certain differences between the two methods. First, in contrast to force detection, the longitudinal component of the magnetization, M_z, an induction coil is sensitive to the transverse magnetization $M_x + iM_y$, that is, to coherences. Measuring the decay of the coherence (i.e., the free induction decay) provides the full spectrum of the excited spin, while the information extractable from the spin-lock decay that is basically the $T1p$ is rather poor. This means that the information content of an MRFM-detected spectroscopy experiment in almost all cases equals that of a 1D-NMR spectrum but requires the measurement time of a 2D-NMR experiment. Second, the MRFM operates in a very inhomogeneous field and includes spins with largely dispersed Larmor frequencies (typically 100 kHz). Reduction of the excitation bandwidth is possible since they are directly proportional to the signal intensity. As a result, broadband pulsing schemes, for example, very hard pulses or adiabatic methods, are strongly preferred. Finally, most experiments require spin temperature alternation to eliminate strong cantilever excitation by rf pulses. A widely applicable scheme, where a single inversion prior to every other acquisition is combined with ± phase cycling of the acquired signal, is used. The single inversion can be shaped to optimize the selected frequency window. The spatial imaging resolution of the MRFM is often much higher along the field axis than for the lateral dimensions. Application to quasi-1D structures such as thin films and interfaces is therefore particularly promising.

To collect a maximum amount of signal from a planar film and to minimize signal contributions from spins outside of the film, the resonant slice must be made as flat as possible. This can be done by optimizing the cylindrical geometry that already possesses rather flat field contours close to the pole. One of several possibilities is shown in Figure 4.13a, using a concave indentation on the lower pole of the cylinder. Thereby, a perfectly flat resonant slice is formed at a fixed distance from the pole, depending mainly on the depth of the indentation. Another possibility is to use a ferromagnetic cone with

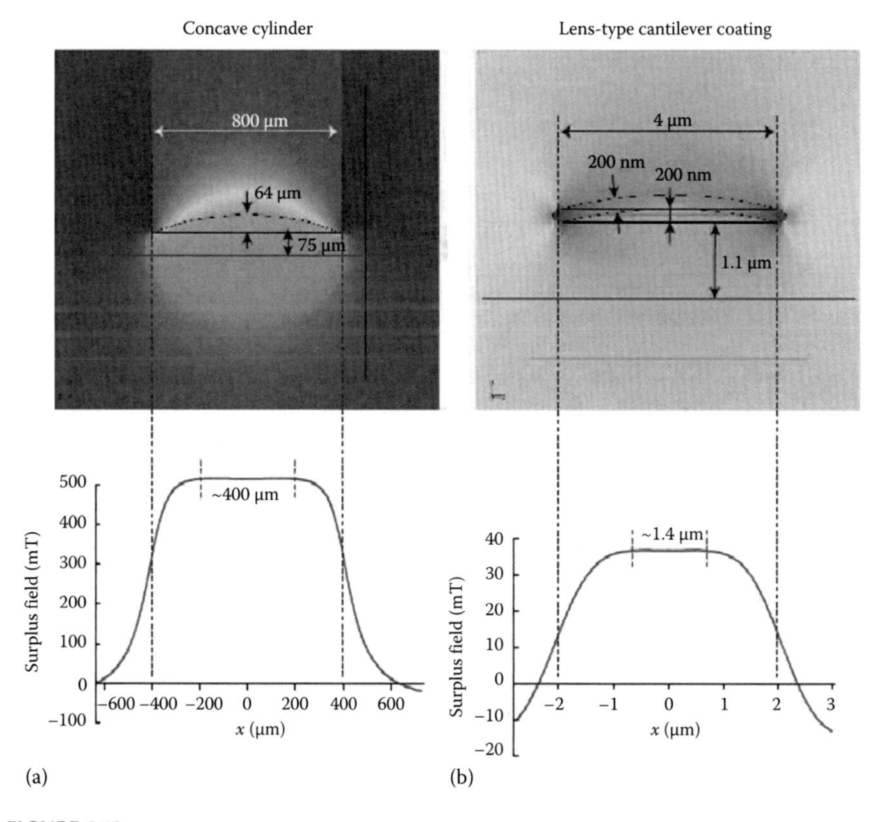

FIGURE 4.13
(See color insert.) Magnet shapes for interface imaging in a (a) sample-on-cantilever and (b) tip-on-cantilever arrangement using a 200-nm thick ferromagnetic coating.

off defined angle and to cut off the tip at a certain width of the cone. One can also integrate the magnet into the cantilever tip, such as the lens-type structure suggested in Figure 4.13b, which can be accurately shaped by epitaxial or vapor deposition techniques.

4.4 Mechanical Characterization of Nanomaterials

Mechanical characterization of 1D nanocomponents is still a very nascent field as the testing methods are not well developed. This has further been complicated by the lack of appropriate analytical tools and models involved in the interpretation of the experimental data obtained at the molecular level [50]. One of the main problems arises from the fact that the techniques used at the macrolevel cannot be implemented at the micro- and nanolevels. For example, tensile testing requires a rigid clamping with a good alignment that is not easy to implement at the nanolevel. Even simple positioning and manipulation of nanostructures or parts is difficult. Among the characterization techniques commonly used for mechanical testing of thin films, microstructures, and nanostructures are nanoindentation- and resonance-based mechanical testing.

Nanoindentation is one of the most effective tools for the study of mechanical properties of micro- or 1D nanomaterials used in various applications. In order to determine the mechanical properties of micro- or nanomaterials, they are placed between two stable edge structures fabricated using standard microfabrication methods. The nanoindenter tip is moved over the chosen location of the sample, where the concerned properties are to be determined. Nanoindenter monitors are used, and the load and displacement of the indenter during indentation with a force resolution of 1 nN or less with a displacement resolution of 2 A are measured where the absolute values of which depend on the instrument. Three-sided pyramidal diamond (Berkovitch) is used as the indenter that provides relatively simpler analysis using the methods of continuum mechanics [50]. During a nanoindentation test, force and displacement values are recorded as the indenter tip is pressed into the test material's surface. Typical load versus displacement curve often seen in a nanoindentation experiment is shown in Figure 4.14. It may be noted that the shape and scales are different for different materials and experiments.

These load–displacement curves not only provide the extent of plastic deformation due to the loading–unloading cycles but also exhibit localized perturbations or discontinuities, which are represented by the characteristic of energy-absorbing or energy-releasing events occurring beneath the tip. Ni and Li showed that ZnO nanobelts failed from brittle fracture in bending, while plastic deformation was indicated in nanoindentation [51].

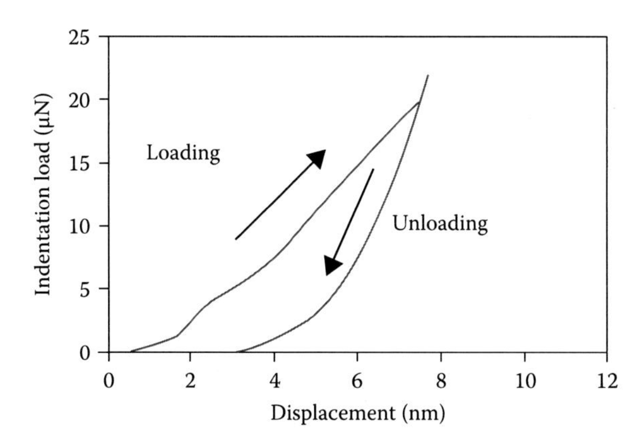

FIGURE 4.14
Nanoindentation load versus displacement curve.

The load–displacement curves obtained in the study through nanoindentation showed a permanent deformation approximately at around 3 nm, after the peak indentation load was removed from ZnO nanobelts with a thickness of 95 nm and a width of 800 nm.

One can determine Young's modulus (E_n) following the linear elastic theory of an isotropic material, knowing the spring constant of the structure to be tested using the formula

$$E_n = \frac{FL^3}{192 d_n I} \tag{4.2}$$

where
- F is the force
- L is the suspended length of the structure or a micro/nano bridge
- (F/d_n) is the slope of the force displacement curve during bending represents the spring constant (K_n)
- I is the area moment of inertia

Nanoindentation hardness is defined as the indentation load divided by the projected contact area of the indentation, as defined by the formula [52,53]

$$H = \frac{P_{max}}{A} \tag{4.3}$$

where
- P_{max} is the peak load
- $A = (\pi/4)\,(S/E\tau)^2$ is the projected contact area
- $E\tau$ is the reduced elastic modulus
- $S = (dP/dH)$ is the slope of the initial portion of the unloading curve

$$\frac{1}{E\tau} = \frac{(1-\upsilon_s^2)}{E_s} + \frac{(1-\upsilon_i^2)}{E_i} \tag{4.4}$$

where E_s, E_i and υ_s, υ_i are Young's moduli and Poisson's ratio of the sample and indenter, respectively. It may be noted that the subscripts s and i refer to the substrate and indenter, respectively. The bending stresses generated in the microbridge (or nanowire) are proportional to the moments generated at the fixed ends. The maximum tensile stress, σ_b, which is the fracture stress, was given by Li and Bhushan [54]:

$$\sigma_b = \frac{F_{\max} l e_1}{8I} \tag{4.5}$$

where
- $F_{\max}$ is the load applied at the instant of failure
- l is the length of the beam
- e_1 is the distance of the top surface from the neutral plane of its cross section

The electric field-induced resonance excitation is used for measuring different mechanical properties of individual nanowire-like structures inside a transmission electron microscope (TEM). This technique has been successfully used to study the properties of carbon nanotubes, silicon nanowires, SiC–SiO_2 composite nanowires, and ZnO nanobelts. The specimen holder used in these measurements has two electrodes and a set of piezomanipulation and translation stages. Typical electrical measurement setup is shown in Figure 4.15. A more detailed representation of the setup with an in situ TEM measurement arrangement is shown by Wang et al. [55,56]. Different voltage configurations have been used to determine the bending modulus of the individual nanotubes, wires, and belts. From the classical elasticity theory, the fundamental resonance frequency is related to the bending modulus and other nanodimensional parameters by

$$\nu_i = \frac{\beta_i^2}{8\pi L^2}\sqrt{\frac{\left(D^2 + D_i^2\right)E_\beta}{\rho}} \tag{4.6}$$

where
- ν_i is the resonance frequency
- β_i^2 is a constant for ith harmonic
- E_β is the bending modulus
- ρ is the density of the tube
- L, D, and D_i are the length, outer diameter, and inner diameter of the tube, respectively

The same relationship can be extended to nanowires or belts representing the diameter of the wire or width of the belt. Individual nanocomponent is

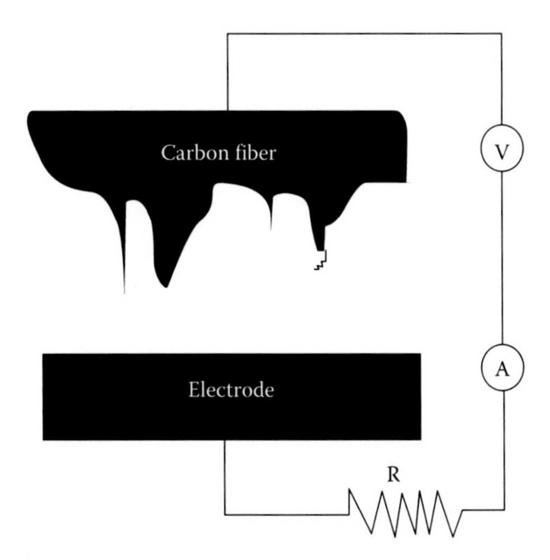

FIGURE 4.15
An experimental setup used during the electric field-induced resonance excitation method.

attached to both the gold electrodes using silver paste, through which the electric contacts were made. A constant or alternating voltage is applied to the two electrodes to induce electrostatic deflection or mechanical resonance. As these tests are performed in TEM, physical observation of changes of the nanoelement features in response to dynamically changing electrical parameters has been added in this method. However, TEM may possibly be replaced by other imaging instruments, which can provide the same information. Due to the electric charge induced at the tip of the 1D nanomaterials, they oscillate at the frequency of the applied voltage. Mechanical resonance results when the applied frequency matches the natural vibration frequency. The resonance frequency should be determined accurately, in order to obtain the bending modulus reliably. Many authors have extended the same technique for determining the bending modulus of ZnO nanobelts, single- and double-walled carbon nanotubes, and other nanowire systems [57–59]. In case of nanobelts that have a rectangular cross section, two fundamental resonance modes corresponding to two orthogonal transverse vibration directions were observed [60]. Bending modulus of ZnO nanobelts was measured to be 52 GPa, which is much lower than the bulk ZnO value, which lies in the range 104–210 GPa. This difference in observed value for single nanobelts was attributed to the scaling effect (with respect to size) and geometrical shape. Further, the anisotropic nature of ZnO may also be another reason showing large differences even in bulk in different directions. This has been one of the very unique techniques, especially measuring the individual single nanotube/wire/belts. Interestingly, these nanowires and belts may be considered similar to nanocantilevers. The values obtained by this technique were in reasonable agreement with those obtained by a nanoindentor [51].

This technique may potentially be extended for determining the fatigue characteristics of the individual 1D nanomaterial components. Most of the earlier mentioned techniques have been used in MEMS-based platforms for studying the mechanical properties of the micro- or nanostructures in the form of bridges or cantilevers where measurements were performed using AFM or in situ TEM. The biggest challenge is mounting or fixing the nanosamples on to the MEMS platforms. Few research groups have been successful in designing and developing the MEMS test beds for mechanical characterization of nanowires, belts, etc. [59,61–65]. The monotonic properties of single- and polycrystalline silicon show that the elastic modulus obtained from miniaturized tensile samples is comparable with those expected for bulk materials. On the other hand, fracture strength may vary significantly from bulk properties. In particular, bending data show significant increase in strength as thickness or cross-section dimensions decrease. The size effect of increasing strength with decreasing cross-section dimensions of brittle Si miniature samples can be attributed to the increased amount of surface energy per unit volume. For bending test there is also high stress gradient that additionally constrains deformation and contributes to the strengthening effect. For metallic materials the most commonly observed size effect in miniature samples is the trend of decreasing ductility with decreasing sample thickness and width. The reason is non-homogeneity of deformation as grain size approaches sample thickness or width. This effect is likely to be enhanced in grains with preferred orientation that results in strain localization and additional stress concentration. Ductility may be reduced further with increasing surface topography and environmental effects. Reduced ductility affects the material strength due to higher sensitivity to porosity, inclusions, and surface roughness. Therefore, the apparent competition between the strengthening effect due to surface energy/stress gradient and weakening effect due to non-homogeneity of deformation for metallic thin films results in inconclusive data reported in the literature. However, further analysis is required to understand the size effects in relation to processing methods (e.g., electroplating, self-assembly) and the testing mode (tensile or bending) used.

4.5 Surface Characterization of Nanomaterials

The mechanisms and dynamics of the interactions of two contacting solids during relative motion, ranging from atomic scale to microscale, need to be understood in order to develop fundamental understanding of adhesion, friction, wear, indentation, and lubrication processes. For most solid–solid interfaces of technological relevance, contact occurs at multiple asperities. Consequently, the importance of investigating single-asperity contacts in studies of the fundamental micro-/nanomechanical and micro-/nanotribological properties of surfaces and interfaces has long been recognized. The recent proliferation of proximal probes, in particular SPMs (the STM and the AFM), the surface force apparatus (SFA), and computational techniques for simulating tip–surface interactions and interfacial properties, has developed systematic investigations of interfacial problems with high resolution as well as ways and means for modifying and manipulating nanoscale structures. These advances have led to the appearance of the new field of nanotribology, which pertains to experimental and theoretical investigations of interfacial processes on scales ranging from the atomic and molecular scale to the microscale, occurring during adhesion, friction, scratching, wear, indentation, and thin-film lubrication at sliding surfaces [65–76]. Proximal probes have also been used for mechanical and electrical characterization, in situ characterization of local deformation, and other nanomechanics studies. Nanotribological and nanomechanics studies are needed to develop the fundamental understanding of interfacial phenomena on a small scale and to study interfacial phenomena in micro-/nanostructures used in magnetic storage devices, nanotechnology, and other applications [68–76]. Friction and wear of lightly loaded micro-/nanocomponents are highly dependent on the surface interactions (few atomic layers). These structures are generally coated with molecularly thin films. Nanotribological and nanomechanics studies are also valuable in the fundamental understanding of interfacial phenomena in macrostructures and provide a bridge between science and engineering. The SFA, STM, AFM, and friction force microscopes (FFMs) are widely used in nanotribological and nanomechanics studies. Typical operating parameters are compared in Table 4.1.

4.5.1 Atomic Force Microscopy

In AFM measurement during surface imaging, the tip comes in intimate contact with the sample surface and leads to surface deformation with finite tip–sample contact area (typically a few atoms). The finite size of the contact area prevents the imaging of individual point defects, and only

TABLE 4.1

Comparison of the Operating Parameters for Different Techniques

Operating Parameter	SFA	STM	AFM
Radius of mating surface/tip (nm)	10	5–100	5–100
Radius of contact area	10–40 nm	N.A	0.05–0.5 nm
Normal load	10–100 mN	N.A	<0.1–500 nN
Sliding velocity (μm/s)	0.001–100	0.02–200	0.02–200
Sample limitations	Optically transparent, smooth surface area	Electrically conducting samples	None

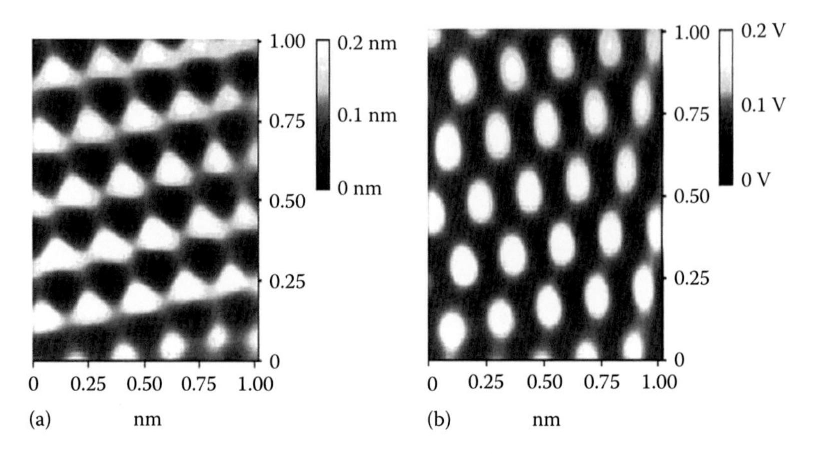

FIGURE 4.16
Grayscale plots of (a) surface topography and (b) friction force maps (2D spectrum filtered), measured simultaneously, of a 1×1 nm^2 area of freshly cleaved highly oriented pyrolytic graphite, showing the atomic-scale variation of topography and friction.

the periodicity of the atomic lattice can be imaged. Figure 4.16 shows the topography image of a freshly cleaved surface of highly oriented pyrolytic graphite [77]. The periodicity of the graphite is clearly observed. Figure 4.17a shows the atomic-scale friction force map (raw data) and Figure 4.16 shows the friction force maps (after 2D spectrum filtering with high-frequency noise truncated) [78]. Figure 4.17a shows a line plot of friction force profiles along some crystallographic direction. The actual shape of the friction profile depends upon the spatial location of the axis of tip motion. Note that a portion of atomic-scale lateral force is conservative. Mate et al. [78] and Ruan and Bhushan [79], Figure 4.17, reported that the average friction force linearly increased with normal load and was reversible with load. Friction profiles were similar during sliding of the tip in either direction. AFM experiments can be generally conducted at relative velocities as high as approximately 100–250 mm/s/K. To simulate applications, it is of interest to conduct friction experiments at higher velocities (upto 1 mm/s/K). Furthermore, high-velocity experiments would be useful to study velocity dependence on friction and wear. One of the approaches has been to mount samples on a shear wave transducer (ultrasonic transducer) and then drive it at very high frequencies (in the MHz range) as reported earlier [80–84]. The coefficient of friction on the nanoscale is estimated based on the contact resonance frequency and requires the solution of the characteristic equations for the tip vibrating in contact with the sample surface. The approach is complex and is dependent upon various assumptions. An alternative approach is to use piezostages with large amplitude (approx. 10–100 mm) and relatively low resonance frequency (few kHz) and measure directly the friction force on the microscale using the FFM signal without any analysis with assumptions used in the previous approaches using shear wave transducers. The commercial AFM setup modified with this approach yields sliding velocities up to 200 mm sK1 [85,86]. It can be argued that for the nanoscale AFM experiments, the asperity contacts are predominantly elastic (with average real pressure being less than the hardness of the softer material) and adhesion is the main contribution to the friction, whereas for the microscale experiments, the asperity contacts are predominantly plastic and deformation is an important factor. It will be shown later that hardness has a scale

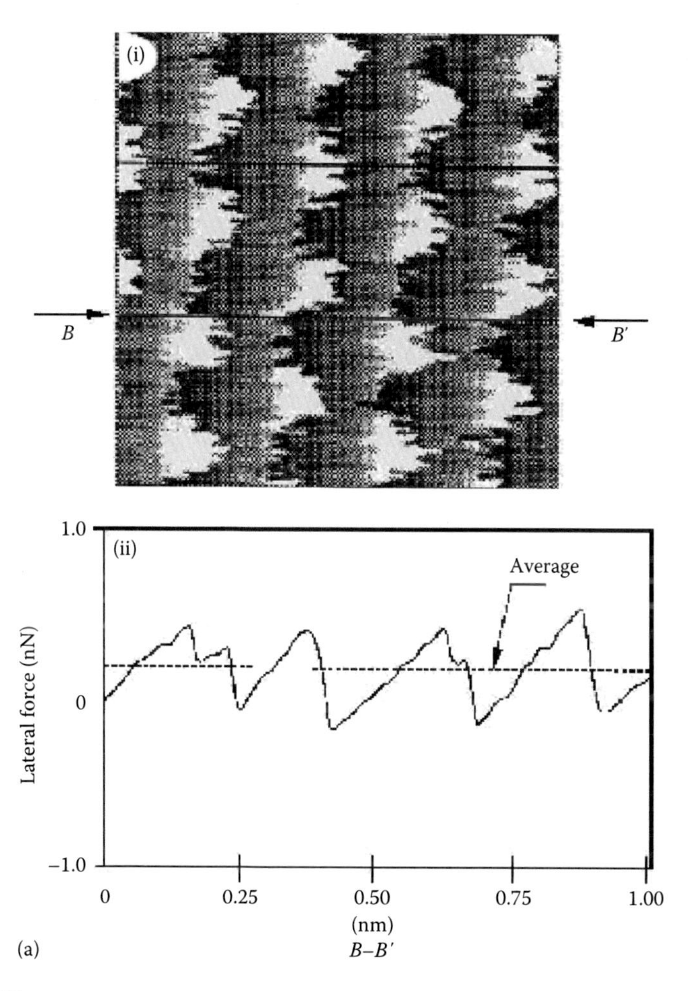

FIGURE 4.17
(a) (i) Grayscale plot of friction force map (raw data) of a 1 × 1 nm^2 area of freshly cleaved HOPG, showing atomic-scale variation of friction force. High points are shown by lighter shading. (a) (ii) Also shown is a line plot of friction force profile along the line indicated by arrows. The normal load was 25 nN and the cantilever normal stiffness was 0.4 N m/K. (From Ruan, J. and Bhushan, B., *J. Appl. Phys.*, 76, 8117, 1994.)

(continued)

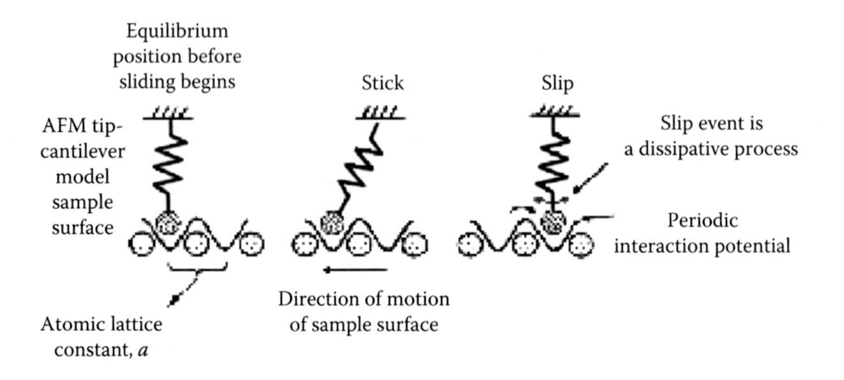

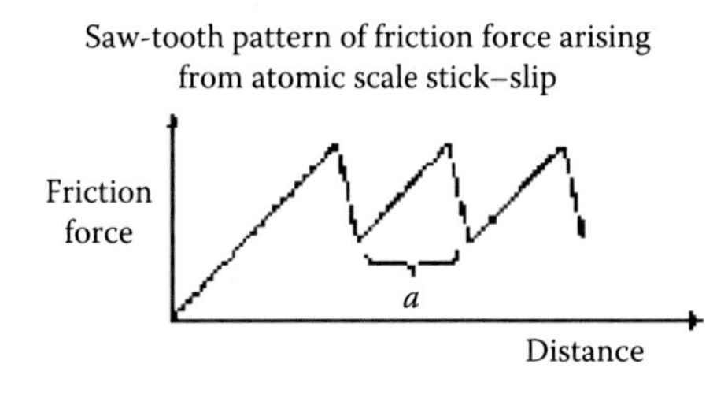

FIGURE 4.17 (continued)
(b) Schematic of a model for a tip atom sliding on an atomically flat periodic surface. The schematic shows the tip jumping from one potential minimum to another, resulting in stick–slip behavior.

effect; it increases with decreasing scale and is responsible for less deformation on a smaller scale. The meniscus effect results in an increase of friction with increasing tip radius. Therefore, third-body contribution, scale-dependent hardness and other properties, transition from elastic contacts in nanoscale contacts to plastic deformation in microscale contacts, and meniscus contribution play an important role [87–88]. The AFM can be used to investigate how surface materials can be moved or removed on micro- to nanoscales, for example, in scratching and wear (where these things are undesirable), and nanofabrication/nanomachining. Scratching can be performed under ramped loading to determine the scratch resistance of materials and coatings. The coefficient of friction is measured during scratching, and the load at which the coefficient of friction increases rapidly is known as the "critical load," which is a measure of scratch resistance. In addition, post-scratch imaging can be performed in situ with the AFM in tapping mode to study failure mechanisms. Figure 4.18 shows data from a scratch test on Si(100) with a scratch length of 25 mm and a scratching velocity of 0.5 mm sK1. At the beginning of the scratch, the coefficient of friction is 0.04, which indicates a typical value for silicon. At approximately 35 mN (indicated by the arrow in the figure), there is a sharp increase in the coefficient of friction, which indicates the critical load. Beyond the critical load, the coefficient of

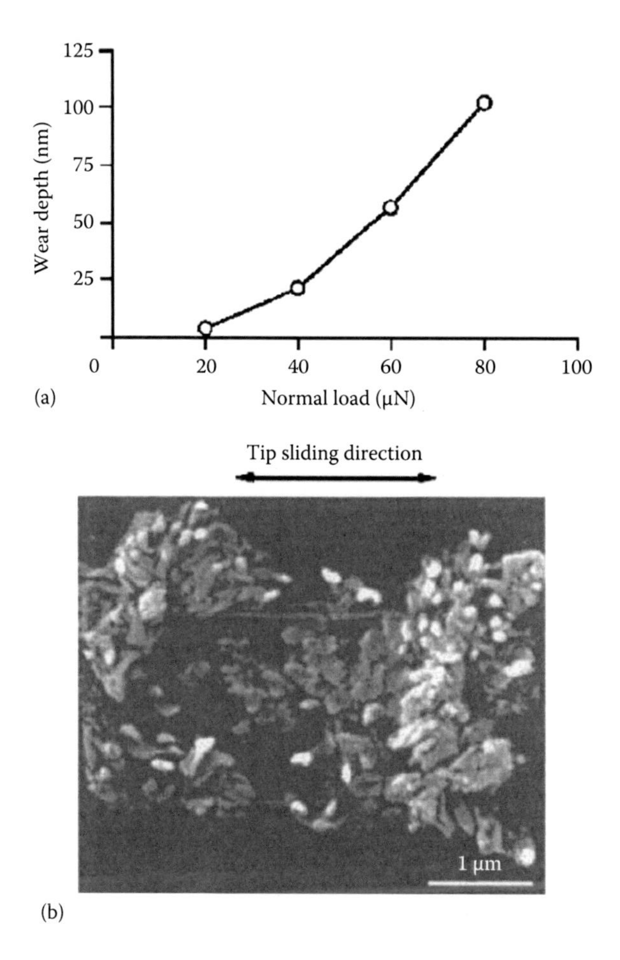

FIGURE 4.18
(a) Wear depth as a function of normal load for Si(100) after one cycle. (From Zhao, X. and Bhushan, B., *Wear*, 223, 66, 1998.) (b) Secondary electron image of wear mark and debris for Si(100) produced at a normal load of 40 mN and one scan cycle.

friction continues to increase steadily. In the post-scratch image, we note that at the critical load, a clear groove starts to form. This implies that Si(100) was damaged by plowing at the critical load, associated with the plastic flow of the material. At and after the critical load, small and uniform debris is observed and the amount of debris increases with increasing normal load. Sundararajan and Bhushan [88] have also used this technique to measure the scratch resistance of diamond-like carbon (DLC) coatings ranging in thickness from 3.5 to 20 nm, Figure 4.19a. By scanning the sample in two dimensions with the AFM, wear scars are generated on the surface. Figure 4.18 shows the effect of normal load on wear depth on Si(100). We note that wear depth is very small below 20 mN of normal load [89]. A normal load of

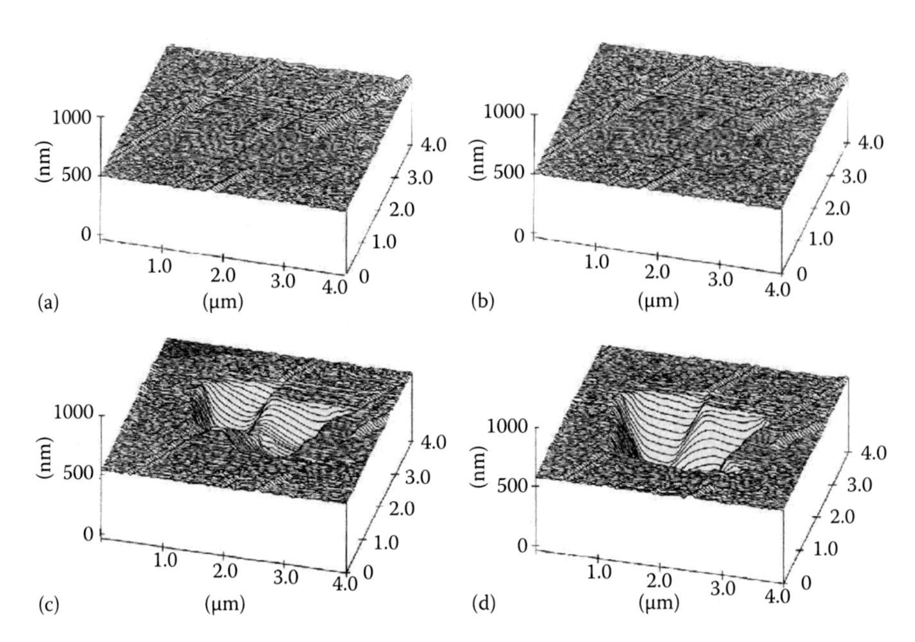

FIGURE 4.19
Surface plots of DLC-coated thin-film disk showing the worn region; the normal load and number of test cycles are (a) 20 mN, 5 cycles; (b) 20 mN, 10 cycles; (c) 20 mN, 15 cycles; and (d) 20 mN, 20 cycles. (From Bhushan, B. et al., *Philos. Mag.*, 74, 1117, 1996.) DLC-coated disk sample.

20 mN corresponds to contact stresses comparable to the hardness of silicon. Primarily, elastic deformation at loads below 20 mN is responsible for low wear [75]. Uniform material removal at the bottom of the wear mark has been reported. An AFM image of the wear mark shows small debris at the edges, probably swiped during the AFM scanning. This indicates that the debris is loose (not sticky) and can be removed during the AFM scanning. Next we examined the mechanism of material removal on a microscale in the AFM wear experiments [90,91]. Figure 4.18 shows a secondary electron image of the wear mark and associated wear particles. The specimen used for scanning electron microscopy (SEM) was not scanned with the AFM after initial wear, in order to retain wear debris in the wear region. To understand wear mechanisms, evolution of wear can be studied using an AFM. Wear is not uniform but is initiated at the nanoscratches. Surface defects (with high surface energy) present at nanoscratches act as initiation sites for wear. Coating deposition also may not be uniform on and near nanoscratches, which may lead to coating delamination. Thus, scratch-free surfaces will be relatively resistant to wear. Wear precursors (precursors to measurable wear) can be studied by making surface potential measurements [92]. The contact potential difference or simply the surface potential between two surfaces depends on a variety of parameters such as electronic work

function, adsorption, and oxide layers. The surface potential map of an interface gives a measure of changes in the work function, which is sensitive to both physical and chemical conditions of the surfaces including structural and chemical changes. Before the material is actually removed in a wear process, the surface experiences stresses that result in surface and subsurface changes of structure and/or chemistry. These can cause changes in the measured potential of a surface. An AFM tip allows mapping of surface potential with nanoscale resolution. Surface height and change in surface potential maps of a polished single-crystal aluminum (100) sample, abraded using a diamond tip at loads of 1 and 9 mN, are shown in Figure 4.20a. It is evident that both abraded regions show a large potential contrast (approx. 0.17 V) with respect to the non-abraded area. The black region in the lower right-hand part of the topography scan shows a step that was created during the polishing phase. There is no potential contrast between the high and low regions of the sample, indicating that the technique is independent of surface height. Figure 4.20b shows a close-up scan of the upper (low load) wear region in Figure 4.20a. While there is no detectable change in the surface topography, there is a large change in the potential of the surface in the worn region. Even in the case of zero wear (no measurable deformation of the surface using AFM), there can be a significant change in the surface potential inside the wear mark, which is useful for the study of wear precursors. It is believed that the removal of the thin contaminant layer including the natural oxide layer gives rise to the initial change in surface potential. The structural changes, which precede generation of wear debris and/or measurable wear scars, occur under ultralow loads in the top few nanometers of the sample and are primarily responsible for the subsequent changes in surface potential. In situ surface characterization of local deformation of materials and thin films is carried out using a tensile stage inside an AFM. Failure mechanisms of coated polymeric thin films under tensile load were studied by [93,94].

The specimens were strained at a rate of 410 K, and AFM images were captured at different strains up to approximately 10% to monitor generation and propagation of cracks and deformation bands. Bobji and Bhushan [93,94] studied various magnetic tapes with thickness ranging from 7 to 8.5 mm. One of these was with acicular-shaped MP coating [70]. They also studied the polyethylene terephthalate substrate with 6 mm thickness. They reported that cracking of the coatings started at approximately 1% strain for all tapes much before the substrate starts to yield at approximately 2% strain. Figure 4.21 shows the topographical images of the MP tape at different strains. At 0.83% strain, a crack can be seen, originating at the marked point. As the tape is further stretched along the direction, as shown in Figure 4.21, the crack propagates along the shorter boundary of the ellipsoidal particle. However, the general direction of the crack propagation remains perpendicular to the direction of the stretching. The length, width, and depth of the cracks increase with strain, and at the same time newer

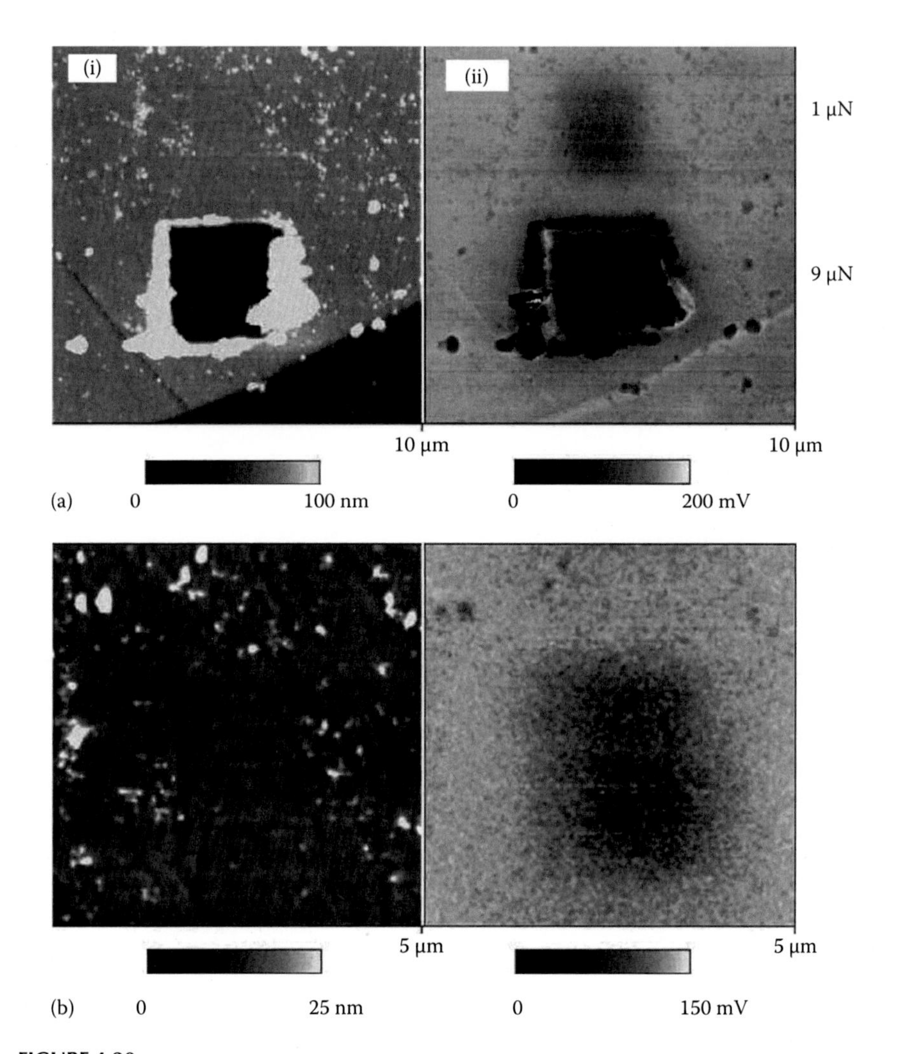

FIGURE 4.20
(a) (i) Surface height and (ii) change in surface potential maps of wear regions generated at 1 and 9 mN on a single-crystal aluminum sample showing bright contrast in the surface potential map on the worn regions. (b) Close-up scan of upper (low load) wear region shown in (a). (From DeVecchio, D. and Bhushan, B., *Rev. Sci. Instrum.*, 69, 3618, 1998.)

cracks keep on nucleating and propagating with reduced crack spacing. At 3.75% strain, another crack can be seen nucleating. This crack continues to grow parallel to the first one. When the tape is unloaded after stretching up to a strain of approximately 2%, that is, within the elastic limit of the substrate, the cracks close perfectly and it is impossible to determine the difference from the unstrained tape. To make accurate measurements of hardness at shallow depths, a depth sensing nano-/picoindentation system is used [70]. Figure 4.22 shows the load–displacement curves at different peak loads

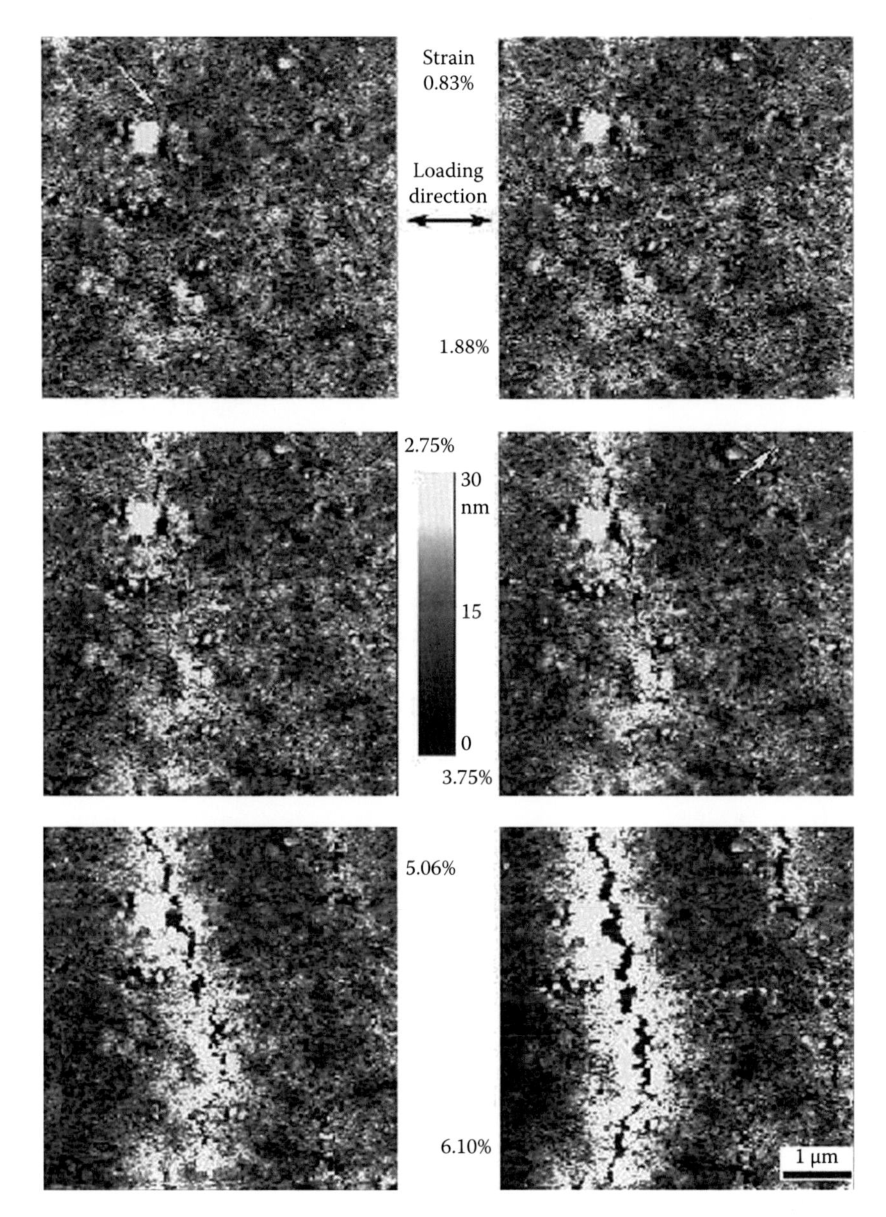

FIGURE 4.21
Topographical images of MP magnetic tape at different strains.

for Si(100). Loading/unloading curves often exhibit sharp discontinuities, particularly at high loads.

Discontinuities, also referred to as pop-ins, occurring during the initial part of the loading part of the curve mark a sharp transition from pure elastic loading to a plastic deformation of the specimen surface and thus correspond

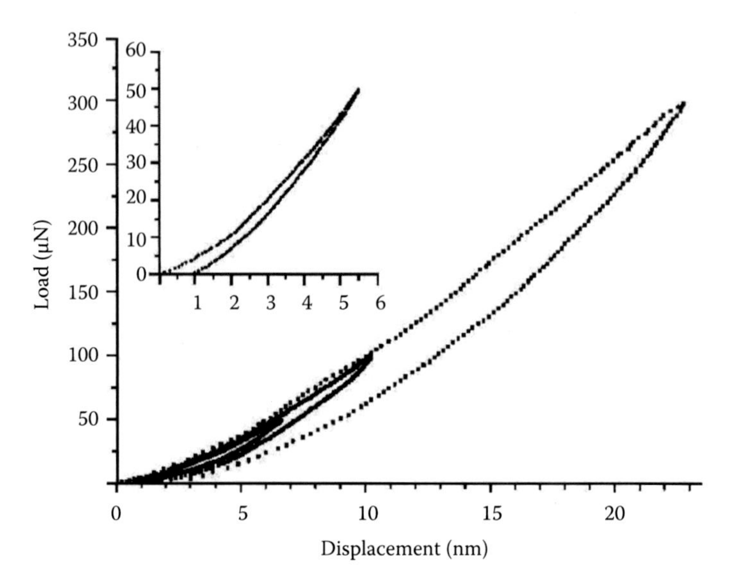

FIGURE 4.22
Load–displacement curves at various peaks for Si(100).

to an initial yield point. The sharp discontinuities in the unloading part of the curves are believed to be due to the formation of lateral cracks that form at the base of the median crack, which results in the surface of the specimen being thrust upward. Load–displacement data at residual depths as low as approximately 1 nm can be obtained. The indentation hardness of surface films has been measured for various materials at a range of loads including Si(100) up to a peak load of 500 mN and Al(100) up to a peak load of 2000 mN by Bhushan et al. [71] and Kulkarni and Bhushan [95–97]. The hardness of single-crystal silicon and single-crystal aluminum at shallow depths of the order of a few nanometers (on a nanoscale) is found to be higher than at depths of the order of a few hundred nanometers or larger (on a microscale), Figure 4.23. Microhardness has also been reported to be higher than that on the millimeter scale by several investigators. The data reported to date show that hardness exhibits a scale (size) effect.

4.5.2 Scanning Probe Microscopy

SPM provides a variety of characterization capabilities with true nanometer-scale resolution of different samples. The types of surfaces that SPM typically investigates range from semiconductor nanostructures to soft polymers, biological tissues, and cells. SPM acquires images point by point, and the throughput gains come either from increasing the scanning speed or from parallelization of imaging process. The speed improves by using shorter cantilevers with higher resonance frequencies or from

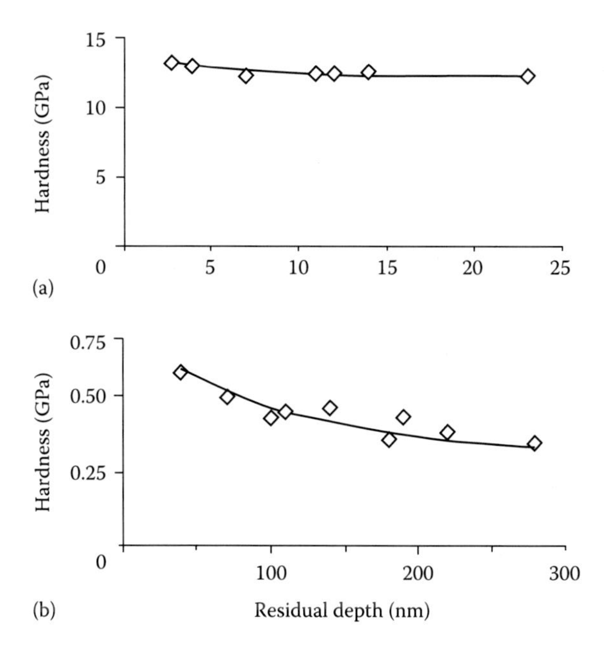

FIGURE 4.23
Indentation hardness as a function of residual indentation depth for (a) Si(100). (From Bhushan, B. et al., *Philos. Mag.*, 74, 1117, 1996.) (b) Al(100). (From Kulkarni, A.V. and Bhushan, B., *Thin Solid Films*, 290–291, 206, 1996.)

other improvements in the AFM bandwidth [98,99]. Another promising approach is the use of parallel cantilevers that scan different parts of the surface at the same time.

SPM measures the topography and forces on the nanoscale, which provides a potential strategy for local dissipation measurement [100–102]. In this, the SPM tip concentrates the probing field to the nanometer level, and the cantilever acts as an energy dissipation sensor. The energy dissipated due to tip–surface interactions is determined using power balance as $P_{diss} = P_{drive} - P_0$, where P_{drive} is the power provided to the probe by an external driving source and P_0 is the sum of intrinsic losses due to cantilever damping by the surroundings and within the cantilever material. The external power can be determined from the cantilever dynamics as $P_{drive} = \langle F_z \rangle$ where F is the force acting on the probe and z is the experimentally measured probe velocity, with the average taken along the probe tip trajectory. The intrinsic losses within the material and due to the hydrodynamic damping by ambient P_0 are determined by calibration at a reference position, $P_{diss} = 0$. The dynamic behavior of the weakly interacting cantilever with the surface in the vicinity of the resonance can be well approximated by a simple harmonic oscillator (SHO) model described by three independent parameters, namely, resonant frequency, ω_0; amplitude at the resonance, A_0; and quality factor, Q, as

$$A(\omega) = \frac{A\max\omega_0^2}{\sqrt{(\omega^2-\omega_0^2)+(\omega\omega_0/Q)^2}} \text{ and}$$
$$\tan(\varphi(\omega)) = \frac{\omega\omega_0/Q}{\omega^2-\omega_0^2} \tag{4.7}$$

From these, ω_0 is related to the tip–surface force gradient, A_0 to the driving force, and Q to the energy dissipation. For constant frequency operation, seminal work by Cleveland et al. [103] and San Paulo and Garcia [102] has related energy loss to the phase shift of a vibrating cantilever. Dissipative tip–surface interactions can be probed via measurement of the amplitude, A, and phase, ϕ, of the cantilever driven mechanically with amplitude, A_d, at a constant frequency, ω, as

$$P_{tip} = \frac{1}{2}\frac{kA^2\omega}{Q_0}\left[\frac{Q_0 A_d \sin\phi}{A} - \frac{\omega}{\omega_0}\right] \tag{4.8}$$

where ω_0 is the resonance frequency of the cantilever with spring constant, k, and the quality factor in free space, Q_0. The emergence of frequency-tracking techniques provides another means to determine dissipation [104,105]. The cantilever is driven at constant amplitude near the resonance frequency, the response amplitude is measured, and by assuming that changes in the signal strength are proportional to the Q-factor, dissipation in the system can be ascertained. In this case, Q is the quality factor in the vicinity of the surface. Experimentally, Q is determined by

$$P_{tip} = \frac{1}{2}\frac{kA^2\omega}{Q_0}\left[\frac{1}{Q_0} - \frac{1}{Q}\right] \tag{4.9}$$

using an additional feedback loop that maintains the oscillation amplitude constant by adjusting the driving amplitude, $Q = A \cdot A_d$. These approaches were implemented by several groups to study magnetic dissipation, electrical dissipation, and mechanical dissipation on atomic and molecular levels [106–110]. Notably, in a standard single-frequency SPM experiment, the number of independent parameters defining the cantilever dynamics (i.e., three SHO parameters) exceeds the number of experimentally observed variables (e.g., amplitude and phase), precluding direct measurement of dissipation. For acoustically driven systems, the constant driving force, $F = \text{constant}$, provides an additional constraint required to determine three independent SHO parameters from two experimentally accessible quantities. However, Equations 4.8 and 4.9 are no longer valid for techniques where the driving signal is position, time, or frequency dependent, $F \neq \text{constant}$. Variations in the signal strength are due to both work function variations and dissipation, and these effects cannot be separated unambiguously. Similarly, in atomic force acoustic microscopy (AFAM) and piezoresponse force microscopy (PFM), which address local mechanical and electromechanical properties, variations in the local response cannot be unambiguously distinguished from dissipation. The nonlinearity in the tip–surface interaction results in the creation of higher harmonics that cause confusion between information about dissipation and other properties [111]. The dissipation measurements are extremely sensitive to SPM electronics. Even small deviations in the phase set point from the resonance condition in frequency-tracking techniques result in major errors in the measured dissipation energy. Hence, the implementation of these techniques requires the calibration of the frequency response of the piezoactuator driving the cantilever. In the absence of such calibration, the images often demonstrate abnormal cantilever-dependent contrast. All these factors contribute to relative paucity on dissipation processes in SPM. This limited applicability of SPM to dissipation measurements is a direct consequence of the fact that traditional SPM excites and samples the response at a single frequency at a time. This allows fast imaging and high signal levels. However, the frequency-dependent response, dissipation and energy transfer, is not probed.

The capability of mapping local energy dissipation on the nanoscale is an enabling technology that will open a pathway toward atomistic mechanism of dissipation and establish relationships between dissipation and structure. This allows the development of strategies to minimize and avoid undesirable energy losses in technologies as diverse as electronics, information technology, and energy storage, transport, and generation. Furthermore, energy dissipation measurements help in understanding energy transformation mechanisms during fundamental physical and chemical processes. Simple estimates suggest that at room temperature the estimated detection limit in band excitation (BE) method as limited by thermomechanical noise is given by

$$P_{tip}^2 = \frac{k_\beta TkA^2\omega_0 B}{Q} \tag{4.10}$$

$2 = 2\omega$, corresponding to ~0.5 fW or ~31 mV/oscillation level for ambient environment (as compared to currently demonstrated 20 fW). In a high-Q environment, detection of single optical phonon generation in the tip–surface junction is possible, providing information on the dissipative processes with broad applicability to nanomechanics and nanotribology. In the BE method, the cantilever is tuned using a standard SPM fast tuning procedure to determine the corresponding resonant frequency, Figure 4.24. The frequency band is chosen such that the resonance corresponds approximately to the center of the band.

A typical example of an excitation and response signal in Fourier and time domains in standard SPM and BE SPM is shown in Figure 4.25. The BE signal is obtained prior to image acquisition and then downloaded to an arbitrary waveform generator and used either to drive the tip electrically

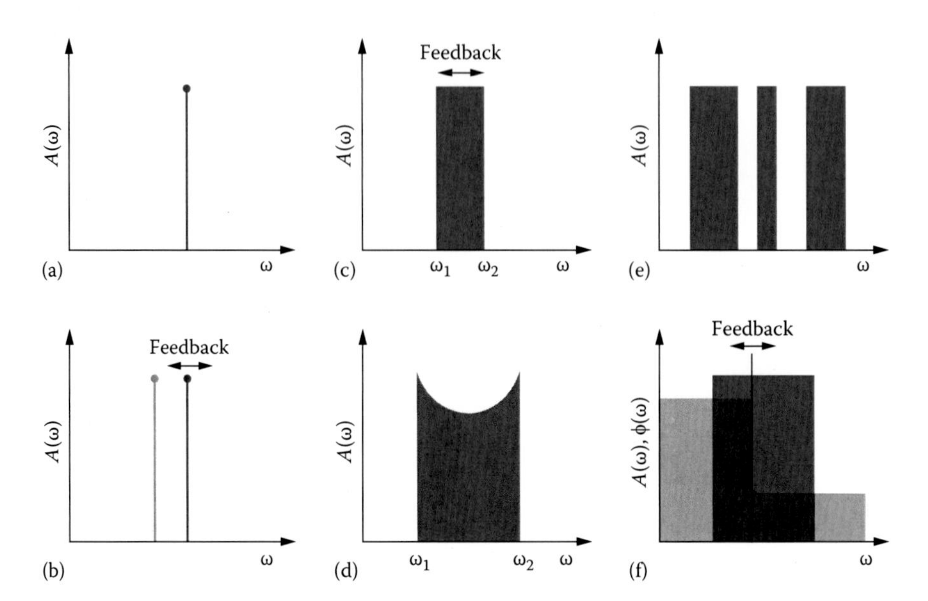

FIGURE 4.24
Frequency spectrum of excitation signal in single-frequency (a) static and (b) frequency-tracking cases. In frequency-tracking methods, the excitation frequency and excitation amplitude are varied from point to point. In the BE method, the response in the selected frequency window around the resonance is excited. The excitation signal can have (c) uniform spectral density or (d) increased spectral density on the tails of resonance peak to achieve better sampling away from the peak. (e) The resonance can be excited simultaneously over several resonance windows. Also, (f) the phase content of the signal can be controlled, for example, to achieve Q-control amplification. The excitation signal can be selected prior to imaging or adapted at each point so that the center of the excitation window follows the resonance frequency (c) or the phase content is updated (f). This is important for, for example, contact mode techniques, when the tip–surface contact area and hence the resonance frequency change significantly with position. Uniform amplitude within the band.

(such as in PFM and Kelvin probe force microscopy [KPFM]) or mechanically (as in tapping mode AFM, magnetic force microscopy [MFM], and electrostatic force microscopy [EFM]) or to drive an external oscillator below the sample (AFAM).

The response signal is acquired using a fast data acquisition card (NI-6115) and Fourier transformed to yield amplitude–frequency and phase–frequency curves. The ratio of the Fourier transforms of the response and excitation signal yields the transfer function of the system within the selected band. The amplitude–frequency and phase–frequency curves in each point are stored as 3D data arrays for subsequent analysis.

The data at each pixel are fitted to the SHO model, Equation 4.7. The fitting yields the local response and maximum dissipation. The fitting can be performed either on amplitude or phase data or simultaneously on both. To ensure adequate weighting, in the latter case, the data are transformed into real and imaginary parts, $A\cos\phi$ and $A\sin\phi$. The derived SHO

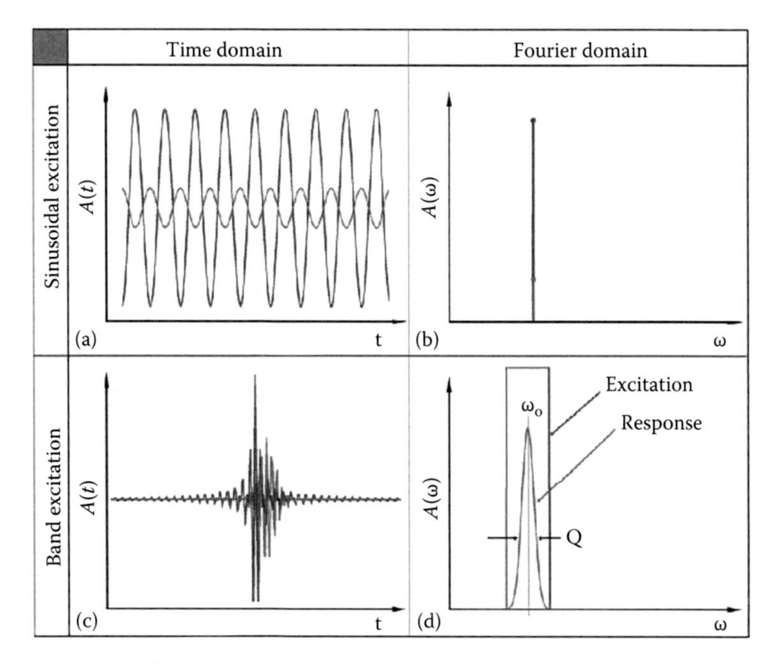

FIGURE 4.25
Excitation and response signals in standard SPM techniques: (a) time domain and (b) Fourier domain, (c) time domain and (d) Fourier domain. In BE, the system response is probed in the specified frequency range (e.g., encompassing a resonance), as opposed to a single frequency in conventional SPMs.

coefficients are plotted as 2D maps. Note that more complex forms of data analysis (using different physical models, 21 statistical fits, wavelet signal transforms) are possible. The BE method for a single point can then be extended to spectroscopy and imaging in the point-by-point and line-by-line modes. In spectroscopic BE measurements, the waveforms are applied to the probe, and the response is measured as a function of a slowly varying external parameter (tip–surface separation, force, or bias) at a single point of the surface to yield 2D spectroscopic response–frequency–parameter maps (spectrograms). Subsequent fitting using SHO model allows 1D response–parameter spectra (e.g., dissipation distance or response–distance curves) to be extracted and compared with the varying parameter (such as force–distance data). In point-by-point measurements, the tip approaches the surface vertically until the deflection set point is achieved. The amplitude–frequency data are then acquired at each point. After acquisition, the tip is moved to the next location. This is continued until a mesh of evenly spaced $M \times N$ points is scanned to yield 3D data array. Subsequent analysis yields 2D maps of corresponding quantities. In line-by-line measurements, the BE signal substitutes the standard driving signal during the interleave line on a commercial SPM (MultiMode NS-IIIA). The topographic information in the main line is collected using standard intermittent contact or contact mode detection. The data are processed using an external data acquisition system and are synchronized with the SPM topographic image to yield BE maps: *iA*, resonant frequency 0*i* ω, and *Q*-factor. The implementation of the BE method at low temperatures holds the promise of an even further increase of sensitivity to the level that a single quasi-particle can be detected, providing insight into the fundamental physics of strongly correlated oxide materials and other systems on the forefront of research. The applicability of the BE approach is demonstrated for mapping energy dissipation in MFM; mechanical and electromechanical probes, including loss processes during ferroelectric domain formation; and the evolution of dynamic behavior of the probes during force–distance curve acquisition. These examples illustrate the universality of the BE method, which can be used as an excitation and control method in all ambient and liquid SPM methods, including standard intermittent mode topographic imaging, magnetic imaging by MFM, electrical imaging by KPFM and EFM, acoustic imaging by AFAM, and electromechanical imaging by PFM. In these techniques, BE allows direct measurement of previously unavailable information of energy dissipation in magnetic, electrical, and electromechanical processes.

The market now offers a wide variety of commercial SPM instruments at different price points and functionality sets. There is also a variety of microfabricated probes available on the market that enable common imaging protocols. In addition, progress in using SPM for semiconductor process metrology resulted in development of the automated probe quality control capabilities and automated tip exchange. The metrological capabilities of SPM were also greatly enhanced by the use of closed-loop scanners that enabled hysteresis-free probe positioning. However, there are still some important challenges. The most important challenge faced by the SPM techniques is that they typically provide no real surface composition information. Other arguably less important challenges include variability in probe size, shape, and lifetime; inability to image high aspect ratio and fragile or very soft structures; and low throughput determined by the limitations of the scanning speed and scanning range.

Characterization of the interactions between nanostructures is at least as important as characterization of their surface morphology and chemical composition. SPM has already been very successful in measuring these interaction forces on truly microscopic scale. A short-term problem is posed by the lack of robust and accurate methods for calibration of the force measurements: standard calibration techniques always produce about 10% error. Perhaps, the first step for measurement standardization could be the establishment of a common molecular scale force standard for probe calibration (perhaps even based on a biological system; one possibility is to use the stretching transition that occurs in DNA). In the longer term, we still need to push the force resolution of the cantilever systems down into the single piconewton regime, all while maintaining adequate cantilever stiffness to avoid jumps. Drastic reductions in the instrument noise level and use of shorter cantilevers should drive progress in that area. We will also need to expand the AFM to probe different loading rates and regimes. Interaction force measurements often require large statistics; therefore, the throughput issues are very important. Most of the throughput increase approaches are similar to the strategies that I have already discussed for imaging. Finally, as a long-term challenge, I should mention the possibility of using SPM for direct mapping of full energy landscapes (perhaps based on using thermal noise-assisted probing). Such capability should then open up a way for a truly rational design of nanoscale assemblies.

Another set of approaches tries to couple SPM with other techniques that typically excel at chemical discrimination. The chief candidate for such technique is optical spectroscopy. NSOM was the first example of such technique; however, NSOM suffers from using a single probe to collect topographical and optical information. As a result, high spatial resolution hurts optical throughput. Another approach combines AFM with optical techniques such as confocal microscopy. In this approach both techniques use their respective strengths: AFM provides topography and confocal microscopy provides single-molecule-level optical signature identification. However, this approach may not resolve optical signatures of the surface features that are located within the diffraction limit. Researchers have proposed placing Raman-enhancing nanoparticles on the SPM probe. Such probes have the potential to provide localized chemical identification of surface features based on Raman signatures. Xie and coworkers proposed the use of local field enhancement by an SPM probe to improve resolution of the fluorescence imaging techniques.

4.5.3 X-Ray Nanodiffraction Technique

The development of x-ray microfocusing optics is based on x-ray microprobe techniques, such as x-ray micro-diffraction, micro-fluorescence, and microspectroscopy, and is applied in materials sciences [112,113], solid-state physics [114] and medical sciences [115], environmental sciences [116], and many other disciplines. However, x-ray scattering characterizations of nanoscale materials have been performed so far on the basis of the information obtained by illuminating a large number of nanomaterials [117–120] as enough photons in an x-ray microbeam are not intercepted by a single nano-object to generate measurable signals. This technique is applicable if high uniformity exists in nanomaterials. Detailed local information, such as nanostructure or structure variation within individual nanomaterials and dimension-dependent structural properties, are lost after averaging the information from an assembly of a large amount of objects with various sizes. Therefore, a tool capable of structural characterization of individual nanomaterials with high sensitivity, high penetrative power, and nondestruction is in high demand. When the dimensions of particles become smaller than the focal spot sizes of currently available x-ray optics, the photon flux density reflects the capacity for a nanosample to intercept photons [121–126]. Kirkpatrick ± Baez mirrors [127] and refractive compound lenses [128] are commonly employed in x-ray focusing optics that are capable of micrometer or submicrometer focusing. Owing to the accuracy of its optical features resulting from the highly mature stage of manufacturing, a zone plate is the most promising optics to provide x-ray beams of the highest photon flux density. Taking full advantage of the brightest synchrotron radiation source and state-of-the-art zone-plate x-ray focusing optics, an x-ray nanodiffraction technique has been developed that allows studies of the material structure of a single nanoscale object. Recently, a new group of semiconducting oxide (tin oxide, zinc oxide, indium oxide, and cadmium oxide) nanostructures that have a belt-like morphology (rectangular cross section) have been synthesized via a thermal evaporation and condensation process [129,130]. The belt-like geometrical morphology is a common structural characteristic of these oxides with different crystallographic structures. Total focal flux is determined by the size of the zone plate and the focal efficiency of the zone plate, which depends on the energy of the x-rays. After the zone plate and the x-ray energy are selected, the focal spot size is measured directly by the photon flux density. The measurement of the focal spot size is performed by measuring the L-line fluorescence profiles of the 100 nm nanobelt held on the sample assembly of the diffractometer scanned across the x-ray beam at the focal plane.

The crystallographic structure of the tin oxide nanobelts has been determined to be triclinic (Sn_2O_3) by x-ray diffraction from individual nanocrystals [131]. The unit-cell parameters of the triclinic tin oxide are $a = 5.457$, $b = 8.179$, $c = 3.714$°A; $\alpha = 93.8$, $\beta = 92.3$, and $\gamma = 90$ [132]. Bragg reflections from atomic planes (030), (011), and $(0\bar{2}1)$ have been measured by rotating the crystal around an axis along the width. The 2θ angles of the reflections match the value of the 2θ angles calculated from the unit-cell parameters at the energy of the measurement, and also the differences of θ angles between any two of the three reflections match with the calculated values [132]. From the orientations of the nanobelts at which the three reflections appeared, it has been determined that the triclinic tin oxide nanobelts grow along [001] and are enclosed by ±(010) and ±(100) crystallographic facets. In Figure 4.26, resolution-limited maps of the tin fluorescence intensity and the diffraction intensity of the (030) reflection against the sample position in the transverse plane on which the nanobelts lie are shown.

The distribution of the diffraction intensity in reciprocal space contains information on the nanostructure within a nanomaterial. In order to further study subgrain structure in Sn_2O_3 nanobelts, a charge-coupled detector (CCD) area detector was used to capture the (030) reflection from Sn_2O_3 nanobelts. In Figure 4.27, the diffraction patterns of the reflection obtained from a 100 nm tin oxide nanobelt are shown. The diffraction patterns clearly show several subspots splitting along the χ direction over a range of degrees, indicating a subgrain with highly textured structure in the nanobelt. The splitting of the

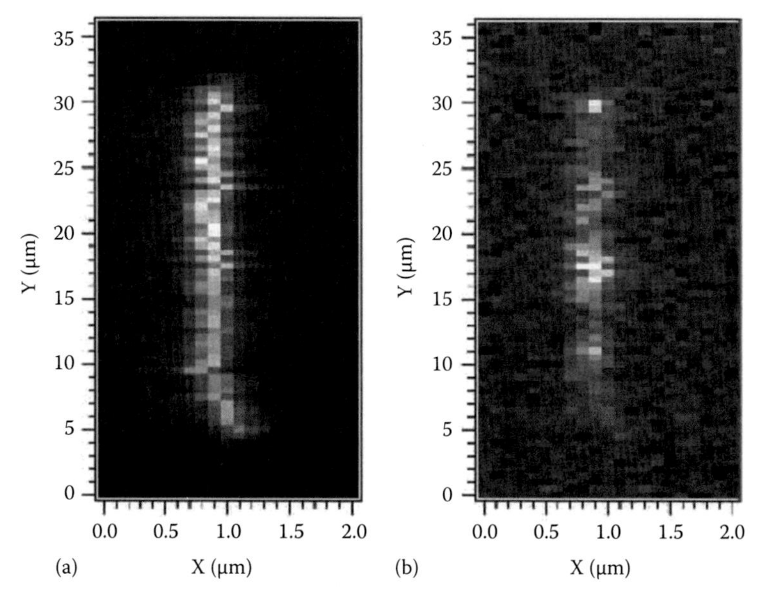

FIGURE 4.26
2D mapping of the tin L-lines fluorescence intensity (a) and the diffraction intensity of the (030) reflection (b) from a 100-nm width tin oxide nanobelt. The maximum intensity per pixel in (a) is 450 counts and that in (b) is 60,000 counts.

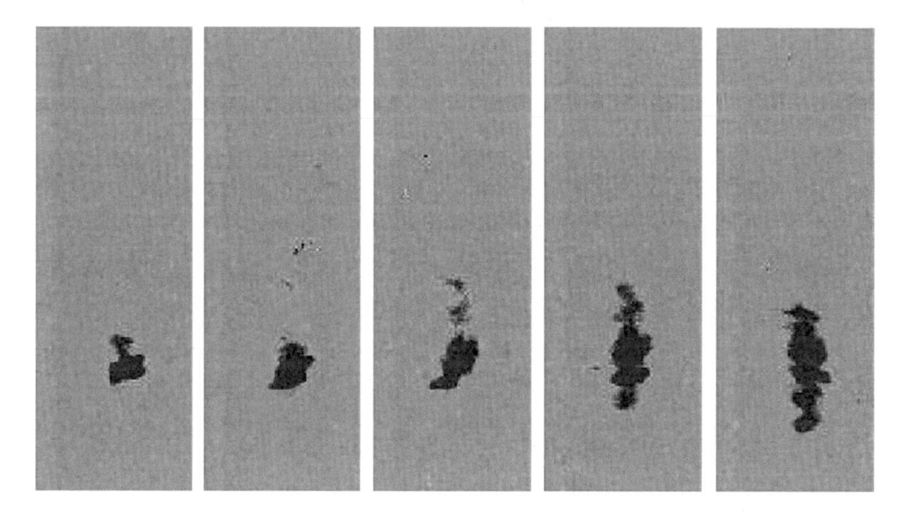

FIGURE 4.27
X-ray diffraction spots of the (030) reflection from a 100-nm Sn_2O_3 nanobelt captured by a CCD detector. Each image was taken after a translation of the microbeam along the length of the nanobelt in steps of 200 nm. The first image is the diffraction spot obtained when the end of the belt was illuminated. The horizontal axis represents the 2θ direction and ranges from 36.5° to 39.5°. The vertical axis represents the χ direction and covers 8.5°.

diffraction spots along the θ direction is much smaller than the splitting along the χ direction, suggesting a relatively small strain variation among grains. Further studies of the grain orientation distribution along the θ direction revealed a three to five times smaller distribution of grain orientation around the θ axis than around the χ axis. When the dimensions of a nanomaterial fall into a region in which the material properties depend on the dimensions, the characterization of the dimensions of a nanomaterial becomes important. The high penetration power of x-rays into the materials is an advantage for embedded nanomaterials. Although SEM is an accurate tool for size characterization of unembedded materials in the plane normal to the electron beam, accurate measurement of the extent of the material along the third dimension sometimes requires better techniques like x-ray scattering. The x-ray nanodiffraction technique allows the study of structures of nanomaterials on a single object. The technique helps in the study of nanostructures and their dependence on material dimensions as in case of nanobelt oxides. Future research into understanding the structure-related electrical, thermal, and optical transport processes in nanoscale materials will benefit from the development of the x-ray nanodiffraction technique and its applications.

4.6 Conclusion and Future Prospects

Advances in the characterization of nanomaterials present both opportunities and challenges. Rapid development of nanodevices with reduced dimensions generated a strong need for an extensive investigation of the size effects in nanomaterials. The insight into the surface to volume characterization needs to be addressed to get an insight into the performance and reliability of nanoscale devices. Techniques of surface characterization such as SPM provide wide opportunities for studying the physical properties and functionality of materials at the nanoscale level. A possibility of manipulation of the characterization techniques will facilitate nanofabrication of devices. Clearly, the future will evidence broad application of these techniques discussed in this chapter for the characterization of nanostructures.

References

1. Courjon, D. and Bainier, C. 1994. Near field microscopy and near field optics. *Rep. Prog. Phys.* 57:989–1028.
2. R Nave. *Hyper Physics: Light and Vision.* http://hyperphysics.phy astr.gsu.edu/hbase/ligcon.html

3. Synge, E.H. 1928. A suggested method for extending microscopic resolution into the ultra-microscopic region. *Philos. Mag.* 6:356–362.
4. Dunn, R.C. 1999. Near-field scanning optical microscopy. *Chem. Rev.* 99:2891–2928.
5. Aston, E. 2001. Scanning probe microscopy: An overview of scanning tunneling and atomic force microscopies, *Short Course on Surface and Colloid Science,* University of Washington, Seattle, WA, July 9–13.
6. Pohl, D.W., Denk, W., and Lanz, M. 1984. Optical stethoscopy: Image recording with resolution $\lambda/20$. *Appl. Phys. Lett.* 44:651–653.
7. Harootunian, A., Betzig, E., Isaacson, M., and Lewis, A. 1986. Super-resolution fluorescence near-field scanning optical microscopy. *Appl. Phys. Lett.* 86:674–676.
8. Lewis, A., Isaacson, M., Betzig, R.E., and Harootunian, A. 1989. Near field scanning optical microscopy, US Patent 4,917,462.
9. Dereux, A., Girard, C., and Weeber, J. 2000. Theoretical principles of near field optical microscopies and spectroscopies. *J. Chem. Phys.* 112:7775–7789.
10. Reddick, R.C., Warmack, R.J., and Ferrell, T.L. 1989. New form of scanning optical microscopy. *Phys. Rev. B* 39:767–770.
11. Courjon, D., Sarayedine, K., and Spajer, M. 1989. Scanning tunneling optical microscopy. *Opt. Commun.* 71:23–28.
12. Turner, D.R. 1989. Etch procedure for optical fibers, US Patent 4,469,554.
13. Hecht, B., Sick, B., Wild, U.P., Deckert, V., Zenobi, R., Martin, O.J., Pohl, D.W. 2000. Scanning near-field optical microscopy and spectroscopy with aperture probes: Fundamentals and applications. *J. Chem. Phys.* 112:7761–7774.
14. Smith, D.A., Webster, S., Ayad, M., Evans, S.D., Fogherty, D., and Batchelder, D. 1995. Development of a scanning near-field optical probe for localized Raman spectroscopy. *Ultramicroscopy* 61:247–252.
15. Jahncke, C.L., Paesler, M.A., and Hallen, H.D. 1995. Raman imaging with near-field scanning optical microscopy. *Appl. Phys. Lett.* 67:2483–2485.
16. Zeisel, D., Dutoit, B., Deckert, V., Roth, T., and Zenobi, R. 1997. Optical spectroscopy and laser desorption on a nanometer scale. *Anal. Chem.* 69:749–754.
17. Webster, S., Smith, D.A., and Batchelder, D.N. 1998. Raman microscopy using a scanning near-field optical probe. *Vib. Spectrosc.* 18:51–59.
18. Thio, T., Lezec, H.J., Ebbesen, T.W., Pellerin, K.M., Lewen, G.D., Nahata, A., and Linke, R.A. 2001. Delay in light transmission through small apertures. *Opt. Lett.* 13:450–452.
19. Bethe, H. 1944. Theory of diffraction by small holes. *Phys. Rev.* 66:163–182.
20. Bouwkamp, C.J. 1950. On Bethe's theory of diffraction by small holes. *Philips Res. Rep.* 5:321–332.
21. Novotny, L., Pohl, D., and Hecht, B. 1995. Scanning near-field optical probe with ultra small spot size. *Opt. Lett.* 20:970–972.
22. Fleischmann, M., Hendra, P.J., and McQuillan, A.J. 1974. Raman spectra of pyridine adsorbed at a silver electrode. *Chem. Phys. Lett.* 26:163–166.
23. Jeanmaire, D.L. and Duyne, R.P.V. 1977. Surface Raman spectroelectrochemistry: Part I. Heterocyclic, aromatic, and aliphatic amines adsorbed on the anodized silver electrode. *J. Electroanal. Chem.* 84:1–20.
24. Albrecht, M.G. and Creighton, J.A. 1977. Anomalously intense Raman spectra of pyridine at a silver electrode. *J. Am. Chem. Soc.* 99:5215–5217.
25. King, F.W., Duyne, R.P.V., and Schatz, G.C. 1978. Theory of Raman scattering by molecules adsorbed on electrode surfaces. *J. Chem. Phys.* 69:4472–4481.

26. Moskovits, M. 1985. Surface-enhanced spectroscopy. *Rev. Mod. Phys.* 57:783–826.
27. Kneipp, K. 2001. Basics of single molecule experiments under ambient conditions. M.I.T Course 6.975, Spring, Cambridge, MA.
28. Campion, A. and Kambhampati, P. 1998. Surface-enhanced Raman scattering. *Chem. Soc. Rev.* 27:241–250.
29. Kneipp, K., Kneipp, H., Itzkan, I., Dasari, R.R., and Feld, M.S. 2002. Surface-enhanced Raman scattering and biophysics. *J. Phys. Condens. Matter* 14:597–624.
30. Binnig, G., Rohrer, H., Gerber, C., and Weibel, E. 1982. Surface studies by scanning tunneling microscopy. *Phys. Rev. Lett.* 49:57–61.
31. Wessel, J.E. 1985. Surface-enhanced optical microscopy. *J. Opt. Soc. Am. B: Opt. Phys.* 2:1538–1541.
32. Stockle, R.M., Suh, Y.D., Deckert, V., and Zenobi, R. 2000. Nanoscale chemical analysis by tip-enhanced Raman spectroscopy. *Chem. Phys. Lett.* 318:131–136.
33. Hayazawa, N., Inouye, Y., Sekkat, Z., and Kawata, S. 2000. Metallized tip amplification of near-field Raman scattering. *Opt. Commun.* 183:333–336.
34. Wang, J.J., Smith, D.A., Batchelder, D.N., Saito, Y., Kirkham, J., Robinson, C., Baldwin, K., Li, G., and Bennett, B. 2003. Apertureless near-field Raman spectroscopy. *J. Microsc.* 210:330–333.
35. Bulgarevich, D.S. and Futamata, M. 2004. Apertureless tip-enhanced Raman microscopy with confocal epi-illumination/collection optics. *Appl. Spectrosc.* 58:757–761.
36. Hartschuh, A., Beversluis, M.R., Bouhelier, A., and Novotny, L. 2004. Tip-enhanced optical spectroscopy. *Philos. Trans. R. Soc. London, Ser. A: Math., Phys. Eng. Sci.* 362:807–819.
37. Bouhelier, A., Hartschuh, A., and Novotny, L. 2005. *Handbook of Microscopy for Nanotechnology*. New York: Springer, pp. 25–54.
38. Anderson, N., Hartschuh, A., Cronin, S., and Novotny, L. 2005. Nanoscale vibrational analysis of single-walled carbon nanotubes. *J. Am. Chem. Soc.* 127:2533–2537.
39. Kerker, M., Wang, D.S., and Chew, H. 1980. Surface enhanced Raman scattering (SERS) by molecules adsorbed at spherical particles. *Appl. Opt.* 19:3373–3388.
40. Kottmann, J.P., Martin, O.J.F., Smith, D.R., and Schultz, S. 2001. Non-regularly shaped plasmon resonant nanoparticle as localized light source for near-field microscopy. *J. Microsc.* 202:60–65.
41. Kottmann, J.P., Martin, O.J.F., Smith, D.R., and Schultz, S. 2001. Plasmon resonances of silver nanowires with a nonregular cross section. *Phys. Rev. B: Condens. Matter Mater. Phys.* 64:235402–235411.
42. Rugar, D., Budakian, R., Mamin, H.J., and Chui, B.W. 2004. Single spin detection by magnetic resonance force microscopy. *Nature (London)* 430:329–332.
43. Chao, S., Dougherty, W.M., Garbini, J.L., and Sidles, J.A. 2004. Nanometer-scale magnetic resonance imaging. *Rev. Sci. Instrum.* 75(5):1175–1184.
44. Alzetta, G., Arimondo, E., Ascoli, C., and Gozzini, A. 1967. Paramagnetic resonance experiments at low fields with angular-momentum detection. *Il Nuovo Cimento B* 52:392–402.
45. Ascoli, C., Baschieri, P., Frediani, C., Lenci, L., Martinelli, M., Alzetta, G., Celli, R.M., and Pardi, L. 1996. Micromechanical detection of magnetic resonance by angular momentum absorption. *Appl. Phys. Lett.* 69(25):3920–3922.

46. Leskowitz, G.M., Madsen, L.A., and Weitekamp, D.P. 1998. Force-detected magnetic resonance without field gradients. *Solid State Nucl. Magn. Reson.* 11:73–86.
47. Madsen, L.A., Leskowitz, G.M., and Weitekamp, D.P. 2004. DP observation of force-detected nuclear magnetic resonance in a homogeneous field. *Proc. Natl. Acad. Sci. USA* 101:12804–12808.
48. Hahn, E.L. 1950. Spin echos. *Phys. Rev.* 80:580–594.
49. Degen, C.L. 2005. Magnetic resonance force microscopy. NMR spectroscopy on the micro- and nanoscale. Christian Lukas Degen. http://dx.doi.org/10.3929/ethz-a-005128707
50. Schuh, C.A. 2006. Nanoindentation studies of materials. *Mater. Today* 9(5):32–40.
51. Ni, H. and Li, X. 2006. Young's modulus of ZnO nanobelts measured using atomic force microscopy and nanoindentation techniques. *Nanotechnology* 17:3591–3598.
52. Bansal, S., Toimil-Molares, E., Saxena, A., and Tummala, R.R. 2005. *Proceedings of the Electronics Components and Technology Conference*, Orlando, FL, May 31–June 3, vol. 1, 71p.
53. Li, X., Gao, H., Murphy, C.J., and Caswell, K.K. 2003. Nanoindentation of silver nanowires. *Nano Lett.* 3:1495–1498.
54. Li, X. and Bhushan, B. 2003. Fatigue studies of nanoscale structures for MEMS/NEMS applications using nanoindentation techniques. *Surf. Coat. Technol.* 163–164:521–526.
55. Wang, Z.L., Poncharal, P., and de Heer, W.A. 2000. Measuring physical and mechanical properties of individual carbon nanotubes by in situ TEM. *J. Phys. Chem. Solids* 61:1025–1030.
56. Wang, Z.L., Poncharal, P., and de Heer, W.A. 2000. Nanomeasurements of individual carbon nanotubes by in situ TEM. *Pure Appl. Chem.* 72:209–219.
57. Wang, Z.L., Dai, Z.R., Bai, Z.G., Gao, R.P., and Gole, J. 2000. Side-by-side silicon carbide–silica biaxial nanowires: Synthesis, structure, and mechanical properties. *Appl. Phys. Lett.* 77:3349–3351.
58. Poncharal, P., Wang, Z.L., Ugarte, D., and de Heer, W.A. 1999. Electrostatic deflections and electromechanical resonances of carbon nanotubes. *Science* 283:1513–1516.
59. Desai, A.V. and Haque, M.A. 2006. Test bed for mechanical characterization of nanowires. *Proc. IMechE: Nanoeng. Nanosys.* 219:57.
60. Bai, X.D., Gao, P.X., Wang, Z.L., and Wang, E.G. 2003. Dual-mode mechanical resonance of individual ZnO nanobelts. *Appl. Phys. Lett.* 82(26):4806–4809.
61. Yi, T. and Kim, C.-J. 1999. Measurement of mechanical properties for MEMS materials. *Meas. Sci. Technol.* 10:706–710.
62. Christofer, H.J. 2004. From micro- to nanosystems: Mechanical sensors go nano. *Micromech. Microeng.* 14:S1–S11.
63. Yu, M., Lourie, O., Dyer, M.J., Moloni, K., Kelly, T.F., and Ruoff, R.S. 2000. Strength and breaking mechanism of multiwalled carbon nanotubes under tensile load. *Science* 287:637–640.
64. Krishnan, A., Dujardin, E., Ebbesen, T.W., Yianilos, P.N., and Treacy, M.M.J. 1998. Young's modulus of single-walled nanotubes. *Phys. Rev. B* 58:14013–14019.
65. Salvetat, J.-P., Bonard, J.-M., Thomson, N.H., Kulik, A.J., Forr`o, L., Benoit, W., and Zuppiroli, L. 1999. Mechanical properties of carbon nanotubes. *Appl. Phys. A* 69:255–260.

66. Singer, I.L. and Pollock, H.M. 1992. *Fundamentals of Friction: Macroscopic and Microscopic Processes*, vol. E220. Dordrecht, the Netherlands: Kluwer Academic Publishers.
67. Persson, B.N.J. and Tosatti, E. 1996. *Physics of Sliding Friction*, vol. E311. Dordrecht, the Netherlands: Kluwer Academic Publishers.
68. Bhushan, B., Israelachvili, J.N., and Landman, U. 1995. Nanotribology: Friction, wear and lubrication at the atomic scale. *Nature* 374:607–616.
69. Bhushan, B., Kulkarni, A.V., Koinkar, V.N., Boehm, M., Odoni, L., Martelet, C., and Belin, M. 1995. Microtribological characterization of self-assembled and Langmuir–Blodgett monolayers by atomic and friction force microscopy. *Langmuir* 11:3189–3198.
70. Bhushan, B., Kulkarni, A.V., Bonin, W., and Wyrobek, J.T. 1996. Nano/picoindentation measurement using a capacitance transducer system in atomic force microscopy. *Philos. Mag.* 74:1117–1128.
71. Bhushan, B., Fuchs, H., and Hosaka, S. 2004. *Applied Scanning Probe Methods*. Heidelberg, Germany: Springer.
72. Bhushan, B., Liu, H., and Hsu, S.M. 2004. Adhesion and friction studies of silicon and hydrophobic and low friction films and investigation of scale effects. *ASME J. Tribol.* 126:583–590.
73. Bhushan, B., Kasai, T., Kulik, G., Barbieri, L., and Hoffman, P. 2005. AFM study of perfluorosilane and alkylsilane self-assembled monolayers for anti-stiction in MEMS/NEMS. *Ultramicroscopy* 105:176–188.
74. Bhushan, B., Hansford, D., and Lee, K.K. 2006. Surface modification of silicon and polydimethylsiloxane surfaces with vapor-phase-deposited ultrathin fluorosilane films for biomedical nanodevices. *J. Vac. Sci. Technol. A* 24:1197–1202.
75. Bhushan, B., Fuchs, H., and Kawata, S. 2007. *Applied Scanning Probe Methods V*. Heidelberg, Germany: Springer.
76. Bhushan, B., Fuchs, H., and Tomitori, M. 2008. *Applied Scanning Probe Methods VIII–X*. Heidelberg, Germany: Springer.
77. Guntherodt, H.J., Anselmetti, D., and Meyer, E. 1995. *Forces in Scanning Probe Methods*, vol. E286. Dordrecht, the Netherlands: Kluwer Academic Publishers.
78. Mate, C.M., McClelland, G.M., Erlandsson, R., and Chiang, S. 1987. Atomic-scale friction of a tungsten tip on a graphite surface. *Phys. Rev. Lett.* 59:1942–1945.
79. Ruan, J. and Bhushan, B. 1994. Frictional behavior of highly oriented pyrolytic graphite. *J. Appl. Phys.* 76:8117–8120.
80. Yamanaka, K. and Tomita, E. 1995. Lateral force modulation atomic force microscope for selective imaging of friction forces. *Jpn. J. Appl. Phys.* 34:2879–2882.
81. Reinstaedtler, M., Rabe, U., Scherer, V., Hartmann, U., Goldade, A., Bhushan, B., and Arnold, W. 2003. On the nanoscale measurement of friction using atomic-force microscope cantilever torsional resonances. *Appl. Phys. Lett.* 82:2604–2606.
82. Reinstaedtler, M., Rabe, U., Goldade, A., Bhushan, B., and Arnold, W. 2005. Investigating ultra-thin lubricant layers using resonant friction force microscopy. *Tribol. Int.* 38:533–541.
83. Reinstaedtler, M., Kasai, T., Rabe, U., Bhushan, B., and Arnold, W. 2005. Imaging and measurement of elasticity and friction using the TR mode. *J. Phys. Appl. Phys.* 38:764–773.
84. Ruan, J. and Bhushan, B. 1994. Atomic-scale and microscale friction of graphite and diamond using friction force microscopy. *J. Appl. Phys.* 76:5022–5035.

85. Tambe, N.S. and Bhushan, B. 2005. A new atomic force microscopy based technique for studying nanoscale friction at high sliding velocities. *J. Phys. D: Appl. Phys.* 38:764–773.
86. Tao, Z. and Bhushan, B. 2006. A new technique for studying nanoscale friction at sliding velocities up to 200 mm/s using atomic force microscope. *Rev. Sci. Instrum.* 71:103705-1–103705-9.
87. Nosonovsky, M. and Bhushan, B. 2005. Scale effects in dry friction during multiple-asperity contact. *ASME J. Tribol.* 127:37–46.
88. Sundararajan, S. and Bhushan, B. 2001. Development of a continuous micro scratch technique in an atomic force microscope and its application to study scratch resistance of ultra-thin hard amorphous carbon coatings. *J. Mater. Res.* 16:75–84.
89. Zhao, X. and Bhushan, B. 1998. Material removal mechanism of single-crystal silicon on nanoscale and at ultralow loads. *Wear* 223:66–78.
90. Koinkar, V.N. and Bhushan, B. 1997. Scanning and transmission electronmicroscopies of single-crystal silicon microworn/machined using atomic force microscopy. *J. Mater. Res.* 12:3219–3224.
91. Bhushan, B., Koinkar, V.N., and Ruan, J. 1994. Microtribology of magnetic media. *Proc. Inst. Mech. Eng. Part J: J. Eng. Tribol.* 208:17–29.
92. DeVecchio, D. and Bhushan, B. 1998. Use of a nanoscale Kelvin probe for detecting wear precursors. *Rev. Sci. Instrum.* 69:3618–3624.
93. Bobji, M.S. and Bhushan, B. 2001. Atomic force microscopic study of the microcracking of magnetic thin films under tension. *Scr. Mater.* 44:37–42.
94. Bobji, M.S. and Bhushan, B. 2001. In-situ microscopic surface characterization studies of polymeric thin films during tensile deformation using atomic force microscopy. *J. Mater. Res.* 16:844–855.
95. Kulkarni, A.V. and Bhushan, B. 1996. Nanoscale mechanical property measurements using modified atomic force microscopy. *Thin Solid Films* 290–291:206–210.
96. Kulkarni, A.V. and Bhushan, B. 1996. Nano/picoindentation measurements on single-crystal aluminum using modified atomic force microscopy. *Mater. Lett.* 29:221–227.
97. Kulkarni, A.V. and Bhushan, B. 1997. Nanoindentation measurement of amorphous carbon coatings. *J. Mater. Res.* 12:2707–2714.
98. Jesse, S., Kalinin, S.V., Proksch, R., Baddorf, A.P., and Rodriguez, B.J. 2007. The band excitation method in scanning probe microscopy for rapid mapping of energy dissipation on the nanoscale. *Nanotechnology* 18(43):435503.
99. Hoefflinger, B. 2012. International technology roadmap for semiconductors. *Chips 2020 The Frontiers Collection*, pp. 161–174.
100. García, R. and Pérez, R. 2002. Dynamic atomic force microscopy methods. *Surf. Sci. Rep.* 47:197–301.
101. Tamayo, J. and Garcia, R. 1998. Relationship between phase shift and energy dissipation in tapping-mode scanning force microscopy. *Appl. Phys. Lett.* 73:2926–2929.
102. San Paulo, A. and Garcia, R. 2001. Tip-surface forces, amplitude, and energy dissipation in amplitude-modulation (tapping mode) force microscopy. *Phys. Rev. B* 64:193411–193414.
103. Cleveland, J.P., Anczykowski, B., Schmid, A.E., and Elings, V.B. 1998. Energy dissipation in tapping-mode atomic force microscopy. *Appl. Phys. Lett.* 72:2613–2615.

104. Tamayo, J. 1999. Energy dissipation in tapping-mode scanning force microscopy with low quality factors. *Appl. Phys. Lett.* 75:3569–3571.
105. Albrecht, T.R., Grütter, P., Horne, D., and Rugar, D. 1991. Frequency modulation detection using high-*Q* cantilevers for enhanced force microscope sensitivity. *J. Appl. Phys.* 69:668–673.
106. Grütter, P., Liu, Y., LeBlanc, P., and Dürig, U. 1997. Magnetic dissipation force microscopy. *Appl. Phys. Lett.* 71:279–281.
107. Proksch, R., Babcock, K., and Cleveland, J. 1999. Magnetic dissipation microscopy in ambient conditions. *Appl. Phys. Lett.* 74:419–421.
108. Kantorovich, L.N. and Trevethan, T. 2004. General theory of microscopic dynamical response in surface probe microscopy: From imaging to dissipation. *Phys. Rev. Lett.* 93:236102–236105.
109. Gauthier, M. and Tsukada, M. 1999. Theory of noncontact dissipation force microscopy. *Phys. Rev. B* 60:11716–11722.
110. Farrell, A.A., Fukuma, T., Uchihashi, T., Kay, E.R., Bottari, G., Leigh, D.A., Yamada, H., and Jarvis, S.P. 2005. Conservative and dissipative force imaging of switchable rotaxanes with frequency-modulation atomic force microscopy. *Phys. Rev. B* 72:125430–1245436.
111. Sebastian, A., Salapaka, M.V., Chen, D.J., and Cleveland, J.P. 2001. Harmonic and power balance tools for tapping-mode atomic force microscope. *J. Appl. Phys.* 89:6473–6481.
112. Soh, Y.A., Evans, P.G., Cai, Z., Lai, B., Kim, C.Y., Aeppli, G., Mathur, N.D., Blamire, M.G., and Isaacs, E.D. 2002. Local mapping of strain at grain boundaries in colossal magnetoresistive films using x-ray micro diffraction. *J. Appl. Phys.* 91:7742–7744.
113. Eastman, D.E., Stagarescu, C.B., Xu, G., Mooney, P.M., Jordan-Sweet, J.L., Lai, B., and Cai, Z. 2002. Observation of columnar microstructure in step-graded $Si_{1-}xGe_x$/Si films using high-resolution x-ray microdiffraction. *Phys. Rev. Lett.* 88:156101–156104.
114. Evans, P.G., Issacs, E.D., Aeppli, G., Cai, Z., and Lai, B. 2002. X-ray microdiffraction images of antiferromagnetic domain evolution in chromium. *Science* 295:1042–1045.
115. Hall, M.D., Dillon, C.T., Zhang, M., Beale, P., Cai, Z., Lai, B., Stamp, A.P.J., and Hambley, T.W. 2003. The cellular distribution and oxidation state of platinum (II) and platinum(IV) antitumour complexes in cancer cells. *J. Biol. Inorg. Chem.* 8:726–732.
116. Yun, W., Pratt, S.T., Miller, R.M., Cai, Z., Hunter, D.B., Jarstfer, A.G., Kemner, K.M., Lai, B., Lee, H.R., Legnini, D.G., Rodrigues, W., and Smith, C.I. 1998. X-ray imaging and micro spectroscopy of plants and fungi. *J. Synchrotron Radiat.* 5:1390–1395.
117. Wang, W., Xu, C., Wang, G., Liu, Y., and Zheng, C. 2002. Synthesis and Raman scattering study of rutile SnO_2 nanowires. *J. Appl. Phys.* 92:2740–2742.
118. Chen, Y., Cui, X., Zhang, K., Pan, D., Zhang, S., Wang, B., and Hou, J. 2003. Bulk-quantity synthesis and self-catalytic VLS growth of SnO_2 nanowires by lower-temperature evaporation. *Chem. Phys. Lett.* 369:16–20.
119. Maniwa, Y., Kataura, H., and Fujiwara, A. 2003. Molecular stumbling inside single-walled carbon nanotubes. *J. Jpn. Soc. Synchrotron Radiat. Res.* 16:296–305.
120. Goldberger, J., He, R., Zhang, Y., Lee, S., Yan, H., Choi, H., and Yang, P. 2003. Single-crystal gallium nitride nanotubes. *Nature (London)* 422:599–601.

121. Yun, W., Vicarro, P.J., Lai, B., and Chrzas, J. 1992. Coherent hard x-ray focusing optics and applications. *Rev. Sci. Instrum.* 63:582–585.
122. Lai, B., Yun, W., Legini, D., Xiao, Y., Chrzas, J., Viccaro, P.J., White, V., Bajikar, S., Denton, D., Cerrina, F., Difabrizio, E., Gentili, M., Grella, L., and Baciocchi, M. 1992. Hard x-ray phase zone plate fabricated by lithographic techniques. *Appl. Phys. Lett.* 61:1877–1879.
123. Arabatzis, I. and Falaras, P. 2003. Synthesis of porous nanocrystalline TiO_2 foam. *Nano Lett.* 3:249–251.
124. Kirkpatrick, P. and Baez, V. 1948. Formation of optical images by x-rays. *J. Opt. Soc. Am.* 38:766–774.
125. Xu, C., Zhao, X., Liu, S., and Wang, G. 2003. Large-scale synthesis of rutile SnO_2 nanorods. *Solid State Commun.* 125:301–304.
126. Raola, O. and Strouse, G. 2002. Synthesis and characterization of Eu-doped cadmium selenide nanocrystals. *Nano Lett.* 2:1443–1447.
127. Wen, X., Zhang, W., Yang, S., Dai, Z., and Wang, Z. 2002. Solution phase synthesis of $Cu(OH)_2$ nanoribbons by coordination self-assembly using Cu_2S nanowires as precursors. *Nano Lett.* 2:1397–1401.
128. Liu, S., Yue, J., and Wehmschulte, R. 2002. Large thick flattened carbon nanotubes. *Nano Lett.* 2:1439–1442.
129. Snigirev, A., Kohn, V., Snigireva, I., and Lengeler, B. 1996. A compound refractive lens for focusing high-energy x-rays. *Nature (London)* 384:49–51.
130. Pan, Z., Dai, Z., and Wang, Z. 2001. Nanobelts of semiconducting oxides. *Science* 291:1947–1949.
131. Xiao, Y., Cai, Z., Wang, Z.L., Lai, B., and Chu, Y.S. 2005. An x-ray nano diffraction technique for structural characterization of individual nanomaterials. *J. Synchrotron Radiat.* 12:124–128.
132. Murken, G. von and Tromel, M. 1973. Uber das bei der Disproportionierung von SnO entstehende Zinnoxid, Sn_2O_3. *Z. Anorg. Allg. Chem.* 397:117–126.

5

Localized Surface Plasmon Resonance Spectroscopy with Applications to Chemistry

Wen-Gang Qu and An-Wu Xu

CONTENTS

5.1 Introduction

Materials that possess a negative real and small positive imaginary dielectric constant are capable of supporting a surface plasmon resonance (SPR), which is a coherent oscillation of the surface conduction electrons excited by electromagnetic radiation. These resonances create sharp spectral absorption and scattering peaks as well as strong electromagnetic near-field enhancements, which led to a broad range of applications such as chemical and biological sensing [1–5] and surface-enhanced spectroscopies [6–13] in the past several decades.

Although there are a rich variety of plasmonic metal nanostructures, they can be differentiated based on the plasmonic modes they support: localized surface plasmons (LSPs) or propagating surface plasmons (PSPs) [14,15]. Figure 5.1 illustrates the difference between PSPs and LSPs [16]. PSPs are supported by structures that have at least one dimension much larger than the excitation wavelength. In such a structure, like a thin metal film for example, SPs propagate in the *x*- and *y*-directions along the metal–dielectric interface back and forth between the ends of the structure. Propagation lengths can be in the tens of micrometers and decay evanescently in the *z*-direction with 1/e decay lengths on the order of 200 nm [17]. This can be described as a Fabry–Perot resonator with resonance condition $l = n\lambda_{sp}$, where l is the length of the metal film, n is an integer, and λ_{sp} is the wavelength of the PSP mode [18]. The PSP waves can be manipulated by controlling the geometrical parameters of the structure [19]. The interaction

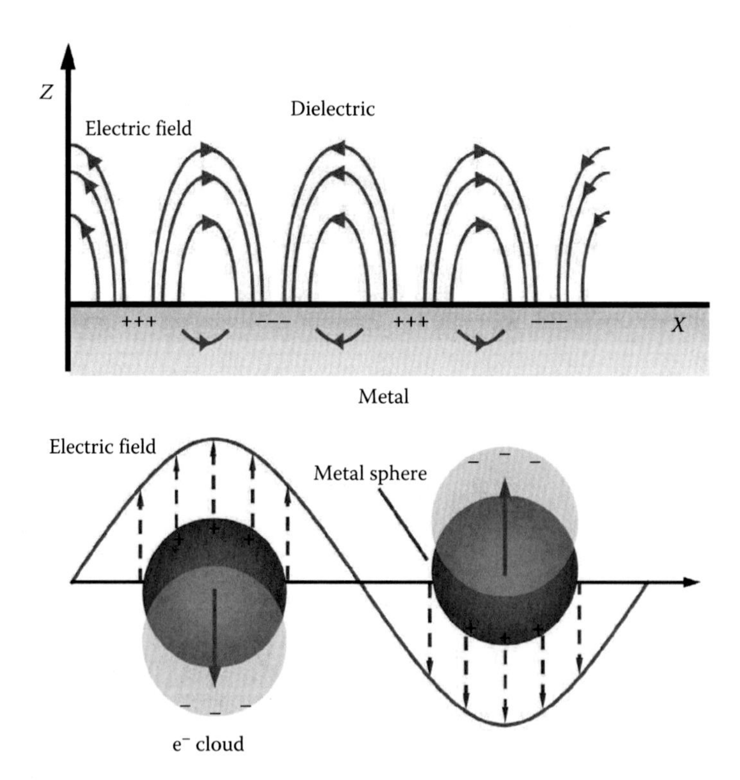

FIGURE 5.1
Schematic of plasmon oscillation for a sphere, showing the displacement of the conduction electron charge cloud relative to the nuclei.

between the metal surface-confined EM wave and a molecular surface layer of interest leads to shifts in the plasmon resonance condition, which can be observed in three modes: (1) angle resolved, (2) wavelength shift, and (3) imaging [17]. In the first two modes, one measures the reflectivity of light from the metal surface as a function of either angle of incidence (at constant wavelength) or wavelength (at constant angle of incidence). The third method uses light of both constant wavelength and incident angle to interrogate a 2D region of the sample, mapping the reflectivity of the surface as a function of position.

For the case of LSPs, light interacts with particles much smaller than the incident wavelength. The oscillating electric field associated with the light (E_0) drives the gas of negatively charged electrons in the conduction band of the metal to oscillate collectively [20]. This is schematically pictured in Figure 5.1. When the electron cloud is displaced relative to the nuclei, a restoring force arises from Coulomb attraction between electrons and nuclei that results in oscillation of the electron cloud relative to the nuclear framework. At a certain excitation frequency (ω), this oscillation will be in resonance with the incident light, resulting in a strong oscillation of the surface electrons, commonly known as a localized surface plasmon resonance (LSPR) mode. The LSPR has two important effects. First, the particle's optical extinction has a maximum at the plasmon resonant frequency, which occurs at visible wavelengths for noble metal nanoparticles. Second, electric fields near the particle's surface are greatly enhanced, this enhancement being greatest at the surface and rapidly falling off with distance.

Generally, the LSPR frequency is dominated by four factors: the density of electrons, the effective electron mass, and the shape and size of the charge distribution. For an individual metallic particle, these four factors are determined by the own characteristics of the particle such as the size, shape, the metal composition, as well as their dielectric environments [20]. Typically researchers studied all of these factors that determine the LSPR wavelength [21]. Whereas these studies provided a fundamental understanding of how plasmons are influenced by local structure and environment, they also suggested the usefulness of plasmons as a sensing modality. Today, plasmon spectroscopy enjoys a reputation as an ultrasensitive method for detecting molecules of both biological and chemical interest, in addition to its continued role in enabling surface-enhanced spectroscopic methods [22–24].

The remainder of this chapter about LSPR spectroscopy is organized as follows: First, the analytical theory of LSPR will be described to explain the physical origin of LSPR. Next a discussion about spectroscopic measurements is performed. Finally, the application of those plasmon resonances is discussed including colorimetric sensing, surface-enhanced Raman scattering (SERS), LSPR-based Rayleigh scattering spectroscopy, and LSPR-controlled fluorescence.

5.2 Theory

In 1908, Gustav Mie developed an analytical solution to Maxwell's equations that describes the extinction spectra (extinction = scattering + absorption) of spherical particles (the radius of the spherical particle is much smaller than the wavelength of incident light λ) [20,25–27]. Although the modern generation of metal nanoparticle science has provided new challenges for theory, Mie's solution remains of great interest to this day. Because Mie's theory is the only simple, exact solution to Maxwell's equations that is relevant to particles. In addition, most of the prepared colloidal particles are approximately spherical, and most of the optical methods for characterizing nanoparticle spectra probe a large ensemble of these particles. All of these lead to results that can be modeled reasonably well using Mie theory [20,25].

For a spherical metal nanoparticle with a size much smaller than the wavelength of incident light, the surface plasmon oscillation is usually called the dipole plasmon resonance (as shown in Figure 5.1). Under this limit, the electric field of the light can be taken to be constant, and the interaction is governed by electrostatics rather than electrodynamics, allowing us to solve Maxwell's equations using a quasistatic approximation [20]. The resulting solution for the E-field outside the particle is given by

$$E_{\text{out}} = E_{\text{o}}\,\hat{x} - \left(\frac{\varepsilon_{\text{i}} - \varepsilon_{\text{o}}}{\varepsilon_{\text{i}} + 2\varepsilon_{\text{o}}}\right) a^3 E_{\text{o}} \left[\frac{\hat{x}}{r^3} - \frac{3x}{r^5}\left(x\hat{x} + y\hat{y} + z\hat{z}\right)\right].$$

Here, ε_i is the dielectric constant of the metal nanoparticle and ε_o is the dielectric constant of the surrounding medium. We note that the first term in Equation 1 is the applied field and the second is the induced dipole field. Because ε_i is strongly dependent on wavelength, the first term in square brackets determines the dielectric resonance condition for the particle. When the dielectric constant of the metal is roughly equal to $-2\varepsilon_o$, the plasmon resonance occurs, thus determining the LSPR frequency. The large optical polarization associated with the LSPR results in a huge local electric field enhancement at the nanoparticle surface as well as strongly enhanced light absorption and scattering by the nanoparticle at the LSPR frequency. For Au and Ag, the resonance condition is fulfilled at visible frequencies, which has important implications for surface-enhanced spectroscopies. The size (a) and external dielectric constant (ε_o) also play key roles in determining the EM field outside the particle, consistent with experimental results.

Plasmon oscillations excited in a metal nanoparticle decay into electron-hole excitations in the metal conduction band [20,26]. This nonradiative pathway of plasmon decay constitutes the absorption of light by the particle. In addition, the oscillating electric field generated by the plasmon can radiate electromagnetic energy at the same frequency as that of the surface plasmon oscillations, which constitutes Rayleigh scattering by the particles [20].

Absorption and scattering together comprise the optical extinction by the particle. According to the Mie theory, for a metal nanosphere with particle size much smaller than the wavelength of light λ (quasistatic and dipole limit), the nanoparticle extinction cross-section is given by [25,27]:

$$C_{\text{ext}} = \frac{24\pi^2 R^3 \varepsilon_{\text{o}}^{3/2}}{\lambda} \frac{\varepsilon_{\text{i}}}{(\varepsilon_{\text{r}} + 2\varepsilon_{\text{o}})^2 + \varepsilon_{\text{i}}^2}.$$

Here, ε_r and ε_i are the real and imaginary components of the metal dielectric function, respectively. Again, we note the wavelength dependence of the metal dielectric function. The optical extinction thus has a band maximum at the resonance condition roughly given by $\varepsilon_r = -2\varepsilon_o$, indicating that the real part determines the wavelength position of the resonance and the imaginary part determines the bandwidth. Both methods allow the evaluation of the extinction of particles of arbitrary shape and size, and the results typically match well with experimental results.

5.3 Spectroscopic Measurements

Figure 5.2 exhibits several approaches to the measurement of nanoparticle LSPR spectra. The most straightforward is ultraviolet-visible spectroscopy, including the transmission and reflection spectroscopy. The transmission spectroscopy measures the extinction spectrum of the nanoparticles by recording the wavelength dependence of the light passing through the sample. For the nontransparent samples, one must use the reflection spectroscopy (Figure 5.2b). Here, a fiber bundle is used both to direct the excitation light to the sample and to collect the light reflected from the surface. Whereas the transmission geometry yields the LSPR wavelength as a maximum value in the extinction curve, the reflected geometry yields the LSPR wavelength as a minimum value (because light absorbed or scattered by the sample is not reflected back). For samples in which small regions or even single nanoparticles are interrogated, dark-field light-scattering measurements are extremely powerful. In this case, the white light is irradiated on the nanoplasmonic particles with an oblique angle by the dark-field condenser lens. The scattering alone is collected by a microscope objective lens with a numerical aperture, and the collected scattering light can be imaged by a true color charge-coupled device (CCD) camera and analyzed by a spectrometer. Figure 5.3c shows an example of this in which a high-numerical aperture condenser brings light to the sample and a low-numerical aperture microscope objective collects the scattered light at low angles. For SERS measurements, the sample is excited by a laser (optimally to the blue of the LSPR of the metal substrate), and the Raman scattered light is passed through a

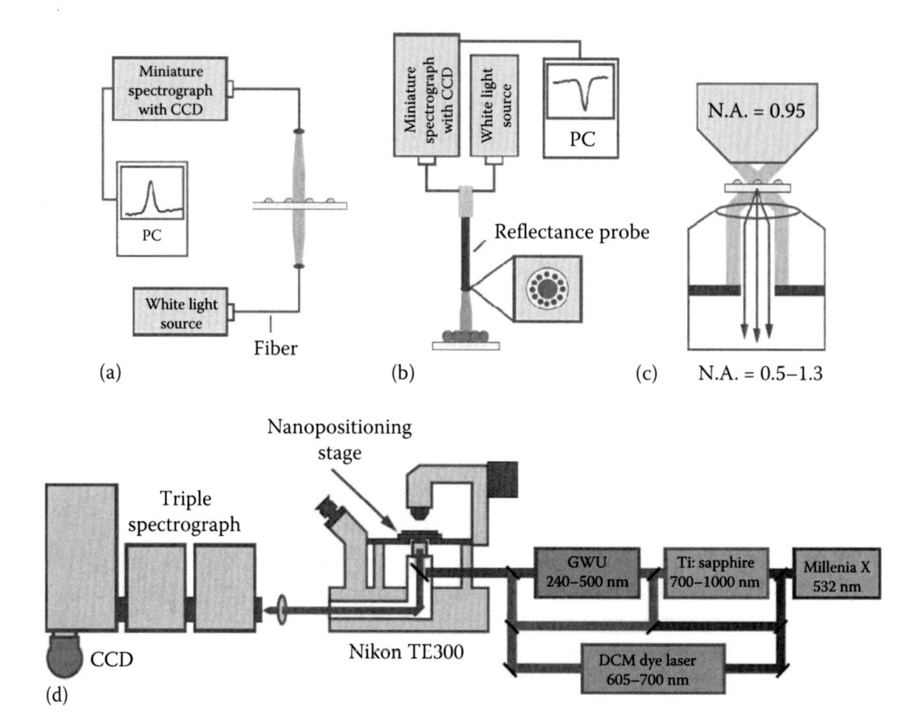

FIGURE 5.2
(a) Transmission and (b) reflectance geometries for measuring extinction spectra of nanoparticle arrays. (c) The detailed experimental configuration of a dark-field microscopy system. Dark-field scattering experimental setup using a high-numerical aperture dark-field condenser and a low-numerical aperture microscope objective for measuring single-nanoparticle scattering spectra. (d) Experimental setup for measuring surface-enhanced Raman scattering in an epi-illumination geometry using a wavelength-scanned laser-excitation system and a triple spectrograph coupled with a charge-coupled device (CCD) camera. (Reproduced from Saha, K., Agasti, S.S., Kim, C., Li, X.N., and Rotello, V.M., *Chem. Rev.*, 112, 2739, 2012. Copyright 2005. With permission from Annul Reviews.)

spectrometer and onto a detector. The excitation light is typically directed to the sample at either a glancing angle or by using a microscope objective in epi-illumination. Figure 5.3d shows the latter configuration that was used for the wavelength-scanned SERS excitation–spectroscopy experiments described in the following section [1].

5.4 Absorption-Based Chemical Sensing

The aggregation of metal nanoparticles of appropriate sizes induces interparticle surface plasmon coupling, resulting in a visible color change [28]. The color change during metal nanoparticle aggregation (or redispersion of

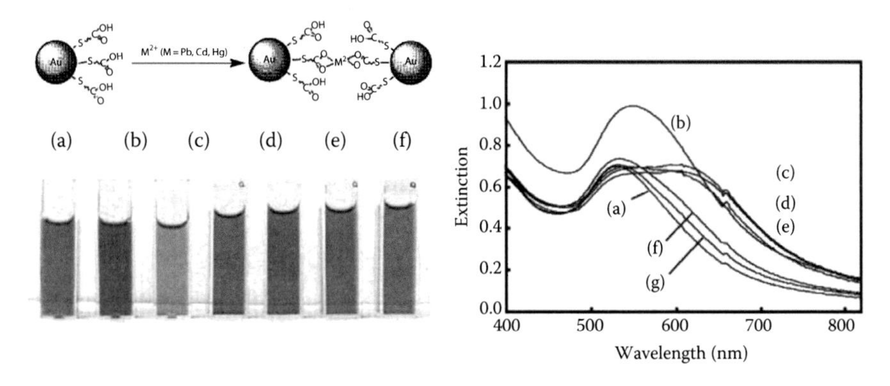

FIGURE 5.3
(See color insert.) Colorimetric responses (a) and corresponding spectral traces (b) from (a) Au-MUA, (b) Au-MUA/Pb^{2+}, and (c–g) Au-MUA/Pb^{2+} and increasing amounts of EDTA. Pb^{2+} concentration in sample (b) is 0.67 mM; EDTA concentrations in samples (c–g) are 0.191, 0.284, 0.376, 0.467, and 0.556 mM. (Reproduced with permission from Kim, Y.J., Johnson, R.C., and Hupp, J.T., *Nano Lett.*, 1, 2001, 165. Copyright 2001, American Chemical Society.)

an aggregate) provides a practical platform for absorption-based colorimetric sensing of variety of target analytes including metal ions, organic molecules, proteins, and nucleic acids, which directly or indirectly trigger the metal nanoparticle aggregation or redispersion [4,29,30].

For example, heavy metal ions such as Pb^{2+}, Cd^{2+}, and Hg^{2+} pose significant public health hazards. Hupp et al. have reported a simple colorimetric technique for the sensing of aqueous heavy metal ions using 11-mercaptoundecanoic acid (MUA)-functionalized 13 nm AuNPs [31]. The color change (red to blue) is driven by a heavy metal ion chelation process where the surface carboxylates act as metal ion receptors. Colorimetric response was observed from Pb^{2+}, Cd^{2+}, and Hg^{2+} ($\geq$400 μM), whereas Zn^{2+} displayed no response to this assay process.

Another scheme for detection of metal ions was developed by Liu and Lu, in which the highly selective lead biosensors using DNAzyme-directed assembly of AuNPs were fabricated, allowing the tuning of sensitivity over several orders of magnitude [32]. In their sensor design, a DNAzyme specific to the Pb^{2+} ion was chosen as the target recognition element and DNA-modified AuNPs were used as the signal transducer element. The Pb^{2+}-specific DNAzyme is comprised of an enzyme strand and a substrate strand. In the presence of Pb^{2+} ion, the enzyme strand carries out catalytic reactions involving hydrolytic cleavage of the substrate strand. When incubated with DNAzyme, the DNA-functionalized AuNPs form blue-colored assemblies through Watson–Crick base pairing as shown in Figure 5.6. The DNAzyme is activated in the presence of Pb^{2+} in the solution. Activated DNAzyme cleaves the substrate strand to dissemble the AuNPs, resulting in a blue-to-red color change. The sensor was capable of detecting Pb^{2+} concentration of 100 nM. More importantly, the sensor also exhibits high selectivity (Figure 5.4).

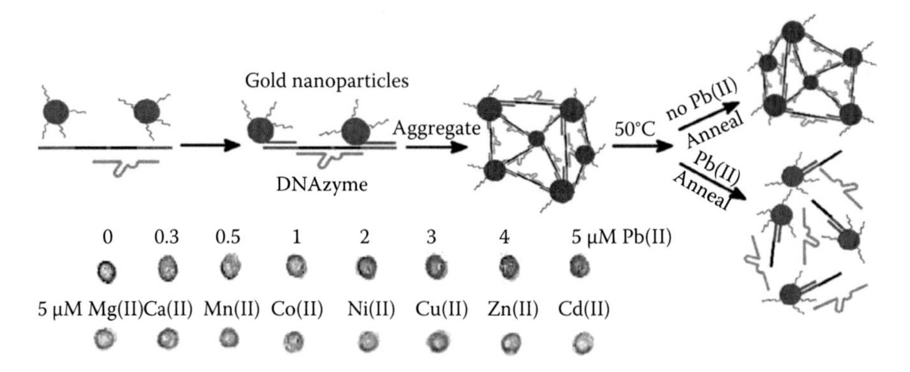

FIGURE 5.4
(See color insert.) Pb^{2+}-directed assembly of AuNPs by the DNAzymes resulting in detection. (Reproduced with permission from Liu, J.W. and Lu, Y., *J. Am. Chem. Soc.*, 125, 2003, 6642. Copyright 2010, American Chemical Society.)

Compared with the detection of metal ions, designing a sensor system for anionic species is more challenging because of their lower charge to radius ratio, pH sensitivity, wide range of geometries, and solvent-dependent binding affinity and selectivity. For example, fast and sensitive detection of CN^- is important for environmental monitoring and the evaluation of food safety. Several strategies for detecting cyanide have been developed. Among them, colorimetric method is particularly attractive because they can be read by the naked eye, and in some cases at the point of use. Although many carefully designed colorimetric sensors for cyanide have been reported, they are limited to their sensitivity, selectivity, and compatibility within a purely aqueous environment. Han et al. reported a colorimetric detection method for cyanide anions in aqueous solution [33]. The detection limit reaches 10^{-5} M in neutral aqueous solution. In this system, adenosine triphosphate-stabilized AuNPs were used as the reporter unit and a Cu^{2+}–phenanthroline complex as the receptor unit. Exposure of CN^- to Cu^{2+}–phenanthroline complex induced a decomplexation process to generate free phenanthroline, which subsequently caused the ATP-stabilized AuNPs to aggregate resulting in color change (Figure 5.5). Besides, Mirkin et al. have utilized the Griess reaction for the AuNP-based colorimetric detection of nitrite [34]. Li et al. have used a "click" reaction coupled with the AuNP probes for the visual detection of ascorbic acid [35]. Han et al. have employed an AuNP-embedded plasticized poly(vinyl chloride) (PVC) membrane for selective detection of iodide anions [36].

Recently, the LSPR-based colorimetric methods also have been used for the detection of organic pollutant. For example, 2,4,6-trinitrotoluene (TNT) is a leading example of a nitroaromatic explosive with significant detrimental effects on the environment and human health [37,38]. As one of the most common explosives, TNT also has been widely applied in terrorist activities. As a result, the detection of trace explosives is a leading priority for the needs of homeland security. Mao et al. have employed cysteamine-modified

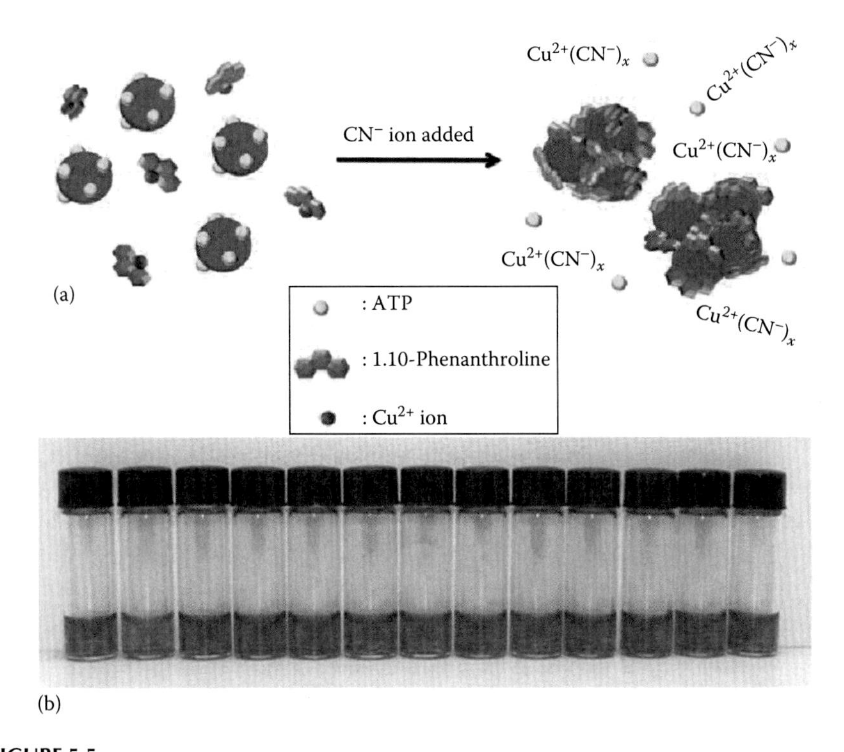

FIGURE 5.5
(See color insert.) (a) A schematic diagram of the cyanide anion sensing ensemble. (b) The color of the solution in the absence and presence of anions: from left to right; no anion, CN^-, F^-, Cl^-, Br^-, ClO_4^-, SO_4^{2-}, HCO_3^-, AcO^-, HPO_4^{2-}, NO_3^-, N_3^-, and $P_2O_7^{4-}$. (Reproduced from Kim, M.H., Kim, S., Jang, H.H., Yi, S., Seo, S.H., and Han, M.S., *Tetrahedron Lett.*, 51, 4712, 2010. Copyright 2010. With permission from Science Direct.)

AuNPs for the colorimetric detection of TNT in real-world matrices [39]. In this study, cysteamine was used both as the primary amine and as the stabilizer for AuNPs to facilitate the donor–acceptor (D–A) interaction between TNT and the primary amine at the AuNP surface for direct visualization of TNT, based on the TNT-induced colorimetric AuNP aggregation phenomenon. Initially, the cysteamine-stabilized AuNPs were well dispersed in distilled water, and the color of the uniform suspension was wine red, because of the strong SPR of the AuNPs. The addition of TNT to the dispersion essentially leads to the aggregation of the cysteamine-stabilized AuNPs as a result of the D–A interaction between TNT and cysteamine (Figure 5.6), and the color of the suspension is accordingly changed from wine red to violet blue. The clear change in the color of the suspension could be used for the direct colorimetric visualization of TNT. This sensor can also be used for the colorimetric detection of melamine [40].

Another important application of the absorption-based colorimetric sensor is the detection of proteins. Many disease states are often associated with the

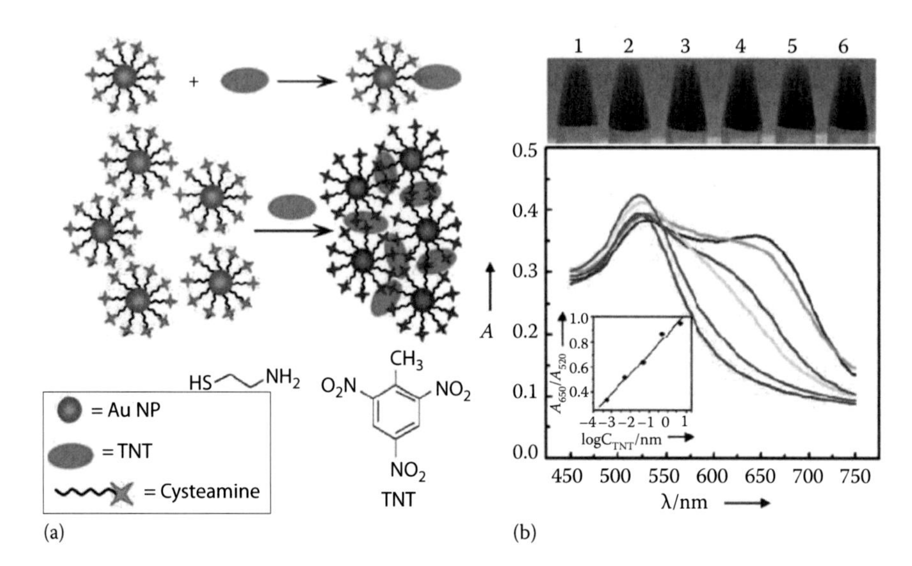

FIGURE 5.6
(See color insert.) (a) Assay for direct colorimetric visualization of TNT based on the electron D–A interaction at the AuNP/solution interface. Colorimetric visualization of TNT by using AuNPs (containing 500 nm cysteamine). TNT concentrations varied from 5×10^{-13} (2) to 5×10^{-9} M (6). (b) UV-vis spectra of the AuNP suspension (10 nm) containing 500 nm cysteamine in the presence of different concentrations of TNT. (From Jiang, Y., Zhao, H., Zhu, N.N., Lin, Y.Q., Yu, P., and Mao, L.Q.: *Angew. Chem. Int. Ed.* 2008. 47. 8601. Copyright Wiley-VCH Verlag GmbH & Co. KGaA. Reproduced with permission.).

presence of certain biomarker proteins or irregular protein concentrations. AuNPs have been successfully applied for colorimetric detection of proteins. A diverse range of functionalized AuNPs have been utilized for the detection of proteins [41–43]. For example, Mirkin et al. have developed a real-time colorimetric screening method for endonuclease activity by using DNA-mediated AuNP assemblies. Aggregates of gold nanoparticles interconnected by DNA duplexes are bluish-purple [44]. Cleavage of the duplexes by deoxyribonuclease I (DNase I) releases the nanoparticles, producing a bluish-purple-to-red color change (Figure 5.7). This method can be used to screen libraries of inhibitors of endonucleases in a high-throughput fashion by using either the naked eye or a simple colorimetric reader.

5.5 Surface-Enhanced Raman Scattering

Raman spectroscopy is a vibrational spectroscopy that provides specific information about molecules. However, the direct application of this technique in sensitive detection and identification of analyte molecules is

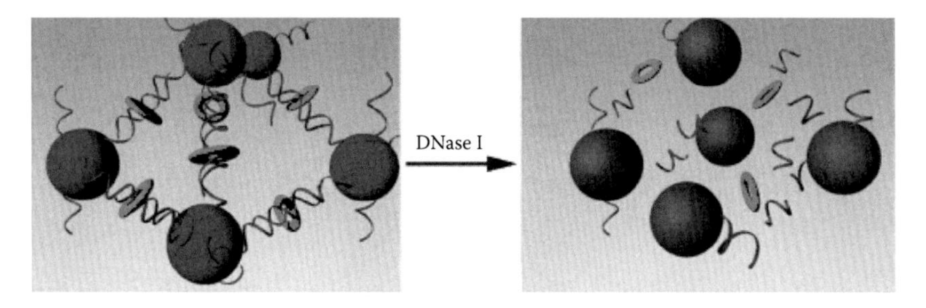

FIGURE 5.7
Illustration of the aggregation and dissociation of the DNA-AuNP probes used in the colorimetric screening of endonuclease inhibitors. (From Xu, X.Y., Han, M.S., and Mirkin, C.A.: *Angew. Chem. Int. Ed.* 2007. 46. 3468. Copyright Wiley-VCH Verlag GmbH & Co. KGaA. Reproduced with permission.)

severely limited due to the low efficiency of inelastic photon scattering by molecules, leading to a weak signal [7–9]. The inherent limitation of low scattering intensity arises from the fact that the Raman scattering cross-sections for molecules are usually small, typically 10^{-30}–10^{-25} cm^2/molecule, 10–15 orders of magnitude smaller than that of fluorescence cross-section. Another major disadvantage is the fluorescence that often accompanies Raman scattering and can sometimes overwhelm the bands in the spectrum, rendering the experiment useless [7–11]. In 1978, it was discovered, largely through the work of Fleischmann, Van Duyne, Creighton, and their coworkers, that molecules adsorbed on specially prepared silver surfaces produce a Raman spectrum that is 10^6 times more intense than expected [19]. This effect was denoted SERS. Then SERS has immediately became the subject of intensive study, with a number of papers published using the technique and investigating the basis behind it growing exponentially year by year. In recent years, SERS has been transformed into a powerful analytic technique due to advances in nanofabrication and an increased understanding of plasmonic properties of nanomaterials [11,45–48]. The two major advantages of using SERS as an analytical technique are its exquisite sensitivity and its molecular specificity. SERS is currently the only method that can simultaneously detect a single molecule and provide its chemical fingerprint [7–11,45–48]. Hence mixtures can be identified without separation, making the surface-enhanced Raman spectroscopy more amenable to more complex analysis than other spectroscopy.

However, even 35 years after its initial discovery, a complete picture of the enhancement mechanism is not available, due to its highly complex nature and required experimental conditions, such as roughened surfaces, nanoparticle junctions and aggregates, and chemical interactions. This problem has become even more apparent recently due to the discovery of single-molecule SERS (SMSERS) that gives rise to the possibility of 10^{14} enhancement factors.

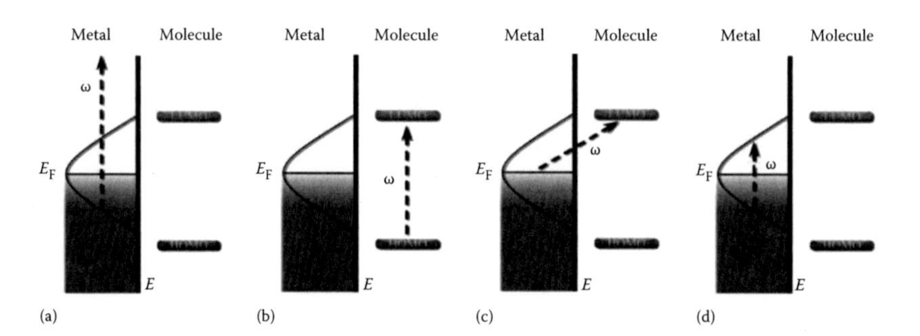

FIGURE 5.8
A cartoon representation of the mechanism of SERS, here using crystal violet on a gold nanotriangle as representative example. (a) CHEM, nonresonant chemical mechanism; (b) RRS, resonance Raman scattering; (c) CT, dynamic charge transfer; and (d) EM, electromagnetic mechanism. (From Jensen, L., Aikens, C.M., and Schatz, G.C., *Chem. Soc. Rev.*, 37, 1061, 2008. Reproduced with permission of The Royal Society of Chemistry.)

Resulting from the experimental observations, two main methods of explaining the larger enhancements of SERS arose, one being an electromagnetic mechanism (EM) [49–51] and the other being a chemical mechanism (CM) [52,53]. Both will contribute to the total SERS enhancement, and it is not possible experimentally to make a clear separation of their individual contributions. However, depending on the specific system, one can identify at least four different situations where the SERS signal is enhanced due to a different mechanism including nonresonant CM, resonance Raman scattering, dynamic charge transfer (CT), and EM. (Figure 5.8).

5.5.1 Electromagnetic Mechanism

EM is a result of SPRs on the nanoparticle surface and is therefore considered to be independent of the nature of the molecule or the chemical bond between the metal and molecule. Since the local field enhancement arising from the plasmon excitation can be very large, the EM is considered as the major contribution to the observed SERS signal. The strong fields lead to the electrodynamic enhancement mechanism, and it is generally assumed that the SERS signal is enhanced by a factor proportional to the fourth power of the electric field enhancement. It can be shown that the enhancement factor due to EM can be written as

$$EF_{EM} = \left|E(\omega)\right|^2 \left|E(\omega_0)\right|^2$$

where

EF_{EM} is the EM enhancement factor
$E(\omega)$ is the frequency-dependent electric field at incident frequency ω
$E(\omega_0)$ is the frequency-dependent electric field at the Stokes-shifted frequency ω_0

However, since the Stokes shift is usually small compared with the wavelength of laser, the enhancement factor is typically simplified as

$$EF_{EM} = |E(\omega)|^4$$

The local electromagnetic field due to plasmon excitations is largest in regions with high local curvature and in particular in the junction between dimers of nanoparticles, the so-called hot spots [50–53]. This is illustrated in Figure 5.9 where Hao and Schatz [54] used discrete dipole approximation (DDA) to simulate the EM enhancements around triangular nanoparticles, which shows that the largest $|E|^2$ values are very similar for triangular prisms, oblate spheroids, or cylindrical rods, with $|E|^2$ always being less than 10^4. For a dimer of nanoparticles, we find $|E|^2$ values of closer to 10^5 for structures where the particle separation is 2 nm. The enhancement is a strong function of separation distance, and it scales with particle size such that larger particles give the same enhancements for larger separations. In addition, the discrete electronic transitions observed in small metal clusters can be used as an analogue for plasmon frequencies [55] and thus serve as a model system for understanding EM. This is illustrated in Figure 5.9b where the electric field enhancement around the Ag_{20} cluster has been calculated using time-dependent density functional theory [53] and is qualitatively similar to results calculated with DDA (Figure 5.9).

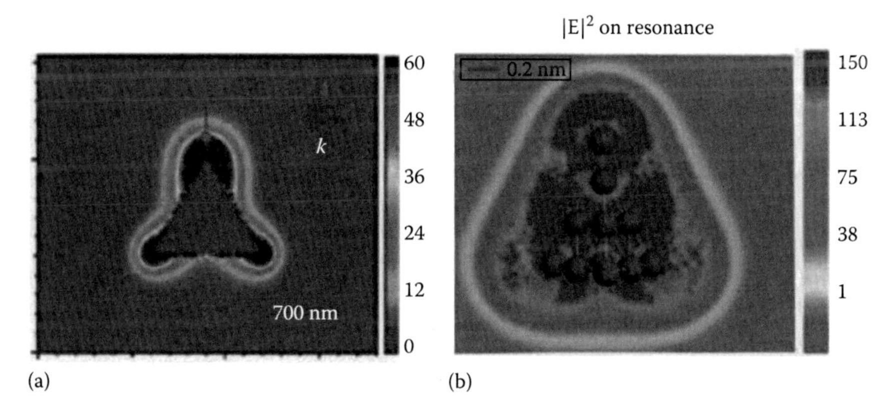

FIGURE 5.9
(See color insert.) (a) DDA calculation of the electric field enhancement $|E|^2$ around a nanotriangle, highlighting the fact that the greatest enhancement is at the tips. (Reprinted with permission from Hao, E. and Schatz, G.C., *J. Chem. Phys.*, 120, 357, 2004, Copyright 2004, American Institute of Physics.) (b) TD-DFT calculation of $|E|^2$ for a tetrahedral Ag_{20} cluster. The two methods yield qualitatively the same results. (From Jensen, L., Aikens, C.M., and Schatz, G.C., *Chem. Soc. Rev.*, 37, 1061, 2008. Reproduced with permission of The Royal Society of Chemistry.)

5.5.2 Chemical Mechanism

CM is also sometimes referred to as the electronic mechanism or first layer effect due to its short-range nature. The CM can be divided into three contributions:

1. A molecular resonance (RRS) mechanism. This is analogous to RRS for a free molecule except that the presence of the surface modifies the enhancements observed. This is not traditionally included as a CM, but since it is in fact altered by the metal, we prefer to include it as a CM. RRS is typically expected to contribute enhancements of 10^3–10^6.
2. A CT mechanism, in which the applied field is in resonance with either a metal–molecule or molecule–metal transition and typically provides enhancements around 10–10^4. It has been proposed that it may be as large as in some situations, based on recent experimental work [56].
3. A nonresonant chemical (CHEM) mechanism, which arises simply from the ground-state interaction of the molecule with the metal. This is the weakest of the mechanisms, only leading to enhancements on the order of ~10–100.

The main features of resonance Raman effect can be explained by assuming a two-state model that neglects vibronic coupling. In this way, the Raman cross-section *I* with respect to normal mode and polarizability derivative can be written as [56–58]

$$I \propto \left(-\frac{\mu}{(\omega_e - \omega - i\Gamma)^2}\right)\left(\frac{\partial \omega_e}{\partial Q_k}\right)^2$$

where
- μ is the transition dipole moment
- ω_e is the energy of the transition
- ω is the laser energy
- Q_k is the normal coordinate of mode k
- Γ is the excited state relaxation rate[4]

Accordingly, at resonance ($\omega = \omega_e$), the resonance Raman intensity is approximately given by

$$I \propto \frac{\mu^2}{\Gamma^4}\left(\frac{\partial \omega_e}{\partial Q_k}\right)^2$$

The CT mechanism is related to RRS since it is a resonance effect between a molecular and a metal electronic state [59,60]. As a result, it can only occur if the molecule is close enough to the metal for the wave functions of the two systems to overlap. CT is undoubtedly the most controversial and least understood mechanism of SERS, because of the difficulty in determining it experimentally and theoretically. Therefore, with both the LSPR and chromophore extending across the wavelength range covered by the Stokes scattered light, the electronic resonance associated with the absorption peak provides a scattering cross-section enhancement of several orders of magnitude in addition to the electromagnetic enhancement.

The nonresonant CM (CHEM, also referred to as the static CM) is very difficult to detect experimentally since it makes the smallest contribution to the overall SERS EF, although there have been recent experimental attempts to quantify it.[61,62] Theoretical studies are expected to be particularly useful in describing this mechanism since it is likely to be dominated by the local environment around the molecule, which can be represented using small metal clusters. For this reason, there are many electronic structure studies of the effect of small metal clusters on the Raman properties of small molecules. In these calculations, the Raman intensities are obtained using static molecular polarizabilities ($\omega = 0$), ensuring no resonances, and therefore the only enhancement mechanism is CHEM.

5.5.3 Single-Molecule Surface-Enhanced Raman Scattering

SMSERS was independently reported in 1997 by Nie and Emory [6] and Kneipp et al. [11], yields richer chemical information and is more broadly applicable, and has the potential to overcome these limitations. Nie and Emory detected Raman scattering from single-rhodamine 6G molecules adsorbed on Ag nanoparticles at concentrations corresponding to zero or one analyte molecule per nanoparticle. Kneipp et al. observed SMSERS of crystal violet on Ag nanoparticles in solution that presented alternative evidence contingent on fluctuations in the intensity domain, using statistical analysis to support the claim of single-molecule behavior. Nie and Emory cited several features of their results that suggest single-molecule behavior: (1) Raman scattering is observed from single nanoparticles for analyte concentrations of less than 10^{-10} M, where each particle is expected to contain mostly zero or one analyte molecule; (2) unlike bulk SERS spectra, the observed single-molecule spectra show sensitivity to polarization; and (3) the position and intensity of vibrational bands exhibit sudden changes as a function of time (Figure 5.10). Additionally, Pettinger and coworkers point out that the SMSERS EF is strongly dependent on the molecule's position in the hot spot, and the variations would make the observation of quantized intensity impossible [63]. Brus and coworkers showed using polarization studies that hot spots formed at the junction of two nanoparticles likely play a major role in SMSERS, which was further supported by atomic force microscopy showing that SMSERS-active structures are aggregates of Ag nanoparticles [64,65]. In addition, SMSERS was shown to depend nonlinearly on excitation power.

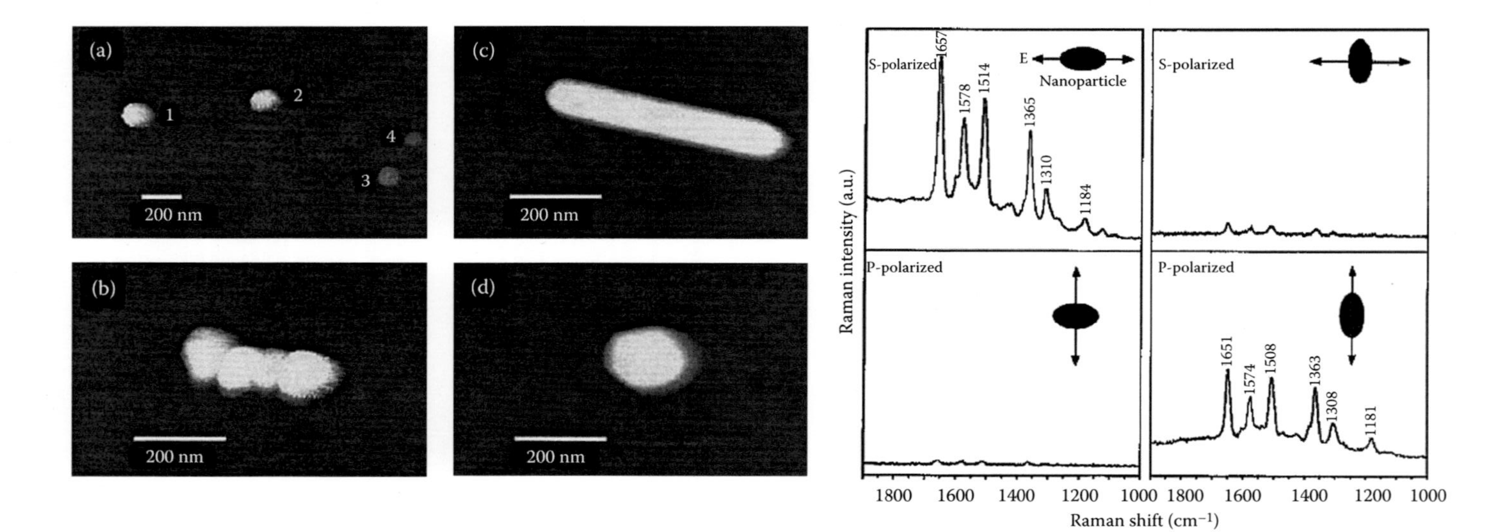

FIGURE 5.10
Surface-enhanced Raman spectra of R6G obtained with a linearly polarized confocal laser beam from two Ag nanoparticles. The R6G concentration was 2 3 10211 M, corresponding to an average of 0.1 analyte molecule per particle. The direction of laser polarization and the expected particle orientation are shown schematically for each spectrum. Laser wavelength, 514.5 nm; laser power, 250 nW; laser focal radius; 250 nm; integration time, 30 s. All spectra were plotted on the same intensity scale in arbitrary units of the CCD detector readout signal. (From Bailo, E. and Deckert, V., *Chem. Soc. Rev.*, 37, 921, 2008. Reproduced with permission of AAAS.)

Etchegoin and coworkers introduced the bianalyte technique that relied on competitive adsorption between two analytes, R6G and benzotriazole [66,67].

For single-rhodamine 6G molecules adsorbed on the selected nanoparticles, the intrinsic Raman enhancement factors were on the order of 10^{14}–10^{15}, much larger than the ensemble-averaged values derived from conventional measurements. This enormous enhancement leads should be resulted from the multiplication of the combination of EM mechanism and resonance Raman mechanism. EM mechanism can only contribute an enhancement factor of 10^{10}–10^{11} [68]. The result part of the SMSERS enhancement should be attributed to the resonance Raman effect. Because all the SMSERS studies of R6G were carried out at laser-excitation wavelengths near the R6G molecular resonance and enhancement from the resonance, Raman effect can contribute an additional factor of 10^2–10^3 [69]. However, the application of this technique, which uses adsorbates with different chemical structures, is complicated due to differences in Raman cross-section, absorption spectra, and surface-binding affinity for the different analytes used. At sufficiently low concentrations where only one molecule is adsorbed to a nanoparticle, SMSER spectra contain the features of only a single isotopologue. Optimization of highly enhancing substrates will enable the general application of SMSERS and provide insight into many chemical problems, including those in heterogeneous catalysis.

5.5.4 Tip-Enhanced Raman Spectroscopy

Tip-enhanced Raman spectroscopy (TERS) is another important variation of SERS, which has emerged as a promising technique for in situ chemical analysis on the nanoscale [70]. In TERS, enhancement arises from a metallic scanning probe microscopy (SPM) tip rather than from the substrate. When an Au or Ag SPM tip is irradiated with visible light, excitation of the LSPR results in an enhanced electromagnetic field that is locally confined around the tip apex. This enhanced electromagnetic field increases Raman scattering from molecules located in the near-field region of the tip by three to six orders of magnitude [70]. The use of an SPM tip also provides nanometer spatial resolution, a significant improvement to the diffraction-limited spatial resolution of SERS which is $\lambda_{ex}/2$. Figure 5.11 shows a schematic diagram of a tip-enhanced Raman scattering setup working in back-reflection mode. An inverted Raman microscope is coupled with an AFM for synchronized use. The microscope is required to illuminate the metal-coated AFM tip. The backscattered Raman signal is collected through the same objective and notch or edge filter is used to block the laser line. After this filter stage, the signal is coupled to a spectrometer equipped with a cooled CCD for spectrally resolved measurements.

In 2000, the Zenobi group reported the first TERS experiment showing among other results a TERS spectrum for brilliant cresyl blue (BCB) with an about 30-fold net increase of Raman scattering upon moving the silverized

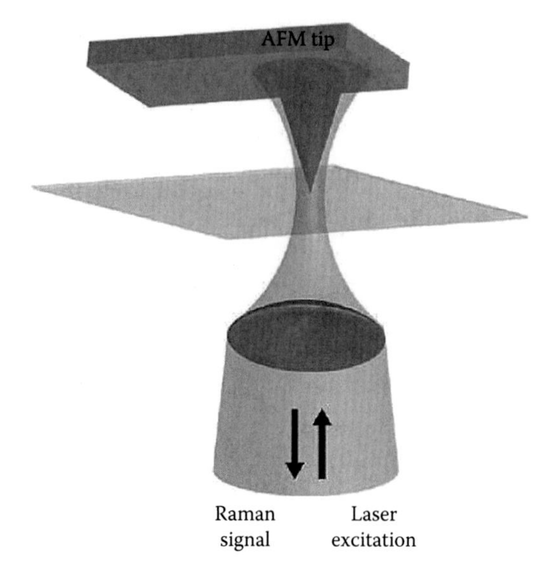

FIGURE 5.11
Schematic diagram of a tip-enhanced Raman scattering setup working in back-reflection mode. (From Bailo, E. and Deckert, V., *Chem. Soc. Rev.*, 37, 921, 2008. Reproduced with permission of The Royal Society of Chemistry.)

AFM tip into the focal region and in contact with a thin dye layer [71]. From then on, this technology has attracted lots of research attention and has been proved a promising tool for investigating nanomaterials and nanostructures. With the development in recent years, it has been applied to dye molecule detection, biological specimen identification, nanomaterial characterization, and semiconductor material determination.

TERS has the potential to be a completely substrate general technique, which overcomes the difficulties and obtains optical resolution beyond the diffraction limit and remarkable signal enhancement in spectral and Raman detection in the nanometer scale. Because the enhancing feature (the tip) is always the same during an experiment, quantitative or semi-quantitative experiments are possible with one tip. Combined spectral and topographical imaging capabilities provide both structural and chemical information about the composition of a surface. Importantly, TERS is a technique compatible with both ambient and ultrahigh vacuum (UHV) environments. Using Raman and AFM, the two techniques combined for TERS, samples can be investigated under conditions relevant to various catalytic processes, as has been demonstrated in the experiment section. Even experiments in liquids (e.g., water) are feasible. Presently the main challenge is a reproducible mass production of the TERS tips. As the tips pick up material during a scan or simply break, this restricts the average lifetime to a few days. Hence, TERS tips have to be exchanged more frequently compared to standard AFM probes. Many groups and joint projects are working on the subject of TERS tip production.

5.5.5 Applications

5.5.5.1 SERS-Based Chemical Sensing

The large enhancement in detectable signal coupled with the unique molecular fingerprints generated has made SERS a powerful tool for the multiplex detection of analytes, with the ability to achieve a detection of single-molecule level. For example, Ray et al. have used AuNPs modified with cysteine as label-free SERS probe for highly selective and sensitive recognition of TNT [72]. Due to the formation of donor–acceptor Meisenheimer complex between TNT and cysteine, gold nanoparticles undergo aggregation in the presence of TNT via electrostatic interaction between Meisenheimer complex-bound gold nanoparticle and cysteine-modified gold nanoparticle. As a result, it formed several hot spots and provided a significant enhancement of the Raman signal intensity by nine orders of magnitude through electromagnetic field enhancements. The resulting "hot spots" for enhancement of the Raman signal provided a sensitivity of 2 pM level (Figure 5.12).

Mirkin et al. have used AuNP probes labeled with oligonucleotides and Raman-active dyes for the SERS-based multiplexed detection of DNA and RNA targets (Figure 5.13) [73]. To detect the presence of specific target DNA strands, a three-component sandwich assay in a microarray format was used. For the assay, dye-labeled AuNP probes were captured by the target oligonucleotide strands, followed by silver enhancement, generating detectable SERS signals exclusively from the Raman dyes immobilized on the particles, with a limit of detection of 20 fM.

As a molecular spectroscopy technique, SERS cannot be used to detect/recognize atomic species directly. Thus, within the 30 years since the discovery of the SERS effect, little research has been carried out toward the analysis of inorganic species in general and inorganic atomic ions in particular. Recently, Alvarez-Puebla et al. demonstrate that SERS spectroscopy can be used for the indirect, quantitative detection of Cl^- [74]. They developed a sensor based on the vibrational changes induced by the interaction of chloride with a Cl-sensitive molecular probe that has a high SERS cross-section. Minute amounts of chloride down to the pM regime were quantitatively detected through the direct comparison of the SERS spectra of the ligand before and after interaction with chloride in aqueous solution (Figure 5.14).

5.5.5.2 TERS-Based Photocatalytic Reaction Monitor

Time-resolved TERS can monitor photocatalytic reactions at the nanoscale. Weckhuysen et al. use a silver-coated atomic force microscope tip both to enhance the Raman signal and to act as the catalyst. The tip is placed in

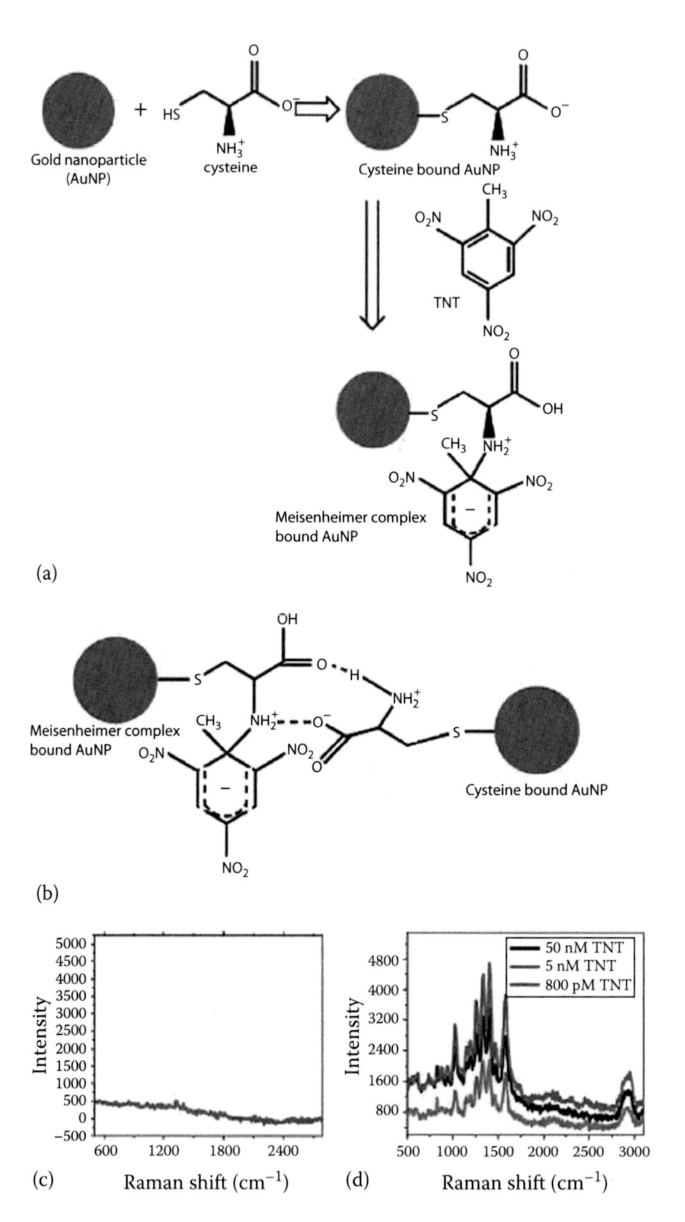

FIGURE 5.12
(a) Schematic representation of the formation of Meisenheimer complex between cysteine-modified gold nanoparticle and TNT. (b) Schematic representation for the possible cross-linking between gold nanoparticle-bound Meisenheimer complex with gold nanoparticle bound cysteine. (c) SERS spectra from gold nanoparticle + TNT (50×10^{-7} M). (d) SERS spectra from cysteine-modified gold nanoparticle + TNT (50 nM–800 pM TNT).

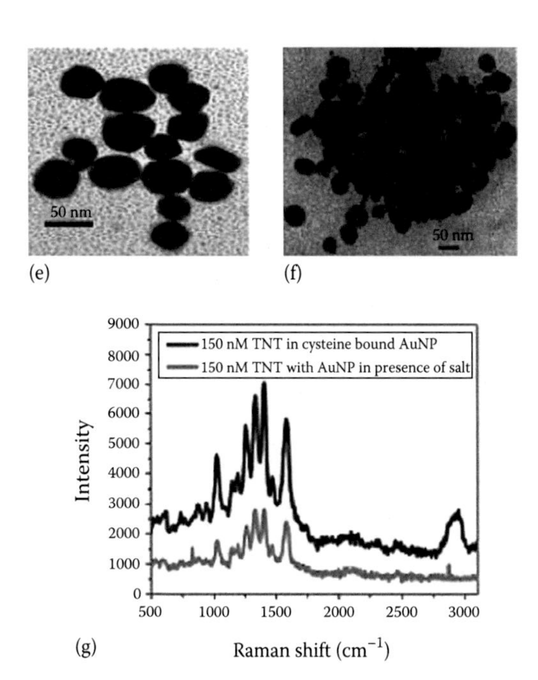

FIGURE 5.12 (continued)
(e) TEM image of cysteine-modified gold nanoparticle. (f) TEM image demonstrating aggregation of gold nanoparticle in the presence of TNT. (g) SERS spectra from gold nanoparticle+TNT (150×10^{-9} M) in the presence of 0.2 M NaCl and SERS spectra from cysteine-modified gold nanoparticle+TNT (150 nM). (Reproduced with permission from Dasary, S.S.R., Singh, A.K., Senapati, D., Yu, H.T., and Ray, P.C., *J. Am. Chem. Soc.*, 131, 2009, 13806. Copyright 2009, American Chemical Society.)

contact with a self-assembled monolayer of p-nitrothiophenol molecules adsorbed on gold nanoplates. A photocatalytic reduction process is induced at the apex of the tip with green laser light, while red laser light is used to monitor the transformation process during the reaction. This dual-wavelength approach can also be used to observe other molecular effects such as monolayer diffusion (Figure 5.15) [75].

5.6 LSPR-Based Rayleigh Scattering Spectroscopy

Recently dark-field microscopy (DMF) and Rayleigh scattering spectroscopy have emerged as complementary technologies for ultrasensitive biological detection and imaging with high spatial and temporal resolution [76]. Plasmonic resonant nanoparticles are key nanoscale probes for these technologies that have enabled single-molecule sensitivity and imaging. Metal nanoparticle has a maximal LSPR extinction (absorption plus scattering) at

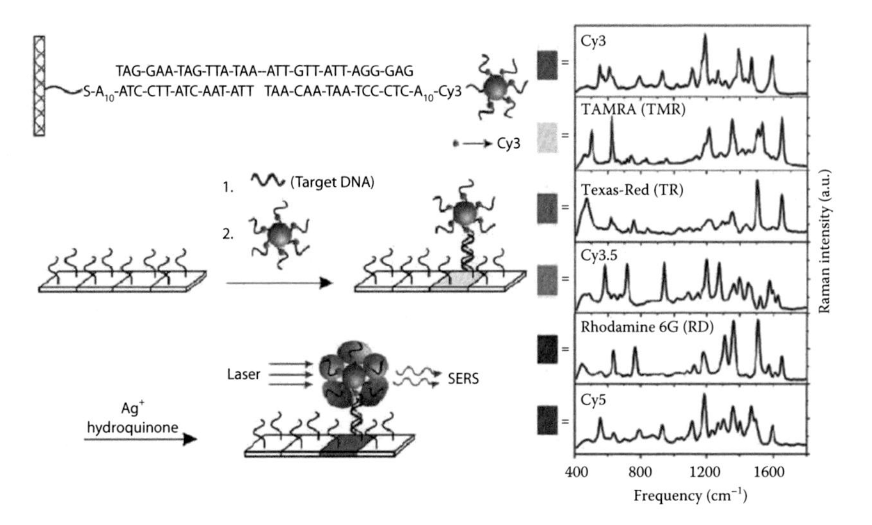

FIGURE 5.13
Schematic representation of three-component sandwich assay for SERS-based oligonucleotide detection (a). The Raman spectra of six dye-labeled nanoparticle probes after Ag enhancing on a chip (b). (From Cao, Y.C., Jin, R., and Mirkin, C.A., *Science*, 297, 1536, 2009. Reproduced with permission of AAAS.)

the SPR frequency, and the elastic collision between particles and photons gives rise to the Rayleigh scattering light without energy loss that can be efficiently collected using DFM. This attribute has attracted considerable attention for applications in biological sensing and single-molecule imaging. Moreover, the scattering light of plasmonic nanoparticles can be simply observed at the single-particle level using DFM, resulting in a new class of sensors with unprecedented detection limits and spatial resolution.

5.6.1 Applications

5.6.1.1 Plasmonic Sensing through Local Changes in Refractive Index

The plasmonic resonance of a metal nanoparticle is dependent on the local refractive index; the scattering spectrum will shift in response to a change in the local surface environment [77]. For example, Van Duyne and coworkers developed an LSPR nanosensor for Alzheimer's disease-associated amyloid b-derived diffusible ligands (ADDLs) [78]. In this work, nanoparticles were lithographically fabricated on a substrate and coupled to antibodies against ADDL, which pull down the antigen to the nanoparticle surface. Binding to ADDL alters the local index of refraction, resulting in a detectable shift in the LSPR spectrum. The adsorption of a second set of antibodies in a sandwich format can further shift the index of refraction and thereby amplify the detected signal. This method also enabled the observation of oligomerization

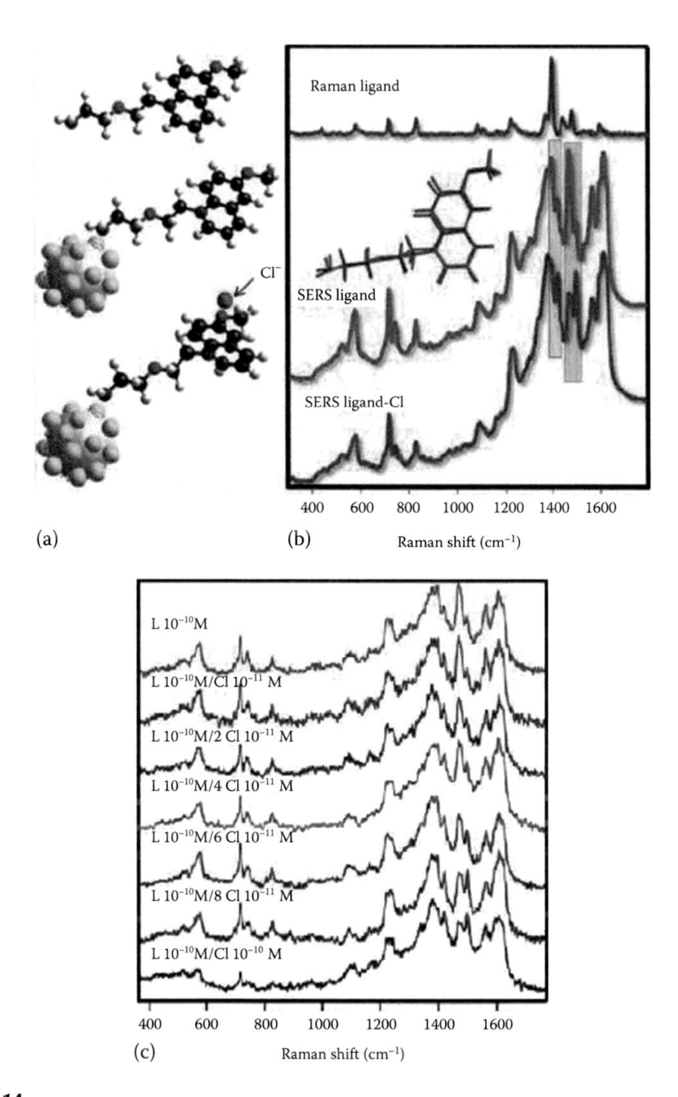

FIGURE 5.14
(a) Schematic representation of the detection process. (b) Raman and SERS spectra of amino-MQAE and SERS spectrum of amino-MQAE in the presence of an equimolar concentration of chloride (10^{-6} M). (c) SERS spectra of amino-MQAE (ligand L) 10^{-10} M with increasing concentrations of chloride (from 10^{-10} to 10^{-11} M). Samples were excited with a 532-nm laser line. (Reproduced with permission from Tsoutsi, D., Montenegro, J.M., Dommershausen, F., Koert, U., Liz-Marzan, L.M., Perak, W.J., and Alvarez-Puebla, R.A., *ACS Nano*, 5, 2011, 7539. Copyright 2011, American Chemical Society.)

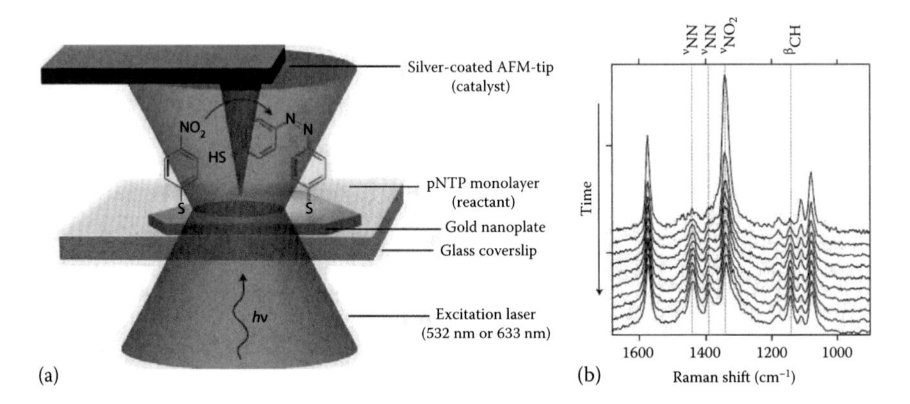

FIGURE 5.15
Schematic overview of the experimental setup (a). Time-dependent TERS measurements of the reduction of pNTP (top) to DMAB (bottom, spectrum mixed with that of pNTP) for a 4 s integration time, taken every 12 s at 532 nm, 24.5 mW (b). (Reproduced with permission from Macmillian Publishers Ltd. *Nat. Nanotechnol.*, van Schrojenstein Lantman, E.M., Deckert-Gaudig, T., Arjan, J.G., Mank, A.J.G., Deckert, V., and Weckhuysen, B.M., 7, 583, Copyright 2012.)

and fibrillogenesis of the antigens and antibodies at very low concentrations relevant to in vivo conditions (Figure 5.16).

5.6.1.2 Plasmonic Sensing through Interparticle Coupling

When multiple plasmonic nanoparticles are in close proximity, their plasmonic oscillations can couple together, resulting in an increase in light-scattering intensity as well as a spectral red shift. Alivisatos and coworkers developed a molecular ruler based on coupling between individual dimers of gold and silver nanoparticles to monitor the kinetics of single DNA hybridization events in real time [79]. Streptavidin-functionalized metal nanoparticles were first attached to a glass surface, and a second nanoparticle conjugated to biotin-functionalized single-stranded DNA (ssDNA) was then added. Binding between streptavidin and biotin resulted in nanoparticle dimerization, inducing plasmonic coupling evident from rapid changes in light scattering. Silver nanoparticles switched from blue to green while gold particles changed from green to orange. Coupling was found to have a much more dramatic effect on the resonance of silver particles compared to gold particles (Figure 5.17). Because plasmonic coupling is distance dependent, these nanoparticle dimers could be used to measure distance changes in a molecule.

5.6.1.3 Plasmon Resonance Energy Transfer

Lee and coworkers observed that nanoparticles coupled to chromophores can exhibit a phenomenon called plasmon resonance energy transfer (PRET) [80]. Nanoparticle plasmon resonance refers to the free-electron

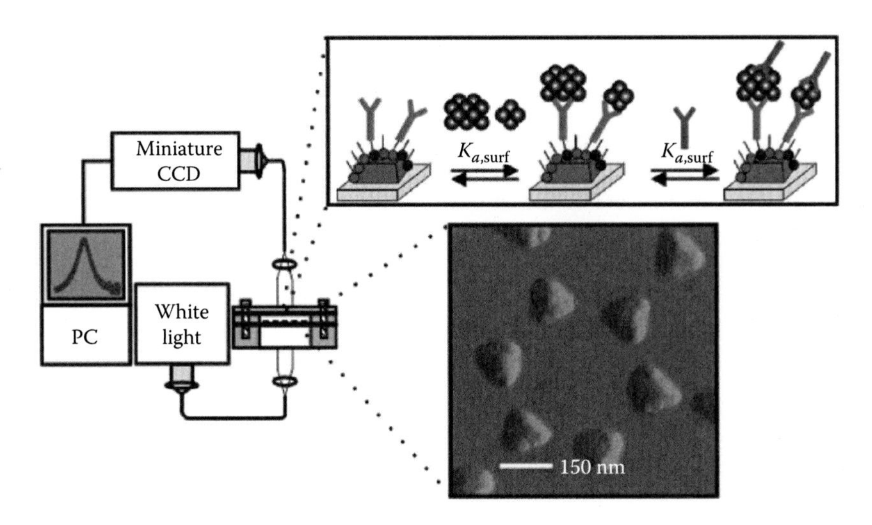

FIGURE 5.16
Design and experimental setup for a biosensor for the detection of amyloid b–derived diffusible ligands using an antibody-based sandwich assay. (Reproduced with permission from Dasary, S.S.R., Singh, A.K., Senapati, D., Yu, H.T., and Ray, P.C., *J. Am. Chem. Soc.*, 131, 2009, 13806. Copyright 2009, American Chemical Society.)

oscillation spatially confined within the physical boundary of metallic nanoparticles. It has been conjectured that the plasmon resonance energy can be transferred to chemical or biological molecules adsorbed on metallic nanostructures. Similar to the donor–acceptor energy matching in fluorescence resonance energy transfer (FRET) between two fluorophores, the PRET process requires that the plasmon resonance peaks of the metallic nanoparticle overlap with the electronic resonance peak positions of the molecule. The quantized energy is likely transferred through the dipole–dipole interaction between the resonating plasmon dipole in the nanoparticle and the molecular dipole. Previous work on the surface plasmon–mediated FRET process, SPR shift owing to redox molecules, and bulk optical extinction spectroscopy of nanoplasmonic particle clusters with conjugated resonant molecules also suggests the possibility of such dipole–dipole interactions (Figure 5.18).

Based on this method, Lee et al. developed a novel sensor for sensitive and selective detection of copper ions. In this scheme, gold nanoparticles were coated with specific ligands that form complexes with copper ion [81]. These were chosen such that the electronic absorption of the metal–ligand complex matched the plasmon resonance of the nanoparticles. This resulted in a quenching of nanoparticle scattering upon copper ion binding for concentrations as low as 1 nM (Figure 5.19).

Later, Xu et al. have shown that the PRET-based spectroscopy can also be utilized for sensitive and selective detection of TNT. The Meisenheimer

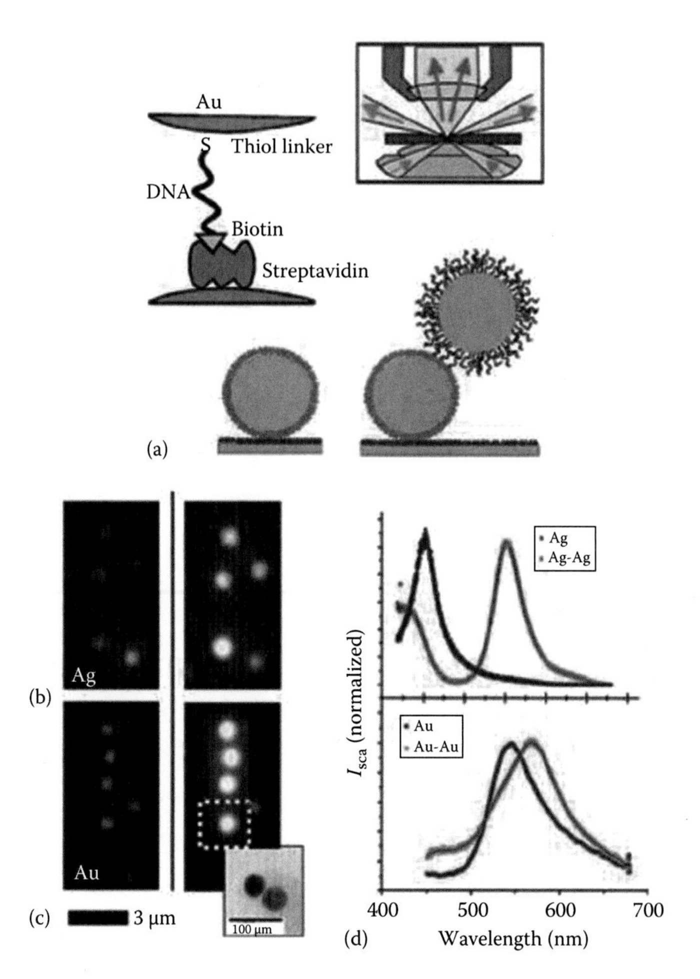

FIGURE 5.17
(See color insert.) Molecular ruler for DNA based on plasmonic coupling between metal nanoparticles. (a) First, nanoparticles functionalized with streptavidin are attached to the glass surface coated with BSA-biotin (left). Then, a second particle is attached to the first particle (center), again via biotinstreptavidin binding (right). Inset: principle of transmission dark-field microscopy. (b) Single silver particles (left) and particle pairs (right). (c) Single gold particles (left) and gold particle pairs (right). Inset: representative transmission electron microscopy image of a particle pair to show that each colored dot comes from light scatted from two closely lying particles, which cannot be separated optically. (d) Representative scattering spectra of single particles and particle pairs for silver (top) and gold (bottom). Silver particles show a larger spectral shift (102 nm) than gold particles (23 nm), stronger light scattering and a smaller plasmon line width. (Reproduced with permission from Macmillian Publishers Ltd. *Nat. Biotechnol.*, Sonnichsen, C., Reinhard, B.M., Liphardt, J., and Alivisatos, A.P., 23, 741, Copyright 2005.)

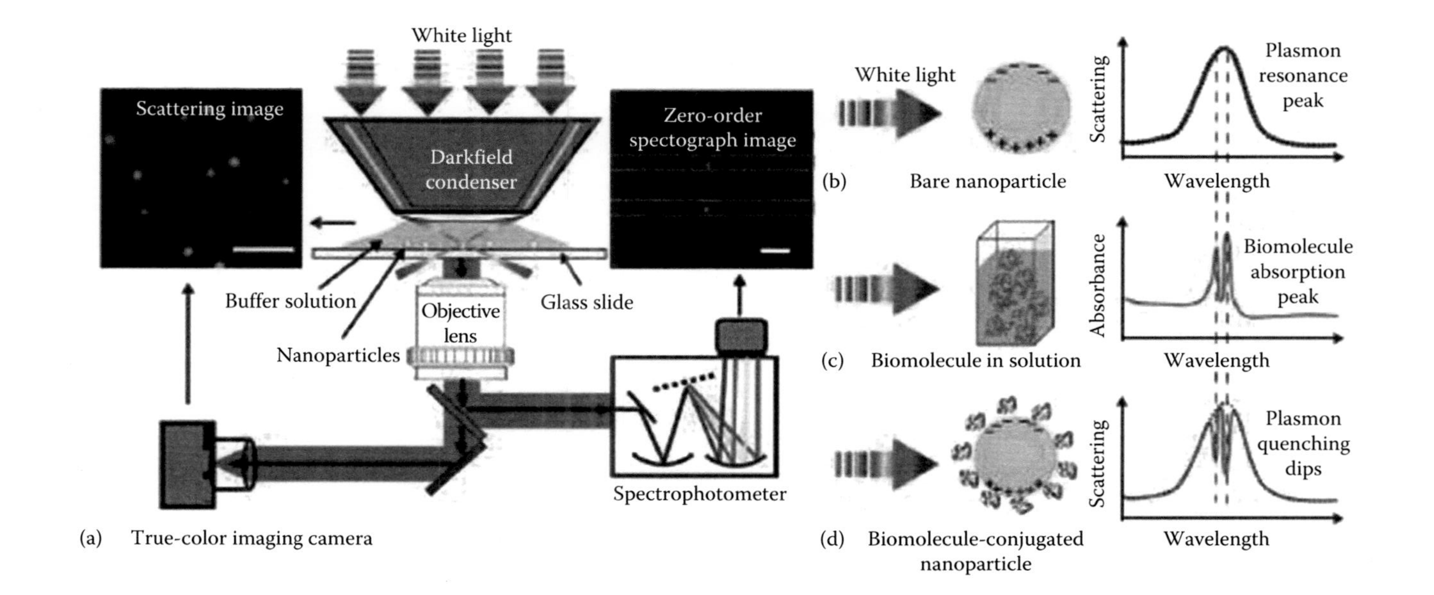

FIGURE 5.18
Schematic diagrams of quantized plasmon quenching dip nanospectroscopy via PRET. (a) Experimental system configuration. (b) A typical Rayleigh scattering spectrum of bare gold nanoparticles. (c) Typical absorption spectra of biomolecule bulk solution. (d) Typical quantized plasmon quenching dips in the Rayleigh scattering spectrum of biomolecule-conjugated gold nanoparticles. Spectra were drawn based on representative data. (Reproduced with permission from Macmillian Publishers Ltd. *Nat. Methods*, Liu, G.L., Long, Y.T., Choi, Y., Kang, T., and Lee, L.P., 4, 1015, Copyright 2007.)

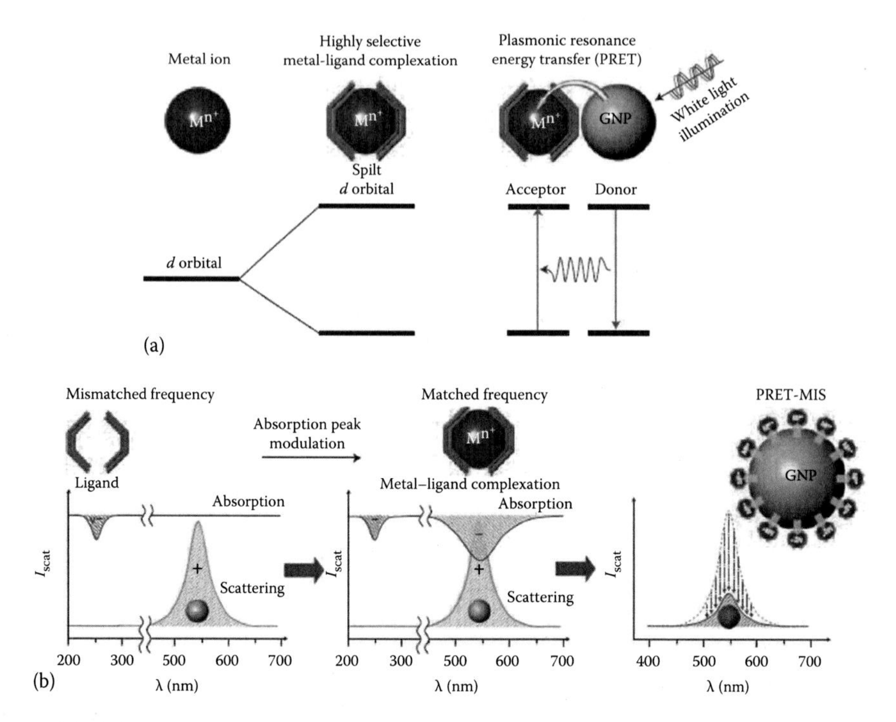

FIGURE 5.19
Plasmonic resonance energy transfer–based copper ion sensing. (a) When the transition metal ion binds with the matching ligand, *d* orbitals are split, which can generate a new absorption band of the metal–ligand complex in the visible range. Owing to the new absorption band, Rayleigh scattering energy from a gold nanoplasmonic probe can be transferred to the metal–ligand complex. (b) There is no spectral overlap between ligands without the metal ion and the gold nanoplasmonic probe. When the electronic absorption frequency of the metal–ligand complex matches with the Rayleigh scattering frequency, the selective energy transfer is induced by this spectral overlap (middle) and the distinguishable resonant quenching on the resonant Rayleigh scattering spectrum is observed (right). (Reproduced with permission from Macmillian Publishers Ltd. *Nat. Nanotechnol.*, Choi, Y., Park, Y., Kang, T., and Lee, L.P., 4, 742, Copyright 2009.)

complex formed from TNT and cysteine exhibits an absorption band centered at 530 nm, which is matched with the scattering peak of AuNPs. Accordingly, AuNP conjugated with cysteine ligands could be used as a TNT probe with high sensitivity and selectivity by providing the quantitative quenching information as a function of the local concentration of the D–A complex around a single nanoscale probe. As shown in Figure 5.14, without addition of TNT, the gold nanoparticle with green color can be clearly seen. After 4 min of exposure to TNT, the Rayleigh scattering exhibits a substantial decrease in intensity but no spectral shift, and the imaging signal in dark field by CCD also decreased (indicated by the white arrows). This phenomenon is more obvious, with the time of exposure to TNT extended to 8 min (Figure 5.3). The scattered light intensity of a single

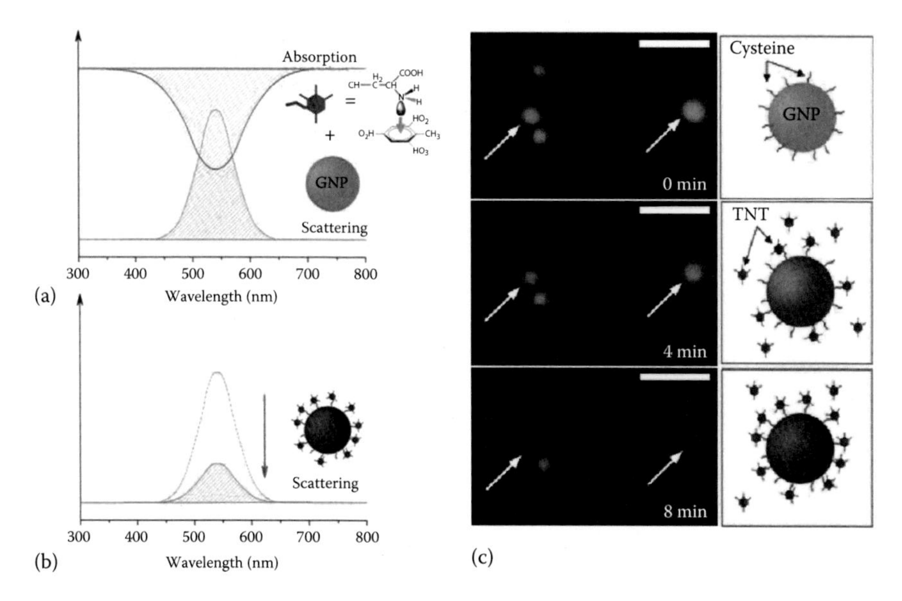

FIGURE 5.20
Schematic representation of TNT detection process based on plasmonic resonance energy transfer (PRET). (a) When the electronic absorption frequency of the donor–acceptor complex overlaps with the Rayleigh scattering frequency of GNP, the selective energy transfer occurs by this spectral overlap. (b) The remarkable resonant quenching on the Rayleigh scattering spectrum is clearly observed. (c) Typical time-dependent true color images of cysteine-modified gold nanoparticles after exposure to 1 mM TNT (left), the corresponding schematics (middle), and the corresponding Rayleigh scattering spectra (right). The white arrow points to a single GNP. The scale bars are 2 μm. (From Qu, W.G. et al., *Chem. Commun.*, 47, 1237, 2009.)

probe decreases rapidly (in ~8 min) to a constant value upon addition of TNT solution. Such a rapid spectral intensity decrease without spectral shift does not cause a local refractive index change adjacent to GNPs or direct optical absorption by the conjugated donor–acceptor complexes. Because direct optical absorption by the donor–acceptor complexes only accounts for less than 0.1% of the intensity loss of scattered light even under the assumption of 100% excitation efficiency (data not shown). The addition of TNT to pure GNPs solution does not lead to a noticeable change in the scattering spectra (Figure 5.20) [82].

5.7 LSPR-Controlled Fluorescence

Metal nanoparticles can also serve as excellent fluorescence quenchers due to their extraordinary high molar extinction coefficients and broad energy

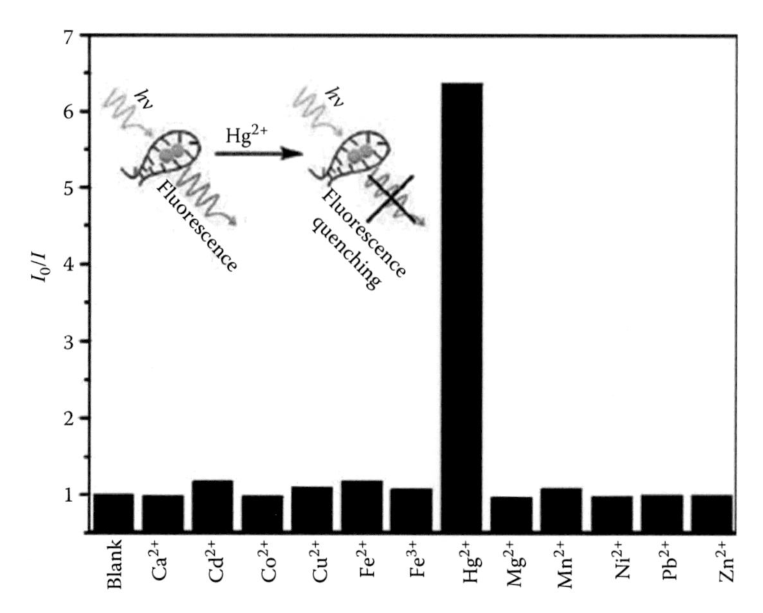

FIGURE 5.21
The "turn-off" fluorescent mechanism for Hg^{2+} sensing with high selectivity. (From Guo, W., Yuan, J., and Wang, E., *Chem. Commun.*, 45, 3395, 2009. Reproduced with permission of The Royal Society of Chemistry.)

bandwidth when the fluorophore is adsorbed directly onto the metal surface, which open rich opportunities for chemical and biological fluorescent sensors design. For example, Wang group used novel and environmentally friendly oligonucleotide-stabilized Ag NCs as fluorescent probes for the determination of Hg^{2+} ions with a low detection limit and high selectivity based on fluorescence quenching mechanism of metal NCs induced by target [83] (Figure 5.21).

Zhang et al. developed a multicolor gold nanoprobe for the simultaneous detection of three analytes—adenosine, potassium ion, and cocaine—which combine the highly specific binding abilities of aptamers with the ultrahigh quenching ability of AuNPs to fluorescence nearby fluorophore-labeled aptamers [84] (Figure 5.22).

However, compared with the fluorescence quencher, the surface plasmon–enhanced fluorescence (SEF) attracted more attention. Fluorescence can be enhanced by the strong local field near a metal surface due to the excitation of surface plasmons. The interactions between the molecule and metal surface alter the decay rates, leading to enhanced fluorescence and the possibility of using radiative decay engineering in designing new sensing systems. This is particularly important for bioanalytical applications since the intrinsic fluorescence of a biological molecule may be amplified so that fluorescent tags are not required.

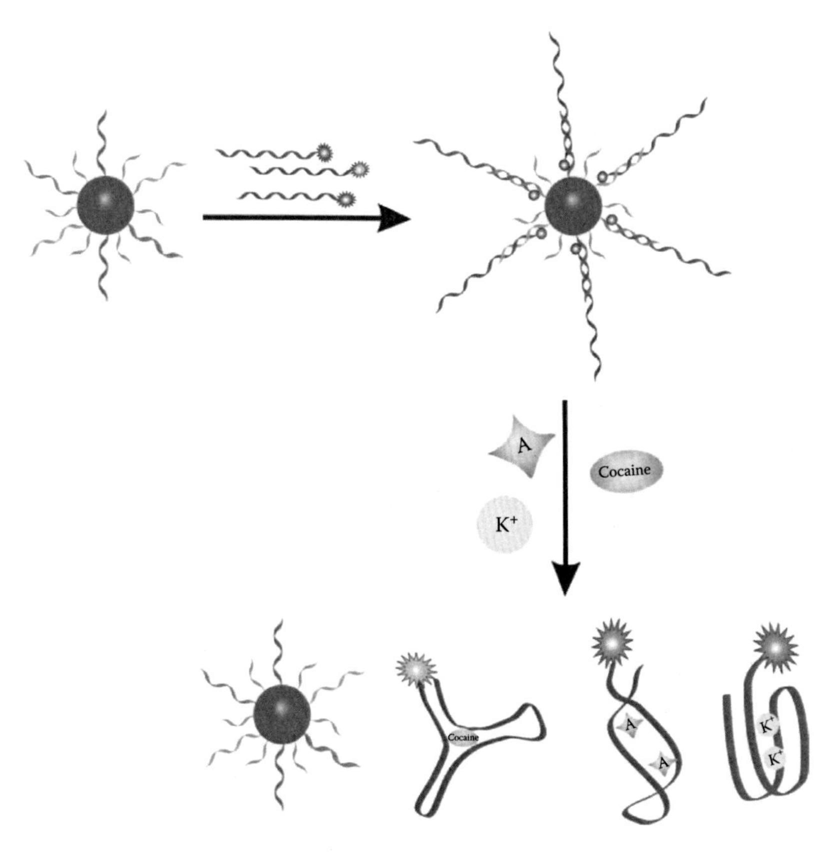

FIGURE 5.22
A multicolor aptamer-based gold nanoprobe for the multiplex detection of adenosine, potassium, and cocaine. (From Zhang, J., Wang, L., Zhang, H., Boey, F., Song, S., and Fan, C.: *Small*. 2010. 6. 201. Copyright Wiley-VCH Verlag GmbH & Co. KGaA. Reproduced with permission.)

Another important application for SEF is the surface plasmon nanolaser. Only SPRs are capable of squeezing optical frequency oscillations into a nanoscopic cavity to enable a true nanolaser. Mikhail Noginov, a physicist at Norfolk State, and his colleagues shine a conventional bluish-green laser beam into a suspension of nanoparticles, each with a gold core surrounded by a layer of sodium silicate and an outer silica shell containing dye molecules [85]. When the gold is excited by the laser photons, its surface starts to dance with collective oscillations of electrons, known as surface plasmons. The plasmons tickle the dye molecules to release more plasmons at the same wavelength, causing the device to emit green laser light (Figure 5.23). To the best of our knowledge, this is the smallest nanolaser reported to date and the first operating at visible wavelength.

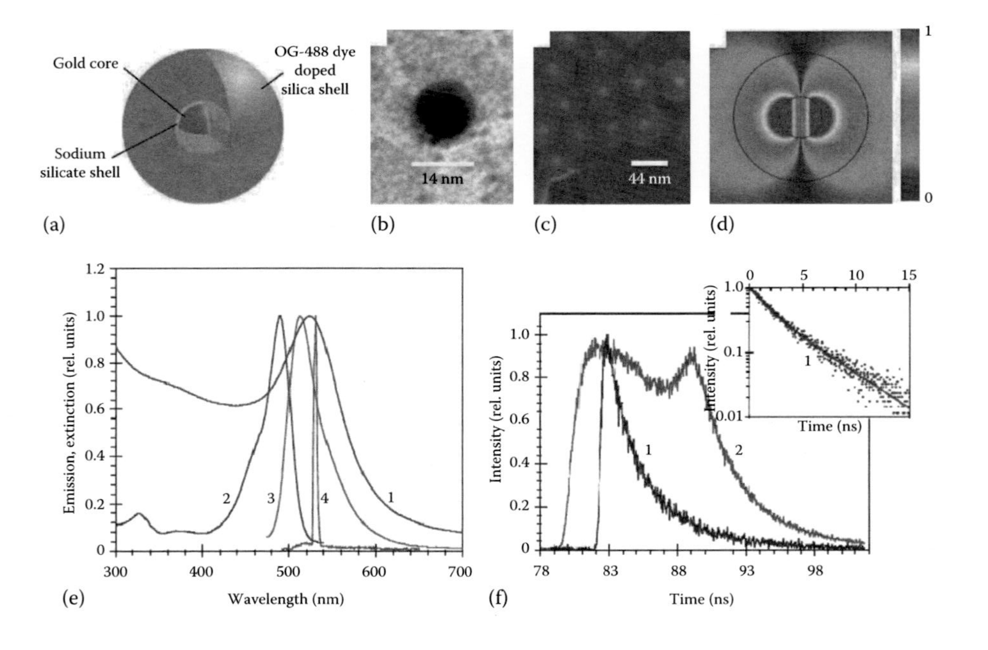

FIGURE 5.23
(See color insert.) (a) Diagram of the hybrid nanoparticle architecture (not to scale), indicating dye molecules throughout the silica shell. (b) Transmission electron microscope image of Au core. (c) Scanning electron microscope image of Au/silica/dye core–shell nanoparticles. (d) Spaser mode (in false color), with $\lambda = 525$ nm and $Q = 14.8$; the inner and the outer circles represent the 14-nm core and the 44-nm shell, respectively. (e) The field strength color scheme is shown on the right. Normalized extinction (1), excitation (2), spontaneous emission (3), and stimulated emission (4) spectra of Au/silica/dye nanoparticles. The peak extinction cross-section of the nanoparticles is 1.1×10^{-12} cm^2. (f) The emission and excitation spectra were measured in spectrofluorometer at low fluence. Main panel: emission kinetics detected at 480 nm (1) and 520 nm (2). Inset, trace 1 plotted in semilogarithmic coordinates (dots) and the corresponding fitting curve. The beginning of each emission kinetic trace coincides with the 90-ps pumping pulse. (Reproduced from Noginov, M.A., Zhu, G., Belgrave, A.M., Bakker, R., Shalaev, V.M., Narimanov, E.E., Stout, S., Herz, E., Suteewong, T., and Wiesner, U., *Nature*, 460, 1110–1113. Copyright 2009. With permission from Nature Progress Group.)

5.8 Conclusion

In this chapter, we offer a general overview of the LSPR spectroscopy. As this topic is widespread and growing, we cannot hope for our review to be comprehensive. However, we have endeavored to describe the major research efforts in the field and to review a wide and varied cross-section of the relevant literature. We first introduced the Mie theory (the analytic theory of LSPR), the physical origin of the LSPR itself and its dependence on the material properties of noble metals, and the surrounding refractive index. Second, we described the basics of nanoparticle-based LSPR sensing, including a description of single-particle and ensemble measurements, a comparison of scattering, absorption, and extinction. Then, we described in detail the theory, application, and the main challenge of four kinds of major LSPR spectroscopies including absorption spectroscopy, SERS, LSPR-based Rayleigh scattering, and LSPR-controlled fluorescence spectroscopy. The applications have shown that a variety of chemically and biologically relevant molecules can be detected—from protein and oligonucleotide to the direct detection of metal ion and anion. In addition, the results of these studies demonstrate the necessity of fundamental spectroscopic studies for guiding experimental design to achieve the largest overall signal or to observe signals at all. As our understanding of the plasmonic response of nanoparticles continues to grow, these LSPR-based sensing experiments will improve as well, leading to higher sensitivity, faster and more reversible responses, and an ever-broadening scope of applicability. The LSPR spectroscopy has generated an exponential increase in their use in chemistry and as our understanding of the plasmonic response of nanoparticles continues to grow, the LSPR-based spectroscopy will improve as well, leading to higher sensitivity and faster and more reversible responses. It will continue to revolutionize the field of spectroscopy for years to come.

Acknowledgments

This work was supported by the National Basic Research Program of China (2011CB933700, 2010CB934700), the 100 Talents program of the Chinese Academy of Sciences, and the National Natural Science Foundation of China (21271165).

References

1. Saha, K., Agasti, S. S., Kim, C., Li, X. N., Rotello, V. M. *Chem. Rev.* **2012**, *112*, 2739.
2. Willet, K. A., Van Duyne, R. P. *Annu. Rev. Phys. Chem.* **2007**, *58*, 267.

3. McFarland A. D., Van Duyne, R. P. *Nano Lett.* **2003**, *3*, 1057.
4. Rosi, N. L., Mirkin, C. A. *Chem. Rev.* **2005**, *105*, 1547–1562.
5. Choi, Y., Kang, T., Lee, L. P. *Nano Lett.* **2009**, *9*, 85–90.
6. Nie, S. M., Emory, S. R. *Science* **1997**, *275*, 1102–1106.
7. Morton, S. M., Silverstein, D. W., Jensen, L., *Chem. Rev.* **2011**, *111*, 3962.
8. Camden, J. P., Dieringer, J. A., Zhao, J., Van Duyne, R. P. *Acc. Chem. Res.* **2008**, *41*, 1653.
9. Moskovits, M. *Rev. Mod. Phys.* **1985**, *57*, 783.
10. Alvarez-Puebla, R. A., Liz-Marzan, L. M., *Chem. Soc. Rev.* **2012**, *41*, 43.
11. Kneipp, K., Yang, W., Kneipp, H., Perelman, L. T., Itzkan, I., Dasari, R. R., Feld, M. S. *Phys. Rev. Lett.* **1997**, *78*, 1667.
12. Jain, P. K., Huang, X., El-Sayed, I. H., El-Sayed, M. A. *Acc. Chem. Res.* **2008**, *41*, 1578.
13. Kawata, S., Inouye, Y., Verma, P. *Nat. Photon.* **2009**, *3*, 388–394.
14. Lal, S., Link, S., Halas, N. J. *Nat. Photon.* **2007**, *1*, 641.
15. Hutter, E., Fendler, J. H. *Adv. Mater.* **2004**, *16*, 1685.
16. Van Duyne, R. P. *Science* **2004**, *306*, 985.
17. Brockman, J. M., Nelson, B. P., Corn, R. M. *Annu. Rev. Phys. Chem.* **2000**, *51*, 41.
18. Yan, R., Gargas, D., Yang, P. *Nat. Photon.* **2009**, *3*, 569.
19. Sanders, A. W., Routenberg, D. A., Wiley, B. J., Xia, Y., Dufresne, E. R., Reed, M. A. *Nano Lett.* **2006**, *6*, 1822.
20. Kelly, K. L., Coronado, E., Zhao, L. L., Schatz, G. C. *J. Phys. Chem. B* **2003**, *107*, 668–677.
21. Xia, Y. N., Xiong, Y. J., Lim, B., Skrabalak, S. E. *Angew. Chem. Int. Ed.* **2009**, *48*, 60–103.
22. Ni, W. H., Ambjörnsson, T., Apell, S. P., Chen, H. J., Wang, J. F. *Nano Lett.* **2010**, *10*, 77–84.
23. Wurtz, G. A., Evans, P. R., Hendren, W., Atkinson, R., Dickson, W., Pollard, R. J., Zayats, A. V., Harrison, W., Bower, C. *Nano Lett.* **2007**, *7*, 1297–1303.
24. Zheng, Y. B., Juluri, B. K., Jensen, L. L., Ahmed, D., Lu, M. Q., Jensen, L., Huang, T. J. *Adv. Mater.* **2010**, *22*, 3603–3607.
25. Mie, G. *Ann. Phys.* (Weinheim, Ger.) **1908**, *25*, 377.
26. Kerker, M. *The Scattering of Light and Other Electromagnetic Radiation*, Academic: New York, **1969**.
27. Bohren, C. F., Huffman, D. R. *Absorption and Scattering of Light by Small Particles*, Wiley Interscience: New York, **1983**.
28. Srivastava, S., Frankamp, B. L., Rotello, V. M. *Chem. Mater.* **2005**, *17*, 487.
29. Liu, R. R., Liew, R. S., Zhou, H., Xing, B. G. *Angew. Chem. Int. Ed.* **2007**, *46*, 8799.
30. Jiang, Y., Zhao, H., Lin, Y. Q., Zhu, N. N., Ma, Y. R., Mao, L. Q. *Angew. Chem. Int. Ed.* **2010**, *49*, 4800.
31. Kim, Y. J., Johnson, R. C., Hupp, J. T. *Nano Lett.* **2001**, *1*, 165.
32. Liu, J. W., Lu, Y. *J. Am. Chem. Soc.* **2003**, *125*, 6642.
33. Kim, M. H., Kim, S., Jang, H. H., Yi, S., Seo, S. H., Han, M. S. *Tetrahedron Lett.* **2010**, *51*, 4712.
34. Daniel, W. L., Han, M. S., Lee, J. S., Mirkin, C. A. *J. Am. Chem. Soc.* **2009**, *131*, 6362.
35. Zhang, Y. F., Li, B. X., Xu, C. L. *Analyst* **2010**, *135*, 1579.
36. Lee, K. Y., Kim, D. W., Heo, J., Kim, J. S., Yang, J. K., Cheong, G. W., Han, S. W. *Bull. Korean Chem. Soc.* **2006**, *27*, 2081.

37. Smith, K. D., McCord, B. R., MacCrehan, W. A., Mount, K., Rowe, W. F. *J. Forensic Sci.* **1999**, *44*, 789.
38. Charles, P. T., Dingle, B. M., Van Bergen, S., Gauger, P. R., Patterson, C. H., Kusterbeck, A. W. *Field Anal. Chem. Technol.* **2001**, *5*, 272.
39. Jiang, Y., Zhao, H., Zhu, N. N., Lin, Y. Q., Yu, P., Mao, L. Q. *Angew. Chem. Int. Ed.* **2008**, *47*, 8601.
40. Liang, X. S., Wei, H. P., Cui, Z. Q., Deng, J. Y., Zhang, Z. P., You, X. Y., Zhang, X. E. *Analyst* **2011**, *136*, 179.
41. Schofield, C. L., Haines, A. H., Field, R. A., Russell, D. A. *Langmuir* **2006**, *22*, 6707.
42. Narain, R., Housni, A., Gody, G., Boullanger, P., Charreyre, M. T., Delair, T. *Langmuir* **2007**, *23*, 12835.
43. Watanabe, S., Yoshida, K., Shinkawa, K., Kumagawa, D., Seguchi, H. *Colloids Surf. B* **2010**, *81*, 570.
44. Xu, X. Y., Han, M. S., Mirkin, C. A. *Angew. Chem. Int. Ed.* **2007**, *46*, 3468.
45. Kneipp, K., Kneipp, H., Itzkan, I., Dasari, R. R., Feld, M. S. *Chem. Rev.* **1999**, *99*, 2957.
46. Anker, J. N., Hall, W. P., Lyandres, O., Shah, N. C., Zhao, J., Van Duyne, R. P. *Nat. Mater.* **2008**, *7*, 442.
47. Prodan, E., Radloff, C., Halas, N. J., Nordlander, P. *Science* **2003**, *302*, 419.
48. Wang, H., Brandl, D. W., Nordlander, P., Halas, N. J. *Acc. Chem. Res.* **2007**, *40*, 53.
49. Birke, R. L., Lombardi, J. R., Gersten, J. I. *Phys. Rev. Lett.* **1979**, *43*, 71.
50. Moskovits, M. *J. Chem. Phys.* **1978**, *69*, 4159.
51. Schatz, G. C. *Acc. Chem. Res.* **1984**, *17*, 370.
52. Lombardi, J. R., Birke, R. L., Lu, T., Xu, J. *J. Chem. Phys.* **1986**, *84*, 4174.
53. Jensen, L., Aikens, C. M., Schatz, G. C. *Chem. Soc. Rev.* **2008**, *37*, 1061.
54. Hao, E., Schatz, G. C. *J. Chem. Phys.* **2004,** *120*, 357.
55. Silverstein, D. W., Jensen, L. *J. Chem. Phys.* **2010**, *132*, No.194302.
56. Fromm, D. P., Sundaramurthy, A., Kinkhabwala, A., Schuck, P. J., Kino, G. S., Moerne, W. E. *J. Chem. Phys.* **2006**, *124*, No. 061101.
57. Hildebrandt, P., Stockburger, M. *J. Phys. Chem.* **1984,** *88*, 5935.
58. Pettinger, B., Krischer, K., Ertl, G. *Chem. Phys. Lett.* **1988**, *151*, 151.
59. Jensen, L., Zhao, L. L., Autschbach, J., Schatz, G. C. *J. Chem. Phys.* **2005**, *123*, No. 174110.
60. Liang, E., Kiefer, W. *J. Raman Spectrosc.* **1996**, *27*, 879.
61. Kim, N.-J., Lin, M., Hu, Z., Li, H. *Chem. Commun.* **2009**, 6246.
62. Maitani, M. M., Ohlberg, D. A. A., Li, Z. Y., Allara, D. L., Stewart, D. R., Williams, R. S. *J. Am. Chem. Soc.* **2009**, *131*, 6310.
63. Domke, K. F., Zhang, D., Pettinger, B. *J. Phys. Chem. C* **2007**, *111*, 8611.
64. Michaels, A. M., Nirmal, M., Brus, L. E. *J. Am. Chem. Soc.* **1999**, *121*, 9932.
65. Le Ru, E. C., Meyer, M., Etchegoin, P. G. *J. Phys. Chem. B* **2006**, *110*, 1944.
66. Michaels, A. M., Jiang, J., Brus, L. *J. Phys. Chem. B* **2000**, *104*, 11965.
67. Xu, H. X., Aizpurua, J., Kall, M., Apell, P. *Phys. Rev. E* **2000**, *62*, 4318.
68. Le Ru, E. C., Blackie, E., Meyer, M., Etchegoin, P. G. *J. Phys. Chem. C* **2007**, *111*, 13794.
69. Hildebrandt, P., Stockburger, M. *J. Phys. Chem.* **1984**, *88*, 5935.
70. Bailo, E., Deckert, V. *Chem. Soc. Rev.* **2008**, *37*, 921.
71. Stockle, R. M., Suh, Y. D., Deckert, V., Zenobi, R. *Chem. Phys. Lett.* **2000**, *318*, 131.
72. Dasary, S. S. R., Singh, A. K., Senapati, D., Yu, H. T., Ray, P. C. *J. Am. Chem. Soc.* **2009**, *131*, 13806.

73. Cao, Y. C., Jin, R., Mirkin, C. A. *Science* **2002**, *297*, 1536.
74. Tsoutsi, D., Montenegro, J. M., Dommershausen, F., Koert, U., Liz-Marzan, L. M., Perak, W. J., Alvarez-Puebla, R. A. *ACS Nano* **2011**, 5, 7539.
75. van Schrojenstein Lantman, E. M., Deckert-Gaudig, T., Arjan, J. G., Mank, A. J. G., Deckert, V., Weckhuysen, B. M. *Nat. Nanotechnol.* **2012**, *7*, 583.
76. Li, Y., Jing, C., Zhang, J., Long, Y. T. *Chem. Soc. Rev.* **2012**, *41*, 632.
77. McFarland, A. D., Van Duyne, R. P. *Nano Lett.* **2003**, *3*, 1057.
78. Haes, A. J., Chang, L., Klein, W. L., Van Duyne, R. P. *J. Am. Chem. Soc.* **2005**, *127*, 2264.
79. Sonnichsen, C., Reinhard, B. M., Liphardt, J., Alivisatos, A. P. *Nat. Biotechnol.* **2005**, *23*, 741.
80. Liu, G. L., Long, Y. T., Choi, Y., Kang, T., Lee, L. P. *Nat. Methods* **2007**, *4*, 1015.
81. Choi, Y., Park, Y., Kang, T., Lee, L. P. *Nat. Nanotechnol.* **2009**, *4*, 742.
82. Qu, W. G., Deng, B., Zhong, S. L., Shi, H. Y., Wang, S. S., Xu, A. W. *Chem. Commun.* **2009**, *47*, 1237.
83. Guo, W., Yuan, J., Wang, E. *Chem. Commun.* **2009**, *45*, 3395.
84. Zhang, J., Wang, L., Zhang, H., Boey, F. Song, S. Fan, C. *Small* **2010**, *6*, 201.
85. Noginov, M. A., Zhu, G., Belgrave, A. M., Bakker, R., Shalaev, V. M., Narimanov, E. E., Stout, S., Herz, E., Suteewong, T., Wiesner, U. *Nature* **2009**, *460*, 1110–1113.

6

Nanostructure Evaluation of Ionic Liquid Aggregates by Spectroscopy

Clarissa P. Frizzo, Aniele Z. Tier, Izabelle M. Gindri, Lilian Buriol, Marcos A. Villetti, Nilo Zanatta, and Marcos A.P. Martins

CONTENTS

6.1 Introduction

Ionic liquids (ILs) are currently defined as liquid organic salts at or close to room temperature (melting point <100°C). They have attracted the attention of the scientific community worldwide due to their physical and chemical properties such as wide liquid range, high conductivity, and good miscibility and/or solubility with organic and inorganic materials and polar, apolar, and nonpolar solvents. These properties have allowed ILs to be used in chemical reactions, electrochemistry, separations, and material synthesis (Martins et al. 2008a,b; Feng et al. 2010). Several properties of ILs such as density, viscosity, polarity, and conductivity can be drastically changed by the presence of small amounts of other substances, and this may lead to significant modifications in, for example, the rate and selectivity of chemical reactions performed in their midst (Bernardes et al. 2011). ILs in a solid state or in solution form aggregates such as clusters, micelles, or particles with long-range order similar to that of surfactants. The aggregation behavior of ILs in aqueous solutions has been shown to be regulated by the alkyl chain length, cationic structure, and anionic type of the IL and by the addition of some inorganic salts (Feng et al. 2010). Aqueous solutions of ILs have become the subject of a variety of potential applications, and this has motivated a significant number of fundamental studies of a theoretical and experimental nature (Bernardes et al. 2011). Aggregates such as micelles, liquid crystals, and microemulsions formed in ILs could open new research lines toward micellar catalysis in IL media, for example (Li et al. 2008). Dicationic ILs consist of a doubly charged cation composed of two singly charged cations linked by an alkyl chain (also called a spacer). Consequently, such ILs have more than one polar and nonpolar region (Bhargava and Klein 2011). Dicationic ILs can be used as solvents and lubricants at high temperatures (Bhargava and Klein 2011). They also find applications in analytical chemistry—particularly in electrospray ionization mass spectrometry (ESI-MS), in the detection of small quantities of anions via gas-phase ion association, and as the stationary phase in gas chromatography columns (Bhargava and Klein 2011). This group of ILs has received very little attention from researchers compared to the monocationic ILs. There are very few experimental studies on these compounds. Computational studies on dicationic ILs are limited to the electronic structure calculations in the gas phase (Bhargava and Klein 2011). In addition, compared with the intensive study on pure ILs, research on ILs in organic solvent is only at an early stage, and the fundamental information about the structural and physicochemical properties has yet to be investigated. Studies on IL groups are of great importance from both an academic and technological point of view (Li et al. 2006; Feng et al. 2010).

Therefore, in this chapter, we will show the importance of studying the organization of pure ILs and ILs in solution by reporting the main results obtained from spectroscopic methods in this research area up until now (monocationic ILs) and the perspectives of these investigations (dicationic ILs). Accordingly, the purpose of this chapter denotes a great deal of material to be covered, and it was necessary to make some limitations to the scope: (1) Only ILs with heterocyclic cations were considered; (2) only papers that related studies of IL aggregates in pure form and in solution with water and other organic solvents and other ILs were considered; (3) only spectrometry methods, namely, mass spectrometry (MS), nuclear magnetic resonance (NMR), light scattering (LS), or photon correlation spectroscopy (PCS) that we consider the most important for IL aggregate detection, would be discussed.

Table 6.1 depicts the papers collected for this chapter. The papers were selected from a search in *Science Finder* using the keywords "ionic liquid aggregates." From these papers, we selected those that are in the scope of this chapter and excluded papers that related only to theoretical studies. Table 6.1 also depicts the spectroscopic methods used in the studies of IL aggregates.

TABLE 6.1

Number of Papers Found at *Science Finder* Using the Keywords "Ionic Liquid Aggregates"

Total papers found	297		
Selected papers	50		
Experimental data papers	40		
Theoretical data papers	11		
Spectroscopic Methods		**Nonspectroscopic Methods**	
DLS	7	Transmission electron microscopy	4
NMR	10	Surface tensiometry	10
Fluorescence	10	Wide- and small-angle x-ray scattering	5
Polarized optical microscopy	3	Other nonspectroscopic methods used to determine the behavior of IL aggregates were cyclic voltammetry, DSC, potentiometry, ultrasound, and speed of sound	
Infrared	3		
UV–vis	2		
Conductivity	14		
X-rays	2		
MS	7		

6.2 Spectroscopic Techniques for Studying Aggregates

Due to their inherent amphiphilicity, ILs form aggregates in solution. The formation of aggregates in amphiphilic molecule solution is expected; however, the nature of the aggregates formed in this solution is rather interesting. The strength of electrostatic and van der Waals interactions in these systems determines which aggregation will be formed. The amphiphilic molecular structure of ILs results in an interesting self-organization behavior when ILs are mixed with water (or organic solvents) at a low or moderate concentration, known as the critical concentration. It is important to note that the critical concentration is defined as the minimum concentration of molecules at which intermolecular hydrogen bonding, micelles, or other aggregates start forming (Wennerström and Lindman 1972). Thus, critical concentrations include the critical hydrogen-bonding concentration (CHC), the critical micelle concentration (CMC), and the critical aggregation concentration (CAC) (Wennerström and Lindman 1972). The CMC is defined as the concentration of amphiphilic molecules above which micelles form spontaneously (Wennerström and Lindman 1972). However, CMC is not a constant that can be affected by many variables. It decreases when increasing the hydrophobic organic chain length of the nonpolar groups, and for ionic amphiphilic molecules, it also depends on the nature and concentration of counterions in solution (Wennerström and Lindman 1972). Since physical properties show different concentration dependences before and after the CMC/CAC, IL aggregate formation in solution can be observed from various experimental methods such as surface tension, LS, fluorescence, heat capacity, conductivity, and chemical shift measurements in NMR spectrometry (Inoue et al. 2007; Blesic et al. 2008; Luczak et al. 2008). Thus, the determination of the CMC is the focus of many experimental studies. The availability of multiple techniques for the same system significantly improves the certainty of the measured CMC value.

Since the spectrometric methods of MS, NMR, LS, and PCS will be discussed in this chapter, we will present a brief introduction about spectrometric principles for each method; how each method can detect the formation of IL aggregates; information that each technique furnishes about aggregate such as size, morphology, aggregation number (N_{agg}), and CMC; and how the graphical results of each method were used to solve a gemini IL performed in our laboratories. Finally, we will show some examples of IL aggregate studies found in the papers selected for this chapter.

6.2.1 Mass Spectrometry Study of IL Aggregation

The ESI-MS is a hyphenated technique applied to the analysis of samples formed by nonvolatile molecules from liquid phase. By using ESI-MS, a sample is impelled through a capillary at a flow rate of 0.1–10 mL/min (Iribarne et al. 1983; Fenn 1993; Cech et al. 2001; Harris 2007), thus forming a spray from the sample. During the nebulization, a high voltage (2–5 kV) coaxial flow of nitrogen gas is applied to the capillary. The applied voltage provides the electric field gradient required to produce charge separation at the surface of the liquid. The charge separation causes the projection of the liquid from the capillary tip and forms what is known as a Taylor cone (Figure 6.1) (Iribarne et al. 1983; Fenn 1993; Cech et al. 2001; Harris 2007). At a certain moment, the solution that covers the Taylor cone reaches the point at which

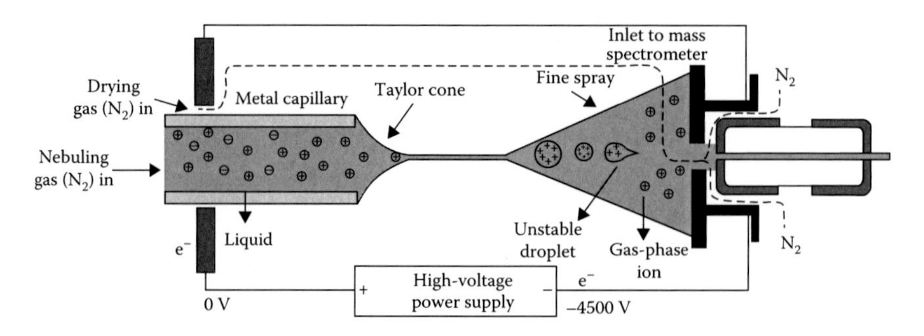

FIGURE 6.1
Schematic representation of the electrospray ionization process.

Coulomb repulsion of the surface charge is equal to the surface tension of the solution (Rayleigh limit). Then, the droplets that contain a charge excess disconnect from their tip (Taflin et al. 1989). These droplets move toward the entrance to the mass spectrometer and generate charged molecules (ions) (Kebarle and Peschke 2000). The amount of charge on the droplets is equal to the amount of charge separation. This charge is sometimes called the excess charge (Enke 2003) to differentiate it from the cations and anions in the droplet that are neutralized by counterions. It is important to distinguish between these excesses and neutralized charges because the neutralized charges are not likely to result in gas-phase ions. The formation of gas-phase ions from neutralized charge would require desolvation energies in excess of the Coulombic forces between the ions and their counterions. Consequently, the maximum rate of vapor phase ion production is equal to the rate of charge separation, and the number of vapor phase ions produced cannot exceed the amount of excess charge introduced into the droplets.

The ESI-MS is a soft ionization technique, which is often used for the analysis of macromolecules, supramolecules, and aggregates because it overcomes the propensity of these systems to fragment during ionization. It also has the capability of transferring liquid-phase noncovalent complexes into the gas phase by minimizing the disruption of the noncovalent interaction (Pramanik et al. 1998; Bordini and Hamdan 1999; Wendt et al. 2003; Vriendt et al. 2004; Anichina and Bohme 2009). Due to this soft ionization method being relatively weak, inter- and intramolecular noncovalent interactions of supramolecules and aggregates can be preserved intact; in other words, they are transported from solution phase to gas phase. Therefore, detailed information on the noncovalent gas-phase molecular complexes from the multiply charged ion peaks generated by ESI-MS represents a promising strategy to probe the intrinsic (Jecklin et al. 2008) properties of aggregates (Wendt et al. 2003) in a solvent-free environment.

Ganem et al. (1991) were the first to demonstrate that noncovalent complexes could be studied by ESI-MS. Most of the gas-phase studies in recent years have focused primarily on protein–ligand complexes and IL aggregates by ionic interaction or hydrogen bonding in solution (Kitova et al. 2004; Sun et al. 2007; Wyttenbac and Bowers 2007; Kitova et al. 2008). These studies have yielded compelling evidence indicating that the solution structures are preserved upon the transfer of the interactions from solution to gas phase (Kitova et al. 2008). Since ILs are totally composed of ions, they may be detected by MS without undergoing further ionization—only the removal of the solvent and separation of the ions in the gas phase are required. The IL aggregates are observed in the ESI-MS spectra by the identification of their molecular weight between fragments. In this manner, the monomer $[C_1A_1]$ is the mass/charge fragment that has a molecule mass that corresponds to ILs formed by a cation $[C]$ and anion $[A]$ in a monocationic IL. In the case of dicationic ILs, the monomer $[C_1A_2]$ is formed by one cation $[C_1]$ and two anions $[A_2]$. Other combinations of cations and anions can be verified in the

ESI-MS spectra, for example, $[C_{(n+1)}A_n]^+$, $[C_{(n+2)}A_n]^{2+}$ in the positive mode and $[C_nA_{(n+1)}]^-$, $[C_nA_{(n+2)}]^{2-}$ in the negative mode in a monocationic IL and $[C_nA_n]^+$, $[C_{(n+1)}A_{(n+1)}]^2$ in the positive mode and $[C_nA_{(n+2)}]^-$, $[C_nA_{(n+3)}]^{2-}$ in the negative mode in a dicationic IL. Typical spectra of the IL aggregates detected by ESI-MS are shown in Figure 6.2.

Cluster abundance distributions are generally considered a direct reflection of their relative stabilities (Echt et al. 1981; Muhlbach et al. 1982; Friedman and Beuhler 1983; Coolbaugh and Garvey 1992). From the ESI-MS spectra with intensity distribution of ionic supramolecules in almost logarithmically decreasing intensities, it is possible to identify the "magic number." The specific aggregate sizes at which the anomalous abundances occur have been termed "magic numbers," which represent clusters with special structural stability (Echt et al. 1981; Muhlbach et al. 1982; Friedman et al. 1983; Coolbaugh and Garvey 1992).

In their pioneering work, Gozzo et al. (2004) proposed the application of Cooks kinetic method (CKM) to data obtained from ESI-MS of IL aggregates.

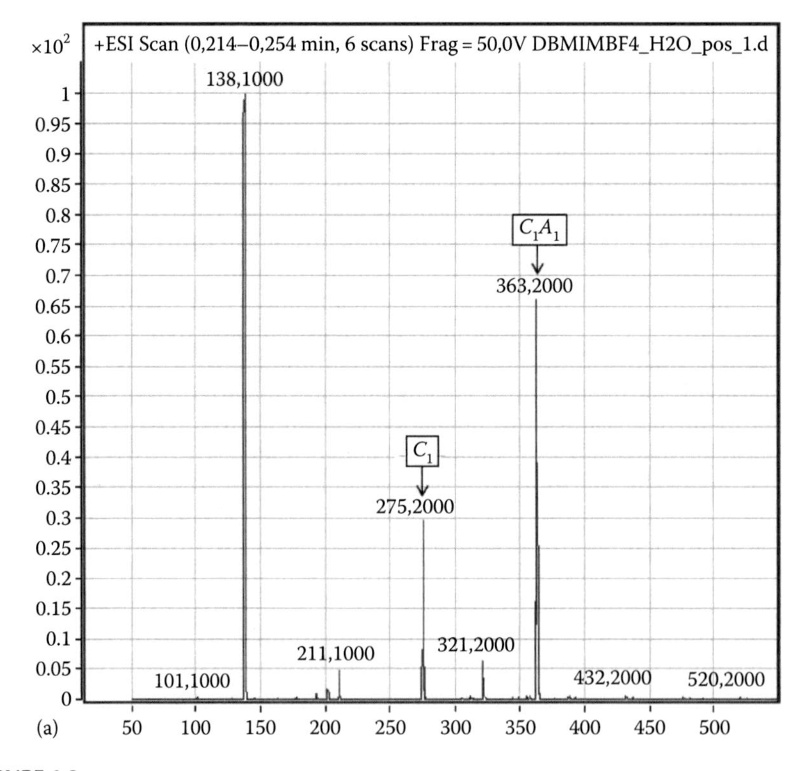

FIGURE 6.2
ESI-MS spectra of the [BisOctMIM][2BF$_4$] in water. (a) Positive-ion mode of the IL [BisOctMIM][2BF$_4$] in water. Aggregates are [C_1] (m/z 275) and [C_1A_1] (m/z 363). Experimental details are given at the end of Section 6.2.1.

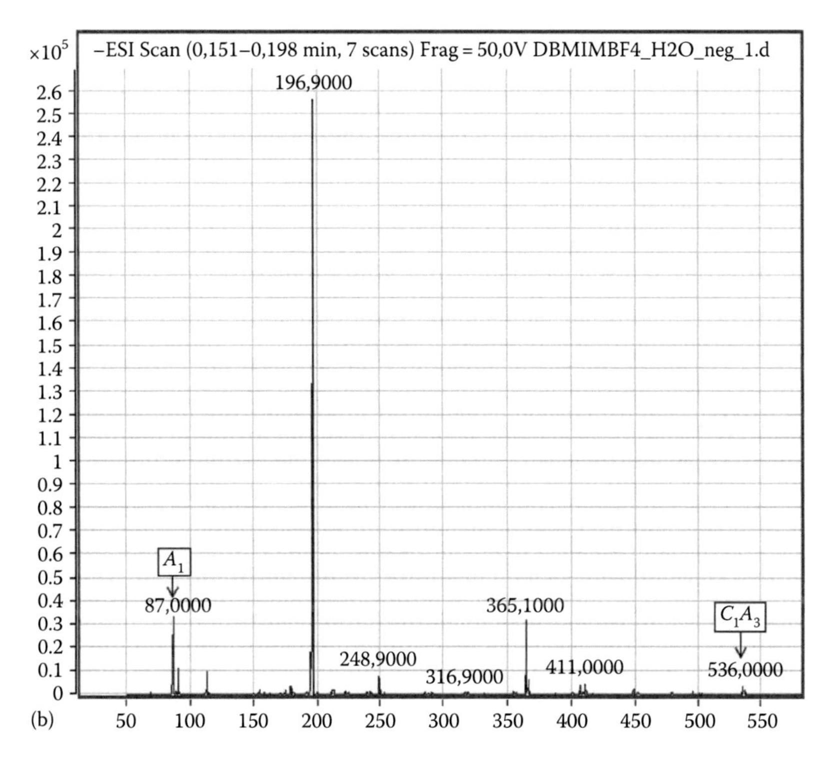

FIGURE 6.2 (continued)
ESI-MS spectra of the [BisOctMIM][2BF_4] in water. (b) Negative-ion mode. Aggregates are [C_1A_2] (m/z 448.9) and [C_1A_3] (m/z 536). Other peaks are associated with a molecular fragment. Experimental details are given at the end of Section 6.2.1.

The kinetic method proposed by Cooks et al. (1994) has been used to determine thermochemical properties based on the rates of competitive dissociations of gaseous mass-selected ionic supramolecules as measured via MS experiments. Because of the ease of use, broad applicability, high sensitivity to small thermochemical differences (typically as small as 0.1 kcal mol^{-1}), and high precision, CKM has found a multitude of applications (Eberlin 1994; Augusti et al. 2002; Wu et al. 2004). In its simplest form, the method relies on the following major assumptions: (1) negligible differences in the entropy requirements for the competitive channels, (2) negligible reverse activation energies, and (3) the absence of isomeric forms of the activated cluster ion. When these conditions are well satisfied, the ratio of the fragment ion abundance, for example, a proton-bound dimer [A_1–H–A_2] as described in Equation 6.1 in which k_1 and k_2 are the rate constants for the competitive dissociations, is related to the proton affinity difference $\Delta(PA)$ of the two anions (A_1 and A_2) by Equation 6.2. T_{eff} is the effective temperature, a thermodynamic quantity (Ervin 2002) apparently related to the internal energy of the dissociating ions, and Δ (Δ*S*) is the reaction entropy difference between the two fragmentation channels. A_1 and A_2 represent the anions (bases) involved in the interactions:

$$A_1H^+ + A_2 \leftarrow k_1 \quad A_1 \cdots H^+ \cdots A_2 \rightarrow k_2 A_1 + A_2H^+ \tag{6.1}$$

$$\ln\left(\frac{k_1}{k_2}\right) = \ln\left(\frac{\left[A_1H^+\right]}{\left[A_2H^+\right]}\right) \approx \frac{PA(A_1) - PA(A_2)}{RT_{eff}} - \frac{\Delta(\Delta S)}{R} \tag{6.2}$$

From Equation 6.2, the CKM reliability and whether or not a significant entropic effect operates are verified by investigating the dissociation of cluster ions formed by a set of reference compounds of known ion affinities (ΔPA for proton affinity) and by plotting $\ln(k_1/k_2)$ versus ΔPA. Thus, considering ΔPA is used to represent the relative $[C_1]$ affinities for anions and the intensity of the anion ratio (ln *IA*/*IA*′) represents $\ln(k_1/k_2)$, a plot can be constructed to furnish the relative intrinsic hydrogen-bond strengths in gaseous IL aggregates. A straight line with a high correlation coefficient intercepting the origin indicates the reliability of the method and the absence of significant entropic effects. T_{eff} is then calculated from its slope. From such a plot, the affinity of an unknown can be determined by forming and dissociating ion clusters of the unknown with reference compounds. For systems where entropy effects are relevant, "extended" CKM versions have been developed (Armentrout 2000). Weakly bonded clusters or supramolecules such as $[H]^-$, $[Cl]^-$, and $[Br]^-$ bound amine dimers and pyridines (McLuckey et al. 1981; Eberlin et al. 1994; Gozzo and Eberlin 2001) usually display a T_{eff} below 700 K, whereas covalently bonded species such as electron-bound dimers display a T_{eff} that is considerably higher, greater than 1500 K (Chen et al. 1997; Meurer et al. 2003). They found that aggregates linked via weak CH hydrogen bonds were efficiently transferred directly from their acetonitrile solutions to the gas phase. They observed this because the mass selection and CID via tandem MS/MS experiments reveal that all of the $[C_nA_{(n+1)}]$ (where $A = [BF_4]^-$) supramolecules dissociate in a characteristic and predictable fashion: owing to their H-bonded nature, they sequentially loose neutral $[C_1]$ $[BF_4]$ molecules. This observation allowed the authors the unprecedented opportunity to form, isolate via mass selection, then gently dissociate mixed gaseous $[C_1 \ldots A \ldots C_2]^+$ supramolecules by low-energy CID, and measure the intrinsic solvent-free relative strengths of their weak, but very relevant, CH hydrogen bonds.

Fernandes et al. (2009) presented the relationship between energy of collision-induced dissociation (CID) and abundance of fragments. To implement the energy-variable CID, the applied collisional activation voltage (E_{lab}) is increased by small increments, while the relative abundances of the precursor and fragment ions are monitored. The energy required to dissociate

50% of the precursor ion is registered as $E_{lab,1/2}$. $E_{1/2}$ is defined here as the value of collision energy in the laboratory frame at which the relative abundance (defined as a percentage of the base peak) of the precursor ion is 50%. In this inelastic collision of the projectile ion with the neutral target, the total available energy for the conversion of translational (or kinetic) to internal (or vibrational) energy of the projectile ion is the E_{cm}, which can be calculated from E_{lab} and from the masses of the neutral target (m_t) and precursor ion (m_p) by Equation 6.3 (McLuckey et al. 1981):

$$E_{cm} = E_{lab}\left(\frac{m_t}{m_p + m_t}\right) \tag{6.3}$$

In one of the first papers that we selected for this chapter, the authors (Alfassi et al. 2009) make an attempt to determine the solubility of ILs (Figure 6.3) by ESI-MS. The most interesting result for this chapter, however, was the observation of IL aggregates. In addition to the parent ions, some fragmentation products and cluster ions were observed. The authors observed that with an increase in cone voltage, the fragmentation products increased and cluster ions decreased. For example, with a cone voltage of 45 V, considerable fragmentation was observed, and there were very few cluster ions. On the other hand, with a cone voltage of 25 V, there was very little fragmentation; however, there were more cluster ions. The cluster ions were composed of the parent cation $[C_n]^+$ or the anion $[A]^-$ with one or more IL molecules observed that formed positive and negative clusters: $[C_{(n+1)}A_n]^+$ and $[C_nA_{(n+1)}]^-$ with $n = 1$–3. The fragmentation under high cone voltage is probably due to the collisions of the accelerated ions under such voltage.

In the chronological sequence, Gozzo et al. (2004) for the first time presented a systematic study of supramolecules of ILs (Figure 6.5) using ESI-MS. The first mixture studied was an acetonitrile solution of $[C_1][BF_4]$, $[C_1][PF_6]$, and $[C_1][CF_3CO_2]$. The authors prepared a mixture of ILs ($[CF_3CO_2{\ldots}C_1{\ldots}BF_4]^-$, $[BF_4{\ldots}C_1{\ldots}PF_6]^-$, and $[CF_3CO_2{\ldots}C_1{\ldots}PF_6]^-$), which were mass selected and then dissociated by 5 eV collisions with argon. The aggregate structures shown in Figure 6.4 were found by the authors in the ESI-MS spectra.

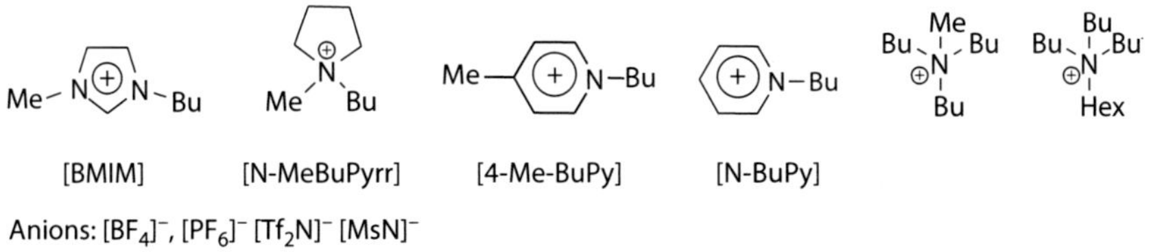

[BMIM] [N-MeBuPyrr] [4-Me-BuPy] [N-BuPy]

Anions: $[BF_4]^-$, $[PF_6]^-$ $[Tf_2N]^-$ $[MsN]^-$

FIGURE 6.3
Chemical structure of cations and anions of ILs studied. (From Alfassi, Z.B. et al., *Anal. Bioanal. Chem.*, 377, 159, 2003, accessed August 2012. http://www.springerlink.com/content/1618-2642/377/1/)

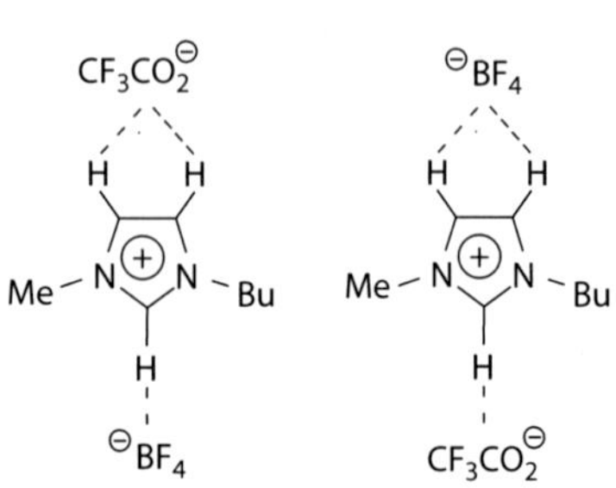

FIGURE 6.4
Aggregate structure found in the ESI-MS spectra for the IL mixture of $[A...C_1...A]^-$. (From Gozzo, F.C. et al., *Chem. A Europ. J.*, 10, 6187, 2004, accessed August 2012. http://onlinelibrary.wiley.com/doi/10.1002/chem.v10:23/issuetoc)

The results showed that the stronger $[CF_3CO_2...C_1]$ hydrogen bond favored the loss of a neutral $[CF_3CO_2...C_1]$ species and, consequently, $[BF_4]^-$ was formed as the main ionic fragment, whereas $[CF_3CO_2]^-$ was of minor abundance. The structure of $[C_1]$, $[C_2]$, and $[C_3]$ and anions is shown in Figure 6.5.

From the results, the authors were able to allocate a qualitative order of intrinsic hydrogen-bond strength to $[C_1]$ for the five anions studied: $[CF_3CO_2]^- > [BF_4]^- > [PF_6]^- > [InCl_4]^- > [BPh_4]^-$. A similar qualitative order of intrinsic hydrogen bond strength was constructed for the cations. Accordingly, the authors prepared three mixed supramolecules from an acetonitrile solution of $[BF_4]$ salts of the $[C_1]$, $[C_2]$, and $[C_3]$ imidazolium ions. They found that upon dissociation for the $[C_2...BF_4...C_3]^+$ mixture, the two imidazolium ions compete for the central $[BF_4]^-$ anion, and the more loosely hydrogen-bonded ion is preferentially expelled as the main ionic fragment. Since $[C_3]$ was formed with an abundance almost twice that of $[C_2]$, the authors concluded that $[C_2]$ was more strongly hydrogen bonded to $[BF_4]$. From these experiments, the authors proposed the following order of H-bond strength for $[BF_4]$: $[C_1] > [C_2] > [C_3]$. The authors believe that this order likely reflects increasing electron-donating or steric effects (or both) of the N-substituents weakening hydrogen bond to the imidazolium ions. In order to verify whether CKM is applicable to measure relative intrinsic hydrogen-bond strengths in gaseous IL supramolecules, the authors plotted $\ln(IA/IA')$ versus

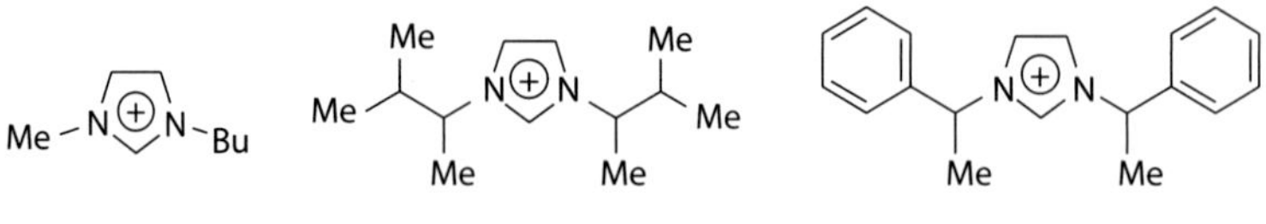

Anions: $[CF_3CO_2]^-$, $[BF_4]^-$, $[PF_6]^-$, $[InCl_4]^-$, $[BPh_4]^-$

FIGURE 6.5
Chemical structure of cations and anions of ILs studied. (From Gozzo, F.C. et al., *Chem. A Europ. J.*, 10, 6187, 2004, accessed August 2012. http://onlinelibrary.wiley.com/doi/10.1002/chem.v10:23/issuetoc)

the relative $[C_1]$ affinities for $[CF_3CO_2]^-$, $[BF_4]^-$, $[PF_6]^-$, and $[BPh_4]^-$ as estimated by B3LYP/6-311+G(d,p) calculations, which used $[CF_3CO_2]^-$ as the reference (zero). The plot obtained by the authors drawing a straight line passing through the origin results in a correlation coefficient as high as 0.998 and a T_{eff} of 430 K. From these results, the authors demonstrate the CKM reliability for ILs and negligible entropic effects, even though there is considerable variation in the nature of the anions $[A]$ investigated.

One of the only examples of studies about the use of the ESI-MS technique to determine the solvent dependence of aggregate formation is that of Dorbritz et al. (2005). In this work, the authors considered ILs as hydrotropes. They show that ILs form solvent-dependent aggregates; however, ILs do not form such large aggregates as surfactants do. The aggregate formation of $[BMIM][BF_4]$ and $[BMIM][Tf_2N]$ (Figure 6.6) was investigated in methanol, 2-propanol, and ethyl acetate. A solution at a concentration of 10^{-4} and 10^{-1} mol/L was injected, and ESI-MS was generated in positive and negative modes. The authors found that the highest in intensity is an aggregate formed of two cations and one anion $[C_2A_1]$, when measured in the positive-ion mode in methanol. In addition, the size and distribution of the aggregates change with different concentrations of IL. For the 10^{-4} mol/L mixture, only the single cation and the aggregate built of two cations and one anion can be found $[C_2A_1]$. For the 10^{-1} mol/L mixture, a complete series of aggregates within the measurable range is detectable $[C_nA_{(n-1)}]^+$. They also found that signal intensity of larger aggregates increases as the concentration increases. The signals (in the amplified region) are easier to recognize as the signal-to-noise ratio improves. In negative mode, $[BMIM][BF_4]$ solutions at 10^{-1} mol/L were analyzed in different solvents, and the authors affirm that the differences in aggregate formation behavior are easily seen. They found that in water, only small aggregates like $[C_1A_2]$ and $[C_2A_3]$ are formed. For methanol, larger aggregates can be found compared to the spectrum obtained for water. With decreasing polarity of the solvent, the authors observed that the formation of higher aggregates was facilitated, giving evidence for all possible aggregates within the range of measurement (2-propanol). They achieved similar results for $[BMIM][BF_4]$ in ethyl acetate. The authors note that the small $[C_1A_2]$ aggregate is not even formed here. Thus, they observed that the IL adapts to its environment: the smaller the $ET_{(30)}$ Reichardt values (Dorbritz et al. 2005) of the solvent used, the larger the aggregates formed,

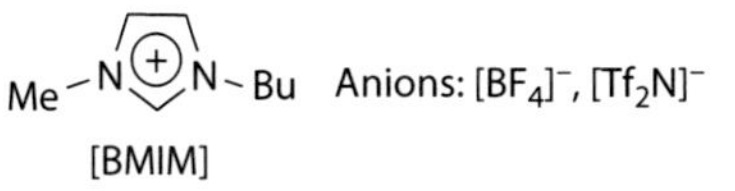

Me–N(+)N–Bu Anions: $[BF_4]^-$, $[Tf_2N]^-$

[BMIM]

FIGURE 6.6
Chemical structure of cations and anions of ILs studied. (From Dorbritz, S. et al., *Adv. Synth. Catal.*, 347, 1273, 2005, accessed August 2012. http://onlinelibrary.wiley.com/doi/10.1002/adsc.v347:9/issuetoc)

and the more distinct the signals from the noise. The authors believe that the aggregate formation, especially in less polar solvents, is probably necessary to minimize the charge density within the ions. All the aforementioned data were described for [BMIM][BF_4]; however, the authors reported similar observations for spectra of the water-nonmiscible [BMIM][Tf_2N]. In this case, the authors assume that the existence of the larger anion minimizes the need for charge delocalization to form aggregates.

In order to extend the work performed by Gozzo et al. (2004) and investigate the possible formation of mixed supramolecular networks formed by different cations coordinated for a selected anion or by different anions bonded to a given cation of composition $[C_1 \ldots A \ldots C_2]^+$ and $[A_1 \ldots C \ldots A_2]^-$, Bini et al. (2007) applied ESI-MS to investigate IL assemblies and to compare and measure the intrinsic "solvent-free" strengths of their hydrogen bonds via CID (Gozzo et al. 2004). The IL structure used in this study is represented in Figure 6.7. The binary mixture solutions of water to methanol (1:1) or acetonitrile at 10^{-3} mol/L concentration were injected, and ESI-MS was generated in both positive and negative modes.

For these cations and anions, generally present in widely used ILs, the authors developed two qualitative "interaction strength" scales on the basis of ESI-MS experiments performed on binary IL mixtures and on pure ILs. The positive-ion mode mass spectrum consisted of the naked cation $[C]^+$, a series of singly charged supramolecules of composition $[C_{(n+1)}A_n]^+$ ($n = 1$–8), and a few doubly charged species $[C_{(n+2)}A_n]^{2+}$ ($n = 1, 3$). Similarly, the negative-ion mode mass spectrum showed the presence of the naked anion $[A]^-$, singly charged anionic aggregates such as $[C_nA_{(n+1)}]^-$, and other doubly and

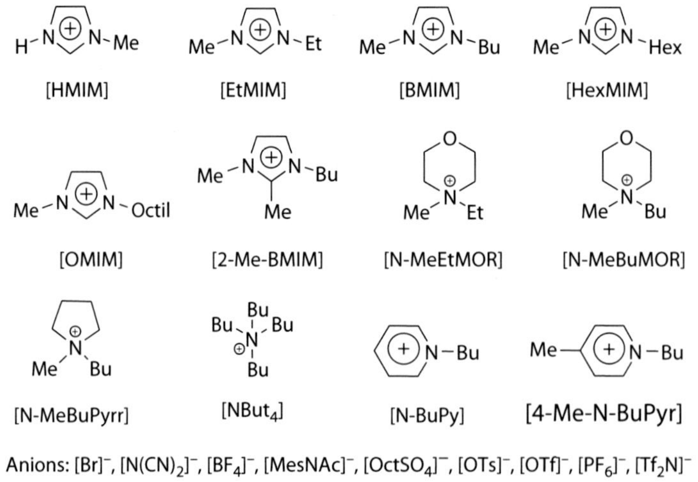

[HMIM] [EtMIM] [BMIM] [HexMIM]

[OMIM] [2-Me-BMIM] [N-MeEtMOR] [N-MeBuMOR]

[N-MeBuPyrr] [$NBut_4$] [N-BuPy] [4-Me-N-BuPyr]

Anions: $[Br]^-$, $[N(CN)_2]^-$, $[BF_4]^-$, $[MesNAc]^-$, $[OctSO_4]^-$, $[OTs]^-$, $[OTf]^-$, $[PF_6]^-$, $[Tf_2N]^-$

FIGURE 6.7
Chemical structure of cations and anions of ILs studied. (From Bini, R. et al., *J. Phys. Chem. B*, 111, 598, 2007, accessed August 2012. http://pubs.acs.org/toc/jpcbfk/111/3)

triply charged species at a low relative intensity, which was in agreement with previous reports (Milman and Alfassi 2000; Tsuzuki et al. 2000; Gozzo et al. 2004; Nohara and Bitoh 2005;). From a distribution of the ESI mass spectrum peaks, the authors showed that whereas the spectrum of pure [EtMIM][Br] was characterized by the presence of clusters $[C_{(n+1)}A_n]^+$ (n up to 10), that of [EMIN][Tf_2N] shows an almost exclusive occurrence of the naked cation peak. An intermediate situation is represented by [EtMIM][TsO]. Thus, the authors could affirm that the distribution of the ESI mass spectrum peaks clearly shows that the $[Br]^-$ anion is much more strongly bounded to the [EtMIM] cation than $[Tf_2N]^-$. The authors also analyzed the mixed supramolecular networks, formed by different cations coordinated to a selected anion or by different anions bonded to a given cation, that is, $[C_1\ldots A\ldots C_2]^+$ and $[A_1\ldots C\ldots A_2]^-$, with the aim of building a scale of the cation–anion interaction strength of the investigated ILs. The ratio between $[C_1]$ and $[C_2]$ in the ESI-MS/MS spectrum estimates the relative intrinsic bond strength for $[A]$. The authors supposed that a stronger bond between the cation $[C_1]$ and $[A]$ accounted for a higher relative intensity of ions $[C_2]^+$ and therefore a higher intrinsic "solvent-free" bond strength of $[C_1]$ for the anionic counterpart. Thus, they established two classes of counteranions: anions tightly coordinated to the cationic moiety that include $[CF_3CO_2]^-$, $[Br]^-$, $[N(CN)_2]^-$, and $[BF_4]^-$ in addition to anions loosely interacting with the alkylimidazolium species, for example, $[TfO]^-$, $[PF_6]^-$, and $[Tf_2N]^-$, this latter one being the least interactive anion among those investigated. The relative sequence of hydrogen-bonding ability that was observed by the authors from these experiments decreases in the order $[CF_3CO_2]^- > [NO_3]^- > [TfO]^- > [ClO_4]^- > [BF_4]^- > [SbF_6]^- > [PF_6]^-$. The authors found $[TfO]^-$ and $[BF_4]^-$ in an unlikely position with respect to their findings. The authors highlight that the solvent used to perform experiments has no influence on the sequence of the relative bond strength but only on the relative intensity of the different peaks in the mass spectrum. In the sequence, the authors performed experiments of gradual dissociation by using increasing collision energies of $[C_1\ldots A\ldots C_2]^+$ and $[A_1\ldots C\ldots A_2]^-$, in addition to aggregates formed and isolated via mass selection. The authors observed a qualitative trend of $[N(CN)_2]^- > [MesNAc]^- \cong [OctSO_4]^- > [BF_4]^- \cong [TsO]^- > [TfO]^- > [PF_6]^-$, which is in reasonable agreement with what was found with binary mixtures of ILs, that is, $[N(CN)_2]^- > [BF_4]^- > [MesNAc]^- > [OctSO_4]^- > [TsO]^- > [TfO]^- > [PF_6]^-$, the only difference being for $[BF_4]^-$. The authors suggest that in virtue of the weaker interaction with the [BMIM] cation, the anion located close to hydrogens H4 and H5 should be considered the departing one. Consequently, the scale arising from the dissociation of the homologous complexes of imidazolium salts gives direct information on the entity of the interaction of the different anions with imidazolium hydrogens H4 and H5.

It is worth mentioning that based on interaction scales, the authors examined the reactivity of ILs as well as their catalyst potential. For example, a strong anion–cation interaction is able not only to reduce the ability of the cation to interact with a substrate or transition state (this latter feature has been evidenced [Aggarwal et al. 2002] in the Diels–Alder reaction) but also to modify the intrinsic ability of the anion to interact with other electrophilic species. An example is the investigation of heterolytic dediazoniation reactions by Bini et al. (2006). In addition, the authors reported that the usual aggregate ion abundance distribution, which is an exponential decay with increasing n, is not observed here, as also reported in the literature for other imidazolium-based ILs (Fernandes et al. 2011).

Two more recent papers describing the systematic studies of IL aggregates were published by Fernandes et al. (2009, 2011). In the first work, the purpose of the authors was to investigate the influence of anion type and alkyl chain length of 1,3-dialkylimidazolium-based ILs on the formation and stability of aggregate dissociation. The structures of the ILs are depicted in Figure 6.8. The solution at concentration of 10^{-4} mol/L in acetonitrile was injected, and ESI-MS was generated in positive and negative modes. The experiments on variable CID were also performed and provided useful information on cluster formation and stability.

The positive-ion mode mass spectra of the ILs studied show $[C_mMIM_{(n+1)}A_n]^+$ cluster ions for n up to 11 with different relative abundances. The authors found that for $[C_mMIM][BF_4]$ ($m = 2$, 4, 8, and 10), the peak distribution clearly shows a decrease in ion abundance for each cluster ion as the alkyl side-chain length increases, demonstrating the expected tendency to decrease aggregate formation with increasing size of the cation.

These variations in peak intensities are best represented using the scaled intensity proposed by Zhang and Cooks where the abundances of individual aggregates are compared to those of the neighbors by plotting $I_2\, s/(I_{(s-1)} \times I_{(s+1)}$ as a function of aggregate size (S), where I is the abundance of the corresponding aggregate ion and the subscript represents the aggregate size (Zhang et al. 2000). The highly symmetric octahedral geometry of the hexafluorophosphate anion compared with other anions might be responsible for the much higher abundance of ion $[C_5A_4]^+$, where $C = [BMIM]$ and $A = [PF_6]$. In this work, the author investigated the dependence of collision energy on the fragment ion abundances for $[BMIM][BF_4]$ clusters. The author observed in this fragmentation pattern a higher abundance of the ion $[C_5A_4]^+$, where $C = [BMIM]$ and $A = [BF_4]$, which corresponds to the monomeric unit of supramolecular aggregates of ILs proposed earlier (Consorti et al. 2005) and indicates extra stability. In order to investigate

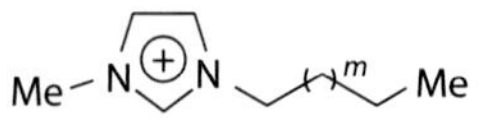

Me–N⊕N (m) Me

$m = 2$, 4, 8, and 10
Anions: $[BF_4]^-$ $[Cl]^-$, $[PF_6]^-$, $[Tf_2N]^-$, $[TfO]^-$

FIGURE 6.8
Chemical structure of cations and anions of ILs studied. (From Fernandes, A.M. et al., *J. Mass Spec.*, 44, 144, 2009, accessed August 2012. http://onlinelibrary.wiley.com/doi/10.1002/jms.v44:1/issuetoc)

further the influence of *N*-alkyl side-chain length and the nature of anions on the kinetic stability of each aggregate ion, a variable-energy CID study was done for the [C_mMIM][BF_4] ($m = 2$, 4, 8, and 10) and [BMIM][*A*] ($A =$ [Cl], [BF_4], [PF_6], [CF_3SO_3], and [$(CF_3SO_2)_2N$]) IL aggregates. The ESI-MS-MS of each aggregate ion, for each IL, was recorded at several collision energies in order to determine $E_{1/2}$. The data from the results found by the authors indicate that, independent of the cation or anion, the energy necessary to dissociate the monomers ($n = 1$) is much higher than for larger aggregates. From the dimer to the octamer, the $E_{cm,1/2}$ values are almost constant, with the exception of the aggregate with $n = 4$, where an increase in energy compared to its neighboring aggregates is observed. For the IL [BMIM][CF_3SO_3], such increased stability for $n = 4$ was not observed. Thus, the authors established an order of decreasing energy for the dissociation of the charged monomers [C_2A_1]$^+$, where $C =$ [BMIM]): $[Cl]^- > [BF_4]^- > [PF_6]^- > [CF_3SO_3]^- > [(CF_3SO_2)_2N]^-$. In the same manner, they observed that the opposite trend is followed by the ionic volumes of the anions (values in Å taken directly from reference [Kobrak 2008]). The energy increase was in the following order: $[Cl]^-$ (47) < $[BF_4]^-$ (73) < $[PF_6]^-$ (107) < $[CF_3SO_3]^-$ (129) < $[(CF_3SO_2)_2N]^-$ (230). From these results, the authors concluded that the Coulomb interaction energy depends, among other factors, on the inter-charge distance that decreases with anion size (mass–energy) necessary to dissociate the monomer. However, its effect on larger aggregates is, in some cases, unnoticeable. In addition, they found that for the monomer [C_2A_1]$^+$ where $C = C_m$MIM and $A =$ [BF_4], the increase in chain length increased the cation–anion distance, thus decreasing the electrostatic energy and the hydrogen-bonding strength.

In the second work of Fernandes et al. (2011), the authors present energy-variable CID measurements for the IL isolated ions, [C_2A_1]$^+$ and [A_2C_1]$^-$, aiming to provide fundamental insights into the relative cation–anion interaction strengths. As a result, the differences in binding energies were evaluated by making use of the structural features of the ionic species, for example, the cation core and ring size, aromaticity, alkyl chain length, anion nature and size, and ion charge densities. The structures of the ILs studied are depicted in Figure 6.9. A solution at concentration of 10^{-4} mol/L in methanol was injected, and ESI-MS was generated in positive and negative modes. From the results, the authors found that the energy required for the separation of the anion from the neutral molecule within the anionic aggregate is practically independent of the alkyl chain length, whereas in the separation of the cation from the neutral molecule within the cationic aggregate, for higher values of X in [CmMIM][Tf_2N], the energy decreases from [DMIM][Tf_2N] to [PentMIM][Tf_2N] toward a constant value. The authors suggest that in the case of [C_2A_1]$^+$ aggregates, the presence of two cations enhances the steric hindrance, which in turn causes an increase in the distance between the ions and therefore leads to a decrease in the relative ionic interactions. The constant energy values for longer alkyl chains

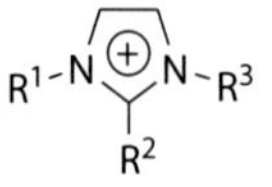

R^1 = Me R^2 = H R^3 = Me [DMIM], R^1 = Et R^2 = H R^3 = Me [EtMIM]; R^1 = Pr R^2 = H R^3 = Me [PrMIM], R^1 = Bu R^2 = H R^3 = Me [BMIM]; R^1 = Pent R^2 = H R^3 = Me [PentMIM], R^1 = Hex R^2 = H R^3 = Me [HexMIM]; R^1 = Heptyl R^2 = H R^3 = Me [HepMIM], R^1 = Octyl R^2 = H R^3 = Me [OctMIM]; R^1 = Decyl R^2 = H R^3 = Me [DecMIM], R^1 = Et R^2 = H R^3 = Et [DEtIM]; R^1 = Bu R^2 = Me R^3 = Me [Bu-2-Me-MIM]
Anions: $[Cl]^-$, $[BF_4]^-$, $[PF_6]^-$, $[OTf]^-$, $[Tf_2N]^-$

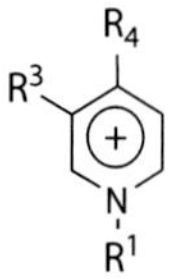

R^1 = Bu R^3 = H R^4 = H [N-BuPy], R^1 = Hex R^3 = H R^4 = H [N-HexPy], R^1 = Octyl R^3 = H R^4 = H [N-OctPy], R^1 = Pr R^3 = Me R^4 = H [N-PrMePy], R^1 = Bu R^3 = Me R^4 = H [N-Bu-3-MePy], R^1 = Bu R^3 = H R^4 = Me[N-Bu-4-MePy] Anions: $[Cl]^-$, $[BF_4]^-$, $[PF_6]^-$, $[OTf]^-$, $[Tf_2N]^-$

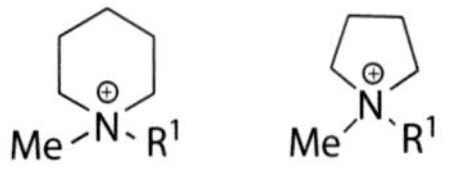

R^1 = Me [DMPip] R^1 = Pr [MePrPip]
R_1 = Me [DMPyrr] R^1 = Pr [MePrPyrr]
Anions: $[Cl]^-$, $[BF_4]^-$, $[PF_6]^-$, $[OTf]^-$, $[Tf_2N]^-$

FIGURE 6.9
Chemical structure of cations and anions of ILs studied. (From Fernandes, A.M. et al., *J. Phys. Chem. B*, 115, 4033, 2011, accessed August 2012. http://pubs.acs.org/toc/jpcbfk/115/14)

at the imidazolium cation are explained by the authors considering that a further extension of the chain away from the ring, where the interaction with the anion predominantly occurs, does not produce any significant effect. The authors also evaluated the influence of the anion on this trend by replacing the bulky $[Tf_2N]^-$ with the $[Cl]^-$ anion. The results for chloride-based ILs displayed the same behavior observed for the $[Tf_2N]$-based counterparts, showing that the anion type/size was not a determinant. Some studies about other structural isomers of the imidazolium-based ILs, which do not involve the presence of alkyl groups at the C2 position of [DMIM] $[Nf_2T]$ and [PrMIM]$[Nf_2T]$, do not show significant differences in their relative ionic interaction strengths. The evaluation of the influence of the cation core on the dissociation energies of $[C_2A_1]^+$ aggregates was investigated using the cation shown in Figure 6.9. The authors concluded that the reduction of the electrostatic strength due to the charge delocalization in aromatic rings, as compared to a localized charge on the nitrogen atom of the saturated rings, is reflected in the relative values of interaction energies. Finally, the authors supposed that aromatic-based ILs (pyridinium and imidazolium based) present lower relative interaction strength values when compared with their saturated counterparts (piperidinium and pyrrolidinium based). The cation size increases from a five- to a six-membered ring in both aromatic and saturated cores, and there is an increase in the distance between charges with delocalization of them as well, resulting in a decrease in the ionic interaction energy. The authors also stated that this effect is similar to that observed for the increase of the cation side alkyl chain length.

From the papers selected to exemplify the use of ESI-MS in the investigation of formation and characterization of IL aggregates, it is possible to show that ESI-MS is an important tool in the spectrometric nanoscale investigation of IL aggregates. Studies of solvent and solution concentration effect on the formation of aggregates showed that in less polar solvents, the aggregates formed were larger, whereas in a more concentrated solution, the aggregates were smaller. However, this effect has been explored very little, and there is an absence of these effects in IL aggregate studies using ESI-MS. Studies about the influence of the alkyl chain in the IL aggregate formation showed that the abundance of aggregates decreases with an increase in the alkyl chain length. The most readily available information is the determination of the hydrogen bond strength between cation and anion in the aggregates, which is obtained from the variation of the CID. Studies to estimate the hydrogen bond strength between cation and anion consist of establishing a relative scale between anions and cations studied in each series. In this manner, it is hard to join all the scales proposed and to compare them. In addition, it was found that the alkyl chain length is not affected by the strength of interaction between cation and anion.

6.2.1.1 Mass Spectrometry Experiments

Mass spectra of the all compounds were acquired with an Agilent 6460 Triple Quadrupole, operating in the positive- and negative-ion mode, equipped with an electrospray source (ESI). The source and desolvation temperatures were 300°C and 350°C, respectively. The ionic liquid solutions in water at concentrations $\sim 10^{-4}$ mol.L^{-1} were introduced at a 5 μL.min^{-1} flow rate. The cone voltage was 50 V. The capillary and the cone voltage were +3500 V to positive mode and −3500 V to negative mode. Nitrogen was used as nebulization gas. The mass spectra recorded were evaluated by the qualitative analysis software from Agilent Technologies. The data were acquired in positive and negative mode MS total ion scan mode (mass scan range m/z 20-1000).

6.2.2 Nuclear Magnetic Resonance Study of IL Aggregation

The aggregated systems that are the focus in this chapter can be investigated by NMR in liquid solution or in solid state. The most important information furnished by NMR is the chemical shift of the ^{1}H nucleus (Blesic et al. 2007; Goodchild et al. 2007; Singh and Kumar 2007; Remsing et al. 2008; Li et al. 2010; Singh et al. 2010; Tariq et al. 2011; Stark et al. 2011; Zhang et al. 2011; Sastry et al. 2012), although investigations on the ^{13}C

(Williams et al. 1973), $^{35/37}Cl$ (Remsing et al. 2008), and ^{19}F (Molinier et al. 2005) have also been performed. Particularly for amphiphilic molecules (focus in this chapter), some specific parameters of the NMR technique can be stated:

1. Variation of NMR longitudinal (T_1) and transverse (T_2) relaxation times. It is determined by some time-dependent interaction involving the nuclear spins. The relaxation rates are given by the interaction strength and a correlation time (τ_c) describing the molecular motion.
2. ^2H in solid experiments and 2D experiments such as nuclear Overhauser effect spectroscopy (NOESY—a method to identify spins undergoing cross relaxation and to measure their cross-relaxation rates (Singh and Kumar 2007)) and rotating frame Overhauser effect spectroscopy (ROESY).
3. Diffusion-ordered spectroscopy (DOSY) with pulsed field gradient spin echo NMR (PFGSE–NMR) to determine diffusion coefficients (Silverstein 2005; Muller 1969).

The proton magnetic relaxation is dominated by inter- and intramolecular magnetic dipole–dipole interactions modulated by the molecular motion. For anisotropic phases, the degree of anisotropy of the molecular motion has a strong influence on the proton spectrum. Particularly for amphiphile molecules, the intermolecular effects are averaged out by rapid translational and rotatory motions. This technique resulted in a sensitive method for determining the size of large micelles. The method can be used to study the evolution of micellar shape and size in relation to amphiphile concentration or solubilization. The proton relaxation of the alkyl chain protons is considerably more effective in the micelles than for the free monomers in aqueous solution (Blesic et al. 2007).

The study of ^2H NMR lies in the electric quadrupole moment of the nucleus (nuclear spin $m_I = 1$). Quadrupolar effects in NMR spectra result from the coupling of the nuclear quadrupole moment with electric field gradients. Studies of ^2H relaxation times for deuterons in an alkyl chain should give quantitative information on chain dynamics, but due to the small chemical shifts, signal overlap (cf. above for ^1H) occurs. From this vast array of NMR experiments, it is possible to extract important information about aggregated systems of amphiphilic molecules, for example, CMC, N_{agg}, diffusion coefficients, mesophase behavior (in connection with other methods), and data about micellar structure organization.

The CMC is determined by varying the concentration of amphiphilic molecules in a solvent and monitoring the ^1H chemical shift of a specific chemical group. Li et al. (2010) proposed that the observed chemical shift for the specific group (δ_{obs}) can be expressed as the sum of the chemical shifts of the monomer (δ_{mon}) and the aggregated form (δ_{mic}). The authors considered that each one is proportional to its respective molar fractions (concentration), where c_{mon}, c_{mic}, and c_{tot} are the concentrations of the amphiphilic molecules existing as monomers, in micelles, and, in total solution, resulting in Equation 6.4:

$$\delta_{obs} = \delta_{mon}\left(\frac{c_{mon}}{c_{tot}}\right) + \delta_{mic}\left(\frac{c_{mic}}{c_{tot}}\right) \tag{6.4}$$

It is assumed that the monomer concentration is constant above the CMC; therefore, the δ_{obs} can be given by Equation 6.5:

$$\delta_{obs} = \delta_{mic}\left(\frac{\text{CMC}}{c_{tot}}\right) + (\delta_{mic} - \delta_{mon}) \tag{6.5}$$

Generally, at low concentrations of the amphiphilic molecule, the chemical shift observed (δ_{obs}) is constant and then changes rapidly and levels off at a higher concentration. This behavior is a consequence of the formation of aggregates. The sudden increase in δ/c corresponds to the aggregate formation (Blesic et al. 2007; Singh and Kumar 2007; Sastry et al. 2012). The CMC can also be found by a plot of δ_{obs} versus $1/c_{tot}$ that should yield two straight lines below and above the CMC. The intersection of these lines corresponds to the CMC (Li et al. 2010). The typical δ_{obs} curve against IL concentration obtained from the ^{1}H NMR experiment is shown in Figure 6.10.

The aggregation number is another piece of information that can be furnished by $\Delta\delta_{obs}$ in the ^{1}H NMR experiments. Sing and Kumar (2007) showed in their work that selecting a chemical group from a system in which chemical shift is not sensitive to solvent effects and applying the mass-action law for the free molecule-aggregate equilibrium result in Equation 6.6:

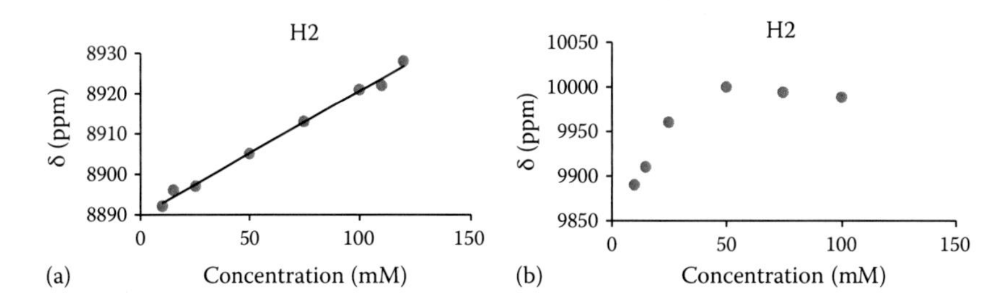

FIGURE 6.10
Monitored ^{1}H NMR chemical shift of C2 proton of [BisOctMIM][2Br] with increase in IL concentration. (a) Water (formation of aggregates was not observed). (b) Chloroform (aggregates formed at 50 mM). Experimental details are given at the end of Section 6.2.2.

$$N_{agg}A_{gg} \leftrightarrows A_{ggN_{agg}} \quad K = \frac{[A_{ggN_{agg}}]}{[Agg_{N_{agg}}]} \tag{6.6}$$

where
N_{agg} is the aggregation number
Agg represents aggregate
K is the equilibrium constant

Thus, the chemical shift can be written as in Equation 6.7 (Wennerström and Lindman 1972):

$$\delta_{obs} = \frac{c_{mic}}{c_{tot}}\delta_m \tag{6.7}$$

where c_{mic} and c_{tot} are the concentration of aggregated IL and the total concentration, respectively, δ_{obs} is the observed chemical shift, and δ_m is the shift of aggregated IL determined by extrapolation to zero IL concentration. It is possible to express the monomer concentration as

$$[A_{gg}] = c_{tot}\frac{\delta_m - \delta_{obs}}{\delta_m} \tag{6.8}$$

and the aggregate concentration as

$$N_{agg}[A_{ggN_{Agg}}] = c_t\frac{\delta_{obs}}{\delta_m} \tag{6.9}$$

then the expression for K may be written as

$$\log\{[c_{tot}](\delta_{mon} - \delta_{obs})\} = N_{agg}\log\{[c_{tot}](\delta_{obs} - \delta_m)\} + \log(N_{agg}K) + (1 - N_{agg})\log(\delta_{mon} - \delta_m) \tag{6.10}$$

The plots of $\log\{[c_{tot}](\delta_{mon} - \delta_{obs})\}$ against $\{[c_{tot}](\delta_{obs} - \delta_m)\}$yield the N_{agg}.

Molinier et al. (2005) showed that by PFGSE–NMR experiments, it is possible to measure self-diffusion coefficients (D_{obs}). This proposal is based on the fact that in liquids, translational diffusion of both solute and solvent molecules occurs in a sufficiently short amount of time to induce observable effects in NMR spin echo experiments (Molinier et al. 2005). This method for measuring D_{obs} can be applied to micellar solutions of surfactants. The authors note this by using the phase-separation model of micelle formation. In other words, the surfactant experiences rapid exchange between the micellar organization and the monomers in solution, and the time that the surfactant molecule remains inside a micelle is far less than the measuring time. Molinier et al. (2005) proposed that the measured D_{obs} can be expressed as a concentration-weighted average of the nonmicellized surfactant diffusion coefficient and the micelle diffusion coefficient. Accordingly, the measured coefficient is given by Equation 6.11:

$$D_{obs} = \left(\frac{c - \text{CMC}}{c}\right) D_{mic} + \left(\frac{\text{CMC}}{c}\right) D_{free} \tag{6.11}$$

where

c is the global concentration of the amphiphilic molecule
CMC is the critical micelle concentration
D_{obs} is the self-diffusion coefficient measured
D_{free} is the self-diffusion coefficient for free amphiphilic molecules
D_{mic} is the micellar self-diffusion coefficient

Constructing D_{obs} as a function of the concentration curve, the typical behavior is a plateau at concentrations lower than the CMC that corresponds to the diffusion of free surfactant. Above the CMC (when the concentration increases), there is an additional surfactant aggregate in micelles, whose diffusion coefficient is about one order of magnitude lower than that of free molecules. The hydrodynamic radius can be calculated from the D_{obs} of micelles by using the Stokes–Einstein equation (see Equation 6.23) (Molinier et al. 2005).

Another important experiment that has been performed to characterize micelle formation and that has been used a few times to study IL aggregates is ^{2}H NMR. ^{2}H NMR spectroscopy was used to explore the formation and identities of the mesophases formed as a function of composition and temperature for IL in solutions (Goodchild et al. 2007). These experiments consider that nuclei with $m_I > 1/2$ have a quadrupole moment that interacts with local electric field gradients in the sample (Khan et al. 1982). If the quadrupolar interaction is anisotropic, the degeneracy of the Zeeman levels is elevated. Following this principle, Khan et al. (1982) presented a method for the identification of lyotropic mesophases in concentrated aqueous surfactant solutions using solid-state NMR spectroscopy. The method considers that a single "horn" in the powder pattern is expected for isotropic phases (cubic and isotropic micellar phases). Powder patterns with "horns" separated in the order of 1 and 2 kHz are expected for the anisotropic hexagonal and lamellar phases, respectively. In other words, the lamellar horn separation is twice that observed for the hexagonal phase.

Although there are several NMR experiments that can furnish information about the structure and formation of micelles (and other aggregates), the most used experiment is the change in chemical shift of ^{1}H nucleus with an increase in amphiphilic molecule concentration in solution.

Particularly in the case of ILs, the changes in chemical shift of ^{1}H are monitored for alkyl chain (terminal methyl) and imidazolinium hydrogens. The change in the chemical shift of the protons of the terminal methyl group in the alkyl chain, as a function of concentration, has been used to determine the aggregation concentration and to provide structural identity of micelles. Protons from the alkyl chain for the ILs generally showed an upfield aggregation shift. There are some facts that can contribute to this upfield shift: aromatic ring-induced shielding of the methylene groups, solvent effects, or structural changes. Structural changes have been associated with increasing importance of the gauche conformation (which causes an upfield shift). In other words, aggregation formation is accompanied by a partial changeover from trans to gauche conformations in the alkyl chain. In addition, this upfield shift has been credited to the fact that alkyl branches squeeze themselves into a nonpolar environment in the core part of the aggregate. ^{1}H nuclei of the methyl group attached directly to the aromatic ring of ILs also generally showed an upfield shift but with a lower magnitude when compared to the protons of the alkyl chain. This change in chemical shift has been ascribed to less shielding of the protons of the methyl group by the aromatic ring. In some cases, these protons change very little or do not change the chemical shift at all, thus showing that the solvent and the aromatic ring have little effect.

The chemical shift of the ^{1}H nucleus attached at the C2, C4, and C5 of imidazolinium has shown that the interaction between them and a solvent is much greater than that experienced by alkyl chains. Protons at the C2 of the imidazolium ring as well as the C2 and C6 of the pyridinium ring are the most acidic due to the vicinity of electronegative nitrogen, and they are most available for hydrogen bonding with water. Thus, changes in the chemical shift due to interaction with hydrogen acceptors (solvents or anions) will be more pronounced in the ^{1}H nucleus.

In most cases, there is competition for the interaction of aromatic protons with the solvent and the anions. A strong interaction between the anion and the aromatic protons results in reduced interaction of the protons with the solvent molecules. When anions interact with water molecules, there is a weakening of the interaction with the aromatic protons, which causes an upfield shift for the aromatic protons in aqueous solutions.

These chemical shift effects will be modulated by anion and solvent basicity and the overall positive charge density of the aromatic ring.

Another comparison that is often made from NMR studies of IL aggregates is a comparison of change in the chemical shift of different protons upon aggregation, relative to that of the monomer, in order to shed some light on aggregate structure. $\Delta\delta_{obs}$ versus carbon number for different ILs to protons of the alkyl chain shows that $\Delta\delta_{obs}$ is the largest for the terminal protons, decreasing progressively toward the aromatic ring. It is accepted that aggregation affects the methyl signal in the same way as the methylene signals and is consistent with the methyl being in the liquid interior of the micelle. The positive value of $\Delta\delta_{obs}$ for protons of the alkyl chain must be due to magnetic shielding of these protons by the aromatic ring. Methyl groups directly attached to the aromatic ring of different ILs also show a positive $\Delta\delta_{obs}$ value in the same order, similar to that of the alkyl chain. The structural changes aforementioned are more commonly observed in IL aggregates studies by NMR, although other alterations to particular IL structures have been observed.

Although CMC correlations offer a straightforward way to relate the structure of a given surfactant with its tendency to self-aggregate, they do not provide a direct quantitative assessment of the energetics involved in the micellization process. From the CMC obtained through NMR data, one can obtain thermodynamic information about micelle formation and its stability. Li et al. (2010) proposed a calculation of the standard Gibbs free energy (ΔG_m°) of aggregate formation for an ionic surfactant with the following equation (Equation 6.12):

$$\Delta G_m^\circ = 2RT\ln_{(\chi CMC)} \tag{6.12}$$

where χ_{CMC} is expressed in molar fraction units. According to the earlier formula, the ΔG_m° for aggregate formation of an IL at different temperatures may be calculated using χ_{CMC}. The authors observed that negative ΔG_m° values suggest that the micellization in aqueous solution is spontaneous. Li et al. (2010) also observed that enthalpy difference (ΔH_m°) decreases with the increase in temperature; thus, the micelle formation process is endothermic at lower temperatures and exothermic at higher temperatures. Similarly, the authors observed that the entropy term (ΔS_m°) increases with an increase in temperature and contributes much to negative values. Thus, the micellization process is entropy driven, as is typical in surface active IL systems. With the help of nonspectroscopic methods such as conductivity, a model describing the aggregation process (generally in aqueous systems), which considers mass-action models, allows the calculation of the standard Gibbs free energy (Equation 6.13) (Villetti 2011):

$$\Delta G_m^\circ = (1-\beta)RT\ln\chi_{CMC} \tag{6.13}$$

In the case of dimeric surfactants, the correct expression is (Equation 6.14) (Zana et al. 1996)

$$\Delta G_m^\circ = (0.5+\beta)RT\ln\chi_{CMC} \tag{6.14}$$

In this brief description about NMR applications in IL aggregation studies, we focus our attention on the NMR experiments that were found in the papers within the scope of this chapter. Therefore, we consider it interesting to describe some of these papers here.

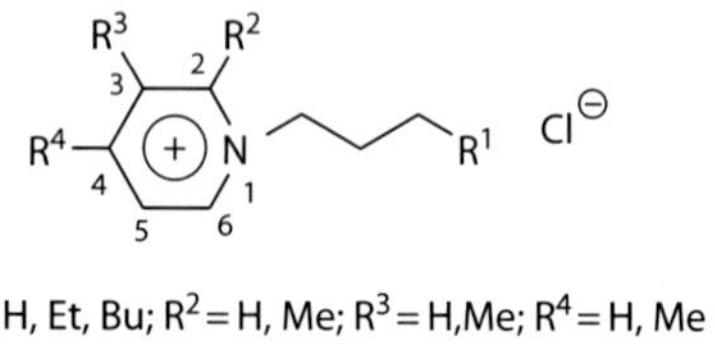

R^1 = H, Et, Bu; R^2 = H, Me; R^3 = H,Me; R^4 = H, Me

FIGURE 6.11
Representative structure of cations and anions of ILs studied. (From Sastry, N.V. et al., *J. Colloid Interface Sci.*, 371, 52, 2012, accessed August 2012. http://www.sciencedirect.com/science/article/pii/S0021979712000082)

In a simple work, Sastry et al. (2012) determined the CAC of the ILs [N-BuPy][Cl], [N-HexPy][Cl], [N-OctPy][Cl], [N-O_{Ct}-2-MePy][Cl], [N-O_{Ct}-3-MePy Py][Cl], and [N-O_{Ct}-4-MePy][Cl] in aqueous solutions, by monitoring the aggregation behavior through ^{1}H NMR chemical shifts. Representative IL structures are shown in Figure 6.11.

In order to reach their objectives, the authors recorded a series of ^{1}H NMR spectra of IL solutions prepared in D_2O, covering the pre- and postmicellar regions. The authors observed a downfield shift of the C2 and C6 protons of 1-octylpyridinium chloride with an increase in the concentration of IL. These trends were expected because the head group protons have water surrounding them even in a typical aggregate structure. The chemical shift of the protons of the terminal methyl was upfield for 1-hexylpyridinium and 1-octylpyridinium chloride, while no appreciable shift was observed for 1-butylpyridinium chloride. As already mentioned, the upfield shifts of the terminal methyl of the alkyl chain are the signature of aggregates arising. The authors believe that the absence of terminal methyl upfield shifts for 1-butyl pyridinium chloride reveals that the nature of the theses aggregates is different from that of hexyl- and octylpyridinium-based ILs. The authors suggest that the clustering of IL molecules ought to occur through an open aggregation process (Sastry et al. 2012).

Remsing et al. (2008) investigated the influence of solvents in the formation of IL aggregates. The authors evaluated binary mixtures of [BMIM][Cl] with water or DMSO, and for each mixture, ^{1}H and $^{35/37}$Cl NMR chemical shift perturbations ($\Delta\delta$) were monitored. The IL structure is shown in Figure 6.12.

The authors monitored the ^{1}H nucleus of an imidazolinium ring (the C2, C4, and C5 protons) and alkyl chain (C6, C7, C8, and C10). With an increase in water content, the authors observed that the chemical shift of the C2 proton (most acidic and forms the strongest hydrogen bonds with the [Cl]$^-$) is the most affected. They reported that increased shielding occurs at this position due to exchange of [Cl]$^-$ ions for D_2O molecules around the five-membered ring. Protons attached to the C4 and C5 form weaker hydrogen bonds with the anion, and changes in the chemical shift were not distinguished. The same happens to protons attached to the C6 and C7. For the H8 and H10,

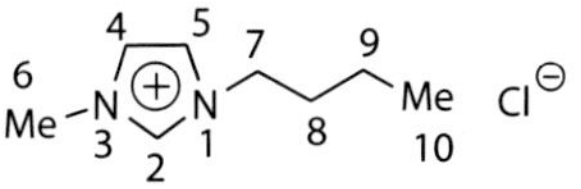

FIGURE 6.12
Chemical structure of [BMIM][Cl] studied. (From Remsing, R.C. et al., *J. Phys. Chem. B*, 112, 7363, 2008, accessed August 2012. http://pubs.acs.org/doi/abs/10.1021/jp800769u)

the authors recorded deshielding at high dilution and an increase in shielding for D_2O concentrations up to 20 wt%. They think that when water is introduced into the system in low quantities, the alkyl groups aggregate and become gradually solvated with an increase in D_2O. The authors highlighted that the binary mixture [BMIM][Cl]/DMSO presents an interesting contrast to what is obtained in an aqueous situation. The change in the chemical shift of the C2 proton was smaller than that observed in water. The authors affirm that the imidazolium ring is not as strongly solvated in DMSO (lower hydrogen bond acceptor). The authors detected that $\Delta\delta$ for protons on the butyl chains are larger in DMSO than in water and proposed that in this medium, the alkyl group becomes solvated before the imidazolium ring. The authors also observed that the aggregation of the nonpolar chains does not appear to occur in the [BMIM][Cl]/DMSO solutions, because there is merely a linear dependence of the C10 proton chemical shifts on solvent content (Remsing et al. 2008). In this work, $^{35/37}Cl$ NMR was used to examine the changes in the environment around the $[Cl]^-$ ion as a function of IL concentration in H_2O and DMSO. The $^{35/37}Cl$ $\Delta\delta$ for the most diluted solution studied was within 0.2 ppm of that of a fully solvated $[Cl]^-$ ion, clearly indicating that water solvates the [BMIM][Cl] anion. $^{35/37}Cl$ $\Delta\delta$ in [BMIM][Cl]/DMSO systems indicates deshielding of the anion as the DMSO concentration increases. This suggests a strengthening of the interionic interactions in this medium. As was the case for the imidazolium ring, an increase in the $^{35/37}Cl$ shielding of the anion indicates that it also becomes solvated at DMSO concentrations above 95 wt% (Remsing et al. 2008).

Zhang et al. (2011) also studied solvent effects in IL aggregate formation. The authors performed ^1H NMR measurements of $[BMIM][CF_3CO_2]/D_2O$ and $[BMIM][CF_3CO_2]/CD_3OD$ mixtures. The IL structure is shown in Figure 6.13.

The authors monitored the ^1H nucleus of an imidazolinium ring (the C2, C4, and C5 protons). In water, the authors noted that the chemical shifts of the C2, C4, and C5 protons had the most significant changes and they moved upfield upon dilution (weakening of the H-bonding interactions involving the C–H groups in the imidazolium ring). The chemical shifts of hydrogen atoms in the alkyl chain are slightly upfield. The authors proposed that this occurs due to the influence of H-bonding cooperativity and the solvent effect of D_2O. For the $[BMIM][CF_3CO_2]/CD_3OD$ system, the authors observed that only the C2 proton shows an upfield shift, whereas the C4 and C5 protons

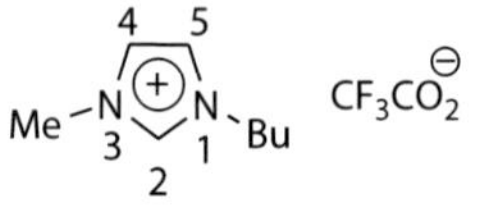

FIGURE 6.13
Chemical structure of [BMIM][CF_3CO_2] studied. (From Zhang, Q.G. et al., *J. Phys. Chem. B*, 115, 11127, 2011, accessed August 2012. http://pubs.acs.org/doi/abs/10.1021/jp204305g)

show a downfield shift with an increase in methanol concentration. Similar to the water mixture, the authors proposed that this occurs due to the influence of H-bonding cooperativity and the solvent effect of CD_3OD.

An interesting work was published by Stark et al. (2011). In this work, the authors established the initial hypothesis that water could tightly solvate the IL up to a water molar fraction of 0.5. This arose from the experimental fact that in some water-sensitive chemical reactions, the IL has been reported to be responsible for deactivating the water by interacting with it. The authors performed a series of experiments; however, of interest in this chapter are the measurement results as a function of composition (generally 0–0.9 water molar fractions), temperature (generally 40°C–85°C) obtained by NMR diffusion measurements, proton chemical shift, and T_1 of cation, anion, and water. In order to reach their objective, the authors studied the binary [EtMIM][Ms]/water system (Figure 6.14).

NMR diffusion measurements were performed using PFGSE–NMR. The self-diffusion was determined by using a method described by other authors (Jerschow and Mueller 1996, 1997; Mao et al. 1997, 1994; Huang et al. 2002). The D_{obs} was measured for water, cation, and anion, since all are observable in proton NMR spectroscopy. In the plot of self-diffusion constants against wt% for all three elements, a nearly linear increase was observed. From the graph, the authors observed that the D_{obs} of the cation and anion are very similar in magnitude, while water diffuses about three times faster. The authors presented a table with D_{obs} for all temperatures and molar fractions investigated, and upon closer inspection of this table, it is seen that they assumed an interesting crossover for the diffusion coefficients of the anion and cation. At low wt%, the self-diffusion of the cation is faster than for the anion, but the reverse is true for high wt%. The composition-dependent proton chemical shifts for the binary water/[EtMIM][Ms] system was also investigated. The authors observed that while the

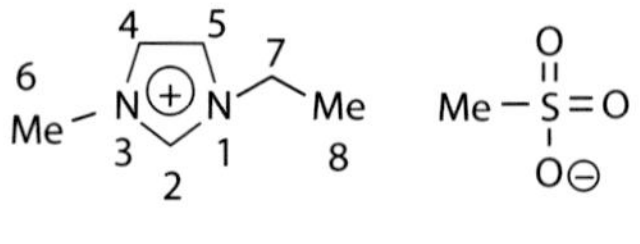

FIGURE 6.14
Chemical structure of [EtMIM][Ms] studied. (From Stark, A. et al., *J. Mol. Liq.*, 160, 166, 2011, accessed August 2012. http://www.sciencedirect.com/science/article/pii/S016773221100073)

water and the anion resonances move downfield with increasing water content, the entire cation resonances shift upfield. The authors plotted chemical shift changes as a function of wt% and observed that the chemical shift changes are linearly related to wt%. In particular, the methyl protons of the anion display a linear relationship between 0 and 10 wt% and also between 10 and 100 wt% but with a different slope. The trend of chemical shifts described here for the data registered at 40°C similarly holds true for the chemical shift data registered at higher temperatures. The authors determined the NMR spin lattice relaxation times (T_1) in conjunction with the diffusion measurements. The plots were constructed of T_1 as a function of wt% for H2, H7, and H8 in addition to the anion methyl protons and the water protons. Overall, simple linear dependencies are observed, especially for the protons of the anion. Thus, in this case, T_1 values were not able to add important information to the dynamic of the IL/water mixture.

Goodchild et al. (2007) used ^{2}H NMR spectroscopy to study the lyotropic mesophases formed in concentrated aqueous solutions of longer chain [C_mMIM][X] ($m=8$, 10 and X = [Cl], [Br]) (Figure 6.15). The ^{2}H NMR spectra obtained by the authors suggest that the major mesophase of the [OctMIM] [Cl]/water system is hexagonal. However, the authors noted that no pure single-phase regions were identified by ^{2}H NMR in the composition ranges studied. Thus, in some cases, the peaks assigned to a lamellar phase are very small relative to the hexagonal horns and could plausibly be artifactual. The authors found that at wt% = 0.262 and $T=263$ K, the pattern from a [OctMIM] [Br]/water mixture of two pairs of horns was observed with separations of 1.2 and 2.3 kHz corresponding to the hexagonal and lamellar phases, respectively. At wt% = 0.406 and 283 K, the pattern from a [DecMIM][Cl]/water mixture of two pairs of horns with separations of 1.4 and 2.8 kHz was seen, and at wt% = 0.187 and $T=298$ K for a [DecMIM][Br]/water mixture, two sets of peaks are again observed with horn separations of 1.0 and 2.0 kHz. In the first two ILs, the relative intensities suggest that the hexagonal phase is present in the greatest proportion, and for the last IL, the lamellar phase is the major phase (Goodchild et al. 2007). The authors also observed that the horn separation of the ^{2}H NMR spectra decreases with increasing chain length for a fixed anion and decreases with increasing anion size for a fixed chain length. Thus, the authors proposed that if a larger horn separation is associated with the D_2O molecules residing in a more ordered environment, this indicates that a more ordered environment is obtained for a shorter alkyl

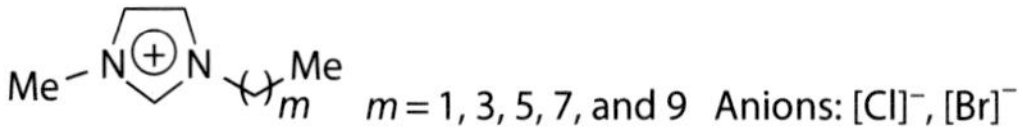

Me–N⊕N–(CH₂)$_m$–Me $m=1, 3, 5, 7$, and 9 Anions: $[Cl]^-$, $[Br]^-$

FIGURE 6.15
Chemical structure of ILs studied. (From Goodchild, I. et al., *J. Colloid Interface Sci.*, 307, 455, 2007, accessed August 2012. http://www.sciencedirect.com/science/article/pii/S0021979706011052)

chain or smaller anion. In this sense, the authors also verified that the anion effect is dominant for the short chain length. When the anions are strongly solvated, the horn separation values suggest that greater order is found in the mesophase. Consequently, a larger anion would be less tightly bound to the micelle surface, which causes enhanced repulsive intermicellar interaction and destabilizes the mesophase.

From the papers selected to exemplify the use of NMR in the investigation of formation and characterization of IL aggregates, it is possible to show that NMR is one of the most important spectroscopic techniques for nanoscale investigation of IL aggregates. Studies of solvent and temperature effects show that in polar protic solvents such as D_2O and CD_3OD, the imidazolium ring is strongly solvated and the alkyl groups aggregate at low D_2O concentration and become gradually solvated with an increase in D_2O. Conversely, in DMSO, the imidazolium ring is not solvated, and in this medium, the alkyl group is solvated before the imidazolium ring. Aggregation of the nonpolar chains does not appear to occur in [BMIM][Cl]/DMSO solutions. Studies of the formation of IL aggregates in nonpolar solvents have not been found, revealing a lack of studies of IL aggregates by NMR. A similar conclusion can be made from studies of temperature effect in the formation of IL aggregates by NMR. There is only one work that has evaluated the temperature effect in the homogeneity of IL aggregate structure, in which it was found that the heterogeneous structure of aggregates is favored at lower temperatures.

6.2.2.1 Nuclear Resonance Magnetic Experiments

^1H spectra were recorded on a Bruker DPX 400 spectrometer (^1H at 400.13 MHz in 5 mm sample tubes at 298 K in D_2O and $CDCl_3$/TMS solutions). The general reproducibility of chemical shift data was estimated to be better than 0.01 ppm. Binary mixtures of IL/water (D_2O) and IL/chloroform ($CDCl_3$) were prepared at concentrations of 10, 15, 25, 50, 75, 100, 110, and 120 mmol/mL. The curve shown in Figure 6.10 was constructed by plotting these concentrations against the chemical shift observed to C-2 proton of imidazolium ring. The IL [BisOctMIM][2BF4] was synthesized in our laboratories according to the literature (Frizzo et al. 2008, 2009a,b, 2011; Martins et al. 2008a,b, 2007, 2009; Buriol, 2011).

6.2.3 Light Scattering Technique Study of Aggregation

Visible (vis) light ($400 < \lambda < 700$ nm) can be used as a nonperturbative probe to obtain information about the structure such as size, size distribution, shape, self-assembly, and the dynamics of particles (translational and rotational diffusion coefficient) from measurements of the LS (Schmitz et al. 1990). When a laser beam passes through a solution or colloidal dispersion, the electric field of the incident light induces an oscillating dipole moment in the particles of the

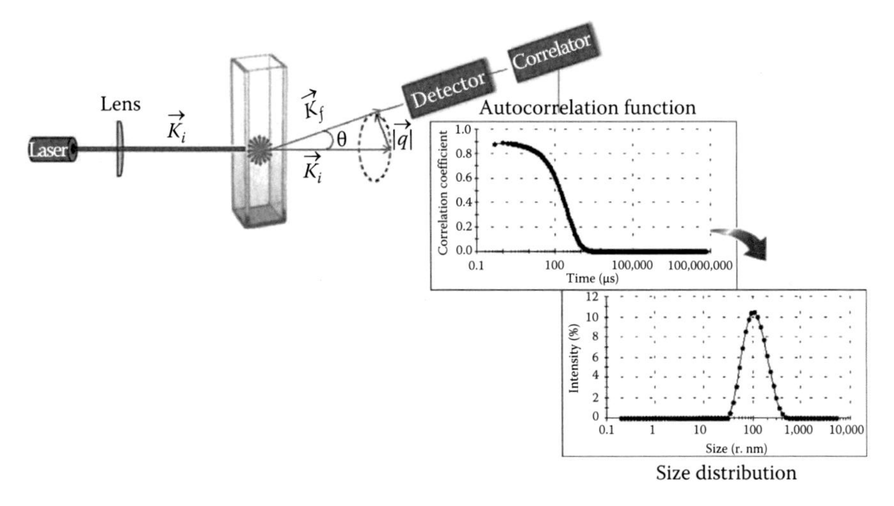

FIGURE 6.16
Schematic diagram of a typical LS experiment. The autocorrelation function and size distribution graph were obtained at 200 mM of [BisDecMIM][2Br]. Experimental details are given at the end of Section 6.2.3.

medium that will radiate light in all directions (Pecora 1985). Figure 6.16 shows a schematic diagram of a typical LS experiment. In the elastic LS, also called Rayleigh scattering, the wavector of incident radiation ($\vec{k}_i = 2\pi n/\lambda_i$) is equal to the wavector of scattered light ($\vec{k}_f = 2\pi n/\lambda_f$), that is, the energy of the incident photon is equal to the scattered photon. If there is an exchange of energy between the incident photon and the particles of the system, then k_i differs from k_f, and the process is known as inelastic LS, as observed in Raman and Brillouin scattering. One of the most important parameters in the scattering techniques is the scattering vector (q) that for elastic scattering is defined as $\vec{K}_f - \vec{K}_i$, which can be seen in Figure 6.16 and is given by Equation 6.15:

$$q = \frac{4\pi n}{\lambda} \sin\left(\frac{\theta}{2}\right) \tag{6.15}$$

where
 θ is the scattering angle
 n is the index of refraction of the medium
 λ is the wavelength of incident light

There is an inverse relationship between q and the dimension of the particle (L) under observation. For the condition $q.L \ll 1$, the scattered radiation contains information about the region of space larger than several particles; however, for the condition $q.L > 1$, it only contains information on the internal relaxations of the particles (see Figure 6.17). Therefore, the modification in

FIGURE 6.17
(See color insert.) Length scales.

the q values, through a change in n, θ, and λ, enables the scattered light to be measured by a set of particles or aggregates or the scattered radiation to be measured by internal segments of a particle.

Experimentally, it is possible to observe time-dependent fluctuations in the scattered intensity $I(t)$ using a suitable detector. Analysis of the scattered light signal can be made in static or dynamic mode. The static light scattering (SLS) technique measures time-averaged scattering intensities at one specific θ but does not consider the fluctuations of the $I(t)$ with time. The intensity of LS for particles larger than $\lambda/20$ is given by Equation 6.16:

$$\frac{K_{opt}}{R(\theta)} = \frac{1}{M_W P(\theta)} + 2A_2.c \tag{6.16}$$

where c and M_w are the concentration and molecular weight of the particle, respectively. A_2 is known as the second virial coefficient, and it is related to the intermolecular interactions between the particles of the medium and the solvent. $R(\theta)$ and K_{opt} are the Rayleigh ratio and the optical constant given by Equations 6.17 and 6.18, respectively:

$$R_\theta = \left(\frac{I_{sample} - I_{solvent}}{I_{standard}}\right) R_{\theta,standard} \tag{6.17}$$

$$K = \frac{4\pi^2 n_0^2}{N_A \lambda_0^4}\left(\frac{dn}{dc}\right)^2 \tag{6.18}$$

where I_{sample}, $I_{solvent}$, and $I_{solvent}$ are the intensity of LS for the sample, the solvent, and the standard (usually toluene), respectively. $R_{\theta,standard}$ is the Rayleigh ratio of the standard, N_A is Avogadro's number, and dn/dc is the refractive index increment. $P(\theta)$ is the particle form factor, which is related to the gyration radius (R_g) of the particle by the Guinier approximation (Equation 6.19):

$$P(\theta) = 1 - \frac{q^2 R_g^2}{3} \tag{6.19}$$

$$\theta \to 0$$

The Zimm method, which employs Equation 6.16, allows simultaneous determination of M_w, A_2, and R_g through a double extrapolation of the plot of $Kc/R(\theta)$ versus $q^2 + k'.c$, for zero angle and zero concentration.

Dynamic light scattering (DLS), also known as photon correlation spectroscopy (PCS), analyzes the fluctuation of LS intensity with time due to local concentration fluctuations caused by Brownian motion of the particles. The temporal fluctuations in the $I(t)$ quantity may be characterized by its autocorrelation function. The time autocorrelation function of the intensity of the scattered light $G^{(2)}(\tau)$ is defined by (Equation 6.20):

$$G^2(\tau) = I(t)I(t+\tau) = \lim_{T \to \infty} \frac{1}{2T} \int_{-T}^{T} I(t)I(t+\tau)dt \tag{6.20}$$

where $I(t)$ and $I(t+\tau)$ are the intensities of LS at some arbitrary time t and $t+\tau$, respectively, τ being the time delay between two counts and $2T$ the total time over which it is averaged. For short-time delays, the correlation is high and over time, as the particles are moving, the correlation diminishes to zero and the exponential decay of the autocorrelation function becomes characteristic of the decay rate, $\Gamma(s^{-1})$. A typical autocorrelation function is also shown in Figure 6.16. Several mathematical methods are used to analyze the autocorrelation function and obtain the distribution of $\Gamma(s^{-1})$, for example, cumulants (Koppel et al. 1972) and inverse Laplace transformation (ILt). The ILt of the autocorrelation function is usually done with the aid of a software such as CONTIN (Provencher 1982a,b). Therefore, a decay rate of a single population or from different populations of the same sample can be determined and is used to produce a size distribution (see Figure 6.16). The $\Gamma(s^{-1})$ is related to the translational diffusion coefficient D of the particles by the relationship (Equation 6.21):

$$\Gamma(s^{-1}) = Dq^2 \tag{6.21}$$

An important feature of Brownian motion is that a population with small particles moves quickly, whereas other populations with large particles move more slowly, this movement being related to the size of the particles.

The relationship between the hydrodynamic radius (R_h) of the particle and its diffusion coefficient is defined by Stokes–Einstein (Equation 6.22):

$$R_h = \frac{k_B T}{6\pi\eta D} \tag{6.22}$$

where

$k_B T$ is the Boltzmann energy
η is the viscosity of the medium

The shape of the particle can be estimated from the parameter ρ defined as the ratio between R_g and R_h (R_g/R_h). The value of ρ leads to an important indication of the topology of the particle in the medium. For example, when ρ is 0.775, 1.0, and 1.9, particles are spherical, spherical shell, and rodlike, respectively (Schurtenberger and Newman 1993).

Thus, from LS techniques (SLS and DLS), it is possible to obtain information about the structure (size and shape) and molecular dynamics of particles in the scattering medium (Berne and Pecora 2000). DLS is becoming a powerful tool for studying the formation of aggregates, and it is being applied to a wide variety of systems such as synthetic and biological aggregates, where the concentration of interest is often much greater than for CAC. Moreover, the information that has been derived from these applications is providing new insights into the structure, thermodynamics, and interactions of aggregate systems and is also yielding valuable clues as to the physiological actions of various biological amphiphiles.

There are few examples of LS being used to study the aggregation behavior of ILs in solution, and we can see in the following description that this method is generally used together with other spectrometric methods and is limited to informing the size distribution of aggregates.

Li et al. (2008) performed a study in order to analyze the aggregation behavior of ILs [$C_m$mim][Br] in [BMIM][BF_4] (Figure 6.18). DLS revealed the existence of spherical aggregates with a size of 70–100 nm, which is larger than traditional micelles. For example, at 100 mM with [TetDecMIM][Br] in [BMIM][BF_4], the aggregates formed were observed to have an average diameter of 76.9 nm.

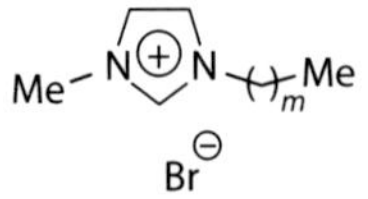

Me N⊕N ( )$_m$ Me
$Br^{\ominus}$

m = 9, 11, 13, and 15

FIGURE 6.18
Structure of ILs studied by [$C_m$mim][Br]. (From Li, N. et al., *Phys. Chem. Chem. Phys.*, 10, 4375, 2008, accessed August 2012. http://pubs.rsc.org/en/content/articlelanding/2008/cp/b807339b)

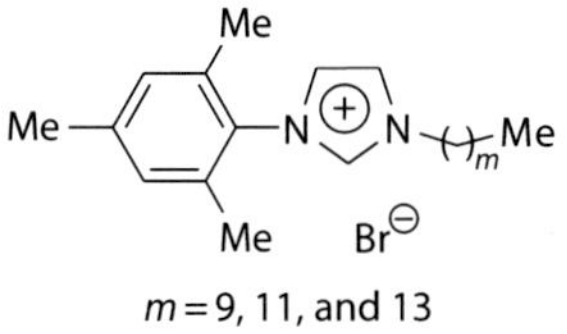

$m = 9, 11$, and 13

FIGURE 6.19
Chemical structure of ILs studied by [C_nPhIM][Br]. (From Shi, L. et al., *J. Phys. Chem. C*, 115, 18295, 2011, accessed August 2012. http:/pubs.acs.org/toc/jpccck/115/37)

Similarly, Shi et al. (2011) investigated the aggregation behavior of 1-(2,4,6-trimethylphenyl)-3-alkylimidazolium bromide [C_mPhIM][Br] ($m = 10$, 12, 14) in [BMIM][BF_4] (Figure 6.19). The authors intended to verify the effect that the incorporation of an N-aryl moiety has on the micellization of IL. DLS results showed that spherical micelles with diameters of 20–40 nm are formed with a concentration of 400 mM. The IL mixtures [DecMIM][Br]/[BMIM][BF_4] and [TetDecPhIM][Br]/[BMIM][BF_4] were detected by DLS forming micelles with an average diameter of 21.6 and 45.1 nm, respectively. The size of micelles formed by [TetDecPhIM][Br] in [BMIM][BF_4] found by Li et al. (2008) was 76.9 nm. It is evident, therefore, that two authors found different-sized micelles for the same IL system. This fact can be attributed to the different concentrations that the systems were submitted to for DLS experiments (100 and 400 mM). It is worth noting that two authors performed TEM analysis that confirms the diameter of micelles.

Chen et al. (2012) studied the behavior in water of ILs with an alkyl chain of several lengths (Figure 6.20) and demonstrated the occurrence of phase separation at both the macroscopic and mesoscopic scale. They established the effect that anion type (Figure 6.20) has on the mesoscopic phase separation and revealed the nature of the mesoscopic phase separation by studying mixtures at 0%–100% weight ratio. They observed that the macroscopic phase is not formed directly by single molecules but rather with a mesoscopic phase as the intermediate. The mesoscopic phase is stable in water mixtures with ILs of shorter alkyl chains. The authors also investigated the filtration of heterogeneous phases, since it is known that filtration is able to remove the heterogeneous structure from the water mixture. However, they observed that the removal of the heterogeneous structure does not change the behavior of solution by DLS. The hypothesis of the authors is

Anions: $[BF_4]^-$, $[CF_3SO_3]^-$, $[NO_3]^-$

$m = 3, 5$, and 7

FIGURE 6.20
Representative structure of cations and anions of ILs studied. (From Chen, G. et al., *Anal. Chem.*, 69, 3641, 1997, accessed August 2012. http://pubs.acs.org/toc/ancham/69/17)

that the level of heterogeneous structure is too small to greatly affect the properties of the mixture. Another explanation given by the authors is that the filtration works as a micromixer and destroys the heterogeneous structure.

The authors evaluated the temperature effect on the heterogeneous structure in both mixtures with and without filtration. They found that the heterogeneous structure is favored at lower temperatures. Thus, the authors concluded that the mesoscopic phase separation of ILs in water is determined by at least four parameters, namely, alkyl chain length of the cation, anion type, concentration, and temperature (all of which affect the miscibility of the IL and water at the molecular level).

Using the DLS method, Wang et al. (2008) studied the aggregate morphology and size of ILs at the IL concentration of 0.45 mol L^{-1}. The apparent hydrodynamic diameter (D_h) can be deduced from the diffusion coefficients. The authors noted that the measured diffusion coefficients were influenced by the repulsion among the highly charged aggregates, which led to the aggregate sizes determined from DLS being smaller than the actual ones. Nevertheless, the DLS experiments performed by Wang et al. (2008) determined that the aggregate size increases with an increase in hydrophobicity of the anions. It was found that the size of the spherical aggregates for the ILs with different anions also increased with the increased hydrophobicity of the anions. This trend was also found for N_{agg} results determined by the fluorescence probe. The aggregate sizes found by D_h values obtained from DLS were supported by TEM analysis. The authors established that anion structures have a very weak effect on morphology, but they affect the aggregate sizes significantly (Figure 6.21).

From the papers selected to exemplify the use of LS in the investigation of formation and characterization of IL aggregates, it is possible to show that LS has been little used for the investigation of IL aggregates. Related results of LS experiments have been limited to informing the size of particles obtained from DLS. This denotes the lack of investigation of IL aggregates using this important tool. Some important information obtained from these studies is that the aggregate size increases with an increase in hydrophobicity of the anions and the anion structures have a

A = $[Cl]^-$, $[BF_4]^-$, $[NO_3]^-$, $[MeCO_2]^-$
$[CF_3CO_2]^-$, $[CF_3SO_3]^-$, $[ClO_4]^-$

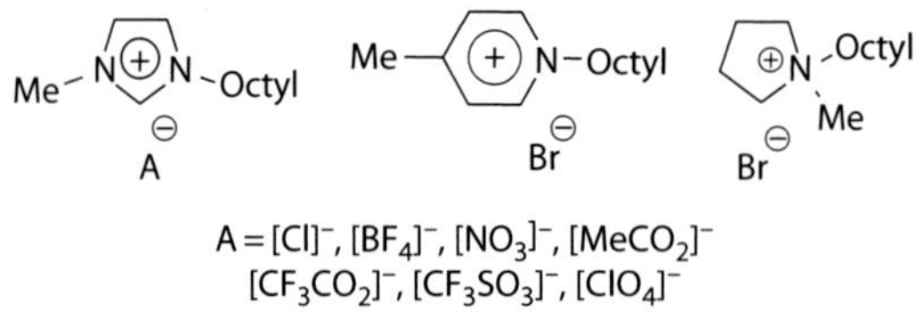

FIGURE 6.21
Chemical structure of the ILs studied. (From Wang, H. et al., *J. Phys. Chem. B*, 112, 16682, 2008, accessed August 2012. http://pubs.acs.org/toc/jpcbfk/112/51)

very weak effect on the morphology, but they do significantly affect the aggregate sizes. Solvents of different polarity, acidity, and basicity were not evaluated using DLS. In addition, just one paper shows a dependence of micelle formation on temperature, indicating that it would be smaller in size at higher temperatures than at lower temperatures. Furthermore, it was found that the heterogeneous structure of IL aggregates is favored at lower temperatures.

6.2.3.1 Dynamic Light Scattering Experiments

The DLS experiments were performed on a Nano ZS, Zetasizer Nanoseries (Malvern Instruments, Worcestershire, UK). The light of $\lambda = 633$ nm from a solid-state He–Ne laser (5.0 mW) was used as the incident beam. All sample solutions were filtered through a 0.45 μm hydrophilic PVDF membrane filter. All measurements were made at 298.2 ± 0.1 K and at 173° scattering angle. At least three measurements were taken for each solution, and the reproducibility of the aggregate sizes was found to be within ±3%. The binary mixtures of IL/water were prepared at concentrations of 0.01, 1, 10, 100, 200, 500, and 800 mM. The size distribution graphic shown in Figure 6.16 was obtained at 200 mM of IL. The [BisDecMIM][2Br] was synthesized in our laboratories according to the literature (Frizzo et al. 2008, 2009a,b, 2011; Martins et al. 2008a,b, 2007, 2009; Buriol, 2011).

6.2.4 Fluorescence Study of IL Aggregation

Fluorescence is the emission of radiation (ultraviolet (UV), vis, or infrared) from any substance and occurs in electronically excited states (Lakowicz et al. 2006). Fluorophore is a fluorescent chemical compound that absorbs energy at a specific wavelength and then reemits it at a different, although equally specific, one. The amount and wavelength of the emitted energy depend on both the fluorophore and the chemical environment of the fluorophore (Sauer et al. 2011). In excited singlet states, the electron in the excited orbital is paired (by opposite spin) to the second electron in the ground-state orbital. Consequently, the return to the ground state is spin allowed and occurs rapidly by emission of a photon (Lakowicz et al. 2006). As a simple rule, the energetically favored electron promotion will be from the highest occupied molecular orbital (HOMO), usually the singlet ground state (S_0), to the lowest unoccupied molecular orbital (LUMO), and the resulting species is called the singlet excited state (S_1). Compounds that absorb in the vis region of the spectrum (these compounds have color) generally have some weakly bound or delocalized electrons and normally comprehend several combined aromatic groups (plane or cyclic molecules) with several π bonds. In these systems, the energy difference between the lowest LUMO and the HOMO corresponds to the quantum energies in the

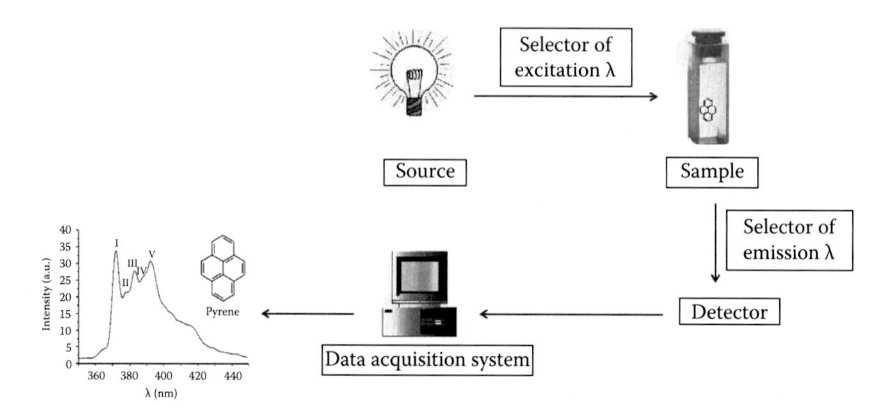

FIGURE 6.22
Schematic representation of fluorescence spectroscopy.

vis region (Sauer et al. 2011). Compounds that emit in the vis region are of interest in the fluorescence spectroscopy. The schematic representation of the fluorescence method is shown in Figure 6.22.

Several probes can be used for the studies of aggregation of ILs, though pyrene is the most widely used fluorescence probe due to its high emission efficiency and hydrophobicity. The emission spectrum of pyrene presents five vibration bands, as can be seen in Figure 6.22. The first (I_1) and the third bands (I_3) are the most important for the aggregation study. It is known that the I_1/I_3 ratio, also called micropolarity index, taken from the first (373 nm, $S_1^{\nu=0} \rightarrow S_0^{\nu=0}$) and third (384 nm, $S_1^{\nu=0} \rightarrow S_0^{\nu=1}$) vibronic peaks in the pyrene emission spectrum, shows linear dependence on the local micropolarity of the medium. Because pyrene preferentially dissolves in hydrophobic regions, the I_1/I_3 ratio may prove the formation of the molecular self-assembly, for example, micelles and aggregates formed by surfactant and IL molecules, respectively (Liu et al. 2011). Fluorescence spectroscopy has been used to determine the CAC in aqueous solutions of ILs. This can be easily done plotting the I_1/I_3 ratio versus the IL concentration. CAC can be taken as the concentration that corresponds to the intersection between the linear extrapolation of the rapidly varied portion of the curve and the relative stabilization portion at higher concentrations (Singh and Kumar 2007; Wang et al. 2007). To illustrate the formation of aggregates, fluorescence spectra of pyrene in aqueous solution at an IL [BisOctMIM][2Br] concentration below and above the CAC were measured and are shown in Figure 6.23a. As can be seen, the band I_1 decreases in intensity above CAC due to aggregation of the IL. The I_1/I_3 ratio as a function of [BisOctMIM][2Br] concentration in aqueous solutions is plotted in Figure 6.23b. The rapid decrease of the I_1/I_3 ratio from 0.1 mM of IL confirms the formation of hydrophobic domains.

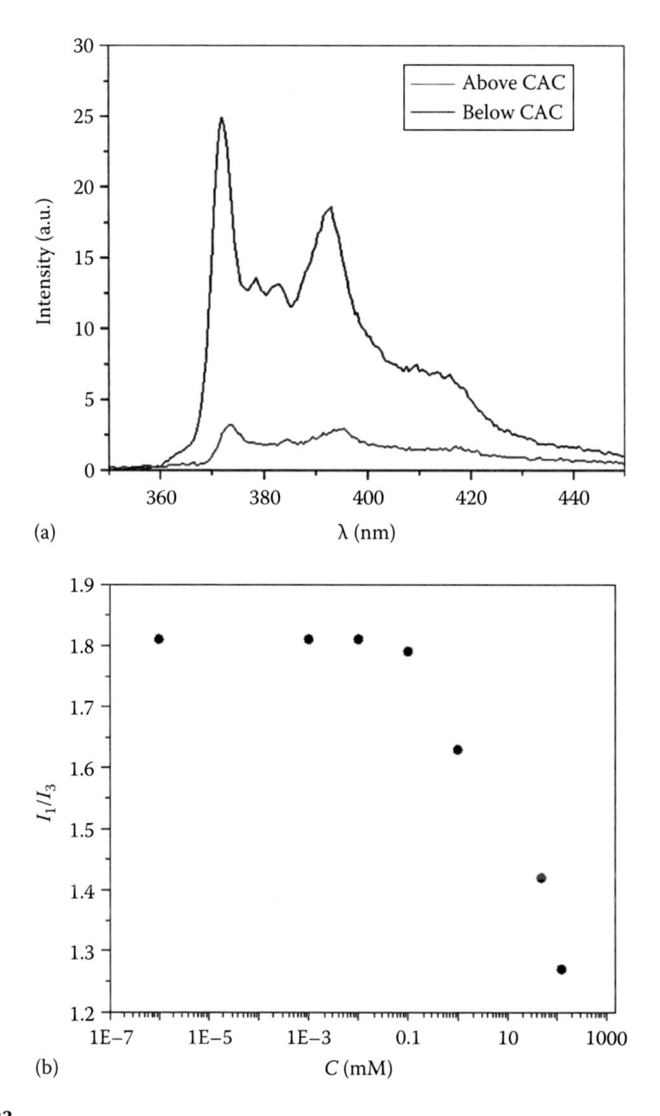

FIGURE 6.23
(a) Representative pyrene fluorescence emission spectra in aqueous [BisOctMIM][2Br] solution at concentrations below and above the CAC. (b) I_1/I_3 ratio of pyrene as a function of [BisOctMIM][2Br] concentration in aqueous solution at 25°C. Experimental details are given at the end of Section 6.2.4.

Other important information about aggregates that one can obtain from fluorescence spectra is the aggregation number (N_{agg}). The aggregation number is the number of monomers in the micelles (aggregate), and it is determined by the fluorescence quenching technique, using pyrene and benzophenone as the probe and quencher, respectively. The aggregation number is determined by the ratio of fluorescence quenching in an aggregated system in Equation 6.23:

$$\ln\left(\frac{I_0}{I}\right) = \frac{N_{agg}c_Q}{c - \text{CMC}} \tag{6.23}$$

where I_0 and I are the pyrene fluorescence intensities in the absence and presence of the quencher, respectively, at a specific wavelength and c_Q and c are the concentrations of quencher and weak fluorescent compound, respectively (Li et al. 2010). Using the slopes of the linear plots of ln (I_0/I) against c_Q, the N_{agg} can be determined (Rao 2011). The optical studies of the pure imidazolium-based or pyridinium-based ILs show that these are weak fluorescent liquids. The analysis of emission spectra from fluorescent probes has provided important details on the structure, dynamics, and transport in organized assemblies. Thus, the dynamics and solvation mechanism of water and various organic solvent ILs have been studied from the steady-state fluorescence responses, and the polarities of these liquids have been examined in this way (Singh and Kumar 2007). Some of these results will now be shown in this chapter.

Wang et al. (2007) studied polarity indexes of pyrene for aqueous solutions of a series of imidazolium-based ILs (Figure 6.24). They showed that the aggregation appeared when the alkyl chain has more than six carbons and the aggregation is strong enough to form micelles above the CMC; however, for the alkyl chains hexyl and butyl, this aggregate formation was not observed. This result indicates that the alkyl chain length of a cation can be tailored to switch the aggregation behavior of ILs. In addition, the author determined the N_{agg} of ILs using the steady-state fluorescence technique. The results showed that the N_{agg} changed with IL concentration; however, at a temperature of 298 K, the N_{agg} determined by fluorescence (35) for [DecMIM][Br] was similar to that determined by Goodchild et al. (2007) and to that predicted by a geometric model proposed by them (38).

Zhao et al. (2011) used fluorescence measurements to determine the micelle N_{agg} in a series of N-alkyl-N-methylpyrrolidinium bromide (C_mMPyrr) ILs (Figure 6.25). The authors concluded that the longer the hydrophobic chains, the larger the N_{agg} values.

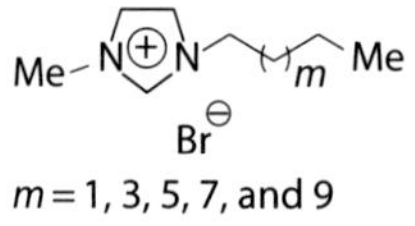

Me–N⊕N–(CH₂)ₘ–Me
Br⊖
$m = 1, 3, 5, 7,$ and 9

FIGURE 6.24
Chemical structure of the ILs studied. (From Wang, J. et al., *J. Phys. Chem. B*, 111, 6181, 2007, accessed August 2012. http://pubs.acs.org/doi/abs/10.1021/jp068798h)

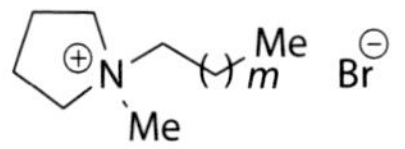

$m = 10, 12$, and 14

FIGURE 6.25
Chemical structure of the ILs studied. (From Zhao, M. et al., *Phys. Chem. Chem. Phys.*, 13, 1332, 2011, accessed August 2012. http://pubs.rsc.org/en/content/articlelanding/2011/cp/c0cp00342e)

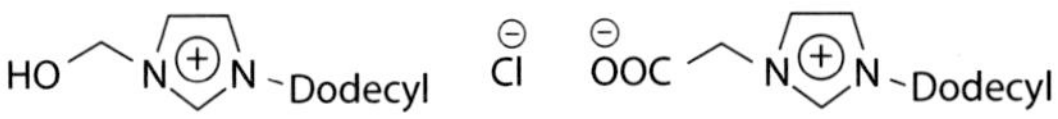

FIGURE 6.26
Chemical structures of ILs studied. (From Liu, X.-F. et al., *J. Surf. Detergents*, 14, 203, 2011a, accessed August 2012. http://www.springerlink.com/content/fhl5037335655574/; Liu, X.-f. et al., *J. Surf. Detergents*, 14, 497, 2011b, accessed August 2012. http://www.springerlink.com/content/f3j108825gn3q210/)

Liu et al. (2011) studied the self-aggregation of 1-hydroxyethyl-3-dodecylimidazolium chloride [OHEtDodecIM][Cl] and 1-carboxymethyl-3-dodecylimidazolium salt [CarboxyMeDodecIM] (Figure 6.26) by fluorescence. The fluorescence measurement was registered at two temperatures, 25° and 35°C. In the case of [OHEtDodecIM][Cl], the authors observed two values for CMC, attributing the first CMC (CMC_1) to the spherical micelles and the second CMC (CMC_2) to the structural transition, that is, from spherical to rodlike ones. The values obtained for CMC_1 were 2.29 mM and 2.67 mM at 25°C and 35°C, respectively. The second breakpoint (corresponding to CMC_2) was observed in the relatively higher-concentration regions. The values obtained at 25°C and 35°C were 14.78 and 15.83 mM, respectively. In the case of the ILs studied, when the IL concentration reaches the CMC_1, the I_1/I_3 ratio is about 1.64, which is higher than for [DodecMIM][Br] and [TetDecMIM][Br]. The authors credited this fact to the side chain (the hydroxyethyl group) in the head group of [OHEtDodecIM][Cl], since the side chain probably produces a larger steric hindrance effect between the head groups, which leads to a relatively looser packing mode within the formed aggregates. The N_{agg} of [OHEtDodecIM][Cl] in aqueous solutions at 35°C was determined by the steady-state fluorescence quenching technique. The values obtained for CMC_1 and CMC_2 were 8 and 63, respectively. For the IL zwitterion [CarboxyMeDodecIM] (Liu et al. 2011), the fluorescence measurements were conducted at 25°C and 35°C. The values obtained for CMC were 1.74 mM (35°C) and 1.25 mM (25°C). The N_{agg} was determined by a steady-state fluorescence quenching method, and the values obtained were 28.3 and 35.8 at 25 ± 0.1°C and 35 ± 0.1°C, respectively. N_{agg} was smaller than that of ammonium salt (about 80–85) (Rosen et al. 2004), which indicates that the [CarboxyMeDodecIM] micelle presents a much looser structure.

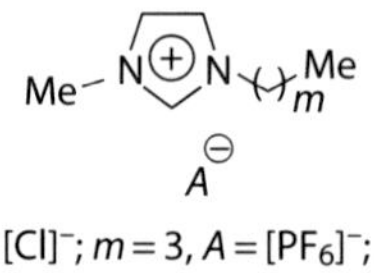

FIGURE 6.27
Representative structure of ILs studied. (From Blesic, M. et al., *Green Chem.*, 9, 481, 2007, accessed August 2012. http://pubs.rsc.org/en/content/articlelanding/2007/gc/b615406a)

Some works selected for this chapter cover more than one spectroscopic method for studying IL aggregation. Blesic et al. (2007), for example, used fluorescence spectroscopy and ^{1}H NMR measurements to monitor the adsorption at the aqueous solution–air interface and self-aggregation behavior of the cations with different linear alkyl chain lengths and different counterions. The IL structure is shown in Figure 6.27.

From fluorescence spectroscopy, the authors observed that although the I_1/I_3 ratio shows the typical increase that indicates the presence of a hydrophobic environment for tails equal to or greater than six carbons, a plateau is reached for $m = 10$, 12, and 14. The authors established that the intersection point between the plateau and the descending part was for the CMC determination of these three ILs ($m = 10$, 12, and 14). The CMC found for ILs using fluorescence was 45, 7, and 3 mM for $m = 10$, 12, and 14, respectively. The authors attributed the absence of a plateau at higher IL concentrations in the case of the other three systems ($m = 4$, 6, and 8) to the progressive dense packing of the individual cation (micelle-like structure). The authors affirm that this produces a fixed increase in the hydrophobicity sensed by pyrene that reflects the continuous increase in the probe response. ^{1}H NMR resonance experiments for the protons [C_mMIM][Cl] showed that only for $m = 10$, 12, and 14 did reasonable shift change occur followed by a plateau region related to the self-aggregation in small micellar aggregates (Blesic et al. 2007).

Singh and Kumar (2008) also report the studies of fluorescence spectroscopy and ^{1}H NMR in aqueous solutions of IL with cations [BMIM], [OctMIM], and [N-Bu-3-MePy] and anions $[BF_4]^-$ and $[Cl]^-$, as shown in Figure 6.28.

Fluorescence spectroscopy data giving I_1/I_3 values of 1.46 and 1.37 for [BMIM][Cl] and [OctMIM][Cl], respectively, indicate that these ILs have similar behavior to that of the classical ionic surfactants where the I_1/I_3 ratio decreases as

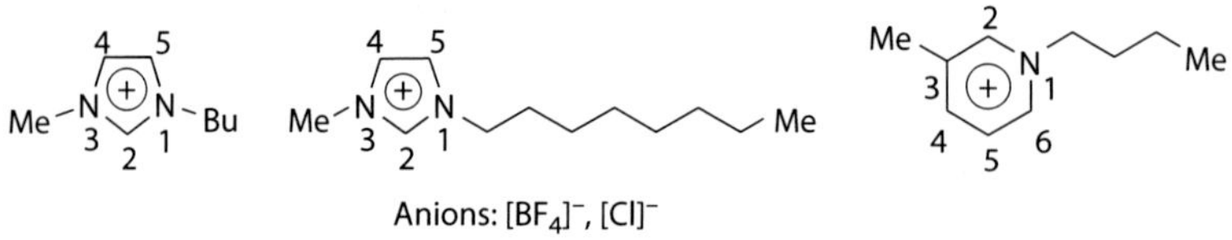

FIGURE 6.28
Chemical structure of ILs studied. (From Singh, T. and Kumar, A., *J. Phys. Chem. B*, 112, 4079, 2008, accessed August 2012. http://pubs.acs.org/doi/abs/10.1021/jp711711z)

the alkyl chain length increases. The CMCs determined by fluorescence spectroscopy were 700, 800, 600, and 101 mM for [BMIM][BF_4], [BMIM][Cl], [BuPy][Cl], and [OctMIM][Cl], respectively. To determine the CMC, the authors also used the change in the ^{1}H NMR chemical shift of the protons of the terminal methyl group in the alkyl chain as a function of concentration. In this chapter, the authors also established the intersection point between the plateau and the descending part for the CMC determination. As expected, protons of the alkyl chain for all the ILs make an upfield shift upon aggregation, showing a partial changeover from trans to gauche conformations in the alkyl chain (Singh and Kumar 2007). The authors used the aforementioned method to determine the N_{agg} by the ^{1}H NMR chemical shift data. The N_{agg} derived by this method were 10, 8, and 13 for [BMIM][BF_4], [BMIM][Cl], and [BuMPy][Cl], respectively (Singh and Kumar 2007). The IL ^{1}H NMR chemical shift of [OctMIM][Cl] at CAC and at higher concentrations resulted in aggregates of different sizes with N_{agg} of 23 and 10. The authors suggest that this decrease in the N_{agg} at higher IL concentration may be due to the penetration of water molecules into the aggregate as a consequence of the bending of the long alkyl chain. The authors also highlighted that information about the aggregate structure can be obtained from the change in the nature of the proton peak (broadening or splitting) upon aggregation. If no change is observed in the shape of the proton peak at the alkyl chain and the aromatic protons upon aggregation, this may be due to formation of a single type of aggregate that is in dynamic equilibrium with the monomers (Singh and Kumar 2007). A small change in the chemical shift of the C2 proton is observed upfield in the case of aqueous solutions of [BMIM][BF_4] and [BuMPy][Cl] with an increase in concentration. Contrary to [BMIM][BF_4] and [BuMPy][Cl], a downfield shift is observed in the aqueous solutions of [BMIM][Cl] and [BMIM][Cl]. Similar behavior is also observed for the other aromatic protons (Singh and Kumar 2007). The authors explain the results by considering that there is a competition for the interaction of aromatic protons for the solvent and the anions. In addition, they attribute the upfield shift in aqueous solutions of [BMIM][BF_4] and [BuMPy][Cl] upon aggregation to possible π–π interactions (ring stacking). The authors suggest that in the case of [BMIM][Cl] and [OctMIM][Cl], hydrogen-bonding interactions of aromatic protons are more important than the π–π interactions, preventing ring stacking upon aggregation (Singh and Kumar 2007).

In addition, Tariq et al. (2011) also studied self-aggregation properties in the aqueous solution of ILs [MeDodecPyrr][Br], [BuDodecPyrr][Br], and [BuOctPyrr][Br] (Figure 6.29) by fluorescence and NMR. NMR experiments were used to determine the CMC and diffusion coefficients by applying the Taylor dispersion method and DOSY (Souza 2002). D_{obs} were measured, and the results showed that all ILs have very similar D_{obs} (1.56 10^{-9} m^2 s^{-1} for [MeDodecPyrr][Br], 1.50 10^{-9} m^2 $s^{-1}$for [BuDodecPyrr][Br], and 1.46 10^{-9} m^2 s^{-1} for [BuOctPyrr][Br].

The CMC values determined by fluorescence and NMR for the aqueous solutions of [MeDodecPyrr][Br], [BuDodecPyrr][Br], and [BuOctPyrr][Br]

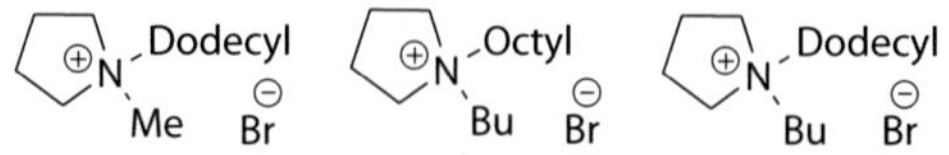

FIGURE 6.29
Chemical structure of ILs studied. (From Tariq, M. et al., *J. Colloid Interface Sci.*, 360, 606, 2011, accessed August 2012. http://www.sciencedirect.com/science/article/pii/S0021979711005315)

TABLE 6.2
CMC for ILs

IL	CMC (mM)	
	Fluorescence	NMR
[MeDodecPyrr][Br]	15 ± 3	10 ± 1
[BuDodecPyrr][Br]	—	5 ± 1
[BuOctPyrr][Br]	150 ± 20	120 ± 20

Source: Tariq, M. et al., *J. Colloid Interface Sci.*, 360, 606, 2011, accessed August 2012. http://www.sciencedirect.com/science/article/pii/S0021979711005315

are listed in Table 6.2. The authors performed the measurements at three temperatures (288, 298, and 323 K); however, they found that the CMC was invariable and expressed itself normally (Table 6.2). The authors noted that the DOSY NMR results yield comparable, albeit slightly lower CMC, values. The authors established that the CMC increases in the order [BuDodecPyrr][Br] < [MeDodecPyrr][Br] < [BuOctPyrr][Br]. This relationship was corroborated by other techniques and justified by the easier micellization process when longer alkyl chains are present. Finally, the authors demonstrated that the nature of the cation has a very minor effect on the thermodynamic quantities of micellization. However, the two alkyl side chains contribute differently to the values of the aggregation properties.

Finally, Li et al. (2010) synthesized a chiral long-chain IL, *S*-3-hexadecyl-1-(1-hydroxy-propan-2-yl)-imidazolium bromide ([HexaDec-OH-PrIM][Br]) (Figure 6.30) and used fluorescence spectroscopy and ^{1}H NMR chemical shifts to determine the CMC of [HexaDec-OH-PrIM][Br] in aqueous solutions.

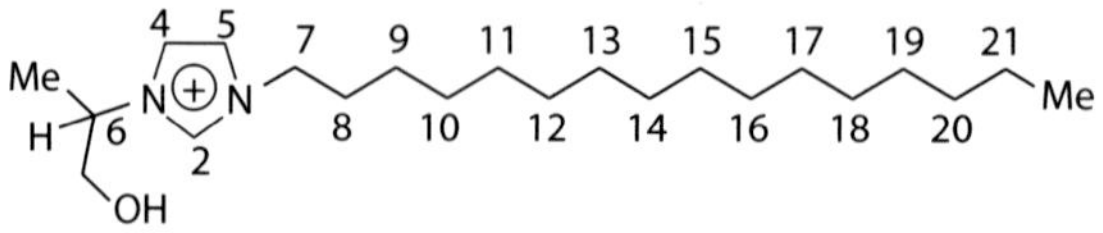

FIGURE 6.30
Structure of IL [HexaDec-OH-PrIM][Br]. (From Li, X.W. et al., *J. Colloid Interface Sci.*, 343, 94, 2010, accessed August 2012. http://www.sciencedirect.com/science/article/pii/S0021979709014519)

The authors found that for the protons in the alkyl chain (C9–C22), the signals shift upfield as the IL concentration increases, while the protons in the imidazolium ring (C2 and C3) or near the imidazolium group (C4–C8) shift downfield.

From this, the authors concluded that the chiral group linked to the imidazolium ring may bend into the hydrophobic region to a small extent, which may result in the C2, C4, C5, and C6 being located in a more nonpolar region of the micelle (Li et al.2010). The authors also investigated the molecular arrangement of [HexaDec-OH-PrIM][Br] in the micelles by 2D ROESY spectroscopic analysis. The spectrum allows adjacent protons in [HexaDec-OH-PrIM][Br] to have identifiable cross peaks, for example, C7/C8, C8/C9–21, C9–21/C22, C4/C5, and C5/C6. In addition, they observed that cross peaks arising from the interactions of C3/C7, C3/C8, and C7/C9–21 suggest that the hydrophilic head of [HexaDec-OH-PrIM][Br] may bend into the hydrophobic region of the micelles. The C2 in the imidazolium ring interacts with the C4 and C5 rather than C6, indicating that the methyl group near the chiral center is located farther from the C2 in the imidazolium ring than the others. The author affirms that the result indicates that the [HexaDec-OH-PrIM][Br] molecules are arranged in a certain order in micelles, which is different from achiral surfactants (Li et al. 2010). An important finding related by the authors is that the CMC obtained for [HexaDec-OH-PrIM][Br] is similar to [HexaDecIM][Br] but lower than for conventional ionic surfactants, demonstrating that long-chain imidazolium ILs have superior capability for micelle formation. The authors also noted that the relatively large head group size of [HexaDec-OH-PrIM][Br] and high electrostatic repulsion between the head groups can lead to a looser packing of the aggregation. Thus, a higher micropolarity and smaller N_{agg} are observed compared to conventional cationic surfactants that have a similar hydrophobic chain length.

Wang et al. (2008) examined the anion effects in the aggregation behavior of ILs based on imidazolinium, pyridinium, and pyrrolidinium cations shown in Figure 6.21. A mixture of IL and water at several molar fractions was submitted to fluorescence and DLS methods. Fluorescence experiments were performed by using a pyrene solvatochromic probe. The variation of the pyrene polarity ratio I_1/I_3 plotted against the IL concentrations shows a specific sigmoidal-type curve with a rapid decrease of I_1/I_3 at concentrations below the CMC and stabilization at concentrations above the CMC, indicating the formation of aggregates. The authors also determined the average N_{agg} of the monomers in the aggregates using the steady-state fluorescence quenching of the micelle-solubilized probe. The authors found that the average N_{agg} generally increases with an increase in hydrophobicity of the anions, indicating that the aggregation is favored by increased hydrophobicity. Therefore, the authors concluded that the anions with higher hydrophobicity can bind the cations more strongly and reduce the repulsive interaction between head groups, thus enhancing the aggregate growth. In addition, this work showed that anions act by altering repulsive head group interactions and altering

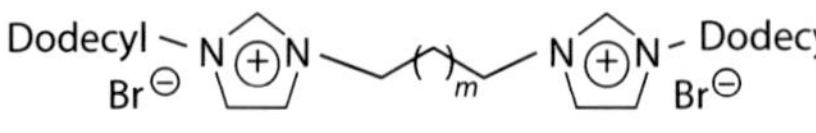

Dodecyl N⊕N N⊕N Dodecyl
Br⊖ m Br⊖

$m = 0$, 2, and 4

FIGURE 6.31
Chemical structure of the IL [Bis-Et[Bu, Hex]DodecIM][2Br] studied. (From Ao, M. et al., *Colloid Polym. Sci.*, 287, 395, 2009, accessed August 2012. http://www.springerlink.com/content/101551/)

the attractive forces that arise from a need to minimize the exposure of the hydrophobic core to water.

Ao et al. (2009) used fluorescence measurements and DLS to study the aggregation behavior and thermodynamic properties for the IL [Bis-Et[Bu, Hex]DodecIM][2Br] with different spacer lengths (Figure 6.31). In accordance with the authors, fluorescence measurements suggest that the micropolarity of micelles increases, but the N_{agg} decreases with an increase in the spacer length of [Bis-Et[Bu, Hex]DodecIM][2Br]. The authors suggested that the reason may be that the longer spacer remains in extended conformation and allows it to aggregate relatively loosely to form smaller aggregates. In addition, the authors observed that with an increase in temperature, there is an increase in the CMC. From DLS experiments, the authors found that the micelle of [Bis-Et[Bu, Hex]DodecIM][2Br] would be smaller in size at higher temperatures than at lower temperatures. As can be seen for [Bis-Et[Bu, Hex]DodecIM][2Br], the apparent hydrodynamic radius decreases from 4.0 nm at 25°C to 2.5 nm at 35°C.

From the papers selected to exemplify the use of fluorescence for investigating formation and characterization of IL aggregates, it is possible to see that fluorescence was essentially used to determine the CMC. Therefore, as well as ESI-MS, it has a complementary role in the investigation of IL aggregates when compared to NMR and LS methods. In some papers, the studies were conducted at different temperatures; however, the CMC and the N_{agg} observed by fluorescence was invariable. This result reveals that fluorescence has not been explored to its full potential for the investigation of IL aggregate formation. The formation of aggregates was found in most of the ILs investigated. The CMC was determined basically by two spectroscopic methods: fluorescence and NMR. We collected these data in a systematic form, as shown in Table 6.3. From Table 6.3, we can see that when the CMC of an IL was determined by NMR and fluorescence, the values are quite close (taking into account experimental errors). In addition, N_{agg} was observed by fluorescence and few times by NMR. However, a comparison between the two methods was not made because the same IL was not used.

Finally, it is worth highlighting that the vast majority of works covered by the scope of this chapter deal with monocationic ILs. Only one experimental

TABLE 6.3

CMC and N_{agg} Determined by NMR and Fluorescence

	CAC (mM) 25°C		N_{agg} 25°C		
ILs	**Fluor**	**RMN**	**Fluor**	**RMN**	**References**
[BMIM][BF_4]	700	952	—	10	Singh et al. (2007)
[BMIM][Cl]	800	935	—	8	Singh et al. (2007)
[N-Bu-4-MePy][Cl][a]	600	609	—	13	Singh et al. (2007)
[OctMIM][Cl]	101	90	—	10/23	Singh (2007)
[OctMIM][Cl][b]	—	200	—	—	Blesic et al. (2007)
[DecMIM][Cl][b]	45	55	—	—	Blesic (2007)
[DodecMIM][Cl][b]	7	13	—	—	Blesic et al. (2007)
[TetDecMIM][Cl][b]	3	4	—	—	Blesic et al. (2007)
[OctMIM][CH3COO]	0.22[c]	—	37	—	Wang et al. (2008)
[OctMIM][Cl]	0.21[c]	—	39	—	Wang et al. (2008)
[OctMIM][NO_3]	0.16[c]	—	65	—	Wang et al. (2008)
[OctMIM][CF_3COO]	0.15[c]	—	68	—	Wang et al. (2008)
[O_{Ct}-4-MePy][Br]	0.13[c]	—	46	—	Wang et al. (2008)
[MeOctPyrr][Br]	0.20[c]	—	74	—	Wang et al. (2008)
[MeDodecPyrr][Br][b]	15 ± 3	10 ± 1	—	—	Tariq et al. (2011)
[BuDodecPyrr][Br][b]	—	5 ± 1	—	—	Tariq et al. (2011)
[BuOctPyrr][Br][b]	150 ± 20	120 ± 20	—	—	Tariq et al. (2011)
[HexaDec-OH-PrIM][Br][b]	0.471	0.523	25	—	Li et al. (2010)
[DodecMIM][Br]	0.010[c]	—	44	—	Wang et al. (2007)
[DecMIM][Br]	0.046[c]	—	35	—	Wang et al. (2007)
[OctMIM][Br]	0.19[c]	—	53	—	Wang et al. (2007)
[HexMIM][Br]	0.88[c]	—	—	—	Wang et al. (2007)
[BMIM][BF_4]	0.96[c]	—	—	—	Wang et al. (2007)
[MeDodecPyrr][Br][b]	—	—	49	—	Zhao et al. (2011)
[MeTetDecPyrr][Br][b]	—	—	55	—	Zhao et al. (2011)
[MeHexDecPyrr][Br][b]	—	—	59	—	Zhao et al. (2011)
[OHEtDodecIM][Cl][b]	2.29	—	8[e]	—	Liu et al. (2011a,b)
	2.67[e]	—	63[d,e]		
	14.78[d]				
	15.83[d,e]				
[CarboxyMeDodecIM][b]	1.25	—	28.3	—	Liu et al. (2011a,b)
	1.74[e]	—	35.8[e]	—	—
[Bis-EtDodecIM][Br_2][b]	0.59	—	21	—	Ao et al. (2009)
[Bis-BuDodecIM][Br_2][b]	0.72	—	16	—	Ao et al. (2009)
[Bis-HexDodecIM][Br_2][b]	0.80	—	14	—	Ao et al. (2009)
[N-BuPy][Cl]	—	900	—	—	Sastry et al. (2012)
[N-HexPy][Cl]	—	800	—	—	Sastry et al. (2012)

(continued)

TABLE 6.3 (continued)

CMC and N_{agg} Determined by NMR and Fluorescence

	CAC (mM) 25°C		N_{agg} 25°C		
ILs	**Fluor**	**RMN**	**Fluor**	**RMN**	**References**
[N-OctPy][Cl]	—	180	—	—	Sastry et al. (2012)
[O_{Ct}-3-MePy][Cl]	—	170	—	—	Sastry et al. (2012)
[O_{Ct}-4-MePy][Cl]	—	175	—	—	Sastry et al. (2012)

[a] Self-excitation: The whole fluorescence spectrum was measured at an excitation wavelength of 320 nm, and the ratio of second peak at ($2\lambda_{excitation}$) to the first peak was used for analysis.
[b] CMC.
[c] $mol.kg^{-1}$.
[d] Second CMC.
[e] Measure at 35°C.

work was found where fluorescence and DLS results were related. No comparison was made with IL monocationic analogue. This indicates that although the study of monocationic ILs has started and is evolving, the same is not true for dicationic ILs. Considering that the existence of imidazolium head groups confers an inherent ionic nature to ILs and that the novel IL dicationics with different spacer lengths would have special properties and potential applications in many areas, such as skin care, medicine, life science, petrochemistry, and the textile industry, this chapter highlights the need for studies on the investigation of properties of new dicationic ILs as gemini surfactants.

6.2.4.1 Fluorescence Experiments

The fluorescence measurements were obtained on a Cary Eclipse (Varian, America) fluorophotometer with cell holder thermostated by a circulating ethylene glycol bath at 25°C ± 0.4°C. For the CAC determination of the IL by using pyrene as a probe, the IL aqueous solution (concentration from 10^{-6} at 1000 mM) was prepared in the presence of 1×10^{-6} $mol.L^{-1}$ pyrene. The fluorescence intensities were measured at 373 nm (band 1) and 383 nm (band 3) with the excitation wavelength of 334 nm. The emission spectra were scanned from 350 to 450 nm, and the excitation and emission band slits were fixed at 2.5 nm. The binary mixtures of IL/water were prepared at concentrations of 1.10^{-6}, 1.10^{-3}, 0.01, 0.1, 1, 50, 125, 175, 500, and 1000 mM. The curves shown in Figure 6.24 were constructed by plotting these concentrations against the fluorescence of pyrene. The [BisOctMIM][$2BF_4$] was synthesized in our laboratories according to the literature (Frizzo et al. 2008, 2009a,b, 2011; Martins et al. 2008a,b, 2007, 2009; Buriol, 2011).

6.3 Conclusion

After having examined the literature on IL aggregate characterization by spectrometric methods including ESI, NMR, LS, and fluorescence, it is possible to conclude that (1) there is a lack of studies on the formation and characterization of dicationic ILs using spectrometric methods, which demonstrates the need for systematic studies of ILs that have high potential as a new soft material; (2) spectrometric methods analyzed in this chapter are complementary in the investigation of formation and characterization of IL aggregates, showing that there is no absolute spectrometric experiment that allows the complete characterization of an IL aggregate. It is evident that NMR and LS are the methods that furnished the most complete and detailed information about aggregated formation and characterization. It is also evident that these same methods have not been exploited to their potential; and (3) there is a deficiency in the investigation of the influence of temperature and solvents in the formation, characterization, and stability of aggregates by the use of the spectrometric methods selected in this chapter. This indicates the great deal of studies that we need to perform in this area to determine the properties of IL aggregates at temperatures and in dissolution media, which are strongly linked to this field of applications.

Finally, in this chapter, we hope to have given a clear idea of the use of important spectrometric tools for the investigation of the formation and characterization of IL aggregates. We conclude with an optimistic view for the future expansion of the development of IL aggregates as a new soft material. This positive view comes from the certainty that the results reported here are the beginning of a great advance in this promising field.

Abbreviations and Symbols

$\partial\delta$	partial derivate of chemical shift
$\Delta(PA)$	proton affinity difference
θ	the scattering angle
η	the viscosity of the medium
λ	the wavelength of incident light
∂c	partial derivate of concentration
ΔG°_{m}	standard Gibbs free energy
ΔH°	standard enthalpy
δ_{mic}	chemical shifts of the aggregated form
δ_{mon}	chemical shifts of the monomer
δ_{obs}	observed chemical shift

ΔS°_m	standard entropy
$[A]$	anion
$[C_{(n+1)}A_n]$	n is the number of anions in the aggregate
$[C_{(n+1)}A_n]^+$	representation of aggregate positively charged
$[C]$	cation
$[C_nA_{(n+1)}]^-$	representation of aggregate negatively charged
A_2	second virial coefficient
B3LYP/6-311+G(d,p)	theoretical level of density functional theory (DFT), where B3LYP is a hybrid functional and represents the Becke, 3-parameter, and Lee–Yang–Parr and 6-311+G (d,p) is the base set
Γ	decay rate
CID	collision-induced dissociation
CKM	Cooks kinetic method
CMC	critical micelle concentration
c_{mic}	concentrations of the amphiphilic molecules existing as micelles
c_{mon}	concentrations of the amphiphilic molecules existing as monomers
c_{tot}	concentrations of the amphiphilic molecules existing as total solution
D_{free}	self-diffusion coefficient of free molecules of amphiphilic
D_h	hydrodynamic diameter
DLS	dynamic light scattering
D_{mic}	self-diffusion coefficient of micelles
DMSO	dimethyl sulfoxide
dn/dc	refractive index increment
D_{obs}	self-diffusion coefficient
DOSY	diffusion-ordered spectroscopy
DSC	differential Scanning calorimetry
$E_{1/2}$	here defined as the value of collision energy, in the laboratory frame, at which the relative abundance (defined as percentage of the base peak) of the precursor ion is 50%
$E_{cm,1/2}$	the energy required to dissociate 50% of the precursor ion
E_{cm}	center of mass energy, which can be calculated from E_{lab}
$E_{lab,1/2}$	defined as the energy required to dissociate 50% of the precursor ion
E_{lab}	applied collisional activation voltage
ESI-MS	electrospray ionization mass spectrometry
$ET_{(30)}$	an empirical solvent polarity measure
$G^{(2)}(\tau)$	time autocorrelation function of the intensity of the scattered light

$I(t)$	intensity of light scattering at some arbitrary time t
I	intensity
I_2	where I is the abundance of the corresponding aggregate ion and the subscript represents the aggregate size
ILs	ionic liquids
ILt	inverse Laplace transformation
K	the equilibrium constant
k'	is an arbitrary constant chosen to provide a convenient spread
k_1 and k_2	rate constants for the competitive dissociations
k_BT	the Boltzmann energy
k_f	wavector of scattered radiation
k_i	wavector of incident radiation
K_{opt}	optical constant
L	dimension of the particle
LS	light scattering
m_l	magnetic quantum number
m_p	mass of the precursor ion
m_t	mass of the neutral target
M_w	molecular weight of the particle
n	index of refraction of the medium
N_A	is Avogadro's number
N_{agg}	aggregation number
NMR	nuclear magnetic resonance
NOESY	nuclear Overhauser effect spectroscopy
$P(\theta)$	particle form factor
PFGSE	pulsed field gradient spin echo
PVDF	polyvinylidene difluoride
q	scattering vector
$R(\theta)$	Rayleigh ratio
R_g	radius of gyration
ROESY	rotating frame Overhauser effect spectroscopy
S	aggregate size
SLS	static light scattering
T_1	longitudinal or lattice relaxation times
T_2	transversal relaxation times
T_{eff}	effective temperature
TEM	transmission electron microscopy
wt%	molar fraction
α	degree of ionization of the aggregates
β	the fraction of counterions that are bound to the aggregate
ρ	defined as the ratio between R_g and R_h (R_g/R_h)

Representation of ILs Found in This Chapter

Representation	Structure	Name
[BisOctMIM]	Me–N⊕N–(CH₂)₆–N⊕N–Me, Br⊖ Br⊖	1,8-bis(3-methylimidazolium-1-yl)
[BisDecMIM]	Me–N⊕N–(CH₂)₈–N⊕N–Me, Br⊖ Br⊖	1,10-bis(3-methylimidazolium-1-yl)
[BMIM]	Me–N⊕N–Butyl	1-butyl-3-methylimadozolium
[PentMIM]	Me–N⊕N–Pentyl	1-pentyl-3-methylimidazolium
DMIM	Me–N⊕N–Me	1,3-dimethylimidazolium
[PrMIM][NTf$_2$]),	Me–N⊕N–Propyl	1-propyl-3-methylimidazolium
[EtMIM]	Me–N⊕N–Ethyl	1-ethyl-3-methylimidazolium
[HexMIM]	Me–N⊕N–Hexyl	1-hexyl-3-methylimidazolium
[HepMIM]	Me–N⊕N–Heptyl	1-heptyl-3-methylimidazolium
[OctMIM]	Me–N⊕N–Octyl	1-octyl-3-methylimidazolium

[DecMIM]	 Me–N ⊕ N–Decyl	1-decyl-3-methylimidazolium
[DEtIM]	 Ethyl–N ⊕ N–Ethyl	1,3-diethylimidazolium
[Bu-2-Me-MIM]	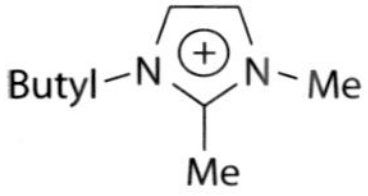 Butyl–N ⊕ N–Me; Me	1-butyl-2-methyl-3-methylimidazolium
[EMIM]	 Me–N ⊕ N–Ethyl	1-ethyl-3-methylimidazolium
[OHEtDodecIM]	 HO N ⊕ N–Dodecyl	1-hydroxyethyl-3-dodecylimidazolium
[DodecMIM]	 Me–N ⊕ N–Dodecyl	1-dodecyl-3-methylimidazolium
[TetDecMIM]	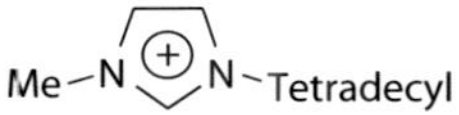 Me–N ⊕ N–Tetradecyl	1-tetradecyl-3-methylimidazolium
[CarboxyMeDodecIM]	 ⊖OOC N ⊕ N–Dodecyl	1-carboxymethyl-3-dodecylimidazolium
[HexaDec-OH-PrIM]	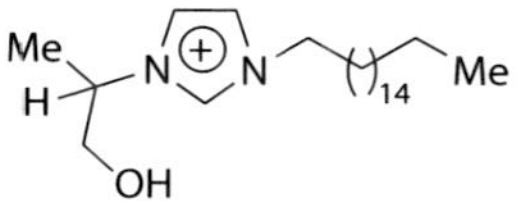 Me, H, N ⊕ N, 14, Me, OH	3-hexadecyl-1-(1-hydroxy-propan-2-yl)-imidazolium

(continued)

Representation of ILs Found in this Chapter (continued)

Representation	Structure	Name
[C_mPhIM] $m = m = 9$, 11, and 13	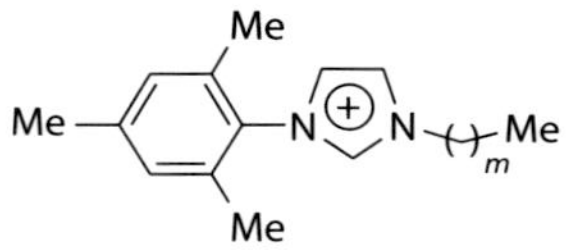	1-(2,4,6-trimethylphenyl)-3-alkylimidazolium
[Bis-Et[Bu, Hex]DodecIM] m = 0, 2, 4	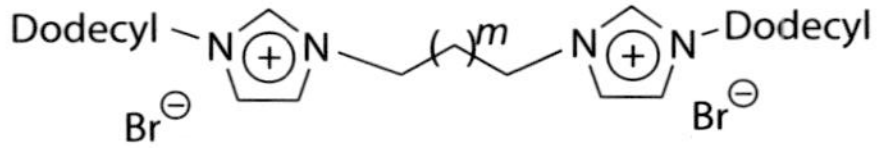	1,2-bis(3-dodecylimidazolium-1-yl) ethane
[N-BuPy]	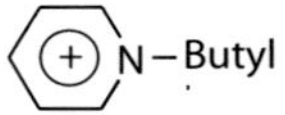	1-butylpyridinium
[N-HexPy]	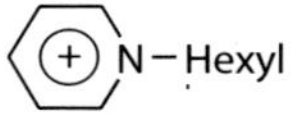	1-hexylpyridinium
[N-OctPy]	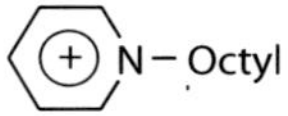	1-octylpyridinium
[N-PrMePy]	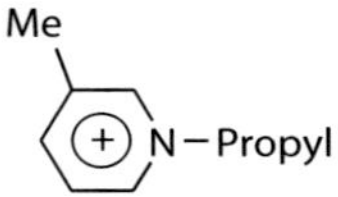	1-propyl-3-methylpyridinium
[N-Bu -3-MePy]	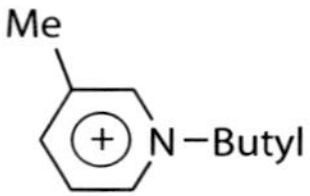	1-butyl-3-methylpyridinium
[N-Bu -4-MePy]		1-butyl-4-methylpyridinium

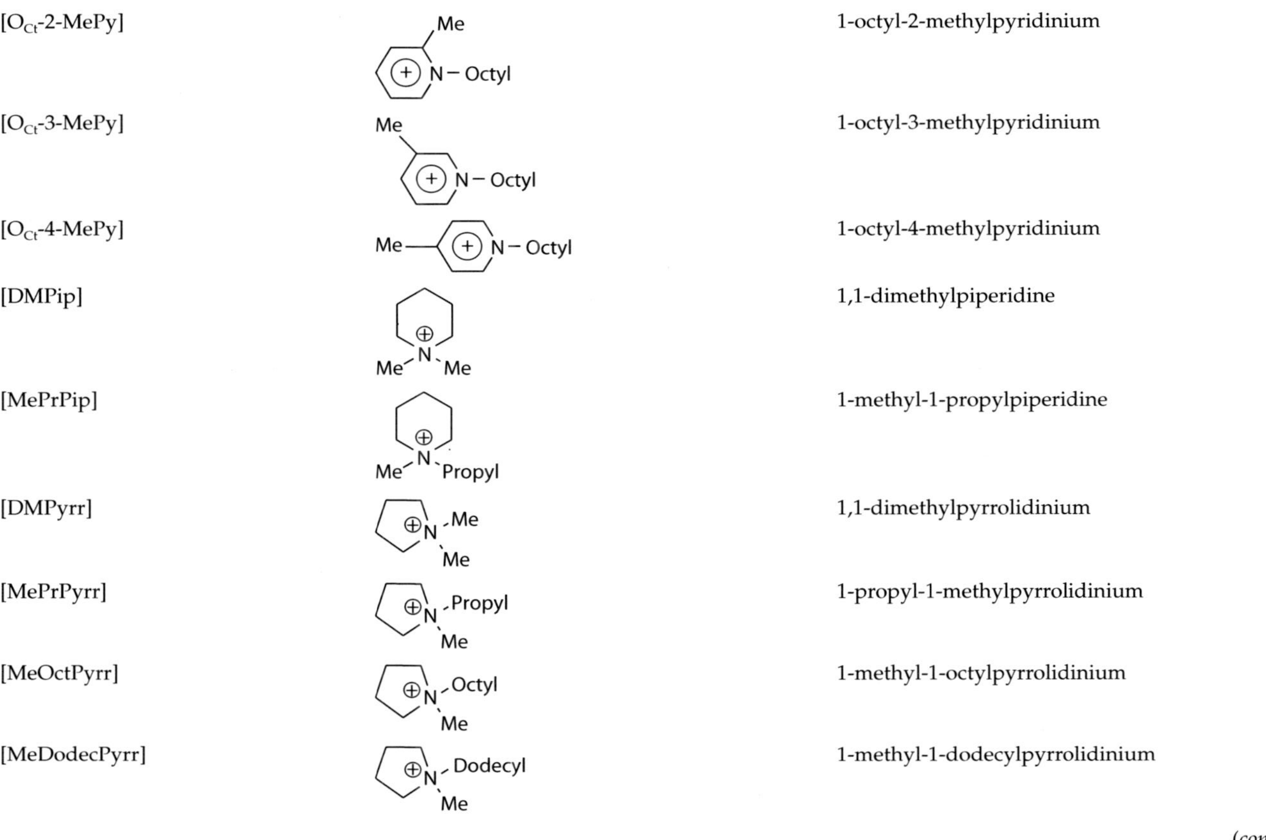

[O_{Ct}-2-MePy]	1-octyl-2-methylpyridinium
[O_{Ct}-3-MePy]	1-octyl-3-methylpyridinium
[O_{Ct}-4-MePy]	1-octyl-4-methylpyridinium
[DMPip]	1,1-dimethylpiperidine
[MePrPip]	1-methyl-1-propylpiperidine
[DMPyrr]	1,1-dimethylpyrrolidinium
[MePrPyrr]	1-propyl-1-methylpyrrolidinium
[MeOctPyrr]	1-methyl-1-octylpyrrolidinium
[MeDodecPyrr]	1-methyl-1-dodecylpyrrolidinium

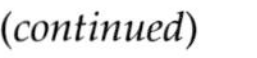
(*continued*)

Representation of ILs Found in this Chapter (continued)

Representation	Structure	Name
[MeTetDecPyrr]	N⊕ -Tetradecyl, Me	1-methyl-1-tetradecylpyrrolidinium
[MeHexDecPyrr]	N⊕ -Hexadecyl, Me	1-methyl-1-hexadecylpyrrolidinium
[BuDodecPyrr]	N⊕ -Dodecyl, Butyl	1-butyl-1-dodecylpyrrolidinium
[BuOctPyrr]	N⊕ -Octyl, Butyl	1-butyl-1-octylpyrrolidinium

References

Aggarwal, A., Lancaster, N., Sethi, A.R., Welton, T. The role of hydrogen bonding in controlling the selectivity of Diels–Alder reactions in room-temperature ionic liquids. *Green Chemistry* 4 (2002): 517–520. Accessed August 2012. http://pubs.rsc.org/en/journals/journalissues/gc#!issueid=gc004005&type=current

Alfassi, Z.B., Huie, R.E., Milman, B.L., Neta, P. Electrospray ionization mass spectrometry of ionic liquids and determination of their solubility in water. *Analytical and Bioanalytical Chemistry* 377 (2003): 159–164. Accessed August 2012. http://www.springerlink.com/content/1618-2642/377/1/

Anichina, J., Bohme, D.K. Mass-spectrometric studies of the interactions of selected metalloantibiotics and drugs with deprotonated hexadeoxynucleotide GCATGC. *The Journal of Physical Chemistry B* 113 (2009): 328–335. Accessed August 2012. http://pubs.acs.org/toc/jpcbfk/113/1

Ao, M., Huang, P., Xu, G., Xiaodeng, Y., Wang, Y. Aggregation and thermodynamic properties of ionic liquid-type gemini imidazolium surfactants with different spacer length. *Colloid & Polymer Science* 287 (2009): 395–402. Accessed August 2012. http://www.springerlink.com/content/101551/

Armentrout, P.B. Entropy measurements and the kinetic method: A statistically meaningful approach. *Journal of the American Society for Mass Spectrometry* 11 (2000): 371–379. Accessed August 2012. http://www.sciencedirect.com/science/journal/10440305/11/5

Augusti, D.V., Carazza, F., Augusti, R., Tao, W.A., Cooks, R.G. Quantitative chiral analysis of sugars by electrospray ionization tandem mass spectrometry using modified amino acids as chiral reference compounds. *Analytical Chemistry* 74 (2002): 3458–3462. Accessed August 2012. http://pubs.acs.org/toc/ancham/74/14

Bernardes, C.E.S., Minas da Piedade, M.E., Lopes, J.N.C. The structure of aqueous solutions of a hydrophilic ionic liquid: The full concentration range of 1-Ethyl-3-methylimidazolium Ethylsulfate and water. *Journal of Physical Chemistry B* 115 (2011): 2067–2074 and references cited therein. Accessed August 2012. http://pubs.acs.org/doi/abs/10.1021/jp1113202

Berne, B.J., Pecora, R. *Dynamic Light Scattering—With Applications to Chemistry, Biology, and Physics*. New York: Dover, 2000.

Bhargava, B.L., Klein, M.L. Nanoscale organization in aqueous dicationic ionic liquid solutions. *Journal of Physical Chemistry B* 115 (2011): 10439–10446 and references cited therein. Accessed August 2012. http://pubs.acs.org/doi/pdf/10.1021/jp204413n

Bini, R., Bortolini, O., Chiappe, C., Pieraccini, D., Siciliano, T. Development of cation/anion "interaction" scales for ionic liquids through ESI-MS measurements. *The Journal of Physical Chemistry B* 111 (2007): 598–604. Accessed August 2012. http://pubs.acs.org/toc/jpcbfk/111/3

Bini, R., Chiape, C., Marmugi, E., Pieraccini, D. The "non-nucleophilic" anion $[Tf_2N]^-$ competes with the nucleophilic Br^-: An unexpected trapping in the dediazoniation reaction in ionic liquids. *Chemical Communications* 8 (2006): 897–899. Accessed August 2012. http://pubs.rsc.org/en/journals/journalissues/cc#!issueid=cc006008&type=current

Blesic, M., Lopes, A., Melo, E., Petrovski, Z., Plechkova, N.V., Lopes, J.N.C., Seddon, K.R., Rebelo, L.P.N. On the self-aggregation and fluorescence quenching aptitude of surfactant ionic liquids. *Journal of Physical Chemistry B* 112 (2008): 8645–8650. Accessed August 2012. http://pubs.acs.org/doi/pdf/10.1021/jp802179j

Blesic, M., Marques, M.H., Plechkova, N.V., Seddon, K.R., Rebelo, L.P.N., Lopes, A. Self-aggregation of ionic liquids: Micelle formation in aqueous solution. *Green Chemistry* 9 (2007): 481–490. Accessed August 2012. http://pubs.rsc.org/en/content/articlelanding/2007/gc/b615406a

Bordini, E., Hamdan, M. Investigation of some covalent and noncovalent complexes by matrix-assisted laser desorption/ionization time-of-flight and electrospray mass spectrometry. *Rapid Communications in Mass Spectrometry* 13 (1999): 1143–1151. Accessed August 2012. http://onlinelibrary.wiley.com/doi/10.1002/%28SICI%291097-0231%2819990630%2913:12%3C%3E1.0.CO;2-D/issuetoc

Buriol, L., Frizzo, C.P., Prola, L.D.T., Moreira, D.N., Marzari, M.R.B., Scapin, E., Zanatta, N., Bonacorso, H.G., Martins, M.A.P. Synergic effects of ionic liquid and microwave irradiation in promoting Trifluoromethylpyrazole synthesis. *Catalysis Letters* 141 (2011): 1130–1135. Accessed August 2012 http://www.springerlink.com/content/7t76u37p53014186/

Cech, N.B., Enke, C.G. Practical implications of some recent studies in electrospray ionization fundamentals. *Mass Spectrometry Reviews* 20 (2001): 362–387. Accessed August 2012. http://www.ncbi.nlm.nih.gov/pubmed/11997944

Chen, G., Wong, P., Cooks, R.G. Estimation of free radical ionization energies by the kinetic method and the relationship between the kinetic method and the Hammett Equation. *Analytical Chemistry* 69 (1997): 3641–3645. Accessed August 2012. http://pubs.acs.org/toc/ancham/69/17

Chen, Y., Ke, F., Wang, H., Zhang, Y., Liang, D., Phase separation in mixtures of ionic liquids and water. *Chemical Physics and Physical Chemistry* 13 (2012): 160–167. Accessed August 2012. http://onlinelibrary.wiley.com/doi/10.1002/cphc.v13.1/issuetoc

Consorti, C.S., Suarez, P.A.Z., Souza, R.F., Burrow, R.A., Farrar, D.A., Loh, W., Silva, L.H.M., Dupont, J. Identification of 1,3-Dialkylimidazolium salt supramolecular aggregates in solution. *Journal of Physical Chemistry B* 109 (2005): 4341–4349. Accessed August 2012. http://pubs.acs.org/toc/jpcbfk/109/10

Cooks, R.G., Patrick, J.S., Kotiano, T., McLuckey, S.A. Thermochemical determinations by the kinetic method. *Mass Spectrometry Reviews* 13 (1994): 287–239. Accessed August 2012. http://onlinelibrary.wiley.com/doi/10.1002/mas.v13:4/issuetoc

Coolbaugh, M.T., Garvey, J.F. Magic numbers in molecular clusters: A probe for chemical reactivity. *Chemical Society Reviews* 163 (1992): 163–169. Accessed August 2012. http://pubs.rsc.org/en/Content/ArticleLanding/1992/CS/CS9922100163

Dorbritz, S., Ruth, W., Kragl, U. Investigation on aggregate formation of ionic liquids. *Advanced Synthesis & Catalysis* 347 (2005): 1273–1279. Accessed August 2012. http://onlinelibrary.wiley.com/doi/10.1002/adsc.v347:9/issuetoc

Eberlin, M.N., Kotiaho, T., Shay, B.J., Yang, S.S., Cooks, G. Gas-phase Cl^+ affinities of pyridines determined by the kinetic method using multiple-stage (MS3) mass spectrometry. *Journal of the American Chemical Society* 116 (1994): 2457–2465. Accessed August 2012. http://pubs.acs.org/toc/jacsat/116/6

Echt, O., Sattler, K., Recknagel, E. Magic numbers for sphere packings: Experimental verification in free xenon clusters. *Physical Review Letters* 47 (1981): 1121–1124. Accessed August 2012. http://adsabs.harvard.edu/abs/1981PhRvL.47.1121E

Enke, C.G. A predictive model for matrix and analyte effects in electrospray ionization of singly-charged ionic analytes. *Analytical Chemistry* 69 (1997): 4885–4893. Accessed, August 2012. http://pubs.acs.org/doi/abs/10.1021/ac970095w

Ervin, K.M. Microcanonical analysis of the kinetic method: The meaning of the "apparent entropy." *Journal of the American Society for Mass Spectrometry* 13 (2002): 435–452. Accessed August 2012. http://www.sciencedirect.com/science/journal/10440305/13/5

Feng, Q., Wang, H., Zhang, S., Wang, J. Aggregation behavior of 1-dodecyl-3-methylimidazolium bromide ionic liquid in non-aqueous solvents. *Colloids and Surfaces A: Physicochemical and Engineering Aspects* 367 (2010): 7–11 and references cited therein. Accessed August 2012. http://www.sciencedirect.com/science/article/pii/S0927775710003298

Fenn, J.B. Ion formation from charged droplets: Roles of geometry, energy and time. *Journal of the American Society for Mass Spectrometry* 4 (1993): 524–535. Accessed August 2012. http://www.sciencedirect.com/science/article/pii/104403059385014O

Fernandes, A.M., Coutinho, J.A.P, Marrucho, I.M. Gas-phase dissociation of ionic liquid aggregates studied by electrospray ionisation mass spectrometry and energy-variable collision induced dissociation. *Journal of Mass Spectrometry* 44 (2009): 144–155. Accessed August 2012. http://onlinelibrary.wiley.com/doi/10.1002/jms.v44:1/issuetoc

Fernandes, A.M., Rocha, M.A.A., Freire, M.G., Marrucho, I.M., Coutinho, J.A.P., Santos, L.M.N.B.F. Evaluation of cation–anion interaction strength in ionic. *The Journal of Physical Chemistry B* 115 (2011): 4033–4041. Accessed August 2012. http://pubs.acs.org/toc/jpcbfk/115/14

Friedman, L., Beuhler, R.J. Magic numbers for argon and nitrogen cluster ions. *The Journal of Chemical Physics* 78 (1983): 4669–4675. Accessed August 2012. http://jcp.aip.org/resource/1/jcpsa6/v78/i7/p4669_s1

Frizzo, C.P., Martins, M.A.P., Moreira, D.N., Marzari, M.R.B., Rosa, F.A., Zanatta, N., Bonacorso, H.G. Synthesis of β-enaminones by ionic liquid catalysis: A one-pot condensation under solvent-free conditions. *Catalysis Communications* 9 (2008): 1375–1378. Accessed August 2012 http://www.sciencedirect.com/science/article/pii/S1566736707005237

Frizzo, C.P., Marzari, M.R.B., Buriol, L., Moreira, D.N., Rosa, F.A., Vargas, P.S., Zanatta, N., Bonacorso, H.G., Martins, M.A.P. Ionic liquid effects on the reaction of β-enaminones and tert-butylhydrazine and applications for the synthesis of pyrazoles. *Catalysis Communications* 10 (2009a): 1967–1970. Accessed August 2012. http://www.sciencedirect.com/science/article/pii/S1566736709002726

Frizzo, C.P., Moreira, D.N., Guarda, E.A., Fiss, G.F., Marzari, M.R.B., Zanatta, N., Bonacorso, H.G., Martins, M.A.P. Ionic liquid as catalyst in the synthesis of N-alkyl trifluoromethyl pyrazoles. *Catalysis Communications* 10 (2009b): 1153–1156. Accessed August 2012. http://www.sciencedirect.com/science/article/pii/S1566736708005621

Frizzo, C.P., Moreira, D.N., Martins, M.A.P. Ionic liquids: Applications in heterocyclic synthesis. In: *Ionic Liquids: Applications and Perspectives*, ed. Prof. Alexander Kokorin, pp. 415–438. InTech: Vienna, Austria, 2011. Accessed August 2012. http://www.intechopen.com/books/ILs-applications-and-perspectives/ILs-applications-in-heterocyclic-synthesis

Ganem, B., Tsyr, Y., Henion, J.D. Detection of noncovalent receptor-ligand complexes by mass spectrometry. *Journal of the American Chemical Society* 113 (1991): 6294–6296. Accessed August 2012. http://pubs.acs.org/toc/jacsat/113/16

Gehlen, M.H., Schryver, F.C.D. Time-resolved fluorescence quenching in micellar assemblies. *Chemical Reviews* 93 (1993): 199–221. Available at: http://pubs.acs.org/doi/abs/10.1021/cr00017a010

Goodchild, I., Collier, L., Millar, S.L., Prokeš, I., Lord, J.C.D., Butts, C.P., Bowers, J., Webster, J.R.P., Heenan, R.K. Structural studies of the phase, aggregation and surface behaviour of 1-alkyl-3-methylimidazolium halide + water mixtures. *Journal of Colloid and Interface Science* 307 (2007): 455–468. Accessed August 2012. http://www.sciencedirect.com/science/article/pii/S0021979706011052

Gozzo, F.C., Eberlin, M.N. Primary and secondary kinetic isotope effects in proton (H+/D+) and chloronium ion (35Cl+/37Cl+) affinities. *Journal of Mass Spectrometry* 36 (2001): 1140–1148. Accessed August 2012. http://onlinelibrary.wiley.com/doi/10.1002/jms.v36:10/issuetoc

Gozzo, F.C., Santos, L.S., Augusti, R., Consorti, C.S., Dupont, J., Eberlin, M.N. Gaseous supramolecules of imidazolium ionic liquids: "Magic" numbers and intrinsic strengths of hydrogen bonds. *Chemistry—A European Journal* 10 (2004): 6187–6193. Accessed August 2012. http://onlinelibrary.wiley.com/doi/10.1002/chem.v10:23/issuetoc

Harris, D.C. *Quantitative Chemical Analysis*. New York: W. H. Freeman and Company, 2007. Accessed August 2012. http://www.sciencedirect.com/science/journal/10440305/13/5

Huang, S.Y., Lin, Y.Y., Lisitza, N., Warren, W.S., Signal interferences from turbulent spin dynamics in solution nuclear magnetic resonance spectroscopy. *Journal of Chemical Physics* 116 (2002): 10325–10337. Accessed August 2012. http://jcp.aip.org/resource/1/jcpsa6/v116/i23/p10325_s1?isAuthorized=no

Inoue, T., Ebina, H., Dong, B., Zheng, L. Electrical conductivity study on micelle formation of long-chain imidazolium ionic liquids in aqueous solution. *Journal of Colloid and Interface Science* 314 (2007): 236–241. Accessed August 2012. http://www.sciencedirect.com/science/article/pii/S0021979707007345

Iribarne, J.V., Dziedzic, P.J., Thomson, B.A. Atmospheric pressure ion evaporation-mass spectrometry. *International Journal of Mass Spectrometry and Ion Physics* 50 (1983): 331–347. Accessed August 2012. http://www.sciencedirect.com/science/article/pii/0020738183870090

Jakes, J. Testing of the constrained regularization method of inverting Laplace transform on simulated very wide quasielastic light scattering autocorrelation functions. *Czech Journal Physics B* 38 (1988): 1305–1316. Accessed August 2012. http://www.springerlink.com/content/t380g54476540j32/

Jecklin, M.C., Touboul, D., Bovet, C., Wortmann, A., Zenobi, R. Which electrospray-based ionization method best reflects protein-ligand interactions found in solution? A comparison of ESI, nanoESI, and ESSI for the determination of dissociation constants with mass spectrometry. *Journal of the American Society for Mass Spectrometry* 19 (2008): 332–343. Accessed August 2012. http://www.sciencedirect.com/science/journal/10440305/19/3

Jerschow, A., Mueller, N., Suppression of convection artifacts in stimulated-echo diffusion experiments: Double-stimulated-echo experiments. *Journal of Magnetic Resonance* 125 (1997): 372–375. Accessed August 2012. http://www.sciencedirect.com/science/article/pii/S109078079791123X

Jerschow, A., Müller, N., 3D diffusion-ordered TOCSY for slowly diffusing molecules. *Journal of Magnetic Resonance* 123 (1996): 222–225. Accessed August 2012. http://www.sciencedirect.com/science/article/pii/S1064185896902417

Kebarle, P., Peschke, M. On the mechanisms by which the charged droplets produced by electrospray lead to gas phase ions. *Analytica Chemica Acta* 406 (2000): 11–35. Accessed August 2012. http://www.sciencedirect.com/science/article/pii/S000326709900598X

Khan, A., Fontell, K., Lindblom, G., Lindman, B. Liquid crystallinity in a calcium surfactant system: Phase equilibriums and phase structures in the system calcium octyl sulfate/decan-1-ol/water. *The Journal of Physical Chemistry* 86 (1982): 4266–4271. Accessed August 2012. http://pubs.acs.org/doi/abs/10.1021/j100218a034

Kitova, E.N., Bundle, D.R., Klassen, J.S. Partitioning of solvent effects and intrinsic interactions in biological recognition. *Angewandte Chemie International Edition* 43 (2004): 4183–4186. Accessed August 2012. http://onlinelibrary.wiley.com/doi/10.1002/anie.v43:32/issuetoc

Kitova, E.N., Seo, M., Roy, P.-N., Klassen, J.S. Elucidating the intermolecular interactions within a desolvated protein–ligand complex. An experimental and computational study. *Journal of the American Chemical Society* 13 (2008): 1214–1226. Accessed August 2012. http://pubs.acs.org/toc/jacsat/130/4

Kobrak, M.N. The relationship between solvent polarity and molar volume in room-temperature ionic liquids. *Green Chemistry* 10 (2008): 80–86. Accessed August 2012. http://pubs.rsc.org/en/journals/journalissues/gc#!issueid=gc010001&type=current

Koppel, D.E. Analysis of macromolecular polydispersity in intensity correlation spectroscopy: The method of cumulants. *Journal Chemical Physics* 57 (1972): 4814–4820. Accessed August 2012. http:/jcp.aip.org/resource/1/jcpsa6/v57/i11/p4814_s1

Lakowicz, J.R. *Principles of Fluorescence Spectroscopy*. New York: Springer, 2006. Accessed August 2012. http://www.springerlink.com/content/978-0-387-46312-4#section=466669&page=1

Li, N., Zhang, S., Zheng, L., Dong, B., Li, X., Yu, L. Aggregation behavior of long-chain ionic liquids in an ionic liquid. *Physical Chemistry Chemical Physics* 10 (2008): 4375–4377 and references cited therein. Accessed August 2012. http://pubs.rsc.org/en/content/articlelanding/2008/cp/b807339b

Li, W., Zhang, Z., Zhang, J., Han, B., Wang, B., Hou, M., Xie, Y. Micropolarity and aggregation behavior in ionic liquid + organic solvent solutions. *Fluid Phase Equilibria* 248 (2006): 211–216 and references cited therein. Accessed August 2012. http://www.sciencedirect.com/science/article/pii/S0378381206003657

Li, X.W., Gao, Y.A., Jie, J., Zheng, L.Q., Chen, B., Wub, L.Z., Tung, C.H. Aggregation behavior of a chiral long-chain ionic liquid in aqueous solution. *Journal of Colloid and Interface Science* 343 (2010): 94–101. Accessed August 2012. http://www.sciencedirect.com/science/article/pii/S0021979709014519

Liu, X.-f., Dong, L.-l., Fang, Y. Synthesis and self-aggregation of a hydroxyl-functionalized imidazolium-based ionic liquid surfactant in aqueous solution. *Journal of Surfactants and Detergents* 14 (2011a): 203–210. Accessed August 2012. http://www.springerlink.com/content/fhl5037335655574/

Liu, X.-f., Dong, L.-l., Fang, Y. A novel zwitterionic imidazolium-based ionic liquid surfactant: 1-carboxymethyl-3-dodecylimidazolium inner salt. *Journal of Surfactants and Detergents* 14 (2011b): 497–504. Accessed August 2012. http://www.springerlink.com/content/f3j108825gn3q2l0/

Luczak, J., Hupka, J., Thöming, J., Jungnickel, C. Self-organization of imidazolium ionic liquids in aqueous solution. *Colloids and Surfaces A: Physicochemical and Engineering Aspects A* 329 (2008): 125–133. Accessed August 2012. http://www.sciencedirect.com/science/article/pii/S0927775708004639

Mao, X., Guo, J., Ye, C.H. Radiation damping effects on spin-lattice relaxation time measurements. *Chemical Physics Letters* 222 (1994): 417–421. Accessed August 2012. http://www.sciencedirect.com/science/article/pii/0009261494003882

Mao, X.A., Ye, C.H. Understanding radiation damping in a simple way. *Concepts in Magnetic Resonance* 9 (1997): 173–187. Accessed August 2012. http://onlinelibrary.wiley.com/doi/10.1002/(SICI)1099-0534(1997)9:3%3C173::AID-CMR4%3E3.0.CO;2-W/abstract

Martins, M.A.P., Frizzo, C.P., Moreira, D.N. An efficient synthesis of 1-cyanoacetyl-5-halomethyl-4,5-dihydro-1 H -pyrazoles in ionic liquid. *Monatshefte für Chemie* 139 (2008a): 1049–1054.

Martins, M.A.P., Frizzo, C.P., Moreira, D.N., Zanatta, N., Bonacorso, H.G. Ionic liquids in heterocyclic synthesis. *Chemical Reviews* 108 (2008b): 2015–2050 and references cited therein. Accessed August 2012. http://pubs.acs.org/doi/pdf/10.1021/cr078399y

Martins, M.A.P., Guarda, E.A., Frizzo, C.P., Moreira, D.N., Marzari, M.R.B., Zanatta N., Bonacorso, H.G. Ionic liquids promoted the C-acylation of acetals in solvent-free conditions. *Catalysis Letters* 130 (2009): 93–99. Accessed August 2012. http://www.springerlink.com/content/r537572653130483/

Martins, M.A.P., Guarda, E.A., Frizzo, C.P., Scapin, E., Beck, P., da Costa, A.C., Zanatta, N., Bonacorso, H.G. Synthesis of 1,1,1-trichloro[fluoro]-3-alken-2-ones using ionic liquids. *Journal of Chemical Catalysis. A, Chemical* 266 (2007): 100–103. Accessed August 2012. http://www.sciencedirect.com/science/article/pii/S1381116906013458

McLuckey, S.A., Cameron, D., Cooks, R.G. Proton affinities from dissociations of proton-bound dimers. *Journal of the American Chemical Society* 103 (1981): 1313–1317. Accessed August 2012. http://pubs.acs.org/toc/jacsat/103/6

Meurer, E.C., Gozzo, F.C., Augustib, R., Eberlin, M.N. The kinetic method as a structural diagnostic tool: Ionized α-diketones as loosely one-electron bonded diacylium ion dimers. *European Journal of Mass Spectrometry* 9 (2003): 295–304. Accessed August 2012. http://www.impublications.com/content/ejms-table-contents?issue=09_4

Milman, B.L., Alfassi, Z.B. Detection and identification of cations and anions of ionic liquids using ESI-MS and MS/MS. *European Journal of Mass Spectrometry* 11 (2005): 35–42. Accessed August 2012. http://www.impublications.com/content/ejms-table-contents?issue=11_1

Molinier, V., Fenet, B., Fitremann, J., Bouchu, A., Queneau, Y. PFGSE–NMR study of the self-diffusion of sucrose fatty acid monoesters in water. *Journal of colloid and interface Science* 286 (2005): 360–368. Accessed August 2012 and references cited therein. http://www.sciencedirect.com/science/article/pii/S0021979704012135

Muhlbach, J., Pfau, S.P., Recknagel, E. Evidence for magic numbers of free lead-clusters. *Physics Letters A* 87 (1982): 415–417. Accessed August 2012. http://www.sciencedirect.com/science/journal/03759601/87/8

Muller, N., Johnson, T.W. Investigation of micelle structure by fluorine magnetic resonance. III. Effect of organic additives on sodium 12,12,12-trifluorododecyl sulfate solutions. *The Journal of Physical Chemistry A* 73 (1969): 2042–2046. Accessed August 2012. http://pubs.acs.org/doi/abs/10.1021/j100726a067

Nohara, D., Bitoh, M. Observation of micelle solution of decyltrimethylammonium bromide by electrospray ionization mass spectrometry. *Journal of Mass Spectrometry* 35 (2000): 1434–1437. Accessed August 2012. http://onlinelibrary.wiley.com/doi/10.1002/10969888%28200012%2935:12%3C%3E1.0.CO;2–5/issuetoc.

Pecora, R. *Dynamic Light Scattering Applications of Photon Correlation Spectroscopy*. New York: Plenum Press, 1985.

Pramanik, B.N., Bartner, P.L., Mirza, U.A., Liu, Y.-H., Ganguly, A.K. Electrospray ionization mass spectrometry for the study of non-covalent complexes: An emerging technology. *Journal of Mass Spectrometry* 33 (1998): 911–920. Accessed August 2012. http://onlinelibrary.wiley.com/doi/10.1002/%28SICI%2910969888%281998100%2933:10%3C%3E1.0.CO;2–5/issuetoc

Provencher, S.W. A constrained regularization method for inverting data represented by linear algebraic or integral equations. *Computer Physics Communication* 27 (1982a): 213–227. Accessed August 2012. http://www.sciencedirect.com/science/journal/00104655/27/3

Provencher, S.W. CONTIN: A general purpose constrained regularization program for inverting noisy linear algebraic and integral equations. *Computer Physics Communication* 27 (1982b): 229–242. Accessed August 2012. http://www.sciencedirect.com/science/journal/00104655/27/3

Rao, K.S., Singh, T., Trivedi, T.J., Kumar, A. Aggregation behavior of amino acid ionic liquid surfactants in aqueous media. *The Journal of Physical Chemistry B* 115 (2011): 13847–13853. Accessed August 2012. http://pubs.acs.org/doi/abs/10.1021/jp2076275

Remsing, R.C., Liu, Z., Sergeyev, I., Moyna G. Solvation and aggregation of N,N′-dialkylimidazolium ionic liquids: A multinuclear NMR spectroscopy and molecular dynamics simulation study. *The Journal of Physical Chemistry B* 112 (2008): 7363–7369. Accessed August 2012. http://pubs.acs.org/doi/abs/10.1021/jp800769u

Rosen, M.J. *Surfactants and Interfacial Phenomena*. Hoboken, NJ: Wiley, 2004. Accessed August 2012. http://pt.scribd.com/doc/52668446/Rosen-M-J-Surfactants-And-Interfacial-Phenomena

Sastry, N.V., Vaghela, N.M., Macwan, P.M., Soni, S.S., Aswal, V.K., Gibaud, A. Aggregation behavior of pyridinium based ionic liquids in water—Surface tension, 1H NMR chemical shifts, SANS and SAXS measurements. *Journal of Colloid and Interface Science* 371 (2012): 52–61. Accessed August 2012. http://www.sciencedirect.com/science/article/pii/S0021979712000082

Sauer, M., Hofkens, J., Enderlein, J. *Handbook of Fluorescence Spectroscopy and Imaging*. Weinheim, Germany: Wiley-VCH, 2011. Accessed August 2012. http://www.wiley-vch.de/books/sample/3527316698_c01.pdf

Schmitz, K.S. *An Introduction to Dynamic Light Scattering by Macromolecules*. New York: Academic Press, 1990.

Schurtenberger, P., Newman, M.E. Characterization of biological and environmental particles using static and dynamic light scattering. In *Environmental Particles*, eds. J. Buffle and H. P. van Leeuwen, pp. 37–115. Lewis Publishers: Boca Raton, FL, 1993.

Shi, L., Li, N., Zheng, L., Aggregation behavior of long-chain *N*-Aryl Imidazolium Bromide in a room temperature ionic liquid. *Journal of Physical Chemistry C* 115 (2011): 18295–18301. Accessed August 2012. http://pubs.acs.org/toc/jpccck/115/37

Singh, T., Drechsler, M., Müeller, A.H.E., Mukhopadhyaya, I., Kumar, A. Micellar transitions in the aqueous solutions of a surfactant-like ionic liquid: 1-butyl-3-methylimidazolium octylsulfate. *Physical Chemistry Chemical Physics* 12 (2010): 11728–11735. Accessed August 2012. http://pubs.rsc.org/en/content/articlelanding/2010/cp/c003855p

Singh, T., Kumar, A. Fluorescence behavior and specific interactions of an ionic liquid in ethylene glycol derivatives. *The Journal of Physical Chemistry B* 112 (2008): 4079–4086. Accessed August 2012. http://pubs.acs.org/doi/abs/10.1021/jp711711z

Singh, T., Kumar, A.J. Aggregation behavior of ionic liquids in aqueous solutions: Effect of alkyl chain length, cations, and anions. *The Journal of Physical Chemistry B* 111 (2007): 7843–7851. Accessed August 2012. http://pubs.acs.org/doi/abs/10.1021/jp0726889

Souza, A.A., Laverde, J.A, Using nuclear magnetic resonance spectroscopy to study molecular diffusion in liquids: The DOSY technique. *Química Nova* 25 (2002): 1022–1026. Accessed August 2012. http://www.scielo.br/scielo.php?script=sci_abstract&pid=S0100404220020006000208&lng=en&nrm=iso&tlng=en

Stark, A., Zidell, A.W., Hoffmann, M.M. Is the ionic liquid 1-ethyl-3-methylimidazolium methanesulfonate [EMIM][MeSO$_3$] capable of rigidly binding water?. *Journal of Molecular Liquids* 160 (2011): 166–179. Accessed August 2012. http://www.sciencedirect.com/science/article/pii/S016773221100073

Stepánek, P. *Dynamic Light Scattering: The Method and Some Application*. Oxford, U.K.: Oxford Science Publications, 1993.

Sun, J., Kitova, E.N., Klassen, J.S. Method for stabilizing protein–ligand complexes in nanoelectrospray ionization mass spectrometry. *Analytical Chemistry* 79 (2007): 416–425. Accessed August 2012. http://pubs.acs.org/toc/ancham/79/2

Taflin, D.C., Ward, T.L., Davis, E.J. Electrified droplet fission and the Rayleigh limit. *Langmuir* 5 (1989): 376–384. Accessed August 2012. http://pubs.acs.org/doi/abs/10.1021/la00086a016

Tariq, M., Podgoršek, A., Ferguson, J.L., Lopes, A., Gomes, M.F.C., Pádua, A.A.H., Rebelo, L.P.N., Lopes, J.N.C. Characteristics of aggregation in aqueous solutions of dialkylpyrrolidinium bromides. *Journal of Colloid and Interface Science* 360 (2011): 606–616. Accessed August 2012. http://www.sciencedirect.com/science/article/pii/S0021979711005315

Trivedi, S., Pandey, S., Baker, S.N., Baker, G.A., Pandey, S. Pronounced hydrogen bonding giving rise to apparent probe hyperpolarity in ionic liquid mixtures with 2,2,2-trifluoroethanol. *The Journal of Physical Chemistry B* 116 (2012): 1360–1369. Available at: http://pubs.acs.org/doi/abs/10.1021/jp210199s

Tsuzuki, S., Tokuda, H., Hayamizu, K., Watanabe, M. Magnitude and directionality of interaction in ion pairs of ionic liquids: Relationship with ionic conductivity. *The Journal of Physical Chemistry B* 109 (2005): 16474–16481. Accessed August 2012. http://pubs.acs.org/toc/jpcbfk/109/34

Villetti, M.A., Bica, C.I.D., Garcia, I.T.S., Pereira, F.V., Ziembowicz, F.I., Kloster, C.L., Giacomelli, C. Physicochemical properties of methylcellulose and dodecyltrimethylammonium bromide in aqueous medium. *The Journal of Physical Chemistry B* 115 (2011): 5868–5876. Accessed August 2012. http://pubs.acs.org/doi/abs/10.1021/jp110247r

Vriendt, K.D., Sandra, K., Desmet, T., Nerinckx, W., Beeumen, J.V., Devreese, B. Evaluation of automated nano-electrospray mass spectrometry in the determination of non-covalent protein–ligand complexes. *Rapid Communications in Mass Spectrometry* 18 (2004): 3061–3067. Accessed August 2012. http://onlinelibrary.wiley.com/doi/10.1002/rcm.v18:24/issuetoc

Wang, H., Wang, J., Zhang, S., Xuan, X. Structural effects of anions and cations on the aggregation behavior of ionic liquids in aqueous solutions. *Journal of Physical Chemistry B* 112 (2008): 16682–16689. Accessed August 2012. http://pubs.acs.org/toc/jpcbfk/112/51

Wang, J., Wang, H., Zhang, S., Zhang, H., Zhao, Y. Conductivities, volumes, fluorescence, and aggregation behavior of ionic liquids [C_4mim][BF_4] and [$C_n$mim]Br (n=4, 6, 8, 10, 12) in aqueous solutions. *The Journal of Physical Chemistry B* 111 (2007): 6181–6188. Accessed August 2012. http://pubs.acs.org/doi/abs/10.1021/jp068798h

Wang, J., Zhang, L., Wang, H., Wu, C. Aggregation behavior modulation of 1-dodecyl-3-methylimidazolium bromide by organic solvents in aqueous solution. *The Journal of Physical Chemistry B* 115 (2011): 4955–4962. Accessed August 2012. http://pubs.acs.org/doi/abs/10.1021/jp201604u

Wendt, S., McCombie, G., Daniel, J., Kienhofer, A., Hilvert, D., Zenobi, R. Quantitative evaluation of noncovalent chorismate mutase-inhibitor binding by ESI-MS. *Journal of the American Society for Mass Spectrometry* 14 (2003): 1470–1476. Accessed August 2012. http://www.sciencedirect.com/science/journal/10440305/14/12

Wennerström, H., Lindman, B. Micelles: Physical chemistry of surfactant association. *Physics Reports (Review Section of Physics Letters)* 52 (1972): 1–86. Accessed August 2012. http://www.sciencedirect.com/science/article/pii/0370157379900875

Williams, C.E., Sears, B., Allerhand, A., Cordes, E.H. Segmental motion of amphipathic molecules in aqueous solutions and micelles: Application of natural-abundance carbon-13 partially relaxes Fourier transform nuclear magnetic resonance spectroscopy. *Journal of the American Chemical Society* 95 (1973): 4871–4873. Accessed August 2012. http://pubs.acs.org/doi/abs/10.1021/ja00796a018

Wu, L., Meurer, E.C., Young, B., Yang, P., Eberlin, M.N., Cooks, R.G. Isomeric differentiation and quantification of α, β-amino acid-containing tripeptides by the kinetic method: Alkali metal-bound dimeric cluster ions. *International Journal of Mass Spectrometry* 231 (2004): 103–111. Accessed August 2012. http://www.sciencedirect.com/science/journal/13873806/240

Wyttenbach, T., Bowers, M.T. Intermolecular interactions in biomolecular systems examined by mass spectrometry. *Annual Review of Physical Chemistry* 58 (2007): 511–533. Accessed August 2012. http://www.annualreviews.org/toc/physchem/58/1

Zana, R. Critical micellization concentration of surfactants in aqueous solution and free energy of micellization. *Langmuir* 12 (1996): 1208–1211. Accessed August 2012. http://pubs.acs.org/doi/abs/10.1021/la950691q

Zhang, D., Cooks, R.G. Doubly charged cluster ions [(NaCl)m(Na)2]2+: Magic numbers, dissociation, and structure. *International Journal of Mass Spectrometry* 195–196 (2000): 667–684. Accessed August 2012. http://www.sciencedirect.com/science/journal/13873806/195-196

Zhang, Q.G., Wang, N.N., Wang, S.L., Yu, Z.W. Hydrogen bonding behaviors of binary systems containing the ionic liquid 1-butyl-3-methylimidazolium trifluoroacetate and water/methanol. *The Journal of Physical Chemistry B* 115 (2011): 11127–11136. Accessed August 2012. http://pubs.acs.org/doi/abs/10.1021/jp204305g

Zhao, M., Zheng, L. Micelle formation by N-alkyl-N-methylpyrrolidinium bromide in aqueous solution. *Physical Chemistry Chemical Physics* 13 (2011): 1332–1337. Accessed August 2012. http://pubs.rsc.org/en/content/articlelanding/2011/cp/c0cp00342e

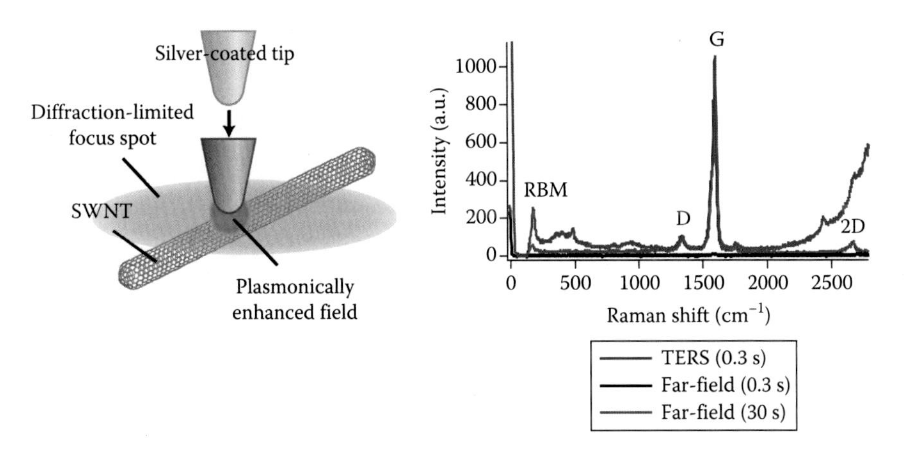

FIGURE 1.7
TERS spectrum of SWNT measured by positioning a silver-coated tip onto SWNT lying in the focus spot. Far-field Raman spectrum, measured with the tip away from the nanotube, is shown as well for comparison. (From Hayazawa, N., Yano, T., and Kawata, S.: Highly reproducible tip-enhanced Raman scattering using an oxidized and metallized silicon cantilever tip as a tool for everyone. *J. Raman Spectrosc.* 2012b. 43(9). 1177–1182. Copyright Wiley-VCH Verlag GmbH & Co. KGaA.)

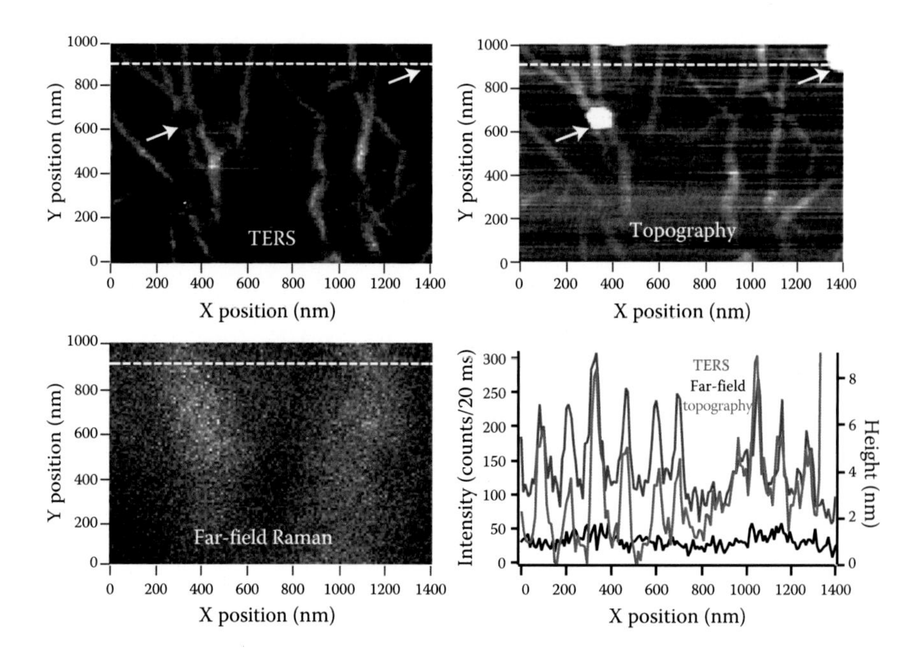

FIGURE 1.8
TERS image of SWNTs on a cover glass as well as simultaneously obtained topographic image. Far-field Raman image at the same area is also shown for comparison. The cross section at the dashed line clearly exhibits a good agreement between TERS and topography. The arrows in the figure indicate the positions of catalysts which are blind in the TERS image. (From Hayazawa, N., Yano, T., and Kawata, S.: Highly reproducible tip-enhanced Raman scattering using an oxidized and metallized silicon cantilever tip as a tool for everyone. *J. Raman Spectrosc.* 2012b. 43(9). 1177–1182. Copyright Wiley-VCH Verlag GmbH & Co. KGaA.)

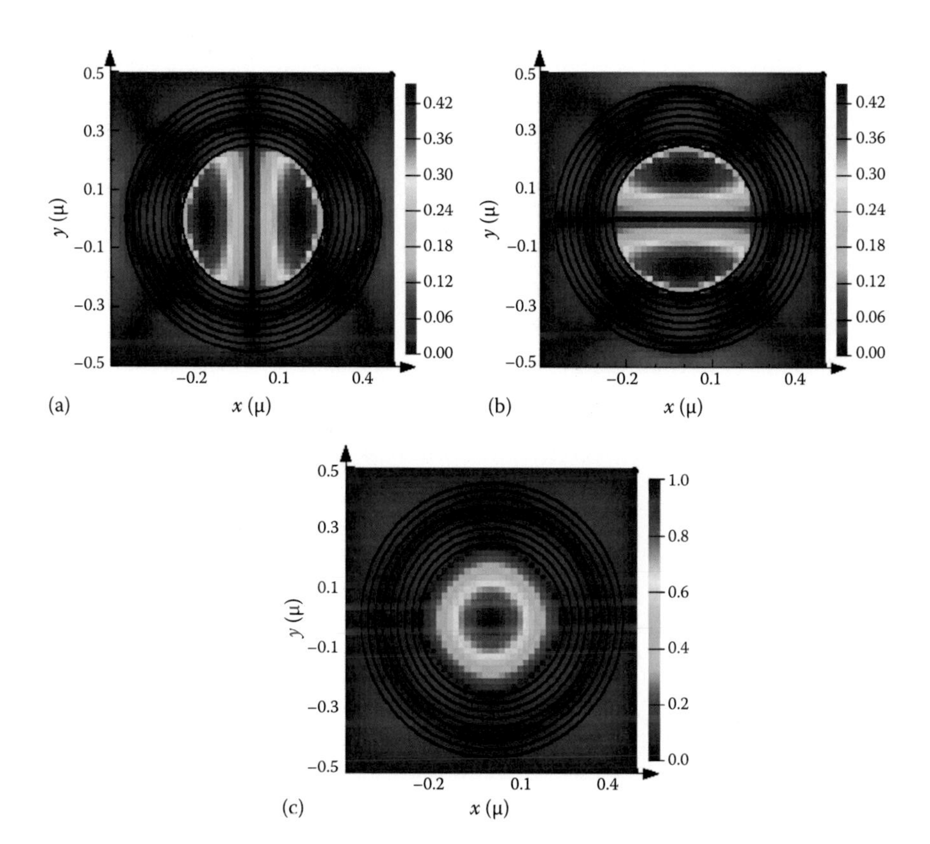

FIGURE 2.6
Electric field distribution of the (a) x-, (b) y-, and (c) z-components.

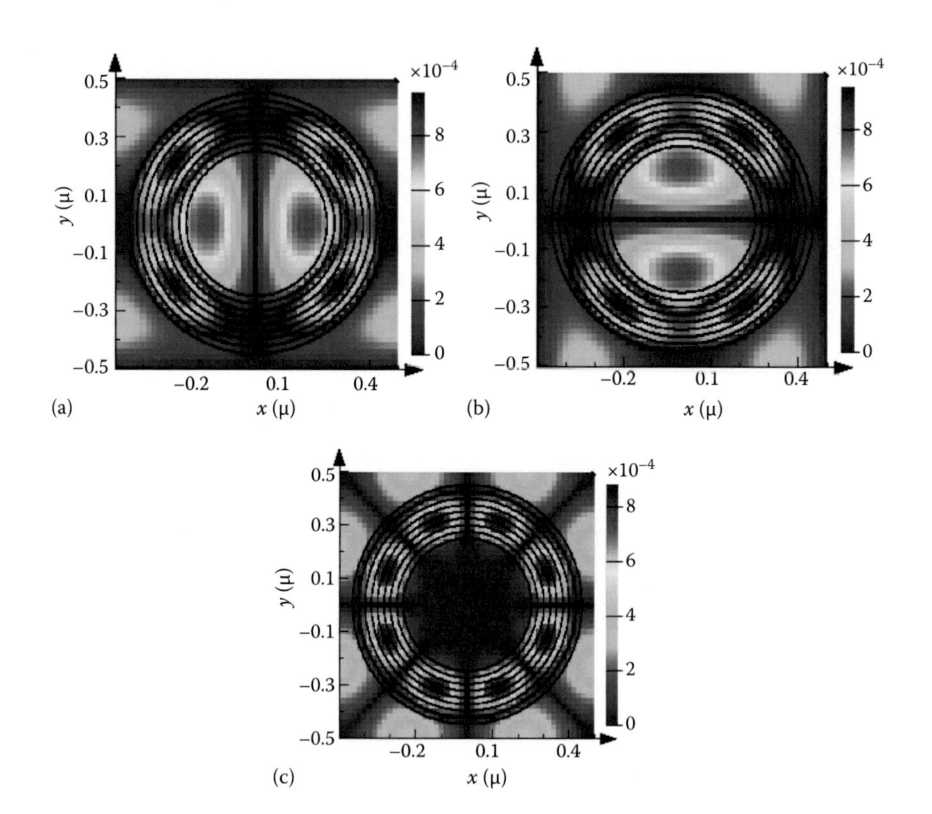

FIGURE 2.7
Magnetic field distribution of the (a) x-, (b) y-, and (c) z-components.

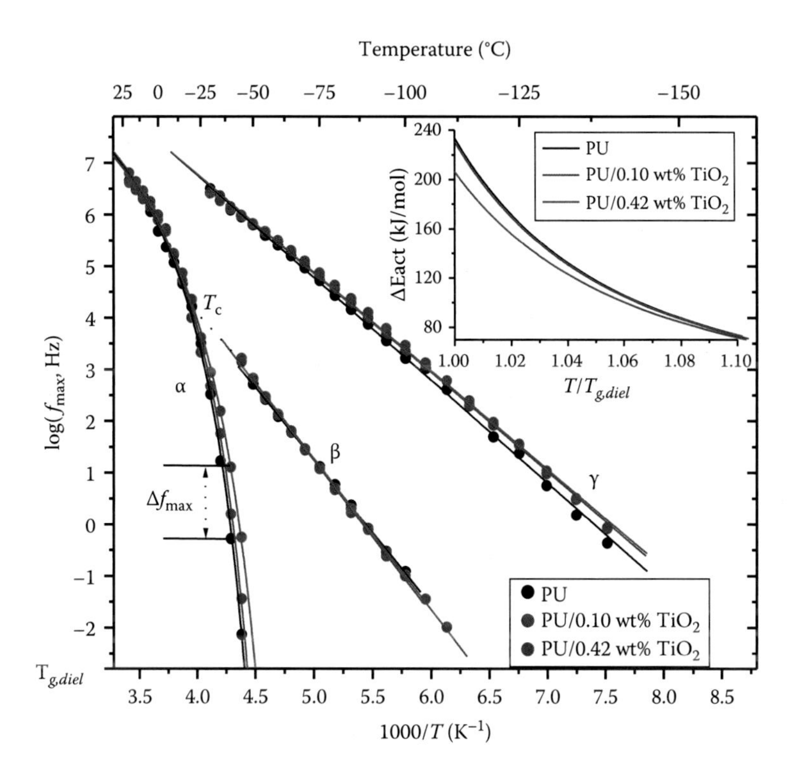

FIGURE 3.3
Summarized Arrhenius plot of the cooperative (α) and local (β,γ) relaxations at temperatures below room temperature along with the fitting lines according to Equations 3.24 and 3.25. T_c corresponds to the merging temperature of the α and β relaxations. Δf_{max} is the frequency shift of the α relaxation (for the 0.42 wt% composite at −40°C) according to Figure 3.2. The dielectric glass transition $T_{g,diel}$, obtained at 1.6 mHz ($\tau = 100$ s), is also indicated on the plot. The normalized (by T_g) apparent activation energies that correspond to the α relaxations were calculated from Equation 3.27 and are shown in the inset plot. (From Polizos, G. et al., *Polymer*, 53, 595, 2012.)

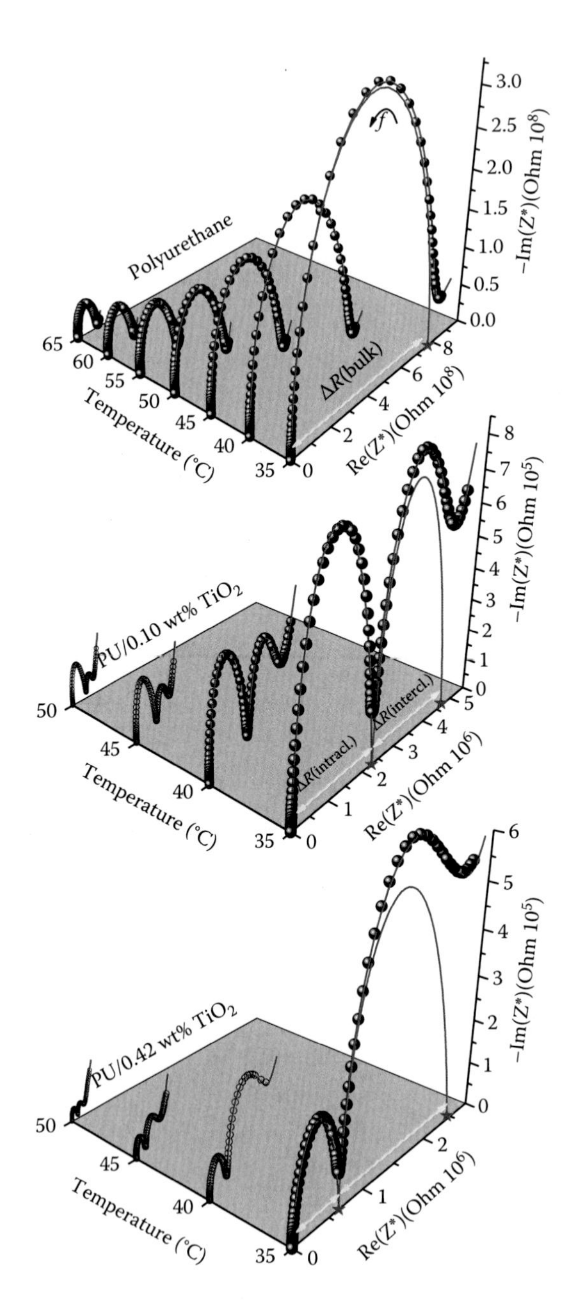

FIGURE 3.4
Complex impedance Nyquist plots for the unfilled polyurethane and the nanocomposites for several temperatures. The fitting curves to the data are according to Equation 3.20. The conductivity process in the unfilled matrix is described by a single relaxation whereas the nanocomposites are characterized by a bimodal relaxation due to the intra- and intercluster conduction processes. (The top two plots are reproduced from Polizos, G. et al., *Polymer*, 53, 595, 2012.)

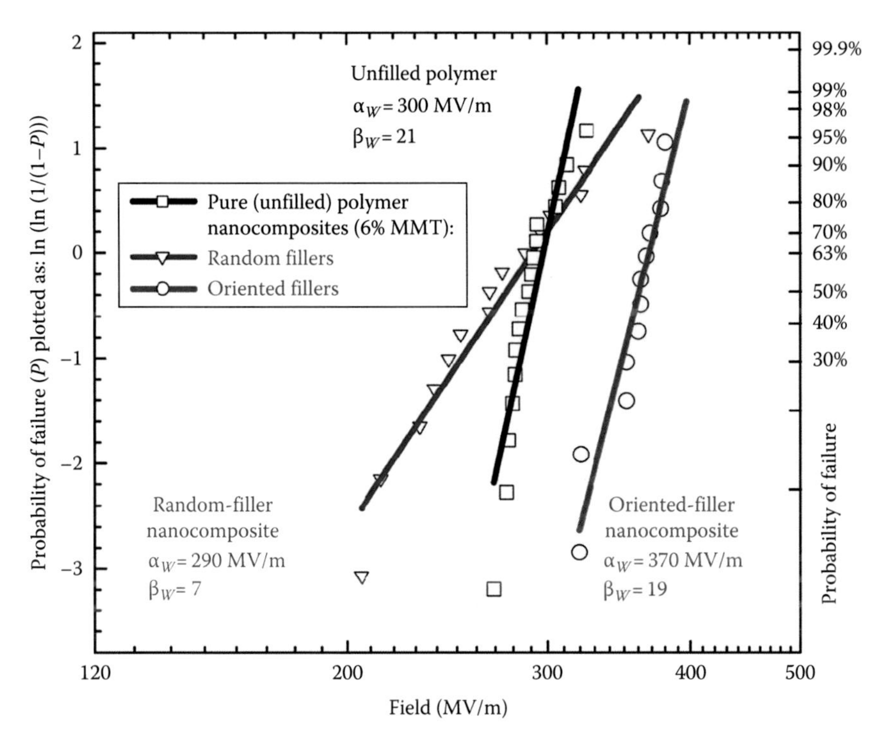

FIGURE 3.19
Dielectric breakdown failure distributions for the unfilled polymer matrix and the nanocomposites based on randomly and aligned layered silicates. The lines are linear fits to the Weibull cumulative probability and the characteristic parameters are indicated on the plot. (Reproduced from Tomer, V. et al., *J. Appl. Phys.*, 109, 074113, 2011.)

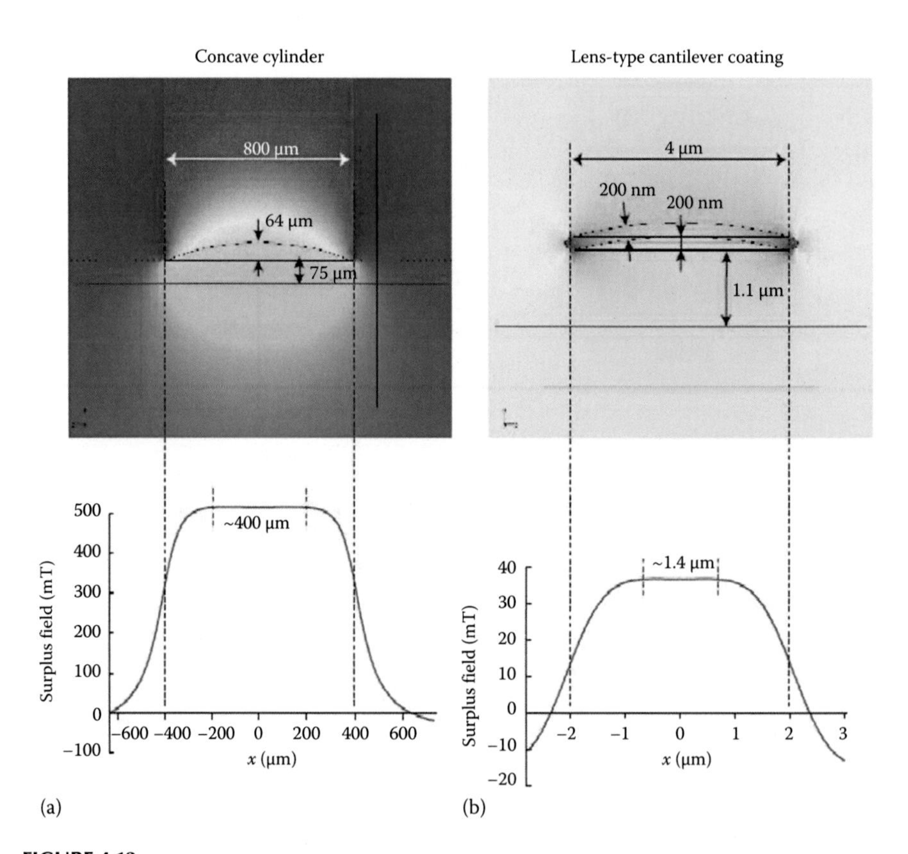

FIGURE 4.13
Magnet shapes for interface imaging in a (a) sample-on-cantilever and (b) tip-on-cantilever arrangement using a 200-nm thick ferromagnetic coating.

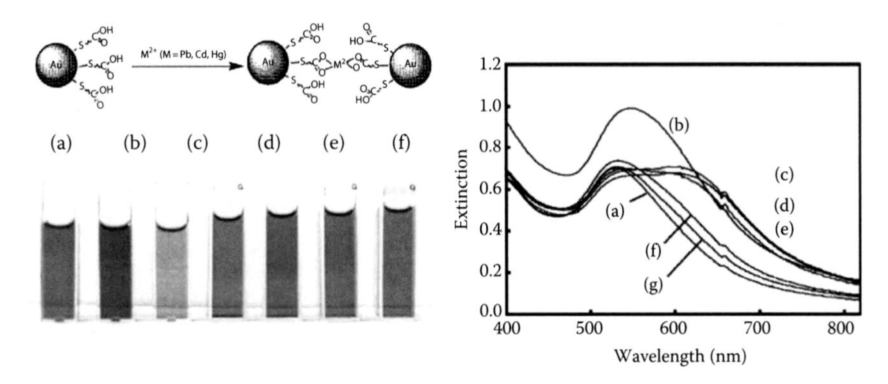

FIGURE 5.3
Colorimetric responses (a) and corresponding spectral traces (b) from (a) Au-MUA, (b) Au-MUA/Pb^{2+}, and (c–g) Au-MUA/Pb^{2+} and increasing amounts of EDTA. Pb^{2+} concentration in sample (b) is 0.67 mM; EDTA concentrations in samples (c–g) are 0.191, 0.284, 0.376, 0.467, and 0.556 mM. (Reproduced with permission from Kim, Y.J., Johnson, R.C., and Hupp, J.T., *Nano Lett.*, 1, 2001, 165. Copyright 2001, American Chemical Society.)

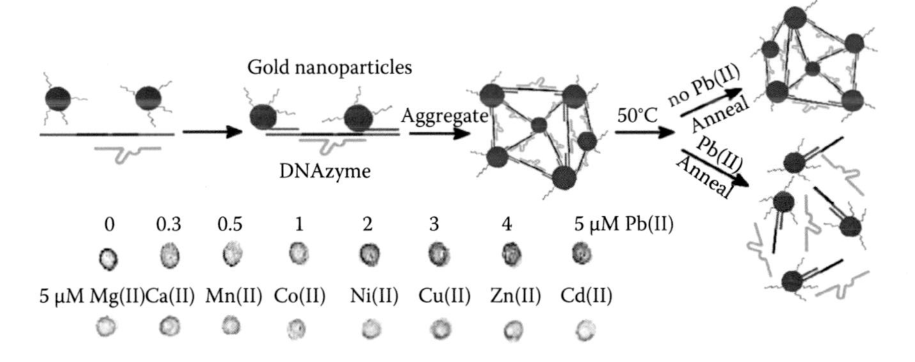

FIGURE 5.4
Pb^{2+}-directed assembly of AuNPs by the DNAzymes resulting in detection. (Reproduced with permission from Liu, J.W. and Lu, Y., *J. Am. Chem. Soc.*, 125, 2003, 6642. Copyright 2010, American Chemical Society.)

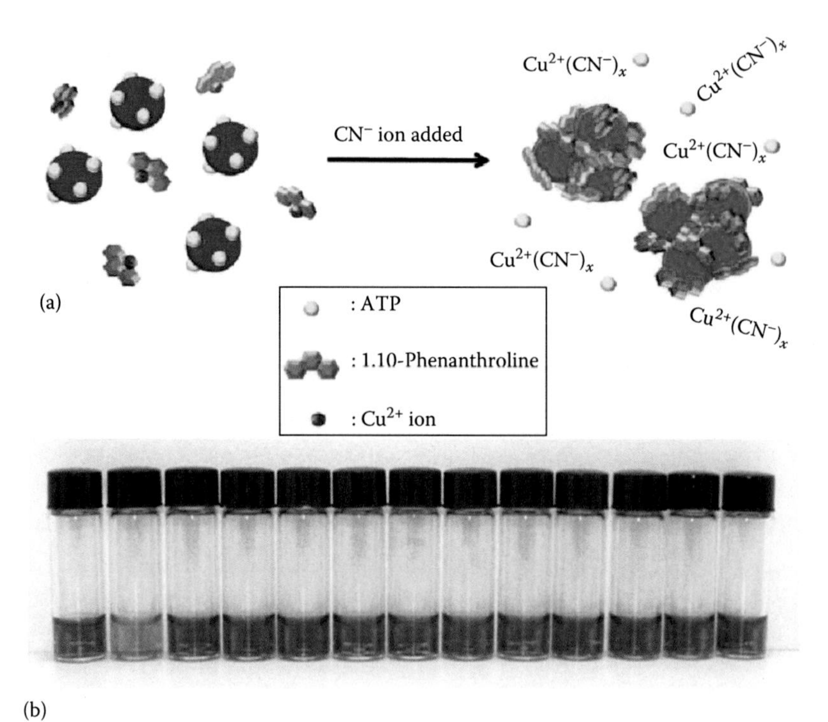

FIGURE 5.5
(a) A schematic diagram of the cyanide anion sensing ensemble. (b) The color of the solution in the absence and presence of anions: from left to right; no anion, CN^-, F^-, Cl^-, Br^-, ClO_4^-, SO_4^{2-}, HCO_3^-, AcO^-, HPO_4^{2-}, NO_3^-, N_3^-, and $P_2O_7^{4-}$. (Reproduced from Kim, M.H., Kim, S., Jang, H.H., Yi, S., Seo, S.H., and Han, M.S., *Tetrahedron Lett.*, 51, 4712, 2010. Copyright 2010. With permission from Science Direct.)

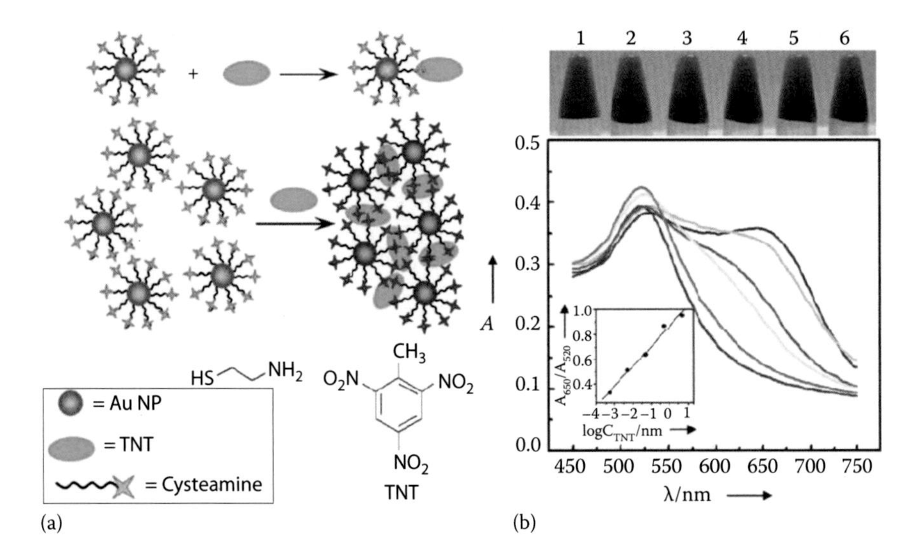

FIGURE 5.6
(a) Assay for direct colorimetric visualization of TNT based on the electron D–A interaction at the AuNP/solution interface. Colorimetric visualization of TNT by using AuNPs (containing 500 nm cysteamine). TNT concentrations varied from 5×10^{-13} (2) to 5×10^{-9} M (6). (b) UV-vis spectra of the AuNP suspension (10 nm) containing 500 nm cysteamine in the presence of different concentrations of TNT. (From Jiang, Y., Zhao, H., Zhu, N.N., Lin, Y.Q., Yu, P., and Mao, L.Q.: *Angew. Chem. Int. Ed.* 2008. 47. 8601. Copyright Wiley-VCH Verlag GmbH & Co. KGaA. Reproduced with permission.).

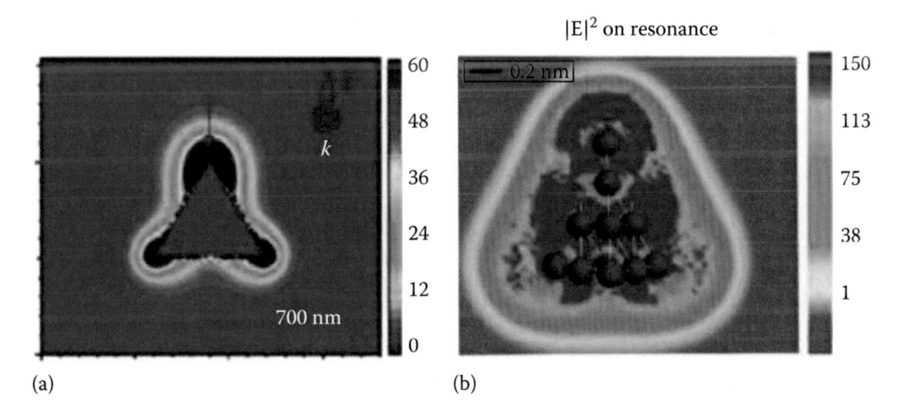

FIGURE 5.9
(a) DDA calculation of the electric field enhancement $|E|^2$ around a nanotriangle, highlighting the fact that the greatest enhancement is at the tips. (Reprinted with permission from Hao, E. and Schatz, G.C., *J. Chem. Phys.*, 120, 357, 2004, Copyright 2004, American Institute of Physics.) (b) TD-DFT calculation of $|E|^2$ for a tetrahedral Ag_{20} cluster. The two methods yield qualitatively the same results. (From Jensen, L., Aikens, C.M., and Schatz, G.C., *Chem. Soc. Rev.*, 37, 1061, 2008. Reproduced with permission of The Royal Society of Chemistry.)

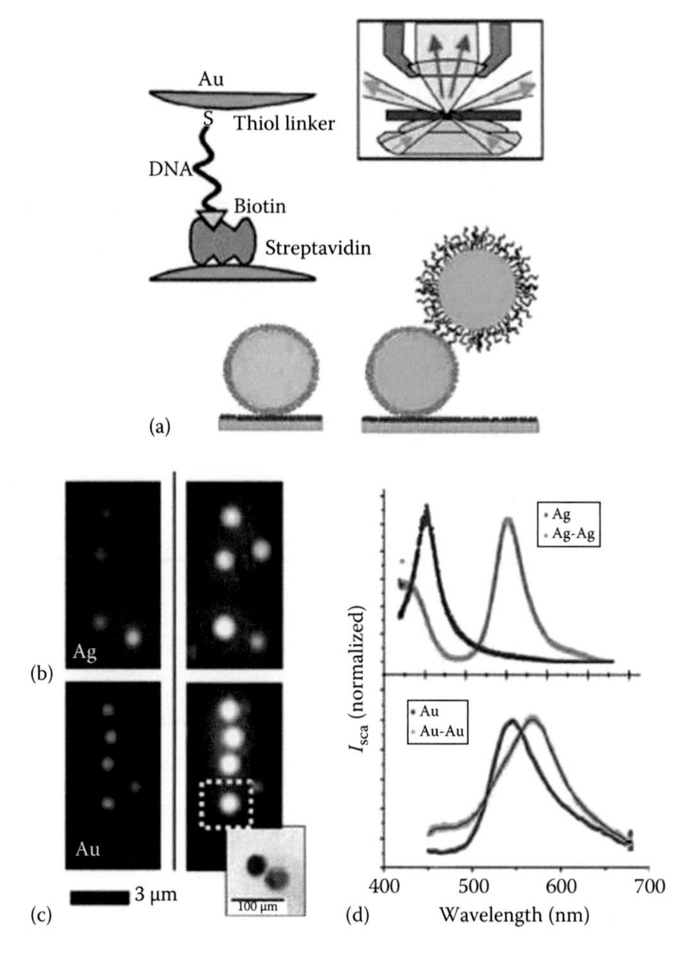

FIGURE 5.17
Molecular ruler for DNA based on plasmonic coupling between metal nanoparticles. (a) First, nanoparticles functionalized with streptavidin are attached to the glass surface coated with BSA-biotin (left). Then, a second particle is attached to the first particle (center), again via biotinstreptavidin binding (right). Inset: principle of transmission dark-field microscopy. (b) Single silver particles (left) and particle pairs (right). (c) Single gold particles (left) and gold particle pairs (right). Inset: representative transmission electron microscopy image of a particle pair to show that each colored dot comes from light scatted from two closely lying particles, which cannot be separated optically. (d) Representative scattering spectra of single particles and particle pairs for silver (top) and gold (bottom). Silver particles show a larger spectral shift (102 nm) than gold particles (23 nm), stronger light scattering and a smaller plasmon line width. (Reproduced with permission from Macmillian Publishers Ltd. *Nat. Biotechnol.*, Sonnichsen, C., Reinhard, B.M., Liphardt, J., and Alivisatos, A.P., 23, 741, Copyright 2005.)

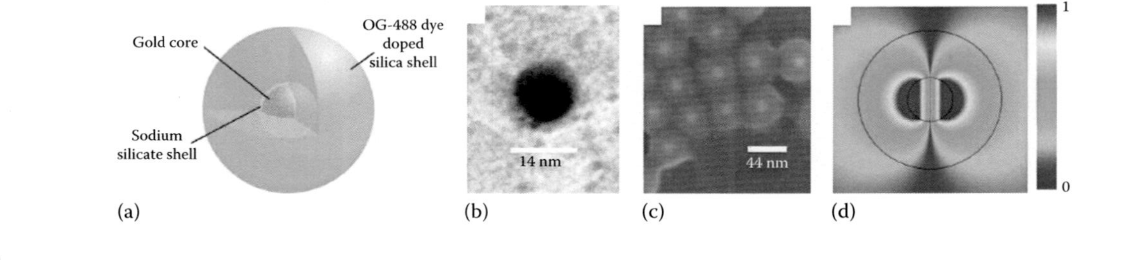

FIGURE 5.23
(a) Diagram of the hybrid nanoparticle architecture (not to scale), indicating dye molecules throughout the silica shell. (b) Transmission electron microscope image of Au core. (c) Scanning electron microscope image of Au/silica/dye core–shell nanoparticles. (d) Spaser mode (in false color), with $\lambda = 525$ nm and $Q = 14.8$; the inner and the outer circles represent the 14-nm core and the 44-nm shell, respectively.

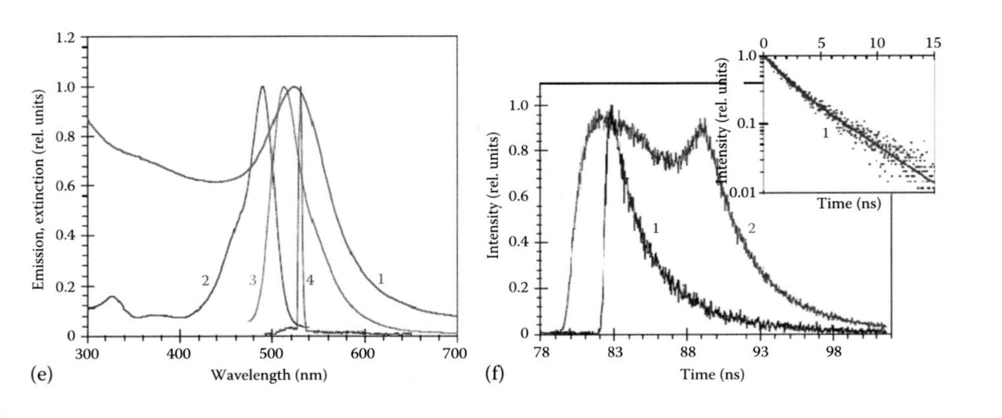

FIGURE 5.23 (continued)
(e) The field strength color scheme is shown on the right. Normalized extinction (1), excitation (2), spontaneous emission (3), and stimulated emission (4) spectra of Au/silica/dye nanoparticles. The peak extinction cross-section of the nanoparticles is 1.1×10^{-12} cm^2. (f) The emission and excitation spectra were measured in spectrofluorometer at low fluence. Main panel: emission kinetics detected at 480 nm (1) and 520 nm (2). Inset, trace 1 plotted in semilogarithmic coordinates (dots) and the corresponding fitting curve. The beginning of each emission kinetic trace coincides with the 90-ps pumping pulse. (Reproduced from Noginov, M.A., Zhu, G., Belgrave, A.M., Bakker, R., Shalaev, V.M., Narimanov, E.E., Stout, S., Herz, E., Suteewong, T., and Wiesner, U., *Nature*, 460, 1110–1113. Copyright 2009. With permission from Nature Progress Group.)

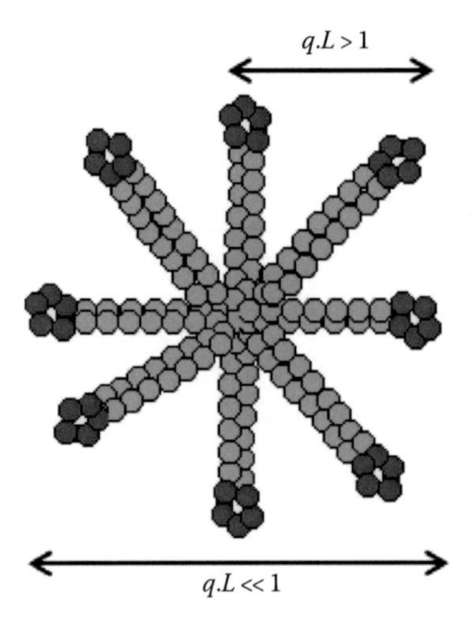

FIGURE 6.17
Length scales.

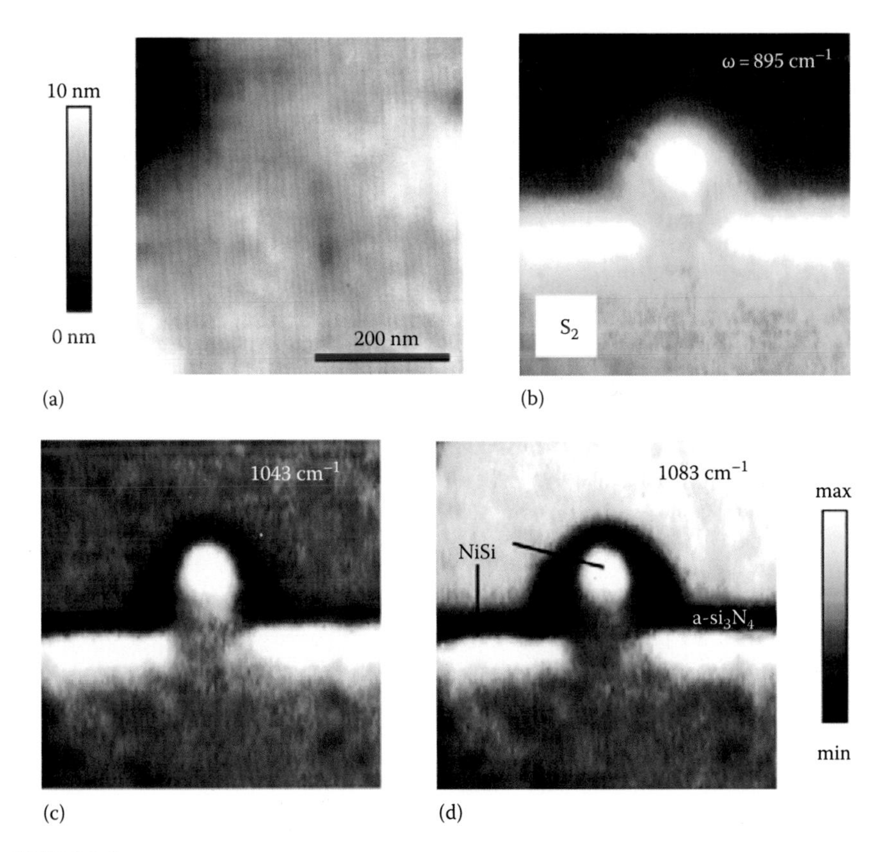

FIGURE 8.6
Despite the fact that the height of the sample shows very little variation (a), the optical signal shows clear variation in these images of a transistor, taken at different wavelengths (b–d). (From Huber, A.J. et al., *Nanotechnology*, 21, 235702, 2010.)

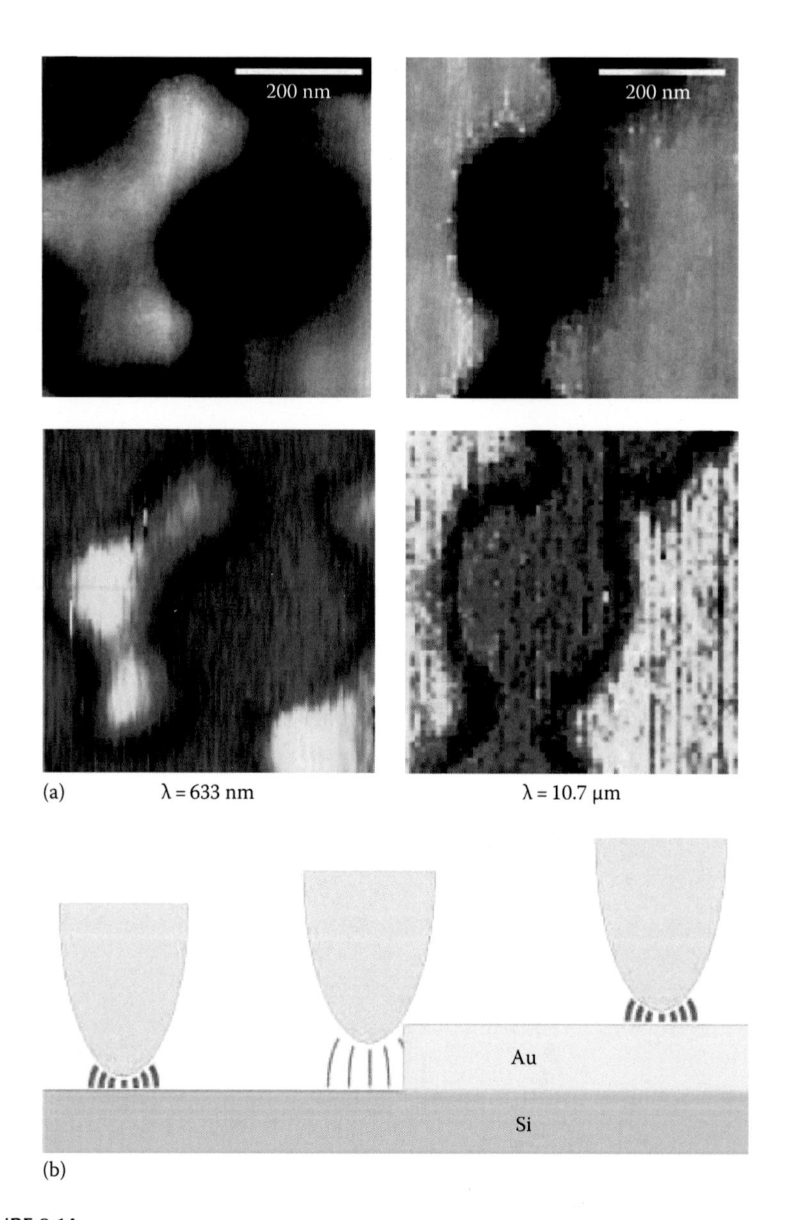

FIGURE 8.14
Topography and optical amplitude images (a) of gold islands on a silicon substrate. The edge darkening effect is clearly visible in the optical images, and a schematic of its mechanism is shown in (b). (From Taubner, T. et al., *J. Microsc. Oxf.*, 210, 311, 2013.)

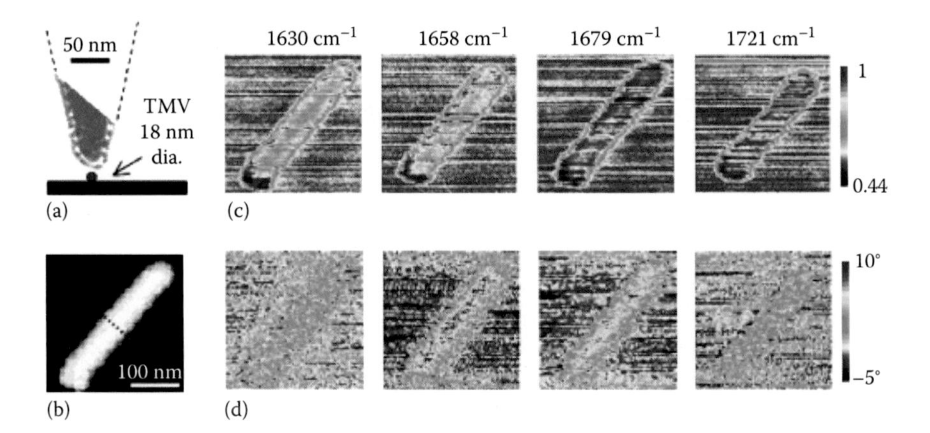

FIGURE 8.15
(a) Scale schematic of tip and virus, (b) topography, (c) and (d) optical amplitude and phase at various wavelengths. (From Brehm, M. et al., *Nano Lett.*, 6, 1307, 2006b.)

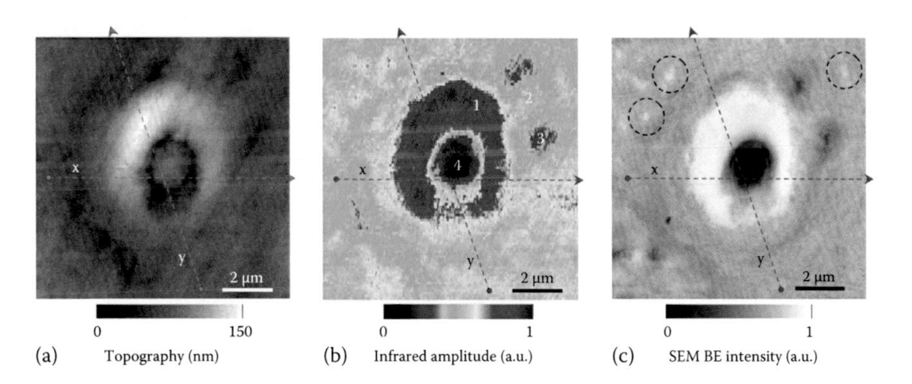

FIGURE 8.18
(a) Topography, (b) spectrally integrated mode near-field image, and (c) a backscattered electron image for comparison. (From Amarie, S. et al., *Beilstein J. Nanotechnol.*, 3, 312, 2012.)

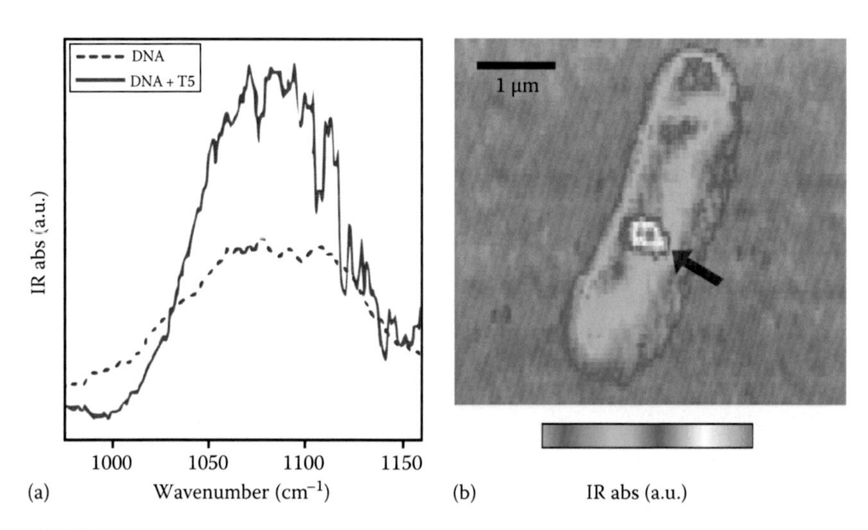

FIGURE 8.23
(a) The nanoscale IR spectra of DNA and DNA with the phage virus present. An IR chemical map at the Amide I band is shown in (b), with the internal phage clearly visible. (Modified from Dazzi, A. et al., *Ultramicroscopy*, 107, 1194, 2008.)

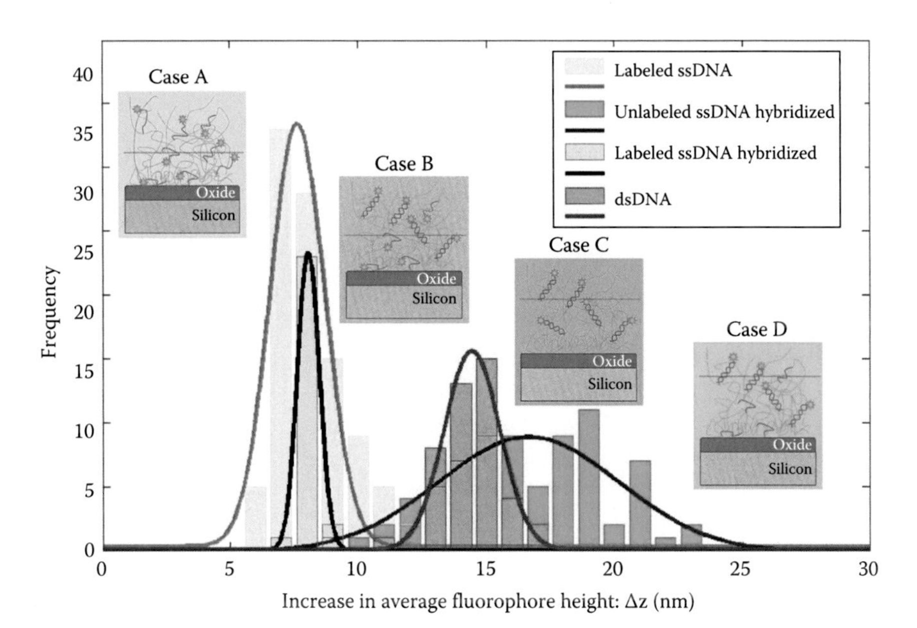

FIGURE 9.9
Histograms summarizing the increase of fluorophore heights measured at spots of labeled ssDNA (Case A), labeled ssDNA hybridized with non-labeled complementary sequences on surface (Case B), labeled dsDNA (Case C), and unlabeled ssDNA hybridized on surface with labeled complementary sequences on surface (Case D). Each histogram contains results of tens of spots and is fitted with a Gaussian curve, the mean values of which indicate the average increase of fluorophore heights. Animated illustration above each histogram shows the experimental conditions of each case. (Reprinted with permission from Yalçın, A., Damin, F., Özkumur, E. et al. Direct observation of conformation of a polymeric coating with implications in microarray applications, *Anal. Chem.*, 81(2), 625–630. Copyright 2009, American Chemical Society.)

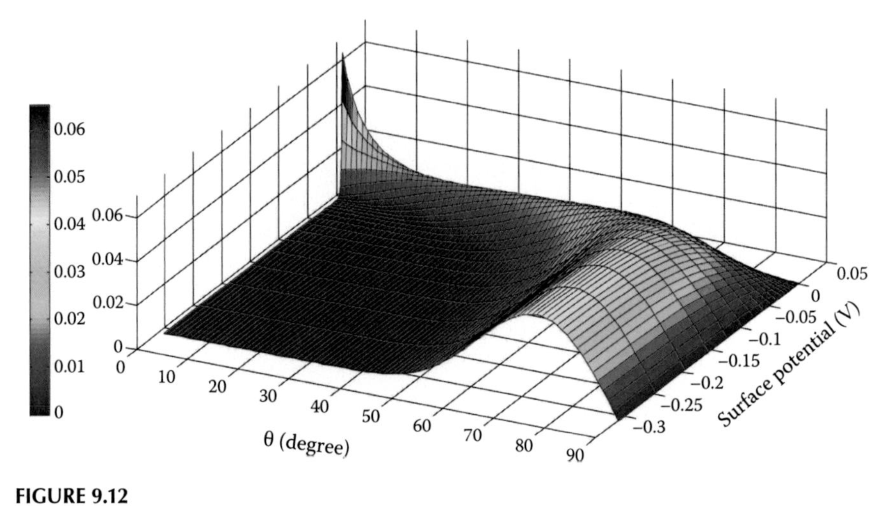

FIGURE 9.12
Probability density functions (PDFs) of dsDNA orientation θ for different surface potentials. Positive potentials pull the dsDNA to the surface, resulting in most dsDNA occupying the lower orientation states. The PDFs are shifted to higher degrees as the surface potential becomes more negative.

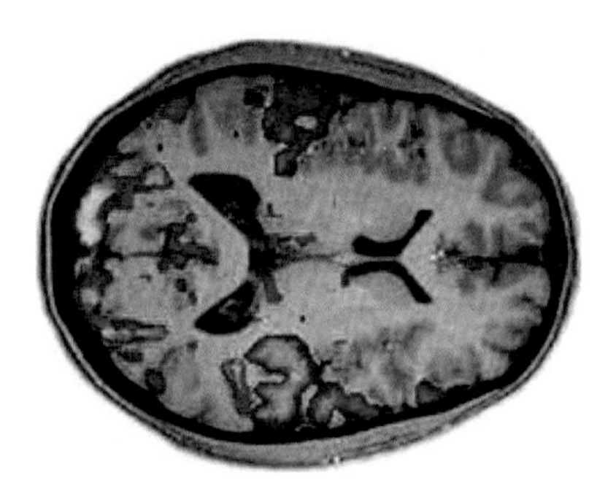

FIGURE 10.18
fMRI scan with the BOLD signal as the contrast. (Image courtesy of M. Ignor, Mind-Body Dualism—Is the mind purely a function of brain?, http://www.godandscience.org/evolution/mind-body_dualism.html)

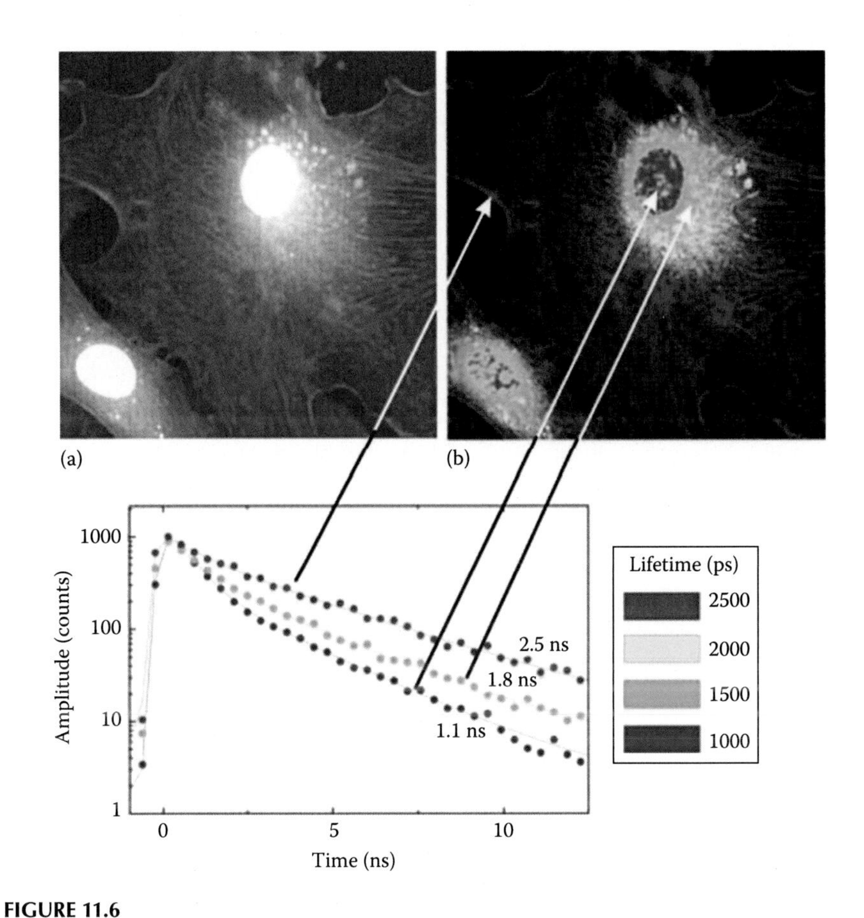

FIGURE 11.6
Intensity (a) and lifetime (b) fluorescence images of bovine artery endothelial cells, stained with three fluorophores. The nuclei were stained with DAPI for DNA (blue), F-actin was stained with Bodipy FL-phallacidin (red), and the mitochondria were stained with MitoTracker Red CMX Ros (green). Two-photon excitation at 800 nm was used. (Reproduced with permission from Springer Science+Business Media: *Principles of Fluorescence Spectroscopy*, 2010, Lakowicz, J.R. Courtesy of Dr. Alex Bergmann, Becker and Hickl GmbH.)

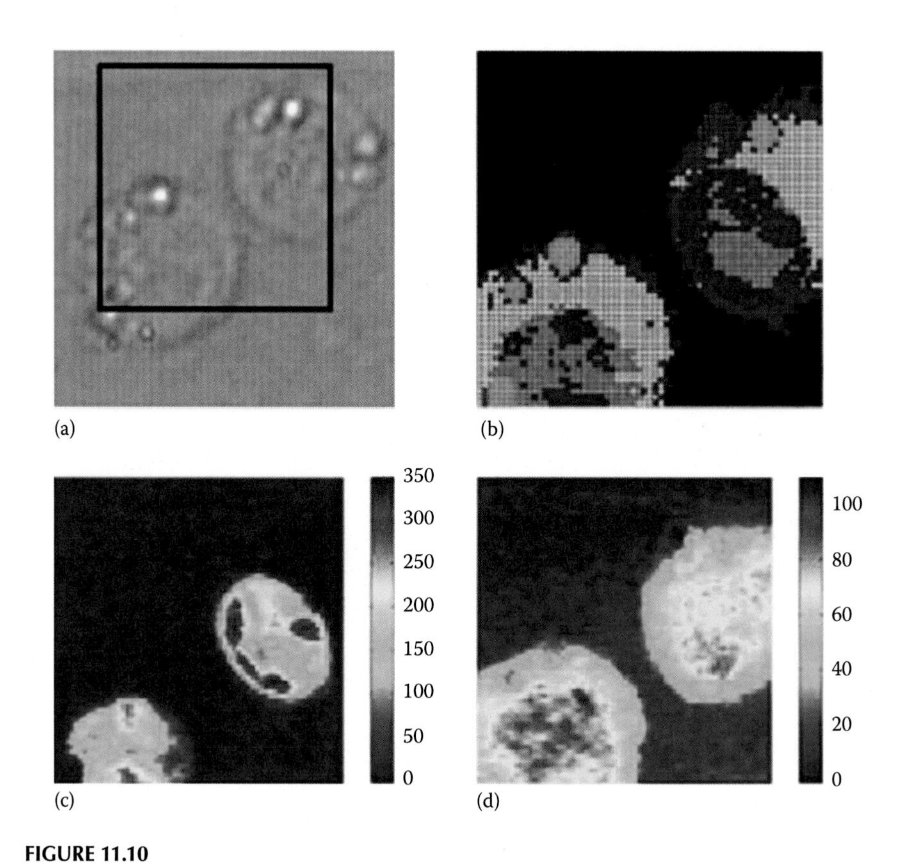

FIGURE 11.10
(a) White light micrograph of freshly extracted bovine chondrocytes seeded on a CaF_2 substrate, (b) corresponding nine-level image showing various organelles in the cell, (c) Raman image for DNA nucleotides at 790 cm^{-1}, (d) Raman image for phenylalanine at 1004 cm^{-1}.

(continued)

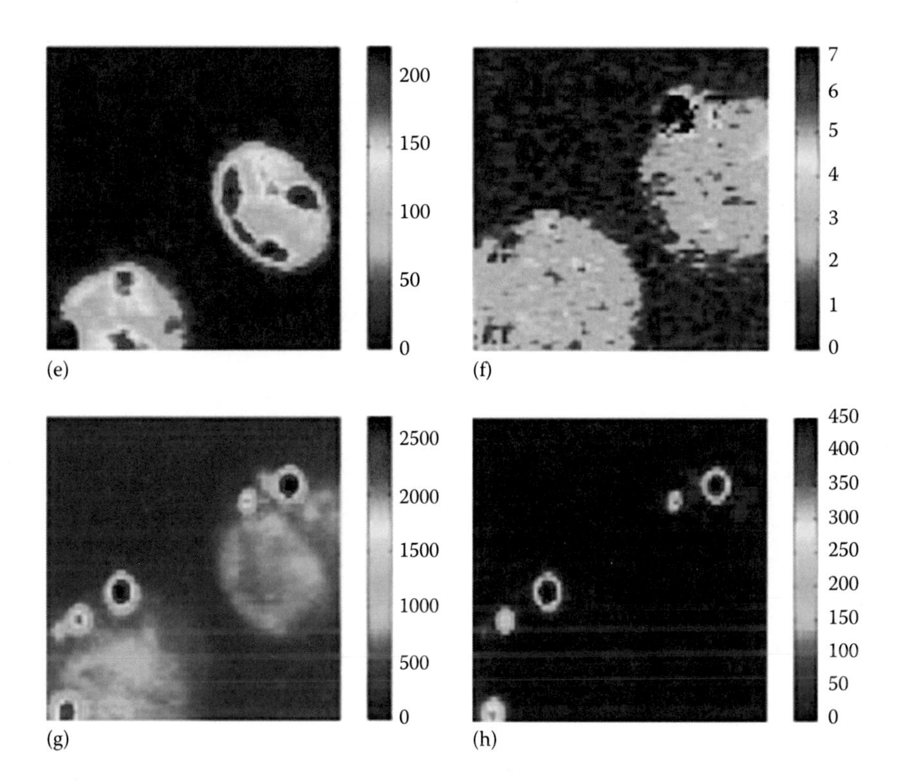

FIGURE 11.10 (continued)
(e) Raman image for DNA O–P–O symmetric stretch at 1094 cm^{-1}, (f) Raman image signifying the distribution of mitochondria at 1602 cm^{-1}, (g) Raman image of lipids and proteins based on the area around 1656 cm^{-1}, and (h) Raman image for ester groups of lipids at 1745 cm^{-1}. (Reproduced from Otto, C. and Pully, V.V., Hyperspectral Raman microscopy of the living cell, In *Applications of Raman Spectroscopy to Biology*, IOS Press, Amsterdam, the Netherlands, pp. 148–173, 2012. With permission from IOS Press.)

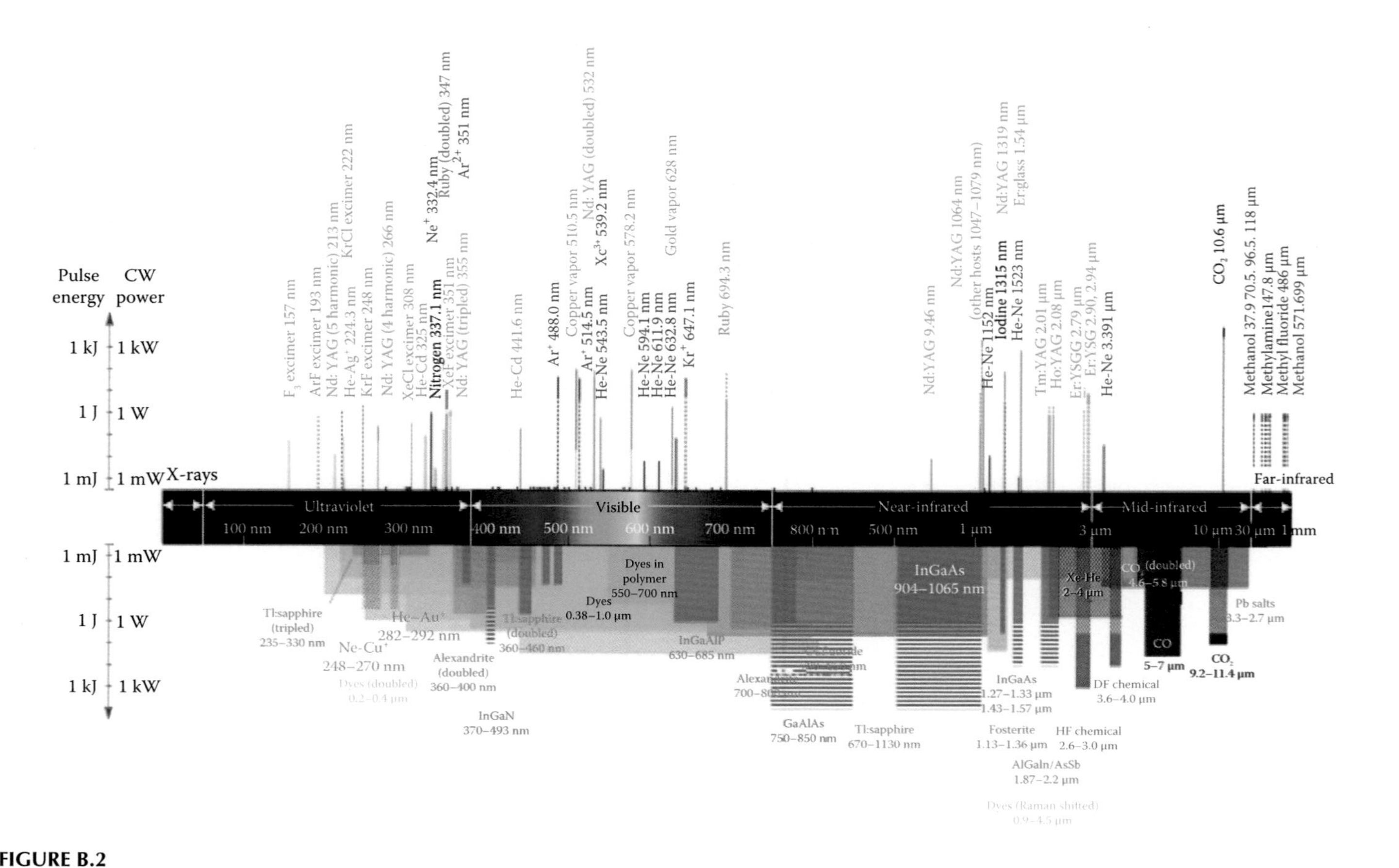

FIGURE B.2
Wavelengths of commercially available lasers. (From Weber, M.J., *Handbook of Laser Wavelengths*, CRC Press, Boca Raton, FL, 1999.)

7

Controlling Reversible Self-Assembly Path of Amyloid Beta Peptide over Gold Colloidal Nanoparticles' Surfaces

Kazushige Yokoyama

CONTENTS

7.1 Introduction

A critical onset process of Alzheimer's disease is the formation of insoluble fibrillar deposits of the amyloid beta peptide (Aβ) as both diffuse and senile amyloid plaque that invades the brain's seat of memory and cognition [1–4]. The major sequences of Aβ associated with the formation of insoluble fibrillar deposits are considered as 42- and 40-residue, that is, $A\beta_{1-42}$ and $A\beta_{1-40}$. The $A\beta_{1-42}$ and $A\beta_{1-40}$ are capable of assembling β-sheet fibrils with 60–100 Å diameter [5,6]. While there are only two sequences of difference between $A\beta_{1-42}$ and $A\beta_{1-40}$, they possess almost opposing property. The $A\beta_{1-42}$ is highly hydrophobic and is significantly associated with the nucleation of amyloid fibril [7]. $A\beta_{1-40}$, on the other hand, is relatively hydrophilic and is

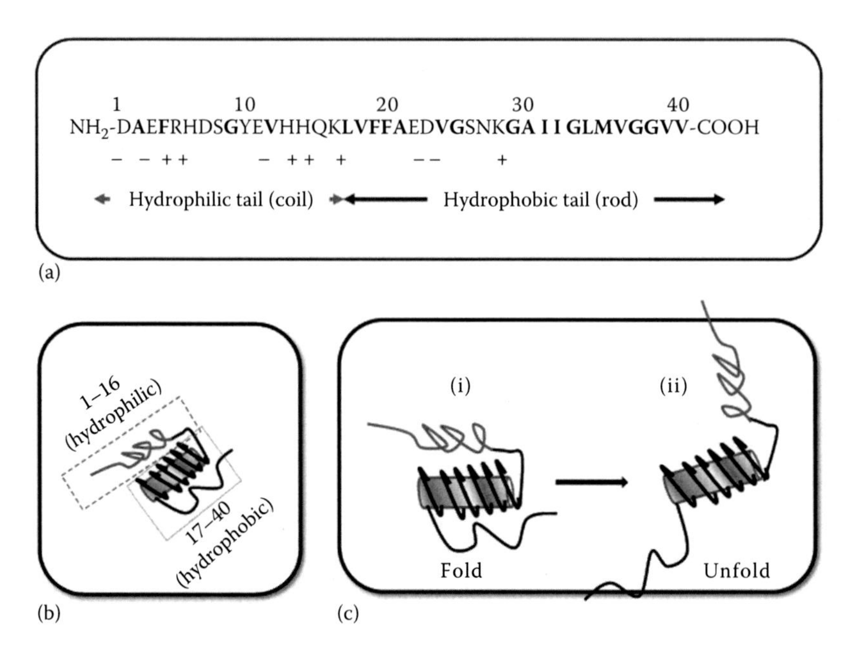

FIGURE 7.1
(a) The entire sequences of $A\beta_{1-40}$. The charges (+ or −) at neutral pHs are given beneath the sequences. Hydrophobic residues are given in bold letters. (b) The sketch of secondary structure of $A\beta_{1-40}$ and portion of hydrophilic ($A\beta_{1-16}$) and hydrophobic ($A\beta_{17-40}$) segments. (c) A schematic of the transition from (i) folded to (ii) unfolded structure of $A\beta_{1-40}$ monomer expected in the solution.

more soluble in aqueous solution. The $A\beta_{1-40}$ is the main form circulating in normal plasma and cerebrospinal fluid (see Figure 7.1) [8,9]. No matter which sequences of Aβ monomers are initially prepared, they are considered to proceed the same steps of a fiber formation (fibrillogenesis). The fibrillogenesis is induced by a nucleation and is regarded as a polymerization process originating from a nucleus unit formed by a limited number of Aβ monomers [10,11]. At an initial stage of the fibrillogenesis, a core aggregate establishes the lattice form of amyloid fibril (Step 1 in Figure 7.2) [12], followed by the elongation of the fibril through the sequential addition of fibril subunits (Step 2 in Figure 7.2). An initial stage of fibrillogenesis, Step 1, is considered to involve the soluble Aβ complex with oligomer form, and this step is a key onset for the subsequent aggregation process [13–15]. Up to now, relatively little is known about the structure of oligomer form. A detection of the Aβ oligomer form and the investigation of associated reversible process are quite important to control fibrillogenesis.

It is plausible to expect that an initial stage of fibrillogenesis may occur under interfacial environment such as blood cell surface or membrane surface. Since Aβ monomer contains both hydrophilic and hydrophobic segments,

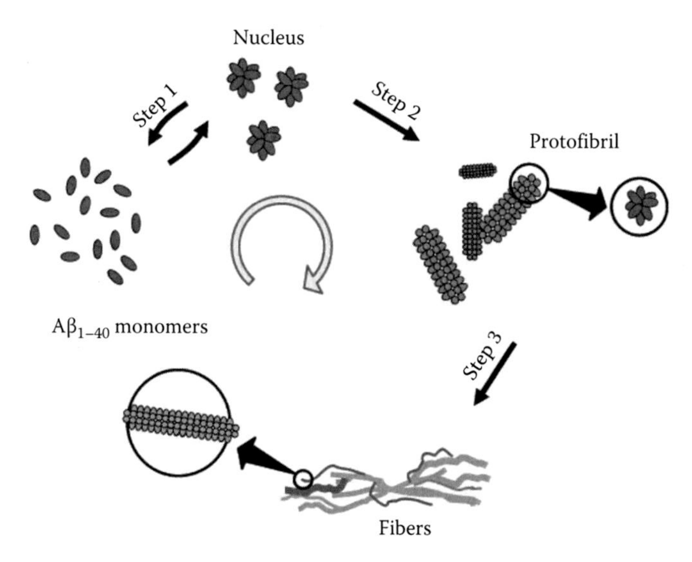

FIGURE 7.2
A model of fibrillogenesis. In Step 1, monomeric Aβ forms nuclei from which protofibrils emanate (Step 2). These protofibrils form fibers (Step 3).

hydrophilic or hydrophobic molecular surface is expected to attract several Aβ monomers to be adsorbed. The adsorption process can help localize a group of Aβ monomers, and this grouping corresponds to an initial stage of the fibrillogenesis (i.e., Step 1 in Figure 7.2). If conformation of Aβ aggregates adsorbed over the surface is investigated, the conformation associated with intermediates in fibrillogenesis can be extracted. The proteins immobilized at an interface were reported to behave differently from their counterparts in bulk solutions [16–19]. The fibrillogenesis that takes place on the surface must progress in a different way than that which takes place in the solution. It is expected to stabilize the oligomer form with the help of the surface potential. Our group's approach is to utilize the spherical surface of gold colloidal nanoparticles (10–100 nm) as a template of a core formation and to locate Aβ over the nanoparticles' surfaces [20,21]. A great advantage of using a metal surface is its high thermal conductivity. The change of ambient temperature can be directly reflected upon the temperature over the surface of the gold colloid. Therefore, it allows us to systematically investigate the effect of temperature change to the initial reversible process (Step 1 in Figure 7.2). The role of the thermal condition in the fibrillogenesis taking place at the interface is considered to be critical, since the variation of the thermal condition must change the surface potential favoring to form an oligomer.

The usage of gold colloidal particles enables us to conduct a spectroscopic (or colorimetric) study to investigate the reversible process that corresponds to the Step 1 of fibrillogenesis. Absorption by the surface plasmons of gold

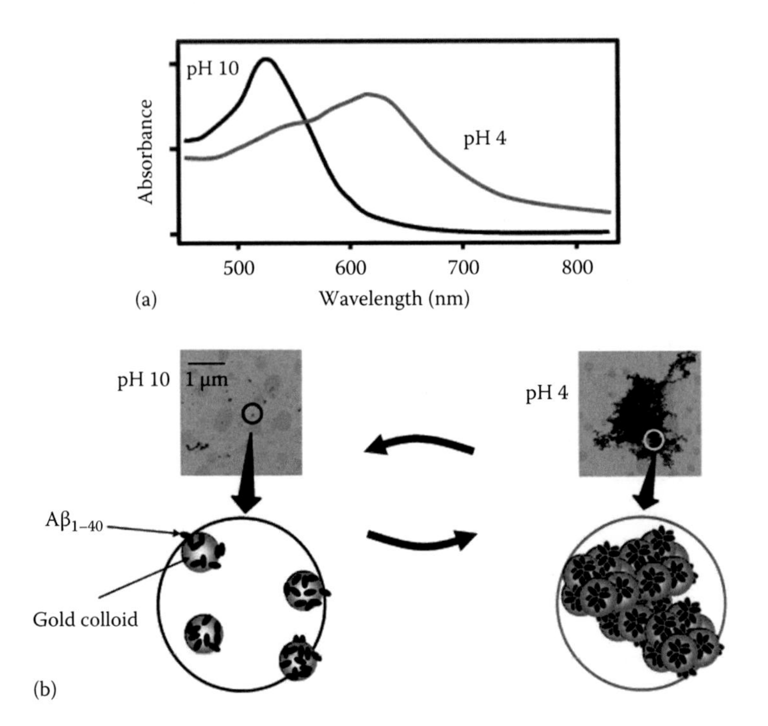

FIGURE 7.3
(a) The absorption spectrum of $A\beta_{1-40}$-coated 20 nm gold colloids observed under pH 10 (black) and pH 4 (gray). (b) The TEM images at each pH condition are shown with the sketches of disperse (at pH 10) or aggregated (at pH 4) conditions of $A\beta_{1-40}$-coated gold colloids. Two arrows imply the observation of reversible process and it corresponds to the reversible process of Step 1 in Figure 7.2. The conformation of $A\beta_{1-40}$ monomer over the colloidal surface also corresponds to that expected in Step 1 of the fibrillogenesis.

colloidal nanoparticles accounts for the observed color [22–24]. As demonstrated in the transmission electron microscopy (TEM) images in Figure 7.3b, Aβ-coated gold colloids that exist dispersely under pH 10 exhibited reddish color. With pH 4, however, the gold colloids showed a bluish color due to the formation of aggregates caused by the networking of Aβ adsorbed on the surface. Since the absorption spectrum that corresponds to each pH condition is distinctively different (see Figure 7.3a), an aggregation and disperse condition can be easily identified. The transition of conformation of Aβ from pH 10 to 4 corresponds to the conformational change expected in the reversible process in the fibrillogenesis. If the size of the core gold colloid and temperature condition were adjusted promptly, the reversible color change can be observed as an external pH changes between pH 4 and 10. This reversible color change (between bluish color and reddish color) reflects the Step 1 of fibrillogenesis, where dispersed monomer reversibly transforms into an oligomer form.

7.2 Investigation of Reversible Self-Assembly

The ultra-pure $A\beta_{1-40}$ (MW: 4329.9 Da) was purchased from American Peptide (Sunnyvale, California) in the form of lyophilized powder (97% by HPLC), and stored at −12°C. The stock solution of 100 μM $A\beta_{1-40}$ was prepared by using double-distilled deionized and filtered water. Gold colloidal nanoparticles were purchased from Ted Pella Inc. (Redding, California) with the diameters of 5 ± 1.0, 9.8 ± 1.0, 15.2 ± 1.5, 19.7 ± 1.1, 30.7 ± 1.3, 40.6 ± 1.1, 51.5 ± 4, 60 ± 1.0, 80 ± 1.0, and 99.5 ± 1.3 nm. The optimum ratio of the particles between the protein and the gold colloidal nanoparticle solution was found to be a 1000:1 [20].

The pH of the solution was repeatedly altered between pH 4 and 10, and the corresponding absorption spectrum was monitored at each pH condition. The original pH of freshly prepared sample was around pH 7. The pH change to pH 4 from pH 7 was completed by dropwise addition of hydrochloric acid (HCl), and the change to pH 10 from pH 4 was by addition of sodium hydroxide (NaOH). The utilized acids (HCl) and bases (NaOH) of various concentrations were prepared in the temperature control cell holder with targeted temperature in order to minimize the temperature variation during the pH adjustment. The average peak position of the surface plasmon resonance (SPR) band of the regions between 400 and 800 nm was monitored as the pH was repetitively changed between pH 4 and 10. The average peak position at given pH, $\lambda_{peak}(pH)$, was calculated based on the following

$$\lambda_{\text{peak}}(\text{pH}) = \sum_i a_i(\text{pH})\lambda_i(\text{pH}) \tag{7.1}$$

where λ_i(pH) and a_i(pH) represent the peak position and fraction of the *i*th component band. The fraction a_i was determined by the fraction of the area (A_i) of the band to the area of the total sum of the entire bands. Most of the bands were analyzed with two components or one component with a large background band. All the absorption bands were fully explained by the fit with Gaussian profile using the peak-fit module of ORIGIN (version 7.0) in the range of 400–800 nm. The initial pH of the bare gold nanoparticle solutions or $A\beta_{1-40}$-coated gold particles was around neutral with reddish color; thus, the SPR band is maximum at 528 nm. When the pH was adjusted to pH 4, the color of the solution that changed into bluish color had bimodal components with absorption band maximum around 600 nm. In our study, the identification of operation to alter pH condition is given by "n." Here, $n = 1$ corresponds to the starting pH condition of solution before acid is inserted and its pH was around pH 7. After $n = 1$, odd numbers of n ($n_{odd} = 3, 5, 7, \ldots$) indicate an operation of base addition leading the pH of solution to be pH 10,

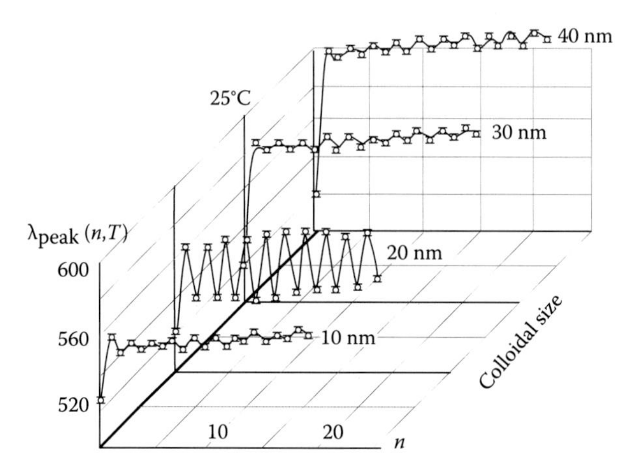

FIGURE 7.4
Collections of peak shift, $\lambda_{peak}(n, T)$, as a function of pH changes (n) for $A\beta_{1-40}$-coated gold colloid particles of 10, 20, 30, 40 nm at 25°C.

whereas even numbers of n ($n_{even} = 2, 4, 6, \ldots$) show an operation of acid addition to decrease the pH of solution to be pH 4. Among all tested sizes of gold colloidal nanoparticles (5–100 nm), only 20 nm gold colloids coated with $A\beta_{1-40}$ exhibited reversible color change at 20°C. (See Figure 7.4. In this figure, $\lambda_{peak}(n, T)$ for $A\beta_{1-40}$-coated 10, 20, 30, and 40 nm gold colloids are represented.) [25]. The color change in the reversible process was not between pure blue and red; rather, it was between purple and red. The peak at pH 10 shifts gradually to ~600 nm (i.e., red shifted) as the repetition number of the pH change, n, increased, while the absorption band at pH 4 appears around 620 nm. Here, it should be noted that the averaged peak position does not necessarily match with the absorption wavelength corresponding to the color of each solution. This is because the peak position is the weighted average of two components as shown in Equation 7.1.

7.3 Temperature/Size Dependence of Reversible Self-Assembly

The reversible color change was further examined at 5°C and 45°C for $A\beta_{1-40}$-conjugated 20, 30, and 40 nm gold colloids as shown in Figure 7.5. All $A\beta_{1-40}$-coated gold colloids exhibited an undulating feature in $\lambda_{peak}(n, T)$ as the pH changed between pH 4 and 10. The reversible process was visually observed for $A\beta_{1-40}$-coated 20 nm gold at 45°C, but not at 5°C. For $A\beta_{1-40}$-coated 30 nm, the amplitudes of $\lambda_{peak}(n)$ were significantly large enough to provide the

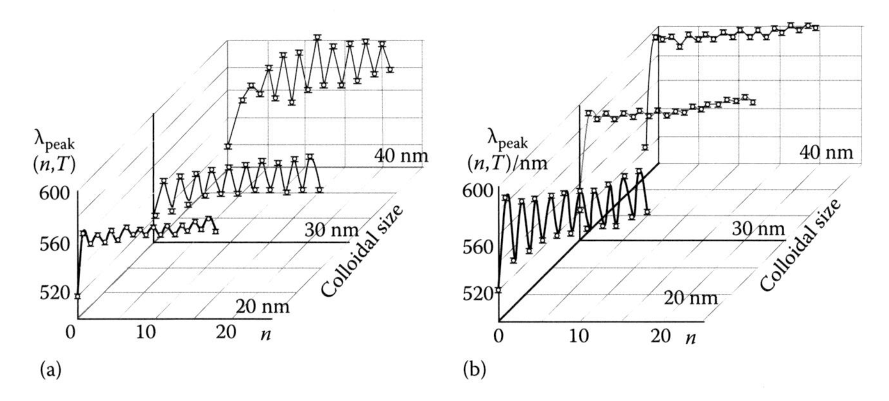

FIGURE 7.5
Collections of peak shift, $\lambda_{peak}(n, T)$, as a function of pH changes (n) for $A\beta_{1-40}$-coated gold colloid particles of 20, 30, 40 nm at (a) 5°C and (b) 45°C.

reversible color change at 5°C. This was true at 5°C for $A\beta_{1-40}$-coated 40 nm gold colloids.

The observed λ_{peak} (n, T) for each gold colloidal size was analyzed by Equation 7.2

$$\lambda_{peak}(n,T) = A_T^d + B_T^d(n-1)^{C_T^d} + D_T^d e^{((n-1)E_T^d)} \cos(n\pi) \tag{7.2}$$

where n indicates the operation of pH change. The parameters B_T^d, D_T^d, and E_T^d were extracted and plotted for each temperature, T (°C), of gold colloids with diameter of d nm and are plotted as a function of colloidal size. In Equation 7.2, an initial peak position at neutral pH (i.e., $\lambda_{peak}[n=1, T]$) is given by $A_T^d - D_T^d$, and the parameters B_T^d and C_T^d show the average wave peak position shift as pH varies between pH 4 and 10. The parameters D_T^d and E_T^d imply the amplitude and damping factor for the repetitive event, and the cosine function was used to indicate the increase and decrease of $\lambda_{peak}(n, T)$ upon change between pH 4 and 10. The visual confirmation of the reversible process was possible when the parameter D_T^d was over 5.5 nm. The values calculated by Equation 7.2 are effective only for each n value, not the values between each n. Thus, the dotted lines shown in Figures 7.4 and 7.5 are given only for the purpose of clarifying the repetitive trend.

Opposed to the trend seen in self-assembly over 20 nm gold colloid, the intermediates over the 30 or 40 nm colloidal surfaces were stably prepared at the lower temperatures. Thus, the kinetics of $A\beta_{1-40}$ nucleation over the 30 or 40 nm colloidal surface must be driven by completely different mechanisms than those for $A\beta_{1-40}$ coated over 20 nm gold. The best-suited concept explains the reversible self-assembly observed for $A\beta_{1-40}$ monomers over

30 or 40 nm gold colloidal surface is *homogeneous nucleation theory* [26]. The Gibbs energy change, ΔG^*, with nucleation radius, R^*, is given by

$$\Delta G^* = \frac{4\pi\sigma R^{*2}}{3} = \frac{16\pi\sigma^3\upsilon^2}{3(kT\ln S)^2} \tag{7.3}$$

Here
- k is the Boltzmann constant
- T is the temperature
- σ is the surface tension of the liquid
- υ is the molecular volume
- S is the super saturation ratio (S) C/C_{sat} where C is the solute concentration
- C_{sat} is the saturation concentration

It should be noted that, for the range of radius focused in this study, the ΔG^* required to form an aggregate from the solution phase is positive. Therefore, the aggregates can be formed only when $S > 1$. The ΔG^* becomes more positive to enhance nucleation more as the temperature is lowered. This explains that the reversible color change was supported at the lower temperature for $A\beta_{1-40}$-coated 30 or 40 nm gold colloid. Utilizing homogeneous nucleation theory, Garai et al. defined the nucleation radius to lie between 5 and 50 nm [27]. The surface tension of the aggregate of the Aβ/water interface is calculated to be more than 4.8 mJ/m^2 at room temperature, and the surface energy barrier to nucleation is estimated to be more than 1.93×10^{-19} J with more than 29 monomers required in the nucleus.

7.4 Investigation of Temperature Shift

Based on the observation of temperature- and size-dependent reversible self-assembly, it allows us to control the self-assembly process by selecting the size and temperature condition. The self-assembly path can be, therefore, well predicted once temperature and core gold colloidal size are determined. For a given size of gold colloidal particle, $\lambda_{peak}(n, T)$ observed under two different temperatures, T_1 and T_2 ($T_1 < T_2$), should show different features (see Figure 7.6a). Generally speaking, the values of $\lambda_{peak}(n, T_2)$ are higher than $\lambda_{peak}(n, T_1)$ if T_2 is greater than T_1. For example, $A\beta_{1-40}$-coated 30 nm gold colloid exhibited non-undulating feature in $\lambda_{peak}(n, 45°C)$ with relatively higher values (see Figure 7.5b). On the other hand, $\lambda_{peak}(n, 5°C)$ showed undulating feature, proving that a reversible self-assembly process

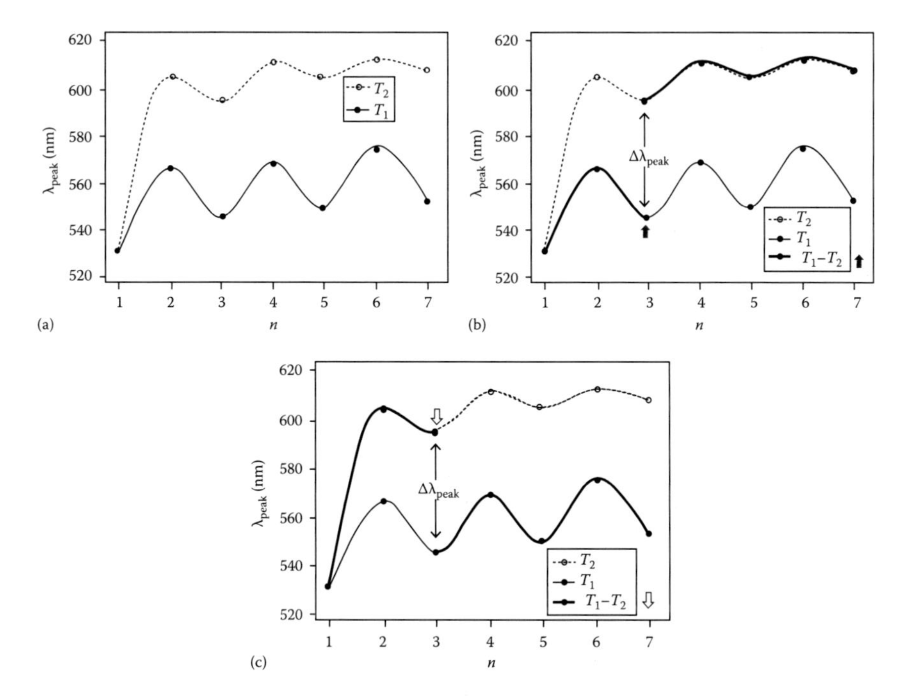

FIGURE 7.6
(a) The representative feature expected for λ_{peak} as a function of repetitive pH change between pH 4 and 10 ($n = 1$–7) at two different temperatures T_1 and T_2 ($T_1 < T_2$). Here, the quasi repetitive color change process is assumed to be observed only at the lower temperature. The bold line shows the expected features of λ_{peak} under temperature change at $n = 3$ are shown for the case when (b) temperature is shifted from the lower (T_1) to the higher (T_2) and (c) temperature is shifted from the higher to the lower. The up or down arrows and discontinuities observed at $n = 3$ in (b) and (c) indicate the temperature change and the expected feature after the temperature change.

takes place at 5°C (see Figure 7.5a). The values of $\lambda_{peak}(n, 5°C)$ are relatively lower than those of $\lambda_{peak}(n, 45°C)$. Therefore, for a given size of colloidal particle, temperature condition controls the reversibility of the self-assembly process. However, it is not known, if temperature change can affect the reversible self-assembly path. For example, as seen in the case of $A\beta_{1-40}$-coated 30 nm gold colloid, a system exhibits a reversible color change at T_1, but not at T_2 ($T_1 < T_2$), as pH changes between pH 4 and 7 (see Figure 7.6a). In Figure 7.6b, the temperature is kept at T_1 from $n = 1$ to 3, and $\lambda_{peak}(n, T_1)$ is observed as indicated by a bold line. Then, temperature is shifted up to T_2 at $n = 3$, and λ_{peak} is observed for $n = 3$–7 under T_2. If the path of the reversible self-assembly heavily depends on a given temperature condition, λ_{peak} value should reproduce $\lambda_{peak}(n, T_2)$ for $n = 3$–7 as given by the bold line. In this case, there should be a jump in λ_{peak} value ($\Delta\lambda_{peak}$) as it shifts from $\lambda_{peak}(n = 3, T_1)$ to $\lambda_{peak}(n = 3, T_2)$. In a similar manner, the same trend should be observed when temperature is shifted down. In Figure 7.6c, the temperature is kept at T_2 from $n=1$ to 3, and $\lambda_{peak}(n, T_2)$ is observed as indicated by a bold line. Then, temperature is shifted down to T_1 at $n=3$, and λ_{peak} is observed for $n=3$–7 under T_1. If the path of the reversible self-assembly heavily depends on a given temperature condition, λ_{peak} value should reproduce $\lambda_{peak}(n, T_1)$ for $n=3$–7 as given by the bold line. In this case, there should be a drop in λ_{peak} value ($\Delta\lambda_{peak}$) as it shifts down from $\lambda_{peak}(n=3, T_2)$ to $\lambda_{peak}(n=3, T_1)$. The difference in wavelength in λ_{peak} after the temperature change at $n=3$ is defined as $\Delta\lambda_{peak}$ in Equation 7.4.

$$\Delta\lambda_{peak} = \lambda_{peak}\left(T_{final}\right) - \lambda_{peak}\left(T_{initial}\right) \tag{7.4}$$

Here, $\lambda_{peak}(T_{final})$ and $\lambda_{peak}(T_{initial})$ indicate the λ_{peak} value before and after the temperature shift at $n=3$, respectively. The $\Delta\lambda_{peak}$ values become positive as the peak is red-shifted or negative as a peak produced a blue shift.

The experimental procedures for the measurement correspond to Figure 7.6b and c and are shown in Figures 7.7a and 7.8a, respectively. As for the case of temperature up-shift (from T_1 to T_2), the procedure is shown in Figure 7.7a. First, absorption spectrum is observed at pH 7 ($n=1$) under temperature T_1. Then, pH is changed to pH 4 and absorption spectrum is measured under T_1 ($n=2$). At $n=3$, pH is changed to pH 10 and spectrum is measured under T_1. In this case, reversible color change is observed as n changes from 1 to 3. Then, temperature up-shift from T_1 to T_2 is conducted while pH is kept pH 10. As temperature is stabilized at T_2, the spectrum is measured ($n=3$ at T_2). Next, pH is changed to pH 4 and spectrum is measured at T_2 ($n=4$). Figure 7.7b shows the plot of $\lambda_{peak}(n=1$–$3, T_1)$ and $\lambda_{peak}(n=3$–$7, T_2)$ as a function of n. While we show examples up to $n=4$, spectrum is obtained for $n=5$–7 under T_2 as pH is alternated between pH 10 and 4. In this example, if we hypothesize that self-assembly process depends on a given temperature, there is no reversible color change observed for the measurements at $n=3$–7.

As for the case of temperature down-shift (from T_2 to T_1), the procedure is shown in Figure 7.8a. First, absorption spectrum is observed at pH 7 ($n=1$) under temperature T_2. Then, pH is changed to pH 4 and absorption spectrum is measured under T_2 ($n=2$). At $n=3$, pH is changed to pH 10 and spectrum is measured under T_2. In this case, reversible color change is not observed for $n=1$–3. Then, temperature is down-shifted from T_2 to T_1 while pH is kept pH 10. As temperature is stabilized at T_1, spectrum is measured ($n=3$ at T_1). Next, pH is changed to pH 4 and spectrum is measured at T_1 ($n=4$). Figure 7.8b shows the plot of $\lambda_{peak}(n=1$–$3, T_2)$ and $\lambda_{peak}(n=3$–$7, T_1)$ as a function of n. While we show examples up to $n=4$, spectrum are obtained for $n=5$–7 under T_1 as pH is alternated between pH 10 and 4. If we hypothesize that self-assembly process depends on a given temperature, reversible color change should be observed for the measurements at $n=3$–7 at T_1.

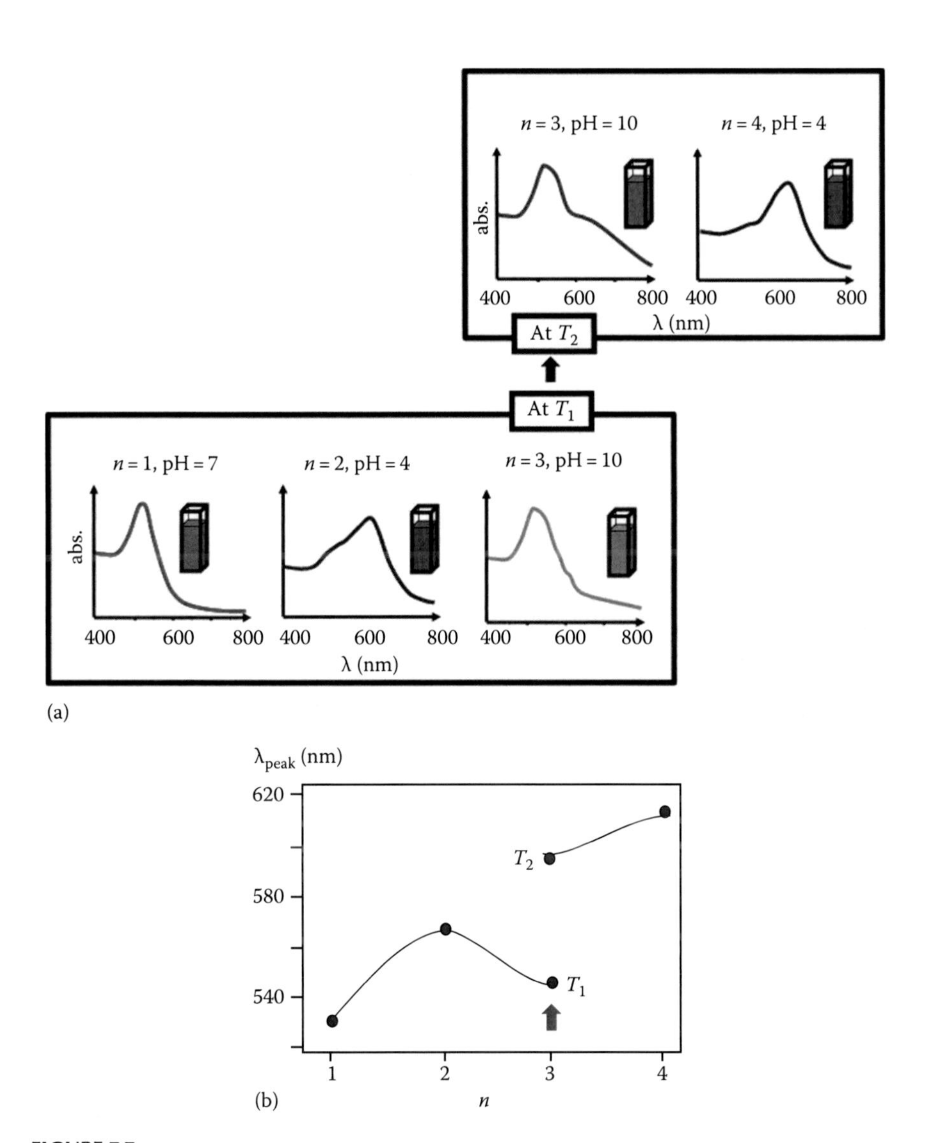

FIGURE 7.7
(a) The schematics showing the procedure of temperature shift from the lower (T_1) to the higher (T_2). The procedure is shown only up to $n=4$, whereas the spectrum was observed up to $n=7$. At $n=3$, temperature is increased from T_1 to T_2 ($T_2>T_1$) as λ_{peak} is measured for each temperature at pH 10. (b) The expected peak position (λ_{peak}) calculated from Equation 7.1 is plotted from $n=1$ to 4. The discontinuity observed at $n=3$ and an upward arrow indicate the temperature change from T_1 to T_2.

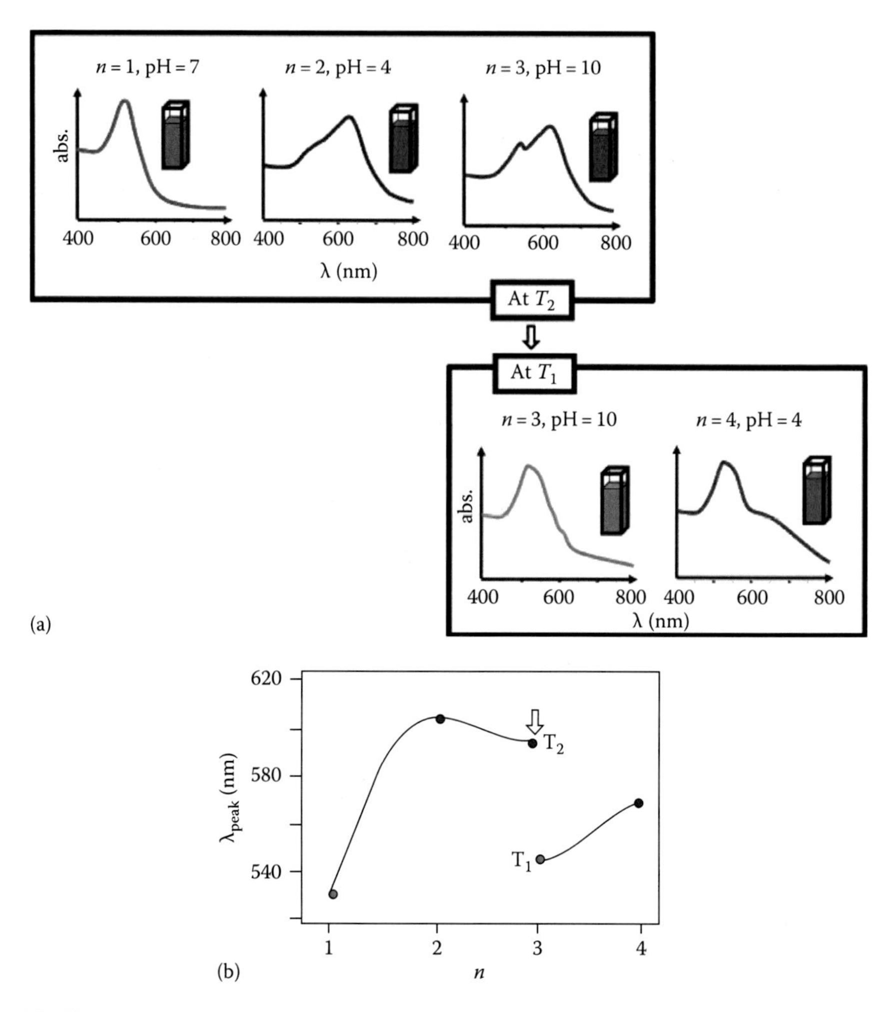

FIGURE 7.8
(a) The schematics showing the procedure of temperature shift from the higher (T_2) to the lower (T_1). The procedure is shown only up to $n=4$, whereas the spectrum was observed up to $n=7$. At $n=3$, temperature is decreased from T_2 to T_1 ($T_2 > T_1$) as λ_{peak} is measured for each temperature at pH 10. (b) The expected peak position (λ_{peak}) calculated from Equation 7.1 is plotted from $n=1$ to 4. The discontinuity observed at $n=3$ and downward arrow indicates the temperature change from T_2 to T_1.

7.5 Effect of Temperature Shift to Nanoscale Reversible Self-Assembly

7.5.1 Temperature Down-/Up-Shift to Reversible Color Change

Following the procedures described in Figures 7.7a and 7.8a, absorption spectrum was observed for $A\beta_{1-40}$-coated gold colloid 20, 30, and 40 nm. The attempted temperature shift was between 5°C and 45°C for $A\beta_{1-40}$-coated gold colloid 20 and 30 nm. As for $A\beta_{1-40}$-coated gold colloid 40 nm, temperature shift was between 3.5°C and 45°C.

20 nm gold colloids coated with $A\beta_{1-40}$: At 45°C, the reversible color change was clearly confirmed, whereas the reversible color change did not occur at 5°C. When temperature was shifted up from 5°C to 45°C, the λ_{peak} ($n=3$, $T=45$°C from 5°C) was not observed at the point expected at 45°C. Instead, the λ_{peak} exhibited a slight red shift. The features after the temperature shift did not follow the position of λ_{peak} ($n=3$, $T=45$°C), whereas an entire peak position was red-shifted by approximately 30 nm. On the other hand, the amplitude of the λ_{peak} between $n=3$ and $n=7$ almost reproduced that of λ_{peak} ($n=4$–7, $T=5$°C) (Figure 7.9a). When temperature was shifted down from 45°C to 5°C, the λ_{peak} ($n=3$, $T=5$°C from 45°C) exhibited a slight red shift and did not exactly overlap with the λ_{peak} ($n=3$, $T=5$°C). The features after the temperature shift did not follow the position of λ_{peak} ($n=3$–7, $T=5$°C); however, the amplitude of the λ_{peak} between $n=3$ and $n=7$ almost reproduced that of λ_{peak} ($n=4$–7, $T=45$°C) (Figure 7.9b).

30 nm gold colloids coated with $A\beta_{1-40}$: A clear sign of reversible color change was observed at 5°C, but not at 45°C. When temperature was shifted up from 5°C to 45°C, the λ_{peak} ($n=3$, $T=45$°C from 5°C) was red-shifted with a significant amount, but not observed at the exact point expected at λ_{peak} ($n=3$, $T=45$°C). The features after the temperature shift almost reproduced an amplitude of λ_{peak} ($n=3$–7, $T=45$°C) with approximately 10 nm of red shift (Figure 7.9c). When temperature was shifted down from 45°C to 5°C, the λ_{peak} ($n=3$, $T=5$°C from 45°C) exhibited a slight blue shift and almost reproduced the feature of λ_{peak} ($n=3$–7, $T=45$°C) with approximately 10 nm of blue shift (Figure 7.9d).

40 nm gold colloids coated with $A\beta_{1-40}$: A reversible color change was observed at 3.5°C after an induction period ($n=1$–4); however, no clear sign of reversible color change was observed at 45°C. When temperature was shifted up from 3.5°C to 45°C, the λ_{peak} ($n=3$, $T=45$°C from 3.5°C) was red-shifted, but still far from the point expected at λ_{peak} ($n=3$, $T=45$°C). As n increases, the features after the temperature shift converged to the feature observed for λ_{peak} ($n=3$–7, $T=45$°C) with red shift (Figure 7.9e). When temperature was shifted down from 45°C to 3.5°C, the λ_{peak} ($n=3$, $T=3.5$°C from 45°C) exhibited only a slight blue shift and did not reach to the point expected at λ_{peak} ($n=3$, $T=3.5$°C). The features after the temperature change still followed the feature of λ_{peak} ($n=3$–7, $T=45$°C) with approximately 10 nm of blue shift (Figure 7.9f).

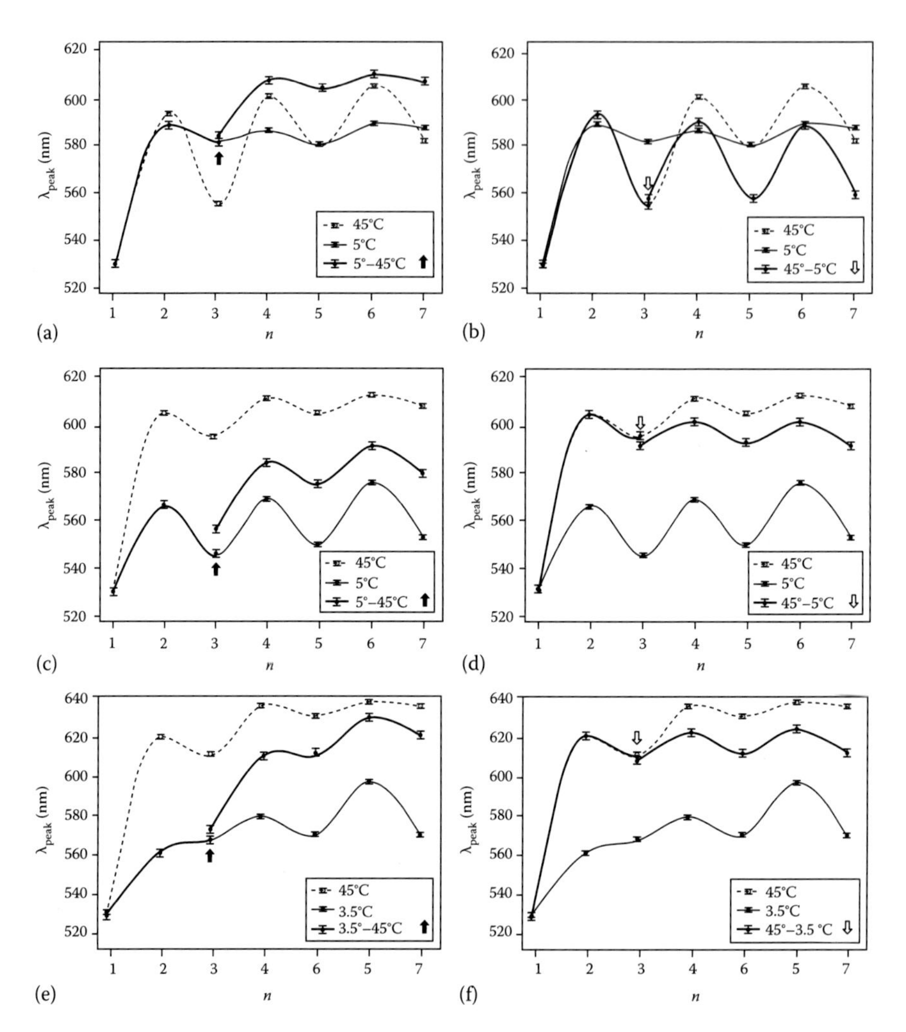

FIGURE 7.9
The peak shift as a function of pH change under constant temperature or as a result of temperature change for $A\beta_{1-40}$ coated (a) 20 nm (5°C, 45°C, and 5°C–45°C), (b) 20 nm (5°C, 45°C, and 45°C–5°C), (c) 30 nm (5°C, 45°C, and 5°C–45°C), (d) 30 nm (5°C, 45°C, and 45°C–5°C), (e) 40 nm (3.5°C, 45°C, and 3.5°C–45°C), and (f) 40 nm (3.5°C, 45°C, and 45°C–3.5°C).

7.5.2 Effect of Temperature Down-/Up-Shift to Reversible Color Change

The observed features in λ_{peak} shown in Figures 7.9a through f were analyzed with Equation 7.2 and extracted parameters are tabulated in Table 7.1. Then, each parameter extracted under temperature shift condition is divided by a corresponding parameter of either upper or lower temperature. For example, based on Table 7.1, parameter A of $A\beta_{1-40}$-coated 20 nm gold colloid at 45°C and 5°C are A(T = 45°C) = 554 ± 7 nm and A(T = 5°C) = 535 ± 4 nm, respectively.

TABLE 7.1

Extracted Parameters B_T^d, D_T^d, and E_T^d and r^2 Values for Equation 7.2 for the Observed pH (4 and 10) Dependent Reversible Peak Shifts under Given Temperature or Temperature Shift Conditions of (a) $A\beta_{1-40}$-Coated 20 nm Gold Colloid, (b) $A\beta_{1-40}$-Coated 30 nm Gold Colloid, and (c) $A\beta_{1-40}$-Coated 40 nm Gold Colloid

(a) 20 nm	***T*=45°C**	**5°C**	**5°C–45°C**	**45°C–5°C**
A_T^{20}	554(7)	535(4)	542 (7)	550(2)
B_T^{20}	17(9)	48(5)	39 (8)	22(3)
C_T^{20}	0.5(2)	0.03(3)	0.32(7)	0.0002(0.05)
D_T^{20}	24(6)	5(3)	12(6)	19(2)
E_T^{20}	0.13(7)	0.2(2)	0.3(2)	0.05(2)
R^2	0.988	0.994	0.990	0.997
(b) 30 nm	***T*=45°C**	**5°C**	**5°C–45°C**	**45°C–5°C**
A_T^{30}	540(2)	541(2)	541(8)	537(3)
B_T^{30}	55(3)	12(3)	13(8)	59(4)
C_T^{30}	0.13(2)	0.4(1)	0.7(3)	0.00003(0.03)
D_T^{30}	11(2)	11(2)	12(6)	5(2)
E_T^{30}	0.23(7)	−0.01(4)	0.08(0.2)	0.03(0.12)
r^2	0.999	0.996	0.953	0.995
(c) 40 nm	***T*=45°C**	**3.5°C**	**3.5°C–45°C**	**45°C–3.5°C**
A_T^{40}	546(5)	531(2)	532(9)	536(2)
B_T^{40}	64(6)	30(3)	29(9)	78(2)
C_T^{40}	0.21(3)	0.36(4)	0.7(2)	0.02(1)
D_T^{40}	15(4)	1.1(5)	5(5)	6(1)
E_T^{40}	0.3(1)	−0.5(1)	−0.2(3)	0.007(5)
r^2	0.998	0.994	0.972	0.999

The parameter A for the $A\beta_{1-40}$-coated 20 nm gold colloid under temperature shifts from 5°C to 45°C is A(T=5°C–45°C)=542±7 nm. Here, either A(T=5°C) or A(T=45°C) is divided by A(T=5°C–45°C) in order to value how much change was caused. For example,

$$A(T=45°C) \div A(T=5°C-45°C) = 554\text{ nm} \div 542\text{ nm} = 1.01$$

$$A(T=5°C) \div A(T=5°C-45°C) = 535\text{ nm} \div 542\text{ nm} = 0.98$$

The list of comparisons calculated by this way are tabulated in Table 7.2 for (a) $A\beta_{1-40}$-coated 20 nm gold colloid, (b) $A\beta_{1-40}$-coated 30 nm gold colloid, and

TABLE 7.2

Comparison of Parameters B_T^d, D_T^d, and E_T^d in Equation 7.2 and $\Delta\lambda$ (nm) in Equation 7.4 Extracted in Temperature Shift Condition to Those under Initial or Final Temperatures for (a) $A\beta_{1-40}$-Coated 20 nm Gold Colloid, (b) $A\beta_{1-40}$-Coated 30 nm Gold Colloid, and (c) $A\beta_{1-40}$-Coated 40 nm Gold Colloid

	T = 5°C–45°C		45°C–5°C	
(a) $A\beta_{1-40}$-Coated 20 nm	**Compared to 5°C (Initial Temp.)**	**Compared to 45°C (Final Temp.)**	**Compared to 45°C (Initial Temp.)**	**Compared to 5°C (Final Temp.)**
A_T^{20}	1.01	0.98	0.99	1.03
B_T^{20}	0.81	2.22	1.26	0.45
C_T^{20}	9.84	0.67	0.0005	0.007
D_T^{20}	2.31	0.50	0.81	3.72
E_T^{20}	1.44	1.93	0.39	0.29
$\Delta\lambda$ (nm)	0.9		−3	
	T = 5°C–45°C		45°C–5°C	
(b) $A\beta_{1-40}$-Coated 30 nm	**Compared to 5°C (Initial Temp.)**	**Compared to 45°C (Final Temp.)**	**Compared to 45°C (Initial Temp.)**	**Compared to 5°C (Final Temp.)**
A_T^{30}	1.00	1.00	0.99	0.99
B_T^{30}	1.11	0.24	1.06	4.87
C_T^{30}	1.97	5.42	0.00023	8.38×10^{-5}
D_T^{30}	1.15	1.20	0.49	0.47
E_T^{30}	−7.98	0.35	0.13	−3.00
$\Delta\lambda$ (nm)	10		−3	
	T = 3.5°C–45°C		45°C–3.5°C	
(c) $A\beta_{1-40}$-Coated 40 nm	**Compared to 3.5°C (Initial Temp.)**	**Compared to 45°C (Final Temp.)**	**Compared to 45°C (Initial Temp.)**	**Compared to 3.5°C (Final Temp.)**
A_T^{40}	1.00	0.97	0.98	1.01
B_T^{40}	0.96	0.45	1.22	2.60
C_T^{40}	1.98	3.39	0.10	0.060
D_T^{40}	3.69	0.27	0.41	5.65
E_T^{40}	0.38	−0.66	0.026	−0.015
$\Delta\lambda$ (nm)	4		−3	

(c) $A\beta_{1-40}$-coated 40 nm gold colloid for each temperature shift measurement. As the values are more deviated from 1.0, the parameter obtained under temperature shift measurement is not regarded as same or close to the parameter of compared temperature.

20 nm gold colloids coated with $A\beta_{1-40}$: Most remarkable feature found in the λ_{peak} plot is that the amplitude of λ_{peak} tends to maintain that at an initial temperature. For example, when temperature up-shift (from 5°C to 45°C) was conducted, the amplitude after the temperature shift almost followed the feature expected at 5°C showing relatively low amplitude implying nonreversible self-assembly process. However, an average peak position is red-shifted by approximately 30 nm. When temperature down-shift (from 45°C to 5°C) was conducted, the amplitude after the temperature shift maintained relatively high amplitude expected at 45°C, implying an occurrence of reversible self-assembly process. The observed $\Delta\lambda_{peak}$ for both temperature up- or down-shift were relatively small, and this is consistent with the feature seen in amplitude where an initial temperature governs the feature of λ_{peak} even after the temperature shift. Generally speaking, a drastic change after temperature shift was mainly observed in a parameter C, which indicates the position of λ_{peak} at given n. This reflects the shift of λ_{peak} observed after temperature shift.

30 nm gold colloids coated with $A\beta_{1-40}$: While there was a significant shift in the average peak positions after the temperature shift, the amplitude of λ_{peak} was found to follow that expected at 45°C. This implies that the $A\beta_{1-40}$ monomers situated over 30 nm gold colloid were denatured at 45°C, and those denatured monomers were not able to fold back even when temperature was lowered. The observed $\Delta\lambda_{peak}$ was relatively large when temperature was up-shifted from 5°C to 45°C, whereas $\Delta\lambda_{peak}$ was almost negligible as temperature decreased from 45°C to 5°C. This trend observed in $\Delta\lambda_{peak}$ supports that $A\beta_{1-40}$ monomers were denatured at 45°C and they did not change denatured conformation when temperature was lowered from 45°C (i.e., negligible value in $\Delta\lambda_{peak}$), whereas conformation was changed into denatured form when temperature was increased to 45°C (i.e., relatively high value in $\Delta\lambda_{peak}$). The parameter C exhibited a remarkable deviation from that at 5°C, implying that observed λ_{peak} can be better reproduced by the parameter obtained at 45°C.

40 nm gold colloids coated with $A\beta_{1-40}$: The features observed for $A\beta_{1-40}$-coated 40 nm gold colloids were similar to those observed in $A\beta_{1-40}$-coated 30 nm gold colloids. After the temperature shift, the amplitude of λ_{peak} was found to follow that expected at 45°C with a significant shift in the average peak positions. Thus, the surface potential prepared by the 40 nm gold colloid is estimated to be similar to that by 30 nm gold colloid. Once $A\beta_{1-40}$ monomers were denatured at 45°C, they were not able to recover to original folded structure even when temperature was lowered to 3.5°C. A similar trend with the case of $A\beta_{1-40}$-coated 30 nm gold colloids was observed in $\Delta\lambda_{peak}$. The $\Delta\lambda_{peak}$ was relatively large when temperature was up-shifted from 3.5°C to 45°C, whereas $\Delta\lambda_{peak}$ was negligible as temperature decreased from 45°C to 3.5°C. The parameter C exhibited a remarkable deviation from that at 3.5°C, implying that observed λ_{peak} can be better reproduced by the parameter obtained at 45°C as it was observed in the case of $A\beta_{1-40}$-coated 30 nm gold colloids.

7.6 Suggested Model for Reversible Self-Assembly

Our observation confirmed the involvement of a particular structure or conformation by Aβ monomers is stable at a particular temperature. Also conformation constricted over the nanocolloidal surface possesses the size dependence. Since the formation of the intermediates can be regarded as the transition state with relatively higher thermal energy, the reversible process is expected to take place at relatively higher temperatures. This assumption is correct for the case of $A\beta_{1-40}$-coated 20 nm gold colloid in which the reversible self-assembly was observed at higher temperature than 5°C. However, $A\beta_{1-40}$-coated 30 and 40 nm gold colloid opened up reversible self-assembly channel at the lower temperature. The nucleation constant on an Aβ monomer is reported to follow the Arrhenius law with an activation energy (E_A) of 311.2 kJ/mol at pH 7.4 over the temperature range from 0°C to 45°C [28]. As for the rate constant of fibril elongation was known to increase by two orders of magnitude by following the Arrhenius law with E_A of 96.2 kJ/mol under pH 1 or 62.8 kJ/mol at pH 3.1 over the temperature range from 4°C to 40°C [29,30], these activation energies imply that there is a significant conformational change in the binding of Aβ monomers to fibril ends. These conformational changes may be associated with peptides or oligomers involved in binding to protofibrils or to a local reorganization of each aggregate [29]. Thus, it is most plausible to consider that protofibrils are formed via a single, no-cooperative, elongation mechanism, and is well described as a linear colloidal aggregation due to diffusion and coalescence of growing aggregates.

7.6.1 Colloidal Size-Dependent Oligomer Formation

Opposed to the fact that reversible self-assembly over 20 nm gold colloid was supported at the higher temperature, the surface potential over the 30 or 40 nm colloidal particles causes $A\beta_{1-40}$ reversibly self-assemble at the lower temperature. This opposite trend can be explained by considering the conformation created over the 20 nm colloidal surface and those created over 30 or 40 nm colloidal surfaces are distinctly different. Thus, the kinetics of $A\beta_{1-40}$ nucleation over the 30 or 40 nm nanocolloidal surface must be involving completely different conformations than those that exist over 20 nm gold colloid's surface. The structure involved in the reversible self-assembly may correspond to the structural change observed in an initial stage of fibrillogenesis, where a cluster of monomeric units form an oligomer form. In this mechanism, only one particular type of oligomer was considered. However, our study revealed at least two different types of oligomers that are critical in the process of fibrillogenesis.

A molecular dynamics simulation performed with $A\beta_{10-35}$ [31] suggests several possible oligomers and their characteristic temperature dependence. While different forms of oligomers (dimer, trimer, tetramer, etc.) can exist in equilibrium, there were specific oligomer forms more preferably stabilized depending on the temperature condition. Essentially, all strands were calculated to exist as a dimer form at the temperature range from 10°C up to 80°C. The maximization of the dimer concentration was predicted at 47°C. On the other hand, the concentration of trimer was calculated to be maximized at 17°C.

By combining the results of molecular dynamics simulation with our study, the dimeric oligomer form of $A\beta_{1-40}$ is dominated over the 20 nm colloidal surface and the trimer form is involved at the 30 and 40 nm gold colloidal surface. The dimer form was calculated to be stable for wider temperature range, and this is consistent with our observation of reversible self-assembly on 20 nm gold colloid at the temperature range between 6°C and 50°C. The simulation implies that the monomer form was supported at temperatures lower than 10°C. Thus, the monomer must be predominantly formed over the dimer form at the lower temperature, and this explains why reversible self-assembly process did not take place at 5°C over the 20 nm gold colloidal surface. We hypothesize that the reversible self-assembly occurs as $A\beta_{1-40}$ monomers construct an organized "cage"-like structure covering over the gold colloidal surface. It is plausible that $A\beta_{1-40}$ monomers construct an organized dimer-unit-based "cage" assembly over 20 nm gold colloidal surface for 6°C–50°C (Figure 7.10a). On the other hand, trimer-unit-based organized "cage" is reversibly constructed over 30 and 40 nm gold colloidal surfaces (Figure 7.10b).

As pH was altered between pH 4 and 10, the internal structure of $A\beta_{1-40}$ monomer is expected to go over the unfolded and the folded structure, respectively. The acidic condition (pH 4) is reported to irreversibly denature $A\beta_{1-40}$ [32,33]. Thus, the structure responsible for the reversible process must minimize this denaturization (i.e., damage to the α-helix portion of the $A\beta_{1-40}$), when $A\beta_{1-40}$ is placed over the gold surface. Over the 20 nm gold colloidal surface, the dimer form consisting of two folded structures of $A\beta_{1-40}$ monomers were prepared at pH 10, and the dimer form consisting of two $A\beta_{1-40}$ monomers with unfolded structure was made at pH 4 (see Figure 7.10a). As for over the 30 or 40 nm gold colloidal surface, the trimer consisting of three folded $A\beta_{1-40}$ must construct a unit of a cage at pH 10 and the trimer unit made out of three unfolded $A\beta_{1-40}$ monomers was stably formed at pH 4 (see Figure 7.10a). For both cases, the unfolded structure is considered to place the hydrophilic portion of $A\beta_{1-40}$ outward from the

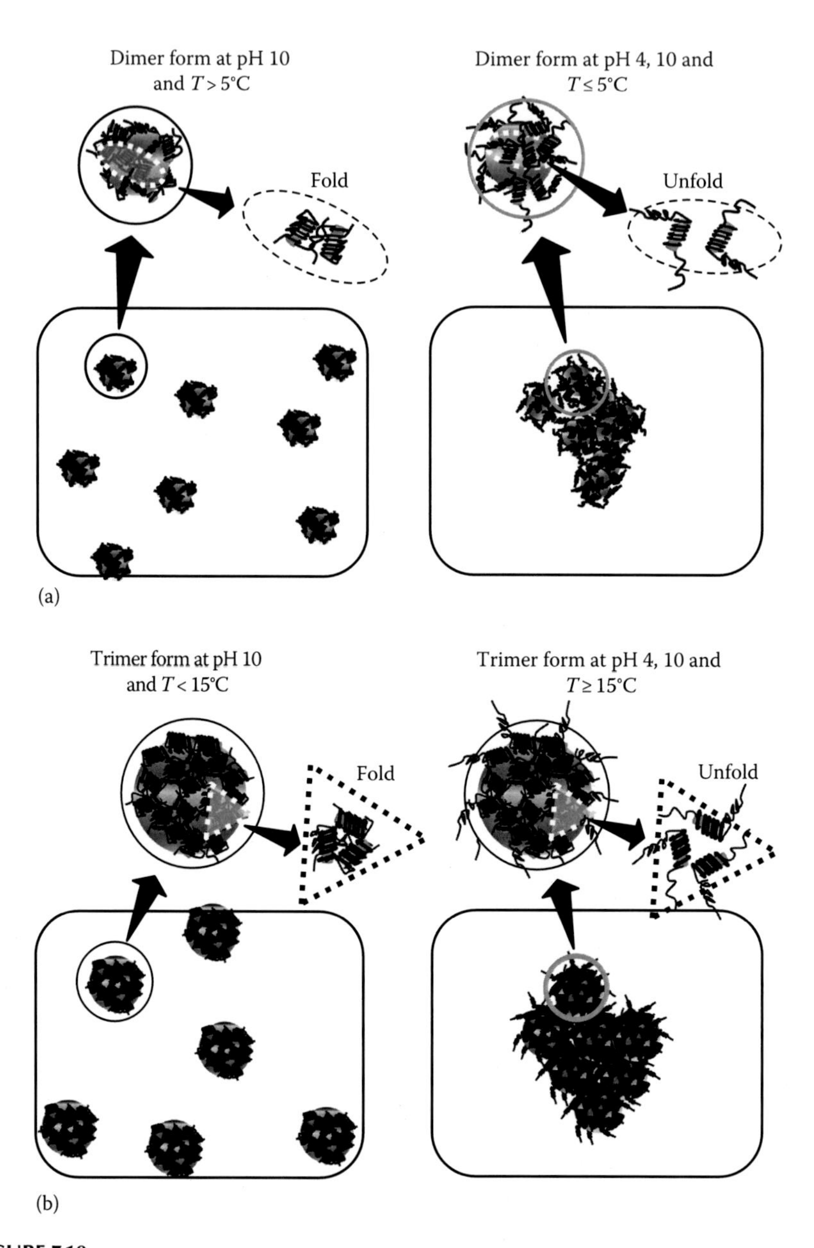

FIGURE 7.10
A schematic sketch showing the stable unit structure of $A\beta_{1-40}$ aggregates constructed at (a) 20 nm and (b) 30 or 40 nm gold colloidal surface. (a) Over the 20 nm gold colloidal surface, the dimer is the unit base and fold structure is supported at the lower temperature. The dimer form with unfold monomer base is supported at the higher temperature. (b) Over the 30 or 40 nm gold colloidal surface, the trimer is the unit base. The trimer consisted of the folded $A\beta_{1-40}$, which is supported at the lower temperature, whereas the trimer made of unfolded $A\beta_{1-40}$ is dominated at the higher temperature. For both cases (a) and (b), the aggregates are networked through hydrophilic segment placed outward from the colloidal surface.

colloidal surface for networking each other. The major structural component observed in the reversible self-assembly process can be simplified as the involvement between β-sheet and α-helices, while the internal bonding between $A\beta_{1-40}$-coated gold colloids is formed at pH 4, or disconnected at pH 10, respectively. The simulation on aggregation of $A\beta_{16-22}$ peptides in water showed a hydrophilic character of aggregation with decrease in density of hydration water, whereas the peptide surface exposed to water became more hydrophobic with increasing aggregate size [15,34]. Thus, it supports the fact that hydrophilic segments are utilized for the aggregation between Aβ monomers. When $A\beta_{1-40}$ monomers are exposed over the gold colloidal surface, the hydrophobic tail must be used as a segment to attach on the colloidal surface, while hydrophilic portion is aligned to be outward.

To enable Aβ to repetitively converge on a corresponding stable structure under acidic or basic conditions as the reversible structure, the conformation of the section of $A\beta_{1-40}$ used for binding to the gold surface must remain relatively unaltered. The segment corresponding to β-pleated sheet formed at pH 4 and α-helical structure formed at pH 10 may be geometrically separated by having the section responsible for binding to gold in between. Because the segments 1–17 are assigned as hydrophilic coil and sequences 18–40 as hydrophobic rod, the hydrophilic segments are considered to be responsible for the formation of α-coil and the hydrophobic component must be responsible for the β-sheet formation. Thus, the central segments may be used to conjugate the gold colloidal surface as a "pivotal point" to transition between the folded and the unfolded structures. Dimers of full-length Aβ indicate that the rate-limiting step involves the formation of a multimeric β-sheet that was found to span the central hydrophobic core sequences between 17 and 21 [35]. Therefore, the mainly hydrophobic segments 17–22 can be considered to be the most plausible section used as a pivotal point. The binding of Aβ to the anionic lipid membrane was reported to be an entropy-driven reversible transition between random coil ↔ β-sheet and the binding enthalpy was found to always be endothermic. The large positive entropy together with the large negative heat capacity suggest a prominent role of hydrophobic interactions in peptide aggregation [36]. An aggregation of soluble Aβ conducted at polar–nonpolar interfaces was reported to produce the aggregates with β-structure-rich conformation, which resembles Aβ protofibrils [37]. Therefore, we estimate that the β-sheet section is aligned outward at the acidic condition and was used for interconnecting gold colloids to each other to form the aggregates.

7.6.2 Effect of Temperature Shift to Reversible Self-Assembly

If temperature is shifted from 5°C to 45°C (or from 45°C to 5°C), the monomers located over the surface of the colloidal surface should attempt to form the most stable form in order to create a new equilibrium. As for the case of 20 nm gold colloidal surface, the dimer form is preferably constructed. Based on the molecular dynamics simulation, the oligomer made out of dimer form stably exists over a wide temperature range including 5°C and 45°C. In our work, the range of the temperature shift was between 5°C and 45°C. This temperature range matches the above temperature condition supporting the formation of the dimer-unit-based oligomer. Therefore, the conformation associated with self-assembly process was maintained even after the temperature shift for the case of Aβ_{1-40} monomers coated over 20 nm gold colloidal surface. With respect to the idea of potential energy of the oligomers, the localized minima of the dimer form must exist with relatively higher transition barrier between 5°C and 45°C for the folded and the unfolded structures. Therefore, once a new equilibrium was made at a given temperature, the conformational change must be conducted within the potential created at the initial temperature condition, since the transition barrier to move to the potential side under other temperature condition. When self-assembly was initiated at 45°C, for the Aβ_{1-40} monomers coated over 20 nm gold colloidal surface, the folded and the unfolded structure-based dimers were stably formed following the potential diagram constructed under 45°C. The potential barrier height between potential under 45°C and that under 5°C is significantly high. Therefore, the self-assembly process has to follow the potential diagram designed for 45°C even after a system is brought to 5°C (see Figure 7.11a).

On the other hand, Aβ_{1-40} monomers coated over 30 or 40 nm gold colloidal surface must possess the relatively lower transition barrier between 5°C and 45°C (or 3.5°C and 45°C) for the folded and the unfolded structures. In this case, however, the potential minimum of the 45°C side must be lower

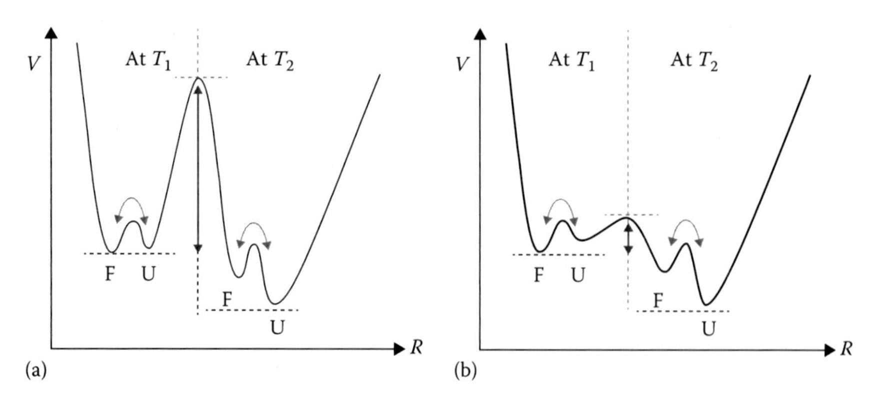

FIGURE 7.11
The potential (V) diagram for (a) dimer-based oligomer formed by Aβ_{1-40} monomers over 20 nm gold colloid surface and (b) trimer-based oligomer formed by Aβ_{1-40} monomers over 30 or 40 nm gold colloidal surface at T_1 and T_2 ($T_2 > T_1$). The reaction coordinate (R) implies that temperature increases as it moves to the right. The curved double-sided arrow indicates the reversible self-assembly process between F and U states. (F: folded structure, U: unfolded structure) The vertical lines with double sided arrow show the potential barrier to overcome from T_1 side to T_2 side.

than that of 5°C. Even a new equilibrium was achieved for reversible self-assembly at the lower temperature side, the conformational change must be conducted at the lower potential side (45°C side) if the temperature was shifted to the higher temperature (45°C) and if a self-assembly equilibrium was created at 45°C, the process follows the potential diagram at 45°C even though temperature is shifted to 5°C, since the potential energy of the folded and the unfolded structure at 45°C side is lower than those at 5°C (or 3.5°C) (see Figure 7.11b).

7.7 Conclusion

In conclusion, two distinctly different self-assembly paths were identified in an initial stage of fibrillogenesis. The potential surface over the 20 nm gold colloid supports an oligomer form with dimer unit. The potential surface for $A\beta_{1-40}$ dimer formation is considered to possess relatively high barrier height for transition of temperature and it conducts a reversible transformation of a conformation, which is resilient to temperature change during its process. The potential surface over the 30 or 40 nm gold colloid, however, supports an oligomer form with trimer unit. The potential surface for $A\beta_{1-40}$ trimer formation is regarded to have a minimum for a denatured form of $A\beta_{1-40}$ with lower barrier allowing conformation converged into a denatured form. The discovery of two different self-assembly exhibited that an initiating core size as well as temperature condition determine the type of oligomer to be constructed and govern a path of an entire self-assembly process. The selection of core size and temperature condition enables us to pinpoint a particular oligomer form, which is significantly associated with a mechanism of Alzheimer's disease.

Acknowledgments

This work was supported by the National Science Foundation under grants NSF-NER 0508240 and NSF 0929615. The Geneseo Foundation is greatly appreciated for their generous contribution toward this project. TEM measurements were conducted under the supervision of Professor Harold Hoops of SUNY Geneseo. The following individuals also should be recognized for their involvement of this project listed here: Nicole B. Gaulin, Nicole M. Briglio, Hyunah Cho, Dewi Sri Hartati, S. M. Winnie Tsang, Winnie W. Eng, Giang T. Nguyen, and Amy L. Tran.

References

1. Selkoe DJ: The molecular pathology of Alzheimer's disease. *Neuron*. 6: 487–498, 1991.
2. Terry RD: Neuropathological changes in Alzheimer disease. *Prog. Brain Res.* 101: 383–390, 1994.
3. Glenner GG and Wong CW: Alzheimer's disease: Initial report of the purification and characterization of a novel cerebrovascular amyloid protein. *Biochem. Biophys. Res. Commun.* 120: 885–890, 1984.
4. Masters CL, Simms G, Weinman NA, Multhaup G, McDonald BL, and Beyreuther K: Amyloid plaque core protein in Alzheimer disease and Down syndrome. *Proc. Natl. Acad. Sci. USA*. 82: 4245–4249, 1985.
5. Rogers J, Cooper NR, Webster S, Schultz J, McGeer PL, Styren SD, Civin WH, Brachova L, Bradt B, and Ward P: Complement activation by beta-amyloid in Alzheimer disease. *Proc. Natl. Acad. Sci. USA*. 89: 10016–10020, 1992.
6. Joachim CL, Mori H, and Selkoe DJ: Amyloid beta-protein deposition in tissues other than brain in Alzheimer's disease. *Nature*. 341: 226–230, 1989.
7. Jarrett JT, Berger EP, and Lansbury Jr, PT: The carboxy terminus of the beta amyloid protein is critical for the seeding of amyloid formation: Implications for the pathogenesis of Alzheimer's disease. *Biochemistry*. 32: 4693–4697, 1993.
8. Shoji M, Golde TE, Ghiso J, Cheung TT, Estus S, Shaffer LM, Cai XD, McKay DM, Tinter R, Frangione B et al.: Production of the Alzheimer amyloid beta protein by normal proteolytic processing. *Science*. 258: 126–129, 1992.
9. Seubert P, Vigo-Pelfrey C, Esch F, Lee M, Dovey H, Davis D, Sinha S, Schlossmacher M, Whaley J, Swindlehurst C et al.: Isolation and quantification of soluble Alzheimer's beta-peptide from biological fluids. *Nature*. 359: 325–327, 1992.
10. Walsh DM, Lomakin A, Benedek GB, Condron MM, and Teplow DB: Amyloid-protein fibrillogenesis. *J. Biol. Chem.* 272: 22364–22372, 1997.
11. Lomakin A, Chung DS, Benedek GB, Kirshner DA, and Teplow DB: On the nucleation and growth of amyloid-protein fibrils: Detection of nuclei and quantitation of rate constants. *Proc. Natl. Acad. Sci. USA*. 93: 1125–1129, 1996.
12. Kirschner DA, Inouye H, Duffy LK, Sinclair A, Lind M, and Selkoe DJ: Synthetic peptide homologous to beta protein from Alzheimer disease forms amyloid-like fibrils in vitro. *Proc. Natl. Acad. Sci. USA*. 84: 6953–6957, 1987.
13. Lambert MP, Barlow AK, Chromy BA, Edwards C, Freed R, Liosatos M, Morgan TE, Rozovsky I, Trommer B, Viola KL et al.: Diffusible, nonfibrillar ligands derived from A 1–42 are potent central nervous system neurotoxins. *Proc. Natl. Acad. Sci. USA*. 95: 6448–6453, 1998.
14. Walsh DM, Klyubin I, Fadeeva JV, Cullen WK, Anwyl R, Wolfe MS, Rowan MJ, and Selko DJ: Naturally secreted oligomers of amyloid beta protein potently inhibit hippocampal long-term potentiation in vivo. *Nature*. 416: 535–539, 2002.
15. Bucciantini M, Giannoni E, Chiti F, Baroni F, Formigli L, Zurdo J, Taddei N, Ramponi G, Dobson CM, and Stefani M: Inherent toxicity of aggregates implies a common mechanism for protein misfolding diseases. *Nature*. 416: 507–511, 2002.

16. Zhu X, Yan D, and Fang Y: In situ FTIR spectroscopic study of the conformational change of isotactic polypropylene during the crystallization process. *J. Phys. Chem. B*. 105: 12461–12463, 2001.
17. Harper SM, Neil LC, and Gardner KH: Structural basis of a phototropin light switch. *Science*. 301: 1541–1544, 2003.
18. Ohba S, Hosomi H, and Ito Y: In situ x-ray observation of pedal-like conformational change and dimerization of trans-cinnamamide in cocrystals with phthalic acid. *J. Am. Chem. Soc*. 123: 6349–6352, 2001.
19. Gupta R and Ahmad F: Protein stability: Functional dependence of denaturational Gibbs energy on urea concentration. *Biochemistry*. 38: 2471–2479, 1999.
20. Yokoyama K: Nanoscale surface size dependence in protein conjugation, in Chen EJ and Peng N (eds.): *Advances in Nanotechnology*, vol. 1, Nova Publishers, New York, 2010.
21. Yokoyama K: Modeling of reversible protein conjugation on nanoscale surface, in Musa SM (Ed.): *Computational Nanotechnology: Modeling and Applications with MATLAB*, CRC Press-Taylor & Francis Group, LLC, Boca Raton, FL, 2011.
22. Link S and El-Sayed M: Spectral properties and relaxation dynamics of surface plasmon electronic oscillations in gold and silver nanodots and nanorods. *J. Phys. Chem. B*. 103: 8410–8426, 1999.
23. Kelly KL, Coronado E, Zhao LL, and Schatz GC: The optical properties of metal nanoparticles: The influence of size, shape, and dielectric environment. *J. Phys. Chem. B*. 107: 668–677, 2003.
24. Jensen TR, Schatz GC, and Van Duyne RP: Nanosphere lithography: Surface plasmon resonance spectrum of a periodic array of silver nanoparticles by ultraviolet-visible extinction spectroscopy and electrodynamic modeling. *J. Phys. Chem. B*. 103: 2394–2401, 1999.
25. Yokoyama K, Briglio NM, Sri Hartati D, Tsang SMW, MacCormac JE, and Welchons DR: Nanoscale size dependence in the conjugation of amyloid beta and ovalbumin proteins on the surface of gold colloidal particles. *Nanotechnology*. 19: 375101–375108, 2008.
26. Huang K: *Statistical Mechanics*. Wiley, New York, 1987.
27. Garai K, Sahoo B, Sengupta P, and Maiti S: Quasi-homogeneous nucleation of amyloid beta yields numerical bounds for the critical radius, the surface tension and the free energy barrier for nucleus formation. *J. Chem. Phys*. 128: 45102, 2008.
28. Sabate R, Gallardo M, and Estelrich J: Temperature dependence of the nucleation constant rate in beta amyloid fibrillogenesis. *Int. J. Biol. Macromol*. 35: 9, 2005.
29. Kusumoto Y, Lomakin A, Teplow DB, and Benedek GB: Temperature dependence of amyloid beta-protein fibrillization. *Proc. Natl. Acad. Sci. USA*. 95: 12277–12282, 1998.
30. Carrotta R, Manno M, Bulone D, Martorana V, and San Biagio PL: Protofibril formation of amyloid β-protein at low pH via a non-cooperative elongation mechanism. *J. Biol. Chem*. 280: 3001, 2005.
31. Jang S and Shin S: Computational study on the structural diversity of amyloid beta peptide (β(10–35)) oligomers. *J. Phys. Chem. B*. 112: 3479, 2008.
32. Barrow CJ, Yasuda A, Kenny PT, and Zagorski MG: Solution conformations and aggregational properties of synthetic amyloid beta-peptides of Alzheimer's disease. Analysis of circular dichroism spectra. *J. Mol. Biol*. 225: 1075–1093, 1992.

33. Wood SJ, Maleeff B, Hart T, and Wetzel R: Physical, morphological and functional differences between pH 5.8 and 7.4 aggregates of the Alzheimer's amyloid peptide Abeta. *J. Mol. Biol.* 256: 870–877, 1996.
34. Brovchenko I, Burri RR, Krukau A, and Oleinnikova A: Thermal expansivity of amyloid β_{16-22} peptides and their aggregates in water. *Phys. Chem. Chem. Phys.* 11: 5035, 1009.
35. Melquiond A, Dong X, Mousseau N, and Derreumaux P: Role of the region 23–28 in β fibril formation: Insights from simulations of the monomers and dimers of Alzheimer's peptides β40 and β42. *Curr. Alzheimer Res.* 5: 244, 2008.
36. Meier M and Seelig J: Length dependence of the coil β-sheet transition in a membrane environment. *J. Am. Chem. Soc.* 130: 1017, 2008.
37. Nichols MR, Moss MA, Reed DK, Hoh JH, and Rosenberry TL: Amyloid-beta aggregates formed at polar–nonpolar interfaces differ from amyloid-beta protofibrils produced in aqueous buffers. *Microsc. Res. Tech.* 67: 164, 2005.

8

Nanoscale Spectroscopy in the Infrared with Applications to Biology

Eamonn Kennedy, Edward Yoxall, and James Rice

CONTENTS

8.1 Introduction

Vision is a remarkable faculty of living organisms. It allows us to quickly extract and interpret data from the enormous number of photons of varying amplitude, phase, and energy that are incident on our eyes per second (Pinker 1997). Our inherent visual and image processing abilities are made more remarkable by the fact that they require no manual effort on our part. Vision is, as the physicist Richard Feynman once described by analogy (Sykes 1989),

> *As if some insect of sufficient cleverness could sit by the corner of a swimming pool and be disturbed by the surface of the waves. And by the nature of the irregularities and bumping of the waves, could figure out what is happening and where at each point across the entire pool.*

However, the human visual system is subject to many limitations. Perhaps the most obvious of these is the finite spectral response to the electromagnetic spectrum exhibited by the cone cells of the human eye shown in gray in Figure 8.1. Motivations to overcome this limitation are abundant and have led to the development of systems that can image from high-energy x-rays, commonly used in medical diagnostics, to lower-than-visible energy infrared (IR) wavelengths, the spectra of which have long been used for chemical identification and purity testing.

Further to expanding the frequency range of detectable light, devices have been invented to exceed the spatial resolution of the human eye—typically just below 0.1 mm at a 0.1 m distance (Russ 2007). However, the frequency of light that an optical system detects and the spatial resolution that the system can achieve are interdependent properties. This interdependence is an inherent physical property caused by the wave nature of light and leads to a fundamental, wavelength-dependent limit to the spatial resolution that can be achieved by all conventional spectroscopic and imaging methods.

This chapter outlines current research into circumventing this so-called far-field diffraction limit by using near-field probe-based approaches.

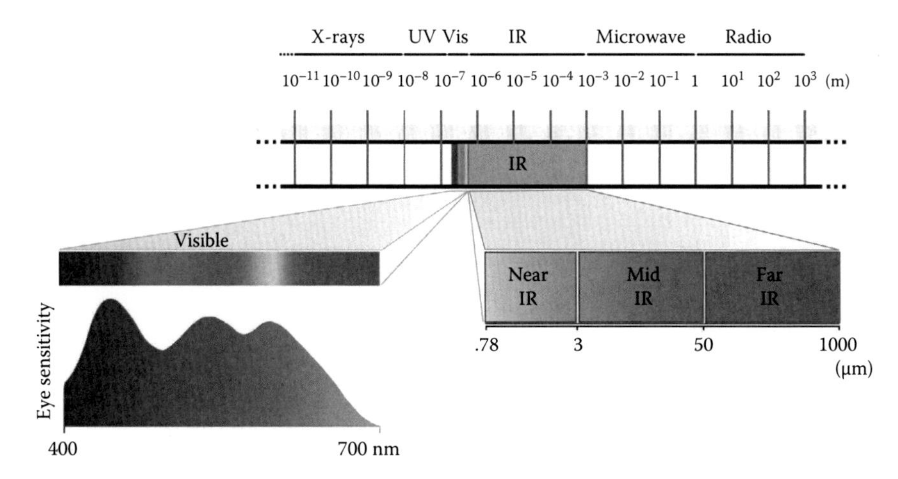

FIGURE 8.1
Location of the visible and IR regions on the electromagnetic spectrum, showing the subdivision into near-, mid-, and far infrared. The human eye cumulative spectral response is shown above the visible region.

We focus on the IR region, for which there are compelling reasons to improve spatial resolution at which point spectra can be obtained.

8.2 Diffraction Limit

The German physicist Ernst Abbe was the first person to present a mathematical formalism for the resolution limit of a microscope (Abbe 1873), which is defined as the minimum distance at which two point sources of light can be resolved. His work concluded that this limit was proportional to the wavelength of light used. Intuitively, we expect that as light waves have finite sizes, they should also have a finite length at which their sources can be resolved; however, defining this limit precisely requires closer inspection.

For this we will use an ideal optical device with no aberrations—a single, circular aperture lens that focuses a parallel source of monochromatic light. The impulse response of such a system to a point source is diffraction limited with a shape that is primarily dependent on the size of the wavelength and aperture radius employed. In the far-field approximation, incident light can be represented as a simple, exponentially time-evolving wave, $E(r) = E_0 e^{i(\mathbf{k}\cdot\mathbf{r})}$, where **k** and **r** are the position and wave vectors of light, respectively. Let us assume that such a wave is incident with a top hat intensity onto our circular lens. The total amplitude of such a wave travelling through our perfect circular aperture on a distant observation plane is therefore the integral of the incident wave's amplitude over the whole aperture surface (Goodman 1996). This process is often physically explained as the reproduction of the original wave at each infinitesimal element of the integral over the aperture's area, $f(r)$. The total amplitude, A, in the observation plane as a function of collection angle can then be calculated by integral transformation.

$$A(\theta)=E_0\int f(r)e^{-i(-\mathbf{k}\cdot\mathbf{r})}dr$$

The working of this integral for a circular aperture is somewhat involved (Hecht 1997). We are primarily interested in the solution, that is, the intensity, $|A(\theta)|^2$ at the observation plane which takes the form

$$I(\theta)=I_0\left(\frac{2J_1 ka\sin(\theta)}{(ka\sin(\theta))}\right)^2$$

where
J_1 is the Bessel function of the first kind of order one
a is the aperture radius

In two dimensions, this function is called an airy pattern. The first minima past the origin for this Bessel function occurs at 3.8317, which corresponds to the first dark ring of the airy disk (Figure 8.2a). When two such patterns' sources overlap, the generally accepted point of minimum resolvable detail is defined as when the minimum of one source point coincides with the

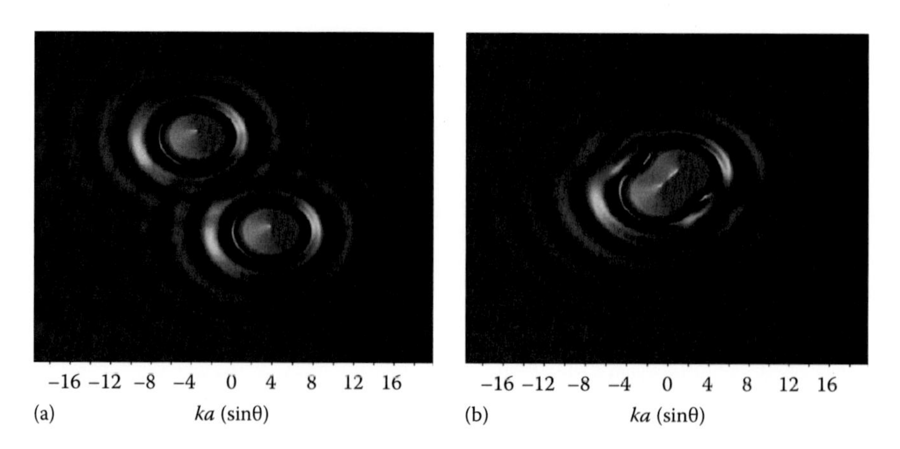

FIGURE 8.2
Intensity profiles of the point spread function from a circular aperture, coded in MATLAB® (a) resolved and (b) at the Rayleigh criterion. The profile has been normalized to unity and false colored to highlight the outer rings.

maximum of another (Figure 8.2b). This relationship follows immediately from the intensity mentioned earlier:

$$ka\sin(\theta)=3.8317$$

Substituting $k=2\pi/\lambda$, we have that for a common microscope which requires that both condenser and objective numerical apertures to be summed, reasonable approximation of numerical apertures yields the spatial resolution:

$$R\approx\frac{1.22\lambda}{NA}\cong\frac{\lambda}{2}$$

Therefore minimum spatial resolution reduces with light energy since $E_p \propto 1/\lambda$. Therefore, in the work presented herein on the IR region, conventional microscopy is extremely limited. For mid-IR wavelengths (3–50 μm), spectroscopy of single, basic biological elements such as a typical 9-μm wide red blood cell cannot be achieved past single-point spectra. Additionally for the IR region in particular, resolution difficulties are further compounded by the thermal heating effect experienced by traditional optics when used for IR applications. This has led to the development of new optical components that are very low IR absorbers (Harbison et al. 1998). However, the energy uptake via absorption of these components is nonlinearly dependent on the intensity of light used, which makes intensity calibration laborious where even possible.

Despite these limitations, microspectroscopic techniques in the IR have been the subject of much research, although mapping spectra at high pixel densities in the IR can take several hours (Miller 2007). Such techniques have yielded a wide range of significant results especially in the life sciences (Peiqiang 2003, Dumas et al. 2004, Miller 2006). The primary advantage of further enhanced techniques in IR microspectroscopy, such as attenuated total reflectance (ATR) methodologies, is the ability to localize the region from which spectra are accumulated, which permits accurate chemical labeling down to the spatial resolution defined by the instrument. Clearly then, there is now a need to surpass this resolution to permit chemical mapping and point spectroscopy of submicron samples, such as nanostructures in cells, which requires the use of new subdiffraction-limited chemical-mapping methodologies.

8.3 Infrared Subdiffraction-Limited Spectral Methods

8.3.1 Photothermal Microspectroscopy

Initial attempts to directly measure thermally based IR spectra below the resolution limit of light were performed by Hammiche et al. (2004).

This technique is called photothermal microspectroscopy (PTMS) and couples a conventional Fourier transform infrared (FTIR) spectrometer with an atomic force microscope (AFM). In PTMS, the probe used is a scanning thermal probe that is connected to an AFM system. The probe is normally a thin platinum wire clad in silver, which permits heat flow from the sample to the tip. A thermal junction can be interfaced with the wire to monitor this heat by means of voltage generation. As well as spectral acquisition, raster scanning can be performed by surface contact with the wire, producing an absorption map of a fixed wavenumber.

8.3.2 Scattering-Type Scanning Near-Field Optical Microscopy

When a metal tip is placed in a beam of light, it has a tendency to confine the fields into the close vicinity of its apex in a phenomenon known as the "lightning rod effect." In this way, the tip can be treated as an extremely small source of light. In one of the many happy accidents of physics, the size of this "nanofocus" does not depend on the wavelength of the illuminating light, but rather on the radius of the tip in question. Any sample that is brought into this very small spot of light will interact with the tip and change the way in which it scatters light into the far field; this change is dependent on the sample's dielectric constant. As such, the tip can be considered both a light source and a light transmitter, which is sensitive to the type of surface beneath. If the scattered light can be collected and analyzed, details about the interaction taking place within the nanofocus can be extracted. Given that in the IR the nanofocus is often several orders of magnitude smaller than the wavelength, scattering-type near-field microscopes are widely used for highly localized nanoscale resolution spectroscopy.

8.3.3 Photothermal-Induced Resonance Approaches

Another more recent technique uses a variation of ATR IR spectroscopy to overcome the diffraction limit for IR spectroscopic measurement (Dazzi et al. 2005). In this methodology, a free electron laser (FEL) is directed through a zinc selenide (ZnSe) prism and totally reflected underneath the sample. Such laser systems typically require large power generation (Sauvage et al. 2011) and are physically large. A small evanescent wave is produced at the boundary of ZnSe prism surface. This evanescent wave penetrates the sample causing thermal heating, which relaxes via mechanical expansion (Mayet et al. 2008). The extent of surface expansion is measured by a cantilever tip placed in contact with the sample surface. In this way, photothermal-induced resonance (PTIR) methods are an indirect, although well-proven method for obtaining pure absorption spectra from point sources on the order of nanometers. PTIR is a rapidly emerging technology that couples the resolution of an AFM with the chemical-mapping ability of IR microspectroscopy.

8.4 Scattering-Type Scanning Near-Field Optical Microscopy

8.4.1 Near Field

All of our visual interaction with the world is with the far field of light—that is the light that scatters or reflects from an object and continues on to our eyes where we can detect it. It is, therefore, quite counterintuitive that there is a significantly different behavior in the field distributions around objects at much smaller length scales. This "near-field" light matter interaction field only exists within several tens of nanometers of a surface; it decays away exponentially with distance. If it can be probed, however, the diffraction limit can be circumvented as it contains all of the information that would have been lost in propagating through any conventional form of microscope.

The idea of using the near field for microscopy was originally conceived of in the late 1920s (Novotny 2007). Irishman Edward Synge proposed raster scanning a hole considerably smaller than the wavelength of light in an opaque screen extremely close to a surface. He realized that the effect of the aperture was to convert the near field into propagating waves that could be measured in the far field (Synge 1928). The amount of light that actually penetrated the hole could then be recorded, and an image relating to the sample absorption could be built up pixel by pixel. The resolution was now dependent on the size of the hole rather than on the wavelength of the light. Interestingly on April 22, 1928, Synge wrote a letter to Einstein in which he states that

> *By means of the method the present theoretical limitation of the resolving power in microscopy seems to be completely removed and everything comes to depend on technical perfection.*

In the 1920s, the technology did not exist to bring his idea to fruition. However, in modern times, the precise positional control offered by piezoelectric crystals offers just such a method (Figure 8.3).

8.4.2 Aperture and Scattering Methods

The key to any kind of SNOM measurement is its use of the near field. This is what sets it apart from traditional forms of microscopy and what allows

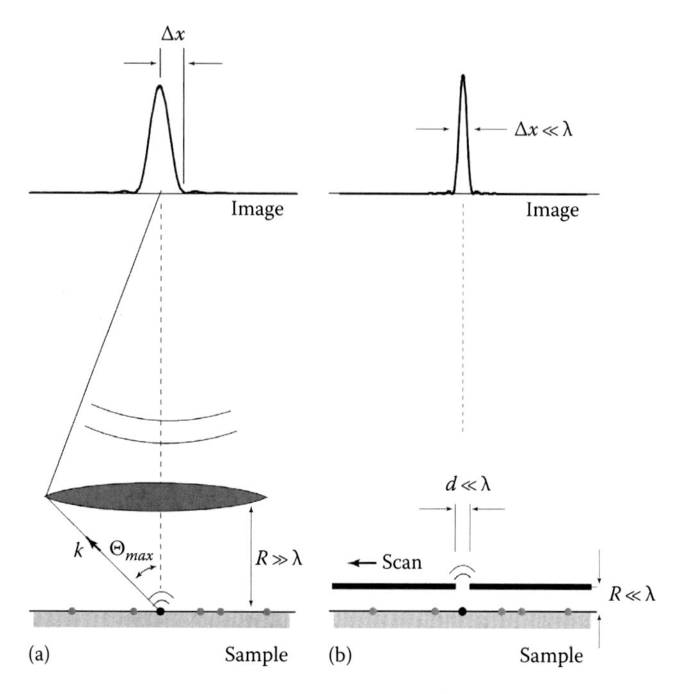

FIGURE 8.3
In traditional microscopy (a), light coming from the sample is collected by a lens and focused at an image plane. The resolution of this image is dictated by the wavelength of light used. With Synge's proposal, however (b), the resolution is limited only by the size of the aperture (and in his view, the ability to control with any degree of finesse the position of the hole). (From Novotny, L., *Progress Opt.*, 50, 137, 2007.)

resolutions higher than the diffraction limit (Hayazawa et al. 2009). There are several ways of collecting near fields—the two most common dictate the different types of SNOM. Aperture SNOM is very similar to Synge's vision of raster scanning a small hole over a surface. Instead of a hole in an opaque screen, aperture SNOM uses a tapered optical fiber that is brought extremely close to a surface to pick up the near field. There are several different modes of operation that simply rely on the way in which the sample is illuminated.

As the more established of the two, aperture SNOM is often referred to as just "SNOM," or occasionally "NSOM" where the word order of the acronym has just been changed. Resolution is determined by the size of the aperture of the fiber. There is a trade-off, however, between the amount of light that can be collected and the resolution; a smaller hole leads to finer image detail but at the cost of illumination intensity. This is also affected by the light's wavelength—the longer the wavelength, the more light is back-reflected by fiber's taper (Figure 8.4).

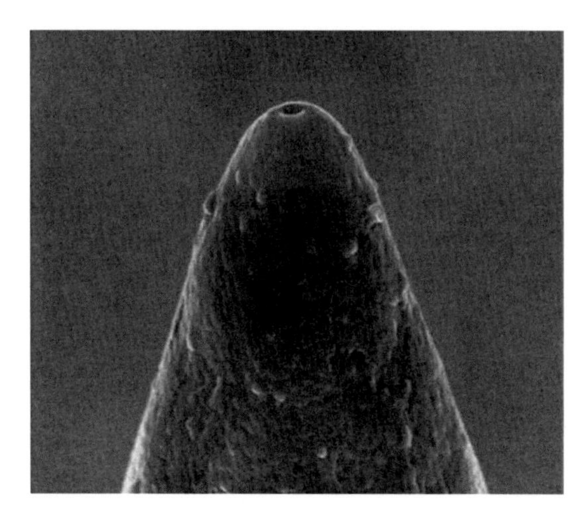

FIGURE 8.4
A scanning electron microscope image of an aperture SNOM probe. The aperture is clearly visible at the apex of the tip.

At wavelengths longer than the near-IR, back-reflection is sufficiently problematic that the technique is not widely used (as well as a less mature industry in mid-IR fiber optics than its shorter wavelength counterpart), although examples do exist where it is felt that fiber delivery of light outweighs the issues.

8.4.3 Scattering SNOM

Given the focus of this chapter on IR spectroscopy, the rest of this section focuses on a technique that is independent of the illumination wavelength. As with many new technologies, some disambiguation in nomenclature is required as different names have been given to the same thing as groups have developed their methods in parallel. Scattering SNOM can equally well be called apertureless SNOM and can be seen in the literature as either "s-SNOM" or "a-SNOM." These authors prefer the former as it prevents any confusion between "aperture" and "apertureless." One further complication is that some authors prefer to use the term s-SNIM when working with IR light, where "optical" has been replaced with "IR."

As the name suggests, this form of SNOM does not rely on a small aperture. Instead, it uses an extremely sharp tip that acts as both a nanolight source and a scatterer of light that contains information about the surface beneath. In a practical sense, the system is run as an AFM with extra optical components attached. A scattering-type scanning near-field optical microscopy (s-SNOM) uses the "tapping" mode where a probe is vibrated

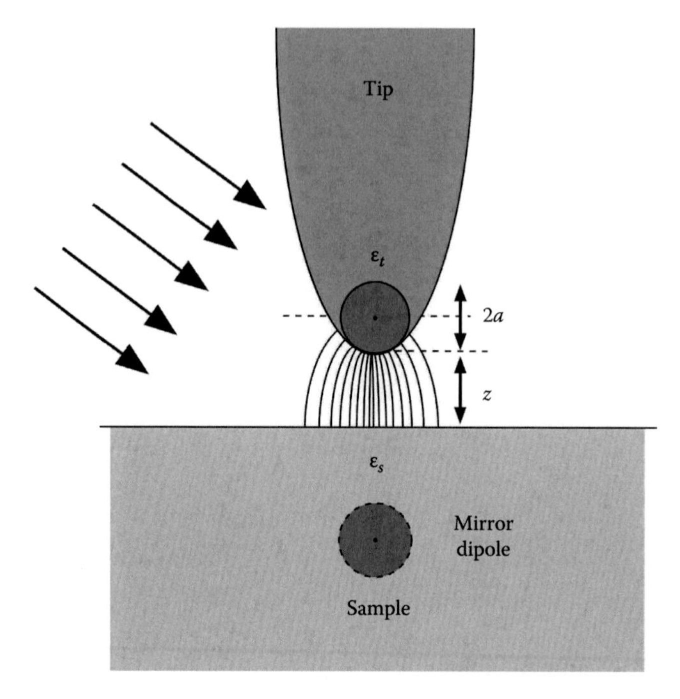

FIGURE 8.5
A simple analytical model for describing the probe–sample interaction is the point dipole model. The presence of the tip is assumed to induce a mirror dipole in the sample, which in turn affects the way in which light is scattered from the probe. (From Keilmann, F. and Hillenbrand, R., *Phil. Trans. R. Soc. Lond. A*, 362, 787, 2004.)

vertically above the sample, just coming into contact with it at the bottom of each oscillation. A feedback loop ensures that the probe's amplitude remains constant by varying the vertical position of the sample; this gives the topography. One of the benefits of s-SNOM is that it keeps all the functionality of an AFM, permitting simultaneous optical and topographical measurement (Figure 8.5).

Several analytical models for this system of varying complexity do exist but will not be described in detail here (Keilmann and Hillenbrand 2004, Cvitkovic et al. 2007). Instead, a qualitative description of the interaction between the tip and the sample is given as follows:

1. Light is focused onto the end of the tip.
2. The "lightning rod effect" confines and amplifies the field in the vicinity of the probe's apex.
3. This field interacts with the surface, which in turn affects the way in which light is scattered from the probe.

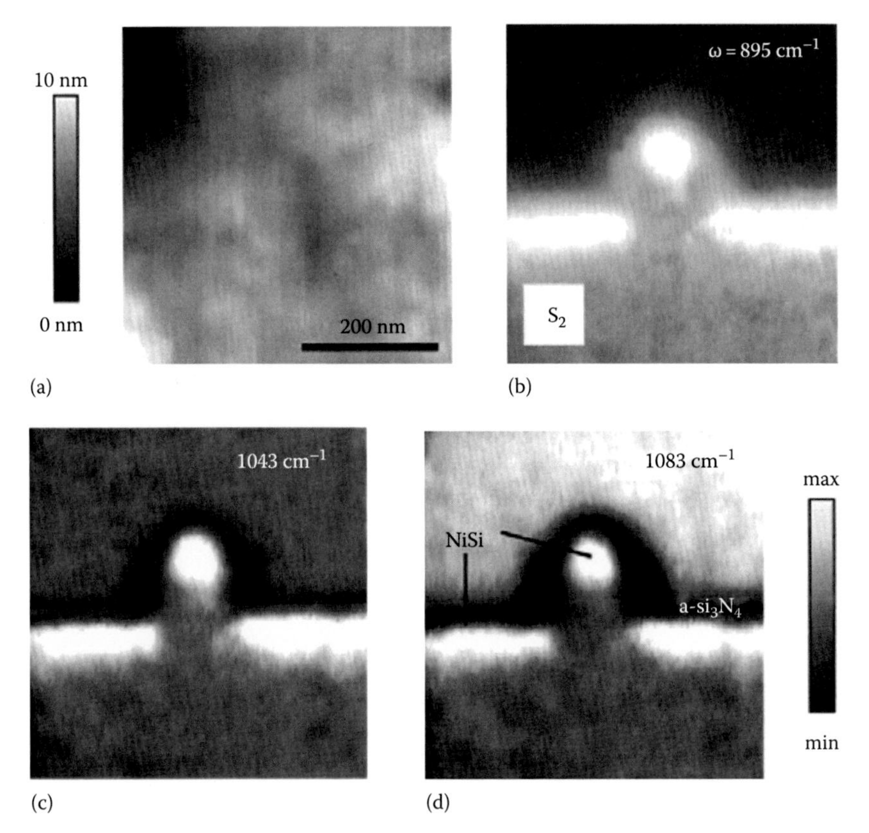

FIGURE 8.6
(See color insert.) Despite the fact that the height of the sample shows very little variation (a), the optical signal shows clear variation in these images of a transistor, taken at different wavelengths (b–d). (From Huber, A.J. et al., *Nanotechnology*, 21, 235702, 2010.)

4. The difference in the scattered light is picked up by the detection system.
5. The interaction depends on the refractive index of the surface, paving the way for surface spectroscopy from each point source (Figure 8.6).

For measurements that aim to distinguish the chemical make-up of a surface, incoming light is polarized along the long axis of the probe. This is in contrast to those that aim to image plasmonic or phonon resonances where light is polarized parallel to the surface to avoid the possibility that the probe itself will affect the resonance. Without the polarization along the long axis, the lightning rod effect is lost, and the near field is not scattered sufficiently so as to be detectable (Figure 8.7).

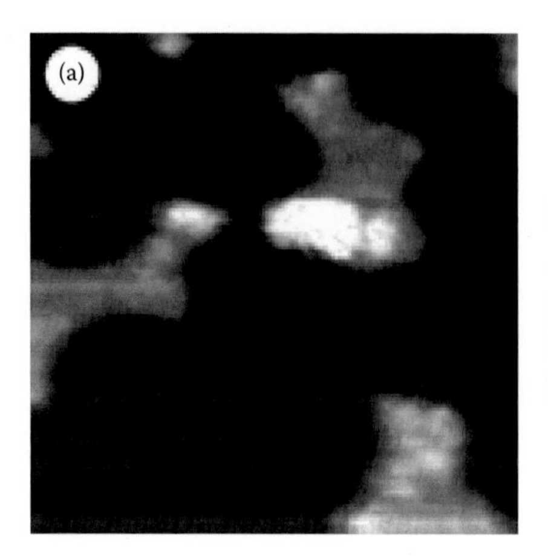

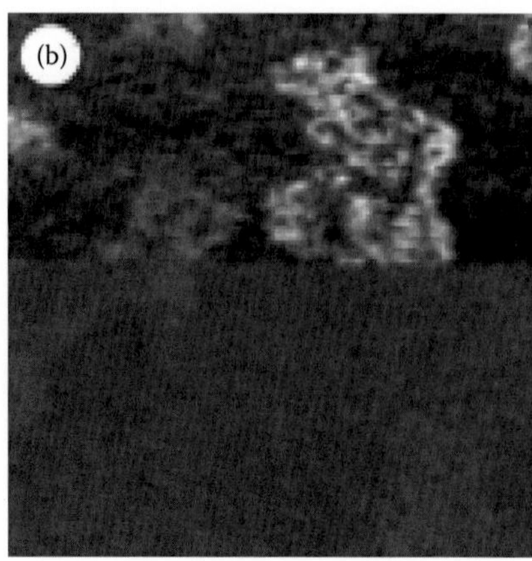

FIGURE 8.7
Topography (a) and IR image (b) of gold islands on a silicon substrate. Halfway through the scan, the polarization is rotated by 90° such that it no longer has any projection onto the axis of the probe, thus destroying the near-field signal. (From Keilmann, F. and Hillenbrand, R., *Phil. Trans. R. Soc. Lond. A*, 362, 787, 2004.)

8.4.4 Comparison of Techniques

The following table gives a succinct, if slightly simplified, comparison of the two major modes of SNOM operation:

	Aperture	Scattering
Wavelength dependence	Backscattering from taper limits to near IR	Can work with source of any IR wavelength
Resolution	Determined by aperture size, typically 100 nm	Determined by the tip radius, typically around 20 nm
Alignment	Fiber delivery is simple and readily repeatable	Focus at apex is challenging. Requires visible tracing
Probes	Expensive taper fibers must be specially prepared	Standard AFM probes can be used

8.5 Signal Detection and Methodology

The previous section gave a brief overview of s-SNOM. We now expand in further detail how it works, with a particular emphasis on how the near-field signal can be extracted. Attention will also be given to two different kinds of light sources—single-wavelength laser sources and broadband thermal sources. The benefits and drawbacks of these are very similar to their far-field counterparts in IR laser imaging and FTIR spectroscopy.

A diffraction-limited focal spot in the mid-IR is perhaps 5 μm in diameter. The apex of the tip—where the near-field interaction with the surface takes place—typically has a radius of 10 nm. The vast majority of light that is scattered from the probe has therefore not been mediated by the near-field interaction. Instead it will be coming from the body of the probe or the sample itself. The challenge is to extract just the near-field signal, coming from the apex of the tip, and exclude everything else—the background signal.

In terms of optical setup, an s-SNOM is designed with as many reflective optical components as possible for simple use at a variety of wavelengths. In fact, the most widely used detection schemes require just one nonreflective component—a beam splitter. For both single-wavelength and thermal light sources, the near-field signal is gathered interferometrically. Before discussing the exact methods by which this is done, it is worth considering the simple electrodynamics of the probe–sample interaction as this gives an insight into what provides contrast in an s-SNOM image.

8.5.1 Scattering Coefficient

From an abstract viewpoint, incoming light is modified in some way by the interaction with the sample. It differs in both amplitude and phase from the outgoing beam. Mathematically, this can be represented by a scattering coefficient σ:

$$E_{out} = \sigma E_{in}$$

where σ is a complex number represented by

$$\sigma = s e^{i\phi}$$

The values of s and ϕ are those that can be measured by an s-SNOM detection system. They are affected by the dielectric constant of the surface beneath and therefore change from material to material. A single s-SNOM scan typically produces three images simultaneously—one topographical image from its operation as an AFM and two optical images—the amplitude and the phase.

8.5.2 Higher Harmonic Detection

As has already been mentioned, the major challenge in s-SNOM measurements is to suppress the background signal—the signal from the probe that would exist without a sample—and to extract the useful near-field signal that contains information about the probe–sample interaction. Fortunately, the differing physics of the two signals provides a method to selectively pick the near field; the background signal varies only very slightly as the tip is retracted several nanometers from the surface whereas the near-field signal is completely lost at these distances. In other words, the background signal is relatively constant with probe height—scattering from the body of the probe or sample does not change. In fact, it varies on length scales of order of the wavelength λ. The near-field signal, however, must necessarily be within the near field to be detected, and this decays away on a length scale related to the tip radius r.

Given that an s-SNOM is based on an AFM operating in tapping mode where the tip is oscillated vertically above a sample, the scattered light from the probe is modulated at its resonance frequency. This affects the background signal very little, but it affects the near-field signal enormously. Therefore, if the signal is demodulated at the harmonics of the probe vibration frequency, one finds that at a sufficiently high harmonic, the contribution from the background signal is negligible and all that remains is the desired near-field signal; the background is suppressed by measurement at higher harmonics. The harmonic at which this occurs varies according to the wavelength being used. Typically, the second harmonic is used in the mid-IR and the third harmonic is necessary for the visible. The following graphs are known as approach curves, and they illustrate the background suppression (Figure 8.8).

At the first harmonic, the signal takes much longer to decay away to zero than at the third harmonic of the probe vibration frequency, showing that at the higher harmonic, only the near field is being measured. Although higher harmonic detection is shared for both single-wavelength and broadband sources, there are subtleties in the experimental setup which set them apart. The following gives a brief overview of what these are.

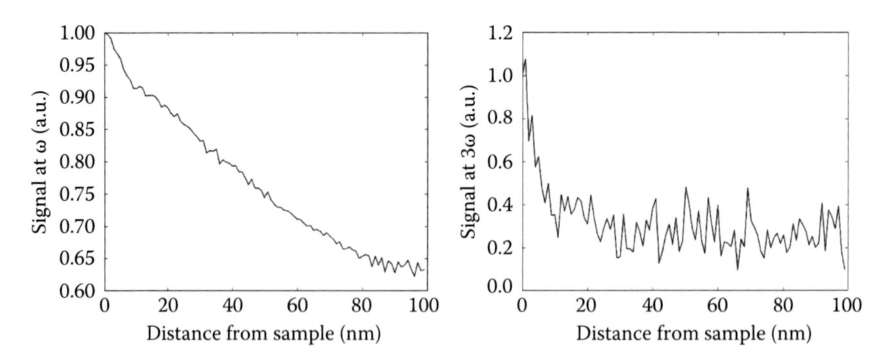

FIGURE 8.8
Approach curves measured at different harmonics of the probe vibration frequency. At the first harmonic of the probe vibration (left), the signal decays away slowly, indicative of the background signal being dominant. At the third harmonic (right), however, the signal decays away within 10 nm—approximately the radius of the tip—showing that the near-field interaction is the dominant contribution.

8.6 Light Sources for s-SNOM

8.6.1 Laser Sources

Since the inception of s-SNOM, three different detection methods have been developed, all of them interferometric. These are the homodyne, heterodyne and pseudoheterodyne schemes. The pseudoheterodyne scheme benefits from not having to introduce extra nonreflective optical components to the setup and is also capable of extracting the amplitude and phase signals simultaneously.

In essence, it works as a Michelson interferometer where one of the arms (the reference beam) is phase modulated by a sinusoidally vibrating mirror (Figure 8.9).

The phase-modulated reference beam has the effect of creating sidebands in the frequency spectrum at the detector (Figure 8.10).

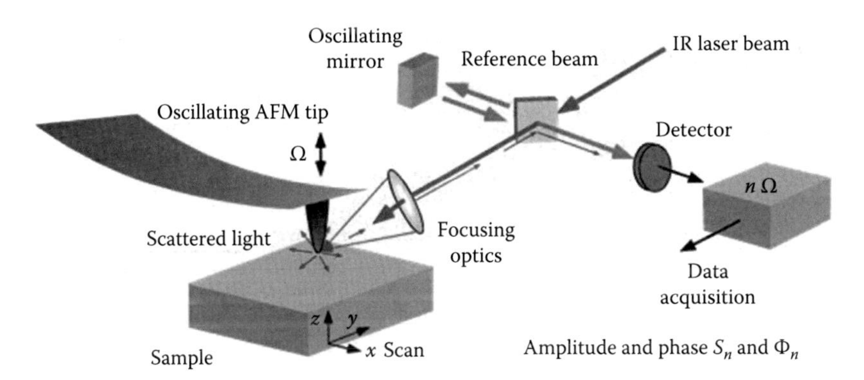

FIGURE 8.9
Experimental setup for single-wavelength s-SNOM measurements using the pseudoheterodyne detection method. This extracts the near-field amplitude and phase (as well as the surface topography) simultaneously in contrast to other techniques. (Retrieved from http://brongersma.stanford.edu/main/InfraredNanooptics.)

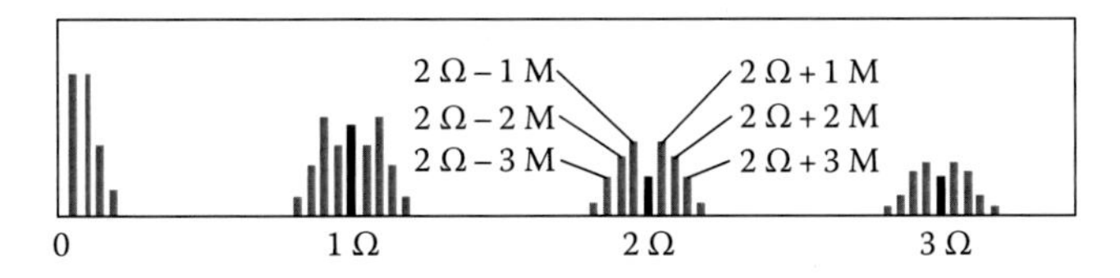

FIGURE 8.10
Effect of the phase-modulated reference beam is to split the harmonics of the signal from the probe (black) into sidebands (gray). The relationship between odd- and even-order sidebands yields the near-field amplitude and phase. (From Ocelic, N. et al., *Appl. Phys. Lett.*, 89(10), 101124, 2006.)

The extraction of the amplitude and phase is rather mathematically involved; so for clarity, their derivation will be left to reference (Ocelic et al. 2006) and their values are stated as

$$s_n = \sqrt{\left|C_{n\Omega+M}\right|^2 + \left|C_{n\Omega+2M}\right|^2}$$

$$\phi_n = \tan^{-1}\frac{\left|C_{n\Omega+M}\right|}{\left|C_{n\Omega+2M}\right|}$$

In an s-SNOM image, these are the variables that give the pixels their value.

8.6.2 Broadband Sources

Broadband sources fall into two main categories—thermal and broadband laser sources. The combination of these and s-SNOM has come to be known as nano-FTIR in recognition of the system similarities with its far-field counterpart. The use of a thermal source for s-SNOM measurements is a relatively new technique having been first demonstrated in 2011 (Huth et al. 2011). Broadband lasers have been used for near-field spectroscopy since 2006 (Brehm et al. 2006a, Amarie et al. 2009). In this case, a femtosecond laser was passed through a nonlinear crystal creating a range of wavelengths. This range, however, was not sufficiently wide to be suited to measurements for the chemical identification of surfaces. More recent incarnations of the laser approach, however, have used more powerful beams with a broader range of wavelengths that are capable of spectroscopically distinguishing materials at a 20-nm resolution. With a broadband source, the mirror in the reference arm of the Michelson interferometer is scanned between two positions resulting in changing optical path differences between the two arms of the interferometer (Figure 8.11).

Just as in standard FTIR, this gives an interferogram that can then be Fourier transformed for the near-field spectrum the difference being that the signal is demodulated at a higher harmonic of the probe vibration frequency to suppress the background (Figure 8.12).

The measured near-field spectrum $S_n(\omega)$ (see previous text) is actually made up of several parts:

$$S_n(\omega) = s_n(\omega) R(\omega) I_{source}(\omega)$$

where

$s_n(\omega)$ is the amplitude of the scattering coefficient

$R(\omega)$ is the spectral responsivity of the optical components in the setup

$I_{source}(\omega)$ is the input spectrum from the light source

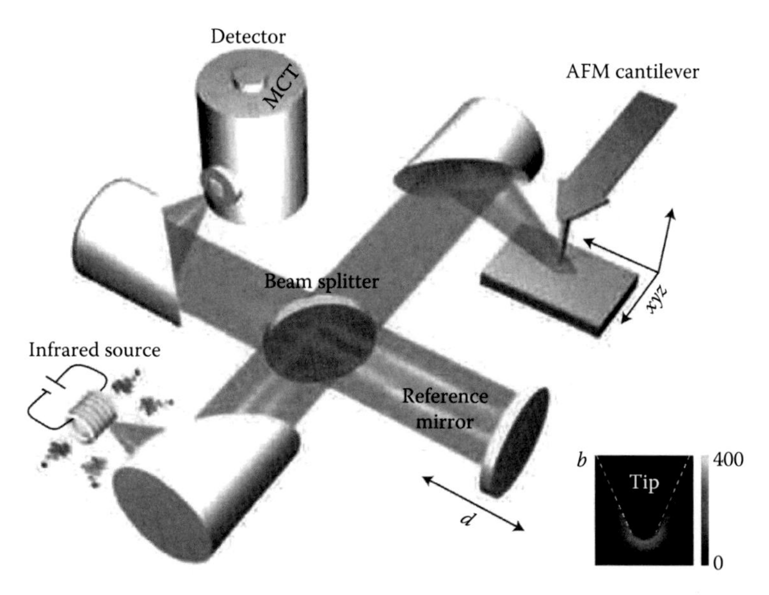

FIGURE 8.11
Diagram showing the experimental setup for broadband s-SNOM measurements. The reference mirror is scanned to create an interferogram that can be Fourier transformed for the spectrum. The inset picture is a numerical simulation of the field enhancement around a metal probe illuminated at 10.7 μm. (From Huth, F. et al., *Nat. Mater.*, 10(5), 352, 2011.)

To compare measurements, it is therefore important to have a reference material. For this silicon is usually chosen due to its flat response in the mid-IR (i.e., $s_n(\omega)$ is constant). The reported spectrum is thus given as a ratio to remove the impact of the optical components and the source:

$$\frac{S_{n,material}(\omega)}{S_{n,Si}(\omega)}$$

One of the key advantages to nano-FTIR is that, for nonpolar materials that do not exhibit any kind of phonon resonance, the imaginary part of the scattering coefficient $Im\{\sigma(\omega)\}$ is directly related to the far-field absorption coefficient $\alpha(\omega)$. Given that s-SNOM extracts both the amplitude and the phase of the scattering coefficient, its imaginary part is simply given by

$$Im\{\sigma_n(\omega)\} = s_n(\omega)\sin[\phi_n(\omega)]$$

Given that $\alpha(\omega) \propto Im\{\sigma_n(\omega)\}$, existing databases containing the spectral responses of different materials can be used without any modification for the direct identification of materials at the nanoscale (Figure 8.13).

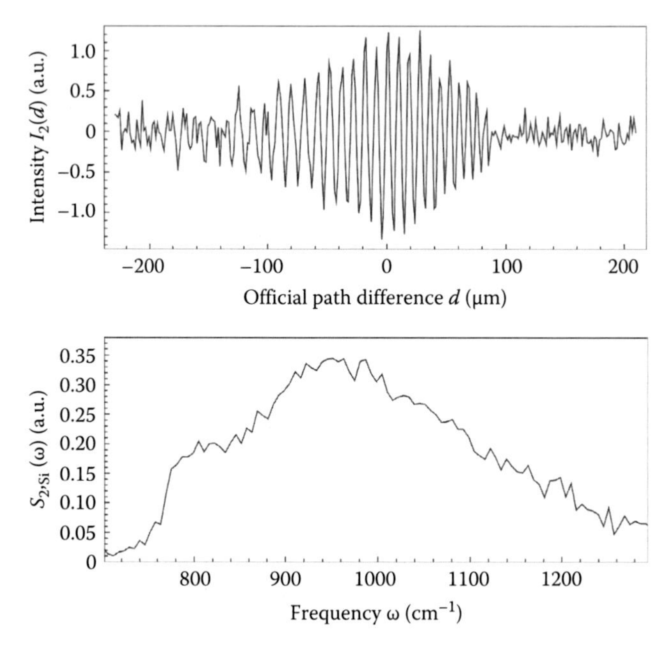

FIGURE 8.12
As the reference mirror is scanned from one point to another, an interferogram is built up (top) that can then be Fourier transformed to yield the near-field spectral response of the material (bottom). (From Huth, F. et al., *Nat. Mater.*, 10(5), 352, 2011.)

8.7 Difficulties with s-SNOM

Until now, we have made no mention of why s-SNOM is not more widespread. S-SNOM has been presented as a tool capable of taking optical images of samples at any wavelength at arbitrarily small resolutions. While this is true, there are restraints that are holding back its further development and that make it unsuitable for certain applications. As such, the following section outlines some of the issues as honestly and frankly as possible.

8.7.1 Wet or Rough Samples

Given its basis in AFM technology, s-SNOM suffers from the same problems that beset AFM measurements. Namely, these are difficulties in working with wet or extremely rough samples. Although it is not impossible to work on wet structures with AFM, the problem is often solved by submerging the whole tip setup in water or other liquids. This would be impossible to implement with s-SNOM, given the requirement to focus light to a point at the

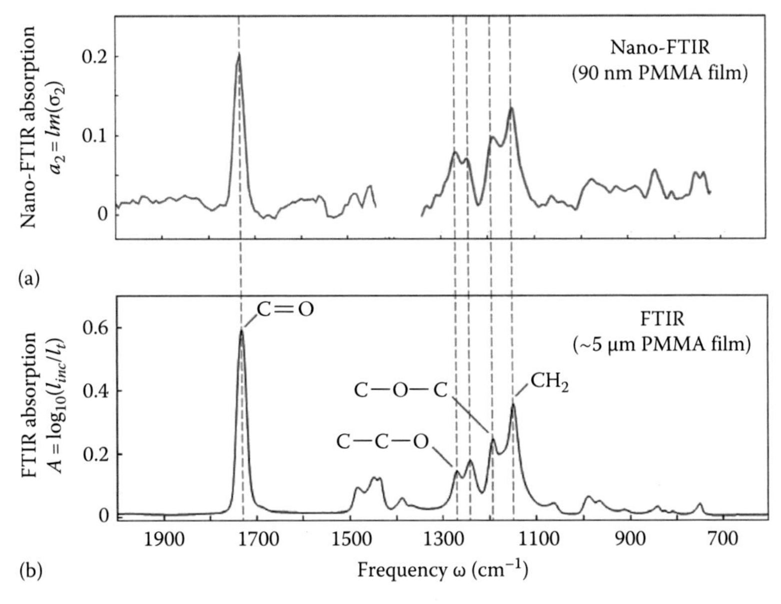

FIGURE 8.13
For nonphononic materials, the imaginary part of the scattering coefficient measured by nano-FTIR compares well with the far-field absorption spectrum, allowing the technique to tap into the existing databases that have been built up for chemical mapping. (From Huth, F. et al., *Nano Lett.*, 12(8), 3973, 2012.)

apex of the probe—as well as the associated alignment difficulties, water is highly absorptive in the IR.

Rough samples present a different set of problems. If a tip is not sufficiently sharp, the topographical image captured by the AFM will not be an accurate representation of the sample. Instead, the peaks and troughs will be "rounded out." A further problem for optical images is that if the apex of the probe is not in the immediate vicinity of the surface due to sharp edges, the signal will not contain the chemical information about the sample, and the "edge darkening" effect is seen (Figure 8.14).

One further problem is that it is impossible to image a sample where the height difference between the minimum and maximum points is greater than the distance the piezoelectric crystals controlling the sample motion can extend vertically. This is typically around several microns and thus, depending on the equipment available, places a requirement on the flatness of a sample.

It should also be noted that s-SNOM is not capable of imaging significantly subsurface structures (at least until nanoscale tomography is practically demonstrated [Sun et al. 2009]). Anything deeper than roughly the wavelength of illuminating light will not be visible (Taubner et al. 2005), rendering s-SNOM very much a surface, rather than volume, imaging technique.

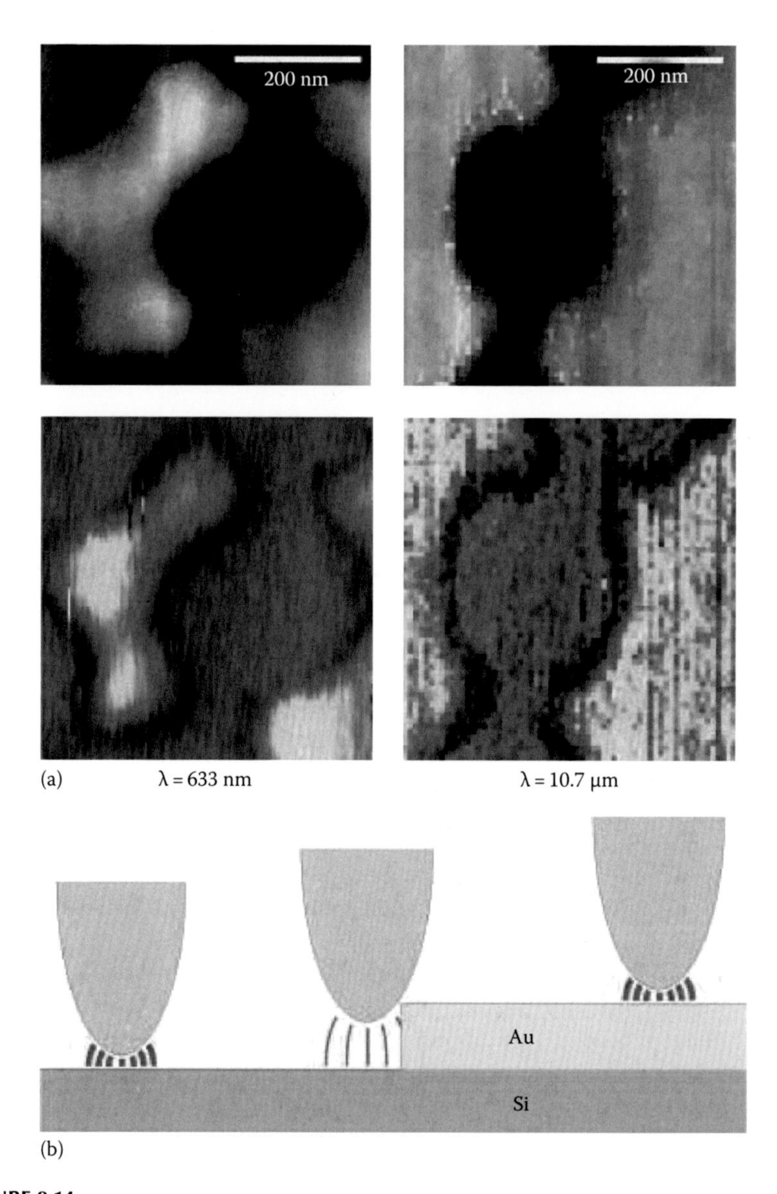

FIGURE 8.14
(See color insert.) Topography and optical amplitude images (a) of gold islands on a silicon substrate. The edge darkening effect is clearly visible in the optical images, and a schematic of its mechanism is shown in (b). (From Taubner, T. et al., *J. Microsc. Oxf.*, 210, 311, 2013.)

8.7.2 Acquisition Speeds

S-SNOM works by raster scanning a sample and building up the image pixel by pixel. One of the inherent drawbacks to this approach is slow data-acquisition times. Although the development of high-speed AFM is starting to solve this (to the extent where video-like acquisition speeds are available), high-speed s-SNOM has yet to be developed and will always be hampered by the need to integrate over a certain number of probe oscillations to extract the near-field signal. Nano-FTIR measurements in particular usually take tens of minutes to obtain due to the necessity of scanning the reference mirror at each pixel.

Coupled with the difficulty in measuring wet samples, slow data-acquisition speeds mean that it is not yet possible to measure biological samples *in vivo*, or indeed any other types of samples that change their configuration over time; temporal resolution in s-SNOM is poor.

8.7.3 Laser Availability

For single-wavelength applications, high-quality lasers are an absolute necessity. To be suitable for use with an s-SNOM system, they must fulfill the following requirements:

1. *Coherence length*—for relative ease of alignment, a coherence length of at least 1 cm is preferable. This is because it is difficult to judge the exact difference in the path length of the two arms in the interferometric setup. Lasers must therefore have a suitably narrow linewidth.
2. *Power*—there must be a sufficient amount of light concentrated at the apex of the probe to extract a near-field signal. This typically means a laser power of at least 1 mW. Although this is a relatively low level, it means that selecting a single wavelength (e.g., by using a prism) from a broadband laser source such as a supercontinuum is not a possibility.
3. *Noise*—the laser must have a very low signal-to-noise ratio (SNR). In other words, the output power must remain constant over time. This is particularly important due to the higher harmonic detection schemes, and for acceptable noise levels in s-SNOM images, SNRs of over 1000 are preferable in the mid-IR. Noise also means that, without suitable measurement synchronization techniques, pulsed lasers are not appropriate.
4. *Broad tuneability*—although not an absolute necessity, a wide selection of possible wavelengths increases the number of possible applications of the system.

Unfortunately, the mid-IR has historically been rather poorly served by lasers meeting the above specifications, and as such the literature only has examples of two types of lasers working in this spectral range—CO_2 or CO lasers and FELs. The former offers a "gold standard" in terms of coherence, usability, and stability but does not offer a particularly wide tuning range, whereas the latter is an extremely expensive tool requiring a team of technicians and a large amount of equipment—it certainly is not a simple bench top option.

The recent commercial growth in the availability of quantum cascade lasers (QCLs), however, looks set to end this lack of choice (Anscombe 2011). These are small, solid-state devices that are broadly tuneable (hundreds of wavenumbers) when used in an external cavity configuration (Hugi et al. 2010). Current commercial technologies are just on the cusp of tuneable, room temperature, CW operation, and their very wide range of possible applications, from manufacturing to security, will likely see enough production volume to push prices down.

8.8 Applications to the Biosciences

Despite the issues outlined in the previous section, there have been some initial applications of s-SNOM to the biosciences. While it is not yet a widely used tool, these proof-of-concept experiments show that—especially once coupled to a mature QCL industry—the future is potentially very bright for biological applications of IR nanoscale imaging and spectroscopy by way of s-SNOM. Here we discuss SNOM applied to three areas of bioscience—Virology, Lipidology, and Osteopathy.

8.8.1 Single Tobacco Mosaic Virus

The first work to be published on a biological sample looked at a tobacco mosaic virus on a silicon substrate (Brehm et al. 2006b). Images were taken at single wavelengths using a CO laser (Figure 8.15).

The tobacco virus is cylindrically shaped, with a diameter of just 18 nm. The clarity of the images proved that IR contrast was indeed a possibility with such small biological particles. The resonance in the phase images is caused by the amide I vibrational resonance in the peptide bonds of the virus shell proteins.

8.8.2 Proteins in Lipid Bilayers

More recently, attention has been given to imaging proteins and how they fit into biological systems (Ballout et al. 2011). A sample was prepared containing surface tethered membrane proteins (cytochrome c oxidase) reconstituted in a lipid bilayer. Gold nanoparticles were also introduced to act as reference points. Using single-wavelength s-SNOM measurements from

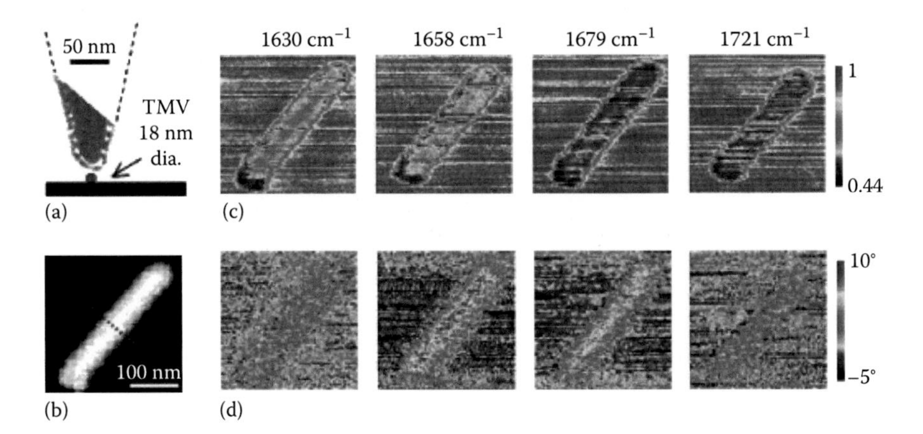

FIGURE 8.15
(See color insert.) (a) Scale schematic of tip and virus, (b) topography, (c) and (d) optical amplitude and phase at various wavelengths. (From Brehm, M. et al., *Nano Lett.*, 6, 1307, 2006b.)

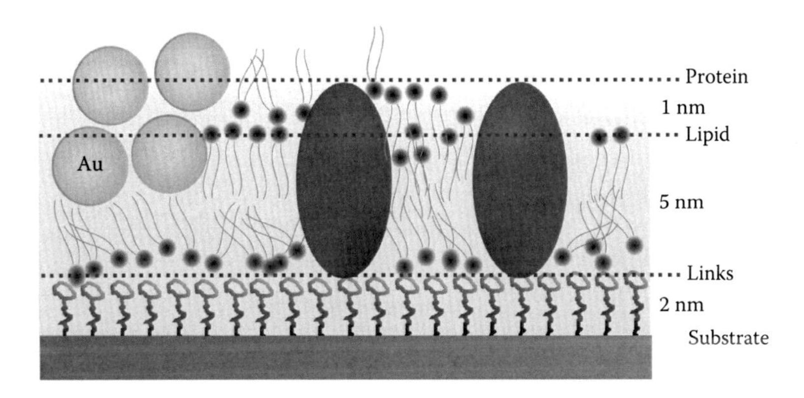

FIGURE 8.16
Based on this work, the sample can be considered as being split into four different areas. From the left: clusters of gold nanoparticles; lipids bound to proteins; a lipid monolayer; aggregated proteins.

a CO laser tuned to the amide I and carboxyl modes of the protein and lipid respectively, images were gathered at a variety of wavelengths. With a spectrum at each pixel and a known surface topography, it was possible to assign a chemical map to the surface as one of three things—gold, proteins, or lipids. From these results, it was also possible to develop a model of the structure of the sample (Figure 8.16).

8.8.3 Minerals in Biological Material

It should be noted that bioimaging is not exclusively the realm of cells and their components. There are many naturally occurring nanostructures that

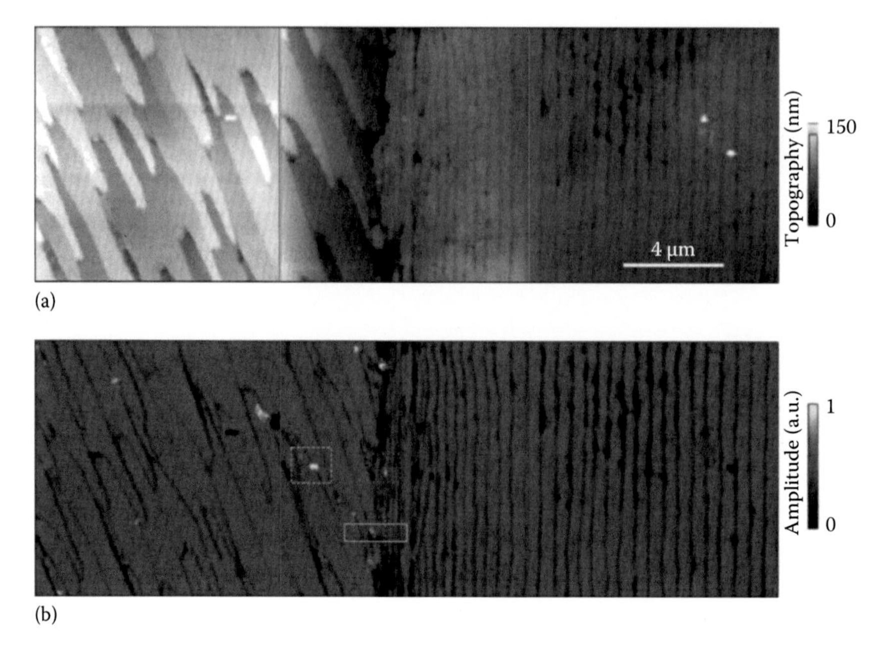

FIGURE 8.17
Topography (a) and near-field amplitude (b) of biocalcite (left) and bioaragonite (right) crystals. The images are viewed at a single wavelength (980 cm^{-1}). The bright spots in the optical image are probably as a result of small phosphate crystals. (From Amarie, S. et al., *Beilstein J. Nanotechnol.*, 3, 312, 2012.)

merit study in this field, among them the mineral particles in shell and bone. In 2012, these formations were imaged by nano-FTIR for the first time on polished sections of *Mytilus edulis* shells, revealing biocalcite and bioaragonite crystals as well as some unexpected particles of a substance resembling phosphate (Figure 8.17).

The same study also introduced a new mode of nano-FTIR operation—spectral integration of the signal. This is essentially a combination of single-wavelength and broadband s-SNOM where the reference mirror is held in a fixed position. The signal at the detector then represents the near-field interaction over a broad spectral range (although without any spectral resolution). This range can be picked to cover a certain absorption peak to ascertain relatively swiftly the amount of any given substance; scan speeds sped up from around 10 s per pixel in nano-FTIR mode to 5 ms per pixel in spectrally integrating mode. The sample beneath is a tubule in human dentin (Figure 8.18).

Different zones are clearly distinguishable that are also readily seen in the backscattered electron image; spectrally integrated nano-FTIR compares well to backscattered electron measurements for chemical mapping and does not require special sample preparation or vacuum conditions.

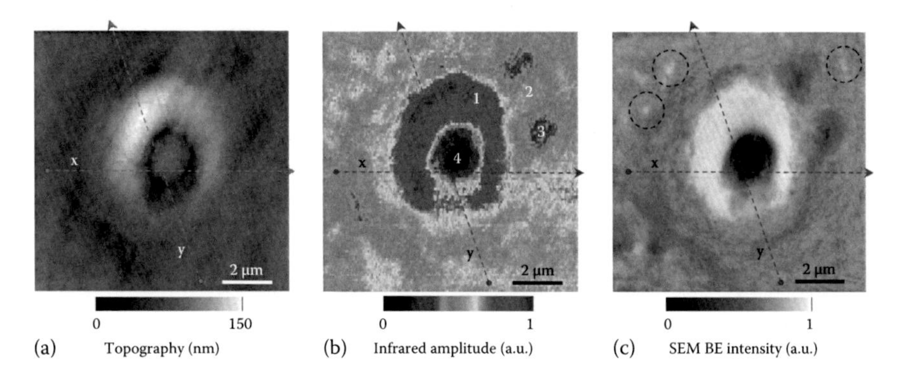

FIGURE 8.18
(See color insert.) (a) Topography, (b) spectrally integrated mode near-field image, and (c) a backscattered electron image for comparison. (From Amarie, S. et al., *Beilstein J. Nanotechnol.*, 3, 312, 2012.)

8.9 Nanoscale Infrared Absorption Spectroscopy with AFMIR

8.9.1 AFMIR

We turn now from scattering-based methods to another approach at nanoscale IR spectroscopy. These methods that come under the umbrella term AFMIR combine an AFM with an IR laser source much as in s-SNOM. They differ in that rather than measuring scattered light, the light–matter interaction is inferred by the deflection of a contact mode probe during photothermal excitation of the sample. First proposed by Alexander Dazzi (2005), it was noted that because the signal output of AFMIR systems is not derived from any light-based surface signal, but rather mechanical expansion, AFMIR is capable of pure absorption measurements, namely those properties that depend on only the imaginary, and not on the real component of the material's dielectric function (Dazzi et al. 2010).

The essential premise of the technique is that a short pulse of IR light incident onto a material will cause thermal heating to an extent that is proportional to the absorption coefficient of the material (Dazzi et al. 2010). This thermal heating quickly relaxes via mechanical expansion of the material volume, which is capable of deflecting a small probe in contact with the sample surface (Figure 8.19). The mechanical force applied to the tip as a result of the material surface expansion is subject to the approximation of Hooke's law; the cantilever will bend upon surface contact. The extent and direction of this deflection is measured using a laser spot reflected from the top surface of the cantilever into a photodiode as in a standard AFM. The cantilever ringdown after each IR excitation pulse is iteratively recorded during scanning or a spectral sweep from the excitation source. Fourier analysis

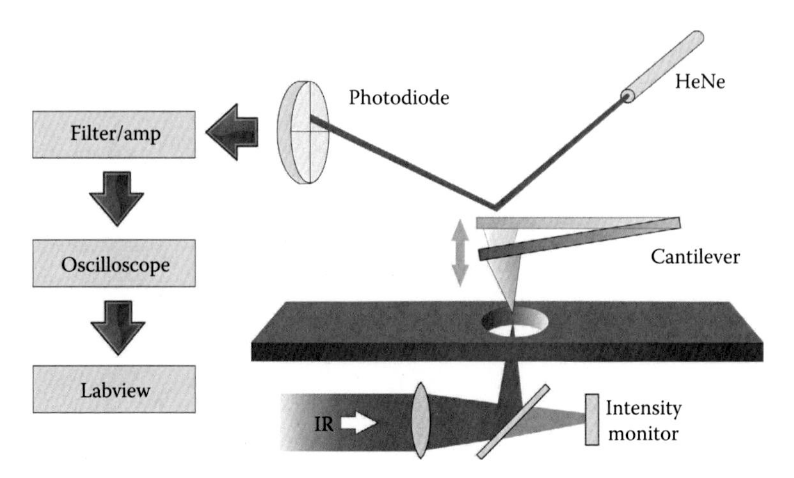

FIGURE 8.19
Experimental arrangement of an AFMIR nanoscope.

of this signal for a given wavelength, laser intensity and spot size can yield the material's absorption coefficient at nanoscale resolutions (Kennedy et al. 2011). We will first explain this method theoretically, followed by a technical overview of a typical AFMIR system. We conclude with a synopsis of work to date on AFMIR methods applied to biological systems.

8.9.2 Theoretical Basis

For particles with sizes smaller than the wavelength of light, Rayleigh scattering and theory provides the best model for light–matter interactions. When the particle size is on or near the order of the wavelength of light, Mie theory provides the most fidelitous calculation of optical properties, such as the absorption cross section (Bohren and Nevitt 1983). Mie solutions to Maxwell's equations account for the scattering of electromagnetic radiation by small particles by modeling the incident plane wave and scattering field as a series of continually growing, radiating spherical vector wave functions. The interaction of this wave of functions at the boundary of the surface provides the absorption coefficients. As one might expect, analytical solutions to such problems are involved and numerical methods are favored.

In this treatment, we will not give a detailed, full account of Mie theory. Instead, we will give a statement of the problem followed by an approximate analytical solution for on-the-fly calculation of the absorption of nanoscale samples. The central premise of AFMIR is that sample surface expansion during photothermal excitation by a short IR pulse is proportional to the power absorbed by the sample. The power absorbed can be related to the wavelength-dependent absorption cross section, $C_{abs}(\lambda)$, by

$$P = C_{abs} I$$

A simple approximation for the absorption cross section of small spheres using Mie theory (Bohren and Nevitt 1983) can be used to demonstrate the proportionality of both the sphere volume V and absorption coefficient α to the absorption cross section

$$C_{abs} = \alpha \frac{9n}{(n^2+2)^2} V$$

The power that such a sphere absorbs can then be found in terms of the incident laser's electromagnetic field E. With the assumption that the light is a monochromatic plane wave with a top hat intensity profile incident orthogonal to the substrate axis, the sinusoidal linearly polarized electromagnetic plane wave of a fixed frequency laser has incident energy flux density

$$I = \overline{(E \times H)} = ce_0 \frac{|E|^2}{2}$$

Substitution with the condition that $\alpha = 4\pi\tilde{n}/\lambda$ yields

$$P_{abs} = \frac{9\kappa n\tilde{n}}{(n^2+2)^2} ce_0 |E|^2 V$$

Thus the absorption cross section of a weakly absorbing small sphere is proportional to its volume, assuming that incident light penetrates uniformly to all parts of the sphere. At the other extreme, the absorption cross section of a highly absorbing sphere is proportional to its area. The method by which absorption is obtained in our experiment is surface expansion, which induces oscillations in a cantilever resting on the sample surface top. Using the calculated power absorbed, we can parameterize this enlargement, dz, which relaxes mechanically in response to thermal excitation (Dazzi et al. 2010)

$$dz(x) = \frac{k' P_{abs}(x)}{12\pi}$$

where k' denotes the ratio of thermal expansion to heat conductivity coefficients. Here we can see that an increase in light absorption causes a linear increase, dz, of surface expansion. This is a significant point as it immediately suggests that the height of the surface deformation as a function of excitation wavelength is an absorption spectrum of the material being probed in arbitrary units. This leads us to the essential definition of the technique:

If a small, sensitive cantilever is placed in contact with the sample surface during photothermal excitation, the cantilever will be displaced with a magnitude proportional to the material absorption coefficient at a given wavelength.

If the cantilever is free standing at one end, such as an AFM tip, we would also expect mechanical relaxation of the tip, such as by the Euler–Bernoulli equation. If the tip is angled from the surface, or the surface is soft, we may also anticipate that at some sufficient laser intensity, higher order vibration modes will be excited by the cantilever.

Some caveats must be noted for this declaration. Since the surface expansion is fast, the influence of velocity effects is also important. However, most of the conditions of concern can be resolved by keeping $P_{abs} \propto \alpha$ or by holding any other λ-dependent properties constant, such as laser intensity. If all these conditions are met, we can then sweep λ while measuring the cantilever displacement from which we can infer an absorption spectrum. From this theoretical basis, we now describe a setup that can measure dz and illustrate how to recover the absorption coefficient.

8.10 AFMIR Technical Overview

In order to understand both the benefits and the limitations of the bottom-up AFMIR method, this section outlines the main technical aspects of a typical AFMIR design. A connectivity diagram of an idealized system is shown in Figure 8.20. A variety of laser types have been successfully employed as sources for AFMIR including optical parametric oscillators (Rice et al. 2010), FELs (Dazzi et al. 2008) and more recently QCLs (Lu and Belkin 2011). Reliable spectral data require careful continual intensity monitoring. This can be achieved by beam splitting the source into a fixed gain intensity monitor that can be iteratively read out and calibrated from the (typically wavelength dependent) detector efficiency and voltage response in order to normalize for the laser intensity variation per pulse. The main beam can then be sized to cover the region of interest, although this step is not necessary when using FEL sources. The beam either can be positioned directly incident onto the tip or can be internally reflected within a ZnSe prism (Rice et al. 2010).

8.10.1 Signal Acquisition

The AFM is engaged with a cantilever above the sample stage and can be brought into feedback when in contact with the surface by the AFM control unit. The raw quadrant detector output of the AFM is routed through a lock-in amplifier that is set to a frequency range that covers the cantilever mode of interest. The time range pulses with the deflection profile immediately

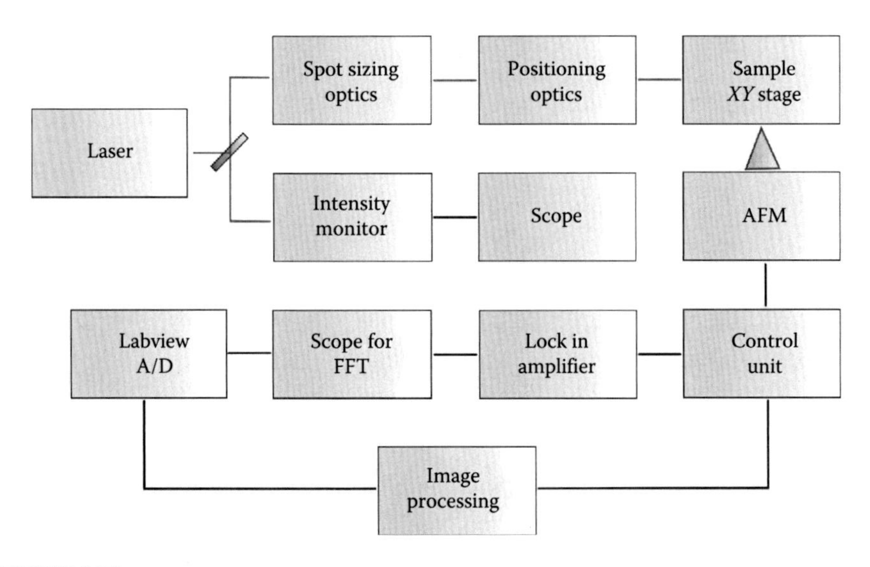

FIGURE 8.20
Connectivity diagram for an idealized AFMIR nanoscope. Gray lines indicate the laser path and black lines represent electrical connection.

after being coaveraged, normally every 128 pulses (Kjoller et al. 2010). The Fourier transform can be taken at this point by the oscilloscope, for example with a Hamming window function. A peak of interest is recorded iteratively. Alternatively, the raw deflection peak to peak maximum V or whole oscillatory response $V(t)$ can be recorded. The magnitude of the signal output and the frequency of the modes present are dependent on the specific tip–sample interaction. However, the cantilever chosen has a significant impact on the efficacy of the system. Often, V-shaped cantilevers with silicon nitride tips are used.

These cantilevers have a strong fundamental oscillatory mode that corresponds to symmetric bending of both arms (Figure 8.21). Another mode often detected is antisymmetric bending of the arms, and several higher order torsional modes are produced on steeply angled surfaces or potentially at the boundary between two media. These modes can be readily assigned using the method of Chang et al., who quantified flexural sensitivity of V-shaped cantilevers as a function of angle between cantilever and sample surface (Chang et al. 2005). During photoexcitation of the sample surface, several of these modes are often observed simultaneously. This is especially prevalent at high laser intensities or when the laser is tuned to output near an absorption band. This is expected as higher energies imparted to the tip will cause higher amplitude oscillation, but only up to a point after which additional impact energy transfer will excite higher resonance modes. Selection of a cantilever of specific size can minimize the observation of simultaneous modes as required by the specific experiment.

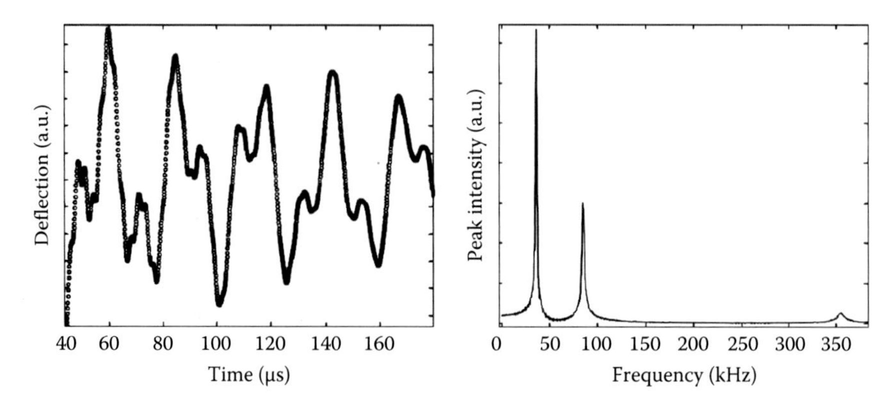

FIGURE 8.21
(Left) A subsection of the decay response from a V-shaped cantilever and (right) the Fourier transform of the whole of this signal, showing a high-frequency torsional mode and the presence of both symmetric and antisymmetric bending.

The viable duration of AFMIR experiments is primarily dependent on the source characteristics. Laser intensity can be kept constant over time scales up to several hours by continually monitoring the laser output intensity and accounting for it by changing the spot size. However, experimentalists should take care to minimize the influence of laser output characteristics that cannot easily be changed during the course of an experiment. These include the consistency of the pulse shape and beam profile, any variation in trigger timing relative to pulse timing and the extent of pulse to pulse intensity variation. Since IR spectroscopy is sensitive to a range of intensity-dependent effects, changes in laser output should ideally be dealt with as they occur rather than by linear calibration after the experiment. The laser spot size can be easily determined from a digital microscope video of the tip, provided the laser contains a visible tracer.

We note that the period of the laser pulse can be tuned to less than the AFM z-piezo feedback response. In this way, the IR data and topology data are in different frequency regions of the signal. Therefore they can each be read simultaneously and without interference. By recording deflection during raster scanning, a matrix can be produced that represents a 2D map of the IR absorption intensity at the wavelength of incidence (Figure 8.22). Alternatively a single point can be chosen within the AFM scan range, and the wavelength can be tuned over a range, yielding an IR spectrum of the sample at the tip apex.

The SNR of the output is not only dependent on the light intensity and material in question, but also on how the data are handled. Pulse averaging is an important factor that greatly adds to the quality of the data. Averaging also increases the efficiency of the software. This is because a certain amount of "dead time" exists between outputs where the scope is processing the FT and thus not reading out data. The three dominant factors to be considered

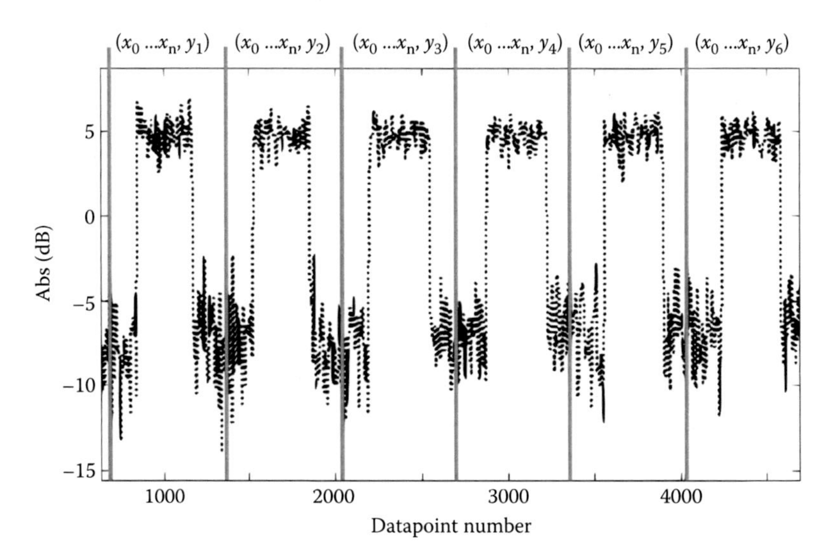

FIGURE 8.22
A raw data stream output from Labview of the FT maximum amplitude in decibels iterated across a protein ledge–glass boundary. The IR absorber is clearly distinguishable at 5 dB from the background level at −7 dB, caused by the spatially varying absorption coefficient. Gray vertical lines indicate column separation in the 2D map.

when deciding on the averaging to be used are the pulses utilized per second, the output sampling frequency desired, and the average percentage error as a function of the pulse averaging employed. With the appropriate choice of averaging and the other experimental restraints described here adhered to, it is now possible to look at the efficacy of the method for real experiments.

8.11 Application of AFMIR to Biosciences

AFMIR has been applied to a number of different biosystems to date. We shall start by reviewing where AFMIR has been applied to study bacterial systems, then moving to cell experiments and then to proteins and lipids.

8.11.1 Bacteria

Single fixed *Escherichia coli* bacteria were studied on an ATR prism substrate (Dazzi et al. 2006, 2008) using AFMIR with a spatial resolution of <100 nm. Localized IR spectra of *E. coli* were recorded using AFMIR and compared with the spectrum of an assembly of bacteria (measured with a far-field FTICR) that showed the same features. The authors of these studies noted that the bacteria are not very well adapted to determine the ultimate resolution and sensitivity because of their rather large size.

AFMIR was applied to study *E. coli* bacteria infected with T5 phage (Dazzi et al. 2008). *E. coli* was grown in LB medium to the exponential growth phase and infected by phages, where the virus infection was studied at various stages using AFMIR. Single T5 phages were studied as dried samples at two different wavelengths, at 1650 cm^{-1} (amide I) characterizing the proteins of the capsid and at 1080 cm^{-1}, the maximum absorption of the DNA band. These studies indicated that IR images of a single T5 phage could only be obtained at 1650 cm^{-1}. IR imaging at the amide I band produced poor IR images as proteins constitute only a small fraction of the phage head. The authors showed that imaging at 1080 cm^{-1} gave blurred IR maps compared to the topography indicating damage from denaturing of the virus following drying that leads to DNA breakdown.

When the T5 phages were prepared infecting the bacteria, IR imaging was performed in resonance with the DNA absorption band in order to detect the presence of the virus as no denaturing of the virus DNA occurs within the bacteria. Dazzi et al. studied three stages of infection. These studies indicated that the virus concentration and location could be identified within the bacteria using AFMIR below $\lambda/60$ resolution (Figure 8.23).

Studies applying AFMIR to spatially map energy storage polymers inside individual bacteria *Rhodobacter capsulatus* have been reported (Mayet et al. 2010). AFMIR studies aimed to chemically map subcellular features with a spatial resolution of <100 nm. The authors used key absorption bands of the energy storage polymer polyhydroxybutyrate (PHB) known from FTIR to

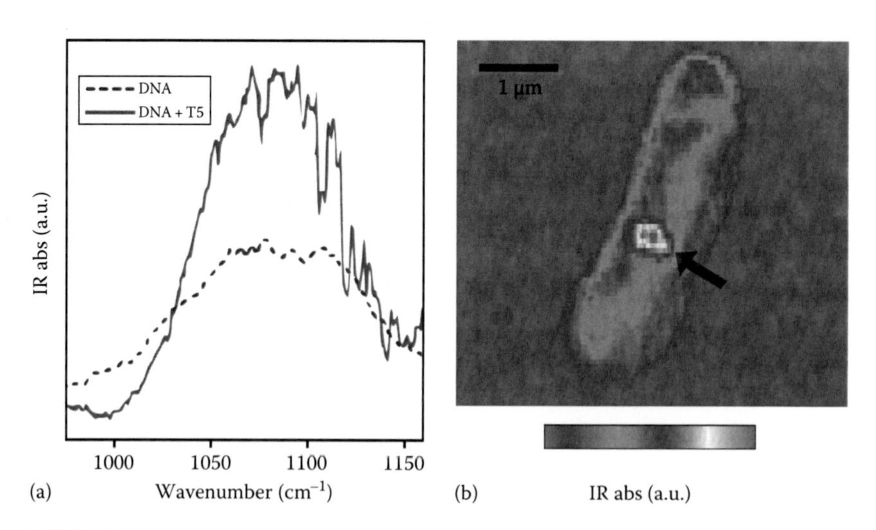

FIGURE 8.23
(See color insert.) (a) The nanoscale IR spectra of DNA and DNA with the phage virus present. An IR chemical map at the Amide I band is shown in (b), with the internal phage clearly visible. (Modified from Dazzi, A. et al., *Ultramicroscopy*, 107, 1194, 2008.)

spatially map the molecular distribution of PHB inside bacteria. The localized IR spectrum of PHB inside the bacteria obtained using AFMIR compared well to the FTIR measurements on bulk PHB. A shift in the location of the carbonyl absorption peak between bulk PHB and PHB inside bacteria was found. The AFMIR studies were supported by finite element analysis to model AFMIR measurements of PHB granules and transmission electron microscopy (TEM) of *R. capsulatus* to determine the size distribution of the PHB granules. Sizes measured by AFMIR correspond well to TEM measurements.

AFMIR was applied to study the production of PHB in *R. capsulatus* and to evaluate the influence of glucose and acetone on the production yield (Mayet et al. 2012). This study demonstrated that AFMIR demonstrated that both glucose and acetone had a positive effect on biopolymer production. AFMIR was shown to be a fast diagnostic for PHB production. The bacteria were sampled directly from the Petri dish or the liquid medium. This suspension was dried on the prism surface. This study compared TEM to AFMIR. It was noted that AFMIR was able to obtain a detailed analysis of the sample after 1 day in comparison to the preparation time for coating samples in epoxy, slicing the resin covered sample and aligning on a copper grid for TEM (1–2 weeks). AFMIR techniques do not need to modify the sample to obtain chemical images. The other advantage of AFMIR was that it directly identifies PHB by spectroscopy (absorption at 1740 cm^{-1}) while the TEM images only show white regions corresponding to the location of the vesicles, which is not a chemically specific signal. By knowing the oscillator force of ester carbonyl and amide I bands, the average production of PHB (with the FTIR measurements) could be quantified by using the PHB volume occupation found from the topography.

8.11.2 Live Cells

AFMIR has been applied to image live cells. Studies of live *Candida albicans* fungi cells were conducted (Mayet et al. 2008) using the ATR excitation alignment. This experimental arrangement reduces IR light propagation into the surrounding water environment, which in turn reduces sample heating and thus minimizing sample perturbation. Blastospore and *Hyphae candida albicans* were studied in water. The spectral region chosen was 800–1200 cm^{-1} in order to assess the water window. IR imaging was performed at the glycogen band centered at 1080 cm^{-1}. The resulting IR images indicated that the glycogen band is distributed uniformly. Water absorption was shown to occur with significant strength from 900 to 800 cm^{-1}, while relatively low water absorption occurred between 900 and 1200 cm^{-1}. Simultaneous AFM images and IR images were recorded enabling cross-confirmation of images profiles. Comparing these images showed that the intensity of the glycogen band signal in the IR image matches the AFM topography image profile for each sample studied. These studies demonstrated that the AFM cantilever vibration in water could be measured and IR imaging was possible.

Effects arising from the water environment on the measured vibrations from the cantilever have been reported, but these effects arising from friction forces did not preclude imaging. The authors showed that Fourier technique analysis enables the contribution of the sample from the surrounding absorbing medium to be resolved. Studies of water absorption were also made showing that the wavelength window of imaging was required to be carefully chosen in order to avoid strong water absorption transitions. The authors reported that the spatial resolution seems identical in liquid than in air at ca. 100 nm as with previous experiments. This study demonstrated that live cell imaging can be performed using AFMIR albeit within a specific spectral region within the water spectral window.

8.11.3 Material Complexes within Cellular Systems

Several studies of mapping materials inside cells using AFMIR have been reported. Studies have applied AFMIR to detect and quantify a metal–carbonyl ($(Cp)Re(CO)_3$) compound in cells (Policar et al. 2011). This study reported to be the first mapping of an exogenous compound in single human cells and in the cell nucleus by AFMIR. Using an IR absorption band originating from the metal–carbonyl, its location was mapped within the cell. The authors used characteristic IR band of metal–tricarbonyl compounds occurring at 1915 cm^{-1} (antisym, e) and at 2017 cm^{-1} (sym, a_1). These bands enabled the presence of the compound to be distinguished from the cell. This study essentially proved that AFMIR enabled the successful detection and quantification of a metal–carbonyl compound in whole cells and the mapping of an exogenous compound in single human cells and in the cell nucleus. The IR signature of the metal–carbonyl unit was used, and the organometallic derivative did not require any further tagging that might modify its physicochemical properties and hence its location. AFMIR was used to localize the nucleus without the use of any nucleus tracker. This study showed that AFMIR is a noninvasive technique that can be applied directly to single cells.

8.11.4 Lipids and Proteins

Several studies have applied AFMIR to study lipids and proteins (Yarrow et al. 2010, Kennedy et al. 2011). Imaging of lipid layers was performed to enable subdiffraction IR signal localization of 1,2-dioleoyl-sn-glycero-3-phosphocholine (DOPC) lipid layers. The multimellar vesicles were made by hydrating a dry film of DOPC lipids with salt solution followed by sonication and deposition onto both cleaved mica and glass substrates. This was significant as it demonstrated the efficacy of the technique without the need of sample deposition on to a ZnSe prism, which is experimentally unfamiliar and even inconvenient for some biologists. The IR absorption was reported not to be fully homogenous throughout the lipid and showed reduced heating at the sample edges relative to the aggregate's interior. This was consistent with computational models of lipid heat diffusion. It was noted that the resolution has a lower bound imposed by the AFM tip–sample interaction area, which varies with the tip used. The AFMIR imaging possessed an image resolution of 37 nm.

Nanoscale IR spectra from single points were produced by holding the AFM tip in position on the surface of a sample, accounting for any thermal drift by regular imaging and moving the tip when necessary, and then iteratively reading the absorption over a frequency range. An ATR FTIR setup was used to record reference IR spectra of the substrate and samples. The AFMIR spectrum of DOPC lipids is shown in Figure 8.24 alongside the lipid's FTIR spectrum. The two spectra had similar features despite the difference in spectral resolution and collection area. The bands at 2851 and 2925 cm^{-1} in the FTIR are assigned to symmetric C–H stretching of the methylene groups, whereas symmetric and asymmetric C–H stretching of the terminal methyl groups are observed at 2875 and 2956 cm^{-1}. The C–H stretching of the oleoyl chain is observed at 3006 cm^{-1}. All these peaks were in agreement with the position of the peaks observed in the AFMIR spectrum in Figure 8.24. An increase in the intensity of the band at 3006 cm^{-1} in the AFMIR spectrum was attributed to the presence of the substrate, owing to the broad mica spectral response near the oleoyl absorption band.

Subdiffraction-limited IR absorption imaging of hemoglobin was performed by AFMIR (Kennedy et al. 2011). Comparisons between the AFM topography and IR absorption images of micron-sized hemoglobin features demonstrated that nanoscale IR spectroscopic analysis of the metalloprotein could be undertaken using AFMIR up to film thickness of 400 nm. Using IR absorption bands for the metalloprotein from the amide B band at 3061 cm^{-1} (hydrogen-bonded N–H stretching) and 2960 cm^{-1} (aliphatic C–H stretching), chemical mapping of hemoglobin aggregates 100 nm height and micron size range in the lateral axial direction were studied using AFMIR. Again, the authors reported some differences between the AFMIR spectrum and the

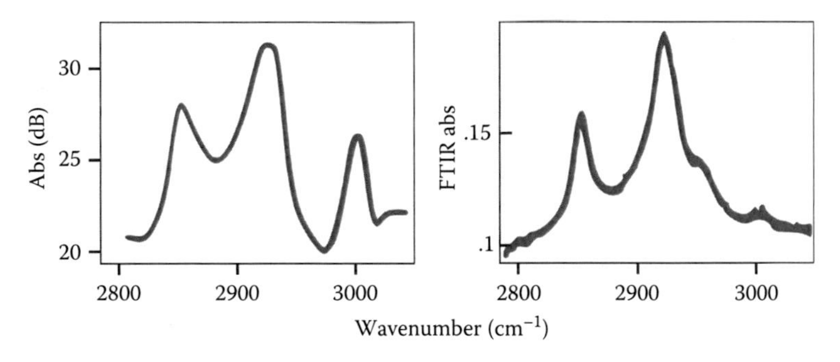

FIGURE 8.24
AFMIR (a) and FTIR (b) IR spectra of DOPC lipids.

FTIR spectrum. However, all of the main absorption bands were the same in both methods. Peak to peak height variation between the FTIR and AFMIR methods is likely attributed to the ratio of the collection areas of the spectra.

FTIR sample collection areas vary, but are often close to 1 mm × 1 mm at the focus. A reasonable estimate of the collection area of the cantilever tip is 50 nm × 50 nm. This gives a ratio for the collection areas on the order of 10^8. Gaining even qualitatively similar spectra from such different detection systems shows that the spectral similarities in the aforementioned studies were not systemic, but rather were measuring a fundamental quantity, in this case, the amount of light energy that is taken up by the sample at varying frequencies. No standard currently exists with which to measure the accuracy of the AFMIR method, as only very few spectra have been taken in this way. As with previous studies, their FTIR spectrum is therefore the best comparison to accuracy.

8.12 Discussion: Applying AFMIR for Bioimaging

A range of methods exist for IR imaging such as photoacoustic microscopy or synchrotron IR microspectrometry; however, the spatial resolution of these approaches is diffraction limited, resulting in the lateral IR image resolution limited to the micron length scale. The use of ATR however provides enhanced vertical spatial resolution by utilizing evanescent waves to probe samples studying samples on the order of the evanescent wave penetration depth (Sohn et al. 2004, Kazarian 2010).

The benefits of nanoscale IR spectroscopy described herein needs to be stated within the context of other vibrational spectroscopic methods. Raman microscopy has been able to provide high spatial resolution spectroscopic imaging with visible and ultraviolet (UV) sources combined with confocal methodologies, matching the spatial resolution of fluorescence microscopy. While in theory it is possible to continuously improve the spatial resolution of Raman microscopy with excitation frequencies into the UV and higher soft x-ray regions, it has been found that the excitation of materials, and specifically biosystems, at above visible excitation energies produces fluorescence that "drowns" the Raman signal. This has effectively resulted in the lateral resolution of Raman microscopy being limited close to the UV Abbe criterion. In contrast, the spatial resolution of AFMIR is ultimately limited by the AFM cantilever profile and geometry. Near-field approaches have been demonstrated that enable IR and Raman spectroscopy imaging to be performed. These approaches utilize optical apertures, plasmon enhancement effects, and/or tip oscillation methods to enable subdiffraction-limited Raman spectroscopy. However, such approaches are not normally able to recover IR absorption information, being scattering techniques, and have signal outputs that are very sensitive to the probe properties.

The application of AFMIR to biological systems has been demonstrated by probing bacterium and living cells as outlined earlier. These studies demonstrated that soft matter can be probed using AFMIR, including living systems. It is noted that the presence of water, which is a strong IR absorber, limits the spectral window in the IR and as a consequence limits the potential of this method to chemically map living biosystems. In contrast to this, other spectroscopic imaging methods such as Raman are not so limited by the presence of water and can potentially probe a wider spectral window.

To date, AFMIR studies of fixed cells have focused on characterizing the occurrence of specific functional groups on the sample surface. These studies have reported homogeneous distribution of specific chemical functional groups such as the amide I band or the glycogen band. However, when probing relatively large features such as viruses within a host organism [jr5], the power of AFMIR becomes apparent, demonstrating that in order to utilize this method to its full potential AFMIR should be applied to systems that have inhomogeneous distributions of material.

Present AFMIR technology has been expanded to include two experimental methodologies, ATR and direct surface excitation, enabling a wide range or samples to be mapped with high lateral resolution on a variety of substrates. It is noted, however, that such mapping is limited in the Z plane and likely has a much larger point spread function along this axis than in the XY plane. Studies of virus in bacteria demonstrated that viruses present beneath the host surface can be spectrally selected, but with undefined resolution (Mayet et al. 2008). Presently, the inability to resolve information in the Z plane is a limit in the spectral localization ability of AFMIR.

Future potential applications of AFMIR could extend to areas such as in mechanical studies of materials. It is noted that AFM-TIRFM (total internal reflection fluorescence microscopy)-based experimental system can observe mechanical processes such as stress transmission in cells (Mathur et al. 2000). In a similar way, AFMIR may in the future be applied to study such processes by varying the AFM cantilevers downward force on the system under study and then recovering the IR data. Other future applications of AFMIR may be combining this method with other techniques to form a single hybrid instrument, thereby achieving potentially enhanced understanding of molecular processes such as morphological or mechanical based processes. This may be achieved by combining AFMIR with other spectroscopic methods—mechanical or electrochemical techniques.

In conclusion, AFMIR enables local spectroscopy and IR chemical mapping with nanoscale resolution while still permitting simultaneous AFM mapping. This nanotool has been applied to a variety of samples with biomaterials being overviewed in this chapter. Interested readers who wish to learn more about applications should note that AFMIR research is also seeing increased popularity in semiconductor research (Sauvage et al. 2011) and pharmaceutical science (Van Eerdenbrugh et al. 2012).

References

Abbe, E. 1873. Beitrage zur theorie des mikroskops und der mikroskopischer wahrnehmun. *Arch. Mikrosk. Anat.* 9(1): 413.

Bohren, C.F. and Nevitt, T.J. 1983. Absorption by a sphere: A simple approximation. *Appl. Opt.* 22 (6) 774–775.

Chang, W. et al. 2005. Flexural sensitivity of a V-shaped cantilever of an atomic force microscope. *Mater. Chem. Phys.* 92: 438332.

Dazzi, A. et al. 2005. Local infrared microspectroscopy with subwavelength spatial resolution with an AFM used as a photothermal sensor. *Opt. Lett.* 30(18): 2388–2390.

Dazzi, A. et al. 2008. Chemical mapping of the distribution of viruses into infected bacteria with a photothermal method. *Ultramicroscopy*, 107: 1194–1200.

Dazzi, A., Prazeres, R., Glotin, F., and Ortega, J.M. 2006. Subwavelength infrared spectromicroscopy using an AFM as a local absorption sensor. *Infrared Phys. Technol.* 49: 113–121.

Dumas, P. et al. 2004. Imaging capabilities of synchrotron infrared microspectroscopy. *Faraday Discuss.* 126: 289–302; discussion 303–311.

Eerdenburgh, B.V. et al. 2012. Nanoscale mid-infrared imaging of phase separation in a drug-polymer blend. *J. Pharm. Sci.* 101(6): 2066–2073.

Goodman, J.W. 1996. *Introduction to Fourier Optics*, 2nd edn. New York: McGraw Hill.

Hammiche, A. et al. 2004. Mid-infrared microspectroscopy of difficult samples using near-field photothermal microspectroscopy (PTMS). *Spectroscopy* 19(14): 2042.

Harbison, B. et al. 1998. Infrared transparent selenide glasses. U.S. Patent 5846889.

Hecht, E. 1997. *Optics*. Reading, MA: Addison-Wesley Publ. Company.

Huber, A.J., Wittborn, J., and Hillenbrand, R. 2010. Infrared spectroscopic near-field mapping of single nanotransistors. *Nanotechnology* 21(23): 235702.

Huth, F. et al. 2012. Nano-FTIR absorption spectroscopy of molecular fingerprints at 20 nm spatial resolution. *Nano Lett.* 12(8): 3973–3978.

Huth, F., Schnell, M., Wittborn, J., Ocelic, N., and Hillenbrand, R. 2011. Infrared-spectroscopic nanoimaging with a thermal source. *Nat. Mater.* 10(5): 352–356.

Kazarian, S.G. and Chan, K.L.A. 2010. Micro- and macro-attenuated total reflection Fourier transform infrared spectroscopic imaging. *Appl. Spec.* 64: 846–846.

Keilmann, F. and Hillenbrand, R. 2004. Near-field microscopy by elastic light scattering from a tip. *Phil. Trans. R. Soc. Lond. A* 362, 787–805, doi:10.1098/rsta.2003.1347.

Kennedy, E. et al. 2011. Nanoscale spectroscopy and imaging of haemoglobin. *J. Biophotonics* 4(9): 588–591.

Kjoller, K. et al. 2010. High-sensitivity nanometer-scale infrared spectroscopy using a contact mode microcantilever with an internal resonator paddle. *Nanotechnology* 21(18): 185705.

Lu, F. and Belkin, M.A. 2011. Infrared absorption nano-spectroscopy using sample photoexpansion induced by tunable quantum cascade lasers. *Opt. Express* 19(21): 19942–19947.

Mathur, A.B. et al. 2000. Atomic force and total internal reflection fluorescence microscopy for the study of force transmission in endothelial cells. *Biophys. J.* 78: 1725–1735.

Mayet, C. et al. 2008. Sub-100 nm IR spectromicroscopy of living cells. *Opt. Lett.* 33(14): 1611–1613.

Mayet, C. et al. 2012. Analysis of bacterial polyhydroxybutyrate production by multimodal nanoimaging. *Biotechnol. Adv.* 31(3): 369–374.

Mayet, C., Dazzi, A., Prazeres, R., Ortega, J.M., and Jaillard, D. 2010. In situ identification and imaging of bacterial polymer nanogranules by infrared nanospectroscopy. *Analyst* 135: 2540–2545.

Miller, L.M. 2006. Infrared microspectroscopy and imaging. *Brookhaven National Laboratory Imaging Workshop*, Brookhaven, New York.

Miller, L.M. et al. 2007. Accretion of bone quantity and quality in the developing mouse skeleton. *J. Bone Miner. Res.* 22: 1037–1045.

Novotny, L. 2007. The history of near-field optics. *Progress Opt.* 50: 137–184.

Ocelic, N., Huber, A., and Hillenbrand, R. 2006. Pseudoheterodyne detection for background-free near-field spectroscopy. *Appl. Phys. Lett.* 89(10): 101124.

Peiqiang, Y. 2003. Chemical imaging of microstructures of plant tissues within cellular dimension using synchrotron infrared microspectroscopy. *J. Agric. Food Chem.*, 52(20): 6062–6067.

Pinker, S.A. 1997. *How the Mind Works*. New York: W.W. Norton & Company.

Policar, C. et al. 2011 Subcellular IR imaging of a metal–carbonyl moiety using photothermally induced resonance. *Angew. Chem. Int.* 50: 860–864.

Rice, J.H. et al. 2010. Nano-infrared surface imaging using an OPO and an AFM. *Eur. Phys. J.*, 51.

Russ, J.C. 2007. *The Image Processing Handbook*, 5th edn. Boca Raton, FL, CRC Press.

Sauvage, S. et al. 2011. Homogenous broadening of the S to P transition in lnGaAs/GaAs quantum dots measured by infrared absorption imaging with nanoscale resolution. *Phys. Rev. B* 83(3): 035302.

Sohn, A.F. et al. 2004. Combined ATR infrared microspectroscopy and SIL Raman microspectroscopy. *Microsc. Microanal.* 10: 1316–1317.

Stanford Materials Science Engineering, Brongersma Lab, accessed August 2012, http://brongersma.stanford.edu/main/InfraredNanooptics with permissions.

Sykes, C. 1989. *The Last Journey of a Genius*. Boston, MA: BBC TV & WGBH Boston.

Van Eerdenbrugh, B. et al. 2012. Nanoscale mid-infrared evaluation of the miscibility behaviour of blends of dextran or maltodextrin with poly(vinylpyrrolidone). *Mol. Pharmaceutics*. 5: 1459–1469.

Yarrow, F. et al. 2011. Subwavelength infrared imaging of lipids. *Biomed. Opt. Express* 2: 37–43.

Some alterative formats for references herein

Amarie, S., Ganz, T., and Keilmann, F. 2009. Mid-infrared near-field spectroscopy. *Opt. Express* 17(24): 21794–21801.

Amarie, S., Zaslansky, P., Kajihara, Y., Griesshaber, E., Schmahl, W.W., and Keilmann, F. 2012. Nano-FTIR chemical mapping of minerals in biological materials *Beilstein J. Nanotechnol.* 3: 312–323.

Anscombe, N. 2011. Quantum leap. *Electro Optics* 2: 16–18.

Ballout, F. et al. 2011. Scanning near-field IR microscopy of proteins in lipid bilayers. *Phys. Chem. Chem. Phys.* 13(48): 21432–21436.

Brehm, M., Schliesser, A., and Keilmann, F. 2006a. Spectroscopic near-field microscopy using frequency combs in the mid-infrared. *Opt. Express* 14(23): 11222–11233.

Brehm, M., Taubner, T., Hillenbrand, R., and Keilmann, F. 2006b. Infrared spectroscopic mapping of single nanoparticles and viruses at nanoscale resolution. *Nano Lett.* 6: 1307–1310.

Cvitkovic, A., Ocelic, N., and Hillenbrand, R. 2007. Analytical model for quantitative prediction of material contrasts in scattering-type near-field optical microscopy. *Opt. Express* 15(14): 8550–8565.

Hayazawa, N., Tarun, A., Taguchi, A., and Kawata, S. 2009. Development of tip-enhanced near-field optical spectroscopy and microscopy. *Jpn. J. Appl. Phys.* 48(8): 08JA02.

Huber, A.J., Wittborn, J., and Hillenbrand, R. 2010. Infrared spectroscopic near-field mapping of single nanotransistors. *Nanotechnology* 21(23): 235702.

Hugi, A., Maulini, R., and Faist, J. 2010. External cavity quantum cascade laser. *Semicond. Sci. Technol.* 25(8): 083001.

Huth, F. et al. 2012. Nano-FTIR absorption spectroscopy of molecular fingerprints at 20 nm spatial resolution. *Nano Lett.* 12(8): 3973–3978.

Huth, F., Schnell, M., Wittborn, J., Ocelic, N., and Hillenbrand, R. 2011. Infrared-spectroscopic nanoimaging with a thermal source. *Nat. Mater.* 10(5): 352–356.

Keilmann, F. and Hillenbrand, R. 2004. Near-field microscopy by elastic light scattering from a tip. *Phil. Trans. R. Soc. Lond. A* 362(1817): 787–805.

Novotny, L. 2007. The history of near-field optics. *Progress Opt.* 50: 137–184.

Ocelic, N., Huber, A., and Hillenbrand, R. 2006. Pseudoheterodyne detection for background-free near-field spectroscopy. *Appl. Phys. Lett.* 89(10): 101124.

Stanford Materials Science and Engineering. 2012. http://brongersma.stanford.edu/main/InfraredNanooptics. Accessed August 2012.

Sun, J., Schotland, J.C., Hillenbrand, R., and Scott Carney, P. 2009. Nanoscale optical tomography using volume-scanning near-field microscopy. *Appl. Phys. Lett.* 95(12): 121108.

Synge, E.H. 1928. A suggested method for extending microscopic resolution into the ultra-microscopic region. *Philos. Mag. Ser. 7* 6(35): 6.

Taubner, T., Hillenbrand, R., and Keilmann, F. 2003. Performance of visible and mid-infrared scattering-type near-field optical microscopes. *J. Microsc. Oxf.* 210: 311–314.

Taubner, T., Keilmann, F., and Hillenbrand, R. 2005. Nanoscale-resolved subsurface imaging by scattering-type near-field optical microscopy. *Opt. Express* 13: 8893–8899.

9

Spectral Self-Interference Fluorescence Microscopy to Study Conformation of Biomolecules with Nanometer Accuracy

Xirui Zhang, Philipp S. Spuhler, David S. Freedman, and M. Selim Ünlü

CONTENTS

9.1 Introduction

Despite the completion of the human genome sequencing, the revolution of personalized medicine still seems years away [1,2]. Although the Human Genome Project provides researchers with enormous amounts of genetic information, the genome is far more complex than the sequences it contains. Part of the reason is that DNA functions through critical interactions with proteins, such as genome packaging, epigenetic modifications, transcription, DNA replication, and DNA repair [3–7]. Since the idealized B-form DNA proposed more than 50 years ago by Watson and Crick, researchers have found that the conformation of DNA is naturally distorted and, depending on the particular sequence, DNA can be curved, tightly bent, or kinked [8–10]. These intrinsic variant conformations of DNA are recognized, stabilized, or even enhanced upon the formation of DNA–protein complexes

[11,12]. DNA–protein complexes bring distant regions of DNA together and can sharply bend or kink the DNA, resulting in conformational changes postulated to play an important role in the recognition of specific binding sites on DNA by proteins [11,13,14]. An understanding of the intrinsic conformation of DNA and deformity of DNA structure to form a specific complex in a native environment is of considerable biological significance.

Conventional tools are available to study DNA–protein binding whose critical size dimensions are on sub-nanometer scales. X-ray crystallography determines structures of biomolecules crystallized under stringent conditions at the atomic level, yet the resulting protein structure from crystallization may not be an accurate representation of its structure in its native environment [15,16]. Nuclear magnetic resonance (NMR) is one of the most precise techniques for imaging biomolecular structures and has the ability to perform dynamic protein structure measurements [17,18]. Electrophoretic mobility shift assay (EMSA) is one of the most inexpensive and easiest ways to investigate DNA conformation. Although EMSA is low throughput and does not allow measurement of protein–DNA dynamics, it is a powerful technique for accurate measurement of changes in DNA secondary structure [19,20]. Supercoiled, kinked, and looped DNA structures and DNA-bending angles in protein–DNA complexes can also be determined with atomic force microscopy (AFM) by scanning DNA (single- or double-stranded) that has been deposited on a very flat, smooth, and stationary substrate to obtain a 2D contour [21–25]. However, the requirement to adsorb the molecules on a surface impacts the conformation of the adsorbed biomolecules, and scanning of a surface by AFM is a slow process. Additionally, the scanning area is usually on a scale of hundreds of nanometers, limiting the throughput of AFM. With the discovery of a variety of synthetic and natural fluorophores and the advancement of specific labeling technologies, fluorescence microscopy has become an essential tool in modern biology and biomedical sciences. Although traditional fluorescence microscopy, like other conventional optical microscopy techniques, cannot provide resolution below the diffraction limit, some new fluorescence imaging techniques have broken the limit. Among these techniques, fluorescence (Förster) resonance energy transfer (FRET) allows investigation of biomolecular structure *in vivo* with a high temporal resolution when combined with stopped flow techniques [26,27] and provides a measurement of the distance between nearby donor and acceptor molecules in a distance range of about 10–100 Å [28]. However, the techniques previously discussed are not high throughput and therefore are not suited to study DNA sequence dependence, one of the most critical parameters influencing protein–DNA interactions. Traditionally, DNA microarray methods are used to study sequence dependence in genetic, epigenetic, transcriptional, and bioinformatics studies [29–35]. Short oligonucleotide probes, that is, 20–80 base pairs (bp), are spotted or synthesized *in situ* via surface engineering to allow covalent attachment of the probes to a solid surface. Detection and quantification of the binding of complementary sequences or protein molecules (targets) are done by measurement of relative fluorescent signals from the labels of the targets. For example, protein-binding microarrays (PBMs) have demonstrated a powerful capability to investigate DNA-binding affinity of proteins at the genome scale [29]. However, there is still an unsolved challenge to accurately quantify conformations of DNA strands and DNA conformational changes induced by specific protein binding in a high-throughput manner under physiological conditions.

Another novel technique of fluorescence microscopy is spectral self-interference fluorescence microscopy (SSFM), which uses optical interference to measure the conformation of surfaced-immobilized biomolecules. From the spectral oscillations emitted by an ensemble of fluorophores located above a reflecting interface (SiO_2–Si interface), the average height of the fluorophores relative to the surface can be determined with sub-nanometer resolution across a broad range from a few nanometers to more than 100 nm [36–39]. Early fluorescence interference microscopy techniques, such as fluorescence interference contrast (FLIC) microscopy, rely on intensity variations of total fluorescent emission. Utilizing FLIC for a high-precision determination of axial distance of fluorophore layers requires multiple measurements and accurate fabrication of spacer steps for use as a reference [40,41]. In contrast, SSFM utilizes spectral information of fluorophore emission on a thick oxide and provides higher precision in one measurement. Measurements of SSFM are immune from factors that can affect the fluorescence intensity, such as the excitation field strength, local fluorophore density, photobleaching, or change of buffer environment. Through the integration of DNA microarray technology and SSFM, we have developed a high-throughput platform that allows precise quantification of intrinsic conformations and conformational changes of surface-immobilized DNA in a physiological-like environment. Such a high-throughput approach increases efficiency and decreases the cost (time and money) required to investigate specific protein–DNA interactions.

In this chapter, we first introduce the fundamentals, principles, and data analysis of SSFM. Then we discuss some of the applications in biosensing and bimolecular studies on the SSFM platform. First, SSFM is used to determine the distance of monolayers of fluorophores relative to a layered substrate with sub-nanometer accuracy. In the next example, SSFM is combined with white light reflectance spectroscopy (WLRS) to measure conformations of surface-immobilized oligonucleotides. The third application demonstrates the usage of SSFM for direct observation of conformational change of a 3D polymeric coating. In the fourth example, the SSFM substrate is functionalized with a novel polymeric surface to achieve control and quantify double-stranded DNA (dsDNA) orientation. In the fifth example, DNA conformational changes upon binding with proteins are measured with SSFM. In the last example, SSFM is developed to accurately measure two axial positions at one location. Finally, we will discuss the future applications of the high-throughput platform in studying DNA–protein interactions.

9.2 Fundamentals of SSFM

9.2.1 Background and Principles

Fluorescence is a process in which an electron from a molecule, atom, or some nanostructures relaxes from an excited state to a ground state by emitting a photon. Molecules that are capable of fluorescence after the electronic transition are often named as fluorescent probes, fluorochromes, dyes, or fluorophores. In this chapter, we use fluorophores to refer to these types of molecules. The fluorescence process mainly consists of three events: excitation from a ground state to an excited state by absorbing an incoming photon, which occurs in femtoseconds; internal conversion, in which the absorbed energy is converted to heat by non-radiative decay from the initial excited state to the lowest vibrational energy level of the excited states in picoseconds; and the return of the molecule to the ground state during which a photon is emitted, which happens in relatively long periods of time, that is, nanoseconds. Since the emitted photons have less energy than the absorbed photon, the fluorescence light is shifted to longer wavelengths. This phenomenon is called the Stokes shift. In fact, the emission spectrum of a fluorophore is often a mirror image of its absorption spectrum, but at lower energies.

Fluorescence microscopy is one of the most powerful tools in modern biological research. The potential of fluorescence as a contrast agent for microscopy was realized by August Köhler in 1904 while he was working on improving the resolution of microscopes by illuminating with short-wavelength light. He realized that biological materials fluoresce when illuminated with ultraviolet (UV) light [42]. The application of a variety of fluorophores has made it possible to identify labeled subcellular components noninvasively and with a high degree of specificity. Like other conventional light microscopy, fluorescence microscopy is constrained by the limitation of spatial resolution due to diffraction. The minimum lateral resolution is proportional to wavelength λ and inversely proportional to numerical aperture (NA), i.e., NA = $n\sin(\theta)$, where n is the refractive index in the object space and θ is the half angle of the largest cone of rays that can enter or leave the optical system. The spatial resolution of conventional high-resolution optical microscopes is not better than about 200 nm laterally and about 600 nm longitudinally. Several methods have achieved spatial resolution beyond the diffraction limit utilizing the specific nature of fluorescence, such as increasing the effective NA (as in 4Pi confocal microscopy) [43–46], introducing spatial variation in the excitation light creating finer spatial features in the image (as in standing wave microscopy) [47–49], using multiple-photon fluorescence absorption or emission mechanisms that lead to nonlinear effects in the light field (as in two-photon microscopy) [50], and selectively quenching the fluorescence from a focal spot to obtain a very small fluorescing volume (as in stimulated emission depletion microscopy) [51–53]. Localization of fluorescent molecules with high accuracy is also of great interest and may provide valuable information not accessible even by high-resolution imaging. Two techniques, stochastic optical reconstruction microscopy (STORM) [54] and photoactivated localization microscopy (PALM) [55], exploit parallel localization of sparse emitters by estimating the center of the emission point spread function of the emitters. Fluorescence emission can also be localized in the axial dimension, and nanometer-scale precision can be achieved by utilizing self-interference of light.

It has been known for more than a century that fluorescence emission is modified by nearby dielectric or metal surfaces [56–58]. In the early 1960s, Drexhage determined the fluorescence decay of an organic dye embedded into a lipid layer as a function of its distance above a silver mirror [59]. Later, Lambacher and Fromherz [41] developed a fluorescent localization technique, FLIC microscopy, based on the change in the total emission intensity as a function of distance from a nearby reflecting substrate. The fluorescently labeled object is within an λ of vertical distance from a reflecting surface, so there is little difference between standing waves for different wavelengths within the emission range. This causes the entire emission spectrum to oscillate as the direct and reflected emitted light undergoes constructive or destructive interference depending on the vertical distance. Careful calibration of fluorescence intensity as a function of monolayer distance from the surface is required to achieve nanometer accuracy [40,60] and to measure cellular membranes [61] and molecular motors [62].

When the separation between a fluorophore and the mirror is larger than 10λ, even at the same height, the interference between the direct optical path of fluorescent emission and the reflected path from the SiO_2–Si interface results in several oscillations or fringes of interference within the fluorophore emission spectrum. The oscillation in the emission spectrum is a unique signature of the optical distance of the fluorophore to the interface. SSFM utilizes this interference-based modulation of the emission spectrum to determine the precise location of fluorophores above a layered surface (SiO_2–Si) with sub-nanometer accuracy. Small height differences produce shifts in the fringes and changes in the period of oscillation, although the latter are less apparent (Figure 9.1). Here, we note again that SSFM determines the axial location of fluorophores from the spectral oscillations, not from fluorescence intensity variations; therefore, experimental conditions that can influence fluorescence intensity do not affect the result.

9.2.2 Physical Model of SSFM

We can model an emitting fluorophore as an oscillating dipole with a random orientation above the SiO_2–Si substrate [36]. All three vectors, the emitter

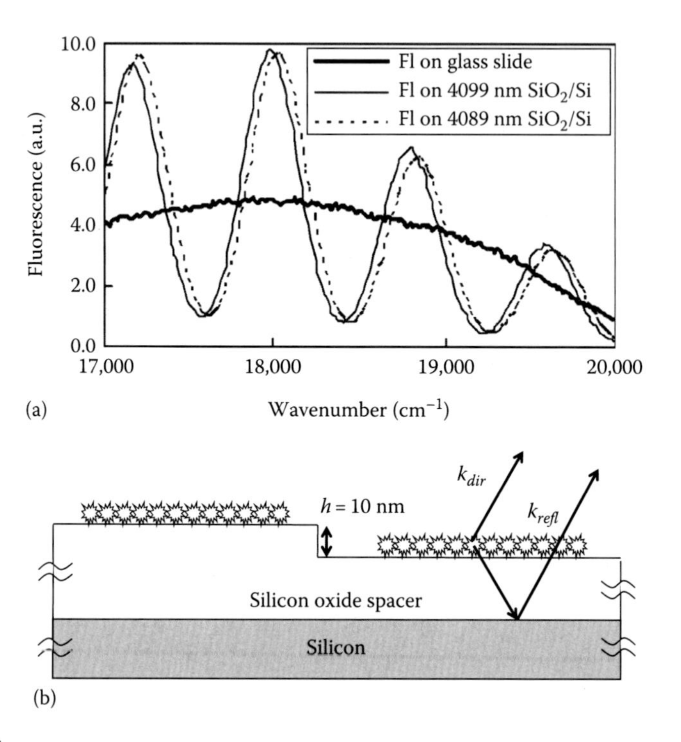

FIGURE 9.1
(a) Emission spectra of monolayer of fluorescein immobilized on a glass slide and on top of a Si–SiO_2 substrate with two different thicknesses of the oxide layer (10-nm difference). (b) Schematic of the Si–SiO_2 substrate (not to scale). (Reprinted with permission from Moiseev, L., Cantor, C.R., Aksun, M.I. et al., Spectral self-interference fluorescence microscopy, *J. Appl. Phys.*, 96(9), 5311–5315. Copyright 2004, American Institute of Physics.)

μ-dipole moment, the wave vector k, and the electric field vector E, lie in the same plane; this is the plane of polarization of the emitted light (Figure 9.2a). If the environmental factors remain constant, the far-field amplitude of the electric field of a fluorescently emitted wave is proportional to the sine of the angle between the dipole and the wave vector. The emission is therefore nonuniform; a 3D picture of the emission profile of a classical dipole has a donut shape (Figure 9.2b).

To describe the emission pattern of a dipole above a reflecting surface, we consider the intensity and polarization of both the direct and reflected waves. Two coherent waves are radiated from each dipole: one goes directly to the observation point (detector) and the other is incident on the mirror and reflected, propagates in parallel with the direct wave in the far field, and arrives at the same spot on the detector. We further describe the model based on two assumptions: First, we ignore near-field radiation because the SiO_2–Si reflecting interface is far away from the dipole. Second, the observation point is in the far field. Therefore, calculation of electrical

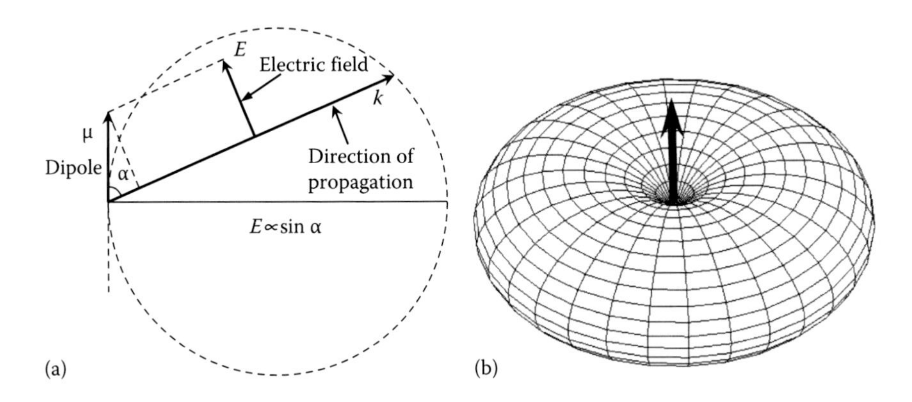

FIGURE 9.2
(a) Intensity and polarization of the electric field emitted by a classical dipole. The circle describes the electrical field amplitude of light emitted at a certain direction. (b) A 3D emission pattern of a classical electric dipole. (From Moiseev, L., Spectral self-interference fluorescence microscopy and its applications in biology, PhD dissertation, Boston University, Boston, MA, 2003.)

fields is sufficient to calculate total intensity, and only one significant plane wave component is used to describe the electrical field in any direction. Reflections of plane waves at the interface are sensitive to polarization; therefore Fresnel reflection coefficients are used to describe reflection for each polarization.

Thus, the direct and reflected fields are deposed into two orthogonal components: transverse electric field E_{TE} and transverse magnetic field E_{TM}. We find the angular dependence of each field by decomposing the dipole moment into three orthogonal components such that each component lies either within or perpendicular to the plane of incidence. E_{TE} is generated by the perpendicular component and E_{TM} is generated by the two in-plane orthogonal components for both direct and incident fields. The reflected fields are incident fields modified by Fresnel reflection coefficients R_{TE} and R_{TM} at the interface.

Figure 9.3 illustrates the angle of the dipole θ as well as the plane of incidence defined by the polar angle θ_{em} and azimuthal angle φ of the observation point and z-axis. The propagation directions of the direct, incident, and reflected waves all lie in the plane of incidence. Therefore, the angular dependence of E_{TE} and E_{TM} can be presented as follows:

$$E_{TE}^{dir} = E_{inc}^{dir} \propto \sin\theta \sin\varphi,$$

$$E_{TE}^{refl} = E_{TE}^{inc} R_{TE} e^{i2\phi},$$

$$E_{TM}^{dir} \propto \cos\theta_{em} \sin\theta \cos\varphi - \sin\theta_{em} \cos\theta,$$

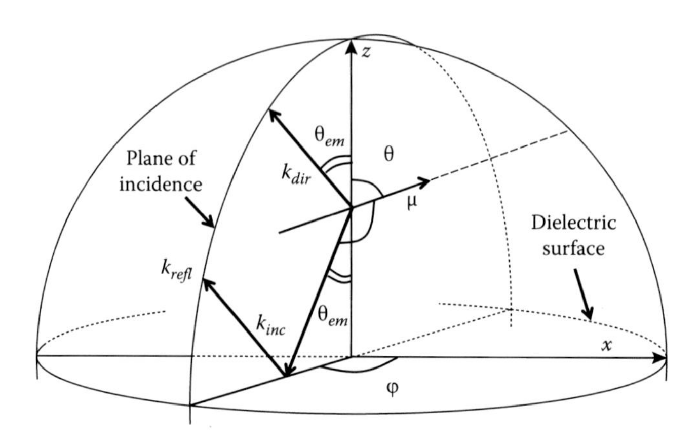

FIGURE 9.3
Dipole emission model showing the direct, incident, and reflected waves k_{inc}, k_{dir}, and k_{ref}. The emitting dipole is located in the x–z plane. (Reprinted with permission from Moiseev, L., Cantor, C.R., Aksun, M.I. et al., Spectral self-interference fluorescence microscopy, *J. Appl. Phys.*, 96(9), 5311–5315. Copyright 2004, American Institute of Physics.)

$$E_{TM}^{inc} \propto \cos\theta_{em}\sin\theta\cos\varphi - \sin\theta_{em}\cos\theta,$$

$$E_{TM}^{refl} = E_{TM}^{inc} R_{TM} e^{i2\phi},$$

where

$\phi = (2\pi n/\lambda) h cos\theta_{em}$

h is the axial position of the dipole

R_{TE} and R_{TM} are Fresnel coefficients for E_{TE} and E_{TM} that take into account the thickness of the SiO_2 spacer layer and the wavelength-dependent refractive indices of Si and SiO_2. In SSFM, since it is the spectral shape of fluorescent emission that is important rather than absolute emission intensity, we only consider the angular dependence of the electrical fields and use the proportionality sign (∝). The total intensity can be represented by the absolute square of the total electrical fields at the observation or detection point in the far field. The intensity is given as

$$I = |E_{TE}|^2 + |E_{TM}|^2,$$

where

$$E_{TM} = E_{TM}^{dir} + E_{TM}^{refl}.$$

The previously mentioned calculation of a dipole emission above a reflecting surface is for a specific direction characterized by θ_{em} and φ. In the applications of SSFM, samples contain thin layers of fluorophores that are assumed to be randomly distributed. Thus, the total emission of a monolayer of random-oriented dipoles should also be integrated over all possible angles of φ and θ. However, the range of polar tilt angles can sometimes be restricted. We should also integrate over $\sin\theta d\theta$ for light collected by a microscope objective with the maximum collection angle θ_{em}^{max}. As a result, the total emission of a monolayer of random dipoles measured with an objective with maximum collection angle θ_{em}^{max} is

$$I_{total} = \int_{\theta=0}^{\pi/2} \int_{\varphi=0}^{\pi} \int_{\theta_{em}}^{\theta_{em}^{max}} I(\theta, \varphi, \theta_{em}) sin\theta_{em} d\theta_{em} d\varphi d\theta.$$

The model is further modified if there are additional layers between the dipole and the microscope objective. The dipole can be considered as residing in a cavity, affecting both the direct and reflected fields in the same way, and the emission of a dipole in such an environment may be modeled in two steps: first, we calculate the total fields without the overhead layers as $E_0 = E_{dir} + E_{refl}$, and then, the resulting field should be multiplied by the "cavity coefficient" caused by multiple reflections from the top and bottom interfaces inside the cavity and the transmission to the objective. The total field at the objective is

$$E = \frac{E_0 T'}{1 - R'R^{i2\phi}}, \quad \phi = \frac{4\pi n}{\lambda} D \cos\theta,$$

where R' and T' are the generalized reflection and transmission coefficients for the layers above the fluorophore in the direction toward the objective, R is the generalized reflection coefficient below the fluorophore, and D and n are the thickness and refractive index of the cavity.

The previously mentioned expression needs to be calculated for each wavelength to express the spectral oscillations. The final SSFM spectrum is composed of the envelope of the free-space fluorescence spectrum and the oscillatory modulation determined by the axial position. We will discuss the interpretation of self-interference spectra of SSFM and the fitting algorithm in Section 9.3.

9.2.3 White Light Reflectance Spectroscopy

WLRS offers great precision for determination of the thickness of a transparent spacer layer on a reflecting surface [63–69]. A broadband light source is focused onto the surface and reflected light is collected. The interference fringes in the collected spectrum are created by waves reflecting from the top and bottom interfaces in contrast to the direct and reflected waves in SSFM (Figure 9.4). The resulting oscillations are based on the total reflectivity of the stack of dielectric layers. With WLRS, 10–20λ thick films of transparent materials can be measured with 1–2 Å precision.

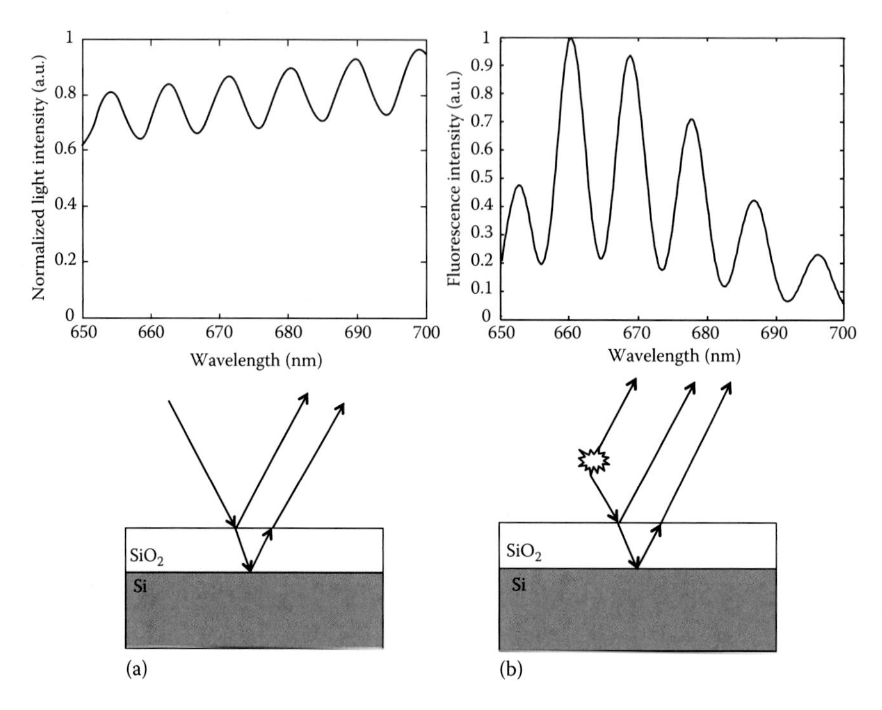

FIGURE 9.4
(a) WLRS is based on spectral variations of reflection from thin transparent films. Interference of light reflected from the top surface and a buried reference surface results in periodic oscillations. (b) SSFM maps the spectral oscillations emitted by a fluorophore located on a layered reflecting surface into a precise position determination.

WLRS is added to the SSFM to determine the amounts of biomass on the surface by measuring the change of the thickness of the spacer layer. For both SSFM and WLRS, we can only determine optical path length, which is the product of refractive index and physical path length. Deposition of DNA and protein molecules on the surface could change the refractive index, and we should regard the change of thickness with caution. Often lower surface density of molecules results in lower refractive index, and for our cases we can assume the refractive index of the biomaterial to be close to that of SiO_2 [70–74]. However, the additional amount of biomaterial on the surface can always be quantified accurately by the change of optical path length.

It needs to be noted that in SSFM, the amplitude of the direct wave and the reflected wave from the SiO_2–Si interface is comparable. However, for WLRS, the reflection from air–SiO_2 or buffer–SiO_2 interface is much less than that of the SiO_2–Si interface, resulting in a reduction of fringe contrasts (Figure 9.4). Since the white light source is external, we can increase light intensity to achieve sufficient spectral fringe contrast not to affect the precision of optical thickness determination. We can quantify the amount of biomass and the

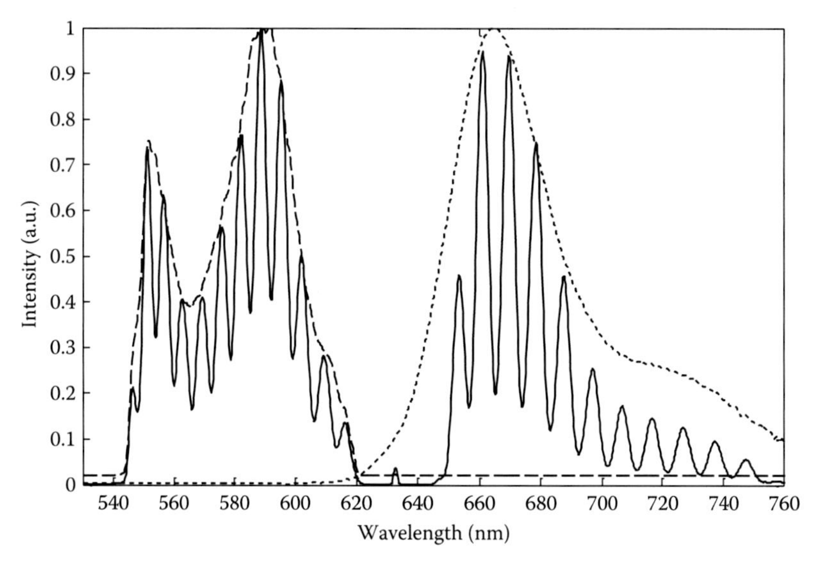

FIGURE 9.5
Interference spectra of white light and fluorescence measured at the same time. Dashed line is spectrum of filtered broadband LED, whereas the dotted line is emission spectrum of Atto647n given from the manufacturer.

height of a monolayer of fluorophores from one spectral measurement by looking at different spectral windows for WLRS and SSFM. For example, we use a broadband LED that is passed through a band-pass filter and a thick spacer layer that results in enough periods of oscillation in each spectrum for data fitting (Figure 9.5).

9.3 Data Acquisition and Analysis

9.3.1 SSFM Setup

SSFM measurements were performed with a system that combines an upright Leica DM/LM microscope and a Renishaw 100B micro-Raman spectrometer. Figure 9.6 is a schematic illustration of the recent SSFM setup built in-house with purchased optical components and instruments. A helium–neon laser with center wavelength of 632.8 nm is used to excite red fluorophores, such as Alexa Fluor 647 and Atto647n, and a diode-pumped solid-state green laser with center wavelength of 532 nm was used to excite green fluorophores, such as Atto532. Dichroics are used in place of beam splitters to maximize excitation and collection efficiency of fluorescence. A 2-axis positioning micro-stage is used to scan samples with a minimum step

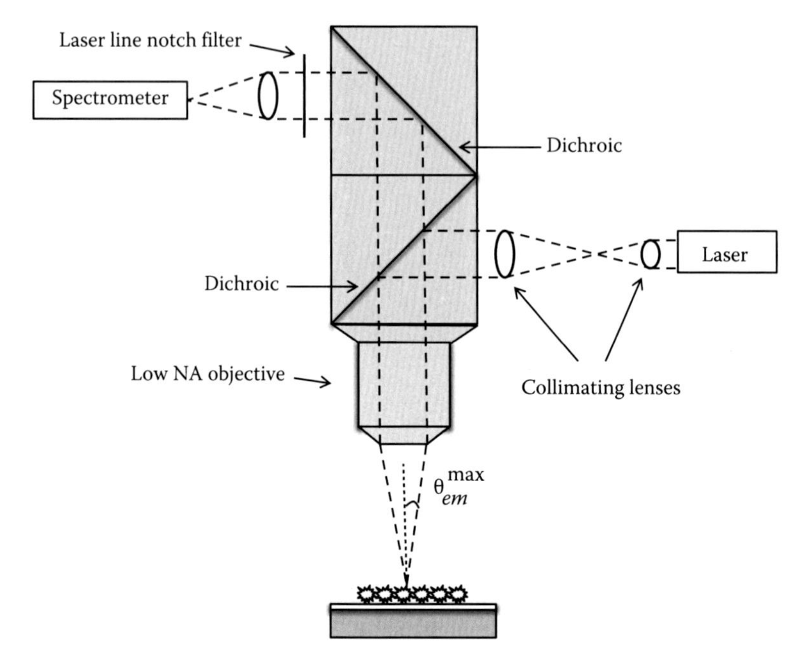

FIGURE 9.6
Microscope setup for SSFM.

size of 190 nm and a scanning range of 2.54 cm for each axis. The emission from the fluorophores is collected with a 5× (NA = 0.13) objective and transmitted through notch filters, which block the excitation laser's wavelength, and is then focused into the slit of the spectrometer, which has an air-cooled CCD camera as the detector. For WLRS measurements, we used normal Köhler illumination with a halogen lamp. This has been replaced with a combination of high-power green–yellow LED and a band-pass filter to allow simultaneous acquisition. Data can be collected manually or automatically using a custom-built application, which controls the lasers, LED, the microstep stage, and the spectrometer.

9.3.2 Fitting Algorithm

Typical SSFM spectra are composed of three parts: the spectral envelope represented by the fluorophore emission profile, the oscillatory interference component, and the noise from the spectrometer. The envelope can be described by a low-order polynomial. Both WLRS and SSFM spectra are fitted with custom algorithms executed with MATLAB®. The algorithms take into account the variation of refractive index of SiO_2 within the wavelength span and the complex reflectivity of the underlying Si. Dipole orientation can also be set as a parameter for fitting. In the fitting procedure, an estimated initial axial position is used to generate an oscillatory curve using the described physical model. Then, the spectrum is divided by the generated curve resulting in the envelope curve, which is fit to a low-order polynomial. If the position is not accurate, the envelope curve is not decoupled from oscillations. The final position is determined iteratively until the envelope spectrum is smooth, free from oscillations, and can be fit to a low-order polynomial with minimum least square error.

The fitting procedure is very fast, which enables real-time feedback within a second during an experiment and allows thousands of spectral measurements to be fit in a short time. In most of our applications, the only variable of interest is the height of a monolayer of fluorophores with respect to the surface. In order to increase the processing speed, the spectral envelope is independently fit with low-order polynomials using linear algebra, leaving the height as the only variable. This procedure dramatically expedited the fitting process and makes the fitting algorithm immune to spectral modifications and any potential quenching of the spectrum.

9.4 Applications

Example 9.1: Determine the Axial Locations of Monolayers of Fluorophores on SiO_2–Si Substrates [36]

To demonstrate the validity of SSFM, the positions of monolayers of fluorescein and quantum dots (QDs) above a silicon mirror were measured. Fluorescein isothiocyanate was immobilized on an aminosilane aminopropyltriethoxysilane (APTES)-covered SiO_2–Si substrate with the thickness of SiO_2 spacer layer being approximately 5 μm. A monolayer of QDs was also prepared on another chip by first silanizing the surface with APTES. QDs were then treated with mercaptoacetic acid to make them hydrophilic and negatively charged at neutral pH, which enables them to electrostatically attach to the aminated surface [75]. The optical thickness of the oxide layer was determined by WLRS. Axial locations of the fluorophores are measured by SSFM regardless of their density. Dipole orientations are assumed to be isotropic in the fitting algorithm. Figure 9.7 shows the schematics of the samples. The WLRS measurement of the chip shows only an additional few angstrom indicating a sparse layer, while SSFM shows the average height of the emitters above the surface to be 3 nm, about half the size of a QD. Compared with smaller molecules such as fluorescein, SSFM determined its position to be within 1 nm from the surface. Hence, axial location of a monolayer of fluorophores can be determined relative to an interface within a few angstrom, even for a sparse layer of fluorophores.

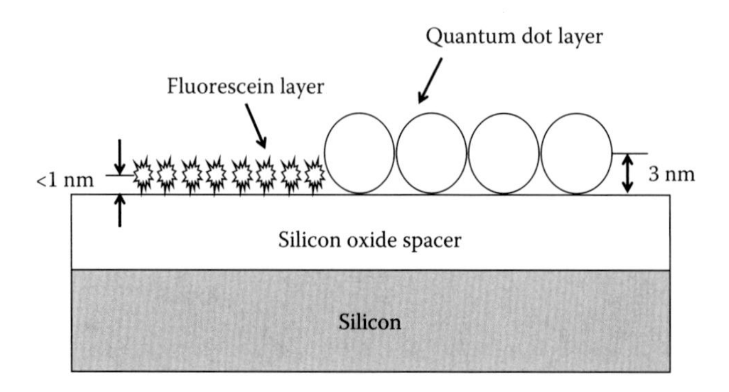

FIGURE 9.7
The average heights of fluorescent emitters above the surface of a Si–SiO_2 substrate. (Figure modified from Moiseev, L., Cantor, C.R., Aksun, M.I. et al., Spectral self-interference fluorescence microscopy, *J. Appl. Phys.*, 96(9), 5311–5315. Copyright 2004, American Institute of Physics.)

Example 9.2: Estimate the Conformation of Surface-Immobilized DNA [39]

DNA microarrays have been widely used in gene expression profiling, biomarker detection, drug discovery, single-nucleotide polymorphism (SNP) detection, and sequencing, all benefiting from the high-throughput capacity of the technology [76–78]. A routinely used detection process of DNA microarrays and other biosensing technologies is the hybridization of surface-immobilized single-stranded DNA (ssDNA) with solution-phase complementary strands or the binding of dsDNA to target protein molecules. The distance of the probe to the solid surface, the surface hydrophobicity and charge, and physical steric hindrance of the probes influence the accessibility of surface-immobilized molecules for contact with target molecules in solution. The closer the probe molecules are to the solid support, the less likely for the target to reach the probe diffusively. Hydrophobic surfaces can act like a shield for bound molecules positioned close to it because of the associated steric factors and lack of diffusion of the bound molecules [79–83]. A charged surface can either repel or attract molecules through electrostatic interactions [84–86]. It has been known that high DNA probe density does affect hybridization kinetics and efficiency, largely due to electrostatic repulsion between DNA strands and steric hindrance [87–89]. Thus, understanding the conformation of surface-bound DNA probes is of great value not only for the future development of DNA microarrays but also for other nanotechnologies that utilize surface-immobilized DNA oligomers as sensing or actuation agents [90–92].

Some methods were proposed to characterize surface-immobilized DNA probes, such as ellipsometry, optical reflectivity, neutron reflectivity, x-ray photoelectron spectroscopy, FRET, surface plasmon resonance (SPR), and AFM [39]. Most of these experimental techniques characterized the DNA layer by studying its thickness or density. Previous studies helped visualize the conformation of surface-immobilized DNA depending on its density and surface charge. The advantage of SSFM combined with WLRS measurements is that it not only measures the optical thickness or the bound mass of DNA layers, but it can also provide insight into the conformations of DNA by measuring axial positions of fluorophores that are tagged to a certain position of the DNA strands.

We studied the conformation of surface-immobilized ssDNA and dsDNA by using 50 and 21 bp long oligonucleotides. The 5′ end of the first strand of the DNA is amine modified and is covalently linked to the SiO_2 surface via a homobifunctional cross-linker. Fluorophores were tagged either at the 3′ end of the immobilized first strand or at the 3′ or 5′ end of the second complementary strand.

WLRS measurements showed that immobilization of 21 and 50 bp long DNA results in additional optical thickness of 1.0–1.5 and 2.0–2.5 nm, respectively, assuming a refractive index of 1.46 for the DNA [70–74]. As we discussed previously, although the absolute thickness depends on the refractive index, WLRS provides an accurate relative measure of the additional biomass by detecting the difference of optical path length, from which we can monitor DNA hybridization. Adding complementary second strands to 50 bp ssDNA results in an increase in the film thickness by approximately 1.0 nm corresponding to a hybridization efficiency of approximately 50%. The optical thicknesses of the layers are used in SSFM measurements to determine the axial location or the height of the fluorophores. Here, we note that the axial locations of the fluorophores are described as the heights of fluorophores above the surface in SSFM data analysis. The average height values measured by SSFM are an indication of different distributions of fluorophore heights within the microscope focal spot. Figure 9.8a shows the incomplete hybridization for the 50 bp ssDNA with the complementary strands labeled either at the 5′ (distal) end or at the 3′ (proximal) end. In principle, the maximum heights of the fluorophores are constrained by the contour length of the dsDNA, which are approximately 7 and 17 nm for 21 and 50 bp long dsDNA fragments, respectively. The average heights of the distal end fluorophores are 5.5 nm for 21 bp and 10.5 nm for 50 bp dsDNA, whereas those of the proximal end are 1 nm for 21 bp and 2.5 nm for 50 bp dsDNA [39].

The persistence length of dsDNA is 50 nm in physiological conditions [93–95], so the short dsDNA fragments in our experiments are considered as rigid rods hinged to the surface. Assuming that these rods can freely rotate on hinges, the average angle of the rods to the surface can be calculated from the average distal height and the length of the rods in this case. A more detailed analysis of the calculation of average angle will be discussed in the next example, where the DNA orientation is additionally influenced by external electrostatic forces. Thus, the angles of the 21 and 50 bp long dsDNA to the surface are approximately 40° and 50°, respectively. We also measured the fluorophore heights tagged on the surface-immobilized strand at the 3′ (distal) end to study the conformation of ssDNA (Figure 9.8b). Unlike dsDNA, ssDNA is very flexible, often described as random coils [96–99], and little is known about its conformation on surfaces. The height of the fluorophore is about 1 nm for 21 bp ssDNA and 5.5 nm for 50 bp ssDNA, which implies a considerably more extended conformation for the 50 bp ssDNA. When a second

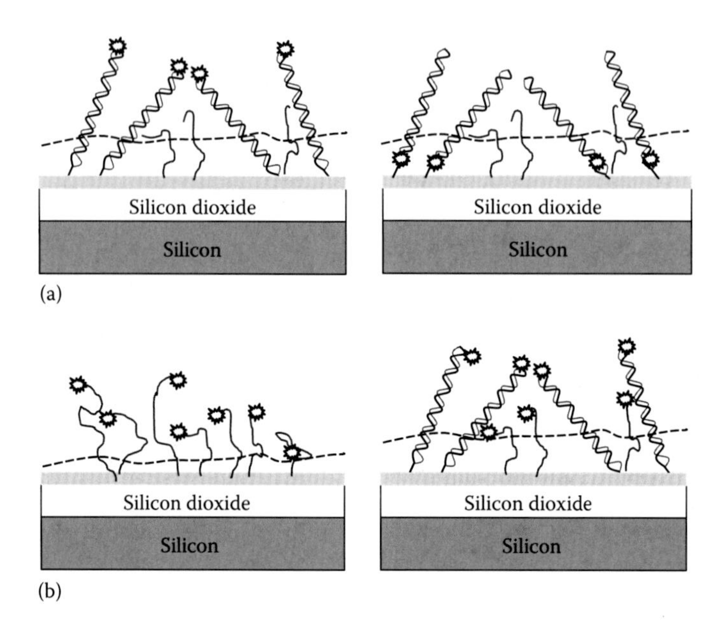

FIGURE 9.8
(a) Schematic illustration of incomplete hybridization of surface-immobilized ssDNA with complementary strand labeled at the 5′ (distal) end (left) and 3′ (proximal) end (right). The SiO_2 surface is functionalized with a saline layer plus phenylene isothiocyanate, which adds about 1.5 nm (light gray layer on top of SiO_2). Dashed lines indicate the thin biomolecule layer whose thickness is obtained by WLRS measurements. (b) ssDNA is labeled at the 3′ (distal) end and covalently bound to the surface (left), and hybridized with non-labeled complementary strand (right). Dashed lines indicate that the thickness of biomolecule layer is increased after hybridization. (Adapted from Moiseev, L. et al., *Proc. Natl. Acad. Sci.*, 103, 2623, 2006.)

unlabeled complementary strand is hybridized, the surface DNA molecules are composed of two species, unhybridized ssDNA and hybridized dsDNA, both with labels at the distal ends, the average height of which should be somewhere between the distal fluorophore heights of ssDNA and dsDNA. The hybridization efficiency can also be estimated from comparing the average height of the first strand or the second strand both labeled at the distal end, which is around 30%–50% in this experiment and is consistent with WLRS measurements. The results demonstrate that SSFM can be used to estimate the conformation of surface-bound ssDNA and dsDNA: ssDNA can be flexibly coiled on the surface, whereas dsDNA is rigid and angled from the surface.

Example 9.3: Direct Observation of Conformational Change of a 3D Polymeric Coating

Surface functionalization is one of the most important components of a solid surface-based biosensor, such as DNA microarrays and SPR biosensors. The direct contact of molecules to the interface as well as other local steric conditions will affect the conformations of the biomolecules, which potentially influence their functions in binding. The ideal surface chemistry for the functionalization of the solid support should have functional groups for probe attachment, minimal nonspecific adsorption, and stability to environmental changes, and for practical applications, it should be low cost, robust, and easily prepared. More importantly, the surface chemistry should keep probe functionality after immobilization for efficient target capture. Among the existing surface chemistries, 3D polymeric surface coatings are the most promising in meeting these criteria. One such 3D polymeric coating, copoly(DMA–NAS–MAPS), has been used in DNA and protein microarray studies [100] and is now commercially available. Copoly(DMA–NAS–MAPS) has improved performance in terms of probe density and hybridization efficiency of DNA microarrays than organosilanization-based microarray surface chemistries [101]. The understanding of the conformation of the surface-adsorbed copoly in both air and solution is extremely important for interpretation and optimization of the observed superior performance.

SSFM was successfully used to quantify the conformation, specifically the swelling of the surface-adsorbed polymer upon hydration [102]. Short DNA oligonucleotides (23 mer), both ssDNA and dsDNA, were used as probes to measure the swelling of the polymer in the axial direction. ssDNA and dsDNA are labeled with fluorophores at one end and modified with amine groups at the other end, which covalently link to the functional groups (NHS esters) in the polymer. An increase in the axial position of the fluorophores measured by SSFM indicates an increase in the length of polymer chains due to hydration (swelling). SSFM and WLRS measurements are performed before and after hydration of the surface. For labeled ssDNA immobilized on the polymer surface, the measured fluorophore height increase upon hydration (Figure 9.9, Case A) is ~7.5 nm, comparing to no significant change (~0.5 nm) for the same probes on non-swelling epoxysilanized surface, indicating that the polymer swells axially upon hydration. For labeled dsDNA immobilized on the surface (Figure 9.9, Case C), a higher difference in averaged fluorophore height (~14 nm) before and after hydration was seen compared to that of labeled ssDNA (Case A, ~7.5 nm). This indicates that before hydration, the immobilized dsDNA is collapsed onto the surface. After hydration, dsDNA is oriented on the surface, and, therefore, the fluorophore height is higher than that of flexible, randomly coiled ssDNA in Case A. When the labeled ssDNA probes are hybridized with complementary strands on surface, the measured average fluorophore height increase upon hydration becomes ~8 nm (Figure 9.9, Case B). Furthermore, unlabeled ssDNA are hybridized with labeled complementary strand on surface (Figure 9.9, Case D). The measured ~17 nm fluorophore height increase in this case is much higher than that of the hybridization of labeled ssDNA (Case B, 8 nm) and also higher than that of the immobilized dsDNA (Case C, ~14 nm). The difference of the fluorophore heights between Case B and Case D indicates that not all ssDNA are hybridized in the experimental condition. In Case B, the signal originates from the spotted labeled ssDNA and the measured fluorophore height is an ensemble average of the ssDNA and hybridized dsDNA. In Case D, fluorescence signal is from the labeled complementary strands, and the average fluorophore height is only of the hybridized

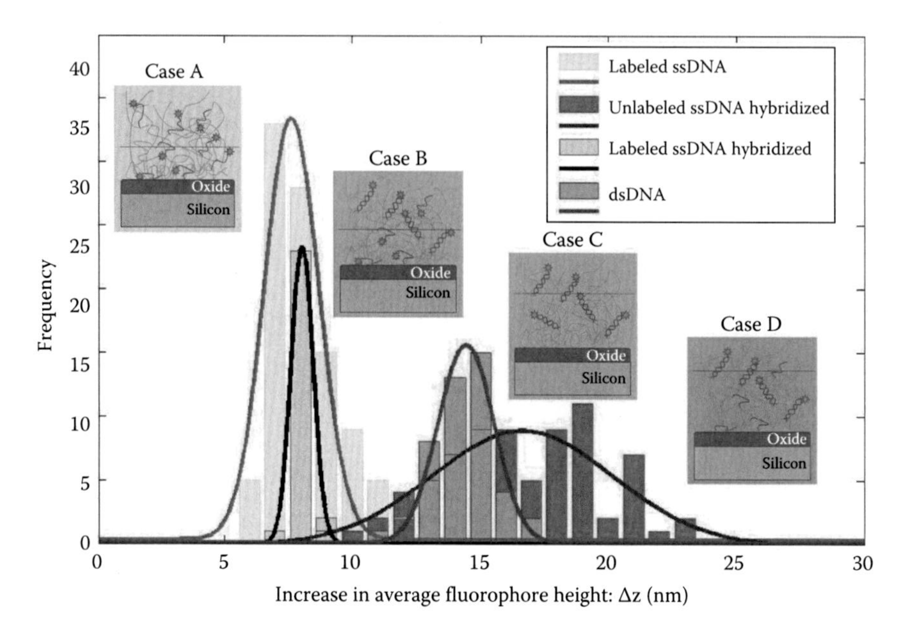

FIGURE 9.9
(See color insert.) Histograms summarizing the increase of fluorophore heights measured at spots of labeled ssDNA (Case A), labeled ssDNA hybridized with non-labeled complementary sequences on surface (Case B), labeled dsDNA (Case C), and unlabeled ssDNA hybridized on surface with labeled complementary sequences on surface (Case D). Each histogram contains results of tens of spots and is fitted with a Gaussian curve, the mean values of which indicate the average increase of fluorophore heights. Animated illustration above each histogram shows the experimental conditions of each case. (Reprinted with permission from Yalçın, A., Damin, F., Özkumur, E. et al. Direct observation of conformation of a polymeric coating with implications in microarray applications, *Anal. Chem.*, 81(2), 625–630. Copyright 2009, American Chemical Society.)

dsDNA. The difference of the measured fluorophore heights between Case C vs. Case D and Case B vs. Case D suggests that the likelihood of hybridization to probes further away from the surface is higher than to the probes located closer to the surface. WLRS technique shows no significant change of the optical thickness upon hydration of the polymer. This is expected because there is no additional mass introduced to the surface during the swelling of the polymeric surface. The information from this example suggests that the control and manipulation of the conformation of the polymeric coating will have a significant impact in improving the hybridization efficiency of microarrays. These results also indicate that the probes are distributed within the hydrated 3D polymer structure in the axial direction. The study of the conformation of 3D polymeric coatings helped understand and optimize the copolymer and strengthened the earlier analysis of this polymeric coating as a high-quality microarray surface chemistry. Characterization of existing surface chemistries and developing new ones with the help of SSFM will enable designing platforms of optimized surface chemistry, probe immobilization, and target detection for DNA microarray and biosensing applications.

Example 9.4: Manipulation and Quantification of dsDNA Orientation

In recent years, researchers have been exploiting surface-immobilized DNA oligomers for a variety of applications in nanotechnology, such as biosensing and DNA switches and motors [90–92], taking advantage of its sequence designability, chemical stability, and automatic complementary base-pairing property. For biosensing and DNA-based molecular devices, sensitive detection and accurate actuation requires ordering and cooperation of the molecules, such as directed movements and simultaneous response of the molecules [90–92,103,104]. In Example 9.2, we discussed the measurement of surface-immobilized dsDNA conformation, in which the DNA is almost randomly oriented to the surface due to stochastic thermal motions. If we want to detect protein-induced conformational changes of dsDNA, such as the binding of integration host factor (IHF) protein, which causes a ~160° kink of the dsDNA at the binding site [105], such disordering of dsDNA molecules on the surface will make it very difficult to accurately quantify conformational changes specific to IHF. Other platforms, such as "switchable DNA" [106], induce ordering of DNA monolayers immobilized on gold electrodes by applying alternating electric fields. On the switched DNA platform, negatively charged short oligonucleotides are switched between lying and upright positions on oscillations of positive and negative surface potentials. The switching dynamics are sensitive to the intrinsic molecular properties of oligonucleotides, such as structural flexibility, and can be characterized and used as a kinetic signature for detection of hybridization and binding events [105,107].

A similar concept of electric-field-induced ordering is adopted with the SSFM platform, but we utilize a novel electromechanical approach to orient dsDNA on the SiO_2 surface. We functionalize the surface with a highly amphoteric polymer that adopts a net negative or positive charge depending on the buffer pH [108–110]. The polymer also has N-hydroxysuccinimide ester (NHS ester) groups that covalently bond with amine-modified oligonucleotides. The isoelectric point of the polymer was tested to be around pH 6 by electroosmotic flow in a capillary coated by a polymer with identical composition. Lower buffer pH results in a net positively charged surface, attracting dsDNA to the surface, whereas higher buffer pH results in a net negatively charged surface, repelling dsDNA to a higher orientation to the surface [111]. An overview of the manipulation of dsDNA orientation on a charged polymer surface through adjustment of buffer pH and ionic strength is shown in Figure 9.10a.

The controlled orientation of dsDNA on the charged polymer can be interpreted by considering the electrical DNA switching on a gold surface previously mentioned, which was studied extensively by Rant and coworkers [112–115]. A diffuse double layer of counterions accumulates at the interface of a charged surface and an electrolyte solution. An intense electric field (~100 kV/cm) results near the ionic buffer–polymer interface due to the high concentration gradient of mobile charges that accumulate. The characteristic length of this electric field (on the scale of nanometers) is inversely proportional to the square root of buffer salt

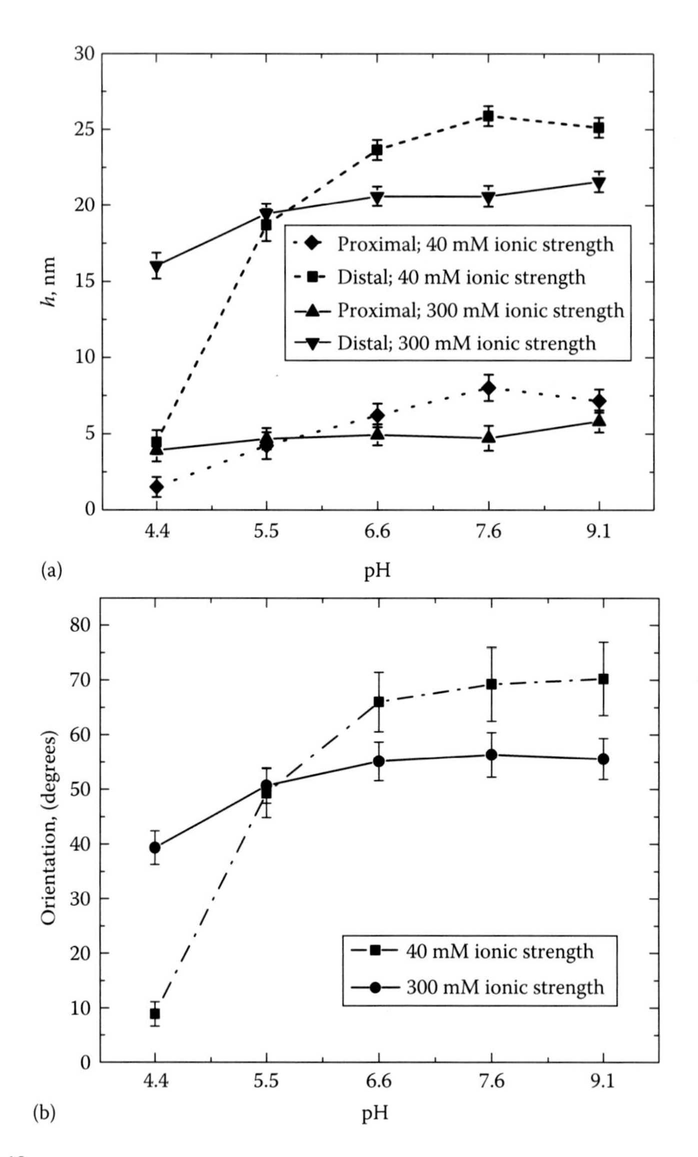

FIGURE 9.10
(a) SSFM measurements of 60 bp dsDNA on charged polymer surface. Surface-proximal labeled dsDNA indicate polymer swelling: positively charged polymer collapses onto the negatively charged oxide surface, and negatively charged polymer is repelled from the oxide surface. Fixed charges within the polymer also repel each other and contribute to polymer swelling. The surface-distal and surface-proximal fluorophore heights allow precise quantification of dsDNA orientation: the dsDNA is oriented in a lying position by the positively charged polymer and in a standing position by the negatively charged polymer. (b) Calculated orientations at each pH for NaCl concentrations at 40 and 300 mM. A low salt concentration allows the electric field to penetrate far from the charged polymer to more effectively orient immobilized dsDNA.

concentration. At low ionic strength (10 mM), for instance, the thickness of the double layer is around 3 nm, whereas at high ionic strength (300 mM), it is about 0.6 nm. Thus, the electric field does not cover the entire length of the dsDNA probes, and the electrostatic interactions are confined to the base of the probes. As a result, the charged polymer surface orients the immobilized dsDNA more effectively in low ionic buffers because the electrostatic force is applied to a larger proportion of the dsDNA (Figure 9.10b). To give a more quantitative view of the control of the orientation of surface-immobilized dsDNA by charged polymer surface, we refer to the model proposed by Rant and coworkers [112,113]. The adaptation of the model helps us understand the relationship between the measured average height and average orientation on the SSFM platform. The model also presents a quantitative analysis of the dsDNA probe orientation over a wide range of salt concentrations, which gives further insight into the behavior of the polymeric system.

First, we regard the polymer surface as a charged plane, the surface potential of which varies with buffer pH. Since dsDNA is inherently negatively charged in electrolyte solution, a negatively charged surface repels dsDNA and a positively charged surface attracts dsDNA. This is the principle of induced ordering of oligonucleotides for both gold and charged polymer surfaces. In electrolyte solution, the surface electrostatic potential is screened by redistributed ions and, according to the Gouy–Chapman theory, reduces as a function of distance to the surface with a characteristic Debye length. The Debye length l_D is defined as

$$l_D^2 = \frac{\varepsilon\varepsilon_0 kT}{2ne^2}$$

where

ε is the dielectric constant of ionic solution
ε_0 is the permittivity of vacuum
k is the Boltzmann constant
T is the temperature
e is the elementary charge
n is the ion density

The Gouy–Chapman equation describes the diffusive potential distribution along the z-axis:

$$\Phi(z) = \frac{2kT}{e}\ln\left(\frac{1+\gamma e^{-\frac{z}{l_D}}}{1-\gamma e^{-\frac{z}{l_D}}}\right), \quad \gamma = \tanh\left(\frac{e\Phi_0}{4kT}\right)$$

where Φ_0 is the surface potential. Since the contour length of short (<80 bp) hybridized oligonucleotides is much shorter than the persistent length of dsDNA (~50 nm) [93–95], they are modeled as rigid rods with equally spaced (0.34 nm) negative point charges, which can freely rotate around the anchor point on the surface (Figure 9.11). A charge of −0.24 e [116] is assumed per point charge along dsDNA to account for the counterion condensation effect [117]. The electrostatic energy of charges along

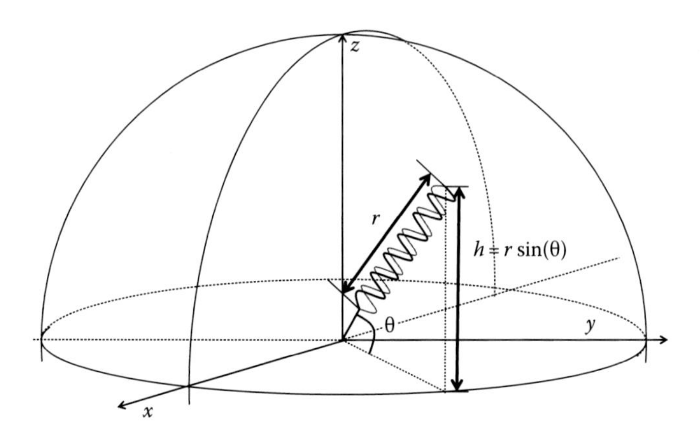

FIGURE 9.11
dsDNA is modeled as a rigid rod with an orientation θ to the surface. *r* is the root-mean-square of the end-to-end distance of dsDNA under experimental conditions. *h* is the distance of surface distal end of dsDNA to the surface along *z*-axis. The half sphere is drawn to illustrate all the possible states occupied by the distal end of dsDNA by rotating around the anchor point on the surface.

the dsDNA depends on their axial location relative to the surface, which is defined as $h = r\sin\theta$, where θ is the orientation of dsDNA to the surface and *r* is the root-mean-square end-to-end length of the DNA fragment, $r = \sqrt{2l^2((L/1) - 1 + e^{-L/l})}$, where *l* is the persistent length of dsDNA and *L* is the contour length of the dsDNA [118].

Thus, the electrostatic energy of each rod is a function of θ and is calculated as the sum of the electrostatic energy of all the charges [113]:

$$E(\theta) = E_{(\Phi_0, l_D)}(\theta) = \sum_i q_{eff}\Phi(z_i)$$

The average orientation of dsDNA is a balance between thermal stochastic 3D rotations and the electrostatic force on the dsDNA.

To calculate the average orientation ⟨θ⟩, without any external electrostatic forces, considering a dsDNA as a rigid rod that rotates freely around the anchor point, we have

$$\langle\theta\rangle = \int_0^{\pi/2} \theta f(\theta)\, d\theta = 33^\circ$$

where $f(\theta)$ is the probability distribution function (PDF) of θ

$$f(\theta) = \frac{\cos(\theta)}{\int_0^{\frac{\pi}{2}} \cos(\theta)\, d\theta} = \cos(\theta), \quad \left(0 \le \theta \le \frac{\pi}{2}\right)$$

In the presence of varying electric potential, the average orientation is calculated by taking the Boltzmann distribution to calculate the probability density of all the energy levels [113]; thus

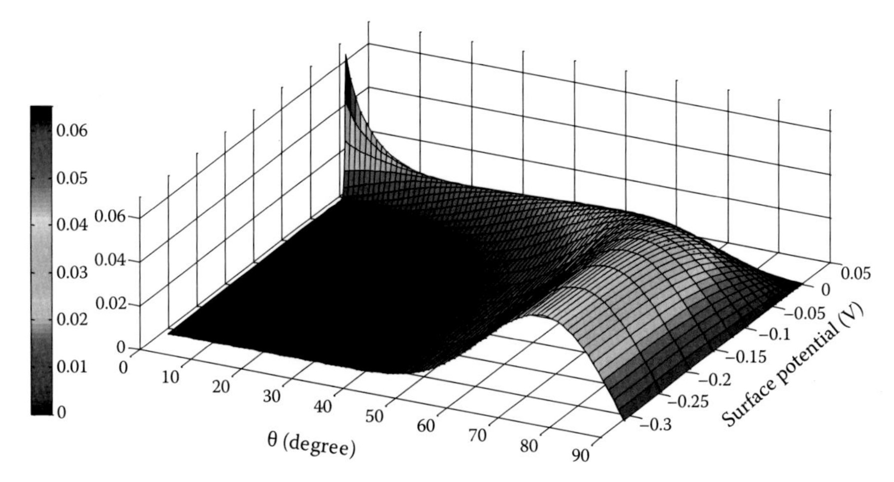

FIGURE 9.12
(See color insert.) Probability density functions (PDFs) of dsDNA orientation θ for different surface potentials. Positive potentials pull the dsDNA to the surface, resulting in most dsDNA occupying the lower orientation states. The PDFs are shifted to higher degrees as the surface potential becomes more negative.

$$\langle\theta\rangle = \frac{\int_0^{\pi/2} \theta\cos(\theta)e^{-\Delta E(\theta)/kT}d\theta}{\int_0^{\frac{\pi}{2}} \cos(\theta)e^{-\Delta E(\theta)/kT}d\theta}$$

$\cos(\theta)$ is the degeneracy factor, representing the stochastic rotations as degenerate microstates, which occupy the same energy level with the same orientation. The potential of a rod for a given orientation is calculated as $\Delta E(\theta) = E(\theta) - E(90°)$, where the standing rod ($\theta = 90°$) is taken as the reference state. Figure 9.12 shows the PDFs of dsDNA orientations for different surface potentials, Φ_0. We can see that when Φ_0 is positive, most of the dsDNA are at lower orientations, and when surface potential is above a critical positive value, almost all of the dsDNA will be pulled down within 2° from the surface with the higher orientation energy states less accessible and unpopulated. When Φ_0 goes from 0 to negative potentials, the PDFs shift to higher orientations with narrower distributions with the low orientation energy states being less probable.

In the experimental data of SSFM, we approximate the mean of dsDNA orientation, $\langle\theta\rangle$, as $a\sin(\langle h\rangle/r)$. To determine the validity of the approximation, we calculate the average of axial height of the dsDNA:

$$\langle h\rangle = \frac{\int_0^{\pi/2} r\sin(\theta)\cos(\theta)e^{-\Delta E(\theta)/kT}d\theta}{\int_0^{\frac{\pi}{2}} \cos(\theta)e^{-\Delta E(\theta)/kT}d\theta}$$

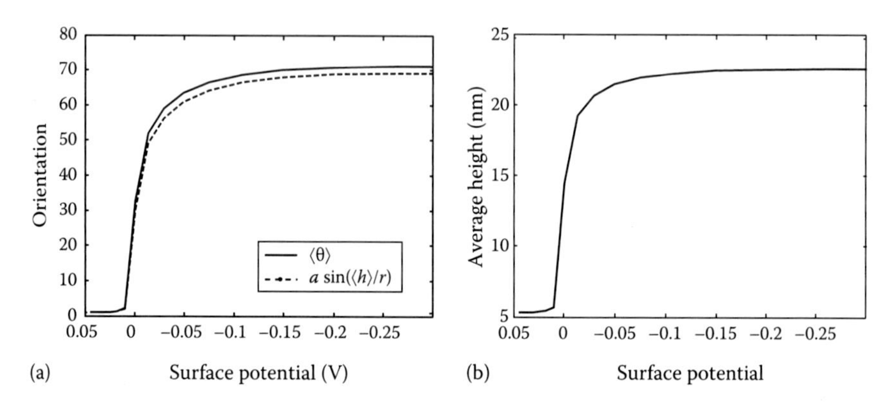

FIGURE 9.13
(a) Plotting mean of θ and calculated θ $\left(\theta_{cal} = a\sin\left(\langle h\rangle / r\right)\right)$ as a function of surface potential for 60 bp dsDNA. When the surface potential equals 0 V, ⟨θ⟩ is 33° corresponding to a stochastically rotating rod model. When the positive surface potential exceeds a particular value, most of dsDNA will be pulled down to the surface with the average orientation of less than 2° The PDF shifts less as the surface potential goes further negative and the mean of θ plateaus. (b) Average height of distal fluorophore of 60 bp dsDNA is plotted as a function of surface potential. The proximal end height is assumed to be 5 nm. Since h equals to $r\sin(\theta)$, it follows the same trend as θ, but plateaus earlier because sin(θ) varies less when θ is at higher degrees.

Figure 9.13a and b plot both theoretical values of ⟨θ⟩, calculated $\langle\theta\rangle_{cal}$, from average height $\left(\langle\theta\rangle_{cal} = a\sin\left(\langle h\rangle/r\right)\right)$ and average height as a function of surface potential, respectively. The analysis shows that our approximation for ⟨θ⟩ underestimates ⟨θ⟩ by 2–3°.

To give a further insight into the working mechanism of the charged polymer surface, we discuss a quantitative analysis and the experimental results of the dsDNA probe orientation over a wide range of salt concentrations. At pH 7.6, when the charged polymer is negatively charged, the measured orientation matches well with the calculated results at low ionic strength, with the maximum orientation exceeding 70° (Figure 9.14). At high ionic strength the effects of negative charge on the polymer become negligible, but the average dsDNA orientation approaches 45°, while the expected average orientation is 33° as calculated previously. The 45° orientation of probes at high ionic strength indicates steric repulsion between the polymer and the base of the dsDNA and between the negatively charged dsDNA themselves. At pH 4.4, when the polymer surface is positively charged, a nearly lying position of dsDNA is induced at low ionic strengths, whereas the orientation once again approaches 45° for high ionic buffer strengths.

In the theoretical model, the steric effects between the polymer scaffold and the dsDNA are not included in the calculations. Thus, the calculated orientations deviate from experimental results and approach 0° for a positively charged surface at low ionic strength and 33° at high ionic strengths. However, the model accurately predicts the ionic strength at which the dsDNA sharply transits from a random orientation to a lying orientation when the polymer is positively charged. This relationship

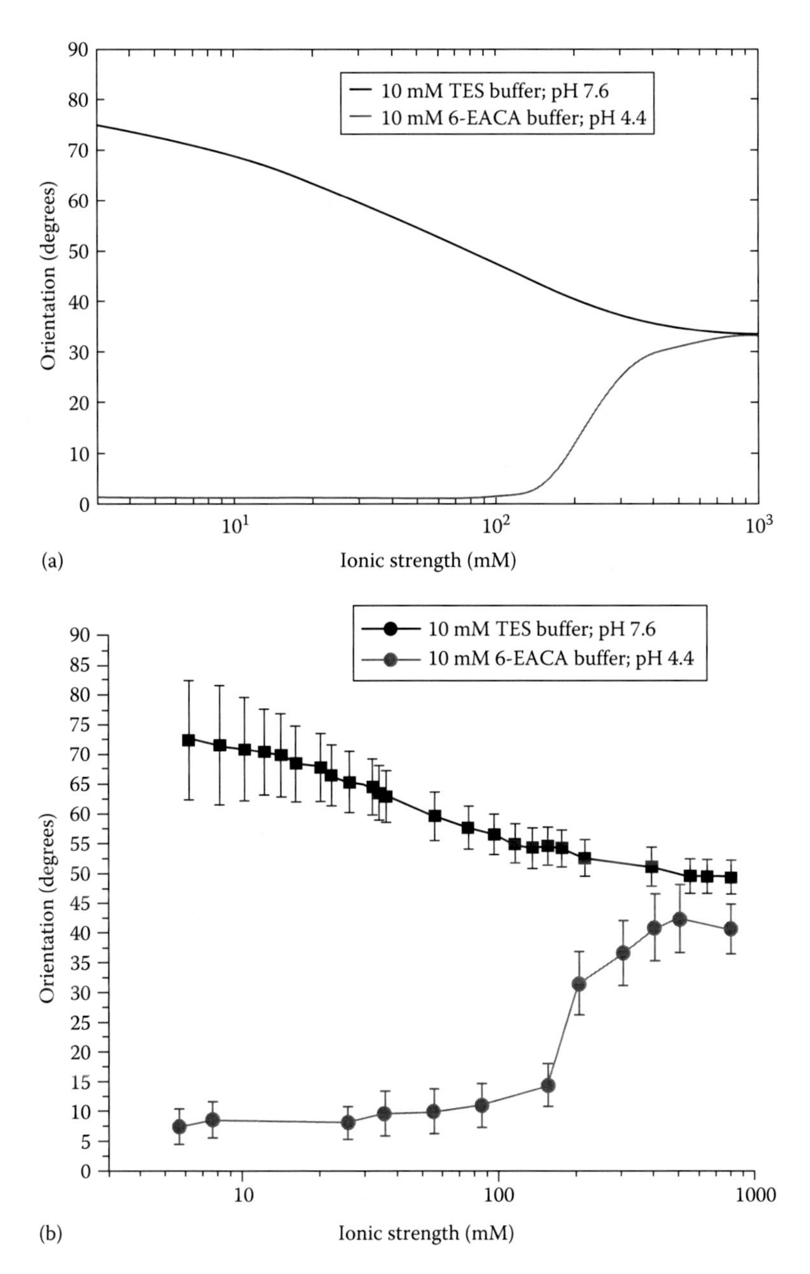

FIGURE 9.14
(a) Calculated orientations of 60 bp dsDNA probes on a charged planar surface as a function of ionic strength. The steric effects between the polymer scaffold and the dsDNA are not included. The surface potential is taken as −100 mV. (b) The measured orientation of 60 bp dsDNA probes for negatively charged (pH 7.6) and positively charged (pH 4.4) polymeric surfaces at a wide range of concentrations of NaCl.

between the probe orientation and the ionic strength reflects the different mechanisms by which DNA are attracted to a positively charged surface as opposed to repelled from a negatively charged surface [115]. As discussed previously, the average orientation measured by SSFM is a balance between the electrostatic force and stochastic thermal motion. On the positively charged polymer surface, when buffer ionic strength transits from high to low, the Debye length is increased and the electrostatic energy is higher for all dsDNA orientations. In the beginning, the attracting electrostatic force does not exceed the thermal motions of dsDNA for most of the orientational states. As the buffer ionic strength goes down further, the dsDNA, whose orientation passes a threshold where the attracting electrostatic force dominates over the thermal motion are pulled down and captured [115]. For SSFM steady-state ensemble measurements, the measured orientation is a weighted average of the orientations of the captured and un-captured probes. As the ionic strength decreases beneath the threshold value, the electrostatic force is large enough to effectively capture a majority of the dsDNA probes. Therefore, the calculated and experimental average orientations are both seen to switch from freely rotating to a horizontal orientation.

Example 9.5: Detection of Protein-Induced Conformational Changes of dsDNA

The ability to control and quantify surface-immobilized dsDNA conformation makes it possible for us to detect specific conformational changes of dsDNA upon binding with proteins in their native environment on SSFM. The advantages of SSFM over other conventional techniques were reviewed in the beginning of this chapter. Here, we use the protein IHF of *Escherichia coli* (*E. coli*) to demonstrate the methodology of SSFM for detecting protein-binding-induced DNA structure distortions.

IHF, a ~20 kDa heterodimeric DNA-binding protein, has been shown to induce a specific bend of dsDNA of more than 140° when estimated by gel mobility assays [119] and approximately 160° by x-ray crystallography [10]. IHF is not only required for the integrative recombination of lambda phage DNA into the bacteria genome, it is also involved in diverse cellular functions, such as DNA replication, transcription regulation, and genome packaging [120–122]. IHF recognizes its binding sequence through recognizing the intrinsic conformation (narrowed minor groove) of dsDNA ("indirect readout") [10,123–125] and functions by causing a sharp bending of dsDNA, bringing separated sequences together [126,127]. It has been shown that IHF binds to its specific sequence with 10^3–10^4 times higher affinity than with nonspecific sequences [125,128,129]. Figure 9.15 shows schematically the detection mechanism on SSFM. We performed a titration experiment and measured the average height of distal fluorophores on 60 bp dsDNA probes with sequentially increasing concentrations of wild-type IHF (Figure 9.16a). Little conformational change was detected for the negative control sequence, and a larger height change was observed in the sequence "IHF 34" versus sequence "IHF 44" (sequences are named by

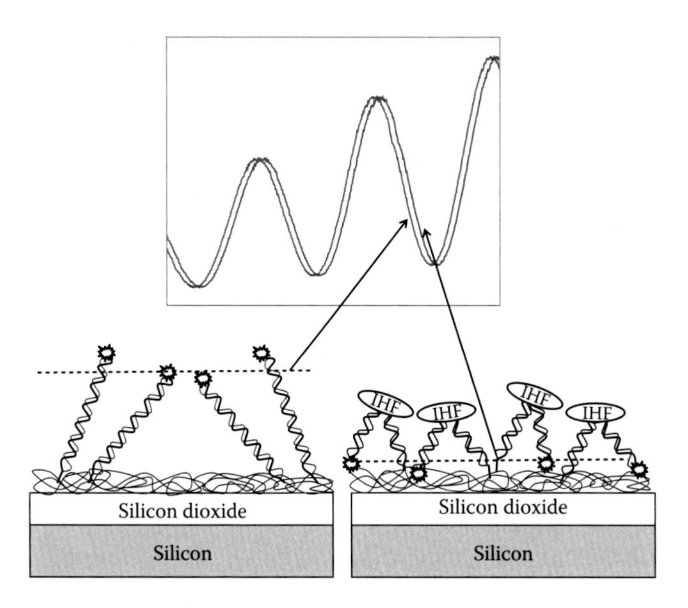

FIGURE 9.15
Schematic illustration of the mechanism for the detection of dsDNA conformational change upon binding with IHF on SSFM. Dotted lines are the estimation of average fluorophore heights. The average fluorophore height changes when dsDNA bends after binding to IHF, resulting in a shift in the fluorescence spectrum.

the position where the IHF consensus binding sequence starts from the surface-proximal end of dsDNA). This is expected because the lower the binding sequence, the larger the change of the fluorophore heights upon binding. Figure 9.16b shows the sequence dependence of distal fluorophore height change after IHF binding, where we can distinguish a two-bp shift of binding locations. Since the charged polymer surfaces are used to covalently link dsDNA and anchor them to the sensor surface and the laser local spot of SSFM is about 10 μm, we can spot or synthesize thousands of different DNA sequences in an array format to study sequence specificity in parallel.

Example 9.6: Apply Dual-Color SSFM to Determine Two Axial Locations of Surface-Immobilized DNA

SSFM utilizes spectral oscillations to locate the distance of a monolayer of fluorophore above a mirror. If the fluorophores are distributed along the *z*-axis, the measured fluorophore height is the average of all the fluorophores within the focal spot. Theoretically, SSFM can be extended to multiple tags in different spectral windows to determine the axial locations of multilayers of fluorophores, each with a different color. The period of spectral oscillations is a function of the thickness of the spacer layer between the fluorophore and the mirror. Thus, we can engineer the thickness of the spacer so that we get sufficient periods of oscillations in

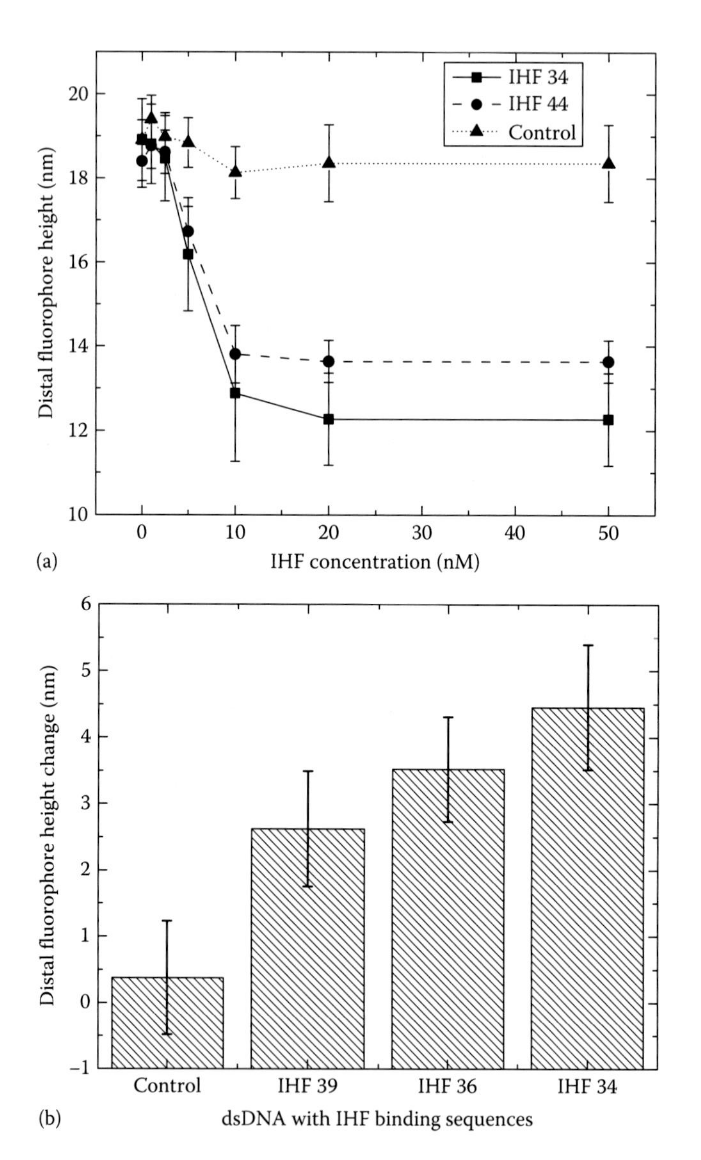

FIGURE 9.16
(a) SSFM measurements of distal fluorophore heights in a titration experiment for two different 60 bp dsDNA. dsDNA is named by the start location of consensus IHF-binding sequence from the bottom of the dsDNA. For example, dsDNA "IHF 34," the IHF-binding sequence, starts at 34 bp away from the bottom. "IHF control" does not have IHF-binding sequence. (b) Distal fluorophore height changes after binding to IHF. Each column is the average height change of distal end of 20 spots of dsDNA spotted at 7.5 μM by Bio-Rad MiniArrayer. The lower localized binding sequence results in a larger height change. SSFM is able to distinguish 2 bp shifts in the binding location.

each spectral window for accurate fitting. Practical application is limited by the availability of thicknesses of SiO_2 (thick thermally grown SiO_2 layers can take months to grow), spectrally separated fluorophores, and multiband dichroics that correspond to the fluorophore spectral bands. As a proof of concept, we demonstrate co-localization of two axial positions of dsDNA using dual-color SSFM.

Figure 9.17a shows the spectra of two fluorophores: Atto532 and Atto647n. With the help of notch filters and multiband dichroics, we can measure the two spectra at the same time and do fitting with same initial estimation of the distance. One problem is that the tail of the spectrum Atto532 falls in the spectral window of Atto647n (shown in Figure 9.17a), potentially affecting the fitting results. More reliable dual-color measurements are taken sequentially. Due to intensity insensitivity, SSFM measurements are not subject to energy transfer and quenching of fluorophores.

Figure 9.17b shows the measurements of height difference between distal and proximal fluorophore heights on dsDNA. Previously, we have used the height difference between surface-distal fluorophore and surface-proximal fluorophore on dsDNA to quantify DNA conformation on polymer surfaces. Due to the swelling of the polymer upon hydration [102], DNA becomes elevated and fluorophores labeled at the proximal end are an indication of the average height of the polymer surface. Therefore, to determine the orientation of dsDNA, we measure fluorophore heights of two separate spots: one with distal fluorophores and the other with proximal fluorophores. Since the conformation of DNA is affected by surface density, if the two spots have different immobilization densities, the results may not be precise. Using dual-color SSFM, we determine the surface-proximal fluorophore height from the spectrum of Atto647n and surface-distal fluorophore height from the spectrum of Atto532. Thus, we can quantify DNA orientation from measurements of one spot. We measured the orientation of dsDNA spots with different immobilization densities on a neutral polymeric surface with dual-color SSFM (Figure 9.18). When the density of dsDNA layer is too high, negatively charged dsDNA molecules repel each other, resulting in higher orientations. At very low surface densities, the measured average orientation is lower than expected for random rotations. This phenomenon on this neutral charged polymer may be due to a slight positive charge resulting from the hydrolysis of some monomers in the polymer, but needs to be investigated further.

It has been well characterized that high surface density in DNA layers results in reduced efficiency for complementary sequences to penetrate the layer to hybridize with probe sequences [130,131]. However, little is known regarding the effect of dsDNA probe density on the binding rates with proteins. SSFM will be used to study the surface density effects on protein binding to surface-immobilized DNA layers, the results of which will be important for all surface-based biosensors. Moreover, the ability to determine the positions of two different fluorophores also extends the application of SSFM in studies of DNA–protein interactions, as we can label the protein with a fluorophore and determine the binding location of the protein on the DNA sequence as well as the conformational change of DNA from a different fluorophore tagged on the distal end of the DNA.

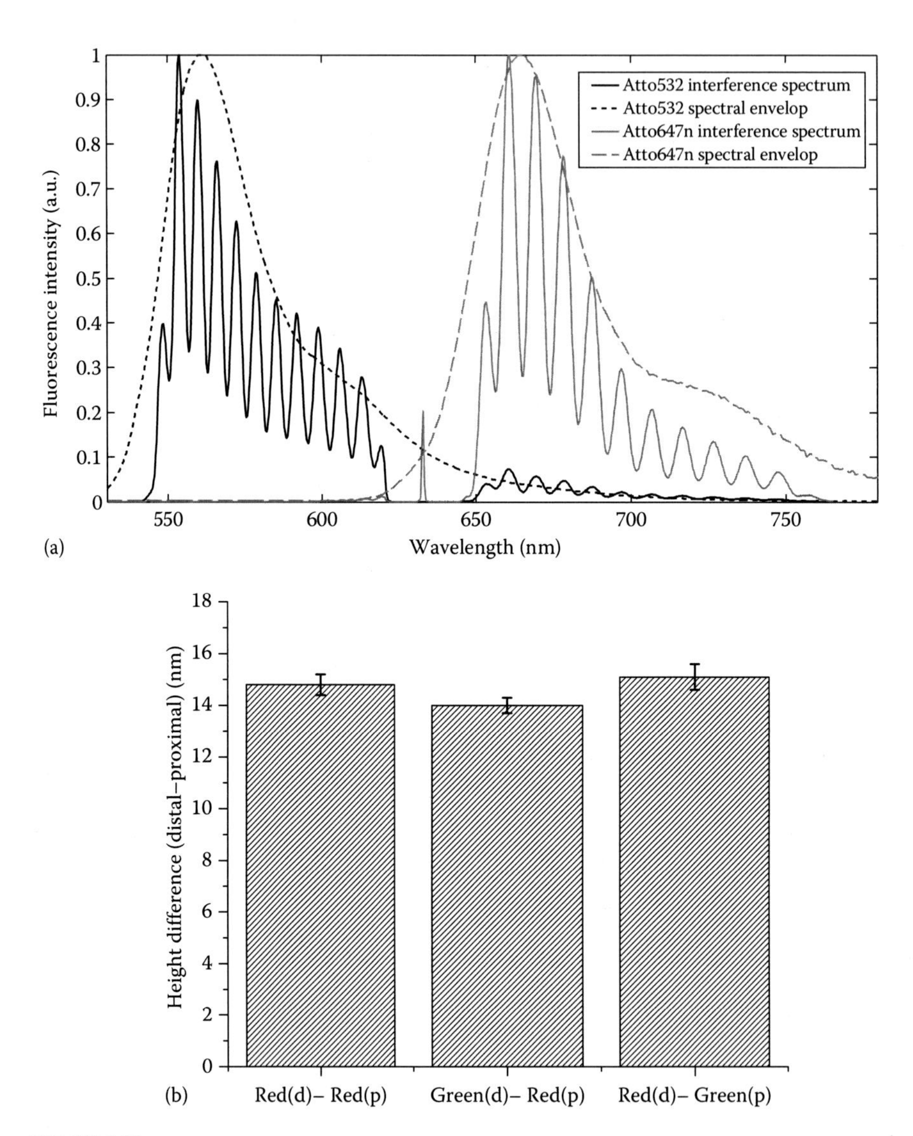

FIGURE 9.17
(a) SSFM measurements of fluorescence spectra of fluorophores Atto532 (green) and Atto647n (red) on a 17.5 μm thick SiO_2. Dotted lines are spectral envelopes given by manufacture. (b) The height difference between distal (d) end and proximal (p) end of 60-bp dsDNA measured by single-color SSFM and dual-color SSFM. For single-color SSFM, the height difference is obtained by measuring the distal fluorophore height (red (d)) of one dsDNA spot and proximal fluorophore (red (p)) height of another dsDNA. For dual-color SSFM, the height difference is obtained by measuring one dsDNA spot that is labeled with different fluorophores at each end, either the distal end with green fluorophores or the proximal end with red fluorophores, or *vice versa*. Each height difference is the average measurement of 15 dsDNA spots.

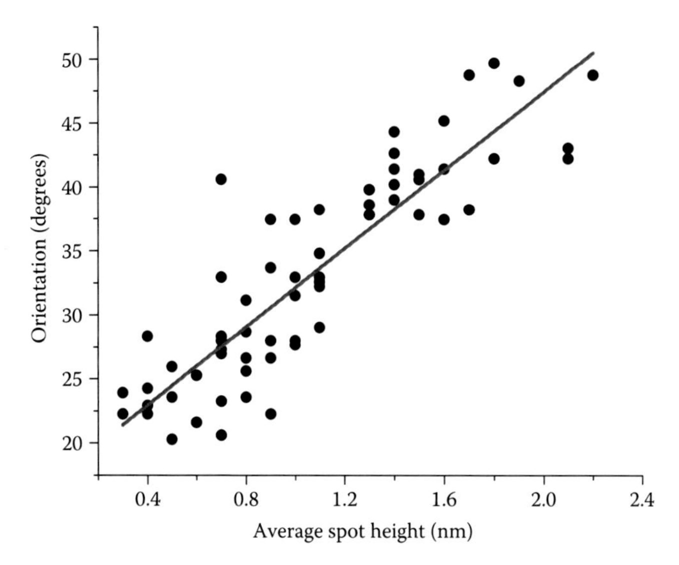

FIGURE 9.18
Measured average orientations of 67 dsDNA spots of different surface densities on a neutral polymeric surface with dual-color SSFM. The heights of optical thickness measurements by WLRS indicate the density of dsDNA layers. Higher surface densities result in higher orientations, showing steric repulsion between negatively charged dsDNA. The line is a guide to the eye.

9.5 Discussions and Conclusion

This chapter presents a novel self-interference method for determining the axial position of fluorescent markers with high precision. In practical applications, the most suitable systems for analysis with SSFM are monolayers of fluorophores on layered surfaces, such as surface-immobilized DNA monolayers tagged with fluorescent labels or monolayers of fluorescein in a lipid membrane or polymer scaffold on a solid substrate [38,39,127,128,132]. For example, in recent years, DNA hairpins and G-quadruplex have been used as nanoswitches for biosensing and molecular motors [133–139], and they are used for the detection of DNA in low concentrations and cancer biomarkers [103,134]. SSFM can be used to confirm the formation of hairpins of DNA microarrays or the detection of its conformational changes due to the occurrence of binding events. G-quadruplex is found to be formed in G-rich telomere repeats, can inhibit telomerase activity [140–143], and is an active target for drug discovery [144–149]. By careful design of the labeling locations of fluorophores, SSFM can be utilized to detect sequence dependence of ssDNA folding into G-quadruplex or the formation of G-quadruplex through binding with small molecules and proteins [150,151]. It should be noted that when labeled DNA is used as probe for detection on SSFM, no labeling is required for the target, which is still in its native state.

DNA conformational changes induced by DNA-binding proteins can be measured noninvasively in their native environment with SSFM. We give another example of a protein that can induce conformational changes in DNA, TATA-binding protein (TBP). TBP is an essential protein required for transcription initiation of all three eukaryote RNA polymerases [152–156]. TBP recognizes its target binding sites (TATA boxes) also by indirect read-out of the DNA conformation or deformation upon forming a DNA–protein complex [157,158]. In co-crystal structures, TBP was found to unwind the bound double helix by about 120° and cause the DNA to bend toward its major groove by 80° [156,159]. The bent angle is sequence dependent, ranging from <34° to 106° for variant "TATA-box" sequences [160], which is correlated with the stability of the complex and transcription activity [160,161]. Some other studies suggest that the flanking sequences of the TATA box also affect the stability of TBP binding to its target sites [158,162,163]. SSFM, for example, can be used to study the different factors that increase or reduce the stability of TBP–DNA complexes by measuring conformational changes of DNA in a high-throughput way. Combining WLRS and SSFM, we can also advance the study of TBP and its cellular function by studying the relationship between binding affinity and specific bending angles for numerous combinations of TATA-box sequences and its flanking sequences. As mentioned in the SSFM setup section, WLRS and SSFM measurements can be taken at the same time. Combined with custom-designed flow cells, SSFM can be used to perform dynamic measurements of both biomass accumulation and DNA conformational changes, making it possible to detect binding and bending simultaneously.

The versatility of the presented platform makes the system an advanced tool to study more complex DNA–protein interactions. For example, in DNA base excision repair (BER) [164–167], a damaged base is specifically recognized and removed by DNA glycosylate to generate an abasic site. Depending on the initial events, the repair patches may be single nucleotide (short patch) or 2–10 nucleotides (long patch). For the short patch repair, the phosphodiester bond at 3′ of the abasic site is cleaved by glycosylate, and the 5′ bond is incised by APE1 endonuclease, which then recruits DNA polymerase to fill the gap that is ligated by a DNA–ligase complex. During the process of BER, the conformation and flexibility of DNA changes with the binding of different enzymes, which kinks the DNA to different angles [168–175]. Through careful design of experiments, the SSFM platform can potentially detect DNA BER through direct detection of DNA conformational changes and protein-binding events. Monitoring DNA repair is an example of applying the SSFM platform to study more challenging DNA–protein interaction processes. The application of polymer-functionalized surfaces and microarray sensor formats adds the capacity of parallel detection of thousands of sequences. SSFM provides critical information required to move DNA interfacial applications forward. What's more, the quantification of DNA conformations and conformational changes, through integration with new surface functionalization techniques and label-free detection, provides critical information to understand DNA–protein interactions in their native environment, allowing the SSFM platform to play a unique and productive role in emerging biotechnological fields.

Acknowledgments

The authors would like to acknowledge National Science Foundation (grant 0933670) for funding support and George Daaboul and Abdulkadir Yurt for their contribution.

References

1. M. J. Khoury. Dealing with the evidence dilemma in genomics and personalized medicine. 2010. *Clin. Pharmacol. Ther.* 87(6): 635–638.
2. E. S. Lander. Initial impact of the sequencing of the human genome. 2011. *Nature* 470: 187–197.
3. G. Li and J. Widom. Nucleosomes facilitate their own invasion. 2004. *Nat. Struct. Mol. Biol.* 11(8): 763–769.
4. A. Portela and M. Esteller. Epigenetic modifications and human disease. 2010. *Nat. Biotechnol.* 28: 1057–1068
5. G. Badis, M. F. Berger, A. A. Philippakis et al. Diversity and complexity in DNA recognition by transcription factors. 2009. *Science* 324(5935): 1720–1723.
6. R. A. Bambara, R. S. Murante, and L. A. Henricksen. Enzymes and reactions at the eukaryotic DNA replication fork. 1997. *J. Bio. Chem.* 272: 4647–4650.
7. A. Sancar, L. A. Lindsey-Boltz, K. Ünsal-Kaçmaz et al. Molecular mechanisms of mammalian DNA repair and the DNA damage checkpoints. 2004. *Annu. Rev. Biochem.* 73: 39–85.
8. C. A. Hunter. Sequence-dependent DNA Structure: The role of base stacking interactions. 1993. *J. Mol. Bol.* 230(3): 1025–1054.
9. P. J. Hagerman. Sequence-directed curvature of DNA. 1986. *Nature* 321(6068): 449–450.
10. K. K Swinger and P. A. Rice. IHF and HU: Flexible architects of bent DNA. 2004. *Curr. Opin. Struct. Biol.* 14(1): 28–35.
11. R. Rohs, X. Jin, S. M. West et al. Origins of specificity in protein–DNA recognition. 2010. *Annu. Rev. Biochem.* 79: 233–269.
12. R. Rohs, S. M. West, A. Sosinsky et al. The role of DNA shape in protein–DNA recognition. 2009. *Nature* 461: 1248–1253.
13. D. B. Starr, B. C. Hoopes, D. K. Hawley et al. DNA bending is an important component of site-specific recognition by the TATA binding protein. 1995. *J. Mol. Bol.* 250(4): 434–446.

14. J. C., Becker, A. Nikroo, T. Brabletz et al. DNA loops induced by cooperative binding of transcriptional activator proteins and preinitiation complexes. 1995. *Proc. Natl. Acad. Sci.* 92: 9727–9731.
15. D. B. Nikolov, H. Chen, E. D. Halay et al. Crystal structure of a human TATA box-binding protein/TATA element complex. 1996. *Proc. Natl. Acad. Sci.* 93(10): 4862–4867.
16. J. Wu, K. M. Parkhurst, R. M. Powell et al. DNA bends in TATA-binding protein-TATA complexes in solution are DNA sequence-dependent. 2001. *J. Biol. Chem.* 276(18): 14614–14622.
17. A. Marcovitz and Y. Levy. Frustration in protein–DNA binding influences conformational switching and target search kinetics. 2011. *Proc. Natl. Acad. Sci.* 108(44): 17957–1796.
18. C. Roll, C. Ketterlé, V. Faibis et al. Conformations of nicked and gapped DNA structures by NMR and molecular dynamic simulations in water. 1998. *Biochemistry* 37(12): 4059–4070.
19. T. E. Cloutier and J. Widom. Spontaneous sharp bending of double-stranded DNA. 2004. *Mol. Cell.* 14: 355–362.
20. S. Geggier and A. Vologodskii. Sequence dependence of DNA bending rigidity. 2010. *Proc. Natl. Acad. Sci.* 107(35): 15421–15426.
21. Y. L. Lyubchenko and L. S. Shlyakhtenko. Visualization of supercoiled DNA with atomic force microscopy *in situ*. 1997. *Proc. Natl. Acad. Sci.* 94(2): 496–501.
22. H. G. Hansma, D. E. Laney, M. Bezanilla et al. Applications for atomic force microscopy of DNA. 1995. *Biophys. J.* 68(5): 1672–1677.
23. H. Yokota, D. A. Nickerson, B. J. Trask et al. Mapping a protein-binding site on straightened DNA by atomic force microscopy. 1998. *Anal. Biochem.* 264(2): 158–164.
24. D. Anselmetti, J. Fritz, B. Smith et al. Single molecule DNA biophysics with atomic force microscopy. 2000. *Single Mol.* 1: 53–58.
25. R. T. Dame, C. Wyman, and N. Goosen. Insights into the regulation of transcription by scanning force microscopy. 2003. *J. Microsc.* 212(3): 244–253.
26. M. J. Jezewska, R. Galletto, and W. Bujalowski. Dynamics of gapped DNA recognition by human polymerase β. 2002. *J. Biol. Chem.* 277: 20316–20327.
27. S. V. Kuznetsov, S. Sugimura, P. Vivas et al. Direct observation of DNA bending/unbending kinetics in complex with DNA-bending protein IHF. 2006. *Proc. Natl. Acad. Sci.* 103(49): 18515–18520.
28. A. Hillisch, M. Lorenz, and S. Diekmann. Recent advances in FRET: Distance determination in protein–DNA complexes. 2001. *Curr. Opin. Struct. Biol.* 11(2): 201–207.
29. N. P. Gerry, N. E. Witowski, J. Day et al. Universal DNA microarray method for multiplex detection of low abundance point mutations. 1999. *J. Mol. Biol.* 292(2): 251–262.
30. M. L. Bulyk. DNA microarray technologies for measuring protein–DNA interactions. 2006. *Curr. Opin. Biotechnol.* 17(4): 422–430.
31. S. Mukherjee, M. F. Berger, G. Jona et al. Rapid analysis of the DNA-binding specificities of transcription factors with DNA microarrays. 2004. *Nat. Genet.* 36(12): 1331–1339.
32. M. F. Berger, A. A. Philippakis, A. M. Qureshi et al. Compact, universal DNA microarrays to comprehensively determine transcription-factor binding site specificities. 2006. *Nat. Biotechnol.* 24(11): 1429–1435.

33. B. Ren, F. Robert, J. J. Wyrick et al. Genome-wide location and function of DNA binding proteins. 2000. *Science* 290(5500): 2306–2309.
34. J. D. Hoheisel. Microarray technology: Beyond transcript profiling and genotype analysis. 2006. *Nat. Rev. Genet.* 7: 200–210.
35. J. D. Lieb, S. Beck, M. L. Bulyk et al. Applying whole-genome studies of epigenetic regulation to study human disease. 2006. *Cytogenet. Genome. Res.* 114(1): 1–15.
36. L. Moiseev, C. R. Cantor, M. I. Aksun et al. Spectral self-interference fluorescence microscopy. 2004. *J. Appl. Phys.* 96(9): 5311–5315.
37. M. S. Ünlü, B. B. Goldberg, A. K. Swan et al. Spectroscopy of fluorescence for vertical sectioning. US Patent. September 2006. #7,110-118.
38. M. Doğan, A. Yalçin, S. Jain et al. Spectral self-interference fluorescence microscopy for subcellular imaging. 2008. *IEEE J. Sel. Top. Quant.* 14(1): 217–225.
39. L. Moiseev, M. S. Ünlü, A. K. Swan et al. DNA conformation on surfaces measured by fluorescence self-interference. 2006. *Proc. Natl. Acad. Sci.* 103: 2623–2628.
40. D. Braun and P. Fromherz. Fluorescence interferometry of neuronal cell adhesion on microstructured silicon. 1998. *Phys. Rev. Lett.* 81(23): 5241–5244.
41. A. Lambacher and P. Fromherz. Fluorescence interference-contrast microscopy on oxidized silicon using a monomolecular dye layer. 1996. *Appl. Phys. A* 63: 207–216.
42. A. Köhler. Microphotographische untersuchungen mit ultraviolettem licht. 1904. *Z. Wiss. Mikrosk.* 21: 129–165, 273–304.
43. S. W. Hell and E. H. K. Stelzer. Properties of a 4Pi confocal fluorescence microscope. 1992. *J. Opt. Soc. Am. A* 9: 2159–2166.
44. S. W. Hell, E. H. K. Stelzer, S. Lindek et al. Confocal microscopy with an increased detection aperture—Type-B 4Pi confocal microscopy. 1994. *Opt. Lett.* 19: 222–224.
45. M. Schrader, K. Bahlmann, G. Giese et al. 4Pi-confocal imaging in fixed biological specimens. 1998. *Biophys. J.* 75: 1659–1668.
46. M. Martinez-Corral. Effective axial resolution in single-photon 4Pi microscopy. 2002. *G.I.T. Imag. Microsc.* 2: 29–31.
47. B. Bailey, D. L. Farkas, D. L. Taylor et al. Enhancement of axial resolution in fluorescence microscopy by standing-wave excitation. 1993. *Nature* 366: 44–48.
48. V. Krishnamurthi, B. Bailey, and F. Lanni. Image processing in 3D standing-wave fluorescence microscopy. 1996. *Proc. SPIE* 2655: 18–25.
49. R. Heintzmann and C. Cremer. Laterally modulated excitation microscopy: Improvement of resolution using a diffraction grating. *Proc. SPIE* 3568: 185–196.
50. P. T. C. So, C. Y. Dong, B. R. Masters et al. Two-photon excitation fluorescence microscopy. 2000. *Ann. Rev. Biomed. Eng.* 2: 399–429.
51. T. A. Klar, S. Jakobs, M. Dyba et al. Fluorescence microscopy with diffraction barrier broken by stimulated emission. 2000. *Proc. Natl. Acad. Sci.* 97: 8206–8210.
52. T. A. Klar, M. Dyba, and S. W. Hell. Stimulated emission depletion microscopy with an offset depleting beam. 2001. *Appl. Phys. Lett.* 78: 393–395.
53. M. Dyba and S. W. Hell. Focal spots of size $\lambda/23$ open up far-field fluorescence microscopy at 33 nm axial resolution. 2002. *Phys. Rev. Lett.* 88(16): 163901.
54. M. J. Rust, M. Bates, and X. Zhuang. Sub-diffraction-limit imaging by stochastic optical reconstruction microscopy (STORM). 2006. *Nat. Methods* 3: 793–796.
55. E. Betzig, G. H. Patterson, R. Sougrat et al. Imaging intracellular fluorescent proteins at nanometer resolution. 2006. *Science* 313: 1642–1645.

56. O. Wiener. Stehende Lichtwellen und die Schwingungsrichtung polarisirten Lichtes.1890. *Ann. Phys.* 276: 203–243.
57. K. H. Drexhage. Monomolecular layers and light. 1970. *Sci. Am.* 6: 108–119.
58. W. Lukosz and R. E. Kunz. Light emission by magnetic and electric dipoles close to a plane dielectric interface: I. Total radiated power. 1977. *J. Opt. Soc. Am.* 67: 1607–1615.
59. K. H. Drexhage. Interaction of light with monomolecular dye layers. 1974. *Progress Opt.* 12: 163–232.
60. D. Braun and P. Fromherz. Fluorescence interference-contrast microscopy of cell adhesion on oxidized silicon. 1997. *Appl. Phys. A* 65: 341–348.
61. J. M. Crane, K. Volker, and L. K. Tamm. Measuring lipid asymmetry in planar supported bilayers by fluorescence interference contrast microscopy. 2005. *Langmuir* 21: 1377–1388.
62. J. Kressemakers, J. Howard, H. Hess et al. The distance that kinesin-1 holds its cargo from the microtubule surface measured by fluorescence interference contrast microscopy. 2006. *Proc. Natl. Acad. Sci.* 103: 15812–15817.
63. E. Hecht. 2002. *Optics*. Reading, MA: Addison-Wesley.
64. J. Greivenkamp. 1995. *Handbook of Optics*. New York: McGraw-Hill.
65. S. V. Baryshev, A. V. Zinovev, C. E. Tripa et al. White light interferometry for quantitative surface characterization in ion sputtering experiments. 2012. *Appl. Surf. Sci.* 258(18): 6963–6968.
66. T. Sandström, M. Stenberg, and H. Nygren. Visual detection of organic monomolecular films by interference colors. 1985. *Appl. Opt.* 24(4): 472–479.
67. R. Jenison, S. Yang, A. Haeberli et al. Interference-based detection of nucleic acid targets on optically coated silicon. 2002. *Nat. Biotechnol.* 19: 62–65.
68. M. Zavali, P. S. Petrou, S. E. Kakabakos et al. Label-free kinetic study of biomolecular interactions by white light reflectance spectroscopy. 2006. *Micro. Nano Lett.* 1(2): 94–98.
69. B. P. Möhrle, K. Köhler, J. Jaehrling et al. Label-free characterization of cell adhesion using reflectometric interference spectroscopy (RIfS). 2006. *Anal. Bioanal. Chem.* 384(2): 407–413.
70. R. K. Bista, S. Uttam, P. Wang et al. Quantification of nanoscale nuclear refractive index changes during the cell cycle. 2011. *J. Biomed. Opt.* 16(7): 070503.
71. K. L. Prime and G. M. Whitesides. Self-assembled organic monolayers: Model systems for studying adsorption of proteins at surfaces. 1991. *Science* 252(5010): 1164–1167.
72. J. Vörös. The density and refractive index of adsorbing protein layers. 2004. *Biophys. J.* 87(1): 553–561.
73. A. Samoc, A. Miniewicz, M. Samoc et al. Refractive-index anisotropy and optical dispersion in films of deoxyribonucleic acid. 2006. *J. Appl. Polym. Sci.* 105(1): 236–245.
74. E. Ozkumur, A. Yalçin, M. Cretich et al. Quantification of DNA and protein adsorption by optical phase shift. 2009. *Biosens. Bioelectron.* 25(1): 167–172.
75. W. C. W. Chan, D. J. Maxwell, X. H. Gao et al. Luminescent quantum dots for multiplexed biological detection and imaging. 2001. *Curr. Opin. Biotechnol.* 13(1): 40–46.
76. S. Drmanac, D. Kita, I. Labat et al. Accurate sequencing by hybridization for DNA diagnostics and individual genomics. 1998. *Nat. Biotechnol.* 16: 54–58.
77. C. Debouck and P. N. Goodfellow. DNA microarrays in drug discovery and development. 1999. *Nat. Genet.* 21: 48–50.

78. C. B. Epstein and R. A. Butow. Microarray technology-enhanced versatility, persistent challenge. 2000. *Curr. Opin. Biotechnol.* 11: 36–41.
79. M. S. Shchepinov, S. C. Case-Green, and E. M. Southern. Steric factors influencing hybridization of nucleic acids to oligonucleotide arrays. 1997. *Nucl. Acids Res.* 25 (6): 1155–1161.
80. E. Southern, K. Mir, and M. Shchepinov. Molecular interactions on microarrays. 1999. *Nat. Genet.* 21: 5–9.
81. K. U. Mir and E. M. Southern. Determining the influence of structure on hybridization using oligonucleotide arrays. 1999. *Nat. Biotechnol.* 17: 788–792.
82. A. Vainrub and B. M. Pettitt. Coulomb blockage of hybridization in two-dimensional DNA arrays. 2002. *Phys. Rev. E* 66: 041905.
83. A. Vainrub and B. M. Pettitt. Surface electrostatic effects in oligonucleotide microarrays: Control and optimization of binding thermodynamics. 2003. *Biopolymers* 68: 265–270.
84. C. F. Edman, D. E. Raymond, D. J. Wu et al. Electric field directed nucleic acid hybridization on microchips. 1997. *Nucl. Acids Res.* 25(24): 4907–4914.
85. R. G. Sosnowski, E. Tu, W. F. Butler et al. Rapid determination of single base mismatch mutations in DNA hybrids by direct electric field control. 1997. *Proc. Natl. Acad. Sci.* 94(4): 1119–1123.
86. C. Gurtner, E. Tu, N. Jamshidi et al. Microelectronic array devices and techniques for electric field enhanced DNA hybridization in low-conductance buffers. 2002. *Electrophoresis* 23(10): 1543–1550.
87. A. W. Peterson, J. R. Heaton, and R. M. Georgiadis. The effect of surface probe density on DNA hybridization. 2001. *Nucl. Acids Res.* 29 (24): 5163–5168.
88. A. W. Peterson, L. K. Wolf, and R. M. Georgiadis. Hybridization of mismatched or partially matched DNA at surfaces. 2002. *J. Am. Chem. Soc.* 124(49): 14601–14607.
89. P. Gong and R. Levicky. DNA surface hybridization regimes. 2008. *Proc. Natl. Acad. Sci.* 105(14): 5301–5306.
90. D. Liu, A. Bruckbauer, C. Abell et al. A reversible pH-driven DNA nanoswitch array. 2006. *J. Am. Chem. Soc.* 128(6): 2067–2071.
91. T. G. Drummond, M. G. Hill, and J. K. Barton. Electrochemical DNA sensors. 2003. *Nat. Biotechnol.* 21: 1192–1199.
92. W. Shu, D. Liu, M. Watari et al. DNA molecular motor driven micromechanical cantilever arrays. 2005. *J. Am. Chem. Soc.* 127(48): 17054–17060.
93. J. Skolnick and M. Fixman. Electrostatic persistence length of a wormlike polyelectrolyte. 1977. *Macromolecules* 10: 944–948.
94. Y. Lu, B. Weers, and N. C. Stellwagen. DNA persistence length revisited. 2002. *Biopolymers* 61(4): 261–275.
95. A. V. Dobrynin. Electrostatic persistence length of semiflexible and flexible polyelectrolytes. 2005. *Macromolecules* 38: 9304–9314.
96. B. Tinland, A. Pluen, J. Sturm et al. Persistence length of single-stranded DNA. 1997. *Macromolecules* 30(19): 5763–5765.
97. C. Bustamante, J. F. Marko, E. D. Siggia et al. Entropic elasticity of lambda-phage DNA. 1994. *Science* 265(5178): 1599–1600.
98. J. Skolnick and M. Fixman. Electrostatic persistence length of a wormlike polyelectrolyte. 1977. *Macromolecules* 10(5): 944–948.
99. N. L. Goddard, G. Bonnet, O. Krichevsky et al. Sequence dependent rigidity of single stranded DNA. 2000. *Phys. Rev. Lett.* 85(11): 2400–2403.

100. G. Pirri, F. Damin, M. Chiari et al. Characterization of a polymeric adsorbed coating for DNA microarray glass slides. 2004. *Anal. Chem.* 76: 1352–1358.
101. R. Suriano, M. Levi, G. Pirri et al. Surface behavior and molecular recognition in DNA microarrays from N,N-dimethylacrylamide terpolymers with activated esters as linking groups. 2006. *Macromol. Biosci.* 6: 719–729.
102. A. Yalçın, F. Damin, E. Özkumur et al. Direct observation of conformation of a polymeric coating with implications in microarray applications. 2009. *Anal. Chem.* 81(2): 625–630.
103. E. Farjami, L. Clima, and K. Gothelf. "Off–on" electrochemical hairpin-DNA-based genosensor for cancer diagnostics. 2011. *Anal. Chem.* 83(5): 1594–1602
104. J. B. Lee, M. J. Campolongo, J. S. Kahn et al. DNA-based nanostructures for molecular sensing. 2010. *Nanoscale* 2: 188–197.
105. P. A. Rice, S. Yang, K. Mizuuchi et al. Crystal structure of an IHF-DNA complex: A protein-induced DNA U-turn. 1996. *Cell* 87(7): 1295–1306.
106. U. Rant, K. Arinaga, S. Scherer et al. Switchable DNA interfaces for the highly sensitive detection of label-free DNA targets. 2007. *Proc. Natl. Acad. Sci.* 104(44): 17364–17369.
107. U. Rant, K. Arinaga, S. Fujita et al. Dynamic electrical switching of DNA layers on a metal surface. 2004. *Nano Lett.* 4(12): 2441–2445.
108. M. Chiari, E. Casale, E. Santaniello et al. Synthesis of buffers for generating immobilized pH gradients. I: Acidic acrylamido buffers. 1989. *Appl. Theor. Electrophor.* 1(2): 99–102.
109. M. Chiari, E. Casale, E. Santaniello et al. Synthesis of buffers for generating immobilized pH gradients. II: Basic acrylamido buffers. 1989. *Appl. Theor. Electrophor.* 1989. 1(2): 103–107.
110. R. G. Righetti. 1990. *Immobilized pH Gradients: Theory and Methodology.* Amsterdam, the Netherlands: Elsevier Science Ltd.
111. P. S. Spuhler, L. Sola, X. Zhang et al. A precisely controlled smart polymer scaffold for nanoscale manipulation of biomolecules. 2012. *Anal. Chem.* 84(24): 10593–10599.
112. W. Kaiser and U. Rant. Conformations of end-tethered DNA molecules on gold surfaces: Influences of applied electric potential, electrolyte screening, and temperature. 2010. *J. Am. Chem. Soc.* 132: 7935–7945.
113. U. Rant, K. Arinaga, S. Fujita et al. Electrical manipulation of oligonucleotides grafted to charged surfaces. 2006. *Org. Biomol. Chem.* 4: 3448–3455.
114. U. Rant, K. Arinaga, S. Fujita et al. Structural properties of oligonucleotide monolayers on gold surfaces probed by fluorescence investigations. 2004. *Langmuir* 20(23): 10086–10092.
115. U. Rant, K. Arinaga, M. Tornow et al. Dissimilar kinetic behavior of electrically manipulated single- and double-stranded DNA tethered to a gold surface. 2006. *Biophys. J.* 90(10): 3666–3671.
116. G. S. Manning. The molecular theory of polyelectrolyte solutions with applications to the electrostatic properties of polynucleotides. 1978. *Q. Rev. Biophys.* 11(2): 179–246.
117. G. S. Manning. Electrostatic free energy of the DNA double helix in counterion condensation theory. 2002. *Biophys. Chem.* 101–102: 461–473.
118. A. Y. Grosberg and A. R. Khokhlov. 1994. *Statistical Physics of Macromolecules.* New York: AIP Press.

119. J. F. Thompson and A. Landy. Empirical estimation of protein-induced DNA bending angles: Applications to λ site-specific recombination complexes. 1988. *Nucl. Acids Res.* 16(20): 9687–9705.
120. H. A. Nash and C. A. Robertson. Purification and properties of the *Escherichia coli* protein factor required for lambda integrative recombination. 1981. *J. Biol. Chem.* 256: 9246–9253.
121. M. Freundlich, N. E. M. Ramani, A. Sirko, and P. Tsui. The role of integration host factor in gene expression in *Escherichia coli*. 1992. *Mol. Microbiol.* 6: 2557–2563.
122. D. I. Friedman. Integration host factor: A protein for all reasons. 1988. *Cell* 55: 545–554.
123. T. W. Lynch, E. K. Read, A. N. Mattis et al. Integration host factor: Putting a twist on protein–DNA recognition. 2003. *J. Mol. Biol.* 330(3): 493–502.
124. P. H. von Hippel. Protein–DNA recognition: New perspectives and underlying themes. 1994. *Science* 263(5148): 769–770.
125. S. Wang, R. Cosstick, J. F. Gardner et al. The specific binding of Escherichia coli integration host factor involves both major and minor grooves of DNA. 1995. *Biochemistry* 34(40): 13082–13090.
126. T. Ellenberger and A. Landy. A good turn for DNA: The structure of integration host factor bound to DNA. 1997. *Structure* 5(2) 153–157.
127. B. S. Parekh and G. W. Hatfield. Transcriptional activation by protein-induced DNA bending: Evidence for a DNA structural transmission model. 1996. *Proc. Natl. Acad. Sci.* 93(3): 1173–1177.
128. S. W. Yang and H. A. Nash. Comparison of protein binding to DNA in vivo and in vitro: Defining an effective intracellular target. 1995. *EMBO J.* 14(24): 6292–6300.
129. C. Murtin, M. Engelhorn, J. Geiselmann et al. A quantitative UV laser footprinting analysis of the interaction of IHF with specific binding sites: Re-evaluation of the effective concentration of IHF in the cell. 1998. *J. Mol. Biol.* 284(4): 949–961.
130. P. Gong and R. Levicky. DNA surface hybridization regimes. 2008. *Proc. Natl. Acad. Sci.* 105(14): 5301–5306.
131. A. W. Peterson, R. J. Heaton, and R. M. Georgiadis. The effect of surface probe density on DNA hybridization. 2001. *Nucl. Acids Res.* 29(24): 5163–5168.
132. L. Moiseev. Spectral self-interference fluorescence microscopy and its applications in biology. 2003. PhD dissertation, Boston University, Boston, MA.
133. H. Du, C. M. Strohsahl, J. Camera et al. Sensitivity and specificity of metal surface-immobilized "molecular beacon" biosensors. 2005. *J. Am. Chem. Soc.* 127(21): 7932–7940.
134. C. M. Strohsahl, B. L. Miller, and T. D. Krauss. Preparation and use of metal surface-immobilized DNA hairpins for the detection of oligonucleotides. 2007. *Nat. Protoc.* 2(9): 2105–2110.
135. Y. Zhang, Z. Tang, J. Wang et al. Hairpin DNA switch for ultrasensitive spectrophotometric detection of DNA hybridization based on gold nanoparticles and enzyme signal amplification. 2010. *Anal. Chem.* 82 (15): 6440–6446.
136. O. Piestert, H. Barsch, and V. Buschmann. A single-molecule sensitive DNA hairpin system based on intramolecular electron transfer. 2003. *Nano Lett.* 3(7): 979–982.

137. Z. S. Wu, C. R. Chen, G. L. Shen et al. Reversible electronic nanoswitch based on DNA G-quadruplex conformation: A platform for single-step, reagentless potassium detection. 2008. *Biomaterials* 29(17): 2689–2696.
138. L. Wang, X. Liu, Q. Yang et al. A colorimetric strategy based on a water-soluble conjugated polymer for sensing pH-driven conformational conversion of DNA i-motif structure. 2010. *Biosens. Bioelectron.* 25(7): 1838–1842.
139. H. Liuw and D. Liu. DNA nanomachines and their functional evolution. 2009. *Chem. Commun.* 19: 2625–2636.
140. K. Paeschke, T. Simonsson, J. Postberg et al. Telomere end-binding proteins control the formation of G-quadruplex DNA structures in vivo. 2005. *Nat. Struct. Biol.* 12: 847–854.
141. G. N. Parkinson, M. P. H. Lee, and S. Neidle. Crystal structure of parallel quadruplexes from human telomeric DNA. 2002. *Nature* 417: 876–880.
142. A. J. Sfeir, W. H. Chai, J. W. Shay et al. Telomere-end processing: The terminal nucleotides of human chromosomes. 2005. *Mol. Cell.* 18(1): 131–138
143. J. R. Willamson. G-quartet structures in telomeric DNA. 1994. *Annu. Rev. Biophys. Biomol. Struct.* 23: 703–730.
144. M. Read, R. J. Harrison, B. Romagnoli et al. Structure-based design of selective and potent G quadruplex-mediated telomerase inhibitors. 2001. *Proc. Natl. Acad. Sci.* 98(9): 4844–4849.
145. N. W. Kim, M. A. Piatyszek, K. R. Prowse et al. Specific association of human telomerase activity with immortal cells and cancer. 1994. *Science* 266(5193): 2011–2015.
146. J. Lopes, A. Piazza, R. Bermejo et al. G-quadruplex-induced instability during leading strand replication. 2011. *EMBO J.* 30(19): 4033–4046.
147. H. Han and L. H. Hurley. G-quadruplex DNA: A potential target for anti-cancer drug design. 2000. *Trends Pharmacol. Sci.* 21(4): 136–142.
148. L. H. Hurleya, R. T. Wheelhouseb, D. Sun et al. G-quadruplexes as targets for drug design. 2000. *Pharmacol. Ther.* 85(3): 141–158.
149. S. Balasubramanian and S. Neidle. G-quadruplex nucleic acids as therapeutic targets. 2009. *Curr. Opin. Chem. Biol.* 13(3): 345–353.
150. S. M. Gowan, J. R. Harrison, L. Patterson et al. A G-quadruplex-interactive potent small-molecule inhibitor of telomerase exhibiting in vitro and in vivo antitumor activity. 2002. *Mol. Pharmacol.* 61(5): 1154–1162.
151. A. J. Zaug, E. R. Podell, and T. R. Cech. Human POT1 disrupts telomeric G-quadruplexes allowing telomerase extension in vitro. 2005. *Proc. Natl. Acad. Sci.* 102(31): 10864–10869.
152. R. J. White and S. P. Jackson. The TATA-binding protein: A central role in transcription by RNA polymerases I, II and III. 1992. *Trends Genet.* 8(8): 284–288.
153. C. R. Wobbe and K. Struhl. Yeast and human TATA-binding proteins have nearly identical DNA sequence requirements for transcription in vitro. 1990. *Mol. Cell. Biol.* 10(8): 3859–3867.
154. N. Hernandez. TBP, a universal eukaryotic transcription factor? 1993. *Genes Dev.* 7(7B): 1291–1308.
155. B. P. Cormack and K. Struhl. The TATA-binding protein is required for transcription by all three nuclear RNA polymerases in yeast cells. 1992. *Cell* 69(4): 685–696.
156. S. K. Burley. The TATA box binding protein. 1996. *Curr. Opin. Struct. Biol.* 6(1): 69–75.
157. M. M. Gromiha, J. G. Siebers, S. Selvaraj et al. Intermolecular and intramolecular readout mechanisms in protein–DNA recognition. 2004. *J. Mol. Biol.* 337(2): 285–294.

158. A. Bareket-Samisha, I. Cohena, and T. E. Haran. Signals for TBP/TATA box recognition. 2000. *J. Mol. Biol.* 299(4): 965–977.
159. D. B. Nikolov, H. Chen, E. D. Halay et al. Crystal structure of a human TATA box-binding protein/TATA element complex. 1996. *Proc. Natl. Acad. Sci.* 93(10): 4862–4867.
160. D. B. Starr, B. C. Hoopes, and D. K. Hawley. DNA bending is an important component of site-specific recognition by the TATA binding protein. 1995. *J. Mol. Biol.* 250: 434–446.
161. A. Grove, A. Galeone, E. Yu et al. Affinity, stability and polarity of binding of the TATA binding protein governed by flexure at the TATA box. 1998. *J. Mol. Biol.* 282(4): 731–739.
162. J. Wu, K. M. Parkhurst, R. M. Powell et al. DNA bends in TATA-binding protein-TATA complexes in solution are DNA sequence-dependent. 2001. *J. Biol. Chem.* 276(18): 14614–14622.
163. B. S. Wolner and J. D. Gralla. TATA-flanking sequences influence the rate and stability of TATA-binding protein and TFIIB binding. 2001. *J. Biol. Chem.* 276(9): 6260–6266.
164. A. Sancar. DNA excision repair. 1996. *Annu. Rev. Biochem.* 65: 43–81.
165. J. H. J. Hoeijmakers. Genome maintenance mechanisms for preventing cancer. 2001. *Nature* 411: 366–374.
166. A. Sancar, L. A. Lindsey-Boltz, K. Unsal-Kaçmaz et al. Molecular mechanisms of mammalian DNA repair and the DNA damage checkpoints. 2004. *Annu. Rev. Biochem.* 73: 39–85.
167. R. Prasad, W. A. Beard, V. K. Batra et al. A review of recent experiments on step-to-step "hand-off" of the DNA intermediates in mammalian base excision repair pathways. 2011. *Mol. Biol. (Mosk)* 45(4): 586–600.
168. M. W. Germann, C. N. Johnson, and A. M. Spring. Recognition of damaged DNA: Structure and dynamic markers. 2012. *Med. Res. Rev.* 32(3): 659–683.
169. M. Garcia-Diaz, K. Bebenek, A. A. Larrea et al. Template strand scrunching during DNA gap repair synthesis by human polymerase. 2009. *Nat. Struct. Mol. Biol.* 16: 967–972.
170. C. Roll, C. Ketterle, V. Faibis et al. Conformations of nicked and gapped DNA structures by NMR and molecular dynamic simulations in water. 1998. *Biochemistry* 37(12): 4059–4070.
171. H. Guo and T. D. Tullius. Gapped DNA is anisotropically bent. *Proc. Natl. Acad. Sci.* 100(7): 3743–3747.
172. J. Ahn, V. S. Kraynov, X. Zhong et al. DNA polymerase β: Effects of gapped DNA substrates on dNTP specificity, fidelity, processivity and conformational changes.1998. *Biochem. J.* 331: 79–87.
173. L. S, D. P. Horning, J. W. Szostak et al. Conformational analysis of DNA repair intermediates by time-resolved fluorescence spectroscopy. 2009. *J. Phys. Chem. A* 113(35): 9585–9587.
174. M. R. Sawaya, R. Prasad, S. H. Wilson et al. Crystal structures of human DNA polymerase beta complexed with gapped and nicked DNA: Evidence for an induced fit mechanism. 1997. *Biochemistry* 36(37): 11205–11215.
175. M. J. Jezewska, R. Galletto, and W. Bujalowski. Dynamics of gapped DNA recognition by human polymerase beta. 2002. *J. Biol. Chem.* 277(23): 20316–20327.

10

fMRI and Nanotechnology

Aditi Deshpande and George C. Giakos

CONTENTS

10.1 Introduction

Medical imaging dates back to the late eighteenth century when Wilhelm Roentgen discovered x-rays, leading to the use of penetrating radiation to detect and study anatomical structures. Since then, x-ray imaging was the basis of the diagnostic techniques implemented to visualize internal structures of the body noninvasively. X-rays were found extremely beneficial in detecting pathologies in bones and, to a certain extent, in soft tissues as well.

Hence, x-ray scans found numerous applications in medical imaging and became the most widely used tool in diagnostic radiography.

Even though x-ray imaging is still a preferred technique in several medical investigations, one of its major flaws is that it is not capable of producing detailed and articulate images of soft tissues as it does of the skeletal system. Hence, not all forms of pathologies or malformations can be detected using x-rays. The images that x-rays offer are all flat and confined to one dimension, which limits the radiologist's understanding of all the aspects of the problem. Apart from this, another major disadvantage of x-ray imaging is the exposure to ionizing radiation. The World Health Organization also has classified x-rays as carcinogenic radiation [1,2]. The physiological effects of ionizing radiation can be determined by the threshold or the level of exposure and the time period for and the area that was exposed [3]. Though low doses are not as damaging as prolonged exposure, x-ray scans do render patients susceptible to harmful effects of radiation, the most prominent of which being cancer.

In the early 1900s, Alessandro Vallebona proposed a method called "tomography" in which we take slices of the object to be scanned. This idea was applied to x-rays, and the technology was advanced forward and extended to computed tomography (CT), which is a technique in which a computer processes the detected x-rays to produce images of "slices" of specific areas of the body. This tomographic scan can also render 3-D images of the body using stacks of 2-D slices. The first commercial CT machine was invented in 1967 by Sir Godfrey Hounsfield. Over the last two decades, CT scans have become very prominent in diagnostic imaging as they provide high-contrast images with good resolution. Modifications such as helical CT also make multiplanar imaging possible. Though CT is a huge improvement over regular x-ray radiography and produces high quality and fast images, the exposure to ionizing radiation still poses the risk of damaging the DNA, which can cause cancer and pose many other risks. Despite this, CT scans are widely used, and this technology is currently among the preferred tools for diagnostic imaging.

The field of medical imaging is one of the most rapidly growing ones. Along with x-ray imaging, other modalities such as ultrasound, nuclear magnetic resonance (NMR) imaging, and radionuclide imaging such as positron emission tomography came into existence. Parallel research was being conducted since many decades to design and implement new techniques of diagnostic imaging that do not employ harmful radiation and can also provide more coherent and lucid images of the soft tissues in the human body. A modality that offered great scope for imaging was ultrasound, which was invented in the late 1940s. This technology utilizes high-frequency acoustic waves to examine the interior of the body. A piezoelectric crystal transducer transmits high energy sound waves into the body, aided by a conducting gel that is applied onto the skin of the patient. The waves travel inside the body and are reflected back to the transducer (after being attenuated by the internal structures in the body), which also acts as a receiver. The time taken for these waves to travel is used to study and construct an image of the area on which the ultrasound waves were incident. Today, ultrasound is the second most employed diagnostic imaging technique [3].

The independent work of Felix Bloch and Edward Purcell involved studying the concept of NMR, for which they were awarded a Nobel Prize in Physics [4]. They performed experiments using chemical compounds [5,6]. NMR is a phenomenon in which, when the magnetic nuclei of certain atoms are exposed to a magnetic field, absorbing and reemitting of electromagnetic (EM) radiation occur. When the external magnetic field is applied, the polarization of spin occurs and a transition between these spin states can be caused by the application of RF waves of appropriate frequency, known as the resonant frequency. This can also cause energy acquisition in certain nuclei spins, which is emitted in the form of RF signal upon switching the incident RF waves off. This emitted signal is captured and can be measured and processed to form an image that gives us the information of the object containing the nuclei that were excited [7]. This is the principle of NMR imaging that was first proposed and experimented upon by Isidor Rabi in 1938 [4]. The current day magnetic resonance imaging (MRI) is based on the previously mentioned phenomenon.

The MRI machine setup consists of a device in which the patient lies with a large powerful magnet surrounding him or her, which provides the external static field described. The images are acquired in MRI in the way described for NMR. This shall be discussed further in the following sections of the chapter. Early MRI scans relied on weak magnets that generated low-quality images [8]. Today, MRI utilizes large and extremely powerful magnets with field in the range of 2–5 T. Undoubtedly, the prime factor behind the boom of medical imaging and the invention of multiple modalities was the revolution of the digital world and the advancement of computer technology [9,10]. These advances in electronics make cutting edge data acquisition and processing possible. The data storage techniques have also improved tremendously due to this. However, the various techniques developed have merits and demerits of their own.

Unlike CT and other scans that use x-rays or other ionizing radiation, MRI does not use any harmful radiation to produce an image of the object scanned and yields detailed physiological information. While CT possesses a greater ability to discriminate two separate structures very close to each other (high spatial resolution), MRI can distinguish two very similar tissues from one another very efficiently (higher contrast). The attenuation of x-rays is the only the basis for contrast in CT, whereas contrast can be introduced and intensified in MRI using a lot of parameters [11]. Both the scans can be enhanced using externally administered contrast agents, which are introduced in the area to be scanned to intensify the contrast in the resulting image. As CT is purely a radiation-based scan, contrast agents, if used, need to be elements with a high atomic number (to cause a large difference in

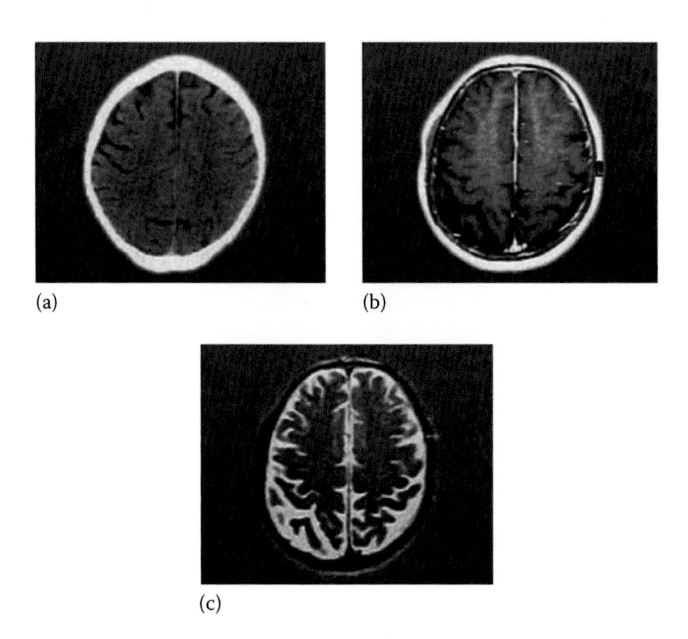

FIGURE 10.1
A comparison of (a) MRI, (b) CT, and (c) contrast-enhanced MRI scans. (From MRI vs CT, The Internet Stroke Center, Dallas, TX, http://www.strokecenter.org/professionals/stroke-diagnosis/mri-compared-to-ct/)

x-ray attenuation and increase contrast), such as barium. In case of MRI, the contrast depends on the magnetic properties of tissues and external contrast agents. Thus, extremely paramagnetic substances such as gadolinium are used. MRI contrast agents modify the relaxation time of the magnetic atoms within the human body, hence altering the data acquisition precision and improving the quality of the final image. These contrast agents can be administered either intravenously or orally (Figure 10.1).

We know that when a particular part of the body is being used, the flow of oxygenated blood to that region increases and so does the blood supply to the part in the brain that controls that action. It has been long established that blood supply and changes in blood oxygenation levels in the brain are synchronous with neural activity [8]. Venous blood (containing deoxygenated hemoglobin [Hb]) exhibits paramagnetism and is much more magnetic than arterial blood (containing oxygenated Hb), which is virtually nonmagnetic or diamagnetic. This difference aids in MRI and acts as a contrasting parameter, distinguishing areas activated by a rushed inflow of oxygenated blood from other regions [8]. Seiji Ogawa [12] was the first to demonstrate the difference in the image of the brain's blood vessels, based on the blood oxygen level, and implement it in MRI scanning. Thus, the "blood-oxygen-level-dependent" (BOLD) contrast came into use in MRI. This concept was the motivation behind the discovery of functional magnetic resonance imaging (fMRI). fMRI is a technique that maps the function of the human brain onto and image by studying the activation of different parts of the brain during different activities.

The procedure adopted in fMRI operates on the principle of change in magnetization between oxygenated and deoxygenated blood. The activated regions of the brain receive an abundant supply of oxygenated blood. This extent of activation is color coded and displayed graphically in the scans. When a region in the brain is activated, the concentration of oxygenated Hb increases rapidly in that area, and the concentration of deoxy-Hb decreases [13]. As deoxy-Hb is paramagnetic in nature, a fall in its concentration decreases its interference with the applied magnetic field and enhances the MR signal. It can be said that fMRI uses neurovascular correlation to encode brain activity into measurable MR signal, which is detected and processed to form an image. The aim of performing fMRI is to correlate the task performed by the subject to the brain activity. It quantifies the brain activation and provides functional information of the brain, unlike regular scans that provide only anatomical information [14]. Specific cognitive stages and activities can be studied and linked to the activity in the brain. Though BOLD fMRI does not directly measure the brain function, the perfusion levels in the brain are used as a proxy to determine the same (Figure 10.2).

Patients with existing pathologies in the brain are much more difficult to scan in this way than the healthy ones, as tumors and malformations

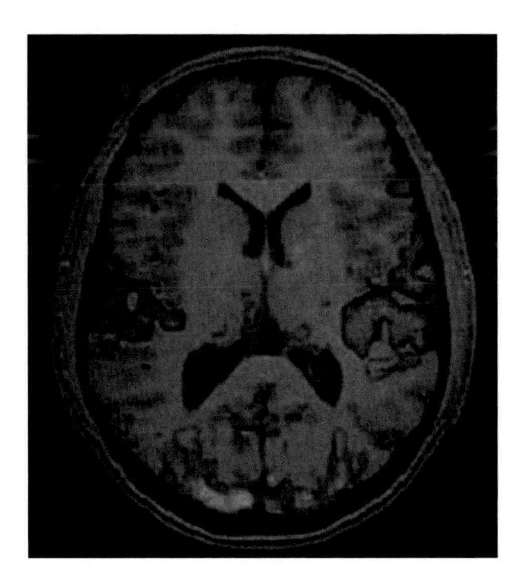

FIGURE 10.2
A simple fMRI scan of a subject being provided visual stimulation. (From H. Devlin, What is functional magnetic resonance imaging (fMRI)? FMRIB Center, Department of Clinical Neurology, University of Oxford, Oxford, U.K., http://psychcentral.com/lib/2007/what-is-functional-magnetic-resonance-imaging-fmri/)

can alter the blood flow in ways unrelated to the neurovascular hemodynamics. Also, fMRI scans are extremely expensive. This limits the use of fMRI in clinical purposes, and hence it is used at a much larger scale in research. Nevertheless, it still is used in a number of applications such as risk assessment for neurosurgery to verify the functioning of the brain during various tasks, in checking for asymmetry corresponding to left–right hemispheres in memory, speech, and other regions; to study the neurocorrelation of a seizure or a stroke; and to examine how the brain recuperates from it. fMRI can also be used to observe the pharmacokinetics of an administered drug and to study its effects on penetration of the blood–brain barrier. The event-related fMRI is a further step in fMRI research, which attempts at modeling the alterations in the fMRI signal due to neural activity by presenting the subject with tasks at certain time intervals, called "events." This allows us to study stimuli responses and cognitive patterns at a higher level.

The basic fMRI scan uses the BOLD as the contrast parameter to image the brain function. New biomarkers are being studied and tested to provide better enhanced contrast [15]. These include temperature and calcium-sensitive agents. As the burning of glucose raises temperature, it can act as a marker of brain activity. We know that action potentials travel through neurons and other cells via calcium channels. It is now possible to track calcium as it flows into the neurons when they activate. Thus, calcium-sensitive compounds can be used as biomarkers for brain function. Researchers at the McGovern Institute for Brain Research at MIT have developed calcium-sensitive contrast agents at the nano-sized level, which the MRI scanner can detect [17]. Their proposed technique of using calcium-sensitive nanoparticles overcomes the two main limitations of conventional fMRI—the time lag between neural activity and corresponding hemodynamic changes and the slightly decreased spatial resolution due to the compact spacing of tiny blood vessels. This is achieved due to the fact that calcium enters neurons almost immediately as they get activated and the potential shoots up. With increased activation levels, the amount of calcium and the rapidness of its inflow increase too. Thus, calcium provides a direct measure of the brain activity. The design of this method shall be elaborated on later in the chapter.

Similar new molecular tools are being developed and studied to enhance the fMRI contrast. Due to their unique properties, nanoparticles are being studied and implemented in MRI contrast agents to enhance the resolution of the scans. Though the design and implementation of nanoparticles in various fields seem to be an extremely modern innovation, the first use of this technology actually dates back centuries. During the middle ages, artisans used to create a film of nanoparticles to coat their creations by superheating copper and silver oxides and salts. In the scientific world, nanoparticles were first described based on their size and optical properties by Michael Faraday in 1857. These ultrafine particles have their diameter between 1 and 100 nm and exhibit superior optical behavior and demonstrate quantum effects due to their unique chemical properties and structure. Various properties of different materials can be altered using nanoparticles, for instance, the melting point of gold, the absorption of solar radiation in photovoltaic cells, surface plasmon resonance in metallic particles, superparamagnetism, and greatly increased surface area. Nanoparticles are linked with biological molecules to act as markers and can be tracked within the body [18]. This nanoparticle-targeted imaging is used to enhance contrast in fMRI. The methods adopted in this technique and various other applications of nanoparticles in medicine shall be discussed in this chapter.

From this point, this chapter is broadly divided into four sections. The section on MRI describes the working principles, current trends, and applications of MRI. We then move onto fMRI—its need, design, working principles, and the contrast agents used—which brings us to the next section that is on nanoparticles. The various properties and fabrication methodologies for nanoparticles along with their numerous advantages and applications have been described. In the final section, we talk about how nanoparticles are implemented in the contrast agents used for fMRI and the ways in which this enhances imaging. Current research and future prospects of this technology have been stated.

10.2 Magnetic Resonance Imaging

Further research led to the fact that the magnetic properties of various nuclei had many potential applications in medical imaging as it is possible to alter the relaxation and resonance properties of such magnetic materials [19]. MRI applies this principle of NMR to the nuclei of the hydrogen atoms in the body based on the fact that the body tissue comprises largely of water, which contains hydronium ions (protons). The difference in water content between various tissues in the body forms the basis of MRI scanning as this difference in the number of hydrogen atoms alters the response to the EM field applied. Since the water content of two dissimilar tissues is different, MRI produces an excellent contrast level, distinguishing the tissues. MRI can penetrate soft and hollow, air-filled structures without much attenuation and also uses nonionizing radiation. These properties give MRI its advantages over many other imaging modalities.

10.2.1 Physical Principles of NMR

The physical phenomenon of NMR is based on the principle that charged particles or certain nuclei possess a characteristic spin associated with their magnetic moment, which produces various effects on the application of an external magnetic field. These particles absorb and reemit EM radiation of a specific frequency value, which is known as the RF for those particles. Magnetic properties of various nuclei are determined by their composition of protons and neutrons. The elements whose nuclei possess an odd number of protons and neutrons are magnetic, and only some elements (or their isotopes) can be utilized in imaging or spectroscopy. The properties of these magnetic nuclei and physical phenomena behind NMR are described in the succeeding text.

10.2.2 Nuclear Magnetic Moment

In case of a magnet, the magnetic moment is a vector entity that governs the amount of force that the magnet exerts on electric currents and the torque produced on it when an external magnetic field is applied.

When an external magnetic field "B" is applied, the magnetic moment "m" possesses a potential energy "U":

$$U = -m \cdot B.$$

The magnetic moment (m) also relates the aligning torque ($\mathcal{T}$) to the applied magnetic field (B) in the following way:

$$\mathcal{T} = m \times B.$$

The magnetic moment is intrinsically related to the angular moment. When an external magnetic field subjects the magnetic moment to a torque, it rotates about the axis of the applied field. This is the intrinsic angular momentum associated with the moment, which, in case of charged particles (atomic particles), is called the "spin."

This spin (angular momentum) of protons and neutrons gives the nucleus its magnetic moment that is nonzero only if the number of neutrons and protons in that nucleus is uneven as protons and neutrons tend to form pairs of opposing magnetic moments. The spin magnetic moment for a subatomic particle can be calculated in the following way:

$$\mu = g \frac{q}{2m} s$$

where

μ is the spin magnetic moment
g is the dimensionless g-factor
q is the charge
m is the mass
s is the angular momentum of the particle

The ratio of a particle's magnetic dipole moment to its angular momentum is called its gyromagnetic ratio, and "g-factor" is the quantity that characterizes the gyromagnetic ratio and the magnetic moment of the particle. Neutrons, protons, and electrons are particles with a ½ spin. For such particles, the energy of the magnetic moment (μ) is in a magnetic field "B," which in some "z" direction is given by

$$E = -\mu_z B$$

where μ_z is the magnetic moment of the particle in the "z" direction.

This energy is the lowest when the applied field and the magnetic moment are aligned together completely and is the highest when they are completely antialigned (Figure 10.3).

The most abundant elements (or their isotopes) in the human body are H^1, C^{12}, N^{14}, and O^{16}. Out of these, the most abundant is hydrogen, and it is the only one that has a magnetic nucleus [16]. As the body consists of a large quantity of water, hydrogen ions (H^+), which are nothing but a single proton, are abundant in the body. Hence, we use these hydrogen ions or protons as the particle of reference/study in NMR imaging.

Naturally, all the magnetic moments in our body are randomized or "jumbled." Hence, the body is nonmagnetic as the net magnetic moment is zero. To study the signals emitted by the protons (after altering their spin using an RF pulse, described later), we first align them in a single direction by applying an extremely strong magnetic field externally.

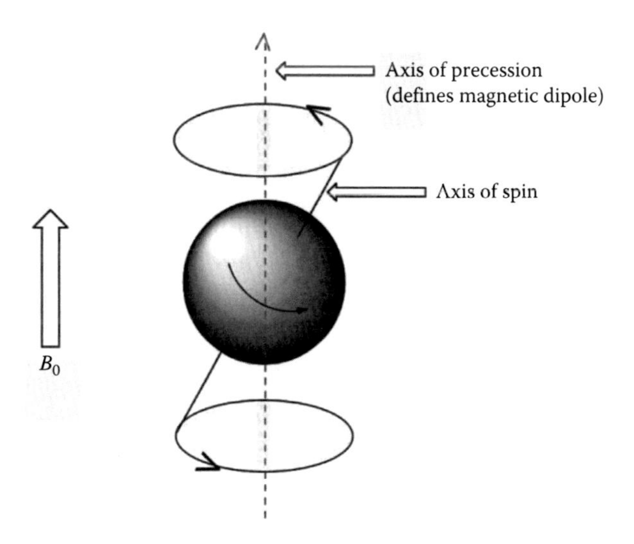

FIGURE 10.3
The spin and magnetization of a precessing proton. (From Soderberg, T., *Organic Chemistry with a Biological Emphasis*, University of Minnesota, Morris, MN, 2010.)

10.2.3 Effect of External Magnetic Field

To align all the protons in the body in one uniform direction, a strong magnetic field of 3 T (3×10^4 gauss) is applied externally. As the proton is a charged particle possessing a spin and a magnetic moment, it behaves like a small bar magnet. When the body is placed in a strong magnetic field, a torque acts on the protons (based on the magnetic moment) and propels them to align themselves along the direction of the field.

However, thermal energy agitates the spin of the nuclei, and not all of them are aligned parallel to the field (Figure 10.4).

The spins are aligned and distributed into two different energy states: parallel to the field and opposite (antiparallel) to the field. As described in the section earlier (nuclear magnetic moment), the protons aligned parallel to the field possess lower energy, and the protons aligned antiparallel to the field possess higher energy. As the applied field strength increases, so does the energy difference between these proton spins. The number of protons with lower energy is much higher in number than the number of protons in the higher-energy state (Figure 10.5).

The difference in energy levels of the parallel and antiparallel protons produces a net magnetization, which is oriented along the direction of the magnetic field. This net magnetization is the "vector sum" of the magnetic moments of the protons under the magnetic field applied. The net magnetization has two components: one component along the direction of the applied field (which is maximum at equilibrium and is known as "longitudinal magnetization") and the other component perpendicular to the applied

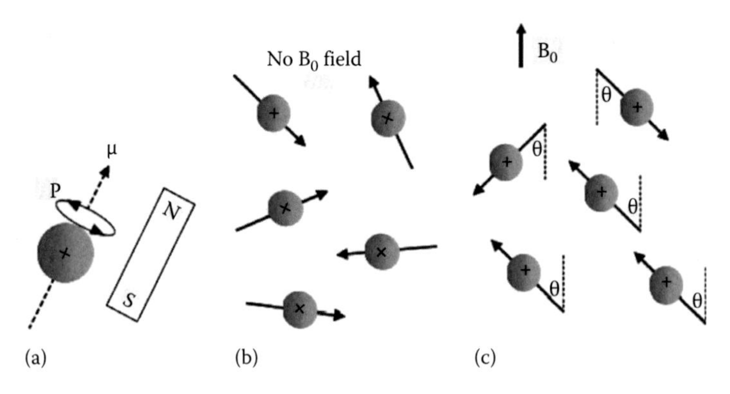

FIGURE 10.4
Representation of the alignment of protons and their spins. (a) The spin and alignment of a single proton, (b) the alignment of various protons in the body at rest, and (c) the alignment of protons in the body after a magnetic field B_0 is applied externally. (From Webb, A. and Smith, N., *Introduction to Medical Imaging: Physics, Engineering, and Clinical Applications*, Cambridge Texts in Biomedical Engineering, 2011.)

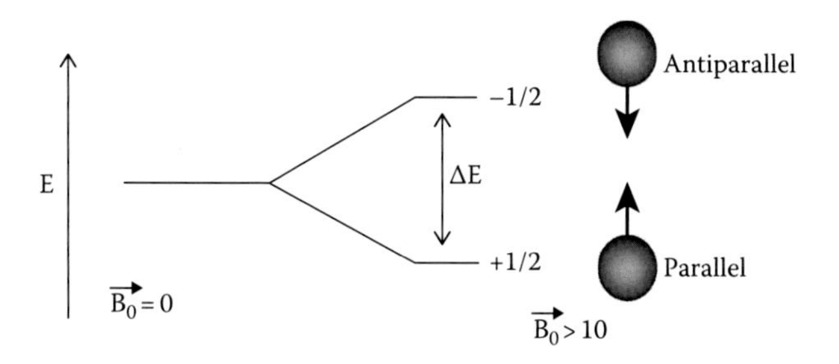

FIGURE 10.5
Difference in energy levels between parallel and antiparallel aligned protons (with respect to the magnetic field B_0). (Image courtesy of Wikipedia [released into the public domain].)

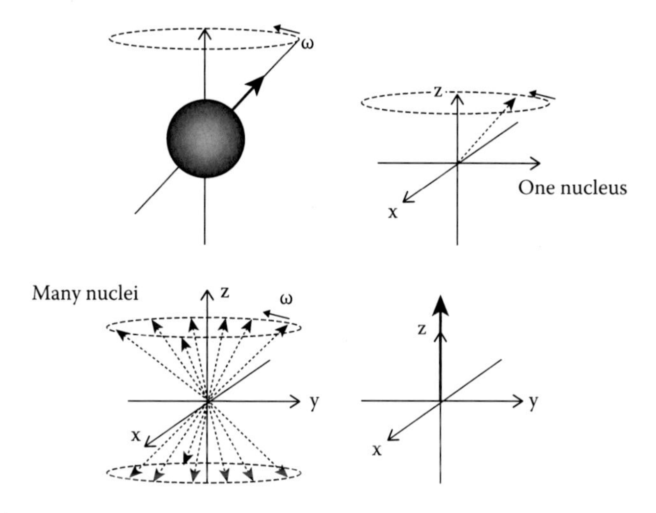

FIGURE 10.6
Spin magnetization vector of a single hydrogen nucleus (proton) as compared to the net spin of many such nuclei; the net magnetization vector points in the z direction. (Image courtesy of Wikipedia [released into the public domain].)

magnetic field (which is zero at equilibrium and is known as "transverse magnetization") (Figure 10.6).

Now, the applied field tries to align the protons with itself, thus producing a torque that is tangential to the direction of the proton's magnetic moment. This causes the proton to rotate or "precess" around the magnetic field axis. This precession is a phenomenon of classical physics, which is the result of the interaction between the spin angular momentum of the nucleus and the externally applied field. Thus, the NMR signal can be seen as a collection of nuclei (behaving as magnetic dipoles) precessing about the direction of the applied magnetic field [20].

10.2.4 Effect of Applying RF Pulse

The importance of nuclear precession is the fact that it renders the nucleus extremely sensitive to radio frequency, which is in resonance with the precessing frequency. All objects, such as radio receivers and musical instruments, are sensitive to their own specific RF. The RF in this case is dependent on the strength of the applied magnetic field and the properties of the nucleus. This is also called the Larmor frequency after the physicist Joseph Larmor. The gyromagnetic ratio described earlier also characterizes this frequency to the applied field.

The NMR signal can be strengthened by increasing the energy difference between the antiparallel aligned protons and by decreasing the temperature (to reduce the randomizing effect of thermal energy). Thus, a stronger NMR signal is produced, on application of a larger magnetic field to increase the energy gap. To stimulate the transition between the energy levels of the protons, the energy is applied externally, which in this case is the EM radiation of radio frequency corresponding to the RF. For a nucleus with net spin of ½, only two energy states exist, separated by an energy gap of $2\ \mu_z B$. A proton with such a magnetic moment in a low-energy state can absorb a photon of energy:

$$\Delta E = hf_0$$

where f_0 is the RF required $\Delta E = 2\ \mu_z B$. Hence, we get

$$f_0 = \frac{2\mu_z B}{h}$$

We can see that the RF is directly proportional to the applied magnetic field and the magnetic moment of the protons. The proton-spin system does not interact with incident radiation of any other frequency other than this RF. So, when an RF pulse of this frequency is applied, the protons in the lower-energy state absorb that energy and get transitioned to the higher-energy state. This flips the spin of the proton from +½ to −½. This causes the net magnetization vector to flip by 90°, and it now precesses about the axis of the applied field. In this excited state, transverse magnetization is maximal, and longitudinal magnetization is minimal (opposite of what the earlier state was). This absorption of radiation (energy) by protons, which leads to flipping of spin, imparts excess energy in them. Thus, they are not in a state of equilibrium. Thus, once we remove the applied RF waves, the nuclei eventually relax and return to their original state of thermal equilibrium. This process is known as relaxation and is of two types—T1 relaxation and T2 relaxation (Figure 10.7).

T1 relaxation, also known as spin–lattice relaxation, is the relaxation of the longitudinal magnetization to its equilibrium state and its realignment along the applied magnetic field. The nuclei dissipate excess energy in the form of

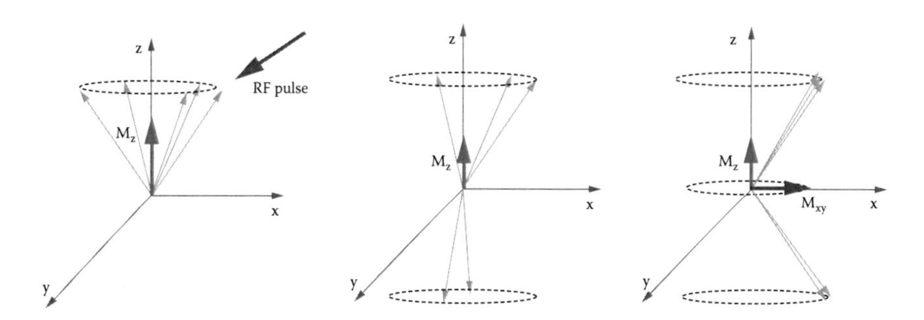

FIGURE 10.7
The effect of applying an RF pulse on the magnetization vector of the applied magnetic field. (Image courtesy of Wikibooks [released into the public domain by Kieran Maher].)

heat to the surroundings, and the spins are reverted to their original state as they were before the application of the RF pulse. This gradually increases the longitudinal magnetization (as it was maximal during equilibrium). This is an exponential process with a time constant T1, that is, T1 is the time taken for 63% of the magnetic field to return to its equilibrium state (Figure 10.8).

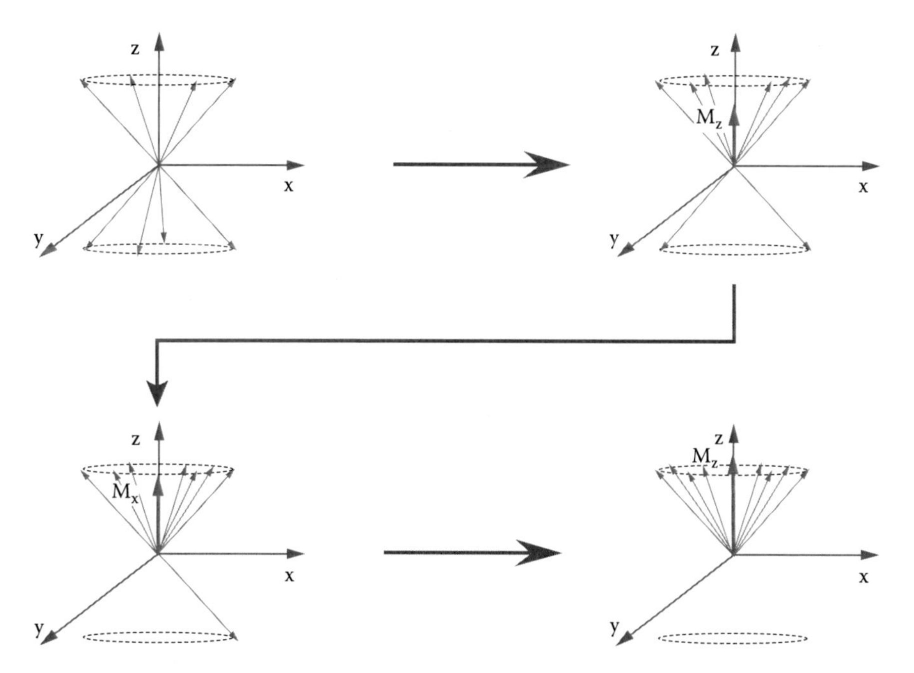

FIGURE 10.8
Process of longitudinal relaxation of the magnetization vector. The magnetization component in the longitudinal z direction becomes maximum, and the transverse magnetization in the x–y plane becomes zero again. (Image courtesy of Wikibooks [released into the public domain by Kieran Maher].)

If M_z is the value of the longitudinal magnetization at a time t and $M_{z.eq}$ is the value of the longitudinal magnetization at equilibrium, then we have

$$M_z = M_{z.eq}(1 - e^{t/T1})$$

T2 relaxation, also known as spin–spin relaxation, is the relaxation of the transverse magnetization from its state of excitation to its equilibrium state (when transverse magnetization was minimal or 0) by way of an exponential decay, which is also called "free-induction decay."

Upon the removal of the RF pulse, the magnetic moments interact with each other, which leads to a decrease in the transverse magnetization; hence, we can say that "decoherence" occurs, and this decaying transverse component produces an oscillating magnetic field that produces an electric current in the receiver coil.

Inconsistent variations and fluctuations in the local magnetic moments result in a loss of phase coherence until there is no magnetization left in the direction perpendicular to the applied field [21]. T2 is the decay for this exponential process (Figure 10.9).

Let M_{xy} be the transverse magnetization at any time t. Let $M_{xy}(0)$ be the transverse magnetization immediately after the application of the RF pulse. This gives us the following:

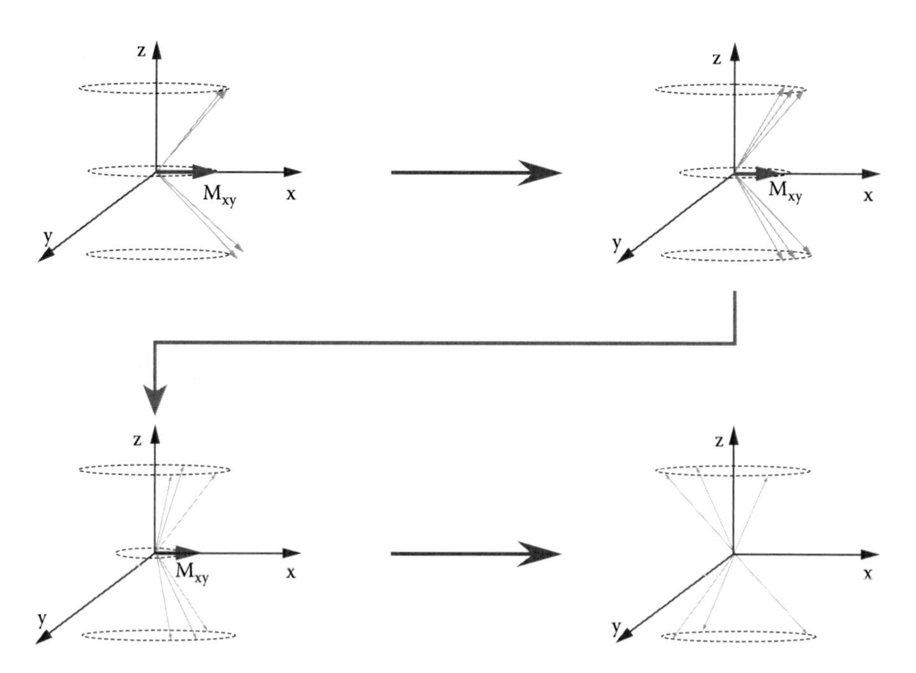

FIGURE 10.9
Transverse relaxation known as T2 decay, illustrated by a progressive loss of coherence. (Image courtesy of Wikibooks [released into the public domain by Kieran Maher].)

$$M_{xy} = M_{xy}(0)e^{-t/T2}$$

Thus, in time T2, the transverse magnetization drops to 37% of its value. Transverse relaxation occurs faster than longitudinal relaxation. The signal that is produced is the measure of the electric signal generated by the resultant magnetic field from the emission of RF energy by protons. As varying tissues in the body contain different water levels (and hence different number of protons), the signal generated from various parts of the body is different. Also, nuclear relaxation times are different for different tissues (healthy and tumorous). This is the basis on which the MR contrast is produced on the spatial encoding of the electric signal [22].

Additionally, the applied static field is varied across the body, causing the different parts of the body to have their protons precess at varied frequencies. This field gradient is pulsed, and there are numerous possible values of the RF pulse sequences. The effect of the relaxation parameters T1 and T2 on the MR signal is one of the most important considerations in the contrast produced between tissues, along with the proton density.

10.2.5 Instrumentation and Image Generation

The basic setup for NMR imaging includes an extremely strong magnet capable of producing field gradients, an RF transmitter and a receiver, a coil (probe), and a computer that processes the data received and generates the final image. The magnet is the prime and the most expensive component in this system; its major qualities that affect the MR imaging system are the field strength and homogeneity (the straightness of the magnetic field lines in the center), temporal stability, and the ability to produce a gradient field [23].

During its early days and until quite recently, either EM or permanent magnets were used to produce the static applied field. As higher and much stronger field was required, superconducting magnets came into use. Permanent magnets were made of ferromagnetic substances and weighed over 100 tons. They were comparatively inexpensive to maintain, but they produced a weaker field compared to EM or superconducting magnets. Also, they could never be "turned off" as they were ferromagnetic; this presented safety issues. Resistive electromagnets are solenoids that use continuous electric energy to produce the desired magnetic field. They produced around 0.1–0.3 T of magnetic field strength and dissipated excessive heat, hence requiring a cooling system. The advantage of this technology over permanent magnets is the ability to turn them off during an emergency. The field strength offered by resistive electromagnets was still found low compared to the needed field for enhanced scanning, and it costs a lot to supply the constant high electric field (Figure 10.10).

Niobium–titanium alloys achieve superconducting properties upon cooling by liquid helium, and these superconducting magnets do not offer any

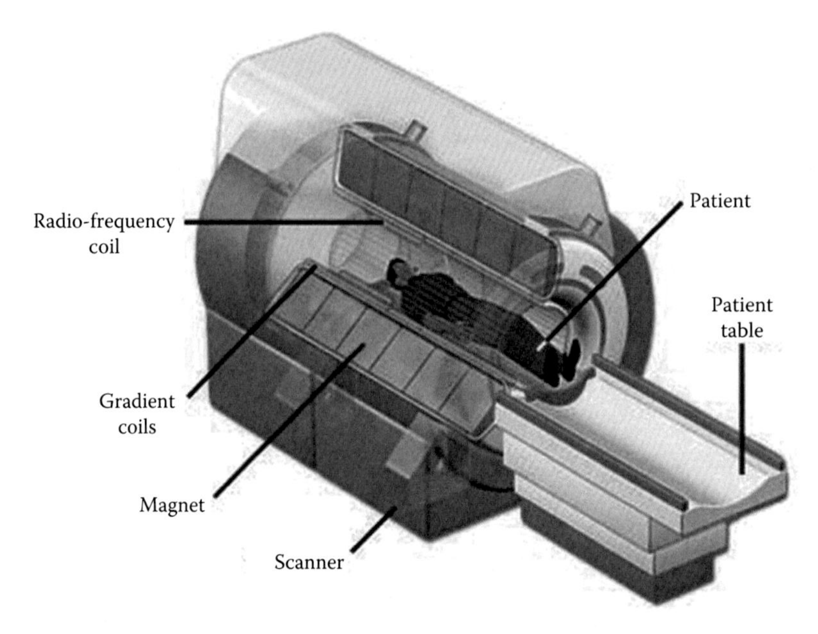

FIGURE 10.10
A basic MRI scanner. (From Coyne, K., MRI: A guided tour, National High Magnetic Field Lab, Florida State University, Tallahassee, FL, http://www.magnet.fsu.edu/education/tutorials/magnetacademy/mri/)

resistance to current flow. They also possess extremely high field strength (0.3–5 T) and high stability along with good homogeneity of the magnetic field. But these magnets are extremely expensive, and their cryostats (for supercooling) are difficult to maintain. Even then, due to the immense advantages and superior quality imaging it provides, this technology has become the most commonly used in MRI.

Magnetic field gradients at room temperature are used to compensate for the inhomogeneities induced by the magnet or to vary the magnetic field across the body, to enhance the contrast in the produced image. This gradient is induced by copper wire loops arranged in the bore of the magnet. These are called "gradient coils." These gradient inducing sets are utilized for space (frequency and phase) encoding and slice selection to produce an image from the NMR signal. These gradient coils are responsible for the banging noise produced during the scan. When the magnetic field changes rapidly, these coils flex on experiencing a torque on them. This generates the noise [20]. "Shim coils" are the ones used to adjust the magnetic field applied by the magnet onto the subject and to correct inhomogeneities (Figure 10.11).

RF waves are applied to the subject using a generator that synthesizes RF pulses, an amplifier, and an RF transmitter coil. This coil should be able to contain the body, apply a uniform and consistent magnetic field, and have

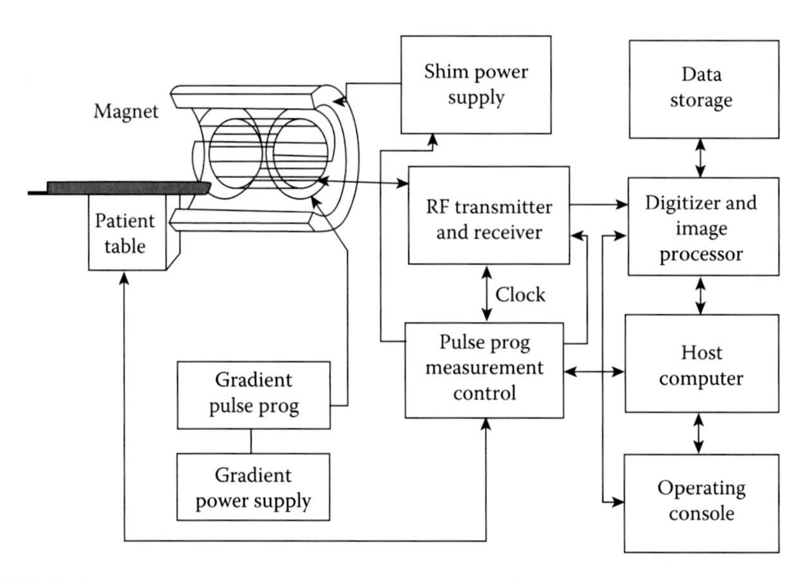

FIGURE 10.11
Block diagram of the complete MRI setup. (From Bushberg, J.T. et al., *The Essential Physics of Medical Imaging*, 3rd edn., North American Edition, Lippincott Williams & Wilkins, Philadelphia, PA, 2011.)

low interference and minimal losses. The same RF coil can be used to transmit and receive the signal, but it is preferred to have separate coils so that the signal-to-noise ratio is not affected. The inductance and capacitance properties of the coil determine its efficiency of resonance at the Larmor frequency, and these RF coils are tuned each time an acquisition of MR signal is to be made. These RF coils are usually classified as either volume coils or surface coils. Volume coils surround the "tissues of interest" and have more or less uniform sensitivity inside them, whereas surface coils are placed over the region of interest and their penetration capability is determined by the coil radius [21]. Surface coils increase sensitivity and offer localized reception of signal from organs of interest, but the field produced by surface coils cannot penetrate deeply [20].

NMR needs a high-power RF field for applying intense RF pulses so as to flip the spin of the protons and alter the alignment of the magnetic field. The power to be transmitted by the RF pulse is adjusted according to the patient's weight. Usually, high-end scanners consist of RF transmitters with peak power up to 30 kW. However, the continuous application of waves of such a high power could be damaging. The radio-frequency EM field produced is a magnetic near field with a very small electric field component associated with it; hence, though a strong EM field is applied (high-intensity waves), the waves are not essentially strong "radio waves" as such. Hence, it does not interfere with the body's produced radio waves. The timing of the applied RF pulse is controlled by a computer. This is an important factor affecting the quality of the signal as the signal-to-noise ratio also depends on the duration of the pulse due to the fact that the signal is proportional to the shift/tilt in the magnetization vector. Also, the bandwidth of the RF radiation depends on the inverse of the pulse duration (frequency). Both the amplitude and phase of these RF waves are computer controlled and can be manipulated to enhance the received signal.

The incoming MR signal (depending on the radio waves emitted by the body) is detected by the receiver system consisting of the receiver coil, an amplifier, a demodulator, and phase-sensitive detectors (which form the signal processing system). After amplification, the signal is demodulated and filtered to remove noise and to remove frequencies that are not near/equal to the RF. This boosts the signal-to-noise ratio of the system. The signal is filtered again to make sure that no aliasing is present in the output signal, and then the signal is digitized fulfilling the Nyquist criterion to avoid aliasing. The resultant output signal is stored in an analog-to-digital converter and fed into the computer system, which is interfaced with the MR scanner for image construction and enhancement.

As ionizing radiation is not made use of in MRI, the procedure is harmless, but there exist several regulations and considerations to ensure the safety of the personnel and the equipment. Any ferromagnetic (or metallic) objects should strictly not come into contact with the magnetic field of the scanner and hence cannot be brought into the room as the applied magnetic field strength of the MRI scanner is tremendously strong. Even nonmetallic implants could heat up due to the rapidly switching field gradient, and, hence, patients with implants should be thoroughly examined and be well aware of the effect of the magnetic torque on the implant material. Inbuilt guards should prevent excessive exposing of the body to high-power RF field as this produces damaging effects and high nerve stimulation. The loud noise generated and the confined space can make the patient uncomfortable and claustrophobic. These issues are now handled using noise cancelling headphones and equipment that creates a virtual surrounding to make the patient comfortable.

10.2.6 Image Construction

The signal that is acquired is processed and coded in space, frequency, and phase domains to finally get an image corresponding to the signal levels received from the different parts of the body. The most commonly used technique for image reconstruction is using the 2-D Fourier transform method of spatial encoding. The RF signals from each "voxel" (a volume element representing a grid in 3-D space, analogous to pixel in 2-D) need to be converted into an intensity level corresponding to a pixel, to obtain an image. To perform this task, we need to create magnetic gradients during every cycle of imaging we perform. The information about the signal is provided along three orthogonal axes, one being the "slice" of the body selected as the region of interest and the other two being the "phase encoded" and the "frequency encoded" signal.

The primary gradient is utilized to specify the thickness and the exact location of the tissue "slice" whose image is to be constructed. The activation of this "slice select gradient" implies that the protons in that tissue "slice" (that particular thin region) have been energized by the RF waves and their spin has been flipped by applying "their" RF. The protons in other regions remain more or less unaffected. This however depends on the amplitude and the time duration of the RF pulse applied along with the dimensions of the selected "slice." This slice gradient becomes the "z-axis." When this gradient is turned off, the protons in this slice return to their original state, emitting RF waves specific to their anatomical position. Hence, this significantly affects the resultant MR signal.

The next step is the phase encoding of the signal, which is also known as the "preparation process." When we apply the RF pulse to the body, the respective spins of the protons are flipped, and the magnetization is now precessing around the originally applied field (a 90° shift is induced). When we turn the field off, they return to their original state, emitting radio waves. As the field applied to an array of selected voxels is constant, the frequency and phase of the emitted signal are the same. If a gradient in RF field is applied along a selected set of voxels, though the magnitude of the emitted pulses by protons from different voxels remains uniform, there is a phase shift induced in them. When this field is turned off, though the protons return to their equilibrium energy level, they still emit signals with the induced phase difference. This phase difference is utilized in the reconstruction process as the second axis along which the information for the image is stored. During each cycle, the gradient is set to different values to create different views [24].

After phase encoding, we move on to frequency encoding. When the frequency encoding gradient is turned on, protons at one end of the "slice" precess at a different rate than protons at the other end. This difference in spinning rate defines a direction to the third axis. Different rows of voxels are encoded with different frequencies of rotation.

The final step is to apply a 2-D Fourier transform to the signal. Using this technique, we can acquire a large data set in one measurement in a short time. These data are later converted into intensity point by using Fourier transform "pairs." The Fourier transform determines the corresponding mathematical representations and points in the signal of the different pixel values. Once all the pixel values and their corresponding signal amplitude levels are obtained, signal averaging is performed reducing noise.

10.2.7 Contrast

The signal density and contrast in the MR image are based on four parameters, namely, the proton density, the T1 relaxation time, the T2 relaxation time, and the blood flow. Longitudinal relaxation time is one of the primary characteristics that can be used to generate contrast in MRI. The extent to which the T1 values contribute to the contrast depends on the repetition time values. The T1-generated contrast is basically the difference between the two magnetization curves at any time. The T1-weighted scans are MRI scans that depict the difference between the longitudinal relaxation times (spin–lattice relaxation) of different tissues. The T1 contrast enhances the difference between water and fat very well.

As mentioned earlier, the variation in proton density throughout the body acts as a great contrast parameter for MR imaging. As the water content in different tissues is different, so is the proton density. This concentration of protons in voxels of the body determines the strength of magnetization after the static field is applied. Even though proton density contrast exists early on in the imaging system, it is smaller than the T1 contrast level. As magnetization increases, so does the proton density contrast. Thus, T1 contrast usually decreases and gives way to the proton density contrast. The variation in proton density between various regions of the body causes a variation in the magnetization levels. Proton density-weighted scans are generated by selecting the repetition time value such that it falls in the later portion of the phase of relaxation of the magnetization.

In certain imaging scans, the difference in T2 relaxation times (spin–spin relaxation) can also be coded into contrast for the resultant image. Say, two different tissues start their transverse relaxation from the same energy level (magnetization value). Even then, this decay continues at different speeds due to a difference in the decay time constant T2 between various tissues. As can be inferred, the tissues that have a longer characteristic rate of T2 decay possess a higher magnetization level at any time. The ratio between the magnetization values of various tissues gives us the T2 contrast [25]. As this contrast depends on the rate of decay of transverse magnetization, there is no T2 contrast at the beginning of the imaging cycle. This contrast builds up and increases with the relaxation of transverse magnetization. Thus, this can be maximized by selecting a long echo time, which is the time between the initial 90° RF pulse and the echo (Figure 10.12).

The MR contrast can also be altered by changing the pulse sequence parameters, which are the specifications of the RF pulse transmitted, that is, the strength, frequency, timing, and the number of RF waves sent. Also, the flow motion of blood and mechanisms associated with it alter the MR signal received and hence change the contrast in the produced image. The increased intensity of the MR signal is related to the flow of blood in the plane of the selected slice of the body, and the extent of contrast enhancement by this is dependent on the velocity of blood flow to the repetition time in the imaging cycle. The blood flow causes a change in the magnetization levels of the blood flowing through the slice, hence changing the MR signal and generating contrast. Dampened signal from within the blood vessels is caused due to very high velocity of flow or because of turbulent flow.

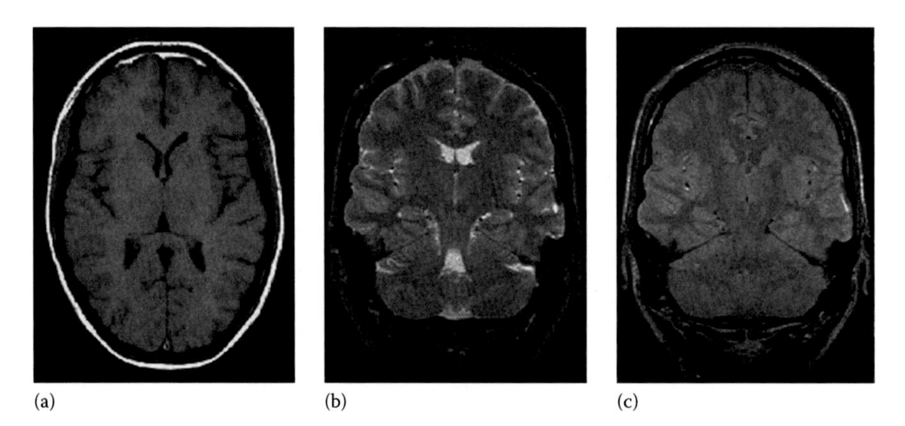

FIGURE 10.12
A comparison of differently weighted scans (a: T1, b: T2, and c: proton density). (Image courtesy of Wikipedia [released into the public domain].)

Sometimes when blood flows through a magnetic field gradient, dephasing of the containing protons can occur, decreasing the MR signal. A variety of these techniques to image the flow of blood in the body are used to visualize the variations in blood vessels in the regions of interest. These principles are employed in MR angiography, which is the imaging of vascular structures in various parts of the body.

We can see that by influencing the parameters such as relaxation times of tissues and proton density, we can enhance the MR contrast in the resultant images. Certain substances are known to enhance the contrast level in the images by doing so. Thus, clinical MRI is often performed with an injected contrast agent, which is magnetic in nature and alters the magnetization level and the rate of relaxation of tissues. These MRI contrast agents may be injected intravenously to enhance the appearance of blood vessels, tumors, or inflammation in the resultant image obtained.

10.3 Contrast Agents for MRI

We know that based on their magnetic properties, materials can be classified as diamagnetic, paramagnetic, or ferromagnetic. Nearly 99% of the tissues in the body are diamagnetic, which is why there is little or no interest in developing contrast agents using diamagnetic materials [19]. Paramagnetic materials possess unpaired electrons and can be used to enhance magnetic relaxation in MRI. These paramagnetic ions bind to the water molecules and tumble (thermally vibrate) in solution, creating a random oscillating magnetic field. This generated field affects the precessing nuclei. When an interaction takes place between the paramagnetic ions and the magnetized protons increase, the rate of relaxation of the neighboring protons and the presence of an additional magnetic field created by these materials increase the dephasing between the protons. In this way, the relaxation time (both T1 and T2) is reduced and therefore enhanced by the use of an external paramagnetic contrast agent. To decrease the toxicity of paramagnetic materials, they are typically chelated with organic ligands before being administered for imaging.

Ferromagnetic materials are those that develop magnetic polarization when an external field is applied. Ferromagnetic materials are said to be magnetized if their "domains" (groups of unpaired electrons) are aligned. The domains of permanent magnets possess this alignment even in the absence of an applied field. As an extremely large number of groups of electrons are required for ferromagnetism, ferromagnetic contrast agents are not administered as solutions of metal ions; instead, they are injected as oxides of ferromagnetic metals and other compounds [26,27].

The most widely used contrast agent is a paramagnetic material called gadolinium, chelated with diethylenetriaminepentaacetic acid (DTPA). The most prominent feature of gadolinium is the high number of unpaired electrons (7). Gadolinium chelates are hydrophilic and do not penetrate the blood–brain barrier, and these are eliminated from the vascular system either into the interstitial spaces or by the kidneys. It is usually administered in small doses (about 0.1 mmol/kg of body weight) for large vessels like aorta, and the concentration can be made stronger for finer blood vessels.

Gadolinium contrast agents can be of various types—blood pool agents (which stay in circulation for a long time and do not get eliminated easily) such as albumin-binding or polymeric gadolinium complexes, extracellular fluid agents (which can be ionic or neutral), and organ-specific agents [28,29]. The gadolinium–DTPA complex had been approved for use in humans and is marketed under the name of Magnevist (in the United States). In the future, the ligand of gadolinium can be coupled to a protein to generate tissue-specific contrast (Figure 10.13) [30].

Superparamagnetic iron oxides (SPIOs) are also used as contrast agents. These are also known as monocrystalline iron oxide nanocompounds (MIONs) as they consist of suspended iron oxide nanoparticles, which reduce the T2 relaxation time of the absorbing tissues. These contrast agents employ nanoparticles and hence shall be discussed in detail in the later sections of this chapter [31].

Porphyrins (a class or aromatic organic compounds) act as markers of various pathological metabolic conditions [24,32]. Metalloporphyrins of manganese and iron demonstrate suitable properties for being MRI contrast agents. Similarly, heme-containing proteins occurring naturally in the body can act as contrast agents as they shorten T2 relaxation time by increasing the dephasing due to an increased magnetic field produced by paramagnetic deoxygenated Hb.

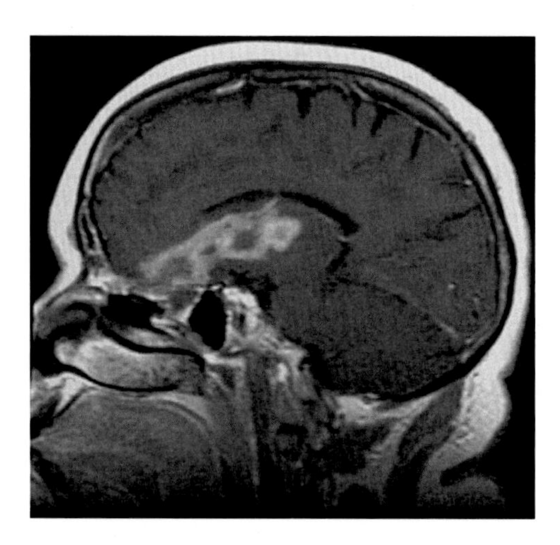

FIGURE 10.13
An MRI scan enhanced using gadolinium as contrast agent. (From Grunsfeld, A.A. and Login, I.S., *BMC Neurol.*, 6, 4, 2006.)

It is difficult to distinguish bowel from abdominal structures in regular imaging techniques, which is the reason that MR scanning is applied limitedly to this area. This can be overcome by the use of gastrointestinal contrast agents. Only one perfluorochemical is currently approved to be used for this purpose. For the detection of lesions within the liver and to assess the functioning of the liver in diseased conditions, hepatobiliary contrast agents are used [33]. They also provide high-resolution images of the gallbladder and the complete biliary network.

MRI can also be weighted using the body's own hemodynamic response as a contrast parameter. This is the concept behind the highly specialized functional MRI. Regular MRI can only provide us with anatomical information regarding the body, but by using fMRI, functional information of the body (especially the brain) can be achieved in the form of an image.

10.4 Functional MRI

As discussed earlier, MRI can only provide anatomical information about the body. A new and improvised technique for performing an MR scan was developed by Seiji Ogawa, which quantifies and locates the brain activity related to various "events" occurring in the brain. This technique is based on the idea that cerebral blood flow could be related to neuronal activity, first experimented by Roy and Sherrington in 1890 [1]. In this specialized scan, the neuronal activity in the brain is mapped to the hemodynamic

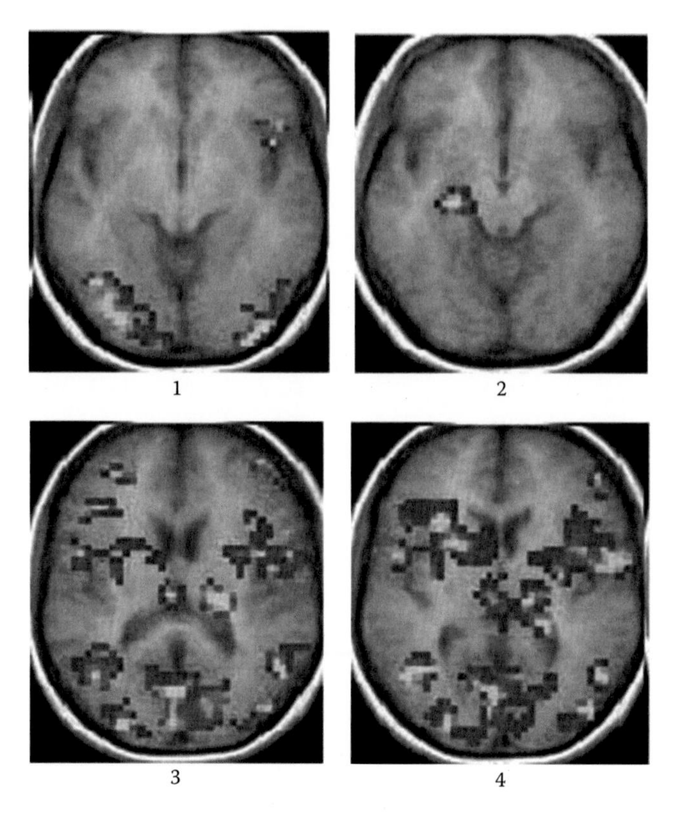

FIGURE 10.14
Four fMRI scans obtained during a visual memory task (each scan represents a different memory). (From Yallapu, M.M. et al., *Biomaterials*, 32, 1890, 2011.)

changes associated with it. The patient is provided with stimuli (tactile or visual), and the MR signal from the brain activity is encoded in space and frequency domains. Using this technique, we can observe the brain function graphically. An example of this is presented in the following image (Figure 10.14).

10.4.1 Working Principles

In 1990, Seiji Ogawa and other researchers at AT&T Bell Laboratories found out that the venous blood oxygen level dependence can be used as a contrast parameter in MRI and can be used to map brain function, based on rat brain studies [34]. This BOLD contrast was soon applied to MRI scans in humans. It is known that the cerebral blood flow is directly related to the glucose metabolism, which in turn is coupled with neuronal activity. Thus, we can relate cerebral blood flow to glucose metabolism. It was assumed that the cerebral metabolism rate of oxygen is also coupled with blood flow and neuronal changes as was the rate of glucose metabolism [35]. It was later reported that the cerebral blood flow changes exceed the rate of oxygen metabolism, which leads to an increase in the venous oxygenation level, providing a new parameter to study and aid in imaging contrast. The basic concept behind the BOLD response and the physiology behind it are described in the succeeding text.

When a task is to be performed or when the brain is stimulated, localized electric neuronal activities are induced in it. Neuronal activity in the brain causes an elevation of blood supply to the surrounding capillary beds (activated regions receive more blood as they need higher levels of glucose and, hence, oxygen). In neural activity, the energy from glucose is required to pump the ions in and out of the neuron membranes. As active cells require more and more energy, they switch to anaerobic glycolysis, which is a comparatively inefficient but much faster method than aerobic glycolysis. This is the reason why the glucose metabolism surpasses the oxygen metabolism during cerebral activation. This hemodynamic response causes a change in the ratio of the amount of oxygenated and deoxygenated Hb present in the blood (Figure 10.15).

Hb in the blood possesses different magnetic properties based on the oxygen molecule bound to it. By nature, deoxygenated Hb is paramagnetic and has much more magnetic properties than oxygenated Hb, which is practically diamagnetic. Thus, deoxy-Hb interferes with the magnetic field and distorts it, causing the excited protons to lose their magnetization

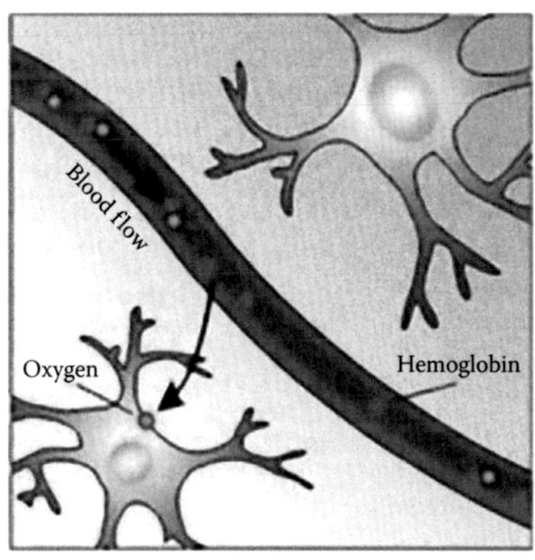

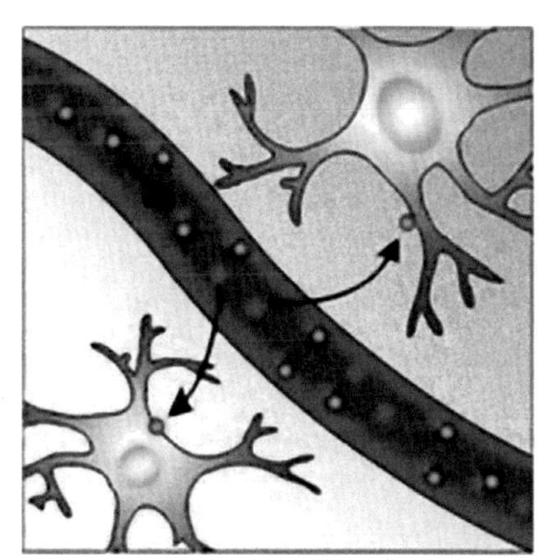

Activated

FIGURE 10.15
Depiction of the change in oxygenation with neuronal activation. (From Devlin, H. What is functional magnetic resonance imaging (fMRI)? FMRIB Center, Department of Clinical Neurology, University of Oxford, Oxford, U.K., http://psychcentral.com/lib/2007/what-is-functional-magnetic-resonance-imaging-fmri/)

faster by speeding up T2 relaxation, whereas oxy-Hb does not interfere with the magnetization produced by the externally applied field. Hence, the MR signal is stronger in the presence of oxy-Hb than deoxy-Hb. As the paramagnetic deoxy-Hb concentration decreases, the dephasing between intravoxel protons decreases, increasing the T2 relaxation time. Thus, as neurons in a region are activated and the blood supply to that region is enhanced (due to need of glucose and oxygen), the MR signal of that region is stronger due to the presence of much higher concentrations of oxy-Hb compared to deoxy-Hb. This is the BOLD contrast that enhances the T2 contrast parameter in MR images [36].

The change in total blood volume is also a good indicator of the change in cerebral blood flow. But we know that venous blood comprises of 75% of the total blood volume, so it is seen that the venous blood volume change is the governing parameter. With respect to BOLD contrast, only venous blood can be a maker for the changes in magnetic susceptibility corresponding to neuronal activation as it contains deoxy-Hb, which is the paramagnetic substance whose change in concentration causes this (Figure 10.16).

Using T2-weighted fMRI, we can depict the changes in magnetic susceptibility corresponding to the effects of the BOLD contrast. The functional MR signal from each voxel in the body is dependent on and defined by the amplitude, phase, and frequency of the proton magnetization and the T2* contrast (T2*-weighted scans are the ones that use a gradient echo and long repetition time with a long echo time). The T2* parameter depends on the spin–spin relaxation (T2) and the inhomogeneities in magnetic field. The BOLD phenomenon also affects the inhomogeneities in the magnetic field. We know that the inhomogeneities in the magnetic field are either caused by the strong magnet used to produce the static field, but they can also arise from the variations in magnetic susceptibility of different tissues in the body. When the concentration of deoxy- and oxy-Hb changes, the magnetic susceptibility of tissues changes, inducing the inhomogeneities in the field. A large echo time and a large repetition time during the imaging cycle help demonstrate this effect of BOLD on the magnetic field inhomogeneities better. Hence, we use T2*-weighted scanning frequently in fMRI.

The repetition time is the parameter that determines how often a "slice" of body tissues is excited (the protons in it and their magnetization) and allowed to relax. In fMRI, the hemodynamic response resulting from brain activity lasts around 1 s, lagging behind the activation by about 3 s (time delay for the vascular system to respond to the need of perfusion in the brain) and peaking at 5–6 s, and ensures continuous firing of the neurons. This peak spreads to form a plateau as the neurons remain active. Once the neurons are no longer active, the BOLD signal falls below its baseline value; this is called undershooting of the signal. One reason behind this undershooting might be the continuous metabolic demands in the body. This BOLD response is a smooth continuous function (Figure 10.17).

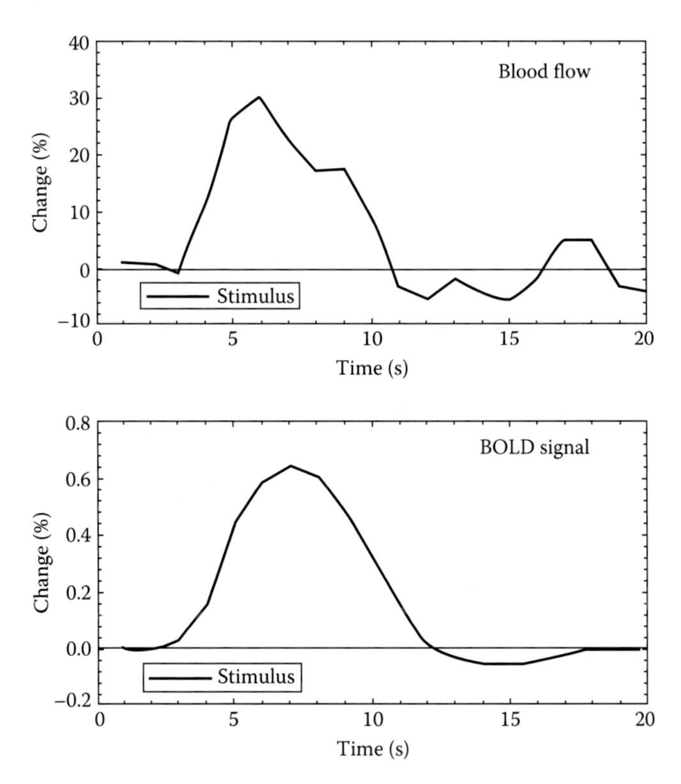

FIGURE 10.16
Relationship of the BOLD signal with time. (Image courtesy of UC San Diego Center for Functional MRI, Copyright 2012, Regents of the University of California, http://fmri.ucsd.edu/Research/whatisfmri.html)

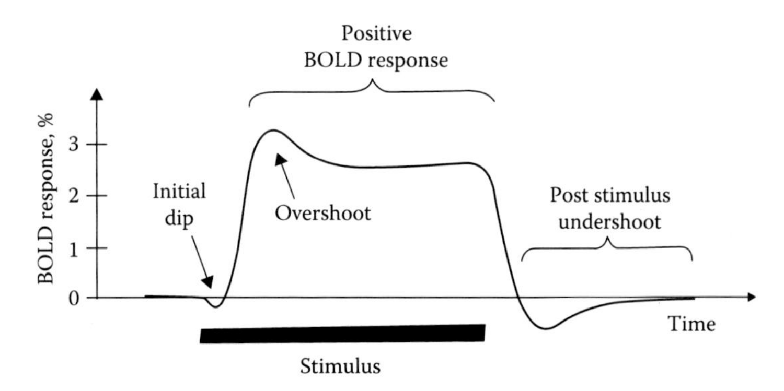

FIGURE 10.17
BOLD response with respect to the stimulation and time.

10.4.2 Resolution

The spatial resolution of fMRI is a measure of its ability to distinguish closely spaced objects. Like in MRI, it is measured by the number and size of voxels in an image (the larger the number of voxels, the finer the image and greater is its spatial resolution). The size range of voxels used is 1–5 mm. Smaller voxels imply lesser number of neurons contained in that voxel; hence, the signal from smaller voxels is weaker. Smaller voxels also take longer to scan as the time required to scan a body slice is proportional to the number of voxels in the slice. For obtaining a high spatial resolution, signals from larger vessels need to be suppressed as they do not correspond to the region of neural activity (they are located further, after blood has been collected from the smaller capillaries where the activity actually was). This is done by increasing the field strength of the static field applied.

Temporal resolution refers to the smallest duration of neural activity accurately distinguished by fMRI reliably. The repetition time is one of the fundamental parameters determining this temporal resolution. This can be enhanced by providing periodic stimulus spaced in time. Thus, by sampling the data trials, higher resolution can be achieved though it also depends on the time needed by the brain to process stimuli and respond to them. Most of the diagnostic study fMRI experiments conducted analyze brain responses lasting 2–10 s at a time with the complete duration of the experiment being about 10 min. Patient movement causes motion artifacts, and changing mental and emotional states of the patient may also affect the signal by increasing blood flow to those centers of the brain.

Multiple activation of the brain also has an effect on the fMRI signal. If the person performs more than one function at once or one after the other (in an overlapping manner), the BOLD response gets added to the previous one in a linear fashion. This might not always be accurate because of the nonlinearity produced by the refractory period of the brain in which the response of the current stimulus suppresses the response being produced for the subsequent stimulus. As the time interval between stimuli becomes shorter, the refractory period is enhanced. This refractory period is different for different regions of the brain.

Experiments were conducted to check the received BOLD signal in two contexts—by implanting electrodes and measuring the detected signal and by measuring the signal from the field potential (which is the electric or magnetic response of the brain activity). Local field potentials, an index of integrated electric activity, form a better correlation with blood flow than the action potentials generated by neuronal shooting activity. The field potential that signifies the internal function in the neuron and the synaptic activity after neuron firing is a better indicator of the BOLD signal. This shows that the BOLD signal is a representation of the input energy to the neuron and its utilization within its body but not of the output potentials that it generates.

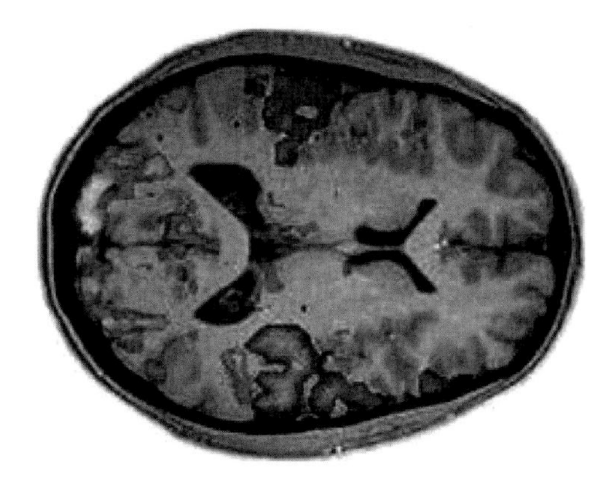

FIGURE 10.18
(See color insert.) fMRI scan with the BOLD signal as the contrast. (Image courtesy of M. Ignor, Mind-Body Dualism—Is the mind purely a function of brain? http://www.godandscience.org/evolution/mind-body_dualism.html)

The BOLD response cannot discriminate between forward and backward mechanisms that take place. Both restraining and stimulating inputs to a neuron from other neurons contribute toward the BOLD response, which is a summation of all the inputs combined. Also, the magnitude of the BOLD signal is not a good indicator of its behavior. An intricate and difficult cognitive task might require high perfusion in the beginning and hence displays a strong BOLD signal (related to good performance by the patient), but the need of excess blood flow might decrease with time as the subject gets better at it. This is because the brain demands lesser perfusion to save energy and reduce wastage of excess of it by channeling the neurons more efficiently. Factors that can alter the hemodynamic response of the body can affect the BOLD response, such as age, medication, anxiety, and diseases (Figure 10.18).

10.4.3 Data Acquisition

To avoid motion artifacts and to enhance the temporal resolution, we try to acquire images as fast as possible as fMRI requires plenty of images. Echo planar imaging and spiral imaging methods were employed to make such a high-speed imaging possible [37,38]. Echo planar imaging is much more readily available and is flexible; hence, it is preferred. In this method, two RF pulses are needed to generate a spin echo. One of these deflects the longitudinal magnetization vector into the transverse plane, whereas the other one rebuilds the phase coherence of the spin that was lost. Say, we have an image with a matrix size of $N \times M$, and we need to acquire information regarding that. For this purpose, a "readout" data set (with spatial information encoded in one direction) consisting of N points is collected and stored M times with a gradual increase in the phase-encoding gradient, which encodes spatial information in a perpendicular direction. Using the 2-D Fourier transform described earlier, we construct an image from this information. By oscillating the readout gradient, we consecutively obtain the readout data set (containing N points), thus taking very less time to acquire a single image. The entire "set" is the collection of the multislice images of the whole brain. For proper analysis and study using fMRI, a large number of data sets need to be collected.

Another mechanism for fMRI image acquisition is the gradient echo formation in which the MR signal is recovered and acquired by reversing the gradients applied and generating an echo gradient. The primary outcome of this echo gradient is the induction of a precession frequency, which is spatially dependent and which makes the isochromatic protons spin at a frequency that can be controlled by the applied gradient field. The method used for encoding the data spatially in gradient echo formation is similar to the one described earlier for spin echo imaging. The net RF energy utilized is decreased significantly, thus allowing us to use more RF pulses stacked closely. The disadvantage of this technique is that it is extremely sensitive to the inhomogeneities in the magnetic field and the susceptibility to magnetic field.

10.4.4 Preprocessing

The data collected from fMRI must be controlled well, and techniques should be implemented to reduce noise from all possible sources as this induces changes in the MR signal from factors we are not interested in. The primary sources of noise are random/irrelevant neural activity, physiological noise, thermal noise, and noise from the system. Another important factor to consider in fMRI is the contrast-to-noise ratio. This is the signal change due to neural activity relative to the random signal fluctuations. It is essential for this parameter to be high in order to obtain a high resolution in the acquired image.

Basically, fMRI tries to qualitatively correlate the task performance of the subject to the activation of specific corresponding regions in his or her brain. It also tries to present an analysis of the cognitive states and behavior of the brain. The BOLD signature of brain activation is slightly weak, and hence the noise needs to be eliminated well in order to receive a clear signal. Hence, preprocessing of fMRI data received is an extremely important step before relevant task-related brain performance can be studied.

The first task to be done while preprocessing the data is "slice-timing correction" also called "slice acquisition correction." Slices acquired from multiple planes assembled together form the data from one repetition time. The scanner collects signals or constructs images from various slices within a given brain volume at various times. This is why the slices together correspond to brain activity at different times. Due to this, an event in the brain that occurs simultaneously in multiple slices (over a large volume in the brain) can be interpreted as different, time-delayed BOLD responses. Thus, a timing correction method is applied so that all the slices are brought to the same time instant reference level.

Another error to be fixed is the one induced by head motion. When the person's head moves in the scanning machine while signal capturing, the neurons inside a designated voxel move (position shift from a slice to another) and might represent another voxel (corresponding to a different brain activity that occurred in the past). Thus, this possible error needs to be corrected, and this correction is usually performed by aligning the signal (and its corresponding image constructed) back to the original image acquired in the beginning of the scanning procedure. The difference between the realigned brain position and the originally positioned brain is minimized using this technique, but this does not completely eliminate the effects of movement on the BOLD response signal, and certain artifacts might still remain.

Due to the inhomogeneities present in the static powerful magnetic field applied, some distortion is usually imparted to the images acquired by fMRI. As described in the previous section (MRI), the use of shimming coils reduces these inhomogeneities. Another method of removing this spatial distortion is to map the magnetic field originally applied by plotting two or more images with different echo times. A uniform field implies uniform differences between the images.

To test an assumption or a hypothesis regarding a particular region in the brain (or a particular function or event in the brain) among a population (group of subjects), we need to recognize the same area in the brain across all the people involved and integrate the results across subjects. One way of doing this is to use a designated common "brain atlas" and adjust and compare all the brains to it and study them as one group. In this technique, we computationally warp or align the anatomical structure of all the subjects' brain to a standard template of the human brain structure. This is termed as spatial normalization. A limitation of this method is the intersubject variability in the anatomy of the brain, which cannot be overcome by aligning to a template. Another way of doing this is to create a "probabilistic map" of the brain structure using the scans as the statistical data from hundreds of individuals.

Another important step in fMRI data processing is temporal filtering in which irrelevant frequencies, which are of no interest, are removed from the signal. The net intensity of the signal from a voxel is the summation of different repeating waves that are plotted (the plot being called the power spectrum) using Fourier transformation. In temporal filtering, we remove certain periodic waves that do not provide any relevant information and then sum the remaining waves back again using inverse Fourier transform. Another processing technique is the spatial filtering or "smoothing," which is done by averaging the various voxel intensities and creating a spatial map of the variations in intensity across the brain regions. This is done to control the rate of false positives being acquired in fMRI and to reduce the number of independent tests that need to be performed.

fMRI is used to assess the risks associated with neurosurgery and also to study the functioning of a brain containing any kind of pathology or injury.

Using fMRI mapping, we can identify and analyze the various regions in the brain associated with different specialized functions. Patients with pathologies in the brain are much harder to scan and diagnose as the hemodynamic response might be altered by the disease and the signals might correspond to different states of the brain and mislead the physician to false positives or false negatives. This is the main reason that the clinical use of fMRI is limited and is mostly restricted to research. The main clinical challenges faced in fMRI are movement artifacts, field inhomogeneities in the region of interest of the brain (which can be induced by the applied field itself or by cerebral pathologies), altered baseline intelligence of the subject (which affects the cognitive behavior of the brain), impaired task compliance or altered mental status, and so on.

Despite these difficulties, fMRI proves extremely helpful in neurosurgeries, mapping activity to specialized brain region, studying the blood–brain barrier penetration by certain drugs, and so on. There are many more future applications, in which fMRI is of prime importance. Some of them include in pain management (to study the cortical area affected by pain therapy), to study and analyze the physiology behind neurological disorders and in the understanding of the physiology behind certain perceptual as well as cognitive events.

Currently, attempts are going on to enhance the contrast in the obtained fMRI images even more by the use of external contrast agents, which are sensitive to certain aspects of specific brain activities. Targeted or "smart" contrast agents are being developed, which are either internalized or "trapped selectively" in the target areas for *in vivo* tracking and imaging by fMRI [39]. The current challenge is to design contrast agents with higher relaxivities that will selectively localize in specific tissues or organs [40]. Biofunctionalized nanoparticles have unique magnetic and diffusion properties, which make them excellent tracers in fMRI. These nano-sized particles are specially designed and implemented for MR imaging and characterization of the anatomy and function of the human body (especially the brain). This new generation of nanoscale contrast agents opens a plethora of imaging options and provides enhanced sensitivity and contrast quality to images acquired through fMRI and regular MRI. The properties and design of nanoparticles including their applications are described in the next section, followed by their implementation as fMRI contrast agents.

10.5 Nanoparticles in Imaging

Nanotechnology is the study and manipulation (designing, separating, consolidating, deforming, etc.) of particles at the atomic or molecular level. The term "nanotechnology" was first defined in 1974 by N. Taniguchi and is derived from the root word "nano," which is the prefix used by the International System of Units for representing one-billionth (10^{-9}) of a meter. To imagine the scale better, the comparison between the lengths of a nanometer to one meter is the same scale as the comparison of a marble to the earth [41]. As mentioned in the introduction, the usage or application of nano-sized particles is not as modern as it seems to be. Artisans and scientists have utilized these nanoscale entities in their work for centuries. What makes nanotechnology a novel science is the utilization of the nanoscale properties of these materials and their application in various diverse fields. This has caused a revolution in the field of science and technology and provided a new platform that offers an abundance of opportunities and opens doors to innovation and numerous new creations. Nanotechnology is now a rapidly growing multidisciplinary platform for research in various fields with the advent of new devices and techniques, which allow a precise control of properties of matter at the nanolevel.

Nanoscience is basically the science of the irregular phenomena and occurrences caused by the nanoscale properties of matter, which are not exhibited by regular macroparticles. The field of nanoscience derives principles and fundamentals from biology, chemistry, and physics. The physics of nanotechnology involves the interaction of nano-sized particles with light, surface, and internal energies and the polarimetric and EM properties of nanoscale elements.

The development of nanotechnology took place over a long period of time, starting in the 1970s. This field saw great development and active research along with commercial interest in the early 2000s. Around the end of the twentieth century, microscopes were being utilized to image biological nanostructures, and various other methods in fluoroscopy and chemiluminescence were being developed for the imaging of nanoscale biological structures [42]. To study the molecular arrangement of nanoscale systems, NMR techniques were being studied by observing the intra- and intermolecular interactions of the nuclear spins of nanostructures to produce a 3-D map. Numerous such tools were being developed to better understand nanomaterials and to master the design and fabrication of such nanoscale systems. Nanotools, as they are referred to, were designed to gather information at nanoresolutions, nanoscale systems were being "simulated," and experiments for the design of such systems were being carried out at a large scale due to the boom in this field since the early 2000s.

Various nanoscale structures such as nanoparticles, nanotubes, and nanofibers were designed to fabricate materials with superior chemical, mechanical, optical, and physical properties than regular macromolecular materials. Various physical phenomena such as quantum mechanical effects and statistical mechanical effects get more pronounced as the size of the basic structural particle decreases to the nanolevel. The quantum size effect alters the electronic properties of solid materials when they are formed of nano-sized particles.

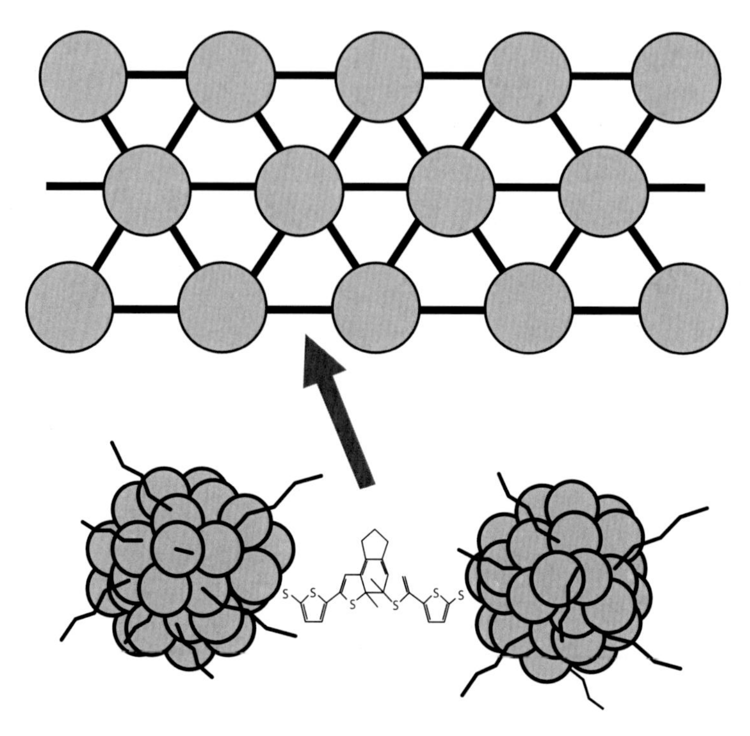

FIGURE 10.19
Schematic picture of a nanoparticle network connected by organic molecules. (Image courtesy of Edwin et al., Studying molecular transport using nanoparticle networks, http://www.physics.leidenuniv.nl/sections/cm/amc/nanoparticle%20networks/nanonetworks.htm)

Another major effect is the prominent increase in the surface area-to-volume ratio of the structure, which also modifies the thermal and catalytic properties of the material. Normally opaque materials exhibit transparency, and the solubility and combustibility of substances are increased, among many other exceptional properties of nanomaterials (Figure 10.19).

10.5.1 Nanoparticles: Properties and Synthesis

A particle is a small structure that behaves as a complete entity in terms of its properties. Nanoparticles are the ultrafine structures that are sized between 1 and 100 nm. Other nanoscale systems include nanoclusters, nanocrystals, and nanopowders. The immense scientific interest in nanoparticles stems from the properties that the small size imparts to these particles, which cannot be found in regular structures. These properties are mostly (though not always) due to the large surface area of the bulk structures formed by nanoparticles as the percentage of atoms at the surface becomes significant, increasing their reactivity.

Nanoparticles exhibit optical properties that are not demonstrated by other macroparticle-based materials by producing "quantum effects" due to their ability to "confine" electrons. The absorption of solar radiation is also enhanced by designing cells made of nanoparticles. They even change color in various media (an example is the deep red appearance of gold nanoparticles in solution). The melting point of gold nanoparticles is much lesser than that of regular gold. Nanoparticles also diffuse very easily, even more so at elevated temperatures, due to the large value of the surface area-to-volume ratio, though agglomeration of nanoparticles can cause complications. Surface plasmon resonance is another property of nanoparticles, which is the oscillation of the electrons in the last shell, of solids that are irradiated with light. Using this phenomenon, the adsorption of materials onto metal surfaces can be measured. Ferromagnetic nanoparticles demonstrate superparamagnetism, which is the state in which the average magnetization appears to be zero (though it actually is not) due to the amount of time taken to measure the "flipping" of magnetization, which exceeds the actual time taken for the magnetization of these particles to flip (called the "Neel" relaxation time). During this state of supermagnetism, it is possible to magnetize these materials like a paramagnet with their magnetic susceptibilities being much higher than those of paramagnets. Thus, superparamagnetic nanomaterials behave like extremely susceptible paramagnets [43]. This is the basic phenomenon that makes nanoparticles suitable contrast agents in MRI and enhances the contrast of the images hence produced.

The synthesis of nanoparticles can be done by following a variety of methods, the most popular being pyrolysis and attrition. In the pyrolysis technique, a vaporous fluid acts as the harbinger and is burned after being transmitted forcefully through an orifice. The result of this is a sooty solid, which is classified in air (elutriated, which is the process of separating lighter particles from heavier ones using a stream of high-speed gas or liquid), and the oxide particles thus formed are retrieved from the gases emitted as by-products. This method of synthesis, many a times, results in aggregates or clusters rather than separate single nanoparticles. The other technique, attrition, dictates that macro- or microparticles (acting as raw materials) are ground finely to reduce their size. This is done in a ball mill grinder, which produces extremely fine powder of the substance (other techniques can also be adopted). Again, air classification is performed to recover nanoparticles.

Another way of synthesizing nanoparticles is to heat microparticles using thermal plasma at temperatures as high as 10,000 K and then cooling them as they leave the plasma region. The main types of plasma generators used to create nanoparticles are DC arc plasma reactors and RF induction plasma torches. In the plasma generators that use RF induction, an induction coil generates an EM field, which provides the heat for the plasma. Many types of gases can be used with this plasma generator—inert, reducing, oxidizing, and so on—and it works in a frequency range of 10 kHz–40 MHz [44].

To obtain complete evaporation, the size of the introduced droplets needs to be small as their residence time is very short. This plasma generator has been applied in the synthesis of many types or nanoparticles such as carbides, oxides, and nitrides of silicon and titanium. In the other kinds of plasma torch, DC arc plasma reactor, the heat energy is provided by an electric arc between the anode and the cathode.

Inert gas condensation is also frequently used to create nanoparticles from metals with a low melting point. The metal is first vaporized and then supercooled to obtain nanoparticles of that metal. Another technique employed is the radiolysis by gamma rays, which creates free radicals in solution that reduce metallic ions to a state of "zero valency." Its reoxidation is prevented by scavenger chemicals. When the metal is in the zero-valance state, their atoms begin to bind into nanoparticles. A chemical surfactant is used to separate and coat these particles and prevent their merging.

Another popular method of fabricating nanoparticles is the wet chemical "sol–gel" technique. This chemical–solution–decomposition process is widely used to process and synthesize materials and utilizes a liquid as the precursor.

For biological processes, the external surface of nanoparticles is coated with agents that allow targeting and control and manipulation of stability and solubility.

10.5.2 Types of Nanoparticles

Nanoparticles can be formed by many kinds of elements or chemical compounds, which prove to be useful. The following are the primary important ones:

1. *Gold nanoparticles*: As gold has always demonstrated superior properties and has been used in medical applications, the development of gold nanoparticles was natural. These nanoscale gold structures can be used as catalysts in chemical reactions (as the reactivity is enhanced) and in targeted drug-delivery mechanisms (attached to site-specific molecules). Gold nanoparticles are also known to convert light of a specific wavelength to heat energy (by absorbing certain wavelengths). For this purpose, nanorods and nanospheres prove to be effective (created from gold) [45].
2. *Iron nanoparticles*: It was found that nanoscale systems constructed with iron and iron oxide are extremely useful. They are magnetic in nature and have a greatly increased surface area. These are useful in medical imaging (due to the magnetic properties) and in cleaning groundwater pollution (Figure 10.20).
3. *Silver nanoparticles*: These particles constitute of a large number of atoms at the surface and are useful in destroying bacteria.

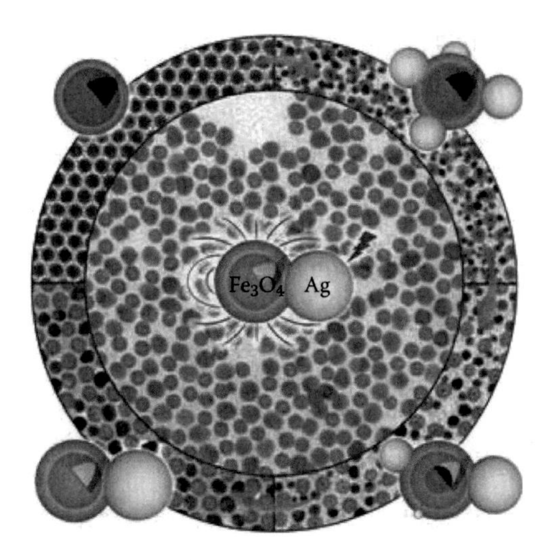

FIGURE 10.20
Evolution pathway from iron nanoparticle coated with thin layers of amorphous iron oxide to hybrid nanoparticles made of solid Ag domain and hollow iron oxide nanoshells. (From Nanoparticles and their Applications, Copyright 2007, Present Hawk's Perch Technical Writing, LLC, http://www.understandingnano.com/nanoparticles.html)

This property is used in wound healing, as an odor repellant, as an antimicrobial agent, etc. It is also used in the design of silver nanowires and carbon nanotubes.

4. *Quantum dots*: Semiconductor nanoparticles are called quantum dots, and they emit a specific color after being emitted by light, which depends on their size, which determines the energy difference between the conduction and the valence band (when a free electron absorbs energy, it is excited to the conduction band; this then loses energy and returns to the valence band, emitting light whose wavelength, hence color, depends on this energy gap).
5. *Titanium dioxide nanoparticles*: These particles are photocatalysts and work very well in sunlight. They emit white light as the light coming from these nanoparticles contains all wavelengths. They are used in targeted drug delivery and as titanium oxide nanotubes.
6. *Silica nanoparticles*: They have poor conductivity of heat and electricity but are of great use in forming aerogels, which are nanotechnology based and prove to be one of the best insulators ever known. Nanospores filled with air are interspersed in silica nanoparticles to form the silica aerogel. Also, attaching these silica nanoparticles with certain molecules that bind to other surfaces, we can functionalize them and utilize them for various purposes. Nanofilms can be made using silicon dioxide (about 1 mm thick) and are used to provide insulation between two devices.

7. *Polymer nanoparticles*: These particles are being developed to directly deliver the chemotherapy drug called docetaxel. The nanoparticles are highly susceptible to a protein present on most of the cancer tumors. Thus, imaging could prove very helpful.

There can be several other types of nanoparticles with numerous applications. Nanoparticles, due to their size and structure and the rare properties induced, find themselves in abundance of applications and uses, with new ones coming up every day. Nanoparticles and nanotechnology as such have had a huge impact on the field of medicine, be it in the area of instrumentation, biological powering systems, medical materials, biomedical imaging, drug delivery, advanced prosthetics, etc. Nanoparticle design and fabrication for contrast enhancement in scanning systems is one of the most popular works of all the ongoing research in this area [46]. This is based on quantum resonance and other properties. The detailed use of nanoparticles in biomedical imaging and drug delivery is explained ahead.

10.5.3 Nanoparticles in Imaging and Drug Delivery

The medical applications of nanoparticles range from functionalized nanobiomolecules acting as probes and therapeutic agents in the body to the development of nanoelectronic biosensors. In the last two decades, various new nanoscale systems have been developed to be used as diagnostic/therapeutic agents. Cells in the precancer stage, disease markers, fragments of viruses, etc., are some of the parameters at the molecular level that can be detected only by using nanoparticles and not by conventional diagnostic tools. The specificity and sensitivity of MR imaging are increased using nanoparticle-based contrast agents. In therapeutics too, nanoparticles offer controlled and targeted administration of drugs.

Along with the chemical composition and structure, the uses of nanoparticles in diagnostics and therapeutics depend on their size and physical properties. Also, the nanoparticles used in medical applications should be biodegradable, biocompatible, and nontoxic. It should be possible for these nanoscale particles to flow inside the bloodstream alongside the blood cells without getting trapped in any microanatomical structure. But, at the same time, in order to perform their function of image enhancement/drug delivery, they should be slightly larger than the size of an atom [42]. The nanoparticles used for medical applications are usually of the size range of 10–100 nm. Their size is optimized to be small enough to avoid blockage and large enough to offer sufficient clearing time (stay in the bloodstream for a good period of time to allow imaging and/or release of drugs at the required site) in comparison to small molecules in the body. We also know that nanoparticles possess superior diffusion properties due to their size; this leads to the easy delivery into areas with tumors or infections, as these abnormalities dilate the tissue lining and blood vessels around them due to inflammation.

For imaging applications, nanoparticles should possess stability in an aqueous medium and demonstrate good absorption and fluorescence. Nanoparticles are not susceptible to optical quenching (reduction in the intensity of luminescence/fluorescence by longer wavelengths) and are thus advantageous and preferable over regular organic dyes and contrast agents [42]. Keeping all these factors in mind, various kinds of nanoparticles can be designed and fabricated using metals, polymers, ceramics, etc. For custom targeting and imaging applications, the surface coating of the nanoparticles should be well designed too. The dominating areas of nanoparticle research in medicine are drug delivery and imaging.

Ever since silver halide was used in photography, nanoparticles were applied in image enhancement. Current technology has been developed to design much more selective and efficient nanoparticles that utilize their advanced optical properties to the fullest. Today, nanoparticles are used to enhance imaging in various modalities such as x-ray, CT, MRI, and ultrasound. For diagnosing diseases (tumors/lesions) in the soft tissues, it is very important to have good contrast between different types of tissues. By using nanoparticle-based agents, this image contrast can be enhanced greatly without the negative effects of radioactive isotopes commonly used.

Nanoparticles can be attached to certain chemicals and functionalized to target specific sites in the body for the detection of certain areas of interest. The contrast in complementary imaging modalities such as MRI and optical imaging can be enhanced by merging various properties of nanoparticles and implementing in one design.

In x-ray and CT imaging scans, regular contrast agents are iodine compounds and radioactive isotopes. These pose possible risks to the body due to their toxicity and also possess a low clearance time, making targeted imaging difficult. Clearance time and targeted imaging can be improved by encapsulating iodine nanoparticles with polymers, but this still poses a risk due to the toxic nature of iodine compounds. Nanoparticles made of heavy metals increase the clearance time; their gold encapsulation makes them biologically inert and stable, eliminating toxicity. Gold particles can also be given a coating of selective binding antigens or target chemicals for certain cells acting as receptors. This boosts the ability of an x-ray scan to detect cancerous cells by a large amount.

In MRI, chelates of gadolinium (paramagnetic in nature) or iron/manganese (supermagnetic in nature) are used to intensify the image and provide better contrast [47]. Metallic and silicon particles that are magnetic in nature are also used as they add to the magnetization and enhance contrast. Contrast agents composed of nanoparticles and coated with polymers are being looked at, as they eliminate the toxic exposure caused by gadolinium or other chelated complexes [48,49]. Silicon nanoparticles of various shapes, after being coated with layers of conducting material greatly, enhance the magnetic interactions with the static field applied in imaging. Nanoparticle-based coating also increases the magnetic susceptibility of materials in implanted devices. Image guidewires can also be coated with these "magnetization-enhancing" nanoparticles for better imaging.

Quantum resonant particles or "dots" are optically powerful and can be used in x-ray and MR imaging for producing an intense signal and also in IR imaging and UV fluorescence. The functionalization of nanoshell particles (spherical nanoparticles with a dielectric core) is done so that they can enhance the detection and imaging of cancer cells. The optical properties of such structures can be modified to suitable levels by adjusting the size and shape of the nanoscale particles/structures. In optical imaging systems like photoacoustic imaging and in optical coherence tomography, nanoparticles with a heightened optical sensitivity are used for contrast enhancement. Various microscopy fields such as fluorescence, confocal, and dark-field microscopy also implement these nanomaterials in their imaging systems [42].

Along with all these imaging applications, nanoparticles are also being utilized to transport and supply energy to targeted areas for treatment (by destroying affected cells, say). The resonant nanoparticles (quantum dots) absorb radiation and transmit it in their surrounding environment in the form of heat. This is a major application of nanoparticles in targeted energy delivery, also used in photothermal and photodynamic therapies.

Cancer cells are released into the bloodstream that are too small (and inflow with normal blood cells) to be detected by conventional imaging methods. Magnetic nanoparticles are attached with specific antigens and functionalized, which separate these cancer cells from the bloodstream and tissue samples, and then concentrate them for enhancing detection. This is an extremely important application as it renders the imaging of even microscopic tumors possible. This can also be done by mass enhancement instead of using magnetic susceptibility, separating the cells based on centrifugation.

As we can see, nanoparticles provide tremendous advantages and applications in medical imaging due to their enhanced optical and magnetic properties, which stem from their size and shape.

Drug delivery is based mainly on the bioavailability of the drug in the body (the presence of drugs in the body when and where they are needed). Nanomedical approaches are being followed to enhance drug delivery by designing and implementing nanoscale systems for targeting and transporting chemicals throughout the body. The pharmacokinetics and the therapeutic properties of drugs being administered can also be improved by implementing nanoparticles in them (Figure 10.21).

There are many advantages of nanoparticle-based drug-delivery systems, such as the increased of drugs (which need to be water soluble), extended life

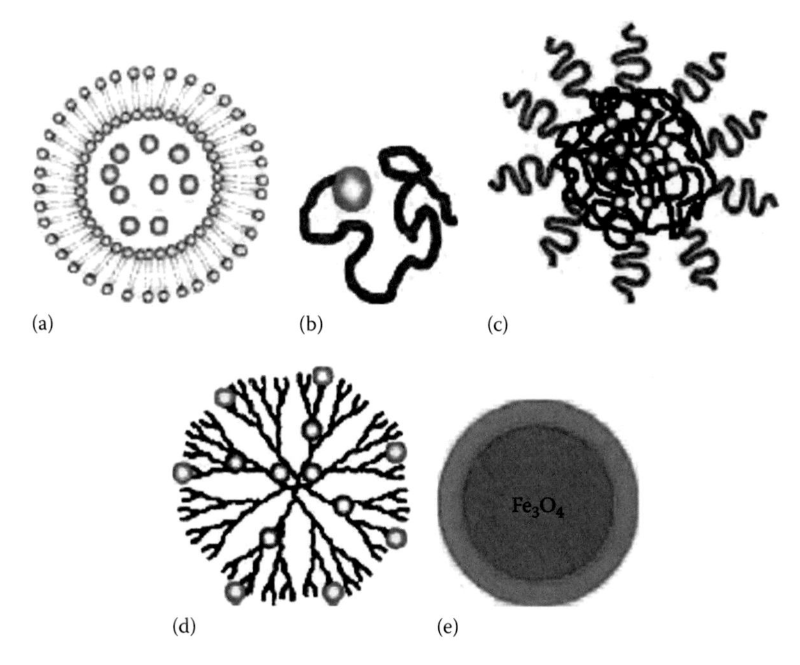

FIGURE 10.21
Schematic illustration of therapeutic nanoparticle platforms in preclinical development: (a) liposome, (b) polymer–drug conjugate, (c) polymeric nanoparticle, (d) dendrimer, and (e) iron oxide nanoparticle. The center dots in (a) and circle in (b) and outer dots in (d) represent hydrophilic drugs, and the center dots in (c) and (d) represent hydrophobic drugs. (From Zhang, L. et al., *Clin. Pharmacol. Ther.*, 83, 761, 2008.)

of the drug in the body (longer time in circulation), control of the rate at which the drug is being administered, and targeted delivery reducing side effects. Thus, many research and commercialization initiatives are being taken up to utilize these properties of nanoparticles into the drug-delivery system.

The straightest way to utilize nanoparticles to target and destroy cancer cells (or other tumor/infections) is to "implant" the nanoparticles (loaded with the drug to be administered) into the diseased site. This method, as is pretty clear, concentrates the drug at the necessary site and does not require it to be circulated through the bloodstream. This technique is often used for localized pathologies in the body tissues. A "molecular recognition function" is added to the nanoparticles so that it targets a specific group of cells in the body, by selectively coupling with them [42]. For this purpose, the basic method followed is the antigen–antibody binding mechanism. Other characteristics are being looked at to better target the nanoparticles to specific tissues/cells.

The most dominant nanoparticle-based drugs are liposomal drugs and polymer drug conjugates [50]. These functionalized nanoparticles are widely being applied in therapeutics worldwide (Figure 10.22).

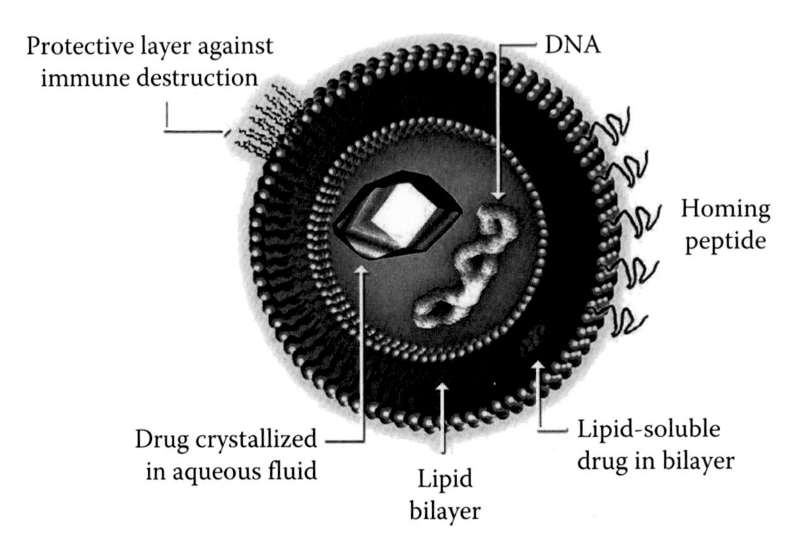

FIGURE 10.22
Liposomes are composite structures made of phospholipids and may contain small amounts of other molecules. Liposomes can vary in size from nanometers to tens of micrometers. (Image courtesy of Wikipedia [released into the public domain].)

10.6 Nanoparticles as Contrast Agents in fMRI

In the previous section, it was briefly explained how nanoscale particles can be used for enhancement of medical imaging techniques. This part of the chapter will focus on the implementation of nanoparticles in functional (and conventional) MRI. We know that gadolinium is the substance of preference in contrast enhancement of MR images; hence, it is but natural that gadolinium-based nanoparticles prove to be excellent tracers in the study of the functional/anatomical information of the human body, acquired by MRI.

These nanoparticles make the study of molecular processes in the body possible, using functional MRI. Molecular imaging (which allows us to observe molecular interactions in healthy and diseased states) is also performed using fMRI by selecting SPIO nanoparticles as the contrast agents, as mentioned briefly earlier.

These agents usually possess a core of magnetite, which is encased in polysaccharides, and have a hydrophilic coating [51]. SPIOs have a strong magnetic susceptibility, and when placed in an external magnetic field, the alignment of their moments increases the magnetic flux and alters the magnetization. This causes a rapid dephasing of protons and leads to a detectable MR signal. Thus, they generate a strong signal, enhancing the contrast quality of the resultant image. SPIO nanoparticles are also used in cell tracking and in magnetic relaxation switching (which includes oligonucleotide detection and proton enzyme detection).

The common methods for synthesizing these magnetite nanoparticles are coprecipitation (of ferrous and ferric salts in an alkaline solution) and microemulsions (which act as bioreactors for the production of nanoparticles) [51]. Recently, in 2010, a technique for the hydrothermal synthesis of magnetite nanoparticles (which were to be used as contrast agents in MRI) was proposed in Malaysia. In this method, Fe_3O_4 is precipitated and treated hydrothermally to produce nano-sized particles of Fe_3O_4 [52]. The surface of these nanoparticles is modified to increase biocompatibility and to avoid aggregation of particles. By using this hydrothermal technique, the cost of production is reduced, highly crystalline particles with controllable size parameters are obtained, no postheat treatment is needed, and we do not need to use any organic reagents. In this work, nanoparticles were functionalized, and T2-weighted MRI scans were performed to analyze the contrast in the image.

Researchers have now developed a contrast agent that is fifteen times more sensitive than the currently used chemicals. This was synthesized by chemically binding gadolinium ions to nanodiamonds (nanoclusters of carbon atoms), which greatly enhances the optical and physical properties of the contrast agent [53]. This increase in sensitivity is greater than what any other nanoparticles offer. These nanodiamonds are also being studied for drug-delivery purposes. These nanodiamonds do not alter the gene sequences in any way (which certain carbon nanoparticles might) (Figure 10.23).

The Center for Medical Image Science and Visualization also has research going on, on the use of biofunctionalized nanoparticles as fMRI

FIGURE 10.23
Nanodiamond structure made with carbon nanoparticles. (From *New Compound Improves MRI Contrast Agents*, MIT Technology Review, India, http://www.technologyreview.in/news/417183/new-compound-improves-mri-contrast-agents/)

contrast agents. They use gadolinium-based nanoparticles biofunctionalized to study microscopic molecular interactions within the body. Monocyte cells were incubated with gadolinium nanoparticles, and the fMRI was performed using these nanoparticles as tracers for watching molecular activity [54].

The researchers at Massachusetts Institute of Technology have developed calcium-sensitive nanoscale contrast agents for functional MRI in an attempt to refine fMRI from just a rough analysis of brain activity to a "fine-tuned analysis" of neuronal activity and its significance [55]. Alan Jasanoff at MIT, along with his colleagues, has synthesized calcium-based nanoparticles, which allows the physician or the researcher to observe and analyze every single element in the neural circuitry as and when the action happens, rather than simply monitoring the behavior of a set of neurons based on the hemodynamics in the brain (Figure 10.24).

Calcium can provide an instantaneous and direct measure of the neural activity as it flows into the neurons almost immediately as they "fire," and this concentration of calcium is directly proportional to the rate of neuron activation. Thus, tracking calcium ion-based nanoparticles in the brain is one technique that allows us to follow the flow of information much better than what the BOLD contrast offers.

It was mentioned earlier that nanoparticles can separate cancer cells out of the bloodstream and be concentrated for better imaging. This multifunctional nanoparticle was recently designed at the University of Washington [56]. These nanoparticles eliminate the background noise from the resultant images and facilitate much more precise medical imaging. This particle combines photoacoustic imaging with the magnetic properties, and a pulsed magnetic field is used to shake the magnetic core of the nanoparticles.

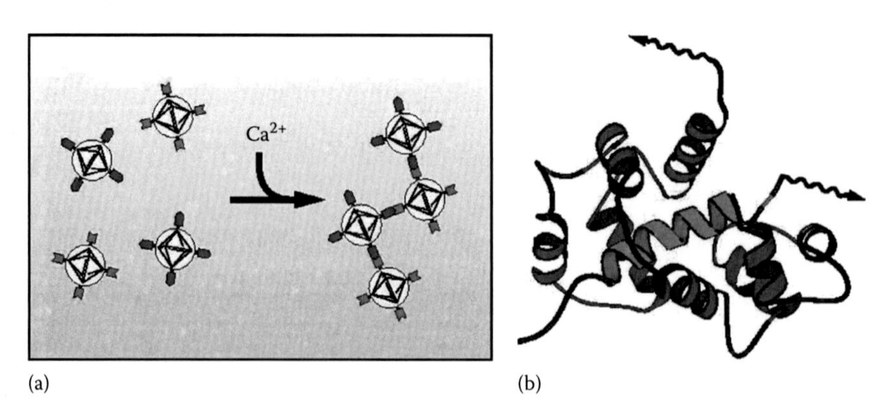

FIGURE 10.24
Calcium-sensitive nanoparticle-based contrast agents. (a) Calcium sensor mechanism: nanoparticles form binary aggregates in the presence of Ca and (b) interaction of nanoparticle polypeptides. (From Bradely, D., Nanoparticles sharpen fMRI, http://www.spectroscopynow.com/details/ezine/sepspec14678ezine/Nanoparticles-sharpen-fMRI.html?tzcheck=1&tzcheck=1&tzcheck=1&tzcheck=1&tzcheck=1&tzcheck=1)

Trials/experiments with synthetic tissue demonstrated that this technique can suppress the noise signal completely out. The size selected for this nanoparticle was around 30 nm, and it was made of a magnetic core consisting of iron oxide and a surrounding shell of gold. This gold shell absorbs IR wavelengths, and they can also be implemented in optical imaging modalities, heat delivery, functionalizing with a biomolecule for cell targeting, etc. [57]. The implementation of such a nanoparticle into the design of contrast agents for fMRI greatly increases its specificity and contrast level.

Another team of researchers has developed an organic nanoparticle (biodegradable and nontoxic) that can be used to find and treat tumors and deliver heat energy at targeted spots. Chlorophyll and lipid were combined to design a versatile nanoparticle, which offers several unique optical and photonic applications [57]. The structure of this nanoparticle is like that of a balloon that is filled with the drug to be delivered at the site. This technology is based on the photothermal effect, which uses light energy to destroy tumors. This can also be used in photoacoustic imaging as well as fMRI. Gold nanoparticles are also being functionalized with gadolinium complexes to improve the contrast and detail of fMRI scans and allow us to study the functional information in the brain at a deeper level [58].

Yallapu et al. [59] have designed a new technique for developing SPIO nanoparticles, which can be dispersed in water. These unique particles can be used for various applications like drug delivery, hyperthermia, and enhancing contrast agents in MRI. To synthesize these, iron particles were precipitated using ammonia as a catalyst and coated with polymers to enhance performance. Due to its chemical structure, this design enables the "encapsulation" of cancer drugs in it and also exhibited higher contrast in MRI images, when used as a contrast agent.

In another experiment, DNA surface modification was performed on gadolinium phosphate nanoparticles to be used as MRI contrast agents as they enhance the relaxation times, improving the image contrast [60]. After the highly water-dispersible and stable $GdPO_4$ nanoparticles were synthesized, their surface chemistry was modified using oligonucleotides. This demonstrates the surface properties of the Gd^{3+} ion. New relaxation times were measured using these nanoparticles as MRI contrast agents. The data confirmed that both T1 and T2 were enhanced [61]. These particles can be biofunctionalized for targeted delivery.

The diagnosis of various liver diseases, studying the growth of cancer, using MRI has been given a tremendous boost by the use of magnetic nanoparticles as contrast agents, due to their unique properties of cell phagocytosis [47,61,62]. Amino-terminal fragment–iron oxide (ATF-IO) nanoparticles have the ability to bind site specifically to breast cancer cells, which is followed by mediated endocytosis. This enhances the MRI contrast in T2-weighted scans [63]. This particular study showed that magnetic nanoparticles, when functionalized with biomarkers, can improve the current state of cancer detection and its specificity.

By engineering functionalized magnetic nanoparticles, different complimentary imaging modalities can be merged, for example, PET and MRI. For this purpose (to combine nuclear imaging with MRI), magnetic nanoparticles are infused with "optical dyes and radioisotope-labeled tracer molecules" [64,65]. Magnetic nanoparticles are also being synthesized for targeted delivery of therapeutic agents using MRI. To functionalize nanoparticles, surface chemistry is altered and various coatings are used that enhance their performance as MRI contrast agents.

When magnetic nanoparticles are used as MRI contrast agents, the relaxation rate per unit concentration of magnetic ions is called relaxivity, and there are two kinds of relaxivities—longitudinal and transverse [60]. The main parameters pertaining to the nanoparticles that determine the relaxivities are the size of the fabricated magnetic nanoparticles, their chemical composition, and their surface properties that can be altered.

It is an established fact that the nanoparticle size affects the relaxation rate in MRI. For smaller nanoparticles, the diffusion of the protons happens faster than the shift in the RF, making it the key parameter contributing to contrast. In larger nanoparticles, this is not the case as the surrounding field is stronger. Hence, larger nanoparticles have a longer magnetization. Also, the magnetic moments of these nanoparticles are also dependent on their size and structural as well as surface properties. The transverse relaxivity is also proportional to the magnetic moment. Thus, the magnetic behavior is directly correlated with the size of the nanoparticles [66,67].

We know that the major factor contributing to the contrast in MRI scans is the variation in the concentration of water in different tissues, under a magnetic field. As the interaction between the magnetic nanoparticles and the water molecules happens on the surface of the nanoparticles, the properties of and the parameters affecting the surface of the nanoparticles should be well studied. To enhance the contrast greatly, hydrophilic substances are used to design the nanoparticle surface, and polymer coating can be applied for increased stability and biocompatibility [67].

The chemical composition of magnetic nanoparticles is another factor that greatly affects the contrast-enhancing capabilities as it can alter the magnetic moments. Manganese ferrous oxide nanoparticles are supposed to have the largest magnetic moment and demonstrate high relaxivity values [68].

10.7 Conclusion

It can be seen that nanoparticles, due to their unique structure and properties that arise from their size, offer a large variety of applications in medicine, more prominently so in targeted drug delivery and medical imaging. For direct drug delivery and as MRI contrast agents, magnetic nanoparticles are the optimal choice as their quality and quantity of biocompatibility and level of toxicity are well understood [60]. Also, these magnetic nanoparticles produce large field gradients (in the presence of an applied field) and shorten both T1 and T2. By altering the surface chemistry and functionalization of these nanoparticles by various processes such as polymer coating, their behavior can be further enhanced. It is very essential to study and analyze the relationship between the size and structure of nanoparticles and the corresponding improvements they cause in MRI relaxation times.

Currently, there is a vast amount of research going on to enhance functional MRI scans by implementing these nanoparticle-based contrast agents into them as this greatly improves the resolution and contrast of the images obtained. Many studies are being conducted to better understand the effect of various parameters on the relaxation times in MRI and to develop new techniques to enhance imaging. This is a relatively new application of nanoparticles (as fMRI itself is currently evolving and being studied greatly) and offers a huge scope for further research.

References

1. C. A. Roobottom, G. Mitchell, and G. Morgan-Hughes, Radiation-reduction strategies in cardiac computed tomographic angiography, *Clinical Radiology* 65(11): 859–867, 2010.
2. National Toxicology Program 2010, U.S. Department of Health and Human Services, Report on Carcinogens—Background document for: Formaldehyde, 11th report on carcinogens, ntp.niehs.nih.gov, Accessed on November 8, 2010.
3. W. G. Bradley, History of medical imaging, *Proceedings of the American Philosophical Society*, 152(3): 349–361, 2008.
4. T. Geva, Magnetic resonance imaging: Historical perspective, *Journal of Cardiovascular Magnetic Resonance* 8: 573–580, 2006.
5. F. Bloch, W. Hanson, and M. Packard, Nuclear infraction, *Physical Review* 69: 127, 1946.
6. E. Purcell, H. Torrey, and R. Pound, Resonance absorption by nuclear magnetic moments in a solid, *Physical Review* 69: 37–38, 1946.
7. J. T. Bushberg et al., *The Essential Physics of Medical Imaging*, 3rd edn., North American Edition, Lippincott Williams & Wilkins, Philadelphia, PA, 2011.
8. S. A. Huettel, A. W. Song, and G. McCarthy, *Functional Magnetic Resonance Imaging*, 2nd edn., Sinauer, Sunderland, MA, 2009. ISBN 978-0-87893-286-3.
9. A. Webb and N. Smith, *Introduction to Medical Imaging: Physics, Engineering, and Clinical Applications*, Cambridge Texts in Biomedical Engineering, Cambridge University Press, Cambridge, U.K.; New York, 2011.
10. W. S. Wong, J. S. Tsukuda, K. E. Kortman, and W. G. Bradley, *Practical Magnetic Resonance Imaging: A Case Study Approach*, Aspen Publishers Inc., New York, 1987.

11. MRI vs CT, The Internet Stroke Center, Dallas, TX, http://www.strokecenter.org/professionals/stroke-diagnosis/mri-compared-to-ct/, Accessed on September 29, 2012.
12. S. Ogawa, T. M. Lee, A. R. Kay, and D. W. Tank, Brain magnetic resonance imaging with contrast dependent on blood oxygenation, *Proceedings of the National Academy of Sciences of the United States of America* 87(24): 9868–9872, 1990.
13. C. S. Roy and C. S. Sherrington, On the regulation of blood supply of the brain, *Journal of Physiology* 11: 85–108, 1890.
14. F. W. Wehrli, J. MacFall, T. H. Newton, and D. G. Potts, *Advanced Imaging Techniques*, Vol. 2, Clavade Press, San Francisco, CA, 1983, pp. 81–118.
15. P. T. Fox, M. E. Raichle, M. A. Mintun, and C. Dence, Nonoxidative glucose consumption during focal physiologic neural activity, *Science* 241: 462–464, 1988.
16. T. Soderberg, *Organic Chemistry with a Biological Emphasis*, University of Minnesota, Morris, MN, 2010.
17. MIT News, Nanoparticles to aid brain imaging, http://web.mit.edu/newsoffice/2006/imaging-neurons.html, Accessed on October 8, 2012.
18. M. E. Akerman, W. C. Chan, P. Laakkonen, S. N. Bhatia, and E. Ruoslahti, Nanocrystal targeting in vivo, *Proceedings of the National Academy of Sciences of the United States of America* 99(20): 12617–12621, 2002.
19. W. R. Hendee and R. E. Ritenour, *Medical Imaging Physics*, 3rd edn., Mosby-Year Book, St Louis, MO, 1992.
20. M. B. Smith, K. Kirk Shung, and B. M. W. Tsui, *Principles of Medical Imaging*, Academic Press, New York, 1992.
21. P. Storey, Introduction to magnetic resonance imaging and spectroscopy, *Methods in Molecular Medicine* 124: 3–57, 2006.
22. F. A. Mettler, L. R. Muroff, and M. V. Kulkarm, *Magnetic Resonance Imaging and Spectroscopy*, Churchill Livingstone, New York, 1986.
23. K. Coyne, MRI: A guided tour, National High Magnetic Field Lab, Florida State University, Tallahassee, FL, http://www.magnet.fsu.edu/education/tutorials/magnetacademy/mri/, Accessed on September 29, 2012.
24. A. Policard, A study on the available aspects of experimental tumours examined by Wood's light, *Comptes Rendus des Séances de la Société de Biologie et de ses Filiales*, 91: 1423–1424, 1924.
25. P. Sprawls Jr., *Physical Principles of Medical Imaging*, 2nd edn., Medical Physics Publishing, Madison, WI, 2005.
26. A. Jedlovszky-Hajdú, E. Tombácz, I. Bányai, M. Babos, and A. Palkó, Carboxylated magnetic nanoparticles as MRI contrast agents: Relaxation measurements at different field strengths, *Journal of Magnetism and Magnetic Materials* 324(19): 3173–3180, 2012.
27. C. C. Berry and A. S. G. Curtis, Functionalisation of magnetic nanoparticles for applications in biomedicine, *Journal of Physics D: Applied Physics*, 36: R198–R206, 2003.
28. F. Soderling et al., Colloidal synthesis and characterization of ultrasmall perovskite $GdFeO_3$ nanocrystals, *Nanotechnology* 19: 085608, 2008.
29. M. Ahrén, L. Selegård, A. Klasson, F. Söderlind, N. Abrikossova, C. Skoglund, T. Bengtsson, M. Engström, P.-O. Käll, and K. Uvdal, Synthesis and characterization of PEGylated Gd2O3 nanoparticles for MRI contrast enhancement, *Langmuir* 26: 5753–5762, 2010.

30. A. A. Grunsfeld and I. S. Login, Abulia following penetrating brain injury during endoscopic sinus surgery with disruption of the anterior cingulate circuit: Case report, *BMC Neurology* 6:4, 2006.
31. Groundbreaking new graphene-based MRI contrast agent, Stony Brook University, Stony Brook, NY (2012, June 7), *ScienceDaily*, http://www.sciencedaily.com-/releases/2012/06/120607175817.htm, Accessed on October 25, 2012.
32. F. Figge et al., Cancer detection and therapy; affinity of neoplastic, embryonic, and traumatized tissues for porphyrins and metalloporphyrins, *Proceedings of the Society for Experimental Biology and Medicine* 68: 640, 1948.
33. G. J. Strijkers, W. J. M. Mulder, G. A. F. van Tilborg, and K. Nicolay, MRI contrast agents: Current status and future perspectives, *Anti-Cancer Agents in Medicinal Chemistry* 7(3): 291–305, 2007.
34. S. Ogawa, T.-M. Lee, A. S. Nayak, and P. Glynn, Oxygenation-sensitive contrast in magnetic resonance image of rodent brain at high magnetic fields, *Magnetic Resonance in Medicine* 14: 68–78, 1990.
35. S.-G. Kim and P. A. Bandettini, Principles of functional MRI, *Springer, Functional Neuroradiology*, 293–303, 2012, DOI:10.1007/978-1-4419-0345-7_16.
36. H. Devlin, What is functional magnetic resonance imaging (fMRI)? FMRIB Center, Department of Clinical Neurology, University of Oxford, Oxford, U.K. http://psychcentral.com/lib/2007/what-is-functional-magnetic-resonance-imaging-fmri/, Accessed on September 29, 2012.
37. P. Mansfield, Multi-planar image formation using NMR spin echoes, *Journal of Physics C* 10: L55–L58, 1977.
38. C. Ahn, J. H. Kim, and Z. H. Cho, High-speed spiral-scan echo planar NMR imaging, *IEEE Transactions on Medical Imaging* MI-5: 2–7, 1986.
39. J. Engelmann et al., Novel contrast agents for MRI, Max Planck Institute of Biological Cybernetics, Baden-Württemberg, Germany, http://www.kyb.tuebingen.mpg.de/?id=141, Accessed on October 8, 2012.
40. K. Raymond et al., MRI contrast agents, Department of Chemistry, University of California, Berkeley, CA, http://www.cchem.berkeley.edu/knrgrp/mri.html, Accessed on October 5, 2012.
41. J. Kahn, Nanotechnology, *National Geographic* June: 98–119, 2006.
42. H. F. Tibbals, *Medical Nanotechnology and Nanomedicine*, CRC Press, Taylor & Francis Group, Boca Raton, FL, 2011.
43. S. P. Gubin, *Magnetic Nanoparticles*. Wiley-VCH, Weinheim, Germany, 2009.
44. M. I. Boulos, The inductively coupled R.F. (radio frequency) plasma, *Pure and Applied Chemistry*, 57(9), 1321–1352, 1985.
45. Nanoparticles and their applications, Present Hawk's Perch Technical Writing, LLC, 2007, http://www.understandingnano.com/nanoparticles.html, Accessed on October 1, 2012.
46. M. Reza Mozafari, *Nanomaterials and Nanosystems for Biomedical Applications*, Springer, New York, 2007.
47. J. W. M. Bulte and D. L. Kraitchman, Iron oxide MR contrast agents for molecular and cellular imaging, *NMR in Biomedicine*, 17: 484–499, 2004.
48. R. M. Petoral Jr. et al., Synthesis and characterization of Tb_{3+} doped Gd_2O_3 nanocrystals: A bifunctional material with combined fluorescent labeling and MRI contrast agent properties, *Journal of Physical Chemistry C* 113: 6913–6920, 2009.

49. A. Klasson et al., Positive MRI contrast enhancement in THP-1 cells with Gd_2O_3 nanoparticles, *Contrast Media and Molecular Imaging* 3: 106–111, 2008.
50. L. Zhang, F. X. Gu, J. M. Chan, A. Z. Wang, R. S. Langer, and O. C. Farokhzad, Nanoparticles in medicine: Therapeutic applications and developments, *Clinical Pharmacology and Therapeutics* 83: 761–769, 2008.
51. D. L. J. Thorek, A. K. Chen, J. Czupryna, and A. Tsourkas, Superparamagnetic iron oxide nanoparticle probes for molecular imaging, *Annals of Biomedical Engineering* 34(1): 23–38, 2006.
52. C. Y. Haw, F. Mohamed, C. H. Chia, S. Radiman, S. Zakaria, N. M. Huang, and H. N. Lim, Hydrothermal synthesis of magnetite nanoparticles as MRI contrast agents, *Ceramics International* 36(4): 1417–1422, 2010.
53. New compound improves MRI contrast agents, *MIT Technology Review*, India, http://www.technologyreview.in/news/417183/new-compound-improves-mri-contrast-agents/, Accessed on October 13, 2012.
54. M. Engstorm, Biofunctionalised nanoparticles—New MRI contrast agents, CMIV, http://www.cmiv.liu.se/research/current-research-projects/biofunctionalised-nanoparticles/, Accessed on October 13, 2012.
55. D. Bradely, Nanoparticles sharpen fMRI, http://www.spectroscopynow.com/details/ezine/sepspec14678ezine/Nanoparticles-sharpen-fMRI.html?tzcheck=1&tzcheck=1&tzcheck=1&tzcheck=1&tzcheck=1&tzcheck=1, Accessed on October 13, 2012.
56. Multifunctional nanoparticle enables new type of biological imaging, University of Washington, Seattle, WA (July 28, 2010), *ScienceDaily*, http://www.sciencedaily.com-/releases/2010/07/100727112831.htm, Accessed on October 26, 2012.
57. Organic nanoparticle uses sound and heat to find and treat tumors, University Health Network (March 21, 2011), *ScienceDaily*, http://www.sciencedaily.com/releases/2011/03/110320164233.htm, Accessed on October 26, 2012.
58. J.-A. Park, P. A. N. Reddy, H.-K. Kim, I.-S. Kim, G.-C. Kim, Y. Chang, and T.-J. Kim, Gold nanoparticles functionalised by Gd-complex of DTPA-bis(amide) conjugate of glutathione as an MRI contrast agent, *Bioorganic & Medicinal Chemistry Letters* 18(23): 6135–6137, 2008.
59. M. M. Yallapu, S. F. Othman, E. T. Curtis, B. K. Gupta, M. Jaggi, and S. C. Chauhan, Multi-functional magnetic nanoparticles for magnetic resonance imaging and cancer therapy, *Biomaterials*, 32(7), 1890–1905, 2011, ISSN 0142-9612, 10.1016/j.biomaterials.2010.11.028.
60. M. F. Dumont, C. Baligand, Y. Li, E. S. Knowles, M. W. Meisel, G. A. Walter, and D. R. Talham, DNA surface modified gadolinium phosphate nanoparticles as MRI contrast agents, *Bioconjugate Chemistry* 23(5), 951–957, 2012.
61. J. Huang, X. Zhong, L. Wang, L. Yang, and H. Mao, Improving the magnetic resonance imaging contrast and detection methods with engineered magnetic nanoparticles, *Theranostics* 2(1): 86–102, 2012, http://www.thno.org/v02p0086.htm
62. D. D. Stark, R. Weissleder, G. Elizondo, P. F. Hahn, S. Saini, L. E. Todd, J. Wittenberg, and J. T. Ferrucci, Superparamagnetic iron oxide: Clinical application as a contrast agent for MR imaging of the liver, *Radiology*, 168: 297–301, 1988.
63. L. Yang, X.-H. Peng, Y. A. Wang, X. Wang, Z. Cao, C. Ni, P. Karna, X. Zhang, W. C. Wood, X. Gao, S. Nie, and H. Mao, Receptor-targeted nanoparticles for in vivo imaging of breast cancer. *Clinical Cancer Research* 15: 4722–4732, 2009.

64. J. Xie, K. Chen, J. Huang, S. Lee, J. Wang, J. Gao, X. Li, and X. Chen, PET/NIRF/MRI triple functional iron oxide nanoparticles, *Biomaterials* 31: 3016–3022, 2010.
65. A. Y. Louie, Multimodality imaging probes: Design and challenges, *Chemical Reviews* 110: 3146–3195, 2010.
66. H. Duan, M. Kuang, X. Wang, Y. A. Wang, H. Mao, and S. Nie, Reexamining the effects of particle size and surface chemistry on the magnetic properties of iron oxide nanocrystals: New insights into spin disorder and proton relaxivity, *The Journal of Physical Chemistry C* 112: 8127–8131, 2008.
67. U. I. Tromsdorf, N. C. Bigall, M. G. Kaul, O. T. Bruns, M. S. Nikolic, B. Mollwitz, R. A. Sperling et al., Size and surface effects on the MRI relaxivity of manganese ferrite nanoparticle contrast agents, *Nano Letters* 7: 2422–2427, 2007.
68. J. H. Lee, Y. M. Huh, Y. W. Jun, J. W. Seo, J. T. Jang, H. T. Song, S. Kim, E. J. Cho, H. G. Yoon, J. S. Suh, and J. Cheon, Artificially engineered magnetic nanoparticles for ultra-sensitive molecular imaging, *Nature Medicine* 13: 95–99, 2007.

11

Review of Nanoscale Spectroscopy in Medicine

Chintha C. Handapangoda, Saeid Nahavandi, and Malin Premaratne

CONTENTS

11.1 Introduction

Widely used biomedical imaging techniques such as radiography, computer-assisted tomography, ultrasound imaging, and magnetic resonance imaging largely focus on structural and anatomical imaging at the tissue or organ levels (Prasad 2003). However, in most of the cases, only information at the molecular and cellular levels can lead to the detection of the early stages of the formation of a disease, such as cancer (Prasad 2003). Nanoscale spectroscopic techniques can be employed to obtain clinically valuable information at the molecular and cellular levels.

Spectroscopy is a major branch of interaction between light and matter that involves the study of transitions of energy between quantized levels in matter (e.g., atoms or molecules) (Prasad 2003). Figure 11.1 schematically shows four different light–molecule interaction processes that are widely encountered in nature (Prasad 2003). As depicted in Figure 11.1, a molecule

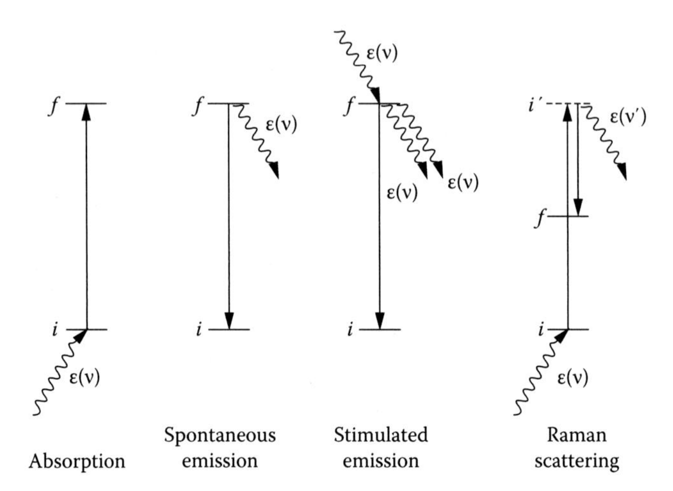

FIGURE 11.1
Schematic diagram of various light–molecule interaction processes. (From Prasad, P.N.: *Introduction to Biophotonics*. 2003. Copyright Wiley-VCH Verlag GmbH & Co. KGaA. Reproduced with permission.)

in the ground state can absorb a photon and be excited to a state with a higher energy. Spontaneous emission involves a molecule randomly relaxing to the ground state from an excited state by emitting a photon. In stimulated emission, a molecule in an excited state absorbs a photon and relaxes to the ground state by emitting an identical photon. Raman scattering involves absorption of a photon and emission of a photon at a different frequency.

Spectroscopy deals with the characterization and applications of such transitions between two or more quantized states of an atom, a molecule, or an aggregate (Prasad 2003). Atoms, ions, and molecules exist in discrete energy states, and their excitation can be electronic, vibrational, or rotational. Figure 11.2 shows the energy level diagram for atomic excitation and emission.

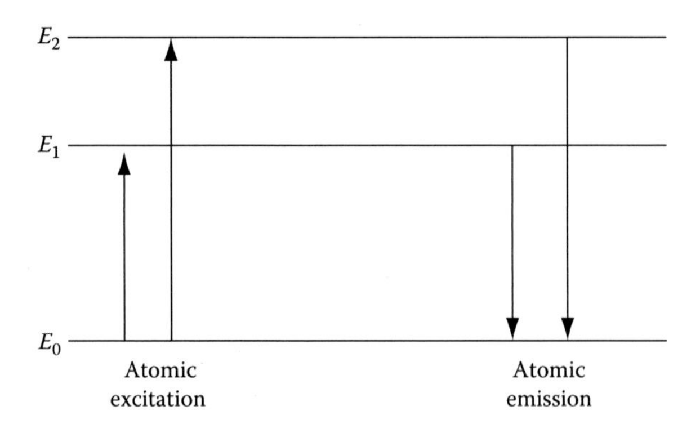

FIGURE 11.2
Energy level diagram depicting atomic excitation and emission.

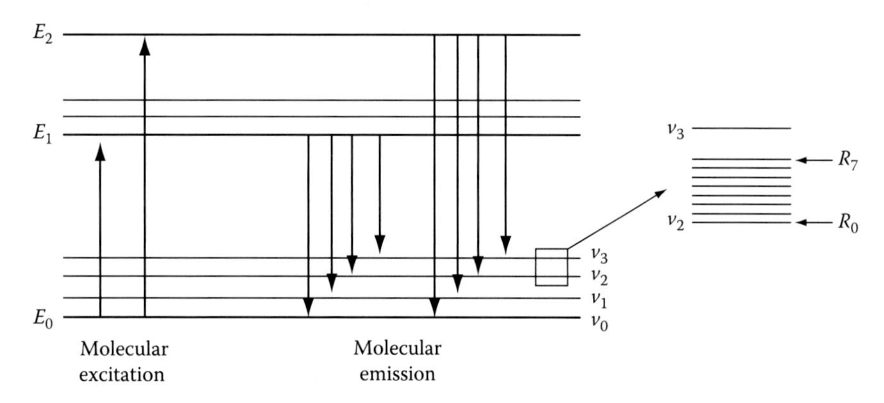

FIGURE 11.3
Energy level diagram depicting molecular excitation and emission.

Atoms exhibit sharp line spectra with few features as they do not have vibrational or rotational energy levels. However, molecules exhibit broad band spectra with many features as they involve electronic as well as vibrational and rotational energy levels. An energy level diagram depicting molecular excitation and emission that involves vibrational states is shown in Figure 11.3.

Each electronic state consists of many vibrational states, and each vibrational state consists of many rotational states, as illustrated by Figure 11.3. Thus, spectroscopy in general can be subdivided into three categories, namely, electronic spectroscopy, vibrational spectroscopy, and rotational spectroscopy. However, quantized levels of biological interest are only electronic and vibrational (Prasad 2003). Therefore, in this chapter, we discuss only electronic and vibrational spectroscopic techniques that are applied in medicine. Nanoscale spectroscopy applications in medicine involve the study of light interaction and propagation through biological tissue (Handapangoda et al. 2008, Premaratne et al. 2005).

A radiative or a nonradiative process can take place following an excitation of a molecule due to the absorption of light. In a radiative process, a photon is emitted to bring the molecule back to the ground state. In a nonradiative process, the excited-state energy is dissipated as heat (e.g., by emission of a phonon) or in producing a chemical reaction (Prasad 2003).

Radiative relaxation includes fluorescence (Vo-Dinh and Cullum 2003) and phosphorescence (Apreleva et al. 2006). Nonradiative relaxation involves many small collisional relaxations and tiny temperature rises of surrounding species. The ground state of most molecules involves paired electrons with a total spin equaling to zero. These states are called singlet states, and they are labeled S_0, S_1, S_2, and so on, in the order of increasing energy (Prasad 2003). A state with a net spin value (i.e., unpaired electrons) is called a triplet state and is denoted by *T*. Figure 11.4 shows the typical spin arrangement in molecular orbitals of the ground state and excited singlet

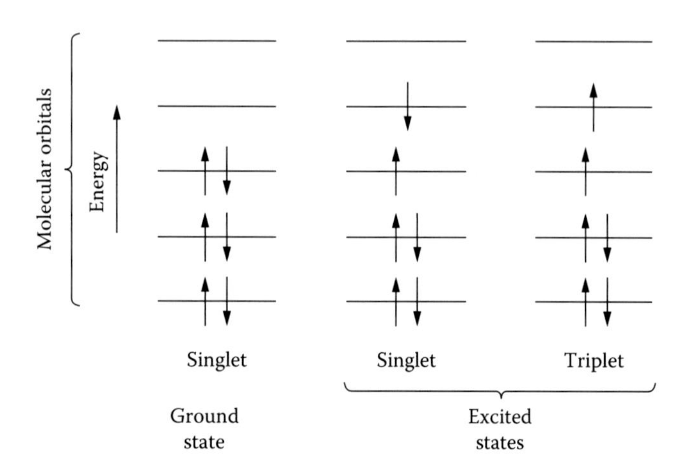

FIGURE 11.4
Schematic diagram of typical spin arrangements in molecular orbitals of the ground, excited singlet, and triplet states. (Reproduced from Vo-Dinh, T. and Cullum, B.M., Fluorescence spectroscopy for biomedical diagnostics, In *Biomedical Photonics Handbook*, CRC Press, Boca Raton, FL, 2003. With permission.)

and triplet states (Vo-Dinh and Cullum 2003). As depicted in Figure 11.4, the excitation of an electron from a paired electron pair of a molecule whose ground state is S_0 can produce either a state where the two electrons are still paired (a singlet state) or a state where the two electrons are unpaired (a triplet state) (Prasad 2003).

This chapter is organized as follows. Section 11.2 discusses nanoscale electronic spectroscopy that is utilized in biomedical applications. This section is subdivided into luminescence spectroscopy and atomic absorption spectroscopy (AAS). In luminescence spectroscopy, we discuss in detail fluorescence spectroscopy and phosphorescence imaging as applied for medical diagnosis. Section 11.3 discusses nanoscale vibrational spectroscopy for biomedical applications. This section is subdivided into Raman spectroscopy, infrared (IR) spectroscopy, and terahertz spectroscopy. The importance of using each of these techniques in medical applications is highlighted. In addition, the basic principles behind each technique and their relative advantages and disadvantages are discussed. Section 11.4 highlights the important points discussed in the chapter.

11.2 Electronic Spectroscopy

Electronic spectroscopy deals with characterization and applications of energy transitions between two quantum electronic states of an atom or a molecule. With the absorption of energy, an electron of an atom or a molecule may be excited to a quantum level with a higher energy. Under specific conditions, it is possible to make such an electron to relax back to its original stable state by releasing energy as photons. Electronic absorption spectroscopy involves one photon absorption per molecule at a time, and the transition between the two electronic states is quantitatively expressed by the Beer–Lambert's law (Prasad 2003)

$$I(\nu)=I_0(\nu)10^{(-\varepsilon(\nu)bc)}, \tag{11.1}$$

where

$I(\nu)$ is the output intensity at distance b from the input
I_0 is the incident intensity at the input
$\varepsilon(\nu)$ is the molar extinction coefficient at frequency ν
c is the molar concentration

Luminescence spectroscopy and AAS are two different branches of nanoscale electronic spectroscopy that can be employed in biomedical applications. This section provides a review of luminescence spectroscopy, mainly focusing on fluorescence spectroscopy, and AAS and their applications and usage in biomedicine.

11.2.1 Luminescence Spectroscopy

Electronic luminescence spectroscopy provides a means of multiparameter analysis using emission spectra, excitation spectra, lifetime of emission, and the polarization characteristics of the emitted light (Prasad 2003). Luminescence is the spontaneous emission of light from any substance and occurs from electronically excited states of a physical system (Bartolo 2011, Lakowicz 2010).

Luminescence can be categorized based on the type of excitation. If excited by light, it is called photoluminescence; if excited by an electric field, it is called electroluminescence; and if excited by a chemical reaction, it is called chemiluminescence (Bartolo 2011). Photoluminescence can be characterized as fluorescence or phosphorescence based on the four physical observables (Vo-Dinh and Cullum 2003):

1. Emission, excitation, and synchronous spectra
2. Quantum yield
3. Lifetime
4. Polarization

Photoluminescence can be divided into two categories—fluorescence and phosphorescence. In fluorescence, an electron from the ground state orbital is excited to an upper orbital where the excited electron is paired

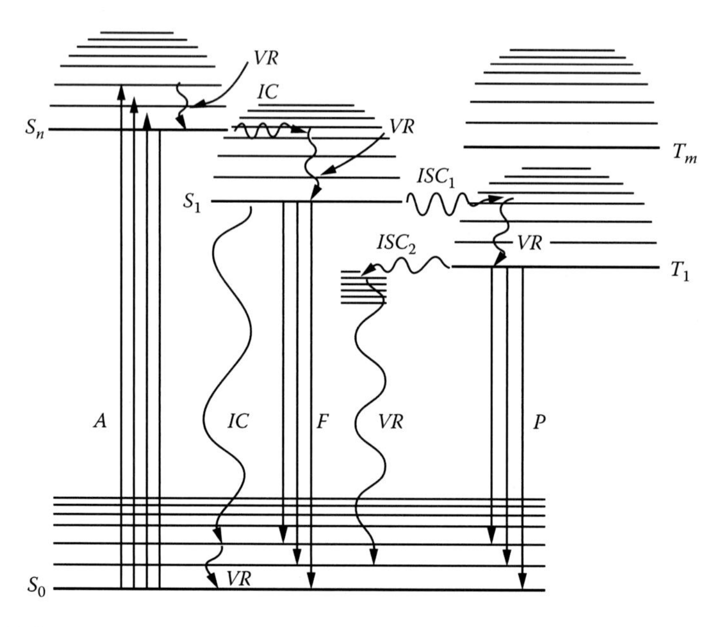

FIGURE 11.5
Jablonski diagram showing different radiative and nonradiative transitions in a molecule upon excitation into a singlet state S_n: S_0, singlet ground state; S_1, first excited singlet state; T_1, first excited triplet state; T_m, excited triplet state; ISC_1 and ISC_2, intersystem crossings; *A*, absorption; *F*, fluorescence; *IC*, internal conversion; *P*, phosphorescence; *VR*, vibrational relaxation. (Reproduced from Vo-Dinh, T. and Cullum, B.M., Fluorescence spectroscopy for biomedical diagnostics, In *Biomedical Photonics Handbook*, CRC Press, Boca Raton, FL, 2003. With permission.)

by opposite spin to the electron in the ground state orbital (i.e., the spin-allowed singlet state) (Bartolo 2011, Lakowicz 2010). In phosphorescence, the electron in the excited orbital and the electron in the ground state orbital have parallel spin (i.e., the spin-forbidden triplet state) (Bartolo 2011, Lakowicz 2010).

Jablonski diagram illustrates the processes that occur between light absorption and emission (Lakowicz 2010). Figure 11.5 shows such a Jablonski diagram (Vo-Dinh and Cullum 2003). In this diagram, singlet states are denoted by S_i and triplet states are denoted by T_i. The ground state of a molecule has no net electron spin, and therefore this state is a singlet state (Vo-Dinh and Cullum 2003). However, depending on the final spin of the electron promoted to the higher orbital, the excited state can be a singlet or a triplet state (Vo-Dinh and Cullum 2003). With the absorption of light, a fluorophore is usually excited to some higher vibrational level of either S_1 or S_2. Generally, molecules in condensed phases rapidly relax to the lowest vibrational level of S_1; and this process is called internal conversion, which occurs generally within a femtosecond or less. Internal conversion is generally complete prior to fluorescence emission (Lakowicz 2010). Return to the ground state via a radiative process typically occurs to a higher vibrational level of the ground state. Then the molecule quickly reaches thermal equilibrium via a nonradiative process (Lakowicz 2010). During an internal conversion process, the excited electron is passed to a lower electronic state without the emission of radiation.

Molecules in the S_1 state can undergo a spin conversion to the first triplet state, T_1, and this conversion from S_1 to T_1 is called intersystem crossing (Lakowicz 2010). Emission from T_1 is called phosphorescence (Lakowicz 2010). During an intersystem crossing process, there is a crossover from a singlet state to a triplet state. This process is mostly observed in molecules with heavy atoms.

Lifetime in luminescence spectroscopy is defined as the time required for the emission to decrease to $1/e$ (where e is the base of the natural logarithm) of its original intensity following an impulse excitation (Vo-Dinh and Cullum 2003). Thus, the lifetime, τ, can be obtained from the relation (Vo-Dinh and Cullum 2003)

$$I(t)=I_0\exp\left(\frac{-t}{\tau}\right), \tag{11.2}$$

where

$I(t)$ is the luminescence intensity at time t

I_0 is the intensity at $t=0$

Lifetime in fluorescence and phosphorescence is defined as the average value of the time that a fluorophore or a phosphor spends in its excited state (Bartolo 2011). The quantum yield is the ratio of the number of emitted photons to the number of absorbed photons (Bartolo 2011).

Luminescence spectroscopy can be further divided as fluorescence spectroscopy and phosphorescence imaging. The ongoing development of fluorescence and phosphorescence optical probes has promising applications in noninvasive diagnostic tissue spectroscopy (Servick-Muraca and Burch 1994). The basic principles behind fluorescence and phosphorescence are similar, the main difference being in the lifetime measurement. Phosphorescence lifetimes are greater than fluorescence lifetimes by several orders. This is because of the spin-forbidden triplet excitation in phosphorescence as opposed to the spin-allowed singlet excitation in fluorescence. In this section, we discuss the principles and applications of nanoscale fluorescence spectroscopy in detail and provide a concise discussion on the clinical applications of phosphorescence imaging.

11.2.1.1 Fluorescence Spectroscopy

Fluorescence spectroscopy is a powerful, highly sensitive analytical technique that has a number of promising applications and that is widely used in biology and medicine (Engels and Wilson 1992, Patterson and Pogue 1994, Schneckenburger 2005). Fluorescence is used extensively in biotechnology, flow cytometry, medical diagnostics, DNA sequencing, forensics, and genetic analysis (Lakowicz 2010). Fluorescence imaging can be used for measurements of intracellular molecules, even at the level of single-molecule detection (Lakowicz 2010). Fluorescence spectroscopy has been applied to the analysis of many different types of samples, ranging from individual biochemical species to live organs (Vo-Dinh and Cullum 2003).

Fluorescence spectroscopy can also be used for the noninvasive diagnosis of the early stages of cancer, atherosclerosis, heart arrhythmia (Vo-Dinh and Cullum 2003) to measure the concentration of various exogenous agents such as the photosensitizers used in photodynamic therapy (Patterson and Pogue 1994) and for the detection of dental caries (Deckelbaum et al. 1987) and many other medical conditions (Vo-Dinh and Cullum 2003). Fluorescence spectroscopy and imaging is useful in the early diagnosis of almost every type of cancer (Vo-Dinh and Cullum 2003).

This technique is also widely used in examining biological objects. Stimulated emission depletion, fluorescence correlation spectroscopy (FCS), and fluorescence lifetime imaging microscopy are some of the new methods of research (Voitovich et al. 2011). Förster resonance energy transfer (FRET) is already being used in the studies of donor–acceptor interactions in biological systems. Fluorescence markers are used in routine procedures in medicine (Voitovich et al. 2011). In vivo real-time diagnosis is one advantage of fluorescence techniques, and fluorescence-based optical diagnoses can provide a more complete examination of organ of interest compared to excisional biopsy or cytology (Vo-Dinh and Cullum 2003).

Fluorescence is the emission of light from a substance following the absorption of radiation. Substances that exhibit fluorescence are called fluorophores (Bartolo 2011). Fluorophores can be divided into two main classes, namely intrinsic fluorophores and extrinsic fluorophores. Fluorescence occurs naturally in systems with intrinsic fluorophores, whereas extrinsic fluorophores are added to systems in which fluorescence do not occur naturally (Bartolo 2011). Fluorescence could serve as a diagnostic probe in medicine due to the visible intrinsic native fluorescence exhibited in many tissues (Tang et al. 1989). In in vivo fluorescence applications, mainly intrinsic fluorescence, which arises from proteins, extracellular fibers, or specific coenzymes, is used (Schneckenburger 2005). Many fluorescent species are present in biological tissues, making it possible to use autofluorescence in diagnostic techniques (Vo-Dinh and Cullum 2003). Encouraging results have been reported when tissue autofluorescence is used to detect a diversity of diseases, such as atherosclerosis in the aorta and the coronary artery and dysplasia in the colon and other tissues (Deckelbaum et al. 1987, Sartori et al. 1987, Wu et al. 1993). Low-power laser radiation induces tissue fluorescence without tissue damage and may be used for diagnostic fluorescence spectroscopy (Deckelbaum et al. 1987).

Distinct different autofluorescent emission spectra can be observed from cells in various disease states as they undergo different rates of metabolism, or have different structures (Vo-Dinh and Cullum 2003). In autofluorescence-based optical biopsy techniques, laser light is used to excite the naturally occurring fluorophores in tissue. Real-time diagnoses without the removal of tissue are possible due to the differences in chemical composition between the various types of tissue (Vo-Dinh and Cullum 2003). Fluorescent probes, such as fluorescein and indocyanine, are used for diagnostic purposes in humans. Such applications include fluorescence angiography, blood volume determination, and photodynamic therapy (Schneckenburger 2005).

Fluorescence microscopy is used for monitoring the spatial distribution of a particular analyte at many different locations (Vo-Dinh and Cullum 2003). Optical detection of malignant or premalignant tissues is the most common type of fluorescence-based biomedical diagnostic procedure used in clinical studies, and the use of fluorescence for the detection of tumors of the skin, urinary bladder, bronchus, gastrointestinal, head and neck, breast, and brain has been investigated (Vo-Dinh and Cullum 2003). The fluorescence and absorption spectra of a system can be used to characterize its physiological state in relation to a normal system (Tang et al. 1989). Fluorescence spectra can be used to obtain detailed information about fluorescent molecules, their conformation, binding sites, and interaction within cells or tissue (Schneckenburger 2005).

Usually, the emission spectrum (fluorescence intensity vs. emission wavelength) is used for fluorescence diagnosis, whereas the excitation spectrum (fluorescence intensity vs. excitation wavelength) is used for extracting additional information about absorption of complex samples or intermolecular energy transfer (Schneckenburger 2005). Emission spectra and intensities of extrinsic probes are often used to determine a probe's location on a macromolecule by exploiting the sensitivity of fluorophores to their surrounding environment (Lakowicz 2010). All of fluorescent characteristics, namely intensity, spectrum, polarization, lifetime, and quantum yield, are used in medical research and diagnostics (Voitovich et al. 2011). Properties such as the quantum yield, the spectral position of the fluorescence emission maximum, spectral line width, and the lifetime are often significantly affected by changes in the microstructure (i.e., the local environment) surrounding the fluorophore (Vo-Dinh and Cullum 2003). Fluorescence spectroscopy can also be used for single-molecule detection due to their high quantum yields and photostability (Lakowicz 2010). A few researchers have explored the use of the fluorescence lifetime as a diagnostic tool (Patterson and Pogue 1994). Lifetime-sensitive fluorescent probes are developed to monitor pH, calcium concentration, glucose, and other metabolites in cellular culture (Burch et al. 1994).

The rate constant of a fluorescence decay is defined as the inverse of the fluorophore's lifetime. The rate constant has two contributions, a radiative decay constant characterized by a radiative lifetime, τ_r, and a nonradiative decay constant characterized by a nonradiative lifetime, τ_{nr}. They are related by

$$\frac{1}{\tau} = \frac{1}{\tau_r} + \frac{1}{\tau_{nr}}. \tag{11.3}$$

Fluorescence quantum efficiency (quantum yield), ϕ, is defined as

$$\phi = \frac{\tau_{nr}}{\tau_{nr} + \tau_r}. \tag{11.4}$$

In fluorescence, the decay of an excited-state singlet to the ground-state singlet occurs by the rapid emission of a photon, and the average time between excitation and return to the ground state (i.e., the lifetime) is around 10 ns (Bartolo 2011, Lakowicz 2010). The same fluorescence spectrum is generally observed independent of the excitation wavelength. This is because luminescence is generally observed from the lowest excited state. This phenomenon is known as the Kasha's rule (Bartolo 2011, Lakowicz 2010). Even though exceptions to Kasha's rule exist, such cases are generally not observed in biological molecules (Bartolo 2011).

If the emission spectrum of a fluorophore (donor) overlaps with the absorption spectrum of another molecule (acceptor), the FRET takes place. The rate of energy transfer is determined by the distance between the donor and the acceptor and by the extent of the spectral overlap (Bartolo 2011). FRET can be used to measure the distance between sites on macromolecules and can provide accurate spatial information about molecular structures in biological systems at distances ranging from 10 to 100 Å. Even geometrical information can be obtained in many cases. The distance between the donor and the acceptor can be calculated from the transfer efficiency (Bartolo 2011). FRET is not the result of emission of light from the donor being absorbed by the acceptor; but the donor and the acceptor are coupled by a dipole–dipole interaction (Lakowicz 2010). There is no intermediate photon involved in FRET (Lakowicz 2010). Resonance fluorescence produces emission at the same frequency as absorption and is common for atoms where there are no vibrational or rotational states. Nonresonance fluorescence that is common in molecules produces emission at lower energy than absorption, due to vibrational relaxation that occurs before fluorescence.

In biological systems, bands of fluorescence and absorption of the systems are often overlapped. Hence, fluorescence intensity measurements of one component or the system as a whole can be distorted (Voitovich et al. 2011). However, overlapping of fluorescence spectrum of a component with absorption spectra of other components in the system does not influence on lifetime measurements, provided that FRET or quenching of fluorescence does not take place. As a result, in many cases, fluorescence lifetime measurements give very important information about positions and interactions of molecules (Voitovich et al. 2011). Fluorescence lifetime measurements down to the subnanosecond range are becoming more and more interesting because they are sensitive to molecular conformations and interactions (i.e., aggregation, binding sites, intermolecular energy transfer, etc.) (Schneckenburger 2005).

Figure 11.6 shows the intensity and lifetime images of bovine artery endothelial cells stained with three fluorophores (Lakowicz 2010). Even though the intensity image shows regions of the cell with different brightness, it does not distinguish between the three fluorophores. However, the lifetime

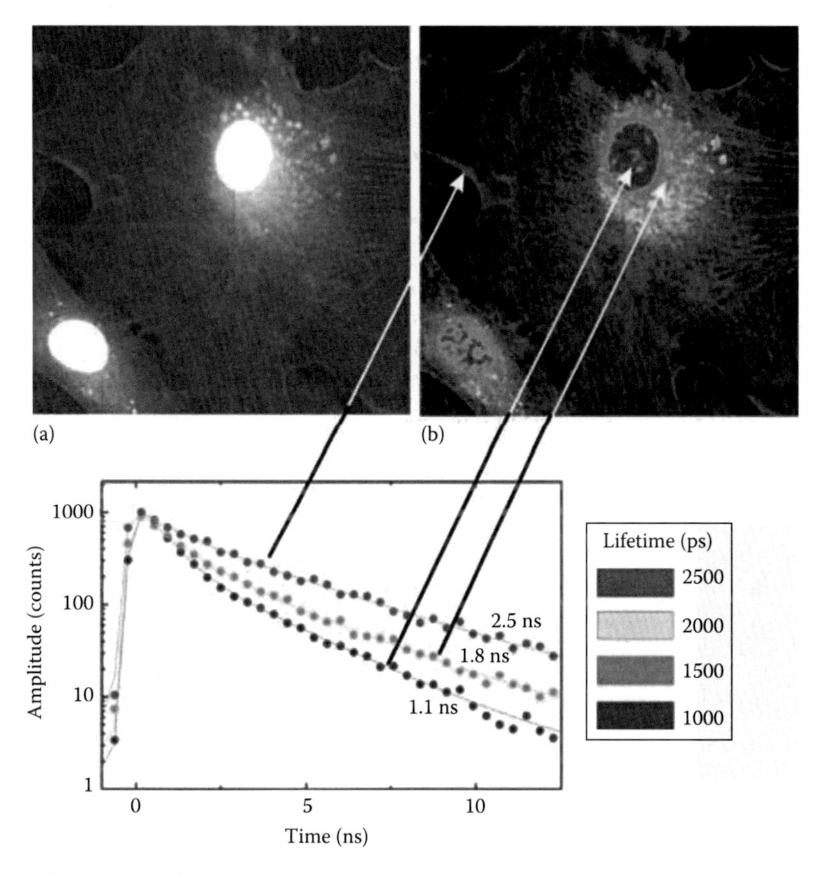

FIGURE 11.6
(**See color insert.**) Intensity (a) and lifetime (b) fluorescence images of bovine artery endothelial cells, stained with three fluorophores. The nuclei were stained with DAPI for DNA (blue), F-actin was stained with Bodipy FL-phallacidin (red), and the mitochondria were stained with MitoTracker Red CMX Ros (green). Two-photon excitation at 800 nm was used. (Reproduced with permission from Springer Science+Business Media: *Principles of Fluorescence Spectroscopy*, 2010, Lakowicz, J.R. Courtesy of Dr. Alex Bergmann, Becker and Hickl GmbH.)

image clearly shows the locations of the probes based on their lifetimes (Lakowicz 2010).

Fluorophores preferentially absorb photons whose electric vectors are aligned parallel to the transition moment of the fluorophore (Lakowicz 2010). They absorb light along a particular direction with respect to the molecular axes. The polarization and anisotropy are determined by the extent to which a fluorophore rotates during its excited-state lifetime (Bartolo 2011). Fluorescence depolarization is a measure of the loss of polarization of fluorescence due to various dynamic effects such as the rotation of the fluorophore (Prasad 2003). Polarization, P, of fluorescence is defined as

$$P = \frac{I_{\parallel} - I_{\perp}}{I_{\parallel} + I_{\perp}}, \tag{11.5}$$

where $I_{\parallel}$ and $I_{\perp}$ are the fluorescence intensities polarized parallel and perpendicular to the polarization of excitation light (Prasad 2003). Fluorescence emission anisotropy,

$$r = \frac{I_{\parallel} - I_{\perp}}{I_{\parallel} + 2I_{\perp}}, \tag{11.6}$$

is a quantity that characterizes the polarization of fluorescence (Prasad 2003). Fluorescence anisotropy measurements provide information on the size and shape of proteins and the rigidity of various molecular environments. These measurements have been used to measure protein–protein associations, fluidity of membranes and for immunoassays of various substances (Lakowicz 2010).

Fluorescence measurements can generally be steady state or time resolved depending on the type of excitation used. For steady-state measurements, continuous-wave excitation is used, whereas for time-resolved measurements pulsed excitation is used. Due to short lifetimes, time-resolved measurements require sophisticated optics and electronics; however, they provide additional important information (Bartolo 2011). Due to the nanosecond time scale of fluorescence, steady state is reached almost immediately after the sample has been exposed to continuous excitation. In pulsed excitation, the light pulse width is typically shorter than the decay time (Bartolo 2011). With continuous excitation, a lot of information available from fluorescence is lost during the time-averaging process. Time-resolved measurements can be used for measuring intensity decays and anisotropy decays (Lakowicz 2010).

Usually, fluorescence is excited by the absorption of a single photon; however, pulse excitation with femtosecond pulse widths can excite fluorophores by two-photon absorption (Lakowicz 2010). Two-photon absorption depends strongly on the light intensity and occurs only at the focal point of the laser beam, provided that the photon wavelength is in a specified region (Lakowicz 2010). Two-photon excitation can be utilized in fluorescence microscopy such that the imaging occurs only at the focal plane; hence the images are free from any distortions arising from fluorescence of planes above and below the focal plane (Lakowicz 2010). One advantage of fluorescence is the very high signal-to-noise ratio that enables one to distinguish spatial distribution of even low concentrations of substances (Prasad 2003).

11.2.1.2 Fluorescence Correlation Spectroscopy

FCS is a technique that can be used to investigate a variety of biological processes such as protein–protein interactions, binding equilibria for drugs, and clustering of membrane-bound receptors (Prasad 2003). In this technique, the sample (either in vitro or in vivo) is illuminated by a light source focused to a very small volume, typically in the order of 1 fL or less. The fluorescence originating from particles diffusing in and out of the detection volume is recorded (Jameson et al. 2009). Therefore, in FCS, spontaneous fluorescence intensity fluctuations in a microscopic volume consisting of only a small number of molecules are monitored as a function of time (Prasad 2003).

While in the illumination volume, the fluorescent particles may be excited more than once and may also undergo chemical or photophysical processes, such as the "blinking" demonstrated by green fluorescent protein and alterations in fluorophore quantum yields upon binding to macromolecules (Jameson et al. 2009). These processes alter fluorescence properties and thus lead to fluctuations in the detected fluorescence signal (Jameson et al. 2009). The fluorescence intensity fluctuations are produced by dynamical processes such as changes in the number of fluorescing molecules due to their diffusion in and out of the observed microscopic volume and change in the fluorescence quantum yield due to processes occurring in the observed volume (Prasad 2003).

It is crucial for FCS to have a strong and stable light source, such as lasers, ultrasensitive detectors, such as avalanche photodiodes, and efficient fluorescence labels with high extinction coefficients and a quantum yield close to 100% (Schwille and Ries 2011). In the case of FCS measurements on fluorescently labeled biomolecules in an aqueous medium, concentrations in the nanomolar range are optimal due to the limiting of the detection volume to less than 1 fL (Schwille and Ries 2011). Additional fluctuations arise in the detected signal due to the chemical or photochemical reactions or conformational changes that may alter the emission characteristics of the fluorophore, during the time a particle spends in the focus (Schwille and Ries 2011).

One way of treating an FCS data stream is using an autocorrelation function, which is the time-dependent decay of the fluorescence intensity fluctuations (Jameson et al. 2009). The autocorrelation function, $G(\tau)$, is the product of the intensity at time t and the intensity at time $t+\tau$ divided by the square of the average fluorescence intensity at time t, mathematically given by (Jameson et al. 2009)

$$G(\tau)=\frac{\langle \delta F(t)\cdot \delta F(t+\tau)\rangle}{\langle F(t)\rangle^2}. \tag{11.7}$$

Even though confocal FCS has become a standard tool in life sciences due to the commercial availability, its application to more complex systems than freely diffusing molecules in buffer solution, such as cells or biological membranes, is difficult due to several limitations of the technique (Schwille and Ries 2011).

Fluorescence cross-correlation spectroscopy that uses multiphoton excitation is an attractive alternative for FCS applications in living cells and tissues because they tolerate near-IR (NIR) radiation relatively well and also there is less autofluorescence and scattering (Schwille and Ries 2011). It is possible to excite two carefully selected spectrally different dyes, with only one IR laser line using two-photon excitation (Schwille and Ries 2011). However, two-photon excitation requires a high intensity, and usually pulsed excitation is used to obtain a high photon density per pulse (Schwille and Ries 2011). Signals from two different fluorophores, associated in some manner, could be followed by cross-correlation (Jameson et al. 2009). In cross-correlation, the signal at one particular time for fluorophore 1 is correlated with the signal at a different time for fluorophore 2 (Jameson et al. 2009). Dual-color cross-correlation spectroscopy requires exact spatial superposition of two laser beams such that the focal volumes overlap, hence making the experimental realization of this technique very demanding (Schwille and Ries 2011).

11.2.1.3 Phosphorescence Imaging

Phosphorescence is the slow emission of light by a substance following absorption of radiation. Phosphorescence emission is very slow relative to fluorescence emission that occurs immediately following absorption. Oxygen is an essential metabolite, and tissue hypoxia is a critical factor in many tissue pathologies such as retinal diseases, brain abnormalities, and cancer (Apreleva et al. 2006, Arden et al. 2005, Evans and Koch 2003, Ferriero 2004, Pena and Ramirez 2005). Imaging techniques for mapping tissue oxygenation range in their accuracy, invasiveness, sensitivity, and temporal and spatial resolution (Apreleva et al. 2006). The development of reliable quantitative methods for tomographic oxygen imaging is very important. One method that can directly quantify oxygen in tissue is based on the ability of oxygen to quench phosphorescence of exogenous probes (Apreleva et al. 2006, Vanderkooi et al. 1987, Wilson and Vinogradov 2003).

The decrease of phosphorescence intensity by some processes is called quenching. Porphyrin compounds that accumulate in tumor tissues may provide a noninvasive means to determine tumor oxygenation accurately for the efficacious choice of chemotherapy or radiotherapy staging (Servick-Muraca and Burch 1994). Phosphorescent lifetime measurements of porphyrin compounds can be used to determine the tumor oxygen concentrations (Servick-Muraca and Burch 1994).

Figure 11.7 shows changes of oxygen distribution in the hepatic microvasculature of rat liver under the course of ischemia and reperfusion, measured by phosphorescence imaging (Lo et al. 2003). Such images can be used to obtain a correlation of morphological features of organs with measured oxygen pressures (Lo et al. 2003).

Even though the theoretical foundation of phosphorescence lifetime imaging closely resembles that of fluorescence lifetime imaging (Handapangoda et al. 2012a,b), the large differences between phosphorescent and fluorescent lifetimes result in significant differences in instrumentation and approaches to lifetime distribution reconstruction (Apreleva et al. 2006, Godavarty et al. 2005, Patterson and Pogue 1994). In phosphorescence, the decay of the excited triplet state to ground state occurs relatively slowly. Phosphorescence lifetimes are typically milliseconds to seconds (Bartolo 2011, Lakowicz 2010). In fluorescence lifetime imaging, it is necessary to account for photon migration due to the comparable rates of photon emission and photon migration (Apreleva et al. 2006). However, in phosphorescence lifetime imaging, the photon migration of excitation and emission light can be considered instantaneous compared to the large triplet lifetimes. Hence, phosphorescence lifetime imaging can be modeled in a much simpler way compared to fluorescence imaging (Apreleva et al. 2006).

11.2.2 Atomic Absorption Spectroscopy

The principle behind AAS is that metal atoms absorb strongly at discrete wavelengths and emit at the same wavelengths (Robinson 1960, Sunderman 1973). In this technique, a small proportion of the metal atoms are thermally excited to emit light, but the majority of the atoms remain in the ground state. These atoms in the ground state are capable of absorbing discrete wavelengths of incident light (Sunderman 1973). Light at a selected wavelength is detected by a photomultiplier (Sunderman 1973).

The recognition of clinical syndromes due to copper and zinc deficiency and human diseases due to abnormalities of chromium and nickel metabolism has expanded the applications of AAS of trace metals in clinical pathology (Sunderman 1973). Atomic absorption techniques had been developed for analyzing a number of metals present in biological components (Sunderman 1973). Clinical applications of AAS include the determination of trace metals such as copper, zinc, lead, and iron, as well as the indirect determination of

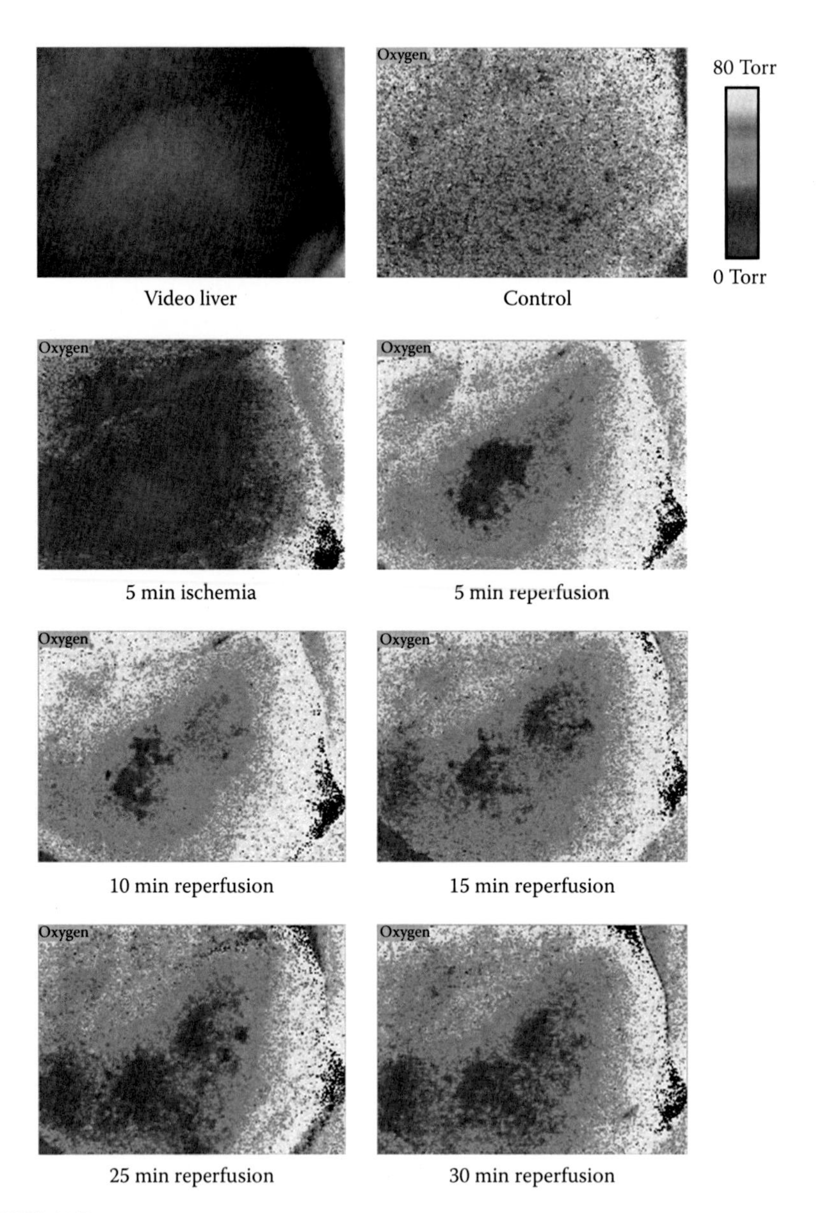

FIGURE 11.7
Changes of oxygen distribution in the hepatic microvasculature of rat liver under the portal triad clamping model of ischemia and reperfusion measured by phosphorescence imaging. (Reproduced from Lo, L.W. et al., *J. Med. Biol. Eng.*, 23, 19, 2003. With permission.)

nonmetallic elements such as sodium, potassium, magnesium, and calcium in biological components (Robinson 1960, Sunderman 1973).

The diagnosis of Wilson disease requires analyses of copper in serum and liver biopsy specimens (Sunderman 1973). Measurements of copper in serum and urine are needed for the diagnosis of acute copper intoxication, which is a complication of renal hemodialysis, neonatal exchange transfusion, and ingestion of copper salts (Sunderman 1973). Copper level in serum is elevated in infections, hyperthyroidism, and leukemia. The effectiveness of antileukemic therapy is routinely monitored by observing serum copper and zinc levels using AAS (Berman 1975). Hypozincemia is an abnormality that is observed in a number of clinical situations such as liver disease, lung disease, acute myocardial infarction, chronic kidney disease with uremia, diseases with increased muscle catabolism, and pregnancy (Sunderman 1973). Diagnosis of acute lead poisoning requires the measurements of lead concentrations in whole blood. Atomic absorption spectrometry of lead in whole blood and urine has become a common analytical task in clinical laboratories due to concerns regarding environmental intoxication (Sunderman 1973). Measurements of serum iron and iron-binding capacity have become routine analyses for the diagnosis of iron deficiency and iron overload (Sunderman 1973).

AAS of trace metals also provides a means of indirect measurement of nonmetallic constituents of clinical interests (Sunderman 1973). By adding an excess of silver nitrate, serum chloride can be precipitated as silver chloride, and then the silver that remains in solution can be measured by atomic absorption (Sunderman 1973). Gold is used to measure urinary bromide, extracted as gold bromide. Barium is used to precipitate urinary sulfate as barium sulfate, which can then be measured by atomic absorption (Sunderman 1973). Copper is used to measure α-amino acids in plasma and urine. Rubidium is used to estimate cardiac glycosides in biological fluids on the basis of their inhibition of rubidium uptake of erythrocytes (Sunderman 1973). Clinical laboratories perform measurements of sodium and potassium in biological fluids routinely (Berman 1975). The sensitivity of detecting sodium and potassium by AAS is 0.005 μg/mL, while the normal range of serum concentration of sodium is 3.10–3.56 mg/mL and that of potassium is 0.14–0.19 mg/mL (Berman 1975). The sensitivity of detecting calcium by AAS is 0.003 μg/mL and the normal range of serum calcium level is 0.09–0.11 mg/mL (Berman 1975).

Determination of the magnesium level of blood serum accurately is crucial and had proven life-saving. However, the measurement of magnesium had been problematic prior to the introduction of atomic absorption instruments into the clinical laboratory (Berman 1975). The normal range of serum magnesium is 10–20 μg/mL, while the sensitivity of detecting magnesium by AAS is 0.0005 μg/mL (Berman 1975). Manganese is another element that is a critical factor of biological interest. It is an activator of various enzymes and is therefore considered an essential trace element. However, excessive manganese is toxic and could result in Parkinson disease (Berman 1975). The manganese content of red cells is elevated in rheumatoid arthritis and also myocardial infarction results in an elevated level of manganese (Berman 1975). Several researchers have shown that AAS could be used to determine the manganese content in serum and urine (Berman 1975).

AAS is free from interference from its environment, which is clearly a great advantage of the technique. It was experimentally shown that the accuracy of this technique is very rarely affected by the presence of other elements, as it is very unlikely that two metals absorb at the same wavelength (Robinson 1960). AAS is preferred in clinical analyses due to the specificity, sensitivity, precision, simplicity, and relatively low cost (Smith et al. 1979).

11.3 Vibrational Spectroscopy

Vibrational spectroscopy deals with the characterization and applications of transition between two vibrational states of a molecule. This technique can be used to analyze chemical composition, molecular structures, conformations, and interactions in a sample (Chan et al. 2005). Vibrational spectroscopy can be employed to obtain information on the vibrational frequencies (or energies) associated with different chemical bonds and bond angles (Prasad 2003).

All molecules are constantly vibrating, and they can absorb energy from an incident photon, which results in an increase of their vibrations. A vibrational transition is considerably weaker than an electronic transition. However, the former is much richer in structures than the latter (Prasad 2003). Hence, vibrational spectroscopy has found wide application in structural characterization of biological materials (Prasad 2003). This technique is ideally suited for monitoring of biological processes as they are accompanied by changes in the intracellular biochemistry, both in compositional and in conformational changes (Chan et al. 2005). Vibrational bands provide a detailed fingerprint of different bonds, functional groups, and conformations of molecules (Prasad 2003).

Molecular vibrations may be excited by two physical mechanisms: the absorption and the inelastic scattering of photons (Siebert and Hildebrandt 2008). The most common types of vibrational spectroscopy are IR spectroscopy and Raman spectroscopy. IR spectroscopy involves absorption of an IR photon to create a vibrational transition, whereas Raman spectroscopy involves a Raman scattering process that generates a vibrational transition (Prasad 2003, Siebert and Hildebrandt 2008). Direct absorption occurs when a molecule is irradiated by a photon whose energy matches the energy difference between the initial and final vibrational states (Siebert and Hildebrandt 2008). The energy differences between the initial and final vibration states are in the order of 0.005–0.5 eV, thus enabling IR light to induce vibrational transitions (Siebert and Hildebrandt 2008). Another type of vibrational spectroscopy that has several potential applications in medicine is terahertz spectroscopy. This technique employs microwave and IR frequencies.

This section provides a review of Raman spectroscopy, IR spectroscopy, and terahertz spectroscopy, focusing on their applications and usage in biomedicine.

11.3.1 Raman Spectroscopy

Raman spectroscopy is the detection of photons that are inelastically scattered by molecular vibrations in a material (Chan et al. 2005). Raman scattering results from the scattering of photons by molecular bonds and thus provides chemical information about the molecular structure and conformations in the sample (Chan et al. 2005). Raman spectroscopy provides chemically specific information about the sample (Talley et al. 2005). It is a technique useful in a wide variety of fields, ranging from forensic analysis to fundamental studies of molecular interactions, due to its ability to provide quantitative and chemically specific information (Talley et al. 2005).

DNA protein interactions, protein folding, and diseased tissues are only a few of the different biological systems that have been studied using Raman spectroscopy (Talley et al. 2005). It has a potential use of rapid noninvasive cancer detection with the goal of developing optical biopsy techniques, and it has the ability of in vivo detection by delivery of laser light and collection of spectral signals using fiber optic probes (Chan et al. 2005). Raman spectroscopy can be used to perform minimally invasive, real-time, in vivo tissue diagnosis in cases where conventional biopsy is difficult to perform—for example in coronary artery disease and Alzheimer disease (Hanlon et al. 2000).

Studies of Raman spectroscopy as a technique for cancer diagnosis have shown that Raman spectra of various tissues (e.g., breast, cervix, lung, skin, and mouth) were able to distinguish between malignant and normal samples with high sensitivity and specificity (Chan et al. 2005, Frank et al. 1995, Gniadecka et al. 1997, Utzinger et al. 2001). Figure 11.8 shows Raman spectra of normal, benign, and malignant breast tissue from a study carried out by Manoharan et al. (1998). Normal breast tissue can be distinguished easily using Raman spectroscopy due to the abundance of fat (Hanlon et al. 2000). As shown in Figure 11.9, the decision plot for Figure 11.8, normal and malignant breast tissue samples are separated with high sensitivity and specificity by Raman spectral features; but benign samples are not well separated (Hanlon et al. 2000, Manoharan et al. 1998).

Images of two closely spaced bovine chondrocyte cells on a CaF_2 slide shown in Figure 11.10 depict the correspondence between univariate and multivariate Raman imaging (Otto and Pully 2012). Raman spectroscopy can

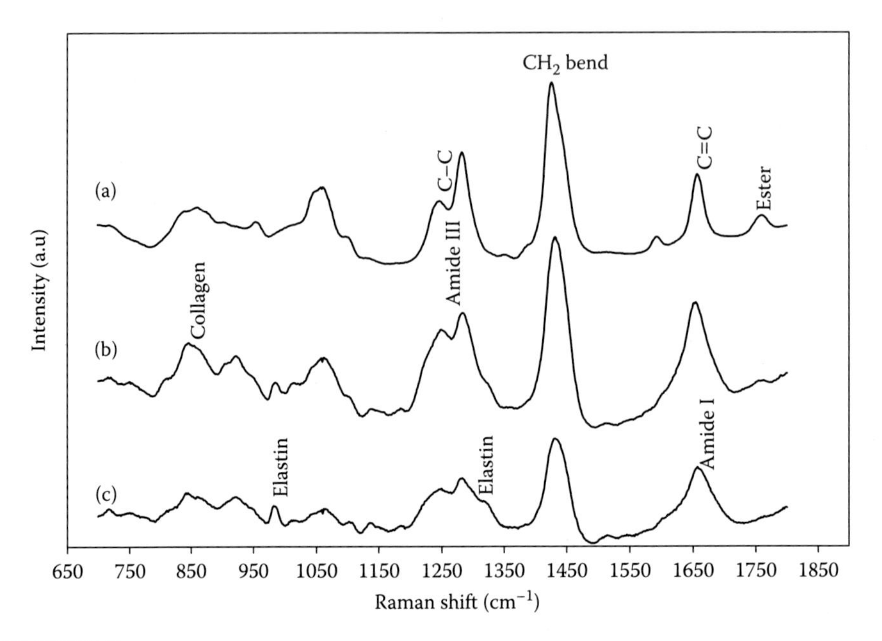

FIGURE 11.8
Raman spectra of (a) normal, (b) benign, and (c) malignant breast tissue. (From Manoharan, R., Shafer, K., Perelman, L., Wu, J., Chen, K., Deinum, G., Fitzmaurice, M., Myles, J., Crowe, J., Dasari, R.R., and Feld, M.S.: Raman spectroscopy and fluorescence photon migration for breast cancer diagnosis and imaging. *Photochem. Photobiol.* 1998. 67. 15–22. Copyright Wiley-VCH Verlag GmbH & Co. KGaA. Reproduced With permission.)

be used to distinguish Alzheimer disease brain tissue from normal brain tissue due to the distinct differences of Raman spectra of each type of these tissues (Hanlon et al. 2000). This technique is also capable of determining concentrations of substances in blood, such as glucose, urea, total protein, albumin, cholesterol, and triglycerides (Hanlon et al. 2000).

Raman scattering is the inelastic scattering of a photon that results in a frequency change in the emitted photon. Stokes Raman scattering produces a photon whose energy (hence frequency) is less than that of the absorbed photon, whereas anti-Stokes Raman scattering produces a photon whose energy (hence frequency) is greater than that of the absorbed photon. Figure 11.11 shows the energy level diagram for elastic Rayleigh scattering and Stokes and anti-Stokes Raman scattering. Approximately 1 in 10^8 photons that are incident on a given bond vibration are Raman scattered (Talley et al. 2005). The low sensitivity of Raman spectroscopy due to the inherent low probability of a photon being Raman scattered has made its applications limited (Talley et al. 2005). Due to this extremely low cross-section, Raman spectroscopy is generally limited to samples that are relatively concentrated and requires high incident power and long integration times to generate Raman spectra of detectable signals (Talley et al. 2005). Biological systems generally

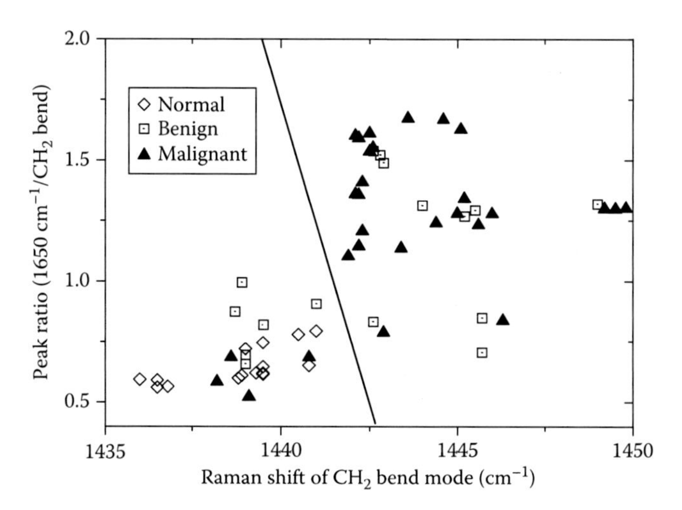

FIGURE 11.9
Decision plot for differentiating normal, benign, and malignant breast tissue. The line shows the separation between normal and malignant samples. (From Manoharan, R., Shafer, K., Perelman, L., Wu, J., Chen, K., Deinum, G., Fitzmaurice, M., Myles, J., Crowe, J., Dasari, R.R., and Feld, M.S.: Raman spectroscopy and fluorescence photon migration for breast cancer diagnosis and imaging. *Photochem. Photobiol.* 1998. 67. 15–22. Copyright Wiley-VCH Verlag GmbH & Co. KGaA. Reproduced with permission.)

involve low concentrations of chemicals such that conventional Raman scattering cannot be used. However, surface-enhanced Raman scattering (SERS) can be used to dramatically increase the Raman scattering cross-section (Talley et al. 2005).

In SERS, the molecule of interest is adsorbed to a metal surface (typically silver or gold), which provides an increase of the observed Raman scattering by many orders of magnitude (Talley et al. 2005). This huge increase of the Raman scattering cross-section in SERS is the result of an electromagnetic enhancement and a chemical enhancement (Talley et al. 2005). The localization of the electric field through surface plasmons provides the electromagnetic enhancement that produces an enhancement of about 10^6 of Raman scattering. The formation of a charge transfer complex that increases the polarizability of the molecule provides the chemical enhancement that produces an increase in Raman scattering by one to two orders of magnitude (Talley et al. 2005). Colloidal nanoparticles have size-dependent plasmon absorption and can be tuned to preferentially adsorb at the desired laser frequency. Hence, these nanoparticles can be used in SERS to obtain Raman spectra from individual molecules (Talley et al. 2005). Gold and silver nanoparticles that are typically used in SERS range from 40 to 100 nm in diameter, and thus can be incorporated into living cells (Talley et al. 2005). SERS spectra of Figure 11.12 show that differences exist in SERS spectra under different polarizations (Feng et al. 2011). These spectra can be used to identify subjects with gastric cancer due to the differences in SERS peaks between healthy subjects and gastric cancer patients (Feng et al. 2011).

Coherent anti-Stokes Raman scattering (CARS) spectroscopy and microscopy is a nonlinear Raman-based technique that is used for noninvasive rapid chemical imaging of live cells (Chan et al. 2005). This technique uses the molecular bonds in the cell as endogenous labels for generating image contrast (Chan et al. 2005). CARS is capable of producing rapid chemical imaging of live cells as a result of the vibrational signals it generates that is of orders of magnitude stronger than spontaneous Raman signals (Chan et al. 2005). The CARS signal is blue shifted with respect to the excitation wavelengths, so that autofluorescence will not generally interfere with the detected CARS signal (Chan et al. 2005).

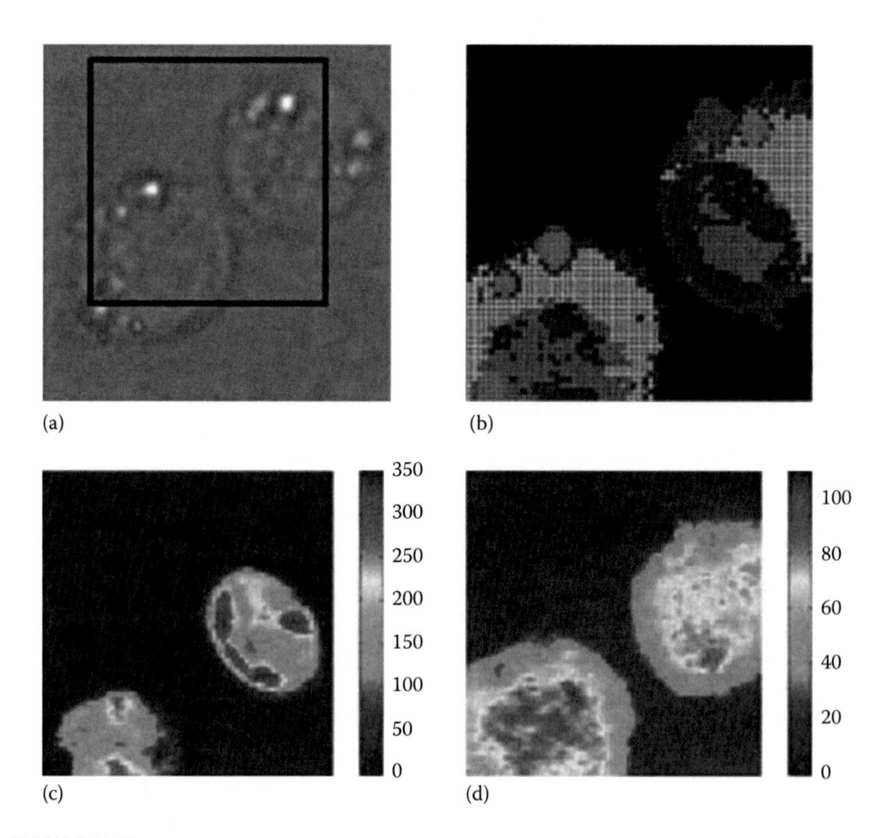

FIGURE 11.10
(**See color insert.**) (a) White light micrograph of freshly extracted bovine chondrocytes seeded on a CaF_2 substrate, (b) corresponding nine-level image showing various organelles in the cell, (c) Raman image for DNA nucleotides at 790 cm^{-1}, (d) Raman image for phenylalanine at 1004 cm^{-1}.

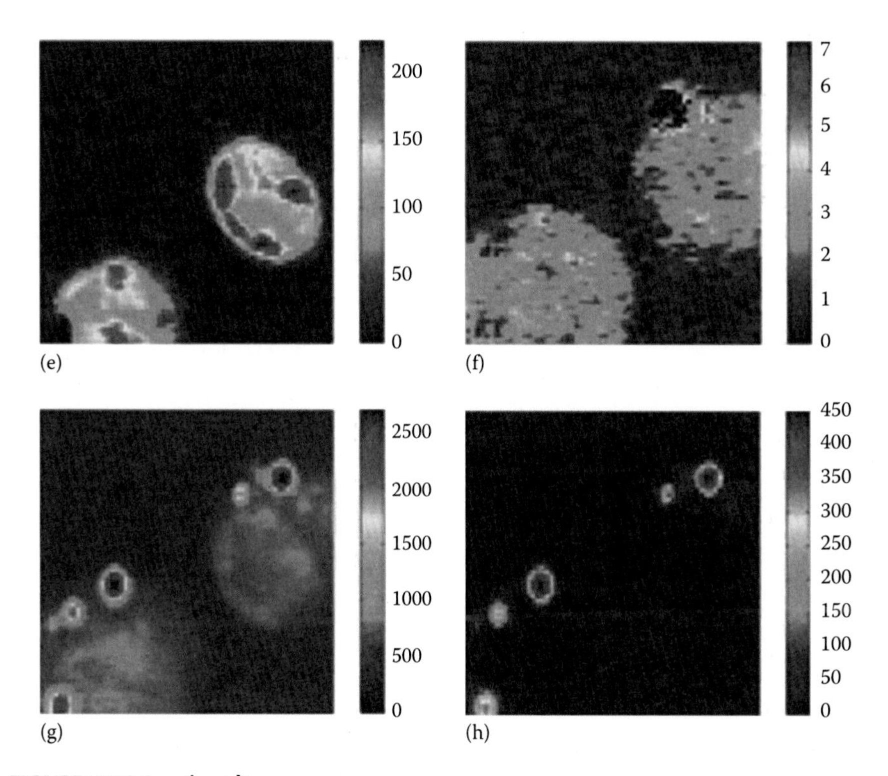

FIGURE 11.10 (continued)
(See color insert.) (e) Raman image for DNA O–P–O symmetric stretch at 1094 cm^{-1}, (f) Raman image signifying the distribution of mitochondria at 1602 cm^{-1}, (g) Raman image of lipids and proteins based on the area around 1656 cm^{-1}, and (h) Raman image for ester groups of lipids at 1745 cm^{-1}. (Reproduced from Otto, C. and Pully, V.V., Hyperspectral Raman microscopy of the living cell, In *Applications of Raman Spectroscopy to Biology*, IOS Press, Amsterdam, the Netherlands, pp. 148–173, 2012. With permission from IOS Press.)

Tissue analysis can be categorized into two distinct approaches—quantitative and qualitative (Hanlon et al. 2000). Qualitative approach deals with disease classification, whereas quantitative approach deals with chemical analysis, such as determining glucose or cholesterol concentrations in blood (Hanlon et al. 2000). Raman spectroscopy can be used for both qualitative and quantitative tissue analyses (Hanlon et al. 2000). This technique can be used in vivo as well as in vitro in a noninvasive and nondestructive manner in different environments (Chan et al. 2005). The nonbleaching property of a Raman signal makes it suitable to be probed for long time periods in living cells for monitoring the dynamic behavior of cells (Chan et al. 2005).

With Raman spectroscopy, there is no need for special sample preparation (Prasad 2003) and also a high spatial resolution can be achieved in subcellular chemical analysis of an individual cell (Chan et al. 2005). Raman spectroscopy can selectively probe a specific molecule or subcellular component

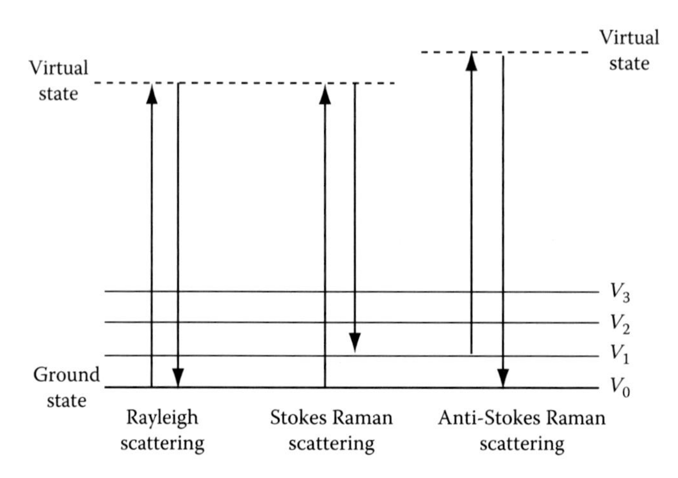

FIGURE 11.11
Energy level diagram depicting Rayleigh scattering, Stokes Raman scattering, and Anti-stokes Raman scattering.

by resonantly enhancing Raman scattering from a desired molecule using an excitation frequency close to its absorption band (Prasad 2003). Due to the very weak Raman scattering of water, Raman spectroscopy has the ability to produce vibrational spectra in an aqueous medium, as opposed to IR spectroscopy in which the IR absorption by water is very strong and hence overwhelms absorption by other cellular constituents (Prasad 2003). In addition, due to this property of weak Raman scattering of water, Raman measurements can be directly made from biofluids (e.g., in vivo measurements from the bladder, prostate, esophagus, skin, cervix, and arteries) (Ellis and Goodacre 2006).

Since Raman signals for visible light excitation suffer from interference from fluorescence, NIR or ultraviolet resonance excitation is generally used in Raman spectroscopy (Hanlon et al. 2000). However, extensive investigations of visible excitation Raman spectroscopy for the study of gallstones and kidney stones have been carried out due to the large Raman scattering cross-section and relatively weak fluorescence from inorganic substances in such stones (Hanlon et al. 2000). Recent advances in Raman spectroscopy have begun to use longer wavelength excitation light (toward NIR) to avoid any potential damage to biological cells, which is a main concern with Raman spectroscopy (Chan et al. 2005). Raman spectroscopy has the potential to diagnose diseases on a chemical level to an extent that no other currently available diagnostic technique is capable of (Hanlon et al. 2000).

11.3.2 Infrared Spectroscopy

With IR spectroscopy, it is relatively easy to obtain spectra of molecules in solution or in gaseous, liquid, or solid states (Stuart 2004). In IR spectroscopy,

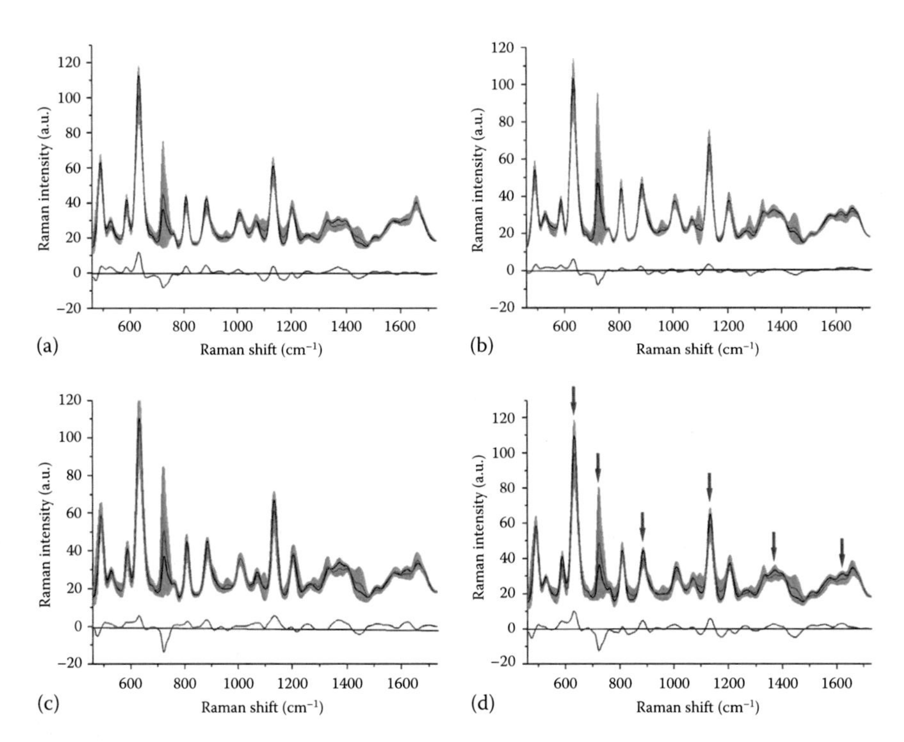

FIGURE 11.12
Comparison of the mean SERS spectra with different polarized laser excitation for the normal blood plasma (darker thin line) versus that of the gastric cancer (lighter thin line), excited by: (a) nonpolarized laser, (b) linear polarization laser, (c) right-handed circularly polarized laser, (d) left-handed circularly polarized laser. At the bottom of each figure, the difference spectra (healthy subject mean spectrum minus that of the cancer group) are shown. The arrows in (d) show the most significant SERS peaks. (Reproduced from *Biosens. Bioelectron.*, 26, Feng, S., Chen, R., Lin, J., Pan, J., Wu, Y., Li, Y., Chen, J., and Zeng, H., Gastric cancer detection based on blood plasma surface-enhanced Raman spectroscopy excited by polarized laser light, 3167–3174, Copyright 2011, with permission from Elsevier.)

mid-IR (MIR) positions are usually reported in units of wavenumbers, whereas NIR positions are reported in units of wavelength (Shaw and Mantsch 2000). MIR range of 400–4000 cm^{-1} and NIR range of 780–2500 nm are used in IR spectroscopy (Shaw and Mantsch 2000). Following the absorption of IR light, the functional groups within the sample vibrate by stretching, bending, or deforming (Ellis and Goodacre 2006). IR spectroscopy has been a widely applied technique in a number of disciplines such as chemical, biochemical, pharmaceutical, and manufacturing industries, for qualitative and quantitative analyses (Chiriboga et al. 1998). The IR absorption spectrum can be directly correlated to biochemical substances (Ellis and Goodacre 2006).

Most researchers focus on using MIR for disease diagnosis (Ellis and Goodacre 2006). Figure 11.13 shows MIR absorption spectra of selected serum

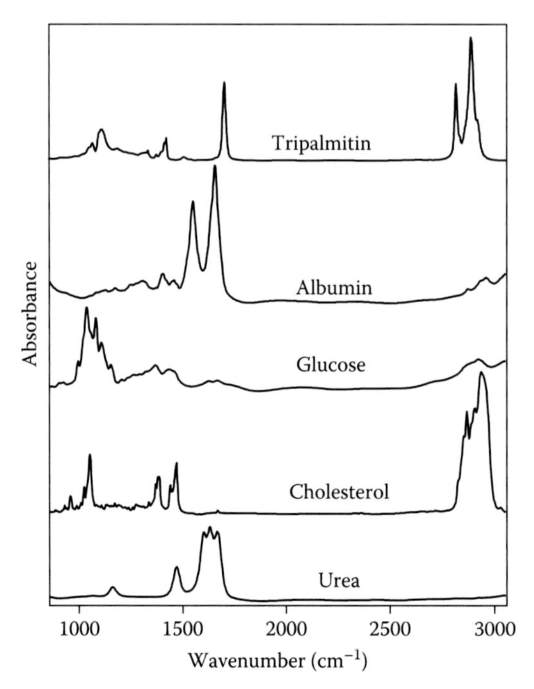

FIGURE 11.13
Mid-infrared absorption spectra of selected serum constituents. (From Shaw, R.A. and Mantsch, H.H.: *Encyclopedia of Analytical Chemistry—Biomedical Spectroscopy*. 1–20. 2000. Copyright Wiley-VCH Verlag GmbH & Co. KGaA. Reproduced with permission.)

constituents (Shaw and Mantsch 2000). Since MIR absorption of water is very intense, MIR spectroscopy requires dehydration of samples or subtraction of the water signal (Ellis and Goodacre 2006). However, unlike with MIR spectroscopy, water does not cause an intense interference in NIR spectroscopy (Ciurczak and Drennen 2002). Another advantage of using NIR radiation is that it can penetrate deep into biological tissue due to the low absorption. Hence NIR spectroscopy has great potential in biomedical applications (Ciurczak and Drennen 2002). The sources used in NIR spectroscopy are intense, and the detectors are sensitive and nearly noise free, resulting in a more precise and accurate spectrum (Ciurczak and Drennen 2002).

In situ glucose measurements is one of the most publicized application of NIR spectroscopy (Ciurczak and Drennen 2002). NIR spectroscopy can also be used to monitor the degree of oxygenation of certain metabolites (Ciurczak and Drennen 2002). NIR radiation can penetrate complex structured matrices of the skin to at least 0.2 mm, and it has been found that this technique may be used to obtain valuable insights into the stratum corneum (Ciurczak and Drennen 2002). NIR spectroscopy can be used to assess the viability of tissue following trauma by identifying tissue regions with poor oxygen supply (Ciurczak and Drennen 2002).

Measurement of body fat in infants is a more accurate measurement of their development than the body weight. Body fat measurements can be easily made by NIR spectroscopy (Ciurczak and Drennen 2002). Several researchers have focused on using NIR spectroscopy to study lung function (Ciurczak and Drennen 2002). This technique can also be used to quantify protein, creatinine, and urea in urine samples (Ciurczak and Drennen 2002). It also has a great potential in cancer research. In a study of cervical cancer detection, it has been shown that NIR spectra of malignant and healthy tissues were distinctly different while abnormal tissues carried spectral features from both sets (Ge et al. 1995). NIR spectroscopy can be used as an alternative to mammograms and offers the advantages of eliminating false-positives and not using harmful radiation as opposed to x-ray–based mammograms (Ciurczak and Drennen 2002). In addition, the use of NIR techniques gives a better understanding of the mass's chemistry, which cannot be achieved with magnetic resonance imaging of breast cancers (Ciurczak and Drennen 2002). This low-cost, noninvasive technique is widely used for measuring tissue oxygen saturation, changes in hemoglobin volume and blood flow, and muscle oxygen consumption (Ferrari et al. 2004).

In IR spectroscopy a polychromatic beam of IR radiation is passed through the sample, and the attenuation of this IR radiation is measured as a function of wavelength (Chiriboga et al. 1998). When a molecule is exposed to IR radiation that matches its vibrational frequency, absorption of IR photons occurs resulting in a characteristic spectrum of the particular molecule (Chiriboga et al. 1998). NIR radiation is highly attenuated by chromophores of variable concentration (e.g., hemoglobin, myoglobin) due to oxygen-dependent absorption and chromophores of fixed concentration (e.g., skin melanine) due to absorption and scattering (Ferrari et al. 2004). Another drawback of this technique is that it is difficult to quantify the NIR signal applied to a biological tissue due to scattering (Ferrari et al. 2004).

Fourier transform-IR (FT-IR) spectroscopy is used for the nondestructive analysis of a wide variety of sample types (Ellis and Goodacre 2006). FT-IR spectroscopy allows for rapid, high-throughput, nondestructive analysis of diseases (Ellis and Goodacre 2006). In FT-IR spectroscopy, the interference of two beams of radiation is used to obtain an interferogram (i.e., a signal produced as a function of the change of pathlength between two beams of light) (Stuart 2004). Figure 11.14 shows a typical FT-IR spectra of white and gray matter of a hamster brain (Naumann et al. 2009).

11.3.3 Terahertz Spectroscopy

In terahertz spectroscopy, microwave and IR frequencies are employed to study the structure, dynamics, and functions of biological systems (Bowen 2011). Terahertz imaging can be used to identify different tissue types, diseased regions, etc. Terahertz radiation is nonionizing, and hence it is safer than using x-rays (Bowen 2011). In many key biological processes, such as

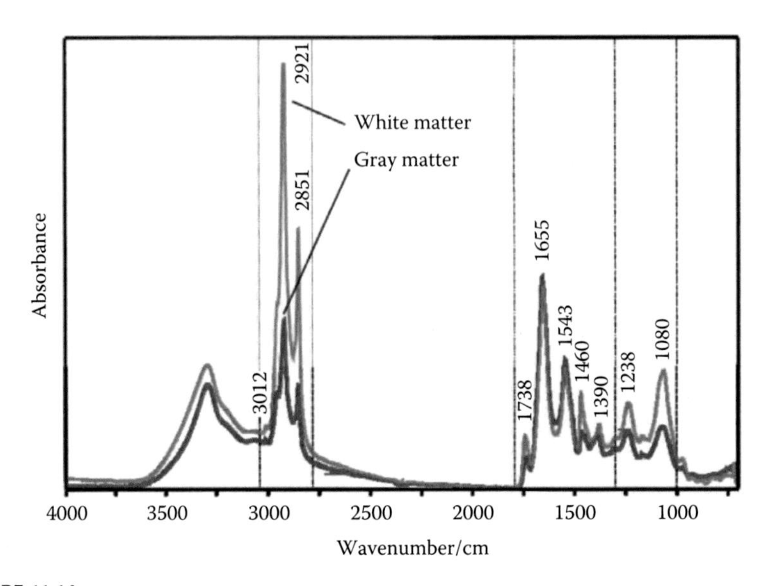

FIGURE 11.14
Typical FT-IR spectra of white and gray matter of a hamster brain. (Reproduced from Naumann, D., Fabian, H., and Lasch, P., FTIR spectroscopy of cells, tissues and body fluids, In *Biological and Biomedical Infrared Spectroscopy, Advances in Biomedical Spectroscopy*, vol. 2, IOS Press, Amsterdam, the Netherlands, pp. 312–354, 2009. With permission from IOS Press.)

protein conformational changes and the collective motion of DNA base pairs along the hydrogen-bonded backbone, the energies involved lie in the terahertz frequency range (Bowen 2011).

Terahertz range of the electromagnetic spectrum covers frequencies from 100 GHz to 10 THz, corresponding to wavelengths from 3 mm to 30 μm (Bowen 2011). Energies corresponding to molecular rotations, hydrogen bonding stretches and torsions, bond vibrations, and crystalline phonon transitions lie within this terahertz range. All of these processes can give rise to distinct spectral signatures (Bowen 2011). Different biomolecules vibrate in different ways, resulting in characteristic signatures. Also, different conformations of the same molecule will have different collective vibrational modes and different spectral signatures in the terahertz range (Bowen 2011).

Cancer imaging at terahertz frequencies shows good agreement with histopathology, and terahertz imaging can be developed as a noninvasive biopsy technique, whereby the full extent of the tumor can be determined prior to carrying out any surgical procedure to remove tissue. Terahertz medical imaging can be used for imaging skin cancer, basal cell carcinoma, breast cancer, colon cancer, and even tooth decay (Bowen 2011, Crawley et al. 2003, Fitzgerald et al. 2006, Woodward et al. 2003).

Nanoscale changes in living cell monolayers can be sensed using terahertz spectroscopy by exploiting its extreme sensitivity to water

(Bowen 2011). Liquid water is a strong absorber of terahertz radiation, the absorption coefficient rising almost linearly from about 125 cm^{-1} at 0.25 THz to around 350 cm^{-1} near 2 THz. Hence, terahertz spectroscopy is a very sensitive probe of water in biological systems (Bowen 2011). Biological processes involving conformational changes of proteins modify both the collective vibrational modes of the protein and its influence on the dynamics of the surrounding water network. Hence, changes occur to the terahertz absorption spectrum, and, therefore, terahertz spectroscopy is useful in understanding protein dynamics and monitoring conformational changes (Bowen 2011). Neural signal transduction along axons involves the flow of potassium and sodium ions out of and into axons through the cell membrane. Potassium ions in solution show a higher terahertz absorbance than sodium ions. Therefore, terahertz frequencies can be used to image neurons and monitor the axonal water flux (Bowen 2011). Terahertz spectrum of a cell is determined by the collective vibrational modes of its molecules and associated hydration water dynamics. Therefore, terahertz spectroscopy can be used to differentiate between cells and to sense changes in cells (Bowen 2011). Figure 11.15 shows parametric images of a wax-embedded melanoma sample.

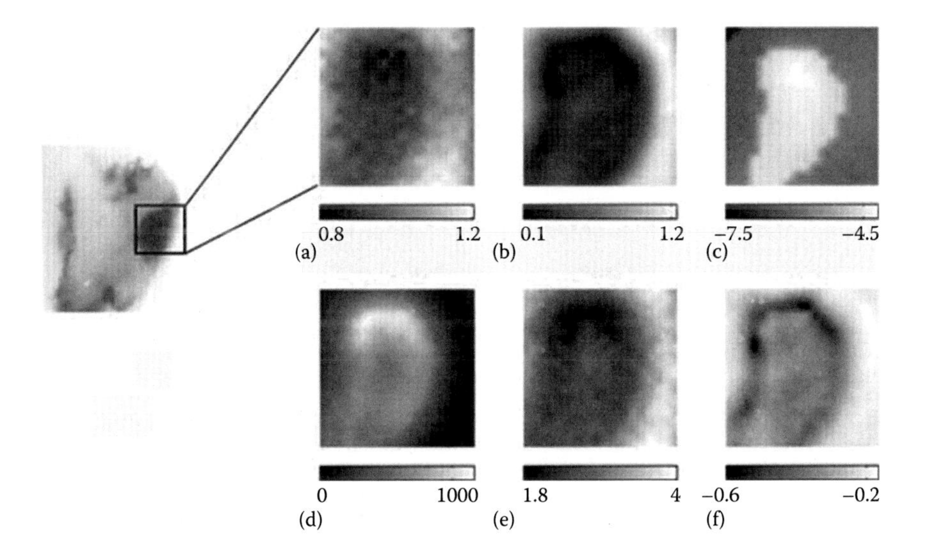

FIGURE 11.15
Parametric images of the wax-embedded melanoma sample (shown in the photograph on the left), which was prepared in the normal histopathological procedure. (a) Transmittance at 0.5 THz, (b) transmittance at 2 THz, (c) phase angle at 1 THz from Fourier transform, (d) absorption coefficient/(refractive index −1), (e) dispersion at 1 THz, and (f) dual-frequency image (transmittance at 1 THz relative to 1.5 THz). (Reproduced from Fitzgerald, A.J., Berry, E., Zinovev, N.N., Walker, G.C., Smith, M.A., and Chamberlain, J.M., An introduction to medical imaging with coherent terahertz frequency radiation, *Phys. Med. Biol.*, 47, R67–R84, 2002. With permission of Institute of Physics Publishing.)

Different imaging parameters, such as the pulse height, pulse width, and time delay, can be used to construct terahertz images. In addition, the time post-pulse, which is the ratio of the electric field amplitude at some time after the pulse minimum to the electric field amplitude minimum, can be used (Woodward et al. 2003). The time at which the electric field is sampled after the pulse minimum can be adjusted to maximize the image contrast (Bowen 2011). Oh et al. (2009) have shown that gold nanorods can be used as nanoparticle contrast agents to increase the terahertz contrast of cancer cells (Bowen 2011). One disadvantage of terahertz imaging is the limited penetration depth into body tissues due to the high water content (Bowen 2011). At 0.5 THz, the penetration depth into skin is around 0.5 mm. However, the penetration depth into fatty tissue (typical of breast tissue) is much higher, around 6.5 mm (Bowen 2011). Nevertheless, an interoperative terahertz probe can be used during surgery to indicate the extent of a tumor, as a high proportion of cancers lie in the outer regions of organs (Ashworth et al. 2008, Bowen 2011). The introduction of image and spectral artefacts is a complication associated with terahertz imaging and spectroscopy of inhomogeneous samples (Bowen 2011).

11.4 Conclusions

In this chapter, we discussed different techniques of nanoscale spectroscopy of clinical importance. These techniques are applied in many different disciplines in biology and medicine. Spectroscopy in general can be categorized as electronic, vibrational, and rotational. However, only electronic and vibrational spectroscopy are of biological interest. Fluorescence and phosphorescence spectroscopy (collectively known as luminescence spectroscopy) and AAS fall into the class of electronic spectroscopy. Raman spectroscopy, IR spectroscopy, and terahertz spectroscopy fall into the class of vibrational spectroscopy.

Fluorescence spectroscopy is used for the noninvasive early diagnosis of various types of cancer, atherosclerosis, and heart arrhythmia and for monitoring many different metabolites in cellular culture such as pH, calcium concentration, and glucose. Due to the very high signal-to-noise ratio, fluorescence spectroscopy enables one to distinguish spatial distribution of even low concentrations of substances. FCS can be used to investigate a variety of nanoscale biological processes such as protein–protein interactions, binding equilibria for drugs, and clustering of membrane-bound receptors. Noninvasive phosphorescence measurements can be used to determine tumor oxygen concentrations.

AAS is applied for the diagnosis and monitoring of Wilson disease, leukemia, hypozincemia, acute lead poisoning, and many more. Clinical applications of AAS include the determination of trace metals such as zinc, copper, lead, and iron as well as the indirect determination of nonmetallic elements such as sodium, potassium, and calcium in biological components. This technique is widely used for the performance of routine analysis of these metallic and nonmetallic elements in clinical laboratories. AAS is free from interference from its environments, and it has been experimentally shown that it is very rarely affected by the presence of other elements.

The application of Raman spectroscopy in medicine ranges from forensic analysis to fundamental studies of molecular interactions. It can be used for rapid noninvasive cancer detection to distinguish Alzheimer disease brain tissue from normal brain tissue and to determine concentrations of substances in blood. SERS is used to enhance the otherwise low Raman scattering cross-section. CARS is another variation of Raman spectroscopy that is used to overcome limitations such as low scattering cross-section and interference from autofluorescence faced by conventional Raman spectroscopy. Raman spectroscopy can be used for both qualitative and quantitative tissue analyses. This technique can be used in vivo as well as in vitro in a noninvasive and nondestructive manner in different environments. Due to the nonbleaching property, Raman signals may be probed for long periods of time in live cells, and there is no need for special sample preparation. A high spatial resolution can be achieved in subcellular chemical analysis of an individual cell.

IR spectroscopy is used for in situ glucose measurements to monitor the degree of oxygenation of certain metabolites, for body fat measurements, to study lung functions, and to quantify various components. IR spectroscopy also has great potential in cancer research. Terahertz spectroscopy can be used for cancer imaging and to image neurons and monitor the axonal water flux.

References

Apreleva, S. V., D. F. Wilson, and S. A. Vinogradov. 2006. Tomographic imaging of oxygen by phosphorescence lifetime. *Applied Optics* 45(33):8547–8559.

Arden, G. B., R. L. Sidman, W. Arap, and R. O. Schlingemann. 2005. Spare the rod and spoil the eye. *British Journal of Ophthalmology* 89:764–769.

Ashworth, P. C., P. O'Kelly, A. D. Purushtham, S. E. Pinder, M. Kontos, M. Pepper, and V. P. Wallace. 2008. An intra-operative THz probe for use during the surgical removal of breast tumors. In *33rd International Conference on Infrared, Millimeter and Terahertz Waves*, pp. 767–769. Pasadena, CA.

Bartolo, B. D. 2011. Fluorescence spectroscopy and energy transfer processes in biological systems. In *Biophotonics: Spectroscopy, Imaging, Sensing and Manipulation*, pp. 107–171. Springer Science + Business Media, Dordrecht, the Netherlands.

Berman, E. 1975. Biochemical applications of flame emission and atomic absorption spectroscopy. *Applied Spectroscopy* 29:1–9.

Bowen, J. W. 2011. Terahertz spectroscopy of biological systems. In *Biophotonics: Spectroscopy, Imaging, Sensing and Manipulation*, pp. 287–303. Springer Science + Business Media, Dordrecht, the Netherlands.

Burch, C. L., J. R. Lakowicz, and E. M. Servick-Muraca. 1994. Biochemical sensing in tissues: Determination of fluorescent lifetimes in multiply scattering media using frequency-domain spectroscopy. *Proceedings of SPIE* 2135:286–299.

Chan, J. W., D. S. Taylor, T. Zwerdling, S. M. Lane, C. E. Talley, C. W. Hollars, and T. Huser. 2005. Raman spectroscopy: chemical analysis of biological samples. In *Advances in Biophotonics*, pp. 148–168. IOS Press, Amsterdam, the Netherlands.

Chiriboga, L., P. Xie, H. Yee, V. Vigorita, D. Zarou, D. Zakim, and M. Diem. 1998. Infrared spectroscopy of human tissue I. differentiation and maturation of epithelial cells in the human cervix. *Biospectroscopy* 4:47–53.

Ciurczak, E. W. and J. K. Drennen. 2002. *Pharmaceutical and Medical Applications of Near-Infrared Spectroscopy*. Marcel Dekker Inc., New York.

Crawley, D., C. Longbottom, V. P. Wallace, B. Cole, D. D. Arnone, and M. Pepper. 2003. Three-dimensional terahertz pulse imaging of dental tissue. *Journal of Biomedical Optics* 8:303–307.

Deckelbaum, L. I., J. K. Lam, H. S. Cabin, K. S. Clubb, and M. B. Long. 1987. Discrimination of normal and atherosclerotic aorta by laser induced fluorescence. *Lasers in Surgery and Medicine* 7:330–335.

Ellis, D. I. and R. Goodacre. 2006. Metabolic fingerprinting in disease diagnosis: Biomedical applications of infrared and Raman spectroscopy. *Analyst* 131:875–885.

Engels, S. A. and B. C. Wilson. 1992. In vivo fluorescence in clinical oncology: Fundamental and practical issues. *Journal of Cellular Pharmacology* 3:66–79.

Evans, S. M. and C. J. Koch. 2003. Prognostic significance of tumor oxygenation in humans. *Cancer Letters* 195:1–16.

Feng, S., R. Chen, J. Lin, J. Pan, Y. Wu, Y. Li, J. Chen, and H. Zeng. 2011. Gastric cancer detection based on blood plasma surface-enhanced Raman spectroscopy excited by polarized laser light. *Biosensors and Bioelectronics* 26:3167–3174.

Ferrari, M., L. Mottola, and V. Quaresima. 2004. Principles, techniques, and limitations of near-infrared spectroscopy. *Canadian Journal of Applied Physiology* 29:463–487.

Ferriero, D. M. 2004. Medical progress—Neonatal brain injury. *New England Journal of Medicine* 351:1985–1995.

Fitzgerald, A. J., E. Berry, N. N. Zinovev, G. C. Walker, M. A. Smith, and J. M. Chamberlain. 2002. An introduction to medical imaging with coherent terahertz frequency radiation. *Physics in Medicine & Biology* 47:R67–R84.

Fitzgerald, A. J., V. P. Wallace, M. Jimenez-Linan, L. Bobrow, R. J. Pye, A. D. Purushotham, and D. Arnone. 2006. Terahertz pulsed imaging of human breast tumors. *Radiology* 239:533–540.

Frank, C. J., R. L. Mccreery, and D. C. B. Redd. 1995. Raman-spectroscopy of normal and diseased human breast tissues. *Analytical Chemistry* 67:777–783.

Ge, Z., C. W. Brown, and H. J. Kisner. 1995. Screening pap smears with near-infrared spectroscopy. *Applied Spectroscopy* 49:432–436.

Gniadecka, M., H. C. Wulf, N. N. Mortensen, O. F. Nielsen, and D. H. Christensen. 1997. Diagnosis of basal cell carcinoma by Raman spectroscopy. *Journal of Raman Spectroscopy* 28:125–129.

Godavarty, A., E. M. Sevick-Muraca, and M. J. Eppstein. 2005. Three-dimensional fluorescence lifetime tomography. *Medical Physics* 32:992–1000.

Handapangoda, C. C., M. Premaratne, and S. Nahavandi. 2012a. A fresh look at the validity of diffusion equations for modelling phosphorescence imaging of biological tissue. *Lecture Notes in Computer Science* 7425:461–468.

Handapangoda, C. C., M. Premaratne, and S. Nahavandi. 2012b. Generalized coupled photon transport equations for handling correlated photon streams with distinct frequencies. *Optics Letters* 37:3444–3446.

Handapangoda, C. C., M. Premaratne, L. Yeo, and J. Friend. 2008. Laguerre rungekuttafehlberg method for simulating laser pulse propagation in biological tissue. *IEEE Journal on Selected Topics in Quantum Electronics* 14(1):105–112.

Hanlon, E. B., R. Manoharan, T.-W. Koo, K. E. Shafer, J. T. Motz, M. Fitzmaurice, J. R. Kramer, I. Itzkan, R. R. Dasari, and M. S. Feld. 2000. Prospects for in vivo Raman spectroscopy. *Physics in Medicine & Biology* 45:R1–R59.

Jameson, D. M., J. A. Ross, and J. P. Albanesi. 2009. Fluorescence fluctuation spectroscopy: Ushering in a new age of enlightenment for cellular dynamics. *Biophysical Reviews* 1:105–118.

Lakowicz, J. R. 2010. *Principles of Fluorescence Spectroscopy*. Springer, New York.

Lo, L. W., S. H. Huang, C. H. Chang, W. Y. Chen, P. J. Tsai, and C. S. Yang. 2003. A phosphorescence imaging system for monitoring of oxygen distribution in rat liver under ischemia and reperfusion. *Journal of Medical and Biological Engineering* 23:19–27.

Manoharan, R., K. Shafer, L. Perelman, J. Wu, K. Chen, G. Deinum, M. Fitzmaurice, J. Myles, J. Crowe, R. R. Dasari, and M. S. Feld. 1998. Raman spectroscopy and fluorescence photon migration for breast cancer diagnosis and imaging. *Photochemistry and Photobiology* 67:15–22.

Naumann, D., H. Fabian, and P. Lasch. 2009. FTIR spectroscopy of cells, tissues and body fluids. In *Biological and Biomedical Infrared Spectroscopy, Advances in Biomedical Spectroscopy*, vol. 2, pp. 312–354. IOS Press, Amsterdam, the Netherlands.

Oh, S. J., J. Kang, I. Maeng, J. S. Suh, Y. Huh, S. Haam, and J. H. Son. 2009. Nanoparticle-enabled terahertz imaging for cancer diagnosis. *Optics Express* 17:3469–3475.

Otto, C. and V. V. Pully. 2012. Hyperspectral Raman microscopy of the living cell. In *Applications of Raman Spectroscopy to Biology*, pp. 148–173. IOS Press, Amsterdam, the Netherlands.

Patterson, M. S. and B. W. Pogue. 1994. Mathematical model for time-resolved and frequency-domain fluorescence spectroscopy in biological tissues. *Applied Optics* 33(10):1963–1974.

Pena, F. and A. M. Ramirez. 2005. Hypoxia-induced changes in neuronal network properties. *Molecular Neurobiology* 32:251–283.

Prasad, P. N. 2003. *Introduction to Biophotonics*. John Wiley & Sons, Inc., Hoboken, NJ.

Premaratne, M., E. Premaratne, and A. J. Lowery. 2005. The photon transport equation for turbid biological media with spatially varying isotropic refractive index. *Optics Express* 13(2):389–399.

Robinson, J. W. 1960. Atomic absorption spectroscopy. *Analytical Chemistry* 32:17A–29A.

Sartori, M., R. Sauerbrey, S. Kubodera, F. Tittel, R. Robert, and P. Henry. 1987. Autofluorescence maps of atherosclerotic human arteries—A new technique in medical imaging. *IEEE Journal of Quantum Electronics* 23:1794–1797.

Schneckenburger, H. 2005. Fluorescence spectroscopy and microscopy. In *Advances in Biophotonics*, pp. 196–209. IOS Press, Amsterdam, the Netherlands.

Schwille, P. and J. Ries. 2011. Principles and applications of fluorescence correlation spectroscopy (FCS). In *Biophotonics: Spectroscopy, Imaging, Sensing, and Manipulation*, pp. 63–85. Springer Science+Business Media, Dordrecht, the Netherlands.

Servick-Muraca, E. M. and C. L. Burch. 1994. Origin of phosphorescence signals reemitted from tissues. *Optics Letters* 19:1928–1930.

Shaw, R. A. and H. H. Mantsch. 2000. Infrared spectroscopy in clinical and diagnostic analysis. In *Encyclopedia of Analytical Chemistry—Biomedical Spectroscopy*, pp. 1–20. John Wiley & Sons Ltd, Chichester, U.K.

Siebert, F. and P. Hildebrandt. 2008. *Vibrational Spectroscopy in Life Sciences*. Wiley-VCH Verlag GmbH & Co, Weinheim, Germany.

Smith, J. C., G. P. Butrimovitz, and W. C. Purdy. 1979. Direct measurement of zinc in plasma by atomic absorption spectroscopy. *Clinical Chemistry* 25:1487–1491.

Stuart, B. 2004. *Infrared Spectroscopy: Fundamentals and Applications*. John Wiley & Sons Ltd, Chichester, U.K.

Sunderman, F. W. 1973. Atomic absorption spectrometry of trace metals in clinical pathology. *Human Pathology* 4:549–582.

Talley, C. E., T. Huser, C. W. Hollars, L. Jusinski, T. Laurence, and S. Lane. 2005. Nanoparticle-based surfaced-enhanced Raman spectroscopy. In *Advances in Biophotonics*, pp. 182–195. IOS Press, Amsterdam, the Netherlands.

Tang, G. C., A. Pradhan, W. Sha, J. Chen, C. H. Liu, S. J. Wahl, and R. R. Alfano. 1989. Pulsed and cw laser fluorescence spectra from cancerous, normal, and chemically treated normal human breast and lung tissues. *Applied Optics* 28:2337–2342.

Utzinger, U., D. L. Heintzelman, A. M. Jansen, A. Malpica, M. Follen, and R. R. Kortum. 2001. Near-infrared Raman spectroscopy for in vivo detection of cervical precancers. *Applied Spectroscopy* 55:955–959.

Vanderkooi, J. M., G. Maniara, T. J. Green, and D. F. Wilson. 1987. An optical method for measurement of dioxygen concentration based on quenching of phosphorescence. *Journal of Biological Chemistry* 262:5476–5482.

Vo-Dinh, T. and B. M. Cullum. 2003. Fluorescence spectroscopy for biomedical diagnostics. In *Biomedical Photonics Handbook*, pp. 1–50. CRC Press, Boca Raton, FL.

Voitovich, A. P., V. S. Kalinov, and A. P. Stupak. 2011. Fluorescence of strongly absorbing multicomponent media. In *Biophotonics: Spectroscopy, Imaging, Sensing and Manipulation*, pp. 173–181. Springer Science+Business Media, Dordrecht, the Netherlands.

Wilson, D. F. and S. A. Vinogradov. 2003. *Handbook of Biomedical Fluorescence*. Marcel Dekker, New York.

Woodward, R. M., V. P. Wallace, R. J. Pye, B. E. Cole, D. Arnone, E. H. Linfield, and M. Pepper. 2003. Terahertz pulse imaging of ex vivo basal cell carcinoma. *Journal of Investigative Dermatology* 120:72–78.

Wu, J., M. S. Feld, and R. P. Rava. 1993. Analytical model for extracting intrinsic fluorescence in turbid media. *Applied Optics* 32(19):3585–3595.

12

Medical Nanoscale Spectroscopy: Concepts, Principles, and Applications

Viroj Wiwanitkit

CONTENTS

12.1 Introduction

Small is the latest trend in science. Almost all sciences and technologies inspire to make things smaller and smaller. The smallest scale available, which measures 10^{-9} m, is called nanoscale. The science that deals with nanoscaled elements is called nanoscience. Nanoscience applications can be useful over various domains in various ways. Today, many new observations and ideas on the subject are being discussed and many nanoscience-powered objects are in use. Nanoscale objects present novel properties that are different from large-sized objects. These unique properties, which include distinctive electrical and biochemical properties, encourage scientists to make new nanoapplications. Nanoobjects can be biological or nonbiological. Hence, nanoscience could be described as a science that is both physical and biological. Nanoscience presents itself as an alternative to solve many challenging and unsolved questions related to nanoscale/size and otherwise, which would have been difficult in earlier days. As a novel science, nanoscience is a perfect convergence of various sciences, which applies the new age concept of integration. Although it is a novel science, nanoscience is widely absorbed and studied in many prestigious higher-education institutions around the world. As an actual multidisciplinary approach, nanoscience can be applied for usage in many aspects. As already noted, nanoscience can be integrated into both physical and biological concerns principle. Many hybrid branches of nanoscience have been introduced to the scientific community over the past few years. Those hybridizations are the result of developments in modern science. A few good examples of hybrid nanosciences are nanomaterial, nanobiology, nanophysics, nanochemistry, nanopharmacology, nanoengineering, nanoradiology, nanodentistry, nanomedicine, etc. Advantages of these hybrids are visible in our day-to-day life.

Among these hybrid branches of nanoscience, nanomedicine is of specific interest. Nanomedicine deals with nanoscaled applications in medical aspects. Nanomedicine is implemented and is used successfully in modern medicine. The challenge today is to make nanomedicine the prominent branch of medical science. Nanomedicine can be used in all the areas of medical science (diagnosis, treatment, and prevention of disease), to carry out any medical procedure successfully, thus making medical services fast, highly accurate, and reliable. Modern medical studies are focusing more on nanomedicine approach, and much research and development is going on. Several diagnostic and therapeutic applications based on nanomedicine are already in use, which includes pathogens, drugs, enzymes, etc. Nanomedicine-based applications can be seen implemented in many modern medical centers. These applications can be in vivo, in vitro, or in silico (computational) [1]. In vivo and in vitro usages are based on the basic concepts in medicine. Examples of nanomedicine used in vivo and in vitro are nanodrugs and nanodiagnostic test kits, respectively. These two usages are expected to become increasingly common in medical society. The third usage, in silico, is comparatively new. It can be an unfamiliar topic for many medical workers. In silico usage of nanomedicine refers to the application of the science for computational-based manipulation. To explain further, it is based on the application of nanoinformatics technique. Using computational informatics, many difficult questions in medical science can be answered. Currently, there are several in silico techniques in use. The main use of this approach is in simulation or imaginary work. Although not a real thing, the simulations are very helpful in medical practice as the results derived from them are acceptable and reliable. In silico approach using nanotechnology can replace the old, classical in vivo and in vitro approaches in medical science. Using this approach, the solutions for advanced medical issues can be derived within a short time. Thus, the advantages of this approach include reduced time, consumption, and cost of study. Further, this approach controls confounding (interference) as it has no interference effects. Computational-simulation technique is presently known as in silico technique in many *omics* sciences [2]. Several new *omics* sciences have been introduced in the past decades and those have become new progressive branches of science. It can be said that we are in the *omics* era today [2]. Similar to other sciences used in medical field, applied computational technology in nanomedicine can be expected to provide solutions to many advanced medical problems. In silico nanoinformatics technique can be useful in solving nanoscale problems on extremely nanoobjects and in predicting the nanomedical phenomena [3,4]. It can provide data useful in modern medicinal science and could be our new hope in fighting diseases and illnesses.

At present, there are many new techniques in nanomedicine research. To select a good technique, knowledge of its basic principles is important. Of several basic techniques, nanoscale spectrometry, a new modality of nanosciene, is discussed here. Spectrometry is a classical technology in science. First, the term "spectrum" should be understood since spectrometry directly deals with spectrum. In science, spectrum is usually mentioned. By definition, spectrum means a condition that shows no boundary. There is no specific limitation. The specific set of values can vary infinitely, or it poses a continuum. Therefore, these are the basic characteristics of spectrum: (a) they are a set of data (values that can be directly measured or not), (b) are specific, (c) a collection all together as a set, (d) a continuum and (e) they really exist. The spectrum is usually mentioned when one talks about light. The specific applied science that deals with the behavior of light is called optics. In nanoscience, there is also a specific branch called nanooptics, which focuses on the behavior of light or optics on the nanometer scale. Applied in medicine, nanophotonics can be useful in many ways such as in ophthalmology and radiology.

Spectroscopy is another scientific approach. In brief, it is a specific assessment of the interaction between object and radiated energy. Classically, spectroscopy has been an important concept in optics. The oldest known work is on visible light. The spectrum of visible light can be dispersed according to its wavelength with the use of a classical scientific tool, the "prism." However, the spectrum is not limited to visible light as already mentioned. The study extends to nonvisible lights (several rays). The interaction is the main assessed thing in spectrometry. Spectroscopic data are useful in representing the interaction, and it is usually written as a simple spectrum. As already mentioned, spectroscopy has these characteristics; (a) focuses on interaction, (b) explores the interrelationship between objects and radiation and (c) is assessed by scientific tools. In a general discussion on radiation, one can imagine the extremely small scale of radiation's property, the wavelength. Generally, the wavelength is usually determined in nanometer scale. Hence, it is not surprising that optics is an actual nanoscience study [5]. In addition, the object that interacts with radiation can be large or small. In case the object is very small, within the nanoscale, the role of nanoscience can be expected. A specific branch of nanoscience, nanophotonics can be useful at this point. In medicine, spectroscopy has been used for a long time, and nanoscale spectroscopy method is a new applied technique. This technique in nanomedicine can be useful in many activities in medicine especially for the study of nanosize cells and particles. In this chapter, the author focuses on the application of nanoscale spectroscopy in medicine with special focus on nanomedicine. Examples of studies based on this technique are discussed in the chapter.

12.2 Overview of Nanoscale Spectroscopy and Its Application in Medicine

Nanoscale spectrometry is a specific new technique in nanoscience. It is widely mentioned in nanophotonics, the new branch of nanoscience. Indeed, nanophotonics is the study of photon or light, which is on the nanoscale. Nanophotonics is an advanced optical science that makes use of optical engineering. The main focus is on optics and interaction between light or photon and objects at nanolevel. In general, nanophotonics can be useful in both basic and applied issues. Focusing on the basic issues, the main themes are fundamentals and principles of light and its interaction with objects at nanoscale. Considering the applied issues, the new nanodevices that deal with opticals is the main focus. The novel nanoengineering can play a role in studying and developing such new nanooptical tools. Indeed, there are many new nanodevices that are good examples of applied nanophotonics tools.

Of several optical tools, spectrometry is a common apparatus. The tool aims at measuring spectrum. Spectrometry or spectrography is mainly the measurement of radiation intensity. The function of wavelength is usually assessed and finalized and referred to as spectrum. There are several measurement devices in spectrometry. Good examples include spectrometers, spectrophotometers, spectrographs, and spectral analyzers. Those tools have been used for a very long time in science. These tools also have a long history in medicine. In discussions on optics in medicine, spectrometry measurement has to be mentioned.

Spectrometry can be applied in several areas including ophthalmology, radiology, and laboratory medicine.

Generally, visualization or the "ability to see" is mainly determined by light. Light is a basic requirement. Without light, visualization cannot be expected. In medicine, visualization at extraordinary small scale is also an important topic. The use of microscope is an example. It leads to the specific medical science, clinical microscopy. This specific science has been employed in the medical society for many years and plays a vital role in modern medical activity. Several new microscopes can be seen at present. It is usually the tool for "seeing" nanoscale objects. The examples are electronic electron microscopy and near-field scanning optical microscopy. There is no doubt that novel nanophotonics techniques can be applied in clinical microscopy [6]. The use of spectrum analysis is important in clinical microscopy. Several new spectrometry tools have recently been developed based on the new knowledge on nanophotonics. Of those new ones, the microscope spectrometer should be mentioned. The microscope spectrometer is a specific microscope aiming at measuring UV–visible–NIR spectrum of microscopic objects (small objects or areas of larger objects) [7]. This kind of microscope is a useful tool that can be applied for nanophotonics study. There are mainly two main kinds of microscope spectrometers. The first one is the basic microspectrometer. It is an integrative microscope that has been built for specific microspectrometry work. The second type is designed as a spectrometer unit. It is not fully integrated, but it is designed as an additional attachment piece to a simple classical open photoport of an optical microscope. The properties of the microspectrometer depend on its composition, strength, and configuration. A microspectrometer helps measure the spectrum of microscopic samples by transmission, absorbance, reflectance, fluorescence, and emission as well as polarization spectrometry. Some specific producers of microspectrometers also design their products for increasing properties of high-resolution digital imaging. The applied computational technology, attaching the new software to a simple microspectrometer, can also increase the ability to measure thin film thickness and colorimetry. As noted, this applied nanophotonics tool can be used as a novel diagnostic tool in laboratory medicine. Extensive applied spectroscopy detection that depends on visualization in laboratory medicine can also be seen in some high-tech investigations such as flow cytometry. Basically, the detection of particles or cells can be done with the use of a flow cytometry system. The main principle is based on detection of light scattering pulse during flowing of the particles. The application of nanophotonics in laboratory medicine can be helpful and already exists in some advanced tools. The new generation of flow cytometry–based tool such as a spectroscopy flow cytometer can be seen. This technique is the basic technology used in several micro- and nanofluidics systems [8]. Conceptually, the nanofabricated impedance spectroscopy flow cytometer can help permit rapid dielectric characterization of a particle population with a simple nanofluidic channel [9]. The measurement of impedance can be done, and this results in laboratory data on particle size, membrane capacitance, and particle's conductivity as a function of frequency [9]. The amplitude, opacity, and phase information can also be derived and can be further used for discrimination or classification of the studied particles into different kinds [9].

In radiology, this medical science also directly deals with radiation. X-ray is also another form of photon. Study by nanophotonics is

TABLE 12.1

Examples of Usages of X-Ray Crystallography

Applications	Details
Hydrocarbon polymers	Oils, fuel, plastic, rubber, textiles
Pharmaceutical products	Drugs, vaccines, antibodies, foodstuffs, cosmetics
Heat-resistant materials	Glass, ceramics
Coatings	Paper, film, polyester, and metals
Wastes	Effluent, cleaning fluids, pools, filters

possible. X-ray spectroscopy is a basic application in radiology [5,9]. In nanoscale, X-ray spectrometry can also be useful. Indeed, the energy-dispersive X-ray fluorescence technology is accepted as a simple, accurate, and economical analytical method for the determination of many chemical compositions (Table 12.1). It can be applied in analysis in material sciences and also be further applied in medicine. A good example is crystallography study of medical molecules. New drugs, antibodies, biomolecules, hormones, or enzymes can be assessed via novel nanophotonics technology. In ophthalmology, the application of nanophotonics can also be useful [5]. Spectrometry can be useful in the assessment of visualization. Spectrometry for retinal assessment is a good example. Functional maps using a chromophore spectrum can be derived from retinal imaging spectroscopy [10,11]. The new snapshot spectral camera is a new ophthalmological tool that provides a new noninvasive way for retinal vessel oximetry mapping [11].

Furthermore, the specific study of the multitude of nanooptical phenomena that is caused by resonant surface plasmons localized in nanosystems by nanoplasmonics can also be useful in medicine [12,13]. Several new devices of nanoplasmonics have been developed and are available for real usage at present. The detection of nanobiochemical molecule with a nanoplasmonics tool is the main application. The application in microscopy should also be mentioned. The Attosecond nanoplasmonic-field microscope is the best example [14]. Finally, flow cytometry–based microfluidics and nanofluidics nanochip can also make use of nanoscale spectroscopy technology [15,16].

As previously mentioned, the use of nanosale spectrometry in medicine can be either simple or advanced. Simple usage is the use of nanoscale spectrometry tool for studying real objects or phenomena. This is the classical assessment in medicine. For more advanced usage, a computational approach can be used. Applied nanophotonics and nanoplasmonics concepts can be useful at this point and can help approach imaginary objects or phenomena. As already mentioned, this is an actual story of in silico approach. Simulating or computational informatics technology can be applied in nanoscale spectrometry and is further discussed in the next section. At this point, the reader might still not be able to imagine the exact application of nanoscale spectrometry in medicine. To help the reader better understand the present applications based on nanoscale spectrometry, some interesting reports on medical nanoscale spectrometry applications are discussed.

1. Application in clinical microscopy
   There is no doubt that nanoscale spectroscopy can be useful for clinical microscopy. This is a useful application in laboratory medicine. The application of nanoscale spectrometry in clinical microscopy is now well accepted. Nanoscale spectrometry is well established at present as a tool for advanced research. There are many new publications on this issue. This implies that the usage is increasing rapidly.
   The following lists important and new interesting publications on applied nanoscale spectrometry.
   a. Fujikawa et al. [17] reported their success in using low energy electron microscope (LEEM) spectromicroscopy for the assessment of surface plasmons. The assessed plasmons might localize on either micro- or nanoscale epitaxial Ag islands [17]. Fujikawa et al. concluded that "LEEM based plasmon spectromicroscopy promises to be a powerful tool for furthering our understanding of nanoplasmonics [17]."
   b. Horiba et al. [18] reported on a new scanning photoelectron microscope for nanoscale 3-D spatial-resolved electron spectroscopy for chemical analysis. This new system has a good capability for pinpoint depth-profile analysis and high-resolution chemical state analysis [18].
   c. Tejedor et al. [19] reported a new in situ molecular analysis technique by live single cell mass spectrometry. In this work, orbitrap mass spectrometer by a nano-electrospray ionization is used as the main analytical tool [19].
   d. Date et al. [20] used video mass spectrometry for assessment of drug metabolism monitoring in a live single hepatic cell. Step by step, nano-spray tip under a video-microscope was first done and then mass spectrometry followed [20].
   e. Hara et al. [21] introduced a new energy-dispersive x-ray spectrometer with a microcalorimeter detector equipped with a transmission electron microscope.
   f. Hodson et al. [22] reported using atomic force microscope as a force spectrometer for assessment of nanomechanical properties of extracellular matrix components.
   g. Sirikatitham et al. [23] reported the use of resin-packed nano-electrospray in combination with video and mass spectrometry for molecular analysis of mast cells. This approach can provide direct and real-time analysis of studied cells [23].

h. Yang et al. [24] reported the use of new nano-ultra high performance liquid chromatography coupling with time-of-flight mass spectrometer (nano-UPLC-TOF-MS) to detect the low molecular weight polypeptides. In this report, nano-UPLC-TOF-MS was used for studying zebrafish embryos [24].
i. Li et al. [25] tried to combine fiber-optical probe and nanoprobe techniques in a scanning electron microscope. This new combination helps provide in situ optical, electrical, and structural characterization of optoelectronic nanomaterials and nanodevices [25].
j. Pettinicchio et al. [26] used light microscope, scanning electron microscope with an energy-dispersive spectrometer, and circularly polarized light microscope for assessment of histological behavior of bone graft materials placed in humans.
k. Hoang et al. [27] reported a new system with high signal-to-noise ratio electron energy spectrometer connecting to the scanning electron microscope. This new tool helps provide mapping specimen surface voltage and atomic number variations on the nanoscale [27].

2. Application in medical instrumentation

In addition to the previously mentioned applications in basic clinical microscopy, nanoscale spectroscopy can be applied in medical instrumentation. New tools such as new chips and nanofluidics analyzers with the use of nanoscale spectroscopy technique can be seen. These new devices are really useful in biomedical science.

Here is a list of important and interesting publications in this area.

a. Christensen et al. [28] reported the use of small unilamellar lipid vesicles in a nanofluidics tool that helps increase the ability of nanoreactors.
b. Batabyal et al. [29] reported a new microfluidics system that combined fluorescence microscopy and femtosecond/picosecond-resolved spectroscopy. This new tool aimed at investigating ultrafast chemical processes in liquid-phase diffusion-controlled reactions [29].
c. Song and Wang [30] reported a new optofluidic differential spectroscopy technique for absorbance detection of sub-nanoliter liquid samples.
d. Prim et al. [31]. reported on an attempt to couple a microfluidic mixer to a Fourier-transform infrared spectrometer for protein-conformation studies. According to the trial, on-line monitoring of protein conformation under varying conditions was successful [31].

e. Greiner et al. [32] reported a new confocal backscattering spectroscopy technique for leukemic and normal blood cell discrimination. Greiner et al. proposed that this technique could be useful for further developing of new nanofluidics system [32].

f. Cecchini et al. [33] reported the use of a new technique based on surface-enhanced resonance Raman spectroscopy with sub-millisecond time resolution. Cecchini et al. proposed that this technique was a powerful detection tool in microdroplet reactors and could be applied in developing the new nanofludics system [33].

g. Ashok et al. [34] first reported on a fiber-based microfluidic Raman spectroscopic detection scheme and called it as "Waveguide Confined Raman Spectroscopy." Ashok et al. said that this new scheme could facilitate reaction monitoring in a microreactor and detection of queried particles in a microdroplet-based nanofluidic system [34].

h. Javanmard et al. [35] presented a new scalable method based on the use of nanofluidics and shear force spectroscopy. Javanmard et al. noted that this method could be used for determining the affinity between molecules [35].

i. Mastrangelo et al. [36] reported the new chip that helped probe protein binding spectra with the use of Fourier microfluidics.

j. Syme et al. [37] first reported on a microstructured trap using Raman spectroscopy. Time-resolved mapping of intracellular nanoparticle labels within living cells could be determined in this nanofluids system [37].

3. Application in clinical oncology

As previously mentioned, nanoscale spectroscopy can be useful in many aspects of medicine. Most of the already mentioned applications in this chapter usually relate to the diagnostic purpose. Applications that can help diagnose in clinical oncology are also available. In addition, nanoscale spectroscopy can also be useful for therapeutic purposes. The application in cancer therapy is an example for therapeutic application. Both diagnostic and therapeutic applications are the new milestones in clinical oncology.

Some important, new, and interesting publications in both diagnostic and therapeutic issues in clinical oncology area are listed here.

a. Birtill et al. [38] described using a photoacoustic spectroscopy system that helped differentiate oxygenated and deoxygenated blood.

b. Bastatas et al. [39] used AFM-based nanomechanics to determine the elastic moduli and the cell-to-substrate adhesion of prostate cancer cells. In this work, fluorescence spectroscopy was also used for assessment of intracellular calcium dynamics [39].

c. Gormley et al. [40] studied the effect of gold nanorods conjugated RGDfK peptide for the treatment of prostate cancer with the use of high-resolution dark field microscopy, inductively coupled plasma mass spectrometry, and transmission electron microscopy.
d. Liu et al. [41] determined gene fragment and PCR amplification products related to chronic myelogenous leukemia with the use of nanoscale fluorescence spectroscopy.
e. Lin et al. [42] reported on real-time detection of β1 integrin expression on cancerous cells using electrochemical impedance spectroscopy. According to this study, evaluation of β1 integrin expression on cell membranes of human osteogenic sarcoma cell line could be successfully performed [42].
f. Semaan et al. [43] reported the use of reversed-phase nano-liquid chromatography coupled to a hybrid linear quadrupole ion trap/Fourier transform ion cyclotron resonance mass spectrometer for determining phosphoprotein levels in breast cancer tissues.

12.3 Nanoscale Spectroscopy for Nanomedicine

Spectroscopy is a useful technique in biomedicine. It is a tool that relates to light or photon at the nanoscale. The photon phenomenon in the nanoscale is not possible to be determined by the naked eye. Despite the simple micro-assessment, it is still difficult. Because classical techniques have limitations, the application of more advanced techniques is needed. The nanoscale spectroscopy technique can be the solution. Several tools have been introduced as previously mentioned. Examples of existing tools include spectrometers, spectrophotometers, and spectral analyzers. The well-known ones in biomedical science are atomic absorption spectrometer analyzer, microscope spectrometer, and x-ray crystallographer. The mentioned tools are based on nanoassessment. Assessing nanolevel medical phenomena can be in either structural or functional aspects (Table 12.2).

In the context of the support offered by computational technology to the use of nanoscale spectroscopy, in silico technique is a new method for manipulation. As noted, an acceptable solution for complex questions is the in silico approach [3,4]. The new nanoinformatics technology can help approach nanoscale phenomena including the nanophotonic ones [44]. Computational approach is required in this approach. Indeed, nanoinformatics helps both structural and functional assessment. Nanoinformatics applied to nanoscale spectroscopy is very interesting. This is a new issue of spectral analysis. The hybrid medical engineering

TABLE 12.2

Application of Computational Nanoscale Spectrometry in Medicine

Applications	Details
Structural approach	This application is the main usage at present. Nanoscale spectrometry is mainly used for determining the compositions of the queried particles. Structural analysis can be done via several analyzers. Accurate results in nanoscale can be derived.
Functional approach	Advanced computational technology can help support the existing spectrometry system. With applied nanoinformatics technology, prediction can be done. The use of nanoinformatics to support nanoscale spectrometry can be useful in simulating for prediction of the function of the system.

technology used in nanoscale spectroscopy helps answer many difficult questions (both structural clarification and functional prediction situation). For sure, there are many advantages of computational nanoscale spectroscopy in medicine at present. Some examples of important reports can be seen next.

12.3.1 Summary of Important Reports on Application of Nanoscale Spectroscopy in Nanomedicine

As noted earlier, nanoscale spectroscopy has many advantages in nanomedicine. It can be in any form (in vivo, in vitro, or in silico). Here, we discuss some important reports on application of nanoscale spectroscopy in nanomedicine (Table 12.3) [45–51].

1. Application for clarification: A good example of application of nanoscale spectroscopy is its use in clarifying the structure of nano-biomaterials. Clarification of the nanostructure of new biomolecules is possible. The advanced microscope, microscope spectrometer, is the best example.
2. Application for prediction: A good example is applied computational nanoscale spectroscopy. It can be useful in designing new nano-biomaterials. Prediction under different conditions can be possible with the use of nanoinformatics techniques.

In summary, nanoscale spectroscopy can be applied in various sections of nanomedicine.

- *In diagnosis (nanodiagnosis)*: Nanoscale spectroscopy in nano-diagnosis is the main application. There are many studies and developments in this area at present. The aim is usually the visualization of the nano-composition of the queried particle as already mentioned.

TABLE 12.3

Some Important Reports on Nanoscale Spectrometry in Nanomedicine

Authors	Details
Lee et al. [45]	In this report [45], a new single nanoparticle spectroscopy for real time in vivo quantitative analysis was reported. Lee et al. used this new technique for monitoring for transportation and toxicity of single nanoparticles in single embryos [45].
Maclaughlin et al. [46]	In this work, dark field imaging, Raman spectroscopy and flow cytometry were used for assessment of the effect of rituximab-conjugated surface-enhanced Raman scattering gold nanoparticles on chronic lymphocytic leukemia cells [46].
Brisebois et al. [47]	In this work, a comparative study of the interaction of fullerenol nanoparticles with eukaryotic and bacterial model membranes was performed [47]. The solid-state NMR and FTIR spectroscopy was used as a key assessor [47].
Tu and Chang [48]	Tu and Chang summarized and discussed on the role of Raman spectroscopy in nanodiagnostics [48].
Janosi et al. [49]	Janosi et al. reported the new computational based theoretical prediction of spectral and optical properties of nanoparticles, bacteriochlorophylls [49]. This is a good example of applied nanoscale spectroscopy in prediction.
Frías et al. [50]	Frías et al. reported the success in developing new neural network system to support analysis by atomic absorption spectrophotometry [50]. This is another good example of applied computational technology for nanoscale spectroscopy.
Sujatha and Chatterji [51]	Sujatha and Chatterji reported the use of atomic absorption spectroscopy to study the enzyme namely *Escherichia coli* RNA polymerase [51]. From the analysis, Sujatha and Chatterji concluded that "the N-terminal domain of the alpha subunit has strong Zn(II) binding ability with no obvious functional implications [51]."

- *In treatment (nanotherapy)*: The use of nanoscale spectroscopy in nanotherapy is also possible. Although it is not a direct therapeutic tool, it can be a supportive tool for treatment. Its use as an investigative and follow-up tool in cancer therapy in clinical oncology as already mentioned is a good example.

12.4 Nanoscale Spectroscopy for Manipulating Biomedical Work

Based on computational technology, a short time is required for the manipulation of biomedical data. To manipulate biomedical data, computational tools are required for fastening the process. A selection of proper tools that fit a specific work is the key to success. It is necessary that practitioners understand the presently available tools. Here, the author briefly summarizes some important available nanophotonics and nanoplasmonics tools. These computational tools can be useful for work in nanomedicine.

12.4.1 Usefulness of Nanoscale Spectroscopy in Nanomedicine

Nanoscale spectroscopy can be useful in many aspects of nanomedicine. As noted, it mainly deals with spectrum. Hence, the nanoscale spectroscopy technique can be applied for any issues relating to spectrum. Using nanoscale spectroscopy to solve a problem or to simulate a case is possible. This can start from basic concepts; (a) observation, (b) generating the query or question, (c) using the tool, nanoscale spectroscopy for answering the question (structure of function), (d) summarizing derived information, and (e) final conclusion to answer the raised query. It can be said that nanoscale spectroscopy brings an effective solution to biomedical research questions.

Focusing on the newest approach, the computation approach, there are some new specific programs that help computational analysis. Several new programs can be selected to support nanoscale spectroscopy. Structural clarification and prediction of phenomena can be possible with the help of those new programs. Here, we discuss the use of the simple program MATLAB® for computational support to nanoscale spectroscopy. MATLAB is considered basic and can be easily available. How MATLAB can be used to support the nanoscale spectroscopy work discussed here.

1. *Creating a graphical model*: MATLAB can be used for creating a graphic model. It can help generate a model for referencing in nanoscale spectroscopy. A 3-D model in several forms (contour, mesh, or surface plots) can be easily created. For example, this is an interesting example of MATLAB usage in creating a model for a contour plot of protein binding identified by nanofluidics analyzers using surface plasmon resonance imaging. The MATLAB code in Command Window is as follows:

   ```
   >> [x, y, z] =peaks;
   >> c=contour (x, y, z, 64);
   >> clabel (c)
   >> title ('2 - D contour plot of protein binding with
      clable');
   ```

   The results of this example of 2-D graphical model in nanoscale spectroscopy created by MATLAB are shown in Figure 12.1.

2. *Solving the problem of differential and integral equation*: Differentiation and integration are usually difficult to solve in mathematical problems. In biomedical information processing, differentiation and

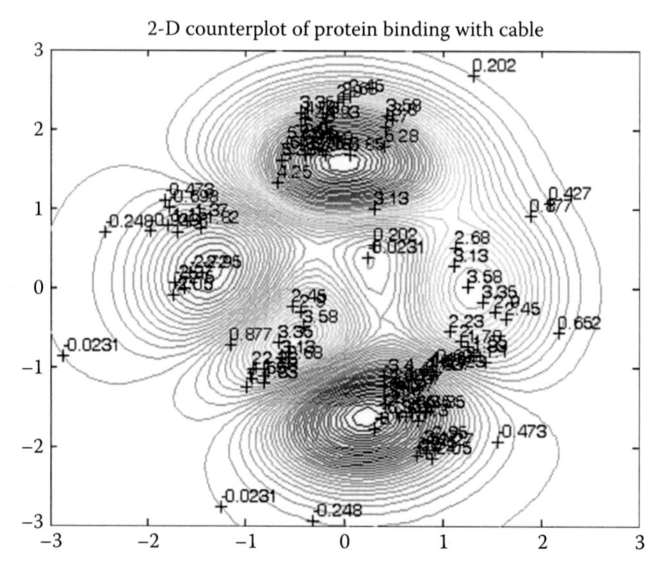

FIGURE 12.1
Two-dimensional graphical model in nanoscale spectroscopy created by MATLAB®.

integration solving is usually required. MATLAB can be the solution for this case. Solving nanoscale spectroscopy differentiation and integration questions by MATLAB is convenient. For example, this is a case of using MATLAB for solving the equation of fluorescence decreasing of carcinoma in situ cells after nanoradiotherapy. An example of code is shown later (Figure 12.2).

```
>> p=- 8; delta=0.08; y(1)=6;
>> k=0
>> for X=[delta: delta: delta: 0.5]
k=k+1;
y (k+1)=y(k)+p*y(k)*delta;
end
>> x=[0: delta: 0.5];
>> y=4*exp (-6*x);
>> y_true=6*exp (-8*x);
>> plot (x, y, '*', x, y_true,'^');
>> legend (' predicted size', ' observed size');
```

12.4.2 Important Nanoscale Spectroscopy Tools for Biomedical Work

As already noted, the use of nanoscale spectroscopy can be both noncomputational (in vivo and in vitro) and computational (in silico). For noncomputational use, there are several basic tools. The examples are as the following:

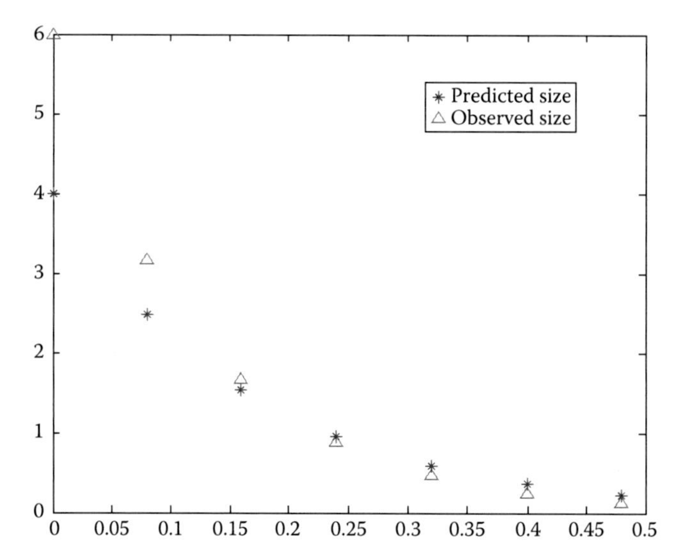

FIGURE 12.2
Solving the equation of fluorescence decreasing of carcinoma in situ cells after nanoradiotherapy using MATLAB®.

- Microscope spectrometer
  This is mainly for visualization of the structure as earlier mentioned.
- Atomic absorption spectrometry
  This is mainly for analysis of the composition of the particle.
- X-ray spectrometry
  This is also mainly for analysis of the composition of the particle.

These examples are the main tools that use noncomputational analysis by nanoscale spectroscopy. However, as already mentioned, another important novel part is the computational analysis by nanoscale spectroscopy. At present, there are several computational nanoscale spectroscopy databases and tools for management of biomedical data. This is based on the concept of bioinformatics. A database is used as a collection of data for basic primary search and the tool is used for further manipulation of the derived data from primary searching. Some important and specific tools are discussed.

1. Tool
   a. RT-PSM [52]
      RT-PSM is a new real-time program [52]. It can help match peptide spectrum with statistical significance [52]. Wu et al. proposed that RT-PSM could compare to more than 7 s per spectrum on average for Sequest and X!Tandem [52].

b. MFPaQ [53]
MFPaQ is a newly developed program that is designed for parsing, validating, and quantifying proteomics data generated by isotope-coded affinity tags and stable isotope labeling with amino acids in cell culture mass spectrometric analyses [53]. This program can be successfully used in proteomics study on proteins and cells [53].

c. EIPeptiDi [54]
The EIPeptiDi tool is designed for peptide discovery [54]. It can support the isotope-coded affinity tags–based nanoscale liquid chromatography-tandem mass spectrometry experiments [54].

d. Corra [55]
Corra is the new software for liquid chromatography-tandem mass spectrometry discovery and targeted mass spectrometry–based proteomics [55]. Corra helps statistical analysis and identifies differentially abundant peptides by tandem mass spectrometry [55].

e. Qupe [56]
Qupe is a specific tool that can help analyze mass spectrometry–based quantitative proteomics experiments [56]. It is an Internet-rich application that provides data management and analysis functions for liquid chromatography-tandem mass spectrometry experiments.

f. APEX [57]
APEX is a new computational tool for quantitative proteomics work [57]. It can estimate protein quantitation from liquid chromatography-tandem mass spectrometry proteomics results [57].

g. IDEAL-Q [58]
IDEAL-Q is a new computational tool that helps label-free quantitation analysis [58]. It uses an efficient peptide alignment approach and spectral data validation in manipulation [58].

h. ScanRanker [59]
ScanRanker is a new program that helps quality assessment of tandem mass spectra via sequence tagging [59]. ScanRanker is available at http://fenchurch.mc.vanderbilt.edu [59].

i. RAId_aPS [60]
RAId_aPS is a new computational program that helps tandem mass spectrometry analysis with the use of multiple scoring functions and spectrum-specific statistics [60]. RAId_aPS is available at http://www.ncbi.nlm.nih.gov/CBBresearch/Yu/raid_aps/index.html [60].

j. VEMS 3.0 [61]
VEMS 3.0 is a set of algorithms and computational tools [61]. VEMS 3.0 helps identify post-translational modifications in proteins based on tandem mass spectrometry [61].

2. Database
   a. Plantmetabolomics.org [62,63]
   Plantmetabolomics.org is a mass spectrometry–based database of Arabidopsis metabolomics [62].
   b. RAId_DbS [64]
   RAId_DbS is a mass spectrometry–based peptide identification database [64]. It is a web server providing data of 17 organisms [64]. It is accessible via http://www.ncbi.nlm.nih.gov/CBBResearch/qmbp/RAId_DbS/index.html [64].
   c. NCBI Peptidome [65]
   NCBI Peptidome is a new computation collecting mass spectrometry proteomics data [65]. NCBI Peptidome is accessible via http://www.ncbi.nlm.nih.gov/peptidome [65].

12.5 Common Applications of Nanoscale Spectroscopy in General Nanomedicine Practice

1. Situation 1: Using atomic absorption spectrometry for measuring trace element in blood samples

   This is an example of a classical nanoscale spectroscopy tool in nanomedicine. It is a diagnostic approach. As already mentioned, most of the classical tools are for diagnosis, and atomic absorption spectrometry is a well-known tool [66–69]. Basically, atomic absorption spectrometry is used in toxicology work. The concept is a quantitative determination of chemical elements [66–69]. The determination is based on tracing absorption of optical radiation by free atoms in the gaseous state [66–69]. Indeed, this kind of analyzer has been in use for many decades. It can help analyze the small quantity of substances, including those in the nanolevel.

   At present, it can be used in monitoring toxic substance exposure. It matches the concept of occupational medicine at present. Risk workers can be monitored. Biological collection for laboratory analysis is the basic practice for predicting the risk of individuals. Atomic absorption spectrometry can be the solution for analysis of the small particles in the collected biological samples.

   In general practice, the blood sample is collected and analyzed. Some recent important publications on this application are as follows:

   a. Paksoy et al. [70] used atomic absorption spectrometry for assessment of blood mercury levels in practicing clinicians, dental students, and dental nurses who took the risk of mercury exposure during clinical practice.

b. Erzen et al. [71] used atomic absorption spectrometry for cadmium measurements in blood and hair of occupationally nonexposed military recruits in Slovenia.
c. Akay et al. [72] used atomic absorption spectrometry for determining serum aluminum levels in glue-sniffer adolescents and in glue containers.

2. Situation 2: Using x-ray spectrometry for finding the structure of enzyme

   This is another example of using the classical nanoscale spectroscopy tool for diagnosis of medical structure. Basically, x-ray spectrometry means spectroscopic techniques for characterization of materials with the use of x-ray excitation. X-ray spectrometry can be divided into two kinds: Energy-dispersive x-ray spectroscopy and wavelength-dispersive x-ray spectroscopy. Indeed, this kind of analyzer has been in use for a long time. It can help analyze the composition of particles, including those at the nanolevel.

   At present, it can be used in monitoring many small biomolecules. It is widely used in high-level biomedical research at present. X-ray spectrometry can provide data on many previously unknown biomolecules. Of several biomolecules, enzymes are the common ones that are investigated. Some recent important publications in this area are as follows:

   a. Sevrioukova and Poulos [73] used x-ray spectrometry for determining the interaction of cytochrome P4503A4 with bromoergocryptine, a type I ligand.
   b. Shiozaki et al. [74] used x-ray spectrometry for studying (1S,2R,3R)-2,3-dimethyl-2-phenyl-1-sulfamidocyclopropane carboxylates, which is a new drug candidate for osteoarthritis.
   c. Begley et al. [75] used x-ray spectrometry for screening for infectious disease drug targets on methyl-D-erythritol-2,4-cyclodiphosphate synthase from *Burkholderia pseudomallei*.

12.5.1 Examples of Nanomedicine Studies Based on Nanoscale Spectroscopy Application

Example 12.1: A Study on Circular Dichroism Spectra of Quantum Dot-Antibody Conjugates

In nanomedicine, quantum dot is accepted as the advanced nanomaterial. It can be used for nanodiagnostic purpose. Quantum dot can be the important composition that possesses nanoproperty and can result in fast diagnosis. Many laboratories develop new quantum dot–based nanodiagnostic tools. For modifying the quantum dot–based nanodiagnostic tool, conjugation is an important step. Several kinds of particles can be used as conjugation that helps upgrade the property of basic

TABLE 12.4

Results from CD Spectra Analysis of Quantum Dot-Fluorescence Dye Conjugation

Random: >0.62
Square Distance: 24.20
Max Error: 0.060

Wavelength	**Original**	**Computed**
300	1.74	0.96
301	−0.82	−1.22
302	−2.44	−3.04
303	−4.82	−4.94
304	−5.76	−6.12
305	−6.14	−6.96

quantum dot. During the conjugation process, the validation step is important, and the circular dichroism (CD) spectra analysis is the key point [76]. This is an applied classical spectrometry. With the use of advanced nanoinformatics technology, this process can be easily done.

Here, the author shows an example of using the analysis tool K2d for predicting the CD of quantum dot-fluorescence dye conjugation. Indeed, this novel conjugation approach is the way to increase the property of bioimaging [77]. This is called luminescent quantum dot technology [77]. Here, in this example, the primary assumption is that the starting condition is the original quantum dot, the predicted condition is the conjugated molecule, and the primary control is both the pH and temperature. The given values of secondary structure are alpha helix = 0.38 and beta strand = 0.22. Based on prediction, the secondary structure of quantum dot–fluorescence dye, the result of conjugation, is still preserved (Table 12.4).

Example 12.2: A Study on Peptide Fragment Mass of Green Pit Viper Protein Sequence

In proteomics, the assessment of protein's structure is an important step. Any new identified protein has to pass the evaluation for its structure and function. There are billions of protein in biomedicine. Some proteins are newly discovered and still require further assessment. Here, we consider an example of a newly discovered protein. The protein is from green pit viper toxin. The mentioned protein was first reported by Soogarun et al [78]. Soogarun et al. used the spectrometry technique for isolating the new proteins and proposed their effects on the coagulation system in human beings [78]. However, more information is still required to enhance the knowledge on this protein.

Here is an example of using the analysis tool, PROWL, for predicting peptide fragment mass information on the green pit viper toxin protein. The basic assumptions in this simulation include (a) all databases, (b) no chemical modification, (c) protein mass 0–3000 kDa, (d) protein pI 1–14,

TABLE 12.5

Results from Using PROWL for Prediction of Peptide Fragment Mass in Green Pit Viper Toxin Protein

Maximum number of proteins in result: 10
Protein mass: 0.0–3000.0 kDa
Protein pI: 0.0–14.0
Enzyme: trypsin, # of incompletes: 2
Fragment mass out of range: 0.00 (allowed range: $0<m<=20,000$)

(e) enzyme = trypsin and (f) monoisotropic mass. According to this work, there is no predicted peptide fragment mass confirming the stability of the identified toxin protein (Table 12.5).

Example 12.3: A Study on Malarial Sporozoite Protein

Malaria is an important tropical mosquito-borne infection. It attacks millions of the world's population each year. Malaria infects red blood cell and causes clinical manifestations (high fever with chill). The development of the malarial parasite within the human body is complex and is the topic for pathophysiology study. The study of the pathological protein expressed in each stage of the parasite is interesting and can be useful information in malariology. An important protein to be mentioned is the sporozoite protein [79,80]. Sporozoite protein might be essential for liver stage development of malaria parasite, which means latency of the infection [79,80].

Here is an example of using the analysis tool, MS-Bridge, for predicting the bridging nature of the sporozoite protein. According to this work, there are 14 predicted bridging points within the studied sporozoite protein (Figure 12.3). Those positions are believed to be the important point playing a role in the pathophysiology of liver infestation.

Example 12.4: A Study on *Mycobacterium avium* Complex Protein

Mycobacterium avium complex is an important group of bacteria that can cause atypical tuberculosis. It is an emerging public health problem for patients with human immunodeficiency virus infection at present [81]. The study of the protein derived from this pathogen can be useful in further specific drug and vaccine development [82]. An important part is the identification of glycopeptides that can serve as the target size in development [82].

Here is an example of using the analysis tool, GlycoMod [83], for predicting the oligosaccharide structures that occur on a protein, namely PPE40. In this study, the allowable missed cleavage is up to 3 and no cutting enzyme. According to this work, there is no identified glycopeptide portion (Figure 12.4). This might imply that the PPE40 should not be the focus for development of new drugs targeting the wall of *Mycobacterium avium*.

Index number: **1**
pI of protein: **8.8**
Protein MW: **12017**
Amino acid composition: **A3 D5 E7 F3 G1 H4 I8 K11 L8 N16 P1 Q8 R3 S7 V7 X3 Y6**

1 INKINLNKPI IENKNNVDVS IKRYNNFVDI ARLSIQKHFE HLSNDQKDSH VNNXEYXQKF VQGLQENRNI SLSKYQENKA
81 VXDLKYHLQK VYANYLSQEE N

FIGURE12.3
Results from using MS-Bridge for prediction of the bridging nature of the sporozoite protein.

Peptide sequence

```
MTAPIFMASP  PELHSALLSS  GPGPASLLAA  AGAWSQLSAE  YASAAEQLST  LLTGVAAGAW
QGVSGESYVG  AHAPYLAWLT  KASADSAAVA  AQHEVAATAY  TTALATMPTL  PELAANHAVH
ATLVATNFFG  VNTVPIAVNE  ADYARMWTQA  ATTMSTYHAV  STAAVASTPQ  AGPAPQIMKS
DASQDDSGNE  DHDPKIDNPF  NDFIANILRN  LGIDWDPAKG  TVNGLDYDAY  TNAGEPIFWV
VRALELLEDF  EQFGYYLVHN  PALAFQYLVQ  LMLFDWPTHI  LEIFMSQPEL  LAPALLLAAA
PFAAVGGFAG  LAGLAALPQP  VAVPAAVAAP  PAPPGLPPAI  AIAPTPVAAA  APVATAPAPA
PAATATTVAG  APPAAPAPVA  PAAGFFPRPV  IGPPGMGTGS  GMSASASSSA  KRKAPEPDSA
AAAAAAAARN  AARSRRRRRA  TRRGHGEEFM  DMNVDVDPDW  GGPAGPESSA  SDRGAGALGF
AGTARKDAAG  AAAGLTTLPG  AELGGPTMPM  MPGSWEPGDD  AGAEHDSPEL  HSALLNXS
```

Maximum number of missed cleavages (MC): 3

Adduct ($[M + H]^+$): 1.00739

None of the peptides obtained by cleavage of your protein sequence contain the motif.

FIGURE 12.4
Results from using GlycoMod for prediction of the glycopeptide portion within a *Mycobacterium avium* protein.

12.6 Conclusion

Spectroscopy is the basic technique is science. This technique can be applied in nanomedicine. With the use of nanoscale spectroscopy, several manipulations can be done. Using nanoscale spectroscopy, the questions on structure and function in nanomedicine can be answered. The application can be in vivo, in vitro, or in silico. There are many basic noncomputational nanoscale spectroscopy tools that can be applied in medicine. In addition, several computational nanoscale spectroscopy tools are also available at present. These advanced tools can be very useful in nanomedicine and are expected to be the widely used tools in the future.

References

1. Gehlenborg N, O'Donoghue SI, Baliga NS, Goesmann A, Hibbs MA, Kitano H, Kohlbacher O, Neuweger H, Schneider R, Tenenbaum D, Gavin AC. Visualization of omics data for systems biology. *Nat Methods*. 2010 March;7(3 Suppl):S56–S68.
2. Haarala R, Porkka K. The odd omes and omics. *Duodecim*. 2002;118(11):1193–1195.
3. Haddish-Berhane N, Rickus JL, Haghighi K. The role of multiscale computational approaches for rational design of conventional and nanoparticle oral drug delivery systems. *Int J Nanomed*. 2007;2(3):315–331.
4. Saliner AG, Poater A, Worth AP. Toward in silico approaches for investigating the activity of nanoparticles in therapeutic development. *IDrugs*. 2008 October;11(10):728–732.

5. Behari J. Principles of nanoscience: an overview. *Indian J Exp Biol.* 2010 October;48(10):1008–1019.
6. Colliex C. From electron energy-loss spectroscopy to multi-dimensional and multi-signal electron microscopy. *J Electron Microsc (Tokyo).* 2011;60(Suppl 1):S161–S171.
7. Meckenstock R. Invited Review Article: Microwave spectroscopy based on scanning thermal microscopy: Resolution in the nanometer range. *Rev Sci Instrum.* 2008 April;79(4):041101.
8. Wu J, Gu M. Microfluidic sensing: state of the art fabrication and detection techniques. *J Biomed Opt.* 2011 August;16(8):080901.
9. Cheung K, Gawad S, Renaud P. Impedance spectroscopy flow cytometry: On-chip label-free cell differentiation. *Cytometry A.* 2005 June;65(2):124–132.
10. Johnson WR, Wilson DW, Fink W, Humayun M, Bearman G. Snapshot hyperspectral imaging in ophthalmology. *J Biomed Opt.* 2007 January–February 12(1):014036.
11. Mordant DJ, Al-Abboud I, Muyo G, Gorman A, Sallam A, Ritchie P, Harvey AR, McNaught AI. Spectral imaging of the retina. *Eye (Lond).* 2011 March;25(3):309–320.
12. Duan H, Fernández-Domínguez AI, Bosman M, Maier SA, Yang JK. Nanoplasmonics: Classical down to the Nanometer Scale. *Nano Lett.* 2012 February 14 [Epub ahead of print].
13. Stockman MI. Nanoplasmonics: past, present, and glimpse into future. *Opt Express.* 2011 October 24;19(22):22029–22106.
14. Lin J, Weber N, Wirth A, Chew SH, Escher M, Merkel M, Kling MF, Stockman MI, Krausz F, Kleineberg U. Time of flight-photoemission electron microscope for ultrahigh spatiotemporal probing of nanoplasmonic optical fields. *J Phys Condens Matter.* 2009 August 5;21(31):314005.
15. Chen HM, Pang L, Gordon MS, Fainman Y. Real-time template-assisted manipulation of nanoparticles in a multilayer nanofluidic chip. *Small.* 2011 October 4; 7(19):2750–2757.
16. Sannomiya T, Vörös J. Single plasmonic nanoparticles for biosensing. *Trends Biotechnol.* 2011 July;29(7):343–351.
17. Fujikawa Y, Sakurai T, Tromp RM. Surface plasmon microscopy using an energy-filtered low energy electron microscope. *Phys Rev Lett.* 2008 March 28;100(12):126803.
18. Horiba K, Nakamura Y, Nagamura N, Toyoda S, Kumigashira H, Oshima M, Amemiya K, Senba Y, Ohashi H. Scanning photoelectron microscope for nanoscale three-dimensional spatial-resolved electron spectroscopy for chemical analysis. *Rev Sci Instrum.* 2011 November;82(11):113701.
19. Tejedor ML, Mizuno H, Tsuyama N, Harada T, Masujima T. In situ molecular analysis of plant tissues by live single cell mass spectrometry. *Anal Chem.* 2011 Dec 20 [Epub ahead of print].
20. Date S, Mizuno H, Tsuyama N, Harada T, Masujima T. Direct drug metabolism monitoring in a live single hepatic cell by video mass spectrometry. *Anal Sci.* 2012;28(3):201–203.
21. Hara T, Tanaka K, Maehata K, Mitsuda K, Yamasaki NY, Ohsaki M, Watanabe K, Yu X, Ito T, Yamanaka Y. Microcalorimeter-type energy dispersive X-ray spectrometer for a transmission electron microscope. *J Electron Microsc (Tokyo).* 2010;59(1):17–26.

22. Hodson NW, Kielty CM, Sherratt MJ. ECM macromolecules: height-mapping and nano-mechanics using atomic force microscopy. *Methods Mol Biol.* 2009;522:123–141.
23. Sirikatitham A, Yamamoto T, Shimizu M, Hasegawa T, Tsuyama N, Masujima T. Resin-packed nanoelectrospray in combination with video and mass spectrometry for the direct and real-time molecular analysis of mast cells. *Rapid Commun Mass Spectrom.* 2007;21(3):385–390.
24. Yang HS, Tang MH, Deng HX, Yang JL. Nano ultra high performance liquid chromatography coupled with time-of-flight mass spectrometer for detecting low molecular weight polypeptides in zebrafish embryos. *Sichuan Da Xue Xue Bao Yi Xue Ban.* 2007 November;38(6):1033–1036.
25. Li C, Gao M, Ding C, Zhang X, Zhang L, Chen Q, Peng LM. In situ comprehensive characterization of optoelectronic nanomaterials for device purposes. *Nanotechnology.* 2009 April 29;20(17):175703.
26. Pettinicchio M, Traini T, Murmura G, Caputi S, Degidi M, Mangano C, Piattelli A. Histologic and histomorphometric results of three bone graft substitutes after sinus augmentation in humans. *Clin Oral Investig.* 2012 February;16(1):45–53.
27. Hoang HQ, Osterberg M, Khursheed A. A high signal-to-noise ratio toroidal electron spectrometer for the SEM. *Ultramicroscopy.* 2011 July;111(8):1093–1100.
28. Christensen SM, Bolinger PY, Hatzakis NS, Mortensen MW, Stamou D. Mixing subattolitre volumes in a quantitative and highly parallel manner with soft matter nanofluidics. *Nat Nanotechnol.* 2011 October 30;7(1):51–55.
29. Batabyal S, Rakshit S, Kar S, Pal SK. An improved microfluidics approach for monitoring real-time interaction profiles of ultrafast molecular recognition. *Rev Sci Instrum.* 2012 April;83(4):043113.
30. Song W, Yang J. Optofluidic differential spectroscopy for absorbance detection of sub-nanolitre liquid samples. *Lab Chip.* 2012 April 7;12(7):1251–1254.
31. Prim D, Crelier S, Segura JM. Coupling of a microfluidic mixer to a Fourier-transform infrared spectrometer for protein-conformation studies. *Chimia (Aarau).* 2011;65(10):815–816.
32. Greiner C, Hunter M, Huang P, Rius F, Georgakoudi I. Confocal backscattering spectroscopy for leukemic and normal blood cell discrimination. *Cytometry A.* 2011 October;79(10):866–873.
33. Cecchini MP, Hong J, Lim C, Choo J, Albrecht T, Demello AJ, Edel JB. Ultrafast surface enhanced resonance Raman scattering detection in droplet-based microfluidic systems. *Anal Chem.* 2011 April 15;83(8):3076–3081.
34. Ashok PC, Singh GP, Rendall HA, Krauss TF, Dholakia K. Waveguide confined Raman spectroscopy for microfluidic interrogation. *Lab Chip.* 2011 Apr 7; 11(7):1262–1270.
35. Javanmard M, Babrzadeh F, Davis RW. Microfluidic force spectroscopy for characterization of biomolecular interactions with piconewton resolution. *Appl Phys Lett.* 2010 October 25;97(17):173704.
36. Mastrangelo CH, Williams LD, Ghosh T. Probing protein binding spectra with Fourier microfluidics. *Conf Proc IEEE Eng Med Biol Soc.* 2010;2010:5318–5321.
37. Syme CD, Sirimuthu NM, Faley SL, Cooper JM. SERS mapping of nanoparticle labels in single cells using a microfluidic chip. *Chem Commun (Camb).* 2010 November 14;46(42):7921–7923.

38. Birtill D, Shah A, Jaeger M, Bamber J. Photoacoustic measurement of the optical absorption spectra of dark or turbid media. *J Acoust Soc Am.* 2012 April;131(4):3478.
39. Bastatas L, Martinez-Marin D, Matthews J, Hashem J, Lee YJ, Sennoune S, Filleur S, Martinez-Zaguilan R, Park S. AFM nano-mechanics and calcium dynamics of prostate cancer cells with distinct metastatic potential. *Biochim Biophys Acta.* 2012 February 16 [Epub ahead of print].
40. Gormley AJ, Malugin A, Ray A, Robinson R, Ghandehari H. Biological evaluation of RGDfK-gold nanorod conjugates for prostate cancer treatment. *Target.* 2011 December;19(10):915–924.
41. Liu A, Sun Z, Wang K, Chen X, Xu X, Wu Y, Lin X, Chen Y, Du M. Molecular beacon-based fluorescence biosensor for the detection of gene fragment and PCR amplification products related to chronic myelogenous leukemia. *Anal Bioanal Chem.* 2012 January;402(2):805–812.
42. Lin CY, Teng NC, Hsieh SC, Lin YS, Chang WJ, Hsiao SY, Huang HS, Huang HM. Real-time detection of β1 integrin expression on MG-63 cells using electrochemical impedance spectroscopy. *Biosens Bioelectron.* 2011 October 15;28(1):221–226.
43. Semaan SM, Wang X, Stewart PA, Marshall AG, Sang QX. Differential phosphopeptide expression in a benign breast tissue, and triple-negative primary and metastatic breast cancer tissues from the same African-American woman byLC-LTQ/FT-ICR masss pectrometry. *Biochem Biophys Res Commun.* 2011 August 19;412(1):127–131.
44. De La Iglesia D, Chiesa S, Kern J, Maojo V, Martin-Sanchez F, Potamias G, Moustakis V, Mitchell JA. Nanoinformatics: new challenges for biomedical informatics at the nano level. *Stud Health Technol Inform.* 2009;150:987–991.
45. Lee KJ, Nallathamby PD, Browning LM, Desai T, Cherukuri PK, Xu XH. Single nanoparticle spectroscopy for real-time in vivo quantitative analysis of transport and toxicity of single nanoparticles in single embryos. *Analyst.* 2012 May 4 [Epub ahead of print].
46. Maclaughlin CM, Parker EP, Walker GC, Wang C. Evaluation of SERS labeling of CD20 on CLL cells using optical microscopy and fluorescence flow cytometry. *Nanomedicine.* 2012 April 25 [Epub ahead of print].
47. Brisebois PP, Arnold AA, Chabre YM, Roy R, Marcotte I. Comparative study of the interaction of fullerenol nanoparticles with eukaryotic and bacterial model membranes using solid-state NMR and FTIR spectroscopy. *Eur Biophys J.* 2012 April 15 [Epub ahead of print].
48. Tu Q, Chang C. Diagnostic applications of Raman spectroscopy. *Nanomedicine.* 2011 October 22 [Epub ahead of print].
49. Janosi L, Kosztin I, Damjanović A. Theoretical prediction of spectral and optical properties of bacteriochlorophylls in thermally disordered LH2 antenna complexes. *J Chem Phys.* 2006 July 7;125(1):014903.
50. Frías S, Conde JE, Rodríguez MA, Dohnal V, Pérez-Trujillo JP. Metallic content of wines from the Canary Islands (Spain). Application of artificial neural networks to the data analysis. *Nahrung.* 2002 October;46(5):370–375.
51. Sujatha S, Chatterji D. Detection of putative Zn(II) binding sites within *Escherichia coli* RNA polymerase: Inconsistency between sequence-based prediction and 65Zn blotting. *FEBS Lett.* 1999 July 2;454(1–2):169–171.

52. Wu FX, Gagné P, Droit A, Poirier GG. RT-PSM, areal-time program for peptide-spectrum matching with statistical significance. *Rapid Commun Mass Spectrom.* 2006;20(8):1199–1208.
53. Bouyssié D, Gonzalez de Peredo A, Mouton E, Albigot R, Roussel L, Ortega N, Cayrol C, Burlet-Schiltz O, Girard JP, Monsarrat B. Mascot file parsing and quantification (MFPaQ), a new software to parse, validate, and quantify proteomics data generated by ICAT and SILAC mass spectrometric analyses: Application to the proteomics study of membrane proteins from primary human endothelial cells. *Mol Cell Proteomics.* 2007 September;6(9):1621–1637.
54. Cannataro M, Cuda G, Gaspari M, Greco S, Tradigo G, Veltri P. The EIPeptiDi tool: Enhancing peptide discovery in ICAT-based LC MS/MS experiments. *BMC Bioinformatics.* 2007 July 15;8:255.
55. Brusniak MY, Bodenmiller B, Campbell D, Cooke K, Eddes J, Garbutt A, Lau H et al. Corra: Computational framework and tools for LC-MS discovery and targeted mass spectrometry-based proteomics. *BMC Bioinformatics.* 2008 December 16;9:542.
56. Albaum SP, Neuweger H, Fränzel B, Lange S, Mertens D, Trötschel C, Wolters D, Kalinowski J, Nattkemper TW, Goesmann A. Qupe— a Rich Internet Application to take a step forward in the analysis of mass spectrometry-based quantitative proteomics experiments. *Bioinformatics.* 2009 December 1;25(23):3128–3134.
57. Braisted JC, Kuntumalla S, Vogel C, Marcotte EM, Rodrigues AR, Wang R, Huang ST et al. The APEX Quantitative Proteomics Tool: Generating protein quantitation estimates from LC-MS/MS proteomics results. *BMC Bioinformatics.* 2008 December 9;9:529.
58. Tsou CC, Tsai CF, Tsui YH, Sudhir PR, Wang YT, Chen YJ, Chen JY, Sung TY, Hsu WL. IDEAL-Q, an automated tool for label-free quantitation analysis using an efficient peptide alignment approach and spectral data validation. *Mol Cell Proteomics.* 2010 January;9(1):131–144.
59. Bais P, Moon-Quanbeck SM, Nikolau BJ, Dickerson JA. Plantmetabolomics.org: mass spectrometry-based Arabidopsis metabolomics—database and tools update. *Nucleic Acids Res.* 2012 January;40(Database issue):D1216–D1220.
60. Alves G, Ogurtsov AY, Yu YK. RAId_aPS: MS/MS analysis with multiple scoring functions and spectrum-specific statistics. *PLoS One.* 2010 November 16;5(11):e15438.
61. Matthiesen R, Trelle MB, Højrup P, Bunkenborg J, Jensen ON. VEMS 3.0: Algorithms and computational tools for tandem mass spectrometry based identification of post-translational modifications in proteins. *J Proteome Res.* 2005 November–December;4(6):2338–2347.
62. Ma ZQ, Chambers MC, Ham AJ, Cheek KL, Whitwell CW, Aerni HR, Schilling B, Miller AW, Caprioli RM, Tabb DL. ScanRanker: Quality assessment of tandem mass spectra via sequence tagging. *J Proteome Res.* 2011 July 1;10(7):2896–2904.
63. Bais P, Moon SM, He K, Leitao R, Dreher K, Walk T, Sucaet Y et al. PlantMetabolomics.org: a web portal for plant metabolomics experiments. *Plant Physiol.* 2010 April;152(4):1807–1816.
64. Alves G, Ogurtsov AY, Yu YK. RAId_DbS: mass-spectrometry based peptide identification web server with knowledge integration. *BMC Genomics.* 2008 October 27;9:505.

65. Ji L, Barrett T, Ayanbule O, Troup DB, Rudnev D, Muertter RN, Tomashevsky M, Soboleva A, Slotta DJ. NCBI Peptidome: A new repository for mass spectrometry proteomics data. *Nucleic Acids Res*. 2010 January;38(Database issue): D731–D735.
66. Hankiewicz J. Atomic absorption spectrophotometry in clinical biochemistry. *Postepy Hig Med Dosw*. 1969 July–August;23(4):465–488.
67. Thamsen J. The atomic absorption spectrophotometer and analysis of metals in solution. An introductory review. *Dan Tidsskr Farm*. 1967 May;41(5):85–99.
68. Sunderman FW Jr. Atomic absorption spectrometry of trace metals in clinical pathology. *Hum Pathol*. 1973 December;4(4):549–582.
69. Willis JB. Recent advances in the analysis of biological materials by atomic absorption techniques. *Endeavour*. 1973 September;32(117):106–111.
70. Paksoy CS, Görgün S, Nalçaci R, Yagbasan A. Assessment of blood mercury levels in practicing Turkish clinicians, dental students, and dental nurses. *Quintessence Int*. 2008 April;39(4):e173–e178.
71. Erzen I, Zaletel Kragelj L. Cadmium measurements in blood and hair of occupationally non-exposed military recruits and in the foods of plant origin produced in Slovenia. *Croat Med J*. 2003 October;44(5):538–544.
72. Akay C, Kalman S, Dündaröz R, Sayal A, Aydin A, Ozkan Y, Gül H. Serum aluminium levels in glue-sniffer adolescent and in glue containers. *Basic Clin Pharmacol Toxicol*. 2008 May;102(5):433–436.
73. Sevrioukova IF, Poulos TL. Structural and mechanistic insights into the interaction of cytochrome P4503A4 with bromoergocryptine, a type I ligand. *J Biol Chem*. 2012 January 27;287(5):3510–3517.
74. Shiozaki M, Maeda K, Miura T, Kotoku M, Yamasaki T, Matsuda I, Aoki K et al. Discovery of (1S,2R,3R)-2,3-dimethyl-2-phenyl-1-sulfamidocyclopropanecarbo xylates: novel and highly selective aggrecanase inhibitors. *Med Chem*. 2011 April 28;54(8):2839–2863.
75. Begley DW, Hartley RC, Davies DR, Edwards TE, Leonard JT, Abendroth J, Burris CA, Bhandari J, Myler PJ, Staker BL, Stewart LJ. Leveraging structure determination with fragment screening for infectious disease drug targets: MECP synthase from *Burkholderia pseudomallei*. *J Struct Funct Genomics*. 2011 July;12(2):63–76.
76. Greenfield NJ. Circular dichroism analysis for protein-protein interactions. *Methods Mol Biol*. 2004;261:55–78.
77. Ishikawa M, Biju V. Luminescent quantum dots, making invisibles visible in bioimaging. *Prog Mol Biol Transl Sci*. 2011;104:53–99.
78. Soogarun S, Sangvanich P, Chowbumroongkait M, Jiemsup S, Wiwanikit V, Pradniwat P, Palasuwan A, Pawinwongchai J, Chanprasert S, Moungkote T. Analysis of green pit viper (Trimeresurus alborabris) venom protein by LC/MS-MS. *J Biochem Mol Toxicol*. 2008 July–August;22(4):225–229.
79. Yuda M. Mechanisms of liver invasion by malaria sporozoites. *Tanpakushitsu Kakusan Koso*. 2009 June;54(8 Suppl):1029–1034.
80. Silvie O, Franetich JF, Rénia L, Mazier D. Malaria sporozoite: Migrating for a living. *Trends Mol Med*. 2004 March;10(3):97–100.
81. Perfect JR. Mycobacterium avium-intracellulare complex infections in the acquired immunodeficiency syndrome. *J Electron Microsc Tech*. 1988 January;8(1):105–113.

82. Chatterjee D. The mycobacterial cell wall: structure, biosynthesis and sites of drug action. *Curr Opin Chem Biol*. 1997 December;1(4):579–578.
83. Cooper CA, Gasteiger E, Packer NH. GlycoMod—a software tool for determining glycosylation compositions from mass spectrometric data. *Proteomics*. 2001 February;1(2):340–349.

13

Nanoscale Spectroscopy for Defense and National Security

Aditi Deshpande, Mohit Agarwal, Suman Shrestha, and George C. Giakos

CONTENTS

13.1 Introduction

The novel science of nanotechnology finds applications in many diverse fields due to the various design prospects of nanoscale systems. The size of their components imparts unique properties to these nanoscale systems, not exhibited by regular-sized particles, and these enhanced properties are utilized in varied sectors such as medicine, defense and security, electronics, and so on. This multidisciplinary development has led to a revolution in the technology used in numerous applications. Varied nanoscale structures such as nanoparticles, nanotubes, and nanofibers were fabricated and implemented in various systems—biological, electronic, optical, etc.

The nanoscale design of these systems alters many optical, mechanical, chemical, and quantum properties of materials, and these differences in properties get more pronounced as the size of the constituting particles approaches the nanolevel. A major change is the ratio of the surface area to the volume of these nanostructured materials. This is a property that imparts enhanced behavior to these structures. The many changes include transparency of usually opaque materials; enhanced absorption of solar radiation; prominent changes in the melting point, solubility, and resistivity of various materials; the exhibition of surface plasmon resonance (SPR) by nanoparticles; superparamagnetism of ferromagnetic nanomaterials, and so on. Various nanostructures such as nanoparticles, nanocrystals, nanoclusters, metallic nanoparticles, and nanofibers can be implemented in designing nanoscale devices. The basic synthesis methods of nanoparticles have been discussed in other chapters.

Prominent increase in terrorism in the past has posed a huge menace to defense and homeland security, which needs to be resolved by designing a wide range of detectors that can perceive the weapons that cause mass annihilation. Nanomaterials necessitate efficient processing and characterization techniques to detect the weapons, and Raman spectroscopy plays a major role in providing information on these materials. This chapter presents a brief insight into the importance of nanoscience and nanotechnology and optimizing new types of material techniques to detect the weapons of mass destruction.

The primary form of terrorist activities in the past has been conventional explosives, and the consequences have been very severe, ending in extensively demolishing the property as well as people (Figure 13.1).

Common peril for military operations is improvised explosive devices that are intricate as the activity is preplanned and may be difficult to divulge as the bomb may be placed to go off as the vehicles pass by. Need of good detectors is essential for defense and military operations, and nanomaterials offer significant importance in designing collectors and detectors [1].

13.2 Applications of Nanotechnology to Defense

The Institute for Soldier Nanotechnologies (ISN) [2] is a research center comprising of a team of MIT, U.S. Army, and industry researchers working in collaboration to discover and implement various nanobased technologies to

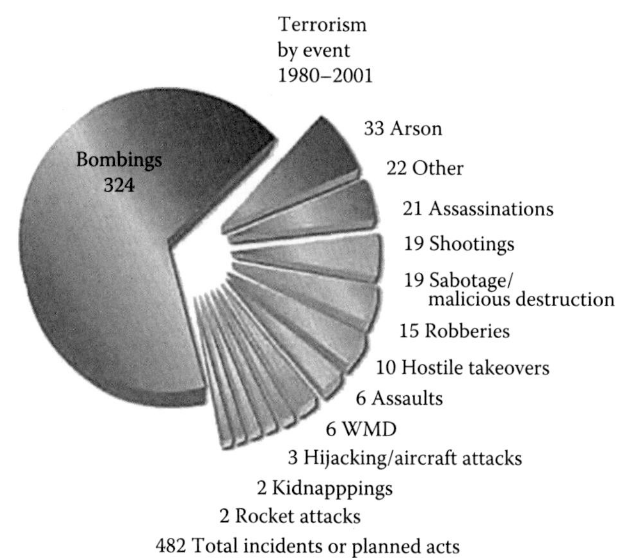

FIGURE 13.1
Chart showing a number of terrorist attacks using different activities. (From U.S. Department of Justice, Federal Bureau of Investigation, Terrorism 2000/2001.)

advance the quality of military surveillance, personnel protection and survivability, defense equipment, and other security issues. Its aim is "unprecedented" advance of the U.S. soldier protection and sustenance capabilities as well as strengthening the defense and offense systems of the nation. The ISN team is carrying out research on a large scale and in many sectors such as the medical care of soldiers (diagnosing and treating effectively), ballistic and blast threats (protective systems against material damage and human injury), detection and destruction of hazardous biological and chemical warfare as mentioned earlier, nanosystems for communication with enhanced security, and speed of transmission [2]. Their research includes using quantum dots to improve high bandwidth communication, nanostructured actuators, films and surfaces for medicine applications on the battlefield, nanoscale designing of materials with enhanced strength and energy absorption, lightweight fibers designed with nanocrystals for blast protection and injury alleviation, and nano-optical electronics, among others.

As can be seen, plenty of applications of nanomaterials exist in the defense and security field, such as coating of military vehicles and other equipment with nanomaterials; various types of sensors that are enhanced and made more efficient, lightweight, and portable using nanostructures; encryption and code breaking using magnetic nanoparticles; lightweight and bulletproof soldier battle-suits [3]; faster communication networks with increased security; nanomaterial-based defense and offense weapons; enhanced fuel efficiency and stealth movement of vehicles using nanoscale systems; enhanced defense and offense weapons constructed by using 3-D arrangement of nanostructures; control of nuclear weapons and nuclear defense platforms using nano-mechanical devices and nanorobots; imaging and military surveillance systems; and many more.

13.3 Detection of Chemical and Biological Warfare

Nanoscale systems can be implemented in a wide variety of defense and security applications. The detection of warfare agents—both chemical and biological—is of prime importance in defense and homeland security, and nanomaterials, including those designed with gold nanoparticles, are being implemented in sensor design. STREM Chemicals have designed chemical resistors comprising of films of gold nanoparticles that are laid on microelectrodes [4]. This sensing system detects harmful vapors and is water resistant. Such nanomaterial-based sensors are lightweight and efficient. Apart from vapor-detecting sensors, microchips that detect other harmful agents are also being designed using nanoparticles. This design uses polymer encapsulation of nanomaterials on microchips followed by a receptor agent that acts as an indicator in the presence of a warfare agent.

Military buildings, vehicles, and other possessions are equipped with these detecting agents. Metal nanocluster resonance technology [4] is also utilized for this detection purpose. In devices that follow this technology, metal nanocluster substrates act as resonators and are used in the transfer and storage of energy. The basic parameters of the sensors used for this purpose that need to be enhanced are the sensitivity, accuracy and the response time, power consumption, size, and portability, which are improved by using nanostructures to design these sensors (Figure 13.2).

Researchers at the U.S. Naval Research Laboratories are also synthesizing semiconductor quantum nanocrystals with high luminescence for use in biosensing. They use negatively charged quantum dots and their electrostatic contact with positively charged proteins. The unique optical properties of these proteins aid in enhanced detection of biological and chemical agents [5]. Some examples of chemical and biological warfare agents that can be used as potential weapons of mass destruction include anthrax, tetrodotoxin, sarin, and other chemicals [6].

Another sensor technology that implements nanostructure design for detection is the concept of ultra-microelectrode arrays. These arrays of microsized electrodes greatly improve the surface area available for detection by implementing 3-D nanostructures [7].

Once detected, these warfare agents also need to be destroyed. This process is also aided by the use of nanoscale structured devices. Nanoparticles

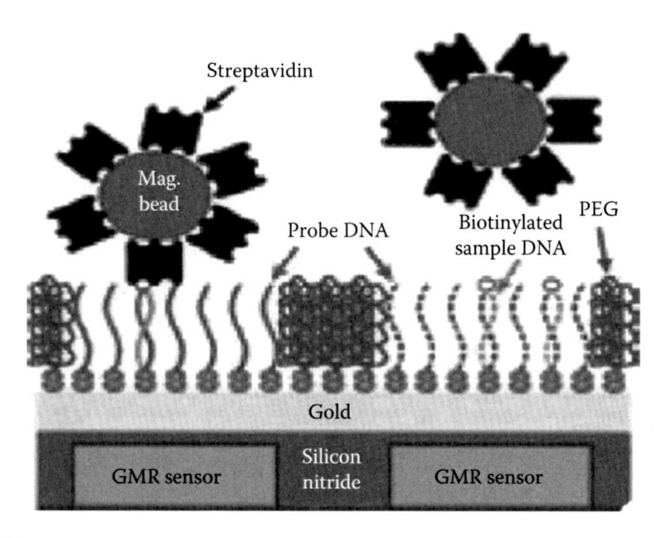

FIGURE 13.2
(See color insert.) Design of chemical and biological sensors using gold nanoparticles. (Courtesy of L.J. Whitman, U.S. Naval Research Laboratory; From Defense and security applications for nanomaterials and nanoparticles, supplier data by STREM Chemicals, Inc., Available: http://www.azonano.com/article.aspx?ArticleID=1337)

with magnetic properties and nanocrystalline metal oxides are used for the destruction of warfare agents and decontamination (detoxification) of exposed areas and personnel. Certain functionalized magnetic nanoparticles are injected, which bind to the toxins present inside and are pulled out by applying a strong magnetic field. This is a currently active research area and shows promising applications in the detoxification process for biological and chemical warfare agents. Apart from these, semiconductor nanomaterials such as ZnO and CdSe exhibit strong photoluminescence (PL) in the visible range of light, and many organic/inorganic molecules present in the biological contaminants absorb this emission [8]. Researchers at the Notre Dame Radiation Laboratory and others have explored the nanosize-dependent sensing and destroying properties of these semiconductor nanomaterials, and these techniques can be applied in the defense field as mentioned earlier (Figure 13.3).

The above-mentioned "tagging and tracking" [4] methodology is also implemented in other military and security applications such as the design of "smart" barcodes that can be magnetically labeled and classified. Gold nanoparticles are blended with the printing ink and radiofrequency identification patterns can be printed. Military equipment can be coded in this manner.

Due to multiple threats to the national security such as bombings using explosives, bacterial or chemical dispersion of harmful agents into the surroundings, targeted firing by armed enemy, and so on, it becomes

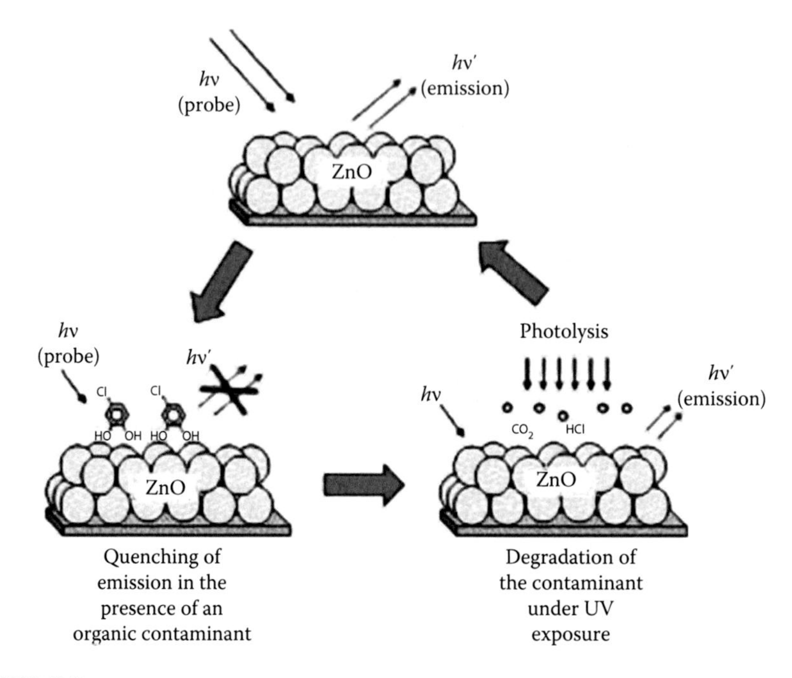

FIGURE 13.3
(See color insert.) Simultaneous detection and degradation process using ZnO in photocatalysis. (From Kamat, P.V. and Meisel, D., *Comp. Rendus Chim.*, 6, 999, 2003.)

extremely essential to devise efficient, flexible, and highly sensitive detection techniques for the defense and security purposes. The two main kinds of detection of explosives are bulk detection, in which explosive substances are detected in the explosive device itself, and detection of the trace of explosives, which is done by identifying and quantifying the trace amount of explosives in the form of particles in the surroundings. As of date, no single detection technology has been enhanced to such an extent that it can cater to the detection requirements in all the above-mentioned fields, but extensive research is going on to develop novel methods for the same. Currently, various multidisciplinary technologies are being studied and implemented for defense purposes, and nanoscale devices and nanomaterials are the frontier in these applications. The properties that are desired in efficient explosive detectors are sensitivity, fast response time, highly selective nature, and excellent accuracy to avoid false-positives. Also, these detection systems should not pose any harm to the persons involved in the scanning process. Nanoscale spectroscopy techniques are of great importance in the detection and imaging of various objects of military importance as Raman spectroscopy is being utilized more intensely in military-based imaging and detection applications such as "bulk detection" of explosives, chemical warfare identification, and so on. Some applications are discussed in detail in this chapter (with a stress on the use of spectroscopy in defense and security applications), with a brief discussion of the future scope of nanostructured materials and devices.

13.4 Spectroscopic Techniques for Nanomaterials

Optical spectroscopic techniques are the basis for studying and analyzing different optical properties of nanomaterials. Devising nanomaterials generally necessitates the binding site to recognize the target of interest. There are several techniques employed currently based on absorption, scattering of light such as electron absorption (UV–vis), PL, infrared (IR) absorption, Raman scattering, transient absorption (TA), and so on. More advanced techniques include single molecular spectroscopy, sum frequency generation, and luminescence up-conversion, and each of these properties holds diverse molecular properties [9].

13.4.1 UV–Visible Spectroscopy

UV–visible spectroscopy is the simplest, cost-effective, and most extensively used optical technique for analyzing different optical properties of nanomaterials. The procedure relies on the measurement of light absorption by using spectrometers (Figure 13.4) [10].

The intensity from the light source is assessed by the detector, which can be a photomultiplier tube or a photodiode with an empty space between the source and the detector. The intensity of transmitted light is reduced if the sample absorbs light at some wavelength, and a plot will give the spectrum

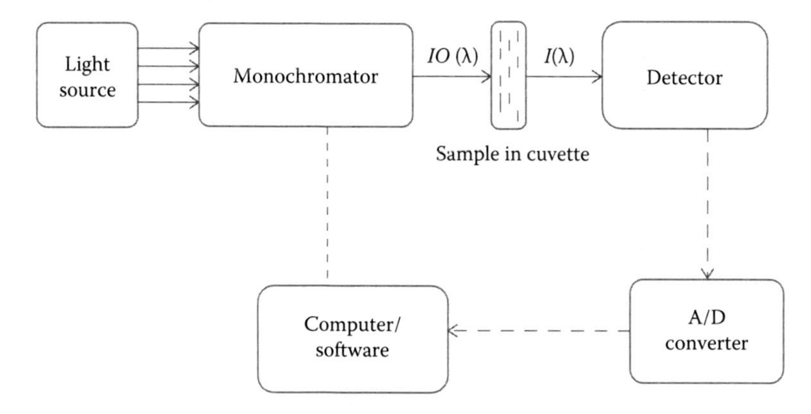

FIGURE 13.4
Schematic of UV–visible spectroscopy. (From Zhang, J.Z., Spectroscopic techniques for studying optical properties of nanomaterials, in *Optical Properties and Spectroscopy of Nanomaterials*, 2009, DOI: 10.1142/9789812836663_0003.)

of the sample absorption. Generally wavelengths in the range of 200–800 nm are preferred, but any wavelength above 800 nm requires entirely different optics [10].

13.4.2 Photoluminescence and Electroluminescence Spectroscopy

PL spectroscopy engages electronic transition of initial and final states coupled by electrical dipole operator. PL is similar to electronic absorption spectroscopy but with a major difference that, the transition between the states in PL is from higher energy level to lower energy level and it does not have any signal in the absence of PL, that is, zero-background experiment. A plot of PL intensity as a function of fixed excitation wavelength is PL spectrum, whereas PLE spectrum is a plot of PL at fixed emission wavelengths (Figure 13.5) [10].

Spectrofluorometer is used for PL measurements, and most of the components are similar to UV–vis spectrometer. Rayleigh scattering is avoided during the detection of PL, and perhaps Raman scattering can be observed in PL spectrum in the case where Raman signal is strong compared to the PL signal. The measurement process involves selection of precise wavelength from light source and directing toward the sample. Light emanated from the sample is gathered by using lens and is dispersed by using another monochromator and detected using a photodetector. The analog signal engendered by photodetector is converted into digital by using analog to digital convertor and is then processed by software on the computer (Figure 13.6).

Principles of PL and Raman play vital roles in extracting useful information among the Raman and PL spectrum. In case of nanomaterials, PL

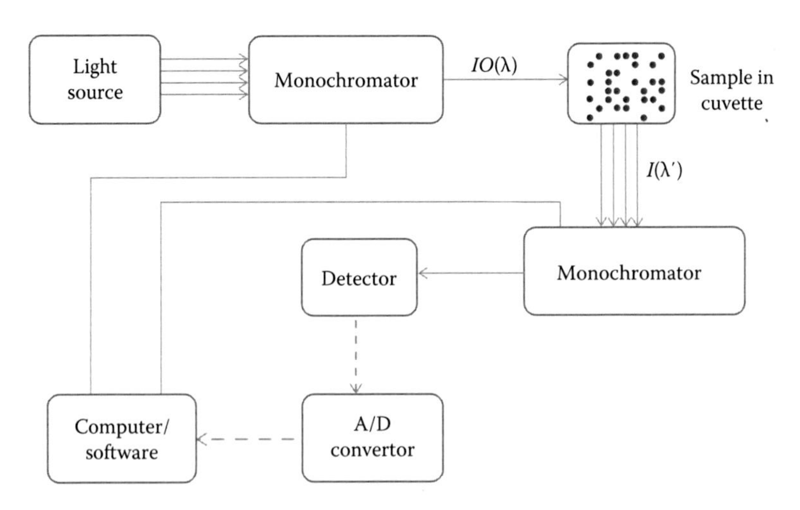

FIGURE 13.5
Schematic of spectrofluorometer. (From Zhang, J.Z., Spectroscopic techniques for studying optical properties of nanomaterials, in *Optical Properties and Spectroscopy of Nanomaterials*, 2009, DOI: 10.1142/9789812836663_0003.)

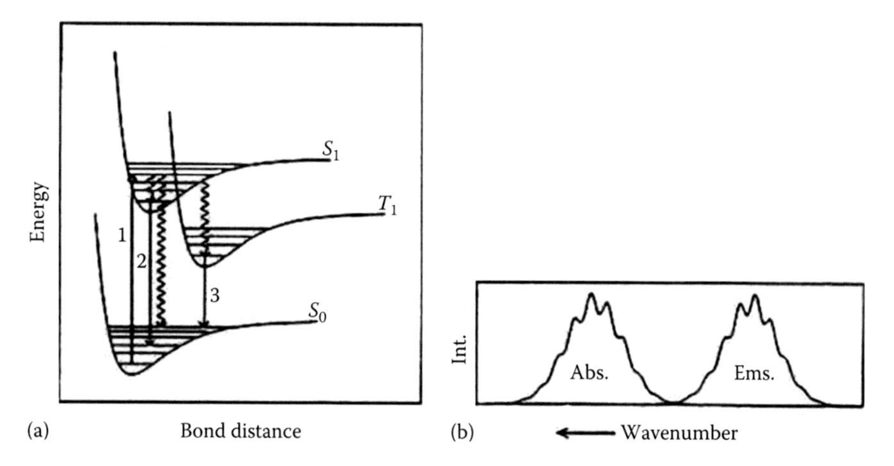

FIGURE 13.6
(a) Photoluminescence spectrum using bound excited state, (b) absorption (Abs.) and emission (Ems.) spectrum for the same. (From Larson, L.J. et al., *J. Am. Inst. Conserv.*, 30, 89, 1991.)

measurement is quite sensitive due to the presence of trap states and provides imperative information on the distribution and density of these states within the bandgap. Thus, it is essential to analyze the properties of nanomaterials thoroughly [6].

13.4.3 Electroluminescence

Electroluminescence (EL) triggers the light-emitting devices (LEDs) and the excitation is electronic, that is, the electrons injected in conduction band and the holes in valence band are injected electrically (Figure 13.7).

EL measurement involves the single-dot LED device being fabricated by focused ion beam (FIB) and molecular beam epitaxy (MBE). EL spectrum consists of sharp emission lines that can be allocated to recombine from a single dot.

13.4.4 Time-Resolved Optical Spectroscopy

Time-domain techniques provide useful information and dynamic properties of nanomaterials. Shortest of lifetime ranging in tens of femtoseconds can be measured quickly using ultrafast lasers. Basic time-resolved laser technique is TA that depends on two short pulses, one to excite and the other to cross-examine the excited-state population.

13.4.5 Time-Resolved Fluorescence

This technique is converse of TA, but it involves assessment of excited-state PL. TRF is receptive as it is zero-background measurement, but has an added

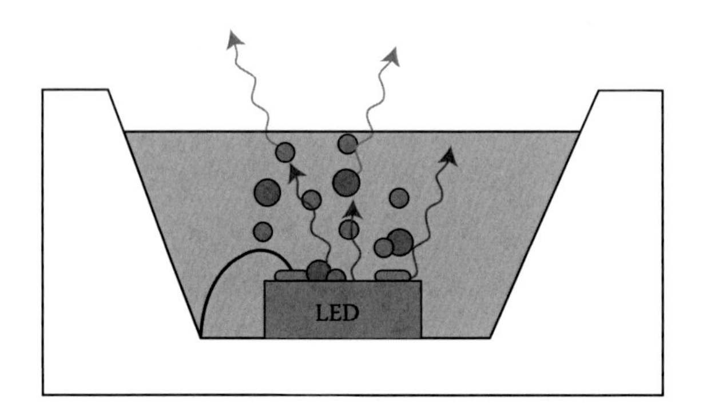

FIGURE 13.7
(See color insert.) Schematic for quantum-dot LED device. (From Pickett, N., Material matters—Commercial volumes of quantum dots: Controlled nanoscale synthesis and micron-scale applications, Nanoco Technologies, Ltd., Manchester, U.K., Available: http://www.nanocotechnologies.com/content/Library/NewsandEvents/articles/Material_Matters_Commercial_volumes_of_quantum_dots__controlled_nanoscale_synthesis_and_micron-scale_applications/45.aspx)

advantage as it uses just the ground state and the excited state of interest. The major drawback of this technique is that the time resolution is low when compared to TA. Fluorescence up-conversion (FUC) technique overcomes this drawback which improves significant amount of time resolution (Figure 13.8).

TRF measurement process uses a short pulse to excite sample, and fluorescence from excited state is combined with second pulse to produce the desired up-converted signal that has the frequency as the sum of fluorescence light and second gating pulse. Excited-state fluorescence decay

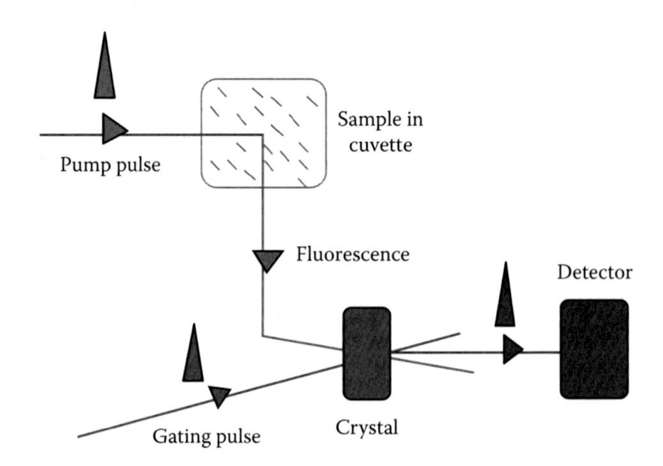

FIGURE 13.8
Schematic of TRF. (From Zhang, J.Z., Spectroscopic techniques for studying optical properties of nanomaterials, in *Optical Properties and Spectroscopy of Nanomaterials*, 2009, DOI: 10.1142/9789812836663_0003.)

profile can be obtained by changing the time delay between gating pulse and up-converted signal.

13.4.6 Enhanced Optical Imaging Using Polarimetry and Optically Active Molecules

Nanostructured particles and optically active molecules can enhance the imaging process greatly. The study carried out by Giakos et al. [11] utilized optically different composite targets immersed in fluids consisting of optically active molecular contrast agents, to test and verify this. This was coupled with polarimetric principles of stoke parameters detection to enhance imaging by increasing the penetration depth of incident light and improving contrast and scatter rejection due to depolarization. As this study was carried out by an author of this chapter, this study and its experimental results are discussed in detail to shed light on this technique that can prove extremely beneficial in defense imaging applications.

Illustration to emphasize the discrete and continuous imaging enhancement by means of optically active molecules, nanostructures, and targets made from optically diverse media immersed into fluids doped with optically active molecules is important and is employed. The conclusion of this study designates that application of Stokes parameters, detection principles with concomitants, administration of fluids containing suitable optically active molecular contrast agents that would enhance the detection and imaging process of internal structures by providing enhanced depth of penetration, high contrast and high depolarized scatter rejection.

Optical imaging presents a detailed depiction of biological tissues and offers significant potential for noninvasive investigation of molecular targets inside the human body. It consent to the categorization of a variety of diseases such as breast cancer, skin cancer, lung cancer, cancer in the bladder, the study of drug treatment effects and the effects on the target pathology, the development of biomarkers and molecular contrast agents analytic of disease and treatment outcomes and the analysis of molecular pathways leading to diseases. Image formation through detection of the polarization states of light offers diverse advantages for a wide range of detection and classification problems and it offers high-specificity and high-contrast images under low-light conditions. On the clinical side, optical polarimetry provides enhanced imaging and spectral polarimetric information regarding the metabolic information of the tissue, drug–cell interaction, single-molecule imaging, molecular mechanism of a biological function, and so on.

Absorption and scattering depends on wavelength and penetration depth. Light scattering from tissue is the limiting factor for imaging of deep biological structures. As a result, the potential to image contrast of deep structures in tissue and distinguish them in the presence of an interfering background is reduced. On the other hand, interrogation of tissue with near-infrared (NIR) wavelengths of light or optical clearing techniques leads to an enhancement of the penetration depth at the outlay of the imaging contrast. Reduced backscattering at molecular level and reduced refractive mismatch across heterogeneous boundaries at macroscopic scale lead to enhanced penetration depth but reduced contrast.

When linearly polarized light is passed through a substance containing optically active asymmetrically arranged chiral molecules or nonchiral molecules, a rotation of the polarization vector occurs. This phenomenon is called optical rotation or optical activity. Glucose and most of the biological molecules such as proteins or enzymes are optically active. The level of polarization preservation and the variation of the rotation of linearly polarized light fraction with glucose concentrations and with the chiral and non chiral molecules in the turbid samples have been detailed in a number of studies. However, most of these research efforts have been concentrated toward the development of noninvasive techniques for glucose monitoring based on reflectance spectroscopy, Raman scattering, or fluorescence rather than on imaging.

The dilemma relies on how to achieve enhanced penetration depth and the imaging contrast at the same time. An immediate question to address is whether utilizing the optical activity associated with both chiral molecules and amplification of the refraction index of the liquid phase, together with Stokes parameters imaging, can alleviate this problem.

Based on this assumption, a novel optical imaging technique has been proposed that relies on the synergistic efforts of doping the background surrounding the target with optical active molecules or high index of refraction molecules in conjunction with the application of efficient polarimetric interrogation techniques. As a result, the proposed technique would provide optical clearing and enhanced contrast capabilities. Minimized refractive index differences between the optically active fluid medium and the target would result in an increase in the degree of polarization (DOP) while also increasing the concentration of the optically active molecules at the expense of the contrast resolution. Due to the intrinsic potential to detect the weakly backscattered linearly polarized radiation, the enhanced image contrast would result using Stokes parameter imaging in the presence of highly backscattered depolarized radiation, while maintaining the original state of polarization of the incident light. Overall, using high refractive index dielectric nanoparticles, the particles disperse within fluids, and it would alter the refractive index of the liquid to volume-weighted average between refractive indices of optically active molecules and the liquid phase and eventually giving rise to an amplification of the index of refraction, yielding to enhanced DOP and DOLP signal-to-background ratio (SBR).

13.4.6.1 Applied Polarimetric Formalism

This research imaging system operates on active polarimetric detection principles performing a Fourier analysis using a rotating quarter-wavelength retarder. The Stokes vector, S', at the input of the detector, is related to the incident on the polarization state analyzer $S = (S_0, S_1, S_2, S_3)^T$ through the Muller matrix M, where, M describes the elements of the analyzer polarization of the phase retarder and the polarizer in front of the detector, including instrumental polarization.

Polarization sensitivity of the detector is given as:

$$M = \mathrm{MR} \cdot \mathrm{Mp} \tag{13.1}$$

where, MR and Mp are the Mueller matrices of the analyzer retarder and polarizer elements, respectively.

Specifically, the Stokes vector of the incident beam is:

$$S = \begin{pmatrix} S_0 \\ S_1 \\ S_2 \\ S_3 \end{pmatrix} \tag{13.2}$$

The Mueller matrix of a rotating quarter-wave retarder is:

$$\mathrm{MR} = \begin{pmatrix} 1 & 0 & 0 & 0 \\ 0 & \cos^2 2\theta & \sin 2\theta \cos 2\theta & \sin 2\theta \\ 0 & \sin 2\theta \cos 2\theta & \sin^2 2\theta & -\cos 2\theta \\ 0 & \sin 2\theta & \cos 2\theta & 0 \end{pmatrix} \tag{13.3}$$

The Mueller matrix of the linear horizontal linear polarizer is:

$$\mathrm{Mp} = \begin{pmatrix} 1 & 1 & 0 & 0 \\ 1 & 1 & 0 & 0 \\ 0 & 0 & 0 & 0 \\ 0 & 0 & 0 & 0 \end{pmatrix} \tag{13.4}$$

The Stokes parameter in the front of the detector, after passing through the retarder-polarizer configuration, is:

$$S' = \frac{1}{2}(S_0 + S_1 \cos^2 2\theta + S_2 \sin 2\theta \cos 2\theta + S_3 \sin 2\theta) \begin{pmatrix} 1 \\ 1 \\ 0 \\ 0 \end{pmatrix} \tag{13.5}$$

But $S_0' = I(\theta)$ is given as:

$$I(\theta) = \frac{1}{2}(S_0 + S_1 \cos^2 2\theta + S_2 \sin 2\theta \cos 2\theta + S_3 \sin 2\theta) \tag{13.6}$$

Reducing Equation 13.6 to a truncated Fourier series

$$I_n(\theta_j) = \frac{1}{2}\left[A + B \sin 2n\theta_j + C \cos 4\theta_j + D \sin 4n\theta_j\right] \tag{13.7}$$

where $\omega t = n\theta_j$, θ_j is the step size of rotation of the analyzer retarder.

And carrying out a Fourier analysis, the four Stokes parameters are calculated through the Fourier series coefficients via the relationship:

$$S_0 = A - C \tag{13.7a}$$

$$S_1 = 2C \tag{13.7b}$$

$$S_2 = 2D \tag{13.7c}$$

$$S_3 = B \tag{13.7d}$$

Once the Stokes parameters are estimated, the DOP was estimated in terms of Stokes parameters, as:

$$\text{DOP} = \frac{(S_1^2 + S_2^2 + S_3^2)^{1/2}}{S_0} \tag{13.8}$$

13.4.6.1.1 Description of the Multi-Index of Refraction for Biological Phantom

The second sets of experiments were performed at high magnification (×40) using a filtered 655 nm light beam from a broadband source. The details of the calibration technique are described specifically; a fiberoptic light illuminator (Fiber-Lite® MI-150W Illuminator, Dolan-Jenner and Boxborough, MA) was used as the broadband light illumination source for the study. An emission filter (Qdot D655/40m, Chroma Technology Corp, San Diego, CA) of central wavelength 655 nm was used with the broadband light source to illuminate the target with visible light (Figure 13.9).

Specifically, the setup includes a broadband light source that emits visible light with the help of a filter of wavelength 655 nm, fiberoptic cable that connects the light source and the light guide, and a visible wavelength polarizer P1 on the polarization generating side. There is no need of an optical attenuator since the broadband light source is equipped with an intensity control knob to control the intensity of the light beam. The polarization analyzing

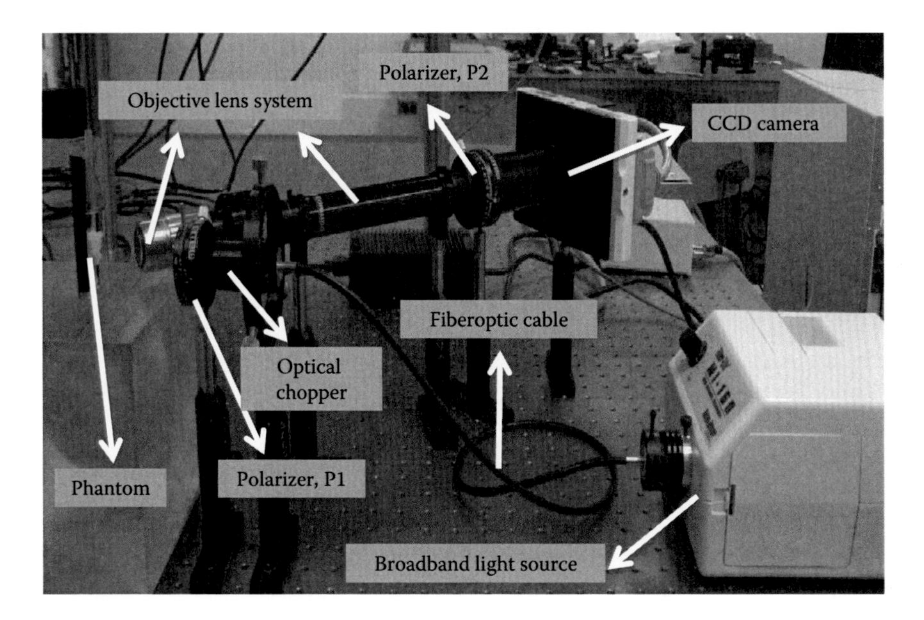

FIGURE 13.9
The high-magnification experimental setup using a 655-nm light source and a CCD camera (University of Akron Surveillance, Imaging, and Nanophotonics Laboratory).

branch contains an infinity-corrected objective lens system followed by an analyzing polarizer P2 and a CCD camera to acquire images.

The light from the broadband laser source is passed through the 655 nm optical filter to allow light of only that particular wavelength into the fiberoptic cable. A light guide is attached to the other end of the fiberoptic cable from which the light is oriented for maximum transmission and is emitted on the polarizer P. The linearly polarized light then enlightens the phantom. The light that backscattered in the direction of the analyzing branch was then detected with the help of the objective lens system that was made to pass onto the CCD camera through an analyzing linear polarizer P2. The images are then acquired by setting the analyzer polarizer P2 parallel (co-polarized) and perpendicular (cross-polarized) to the generating polarizer P1. Thus, pair of images is acquired through this method, which is then processed to obtain the corresponding DOLP images, according to:

$$\text{DOLP} = \frac{I_{\text{copolarized}} - I_{\text{cross polarized}}}{I_{\text{copolarized}} + I_{\text{cross polarized}}} \tag{13.9}$$

A fiberoptic light guide (Edmund Optics, Inc., Barrington, NJ) of bundle diameter 0.0625 in. and length 36 in. is used to illuminate the phantom. The spectrum of the filtered broadband source is recorded by means of a spectrometer (Ocean Optics OOIBase32 Spectrometer). An infinity-corrected long working distance objective lens system (Mitutoyo 46-144 M-Plan Apo Objective combined with Edmund Optics 54-774 MT-1 Tube Lens) is used for the second set of experiments (655 nm).

The biological phantom consists of a cluster of polypropylene spheres (refractive index $n = 1.49$) with diameter 2 mm and high-density polyethylene spheres (refractive index $n = 1.54$) with diameter 3.5 mm bonded with epoxy adhesive (refractive index $n = 1.65$) immersed in water solution. The above indexes of refraction were chosen to emulate biological tissues, namely calcified structures ($n = 1.53$), hydrated collagen ($n = 1.47$), and highly calcified mineralized structures ($n = 1.65$). Typically, microcalcifications depend on their shape, geometry, and composition, and they can be classified as precursors of malignancies in breast mammography.

Indeed, systematic differences between hydrated collagen in the intensities between the collagen of malignant, benign, and normal tissue groups appear to be due to a significantly lower structural order within the malignant tissues. The entire cluster is glued to a polystyrene cylindrical fixture that is fixed to the bottom of the test tube for a firm support. The distance between the cluster surface and the wall of the test tube was 3 mm ± 0.5 mm. The inner diameter of the test tube was 22 mm and the length of the test tube was 95 mm.

To perform the experiment with glucose, first an aqueous solution is prepared by adding up 12.5 g of glucose to 100 mL of water. The glucose solution was then added to the phantom that contains 16.8 mL of water in concentration increments of 10.42 mg/mL up to 62.50 mg/mL. For each concentration, increment the glucose solution, the co-polarized and cross-polarized images along with the DOLP images that were obtained. Thus, a total of seven pairs of co-polarized and cross-polarized images are obtained for this experiment.

By studying the optical phantom of the system in Figure 13.10, images of the degree of linear polarization at a ×40 magnification and varying glucose concentrations were obtained, and this is represented in Figure 13.11. It shows two parts of area with the index of refraction of the material $n = 1.49$ which are separated by a strip of area with the index of refraction $n = 1.65$.

The Canny edge detection algorithm [12] was used to study the edge densities and quantify the image contrast, using a threshold range of 0.05–0.125 (this range was selected after being obtained by applying the Canny operator on the degree of linear polarization images for 16.8 mL water instead of glucose solution, as a calibration). The Canny edge detector algorithm applied is implemented in several steps. First, the image is smoothed to remove any noise using a Gaussian filter and then areas with high spatial derivatives are highlighted using the well-known Sobel operator used in image processing (this is done by measuring 2-D gradients across the image). The next step is to check along gradients that are increasing in order and eliminate the pixels that do not correspond to any local maxima. Thresholding is done to further eliminate unwanted points. For each different glucose concentration,

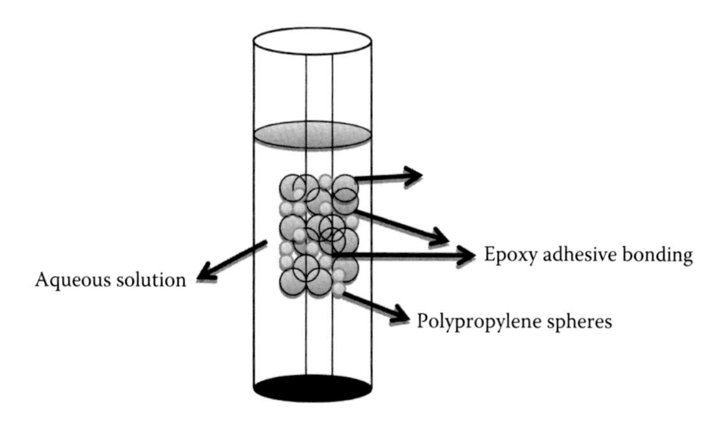

FIGURE 13.10
The multi-index of refraction phantom.

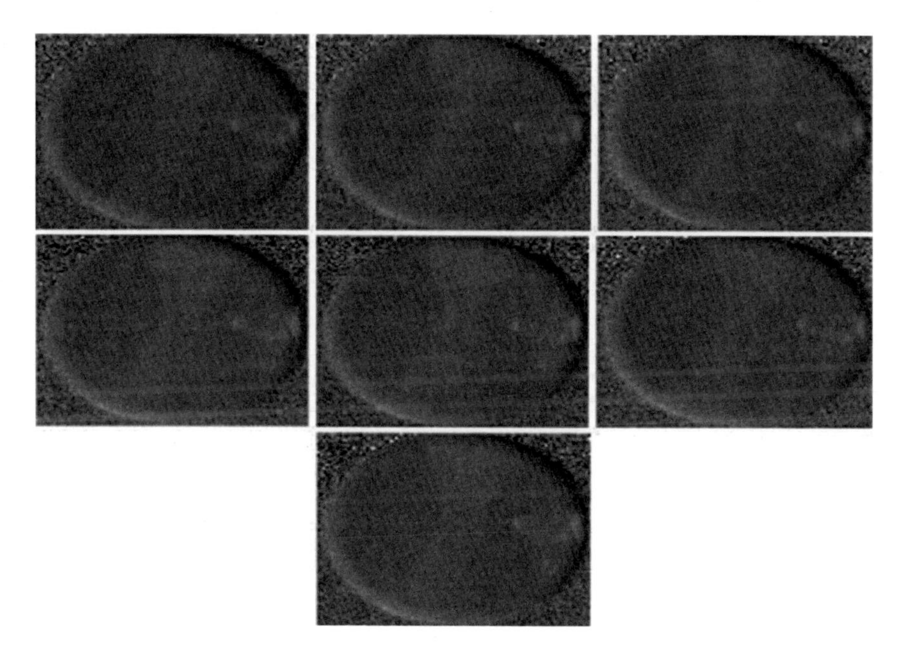

FIGURE 13.11
Images of degree of linear polarization for an aqueous glucose solution at ×40 magnification using a light source at 655 nm.

the "edge pixels" are determined and the number of edge pixels is plotted against concentration. The corresponding confidence intervals are also computed for each measurement (Figure 13.12).

For each optical molecular contrast agent, there are three sets of data values as the experiments were run thrice for each one of them. The number of edge pixels found was treated as the dependent variable and the concentration

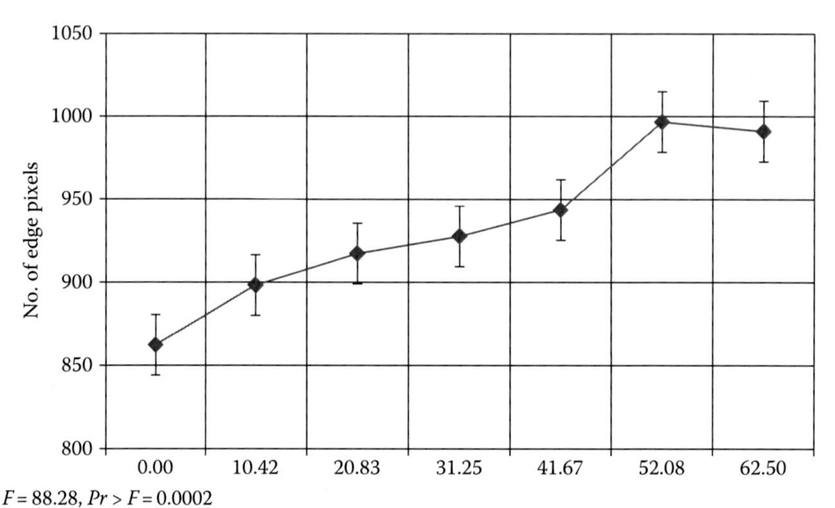

FIGURE 13.12
Number of edge pixels plotted against concentration for various glucose solutions.

TABLE 13.1

Regression Analysis for the Glucose Solution Experiment

Source	DF	Sum of Squares	Mean Square	*F* Value	*Pr*>*F*
Regression	1	13,245.35	13,245.35	88.28	0.0002
Unexplained	5	750.17	150.034		
Total	6	13,995.52			

Regression coefficient (R^2) 0.94.

was taken as the independent variable and a regression analysis was performed for this data. The results are summarized in Table 13.1.

It can clearly be seen that, for different concentrations of aqueous glucose solutions, there is a significant difference in the number of edge pixels computed for the degree of linear polarization images.

The number of edge pixels represents a measure of the contrast of the image and can be affected by noise, influencing and changing the contrast-specific effects. Hence, a different way to quantify the image contrast with respect to the concentration was determined by designing a statistical model for the degree of linear polarization images. At this wavelength of light and magnification level being used, it is seen that the image contrast information can be classified using intensity levels. The local variance of intensity is not useful for region differentiation as the "texture" of subregions is similar everywhere. But, there is a difference in the level of illumination within the various regions of interest as the absorption characteristics vary.

Further, a moving average filter was applied to obtain local mean images. This moving average operator smoothens the abnormal intensities of various regions and provides us with a much more even and compact histogram, which is better for statistical modeling. The mean images obtained after filtering show two different but symmetric distributions that can be efficiently modeled using normal distributions.

A sample set of an original and its corresponding mean image along with their histograms has been depicted in Figure 13.13. The structural variations between regions are represented by the distance of the distribution. From this, it was proposed that the difference of model modes can be used as a quantifying measure of images for varying concentrations.

The results extracted for the glucose aqueous solution experiment in different concentrations are illustrated in Figure 13.14, clearly revealing a contrast improvement over increased concentration levels.

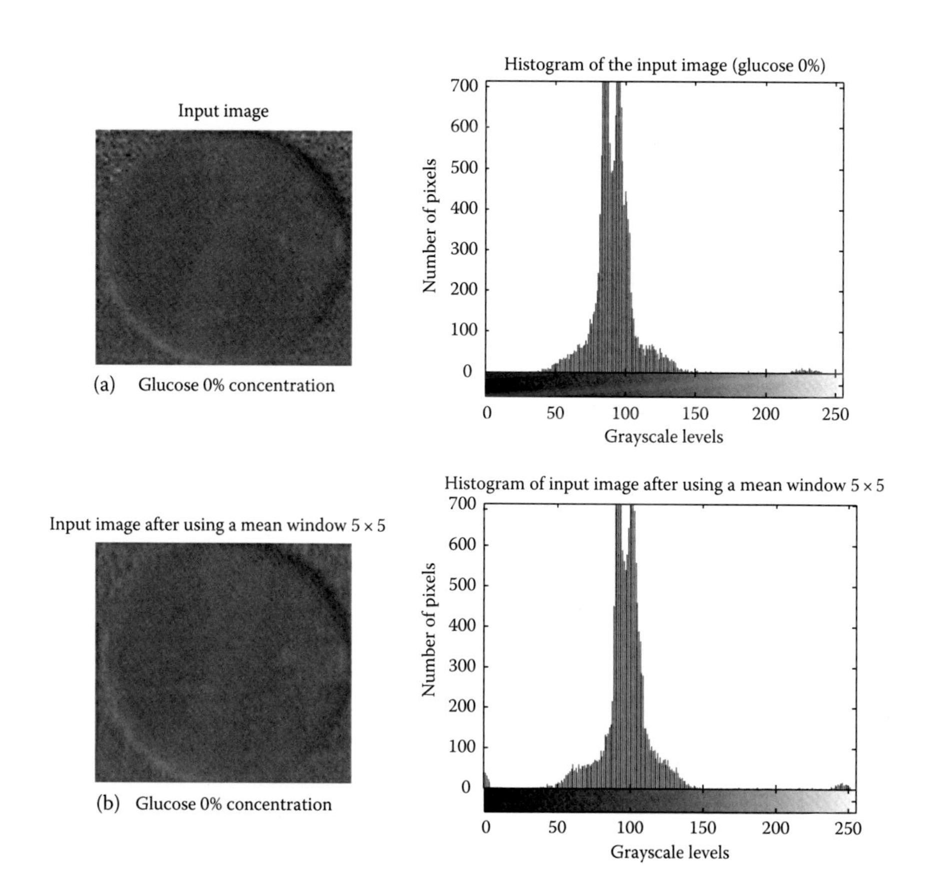

FIGURE 13.13
(a) Original and (b) mean processed DOLP images along with their corresponding histograms.

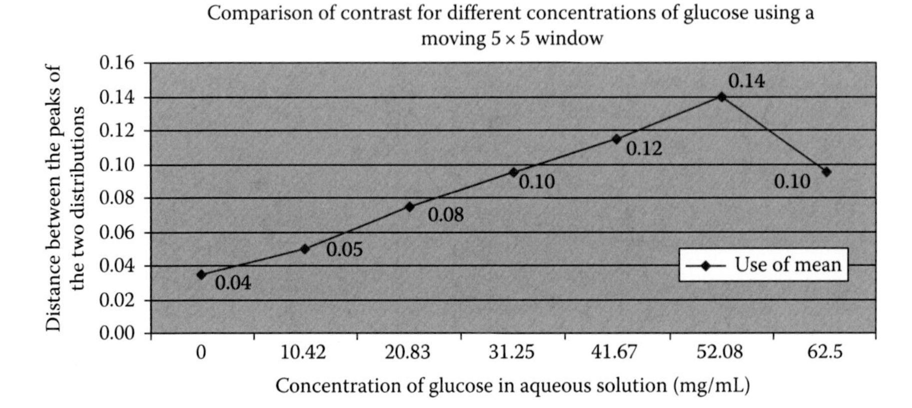

FIGURE 13.14
Contrast comparison for different concentrations using moving average filtering.

Thus, an image-enhancing technique was developed, which increases the penetration depth and contrast in images. Along with Stokes parameter imaging, if we combine fluid solutions that are doped with optically active molecular contrast agents, optically enhanced images can be obtained. Further statistical analysis of these images for different concentration of optically active molecules indicated that higher concentration of these agents increases the DOP and enhances contrast. This technique described above can be applied in defense imaging settings in various military fields.

13.5 Surface-Enhanced Raman Spectroscopy for Defense Purposes

We know that when photons are scattered from an object upon interaction with it, most of the photons undergo elastic scattering, that is, the energy of the scattered photons is the same as the incident photons. However, there is a portion of photons, which are scattered with a frequency (thus, energy level) different than that of the incident photons. This energy may be higher or lower than the original energy level of the incident photons. This phenomenon of inelastic scattering of photons is called the Raman effect or Raman scattering. The incident light interacts with the "molecular vibrations" in the object, and these vibrations shift the energy of the emitted photons either up or down. If the energy of the scattered photons is lower than the incident ones, this is termed as "Stokes" Raman scattering and if the energy of the scattered photons is greater than the incident ones, this is called "anti-Stokes" Raman scattering. The energy difference between the incident and

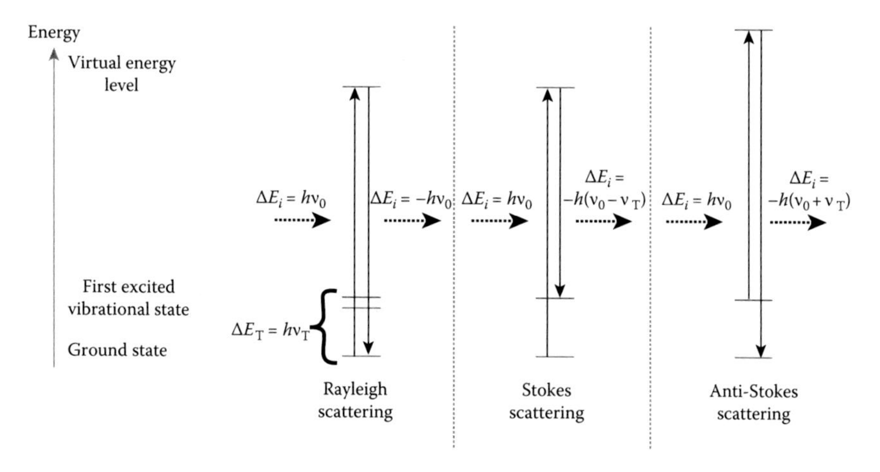

FIGURE 13.15
Energy level depiction of different possible scattering processes. (Image was released into the public domain by Wikipedia.)

the scattered photon is representative of the energy difference between the two resonant states of the material of interaction. The spectrum of Raman effect is called the Raman spectrum, and it corresponds to the different energy levels of the absorbed photons (Figure 13.15).

It is important to distinguish Raman scattering from the process of fluorescence, in which the system completely absorbs the incident light and is shifted to a higher energy level and stays in this state for a specific time before returning to various possible lower energy levels. Raman scattering is sensitive to the various vibrational modes of the particles of the object, whereas in fluorescence, relaxation of vibration occurs. Thus, Raman scattering is also known as the "fingerprint of the molecules" [13].

Raman spectroscopy is a technique that allows us to study the vibrational modes of a system by measuring the Raman scattering. Generally, a sample object is irradiated with a laser beam, usually in the visible or NIR range, and this incident laser light interacts with the molecular vibrations in the sample object and produces scattered photons that are at a higher or lower level of energy compared to the incident light (inelastic scattering). Other type of scattering that occurs is Rayleigh scattering (elastic) in which the energy of the scattered photons remains the same as the incident ones. To measure Raman scattering, light is collected using a lens and is sent through a monochromator. The wavelengths nearer to the incident laser beam (produced due to elastic Rayleigh scattering) are filtered out and only the inelastic component of the scattered light is sent to the detector. This signal is weak and difficult to detect as maximum number of photons undergo elastic scattering and very few undergo Raman scattering. Multiple gratings and dispersion techniques are used to reject as much elastically scattered light as possible (Figure 13.16).

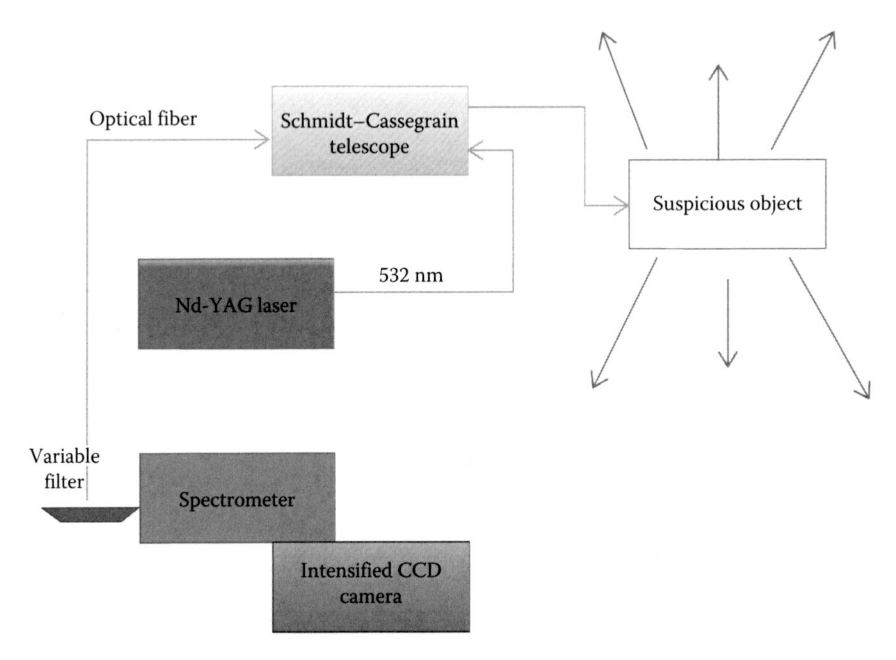

FIGURE 13.16
Raman spectroscopic setup. (From Brandner, B. et al., *Explosives Detection by LIMS and Resonance Raman Spectroscopy*, FOI, Defense Research Agency, Ministry of Defense, Stockholm, Sweden.)

Surface-enhanced Raman spectroscopy (SERS) is a technique to greatly enhance Raman scattering using surface phenomenon of the material such as SPR by "Raman-active" molecules adsorbed or layered on rough metallic surfaces. Various kinds of surface phenomenon (occurring due to the special properties of the surface molecules) enhance the Raman scattering up to 10^{10} times; this level of signal intensity makes it possible to detect even single molecules. SERS greatly improves the selectivity and sensitivity of Raman spectroscopy [14]. There are two main kinds of processes that modify the surface behavior and enhance spectroscopy—electromagnetic and chemical—the primary one being the electromagnetic enhancement that is due to the roughness and the increased electrical activity of the metallic surface specially designed for SERS. Localized surface plasmons are created when the light is incident on the surface of the material being used. For SERS, we mostly use light of wavelength in the visible or NIR range. If the frequency of the plasmon oscillations is in resonance with the incident light, the Raman scattering is the strongest, that is, when SPR occurs. Also, these oscillations must be in a plane normal to the plane of incidence, that is, perpendicular to the surface, for scattering to occur. This SPR perpendicular to the surface is induced/enhanced by using rough metallic surfaces or, even better, nanoparticles. Surfaces composed of nanostructured materials offers the incident light a large change in the vibrational state and greatly intensifies Raman scattering by increasing SPR.

This phenomenon of SERS is mostly observed when the surface is composed of metals such as gold, silver, or copper or of alkaline metals such as lithium, sodium, or potassium as SERS is highly surface selective. Also, this enhancement in the Raman scattering signal is not the same for all frequencies of incident light. Au and Ag are preferred for these surfaces as their frequency values during SPR are in accordance with the wavelengths of visible and NIR rays. Earlier, SPR was studied using roughened silver or gold surfaces. Advancements led to the use of metallic nanoparticles for this purpose. The nanosize of the particles greatly increases the surface area to volume ratio and enhances SPR and increases scattering. The Raman signal is directly proportional to the electromagnetic field incident on the surface, which is a summation of the electromagnetic field induced by regular Raman scattering of a nonroughened surface and that by a surface roughened with metallic particles. At the joining between the particle and the medium, the electromagnetic radiation that is incident gets coupled into a surface plasmon, and this gives rise to the absorption band of metallic nanoparticles [15]. These confined electrons undergo oscillations collectively that causes SPR, which is dependent on the size of each particle [16].

As metallic nanoparticles exhibit quantum confinement and the surface electrons are confined to small-sized particles, SPR is highly enhanced by the metallic nanoparticles causing roughness. Also, the shape and the alignment of the particles also determine the extent of perpendicular oscillation causing resonance and scattering. These nanoparticle-coated surfaces greatly increase the Raman spectroscopy signal and are hence very useful in sensing and detecting. Thus, spherical metallic nanoparticles (with a preference to gold) are utilized for this purpose (Figure 13.17).

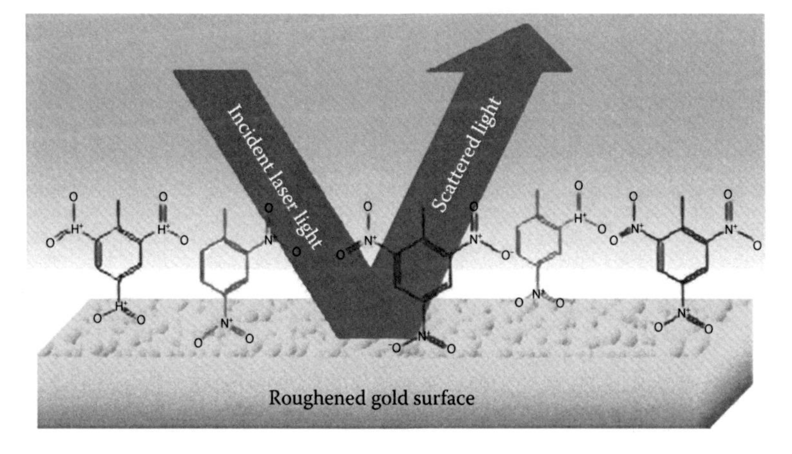

FIGURE 13.17
(See color insert.) Raman scattering, enhanced with gold nanoparticles. (From EIC Laboratories, Instrument development and applications of Raman spectroscopy, Available: http://www.eiclabs.com/resspec.htm)

FIGURE 13.18
SERS probe for mine detection. (From Kawai, N.T. et al., Raman spectroscopy for homeland defense applications, InPhotonics, Inc., Norwood, MA, 2004.)

This technique of Raman spectroscopy enhanced by gold nanoparticles is extremely useful as a stand-off detection technique for explosive detection in defense applications.

A novel SERS application in defense and security is the ongoing development of a Raman fiberoptic probe with an activated gold substrate to be used for landmine detection [17]. This is being done by measuring the explosive signature of DNT. SERS is also being used as a chemical warfare detection agent as it aids in the detection of cyanide, toxins, nerve agent degradation products, and so on (Figure 13.18) [18].

13.6 Bulk Detection Techniques for Explosives

Explosives can be detected in bulk by detecting certain elements (such as nitrogen, oxygen, carbon, and Hydrogen) that are present in almost all types of explosives (trinitrotoluene, RDX, HMX, also called octogen). Even though various kinds of explosives contain more or less the same elements, their chemical composition varies greatly and an accurate detection of that can be useful in the identification of explosive. More the number of types of elements that can be detected, higher the accuracy of our detecting system as it gives a very less probability of false-positives or false-negatives.

13.6.1 X-Ray Detection Techniques

The contents inside any kind of bags or packaging can be detected easily by using x-ray scattering imaging techniques as x-rays easily penetrate most of

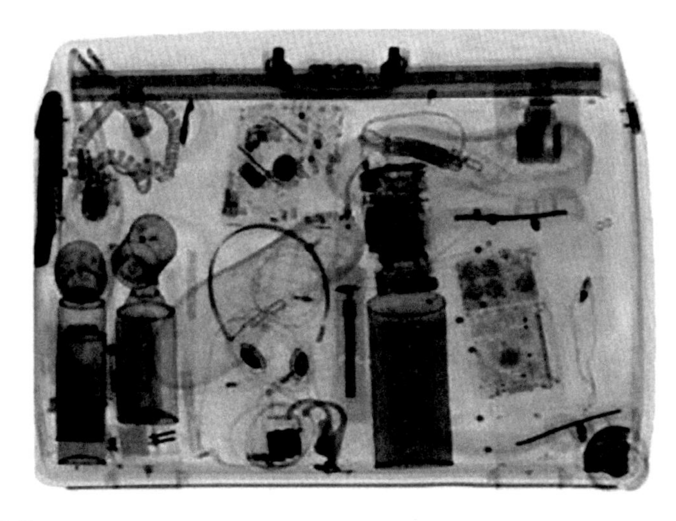

FIGURE 13.19
(See color insert.) A typical x-ray scan of a bag at the airport security screening. (From Tyson, J. and Grabianowski, E., How airport security works, Available: http://science.howstuffworks.com/transport/flight/modern/airport-security4.htm)

the materials. X-ray imaging is a transmission technique, which means that the source, the object, and the detector are in a straight line, with the source and the detector opposite to each other (scattering) or the source and the detector on the same side (backscattering) (Figure 13.19).

X-rays undergo both the different kinds of scattering—Compton scattering (which is inelastic) and elastic scattering. In Compton scattering, we study the scatter density that depends on the local electron density (as it determines the energy loss of the scattered light), which is in accordance with the physical density for majority of the elements with a low atomic number, except hydrogen [1]. In elastic scattering, the energy remains constant but the angle of scattering varies a lot, which can be measured by fixing the detector angle constant. This technique is more applicable to the detection of elements with a higher atomic number, as contained by the detonators of explosives.

Apart from scattering techniques, x-ray fluoroscopy is also performed routinely to examine and detect harmful explosives and their detonator components.

13.6.2 Magnetic Resonance Detection Technique

Nuclear magnetic resonance can distinguish between different chemicals based on their behavior in a strong magnetic field due to the different content of the contrast parameter in them, the most important being the proton density, or the concentration of the hydrogen ion. The biggest disadvantage of this technique is the difficulty in applying a large, constant, homogenous field uniformly over the surface and objects to be detected and the high cost.

13.6.3 Neutron- and Gamma-Ray-Based Detection Techniques

Though gamma rays and neutrons provide good screening techniques as they penetrate very deep into objects, they pose health hazards, making this technique useful only for cargo purposes. Neutrons interact with the object and produce characteristic gamma ray emission by a chain of several nuclear reactions. The energy of the neutrons to be used depends on the various chemical elements that need to be detected.

13.6.4 Millimeter Wave and THz Spectroscopy

In electromagnetic radiation with a wavelength greater than 300 μm, clother and other similar materials become transparent. Thus, this technique of using terahertz (THz) waves (which lie between the mid-IR and microwave bands) provides a method for efficient explosive detection without any health hazards. This technique can also be used to extract structural and spectroscopic information, and it is based on the vibration of molecules at different frequency levels.

Free electron lasers are the strongest THz radiation sources available and they produce both continuous and pulsed THz waves [19]. This spectroscopic technique produces and detects THz pulses and allows the characterization of dielectric properties of materials (Figure 13.20).

These THz waves are excellent for biological and chemical sensing and are being used extensively in explosive detection like TNT, RDX, HMX, and PETN. As most targets are bulky and it is hard to image them in transmission mode, reflection measurements are preferred.

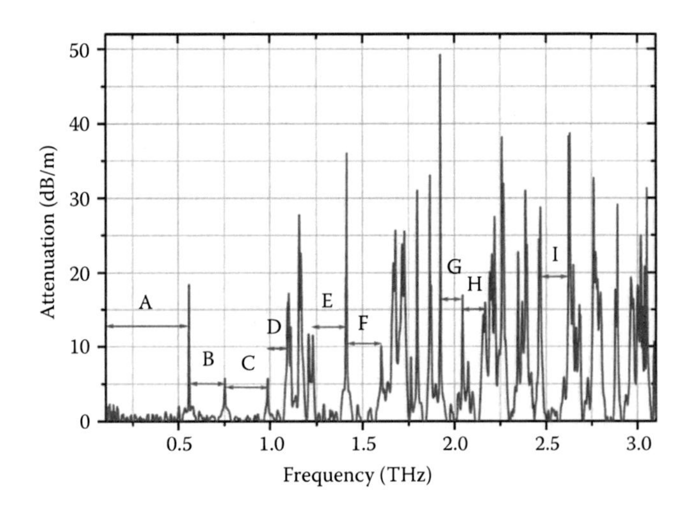

FIGURE 13.20
Atmospheric attenuation of THz waves in the nine major THz propagation bands. (From Liu, H.-B. et al., *Proc. IEEE*, 95, 1514, 2007.)

Millimeter wave imaging allows us to look through materials. Reflected or emitted waves from objects, at a wavelength of about 3 mm, are detected by sensors. This spectroscopy technique can be operated in either an active mode or a passive one.

13.7 Trace Detection Methods

These techniques employ the detection of the trace amounts of explosives present in the atmosphere (in the gas or around an object or on its own packaging material). Particles and vapors are sampled properly for effective tracing. Various methods are implemented for detecting the trace residual amount of explosives. Chemiluminescence, desorption electrospray ionization, matrix-assisted laser desorption/ionization mass spectroscopy, and electronic noses are some of the techniques utilized for this purpose.

The laser ionization mass spectroscopy (LI-MS) techniques being developed at the FOI, Sweden, include the ESSEX (extremely sensitive and selective explosive detector) system. A very sensitive detection method is obtained by merging three different methods—optical spectroscopy, mass spectrometry, and the jet expansion of gases [20]. This is called jet-REMPI (resonant enhanced multiphoton ionization), which is a special case of laser ion mass spectrometry (LI-MS) and gives us properties specific to the molecules comprising the object. As mentioned, this REMPI technique combines laser ionization along with mass spectroscopy. Upon irradiation by a laser, molecules get excited to a known higher level of energy and when it absorbs another photon from the laser, the molecule gets ionized. Using a time-of-flight mass spectrometer, the mass of this ion is determined [20]. As the ionization process is highly selective, only those molecules get ionized for which the system is being tested right now, eliminating any interferences. This technique is useful in analyzing combustion products and in detecting explosives. The measured data using jet-REMPI are obtained in two dimensions, yielding high selectivity and sensitivity, allowing REMPI to detect concentrations of contaminants at a very low level. Also, this technique can be extended to a large number of substances, more than any of the current systems can. This is also a very fast system as there is no need of any preparation for testing of the sample and it can directly be taken in its available form from air. Various processes are being developed to mobilize the REMPI system and make it easily transportable by reducing the size of both the laser and the spectrometer.

As can be clearly seen, over the past decade, numerous techniques have been developed and implemented using nanoscale systems to aid in defense and military applications. The various spectroscopic techniques and homeland security applications mentioned in this chapter represent the current trends in research in this area. This application of nanomaterials offers exceptional scope and promise for future research and development.

References

1. A. Pettersson, S. Wallin, B. Brandner, C. Eldsater, E. Holmgren, *Explosives Detection: A Technology Inventory*, FOI, Defense Research Agency, Ministry of Defense, Stockholm, Sweden.
2. Institute for Soldier Nanotechnologies, Collaboration of the U.S. Army and MIT. Available: http://web.mit.edu/isn/index.html (accessed on October 12, 2012).
3. Defence, weapons and the use of nanotechnology in modern combat equipment and warfare systems, Azonano.com: The A to Z of nanotechnology. Available: http://www.azonano.com/article.aspx?ArticleID=1818 (accessed on October 18, 2012).
4. Defense and security applications for nanomaterials and nanoparticles, supplier data by STREM Chemicals, Inc. Available: http://www.azonano.com/article.aspx?ArticleID=1337 (accessed on October 18, 2012).
5. J. M. Mauro, H. Mattoussi, I. L. Medintz, E. R. Goldman, P. T. Tran, and G. P. Anderson, Receptor protein-based bioconjugates of highly luminescent CdSe-ZnS quantum dots: Use in biosensing applications, *Defense Applications of Nanomaterials* 891, 16–30, 2005.
6. J. G. Reynolds and B. R. Hart, Nanomaterials and their applications to defense and homeland security, *JOM*, 56, 36–39, 2004.
7. J. Morse, Nanostructured sensors for homeland security applications, InterNano, National NanoManufacturing Networks. Available: http://www.internano.org/content/view/322/251/ (accessed on October 18, 2012).
8. P. V. Kamat, D. Meisel, Nanoscience opportunities in environmental remediation, *Comptes Rendus Chimie* 6(8–10), 999–1007, 2003. ISSN 1631-0748, 10.1016/j.crci.2003.06.005.
9. J. Z. Zhang, Introduction, in *Optical Properties and Spectroscopy of Nanomaterials*, July 21, 2009, DOI: 10.1142/9789812836663_0001.
10. J. Z. Zhang, Spectroscopic techniques for studying optical properties of nanomaterials, in *Optical Properties and Spectroscopy of Nanomaterials*, July 21, 2009, DOI: 10.1142/9789812836663_0003.
11. G. C. Giakos, K. Valluru, K. Ambadipudi, S. Paturi, P. Bathini, M. Becker, P. Farajipour, S. Marotta, J. Paxitzis, B. Mandadi, Stokes parameters imaging of multi-index of refraction biological phantoms utilizing optically active molecular contrast agents, *Measurement Science and Technology* 20(2209), 1–12, 2009.
12. J. F. Canny, A computational approach to edge detection, *IEEE Transactions on Pattern Analysis and Machine Intelligence* 8(6), 679–698, 1986.
13. S. Jiang, Surface enhanced Raman scattering spectroscopy, Term paper for Physics 598 OS. Available: http://mit.edu/sjiang2/www/Resources/Term%20Paper_SERS.pdf (accessed on October 18, 2012).

14. C. J. Hicks, Surface enhanced Raman spectroscopy, MSU CEM 924, 2001. Available: http://www.cem.msu.edu/~cem924sg/ChristineHicks.pdf (accessed on October 18, 2012).
15. M. Alvarez, J. T. Khoury, G. Schaaff, M. N. Shafigullin, I. Vezmar, R. L. J. Whetten, Optical absorption spectra of nanocrystal gold molecules, *Physical Chemistry B* 101(19), 3706–3712, 1997.
16. D. B. Pederson, E. J. S. Duncan, *Surface Plasmon Resonance Spectroscopy of Gold Nanoparticle Coated Substrates*, Defense Research and Development, Ottawa, Ontario, Canada.
17. EIC Laboratories, Application summary: Buried landmine detection with SERS. Available: http://www.eiclabs.com/Detection_of_Landmines.pdf (accessed on October 21, 2012).
18. EIC Laboratories, Application summary: Chemical warfare agent detection using SERS. Available: http://www.eiclabs.com/CW_Agent_Detection_via_SERS.pdf (accessed on October 21, 2012).
19. H.-B. Liu et al., Terahertz spectroscopy and imaging for defense and security applications, *Proceedings of the IEEE* 95(8), 1514–1527, 2007.
20. B. Brandner et al., Explosives detection by LIMS and resonance Raman spectroscopy, FOI, Defense Research Agency, Ministry of Defense, Stockholm, Sweden.
21. U.S. Department of Justice, Federal Bureau of Investigation "Terrorism 2000/2001."
22. L. J. Larson et al., Photoluminescence spectroscopy of natural resins and organic binding media of paintings, *Journal of the American Institute of Conservation* 30(1), 89–104, 1991.
23. N. Pickett, Material matters—Commercial volumes of quantum dots: Controlled nanoscale synthesis and micron-scale applications. Nanoco Technologies, Ltd., Manchester, U.K. Available: http://www.nanocotechnologies.com/content/Library/NewsandEvents/articles/Material_Matters__Commercial_volumes_of_quantum_dots__controlled_nanoscale_synthesis_and_micronscale_applications/45.aspx (accessed on October 23, 2012).
24. EIC Laboratories, Instrument development and applications of Raman spectroscopy. Available: http://www.eiclabs.com/resspec.htm (accessed on October 21, 2012).
25. N. T. Kawai et al., Raman spectroscopy for homeland defense applications, InPhotonics, Inc., Norwood, MA, 2004.
26. J. Tyson, E. Grabianowski, How airport security works. Available: http://science.howstuffworks.com/transport/flight/modern/airport-security4.htm (accessed on October 18, 2012).

Appendix A: Material and Physical Constants

A.1 Common Material Constants

TABLE A.1

Approximate Conductivity at 20°C

Material	Conductivity (S/m)
1. Conductors	
Silver	6.3×10^7
Copper (standard annealed)	5.8×10^7
Gold	4.5×10^7
Aluminum	3.5×10^7
Tungsten	1.8×10^7
Zinc	1.7×10^7
Brass	1.1×10^7
Iron (pure)	10^7
Lead	5×10^7
Mercury	10^6
Carbon	3×10^7
Water (sea)	4.8
2. Semiconductors	
Germanium (pure)	2.2
Silicon (pure)	4.4×10^{-4}
3. Insulators	
Water (distilled)	10^{-4}
Earth (dry)	10^{-5}
Bakelite	10^{-10}
Paper	10^{-11}
Glass	10^{-12}
Porcelain	10^{-12}
Mica	10^{-15}
Paraffin	10^{-15}
Rubber (hard)	10^{-15}
Quartz (fused)	10^{-17}
Wax	10^{-17}

TABLE A.2

Approximate Dielectric Constant and Dielectric Strength

Material	Dielectric Constant (or Relative Permittivity) (Dimensionless)	Strength, E (V/m)
Barium titanate	1200	7.5×10^6
Water (sea)	80	—
Water (distilled)	8.1	—
Nylon	8	—
Paper	7	12×10^6
Glass	5–10	35×10^6
Mica	6	70×10^6
Porcelain	6	—
Bakelite	5	20×10^6
Quartz (fused)	5	30×10^6
Rubber (hard)	3.1	25×10^6
Wood	2.5–8.0	—
Polystyrene	2.55	—
Polypropylene	2.25	—
Paraffin	2.2	30×10^6
Petroleum oil	2.1	12×10^6
Air (1 atm)	1	3×10^6

TABLE A.3

Relative Permeability

Material	Relative Permeability, μ_r
1. Diamagnetic	
Bismuth	0.999833
Mercury	0.999968
Silver	0.9999736
Lead	0.9999831
Copper	0.9999906
Water	0.9999912
Hydrogen (STP)	$\simeq$1.0
2. Paramagnetic	
Oxygen (STP)	0.999998
Air	1.00000037
Aluminum	1.000021
Tungsten	1.00008
Platinum	1.0003
Manganese	1.001
3. Ferromagnetic	
Cobalt	250
Nickel	600
Soft iron	5000
Silicon iron	7000

TABLE A.4

Approximate Conductivity for Biological Tissue

Material	Conductivity (S/m)	Frequency
Blood	0.7	0 (DC)
Bone	0.01	0 (DC)
Brain	0.1	10^2–10^6 Hz
Breast fat	0.2–1	0.4–5 GHz
Breast tumor	0.7–3	0.4–5 GHz
Fat	0.1–0.3	0.4–5 GHz
	0.03	10^2–10^6 Hz
Muscle	0.4	10^2–10^6 Hz
Skin	0.001	1 kHz
	0.1	1 MHz

TABLE A.5

Approximate Dielectric Constant for Biological Tissue

Material	Dielectric Constant (Relative Permittivity)	Frequency
Blood	10^5	1 kHz
Bone	3,000–10,000	0 (DC)
Brain	10^7	100 Hz
	10^3	1 MHz
Breast fat	5–50	0.4–5 GHz
Breast tumor	47–67	0.4–5 GHz
Fat	5	0.4–5 GHz
	10^6	100 Hz
	10	1 MHz
Muscle	10^6	1 kHz
	10^3	1 MHz
Skin	10^6	1 kHz
	10^3	1 MHz

A.2 Physical Constants

Quantity	Best Experimental Value	Approximate Value for Problem Work
Avogadro's number (kg/mol)	6.0228×10^{26}	6×10^{26}
Boltzmann constant (J/K)	1.38047×10^{-23}	1.38×10^{-23}
Electron charge (C)	-1.6022×10^{-19}	-1.6×10^{-19}
Electron mass (kg)	9.1066×10^{-31}	9.1×10^{-31}
Permittivity of free space (F/m)	8.854×10^{-12}	$\frac{10^{-9}}{36\pi}$
Permeability of free space (H/m)	$4\pi \times 10^{-7}$	12.6×10^{-7}
Intrinsic impedance of free space (Ω)	376.6	120π
Speed of light in free space or vacuum (m/s)	2.9979×10^{8}	3×10^{8}
Proton mass (kg)	1.67248×10^{-27}	1.67×10^{-27}
Neutron mass (kg)	1.6749×10^{-27}	1.67×10^{-27}
Planck's constant (J s)	6.6261×10^{-34}	6.62×10^{-34}
Acceleration due to gravity (m/s^2)	9.8066	9.8
Universal constant of gravitation ($m^2/kg\ s^2$)	6.658×10^{-11}	6.66×10^{-11}
Electron volt (J)	1.6030×10^{-19}	1.6×10^{-19}
Gas constant (J/mol K)	8.3145	8.3

Appendix B: Photon Equations, Index of Refraction, Electromagnetic Spectrum, and Wavelength of Commercial Laser

TABLE B.1

Photon Energy, Frequency, Wavelength

Photon energy (J)	Planck's constant × frequency
Photon energy (eV)	$\frac{\text{Planck's constant} \times \text{frequency}}{\text{Electron charge}}$
Photon energy (cm^{-1})	$\frac{\text{Frequency}}{\text{Speed of light in vacuum}}$
Photon frequency (Hz)	$\frac{1\text{(cycle)}}{\text{Period (s)}}$
Photon wavelength (μm)	$\frac{\text{Speed of light in free space}}{\text{Frequency}}$

TABLE B.2

Index of Refraction for Common Substances

Substance	Index of Refraction
Air	1.000293
Diamond	2.24
Ethyl alcohol	1.36
Fluorite	1.43
Fused quartz	1.46
Crown glass	1.52
Flint glass	1.66
Glycerin	1.47
Ice	1.31
Polystyrene	1.49
Rock salt	1.54
Water	1.33

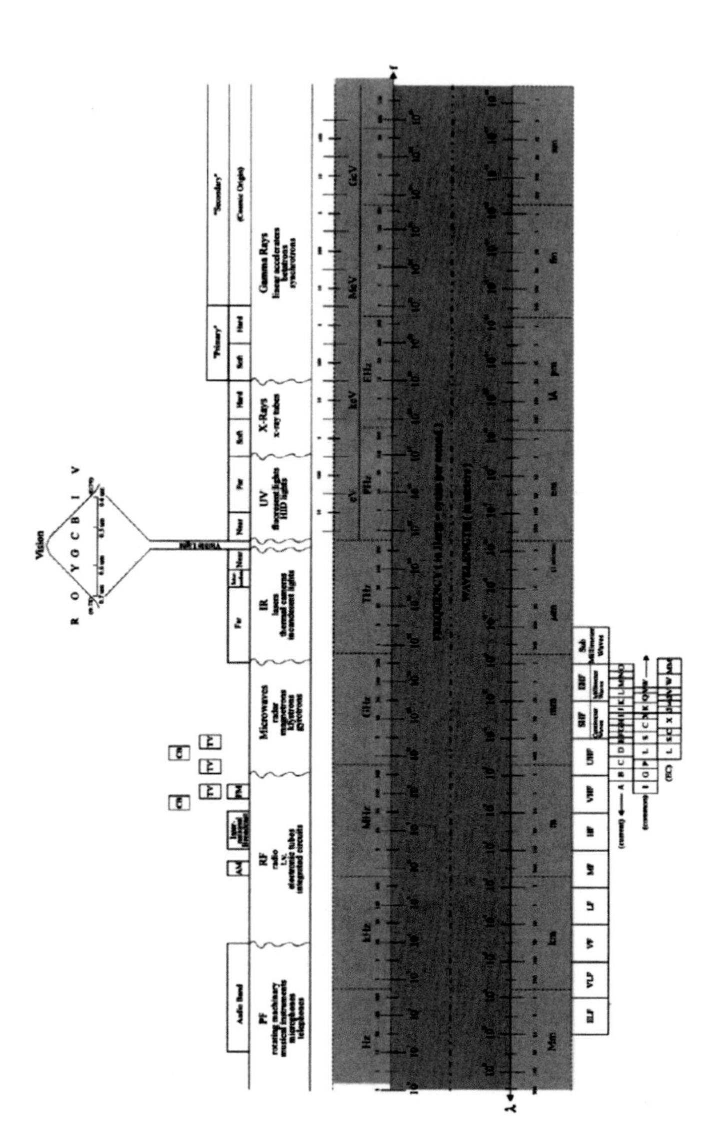

FIGURE B.1
Simplified chart of the electromagnetic spectrum. (From Whitaker, J.C., *The Electronics Handbook*, CRC Press, Boca Raton, FL, 1996.)

TABLE B.3

Approximate Common Optical Wavelength Ranges of Light

Color	Wavelength
Ultraviolet region	10–380 nm
Visible region	380–750 nm
Violet	380–450 nm
Blue	450–495 nm
Green	495–570 nm
Yellow	570–590 nm
Orange	590–620 nm
Red	620–750 nm
Infrared	750 nm–1 mm

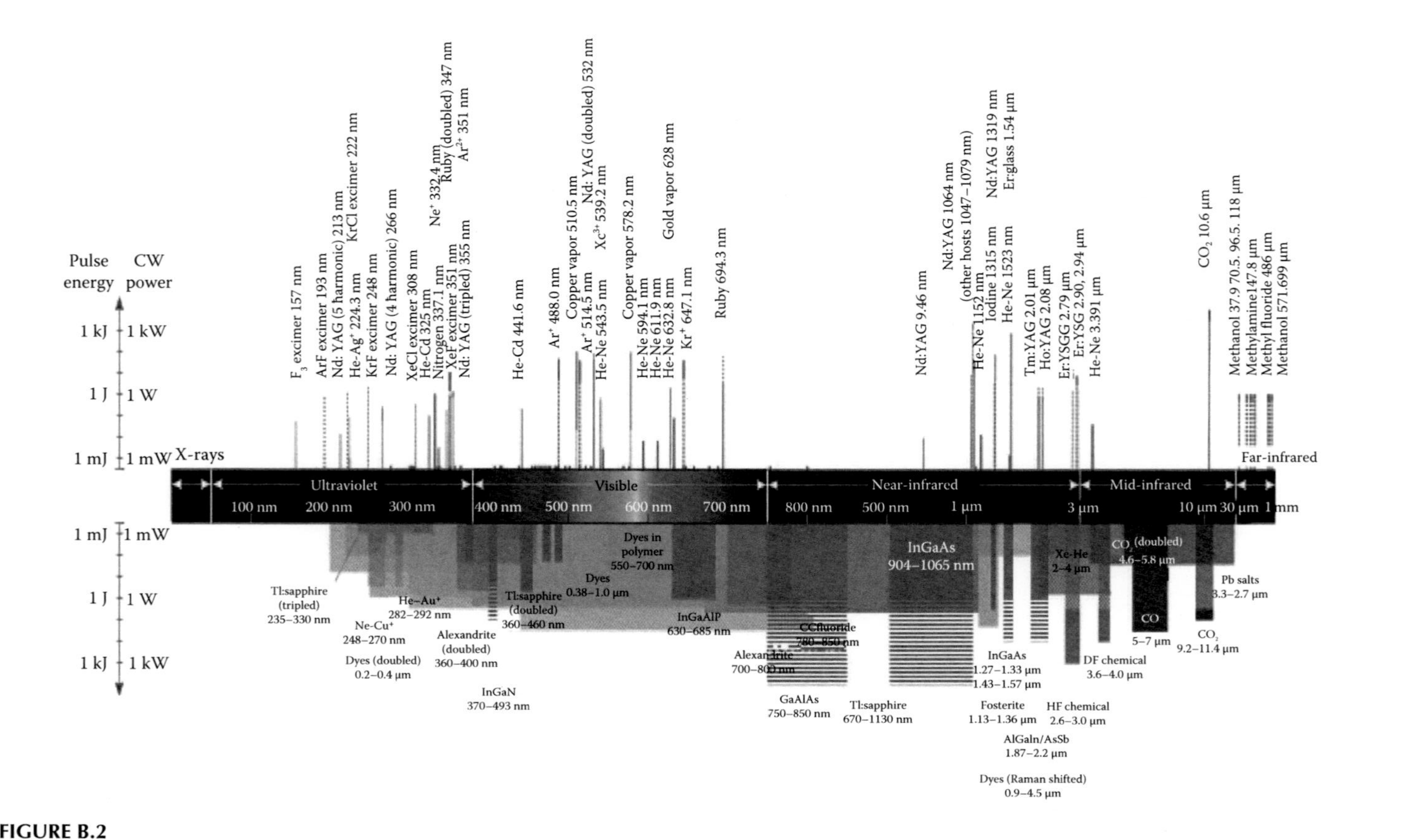

FIGURE B.2
(See color insert.) Wavelengths of commercially available lasers. (From Weber, M.J., *Handbook of Laser Wavelengths*, CRC Press, Boca Raton, FL, 1999.)

References

Weber, M.J. *Handbook of Laser Wavelengths*, CRC Press, Boca Raton, FL, 1999.
Whitaker, J.C., *The Electronics Handbook*, CRC Press, Boca Raton, FL, 1996.

Appendix C: Symbols and Formulas

C.1 Greek Alphabet

Uppercase	Lowercase	Name
Α	α	Alpha
Β	β	Beta
Γ	γ	Gamma
Δ	δ	Delta
Ε	ε	Epsilon
Ζ	ζ	Zeta
Η	η	Eta
Θ	θ, ϑ	Theta
Ι	ι	Iota
Κ	κ	Kappa
Λ	λ	Lambda
Μ	μ	Mu
Ν	ν	Nu
Ξ	ξ	Xi
Ο	ο	Omicron
Π	π	Pi
Ρ	ρ	Rho
Σ	σ	Sigma
Τ	τ	Tau
Υ	υ	Upsilon
Φ	ϕ, φ	Phi
Χ	χ	Chi
Ψ	ψ	Psi
Ω	ω	Omega

C.2 International System of Units (SI) Prefixes

Power	Prefix	Symbol	Power	Prefix	Symbol
10^{-35}	Stringo	—	10^{0}	—	—
10^{-24}	Yocto	y	10^{1}	Deka	da
10^{-21}	Zepto	z	10^{2}	Hecto	h
10^{-18}	Atto	a	10^{3}	Kilo	k
10^{-15}	Femto	f	10^{6}	Mega	M
10^{-12}	Pico	p	10^{9}	Giga	G
10^{-9}	Nano	n	10^{12}	Tera	T
10^{-6}	Micro	μ	10^{15}	Peta	P
10^{-3}	Milli	m	10^{18}	Exa	E
10^{-2}	Centi	c	10^{21}	Zetta	Z
10^{-1}	Deci	d	10^{24}	Yotta	Y

C.3 Trigonometric Identities

$$\cot\theta = \frac{1}{\tan\theta}, \quad \sec\theta = \frac{1}{\cos\theta}, \quad \csc\theta = \frac{1}{\sin\theta}$$

$$\tan\theta = \frac{\sin\theta}{\cos\theta}, \quad \cot\theta = \frac{\cos\theta}{\sin\theta}$$

$$\sin^2\theta + \cos^2\theta = 1, \tan^2\theta + 1 = \sec^2\theta, \cot^2\theta + 1 = \csc^2\theta$$

$$\sin(-\theta) = -\sin\theta, \ \cos(-\theta) = \cos\theta, \ \tan(-\theta) = -\tan\theta$$

$$\csc(-\theta) = -\csc\theta, \ \sec(-\theta) = \sec\theta, \ \cot(-\theta) = -\cot\theta$$

$$\cos(\theta_1 \pm \theta_2) = \cos\theta_1 \cos\theta_2 \pm \sin\theta_1 \sin\theta_2$$

$$\sin(\theta_1 \pm \theta_2) = \sin\theta_1 \cos\theta_2 \pm \cos\theta_1 \sin\theta_2$$

$$\tan(\theta_1 \pm \theta_2) = \frac{\tan\theta_1 \pm \tan\theta_2}{1 \mp \tan\theta_1 \tan\theta_2}$$

$$\cos\theta_1 \cos\theta_2 = \frac{1}{2}\left[\cos(\theta_1 + \theta_2) + \cos(\theta_1 - \theta_2)\right]$$

$$\sin\theta_1 \sin\theta_2 = \frac{1}{2}\left[\cos(\theta_1 - \theta_2) + \cos(\theta_1 + \theta_2)\right]$$

$$\sin\theta_1 \sin\theta_2 = \frac{1}{2}\left[\sin(\theta_1 + \theta_2) + \sin(\theta_1 - \theta_2)\right]$$

$$\cos\theta_1 \sin\theta_2 = \frac{1}{2}\left[\sin(\theta_1 + \theta_2) + \sin(\theta_1 - \theta_2)\right]$$

$$\sin\theta_1 + \sin\theta_2 = 2\sin\left(\frac{\theta_1 + \theta_2}{2}\right)\cos\left(\frac{\theta_1 - \theta_2}{2}\right)$$

$$\sin\theta_1 - \sin\theta_2 = 2\cos\left(\frac{\theta_1 + \theta_2}{2}\right)\sin\left(\frac{\theta_1 - \theta_2}{2}\right)$$

$$\cos\theta_1 + \cos\theta_2 = 2\cos\left(\frac{\theta_1 + \theta_2}{2}\right)\cos\left(\frac{\theta_1 - \theta_2}{2}\right)$$

$$\cos\theta_1 - \cos\theta_2 = 2\sin\left(\frac{\theta_1 + \theta_2}{2}\right)\sin\left(\frac{\theta_1 - \theta_2}{2}\right)$$

$$a\sin\theta - b\cos\theta = \sqrt{a^2 + b^2}\,\cos(\theta + \phi), \quad \text{where } \phi = \tan^{-1}\left(\frac{b}{a}\right)$$

$$a\sin\theta - b\cos\theta = \sqrt{a^2 + b^2}\,\sin(\theta + \phi), \quad \text{where } \phi = \tan^{-1}\left(\frac{b}{a}\right)$$

$$\cos\,(90° - \theta) = \sin\,\theta,\ \sin\,(90° - \theta) = \cos\,\theta,\ \tan\,(90° - \theta) = \cot\,\theta$$

$$\cot\,(90° - \theta) = \tan\,\theta,\ \sec\,(90° - \theta) = \csc\,\theta,\ \csc\,(90° - \theta) = \sec\,\theta$$

$$\cos\,(\theta \pm 90°) = \mp\sin\,\theta,\ \sin\,(\theta \pm 90°) = \pm\sin\,\theta,\ \tan\,(\theta \pm 90°) = -\cot\,\theta$$

$$\cos\,(\theta \pm 180°) = -\cos\,\theta,\ \sin\,(\theta \pm 180°) = -\sin\,\theta,\ \tan\,(\theta \pm 180°) = \tan\,\theta$$

$$\cos\,2\theta = \cos^2\theta - \sin^2\theta,\ \cos\,2\theta = 1 - 2\sin^2\theta,\ \cos\,2\theta = 2\cos^2\theta - 1$$

$$\sin 2\theta = 2\sin\theta\cos\theta, \quad \tan 2\theta = \frac{2\tan\theta}{1 - \tan^2\theta}$$

$$\cos\,3\theta = 4\,\cos^3\theta - 3\,\sin\,\theta$$

$$\sin\,3\theta = 3\,\sin\,\theta - 4\,\sin^3\theta$$

$$\sin\frac{\theta}{2} = \pm\sqrt{\frac{1-\cos\theta}{2}}, \quad \cos\frac{\theta}{2} \pm \sqrt{\frac{1+\cos\theta}{2}},$$

$$\tan\frac{\theta}{2} = \pm\sqrt{\frac{1-\cos\theta}{1+\cos\theta}}, \quad \tan\frac{\theta}{2} = \frac{\sin\theta}{1+\cos\theta}, \quad \tan\frac{\theta}{2} = \frac{1-\cos\theta}{\sin\theta}$$

$$\sin\theta = \frac{e^{j\theta}-e^{-j\theta}}{2j}, \quad \cos\theta = \frac{e^{j\theta}+e^{-j\theta}}{2}\left(j=\sqrt{-1}\right), \quad \tan\theta = \frac{e^{j\theta}+e^{-j\theta}}{j\left(e^{j\theta}+e^{-j\theta}\right)}$$

$$e^{\pm j\theta} = \cos\,\theta \pm j\sin\,\theta \;\left(\text{Euler's identity}\right)$$

$$1\text{ rad} = 57.296°$$

$$\pi = 3.1416$$

C.4 Hyperbolic Functions

$$\cosh x = \frac{e^{x}+e^{-x}}{2}, \quad \sinh x = \frac{e^{x}-e^{-x}}{2}, \quad \tanh x = \frac{\sinh x}{\cosh x}$$

$$\cosh x = \frac{1}{\tanh x}, \quad \text{sech}\, x = \frac{1}{\cosh x}, \quad \text{csch}\, x = \frac{1}{\sinh x}$$

$$\sin jx = j\sinh x, \; \cos jx = \cosh x$$

$$\sinh jx = j\sin x, \; \cosh jx = \cos x$$

$$\sin\,(x \pm jy) = \sin x\,\cosh y \pm j\cos x\,\sinh y$$

$$\cos(x \pm jy) = \cos x\,\cosh y \pm j\sin x\,\sinh y$$

$$\sinh\,(x \pm y) = \sinh x\,\cosh y \pm \cosh x\,\sinh y$$

$$\cosh\,(x \pm y) = \cosh x \cosh y \pm \sinh x\,\sinh y$$

$$\sinh\,(x \pm jy) = \sinh x\,\cos y \pm j\cosh x\,\sin y$$

$$\cosh\,(x \pm jy) = \cosh x\,\cos y \pm j\sinh x\,\sin y$$

$$\tanh(x \pm jy) = \frac{\sinh 2x}{\cosh 2x + \cos 2y} \pm j \frac{\sin 2y}{\cosh 2x + \cos 2y}$$

$$\cosh^2 - \sinh^2 x = 1$$

$$\text{sech}^2 + \tanh^2 x = 1$$

C.5 Complex Variables

A complex number can be written as

$$z = x = jy = r\angle\theta = re^{j\theta} = r(\cos\theta + j\sin\theta)$$

where

$x = \text{Re}\ z = r\cos\theta,\ y = \text{Im}\ z = r\sin\theta$

$$r = |z| = \sqrt{x^2 + y^2}, \quad \theta = \tan^{-1}\left(\frac{y}{x}\right)$$

$$j = \sqrt{-1}, \quad \frac{1}{j} = -j, \quad j^2 = -1$$

The complex conjugate of

$z = z^* = x - jy = r\angle -\theta = re^{-j\theta} = r(\cos\theta - j\sin\theta)$

$\left(e^{j\theta}\right)^n = e^{jn\theta} = \cos n\theta + j\sin n\theta$ (de Moivre's theorem)

If $z_1 = x_1 + jy_1$ and $z_2 = x_2 + jy_2$, then only if $x_1 = x_2$ and $y_1 = y_2$.

$$z_1 \pm z_2 = (x_1 + x_2) \pm j(y_1 + y_2)$$

$$z_1 z_2 = (x_1 x_2 - y_1 y_2) + j(x_1 y_2 + x_2 y_1) = r_1 r_2 e^{j(\theta_1 + \theta_2)} = r_1 r_2 \angle\theta_1 + \theta_2$$

$$z_1 z_2 = (x_1 x_2 - y_1 y_2) + j(x_1 y_2 + x_2 y_1) = r_1 r_2 e^{j(\theta_1 + \theta_2)} = r_1 r_2 \angle\theta_1 + \theta_2$$

$$\frac{z_1}{z_2} = \frac{(x_1 + jy_1)}{(x_2 + jy_2)} \cdot \frac{(x_2 + jy_2)}{(x_2 + jy_2)} = \frac{x_1 x_2 + y_1 y_2}{x_2^2 + y_2^2} + j\frac{x_2 y_1 - x_1 y_2}{x_2^2 + y_2^2} = \frac{r_1}{r_2} e^{j(\theta_1 + \theta_2)}$$

$$= \frac{r_1}{r_2} \angle\theta_1 - \theta_2$$

$$\ln\left(re^{j\theta}\right) = \ln r + \ln e^{j\theta} = \ln r + j\theta + j2m\pi \quad (m = \text{integer})$$

$$\sqrt{z} = \sqrt{x+jy} = \sqrt{r}(e^{j\theta/2}) = \sqrt{r}\angle\theta/2$$

$$z^n = (x+jy)^n = r^n e^{jn\theta} = r^n\angle n\theta \quad (n = \text{integer})$$

$$z^{\frac{1}{n}} = (x+jy)^{\frac{1}{n}} = r^{\frac{1}{n}} e^{\frac{j\theta}{n}} = r^{\frac{1}{n}}\angle\theta/2 + 2\pi m/n \quad (m = 0,1,2,\ldots,n-1)$$

C.6 Table of Derivatives

$y =$	$\frac{dy}{dx} =$
c (constant)	0
cx^n (n any constant)	cnx^{n-1}
e^{ax}	ae^{ax}
a^x $(a>0)$	$a^x \ln a$
$\ln x$ $(x>0)$	$\frac{1}{x}$
$\frac{c}{x^a}$	$\frac{-ca}{x^{a+1}}$
$\log_a x$	$\frac{\log_a e}{x}$
$\sin ax$	$a \cos ax$
$\cos ax$	$-a \sin ax$
$\tan ax$	$-a\sec^2 ax = \frac{a}{\cos^2 ax}$
$\cot ax$	$-a\csc^2 ax = \frac{-a}{\sin^2 ax}$
$\sec ax$	$\frac{a\sin ax}{\cos^2 ax}$
$\csc ax$	$\frac{-a\cos ax}{\sin^2 ax}$
$\arcsin ax = \sin^{-1} ax$	$\frac{a}{\sqrt{1-a^2x^2}}$
$\arccos ax = \cos^{-1} ax$	$\frac{-a}{\sqrt{1-a^2x^2}}$
$\arctan ax = \tan^{-1} ax$	$\frac{a}{1+a^2x^2}$
$\text{arccot}\, ax = \cot^{-1} ax$	$\frac{-a}{1+a^2x^2}$
$\sinh ax$	$a \cosh ax$
$\cosh ax$	$a \sinh ax$

$\tanh ax$	$\frac{a}{\cosh^2 ax}$
$\sinh^{-1} ax$	$\frac{a}{\sqrt{1-a^2x^2}}$
$\cosh^{-1} ax$	$\frac{a}{\sqrt{a^2x^2-1}}$
$\tanh^{-1} ax$	$\frac{a}{1-a^2x^2}$
$u(x)+v(x)$	$\frac{du}{dx}+\frac{dv}{dx}$
$u(x)v(x)$	$u\frac{du}{dx}+v\frac{dv}{dx}$
$\frac{u(x)}{v(x)}$	$\frac{1}{v^2}\left(v\frac{du}{dx}-u\frac{dv}{dx}\right)$
$\frac{1}{u(x)}$	$\frac{-1}{u^2}\frac{dv}{dx}$
$y(v(x))$	$\frac{dy}{du}\frac{du}{dx}$
$y(v(u(x)))$	$\frac{dy}{dv}\frac{dv}{du}\frac{du}{dx}$

C.7 Table of Integrals

$$\int a dx = ax + c \quad (c \text{ is an arbitrary constant})$$

$$\int x dy = xy - \int y \, dx$$

$$\int x^n dx = \frac{x^{n+1}}{n+1} + c \quad (n \neq -1)$$

$$\int \frac{1}{x} dx = \ln|x| + c$$

$$\int e^{ax} dx = \frac{e^{ax}}{a} + c$$

$$\int a^x dx = \frac{a^x}{\ln a} + c \quad \text{for } (a > 0)$$

$$\int \ln x \, dx = x \ln x - x + c \quad \text{for } (x > 0)$$

$$\int \sin ax \, dx = \frac{-\cos as}{a} + c$$

$$\int \cos ax\, dx = \frac{\sin ax}{a} + c$$

$$\int \tan ax\, dx = \frac{-\ln|\cos ax|}{a} + c$$

$$\int \cot ax\, dx = \frac{-\ln|\cos ax|}{a} + c$$

$$\int \sec ax\, dx = \frac{-\ln(1-\sin ax/1+\sin ax)}{2a} + c$$

$$\int \csc ax\, dx = \frac{-\ln(1-\cos ax/1+\cos ax)}{2a} + c$$

$$\int \frac{1}{x^2+a^2}\, dx = \frac{\tan^{-1}(x/a)}{a} + c$$

$$\int \frac{1}{x^2-a^2}\, dx = \frac{\ln(x-a/x+a)}{2a} + c \text{ or } \frac{\tanh^{-1}(x/a)}{a} + c$$

$$\int \frac{1}{a^2-x^2}\, dx = \frac{\ln(x+a/x-a)}{2a} + c$$

$$\int \frac{1}{\sqrt{a^2-x^2}}\, dx = \sin^{-1}\left(\frac{x}{a}\right) + c$$

$$\int \frac{1}{\sqrt{a^2-x^2}}\, dx = \frac{\sinh^{-1}(x/a)}{a} + c \text{ or } \ln\left(x+\sqrt{x^2+a^2}\right) + c$$

$$\int \frac{1}{\sqrt{x^2-a^2}}\, dx = \ln\left(x+\sqrt{x^2+a^2}\right) + c$$

$$\int \frac{1}{x\sqrt{x^2-a^2}}\, dx = \frac{\sec^{-1}(x/a)}{a} + c$$

$$\int xe^{ax} dx = \frac{(ax-1)e^{ax}}{a^2} + c$$

$$\int x\cos ax dx = \frac{\cos ax + ax\sin ax}{a^2} + c$$

$$\int x\sin ax dx = \frac{\sin ax + ax\cos ax}{a^2} + c$$

$$\int x\ln x dx = \frac{x^2}{2}\ln x - \frac{x^2}{4} + c$$

$$\int xe^{ax} dx = \frac{e^{ax}(ax-1)}{a^2} + c$$

$$\int e^{ax}\cos bx\, dx = \frac{e^{ax}(a\cos bx + b\sin bx)}{a^2+b^2} + c$$

$$\int e^{ax}\sin bx dx = \frac{e^{ax}\left(-b\cos bx + a\sin bx\right)}{a^2+b^2} + c$$

$$\int \sin^2 x dx = \frac{x}{2} - \frac{\sin 2x}{4} + c$$

$$\int \cos^2 x dx = \frac{x}{2} - \frac{\sin 2x}{4} + c$$

$$\int \tan^2 x dx = \tan x - x + c$$

$$\int \cot^2 x dx = \cot x - x + c$$

$$\int \sec^2 x dx = \tan x + c$$

$$\int \csc^2 x dx = -\cot x + c$$

$$\int \sec x \tan x dx = \sec x + c$$

$$\int \csc x \cot x dx = -\csc x + c$$

C.8 Table of Probability Distributions

Discrete Distribution	Probability, $P(X=x)$	Expectation (Mean), μ	Variance, σ^2
Binomial $B(n,p)$	$\binom{n}{r}p^r(1-p)^{n-r} = \frac{n!p^r q^{n-r}}{r!(n-r)!'}$ $r = 0,1,\ldots,n$	np	$np(1-p)$
Geometric $G(p)$	$(1-p)^{r-1}p$	$\frac{1}{p}$	$\frac{1-p}{p^2}$
Poisson $p(\lambda)$	$\frac{\lambda^n e^{-\lambda}}{n!}$	λ	λ
Pascal (negative binomial) $NB(r,p)$	$\begin{pmatrix} x & -1 \\ r & -1 \end{pmatrix}p^r(1-p)^{x-r}$, $x = r, r+1, \ldots$	$\frac{r}{p}$	$\frac{r(1-p)}{p^2}$
Hypergeometric $H(N,n,p)$	$\frac{\binom{Np}{r}\binom{N-Np}{n-r}}{\binom{N}{n}}$	np	$np(1-p)\frac{N-n}{N-1}$

Continuous Distribution	Density, $f(x)$	Expectation (Mean), μ	Variance, σ^2
Exponential $E(\lambda)$	$\begin{cases}\lambda e^{-\lambda x}, & x \ge 0 \\ 0, & x < 0\end{cases}$	$\frac{1}{\lambda}$	$\frac{1}{\lambda^2}$
Uniform $U(a,b)$	$\begin{cases}\frac{1}{b-a}, & a < x < b \\ 0, & \text{elsewhere}\end{cases}$	$\frac{a+b}{2}$	$\frac{(b-a)^2}{12}$
Standardized normal $N(0,1)$	$\varphi(x) = \frac{e^{\frac{-x^2}{2}}}{\sqrt{2\pi}}$	0	1
General normal	$\frac{1}{\sigma}\varphi\left(\frac{x-\mu}{\sigma}\right)$	μ	σ^2
Gamma $\Gamma(n,\lambda)$	$\frac{\lambda^n}{\Gamma(n)} x^{n-1} e^{-\lambda x}$	$\frac{n}{\lambda}$	$\frac{n}{\lambda^2}$
Beta $\beta(p,q)$	$a_{p,q}xp-1(1-x)q-1, 0 \le x \le 1$ $a_{p,q} = \frac{\Gamma(p+q)}{\Gamma(p)\Gamma(q)}, p > 0, q > 0$	$\frac{p}{p+q}$	$\frac{pq}{(p+q)^2(p+q+1)}$
Weibull $W(\lambda, \beta)$	$\lambda^\beta \beta x^{\beta-1} e^{-(\lambda x)\beta}, x \ge 0$ $F(x) = 1 - e^{-(\lambda x)\beta}$	$\frac{1}{\lambda}\Gamma\left(1+\frac{1}{\beta}\right)$	$\frac{1}{\lambda^2}(A-B)$
			$A = \Gamma\left(1+\frac{2}{\beta}\right)$
			$B = \Gamma^2\left(1+\frac{1}{\beta}\right)$
Rayleigh $R(\sigma)$	$\frac{x}{\sigma^2} e^{\frac{-x^2}{2\sigma^2}}, x \ge 0$	$\sigma\sqrt{\frac{\pi}{2}}$	$2\sigma^2\left(1-\frac{\pi}{4}\right)$

C.9 Summations (Series)

1. Finite element of terms

$$\sum_{n=0}^{N} a^n = \frac{1-a^{N+1}}{1-a}; \quad \sum_{n=0}^{N} na^n = a\left(\frac{1-(N+1)a^N + Na^{N+1}}{(1-a)^2}\right)$$

$$\sum_{n=0}^{N} n = \frac{N(N+1)}{2}; \quad \sum_{n=0}^{N} n^2 = \frac{N(N+1)(2N+1)}{6}$$

$$\sum_{n=0}^{N} n(n+1) = \frac{N(N+1)(N+2)}{3};$$

$$(a+b)^N = \sum_{n=0}^{N} NC_n a^{N-n} b^n, \quad \text{where} \quad NC_n = NC_{N-n} = \frac{NP_n}{n!} = \frac{N!}{(N-n)!n!}$$

2. Infinite element of terms

$$\sum_{n=0}^{\infty} x^n = \frac{1}{1-x}, (|x|<1); \quad \sum_{n=0}^{\infty} nx^n = \frac{1}{(1-x)^2}, (|x|<1)$$

$$\sum_{n=0}^{\infty} n^k x^n = \lim_{a\to 0}(-1)^k \frac{\partial^k}{\partial a^k}\left(\frac{x}{x-e^{-a}}\right), (|x|<1);$$

$$\sum_{n=0}^{\infty} \frac{(-1)^n}{2n+1} = 1 - \frac{1}{3} + \frac{1}{5} - \frac{1}{7} + \cdots = \frac{1}{4}\pi$$

$$\sum_{n=0}^{\infty} \frac{1}{n^2} = 1 + \frac{1}{2^2} + \frac{1}{3^2} + \frac{1}{4^2} + \cdots = \frac{1}{6}\pi^2$$

$$e^x = \sum_{n=0}^{\infty} \frac{x^n}{n!} = 1 + \frac{1}{1!}x + \frac{1}{2!}x^2 + \frac{1}{3!}x^3 + \cdots$$

$$a^x = \sum_{n=0}^{\infty} \frac{(\ln a)^n x^n}{n!} = 1 + \frac{(\ln a)x}{1!} + \frac{(\ln a)^2 x^2}{2!} + \frac{(\ln a)^3 x^3}{3!} + \cdots$$

$$\ln(1\pm x) = \sum_{n=1}^{\infty} \frac{(\pm 1)^n x^x}{n} = \pm x - \frac{x^2}{2} \pm \frac{x^3}{3} - \cdots, \ (|x|<1)$$

$$\sin x = \sum_{n=0}^{\infty} \frac{(-1)^n x^{2n+1}}{(2n+1)!} = x - \frac{x^3}{3!} + \frac{x^5}{5!} - \frac{x^7}{7!} + \cdots$$

$$\cos x = \sum_{n=0}^{\infty} \frac{(-1)^n x^{2n}}{(2n)!} = 1 - \frac{x^2}{2!} + \frac{x^4}{4!} - \frac{x^6}{6!} + \cdots$$

$$\tan x = x + \frac{x^3}{3} + \frac{2x^5}{15} + \cdots, \ (|x|<1)$$

$$\tan^{-1} x = \sum_{n=0}^{\infty} \frac{(-1)^n x^{2n+1}}{(2n+1)} = x - \frac{x^3}{3} + \frac{x^5}{5} - \frac{x^7}{7} + \cdots, \ (|x|<1)$$

C.10 Logarithmic Identities

$$\log_e a = \ln a \text{ (natural logarithm)}$$

$$\log_{10} a = \log a \text{ (common logarithm)}$$

$$\log ab = \log a + \log b$$

$$\log \frac{a}{b} = \log a - \log b$$

$$\log a^n = n \log a$$

C.11 Exponential Identities

$$e^x = 1 + x + \frac{x^2}{2!} + \frac{x^3}{3!} + \frac{x^4}{4!} + \cdots, \quad \text{where} \quad e \simeq 2.7182$$

$$e^x e^y = e^{x+y}$$

$$\left(e^x\right)^n = e^{nx}$$

$$\ln e^x = x$$

C.12 Approximations for Small Quantities

If $|a| \ll 1$, then

$$\ln(1+a) \simeq a$$

$$e^a \simeq 1 + a$$

$$\sin a \simeq a$$

$$\cos a \simeq 1$$

$$\tan a \simeq a$$

$$(1 \pm a)^n \simeq 1 \pm na$$

C.13 Matrix Notation and Operations

1. Matrices

A *matrix* is a rectangular array of elements arranged in rows and columns. The array is commonly enclosed in brackets. Let a matrix A (expressed in boldface as **A** or in bracket as $[A]$) have m rows and n columns, and then the matrix can be expressed by

$$\mathbf{A} = [A] = \begin{bmatrix} a_{11} & a_{12} & \cdot & \cdot & \cdot & a_{1j} & \cdot & \cdot & \cdot & a_{1n} \\ a_{21} & a_{22} & \cdot & \cdot & \cdot & a_{2j} & \cdot & \cdot & \cdot & a_{2n} \\ \cdot & \cdot & \cdot & \cdot & \cdot & \cdot & \cdot & \cdot & \cdot & \cdot \\ \cdot & \cdot & \cdot & \cdot & \cdot & \cdot & \cdot & \cdot & \cdot & \cdot \\ \cdot & \cdot & \cdot & \cdot & \cdot & \cdot & \cdot & \cdot & \cdot & \cdot \\ a_{i1} & a_{i2} & \cdot & \cdot & \cdot & a_{ij} & \cdot & \cdot & \cdot & a_{in} \\ \cdot & \cdot & \cdot & \cdot & \cdot & \cdot & \cdot & \cdot & \cdot & \cdot \\ \cdot & \cdot & \cdot & \cdot & \cdot & \cdot & \cdot & \cdot & \cdot & \cdot \\ \cdot & \cdot & \cdot & \cdot & \cdot & \cdot & \cdot & \cdot & \cdot & \cdot \\ a_{m1} & a_{m2} & \cdot & \cdot & \cdot & a_{mj} & \cdot & \cdot & \cdot & a_{mn} \end{bmatrix}$$

where the element a_{ij} has two subscripts, of which the first denotes to the row ith and the second denotes to the column jth that the element locates in the matrix. A matrix with m rows and n columns, $[A]$, is defined as a matrix of order or size $m \times n$ (m by n), or an $m \times n$ matrix. A vector is a matrix that consists of only one row or one column.

Location of an element in a matrix:

$$\text{Let } A = \begin{bmatrix} a_{11} & a_{12} & a_{13} & a_{14} \\ a_{21} & a_{22} & a_{23} & a_{24} \\ a_{31} & a_{32} & a_{33} & a_{34} \\ a_{41} & a_{42} & a_{43} & a_{44} \end{bmatrix} \text{is matrix with size } 4 \times 4$$

where

a_{11} is element a at row 1 and column 1
a_{12} is element a at row 1 and column 2
a_{32} is element a at row 3 and column 2

2. Special common types of matrices
   a. If $m \neq n$, then the matrix $[A]$ is called *rectangular matrix*.
   b. If $m = n$, then the matrix $[A]$ is called *square matrix of order n*.
   c. If $m = 1$ *and* $n > 1$, then the matrix $[A]$ is called *row matrix or row vector*.

d. If $m>1$ *and* $n=1$, then the matrix $[A]$ is called *column matrix or column vector.*

e. If $m=1$ *and* $n=1$, then the matrix $[A]$ is called a scalar.

f. A *real matrix* is a matrix whose elements are all real.

g. A *complex matrix* is a matrix whose elements may be complex.

h. A *null matrix* is a matrix whose elements are all zero.

i. An *identity* (or *unit*) *matrix*, $[I]$ or **I**, is a square matrix whose elements are equal to zero except those located on its *main diagonal* elements, which are unity (or one). *Main diagonal* elements have equal row and column subscripts. The main diagonal runs from the upper left corner to the lower right corner. If the elements of an identity matrix are denoted as e_{ij}, then

$$e_{ij} = \begin{cases} 1, & i = j \\ 0, & i \neq j \end{cases}$$

j. A *diagonal matrix* is a square matrix that has zero elements everywhere except on its main diagonal. That is, for diagonal matrix, $a_{ij}=0$ when $i \neq j$ and not all a_{ii} are zero.

k. A *symmetric matrix* is a square matrix whose elements satisfy the condition $a_{ij}=a_{ji}$ for $i \neq j$.

l. An *antisymmetric (or skew symmetric) matrix* is a square matrix whose elements $a_{ij}=-a_{ji}$ for $i \neq j$, and $a_{ii}=0$.

m. A *triangular matrix* is a square matrix whose all elements on one side of the diagonal are zero. There are two types of triangular matrices: first, an upper triangular **U**, whose elements below the diagonal are zero, and second, a lower triangular **L**, whose elements above the diagonal are all zero.

n. A *partitioned (or block) matrix* is a matrix that is divided by horizontal and vertical lines into smaller matrices called *submatrices* or blocks.

3. Matrix operations

a. Transpose of a matrix

The *transpose* of a matrix $\mathbf{A}=[a_{ij}]$ is denoted as $\mathbf{A}^T=[a_{ji}]$ and is obtained by interchanging the rows and columns in matrix **A**. Thus, if a matrix **A** is of order $m \times n$, then $\mathbf{A}^T$ will be of order $n \times m$.

b. Addition and subtraction

Addition and subtraction can only be performed for matrices of the same size. The addition is accomplished by adding corresponding elements of each matrix. For addition, $\mathbf{C}=\mathbf{A}+\mathbf{B}$ implies that $c_{ij}=a_{ij}+b_{ij}$.

Now, the subtraction is accomplished by subtracting corresponding elements of each matrix. For subtraction, $\mathbf{C}=\mathbf{A} - \mathbf{B}$

implies that $c_{ij} = a_{ij} - b_{ij}$ where c_{ij}, a_{ij}, and b_{ij} are typical elements of the **C**, **A**, and **B** matrices, respectively.

Both **A** and **B** matrices are in the same size $m \times n$. The resulting matrix **C** is also of size $m \times n$.

Matrix addition and subtraction are associative:

$$\mathbf{A}+\mathbf{B}+\mathbf{C}=(\mathbf{A}+\mathbf{B})+\mathbf{C}=\mathbf{A}+(\mathbf{B}+\mathbf{C})$$

$$\mathbf{A}+\mathbf{B}-\mathbf{C}=(\mathbf{A}+\mathbf{B})-\mathbf{C}=\mathbf{A}+(\mathbf{B}-\mathbf{C})$$

Matrix addition and subtraction are commutative:

$$\mathbf{A}+\mathbf{B}=\mathbf{B}+\mathbf{A}$$

$$\mathbf{A}-\mathbf{B}=-\mathbf{B}+\mathbf{A}$$

c. Multiplication by scalar

A matrix is multiplied by a scalar by multiplying each element of the matrix by the scalar. The multiplication of a matrix **A** by a scalar c is defined as

$$c\mathbf{A}=\left[ca_{ij}\right]$$

The scalar multiplication is commutative.

d. Matrix multiplication

The product of two matrices is $\mathbf{C}=\mathbf{AB}$ if and only if the number of columns in **A** is equal to the number of rows in **B**. The product of matrix **A** of size $m \times n$ and matrix **B** of size $n \times r$ results in matrix **C** of size $m \times r$. Then, $c_{ij} = \sum_{k=1}^{n} a_{ik} b_{kj}$. That is, the ($ij$)th component of matrix **C** is obtained by taking the dot product:

$$c_{ij} = (i\text{th row of } \mathbf{A})\ (j\text{th column of } \mathbf{B})$$

Matrix multiplication is associative:

$$\mathbf{ABC}=(\mathbf{AB})\mathbf{C}=\mathbf{A}(\mathbf{BC})$$

Matrix multiplication is distributive:

$$\mathbf{A}\ (\mathbf{B}+\mathbf{C})=\mathbf{AB}+\mathbf{AC}$$

Matrix multiplication is not commutative:

$$\mathbf{AB} \neq \mathbf{BA}$$

e. Transpose of matrix multiplication

The transpose of matrix multiplication is usually denoted $(\mathbf{AB})^T$ and is defined as

$$(\mathbf{AB})^T=\mathbf{B}^T\mathbf{A}^T$$

f. Inverse of square matrix

The inverse of a matrix $\mathbf{A}$ is denoted by $\mathbf{A}^{-1}$. The inverse matrix satisfies

$$\mathbf{A}\mathbf{A}^{-1} = \mathbf{A}^{-1}\mathbf{A} = \mathbf{I}$$

A matrix that possesses an inverse is called a *nonsingular matrix (or invertible matrix)*. A matrix without an inverse is called a *singular matrix*.

g. Differentiation of a matrix

The differentiation of a matrix is differentiation of every element of the matrix separately. To emphasize, if the elements of the matrix $\mathbf{A}$ are a function of t, then

$$\frac{d\mathbf{A}}{dt} = \left[\frac{da_{ij}}{dt}\right]$$

h. Integration of a matrix

The integration of a matrix is integration of every element of the matrix separately. To emphasize, if the elements of the matrix $\mathbf{A}$ are a function of t, then

$$\int \mathbf{A}dt = \left[\int a_{ij}dt\right]$$

i. Equality of matrices

Two matrices are equal if they have the same size and their corresponding elements are equal.

4. Determinant of a matrix

The determinant of a square matrix $\mathbf{A}$ is a scalar number denoted by $|\mathbf{A}|$ or det $\mathbf{A}$.

The value of a second-order determinant is calculated from

$$\det\begin{bmatrix} a_{11} & a_{12} \\ a_{21} & a_{22} \end{bmatrix} = \begin{vmatrix} a_{11} & a_{12} \\ a_{21} & a_{22} \end{vmatrix} = a_{11}a_{22} - a_{12}a_{21}$$

By using the sign rule of each term, the determinant is determined by the first row in the diagram $\begin{vmatrix} + & - & + \\ - & + & - \\ + & - & + \end{vmatrix}$.

The value of a third-order determinate is calculated from the following:

$$\det\begin{bmatrix} a_{11} & a_{12} & a_{13} \\ a_{21} & a_{22} & a_{23} \\ a_{31} & a_{32} & a_{33} \end{bmatrix} = \begin{vmatrix} a_{11} & a_{12} & a_{13} \\ a_{21} & a_{22} & a_{23} \\ a_{31} & a_{32} & a_{33} \end{vmatrix}$$

$$= a_{11}\begin{vmatrix} a_{22} & a_{23} \\ a_{32} & a_{33} \end{vmatrix} - a_{12}\begin{vmatrix} a_{21} & a_{23} \\ a_{31} & a_{33} \end{vmatrix} + a_{13}\begin{vmatrix} a_{21} & a_{22} \\ a_{31} & a_{32} \end{vmatrix}$$

C.14 Vectors

1. Vector derivative

   a. Cartesian coordinates

Coordinates	(x,y,z)
Vector	$\mathbf{A} = A_x a_x + A_y a_y + A_z a_z$
Gradient	$\nabla \mathbf{A} = \frac{\partial A}{\partial x} a_x + \frac{\partial A}{\partial y} a_y + \frac{\partial A}{\partial z} a_z$
Divergence	$\nabla \cdot \mathbf{A} = \frac{\partial A_x}{\partial x} + \frac{\partial A_y}{\partial y} + \frac{\partial A_z}{\partial z}$
Curl	$\nabla \times \mathbf{A} = \begin{vmatrix} a_x & a_y & a_z \\ \frac{\partial}{\partial x} & \frac{\partial}{\partial y} & \frac{\partial}{\partial z} \\ A_x & A_y & A_z \end{vmatrix}$
	$= \left(\frac{\partial A_z}{\partial y} - \frac{\partial A_y}{\partial z}\right) a_x + \left(\frac{\partial A_x}{\partial z} - \frac{\partial A_z}{\partial x}\right) a_y + \left(\frac{\partial A_y}{\partial_x} - \frac{\partial A_x}{\partial y}\right) a_z$
Laplacian	$\nabla^2 \mathbf{A} = \frac{\partial^2 A}{\partial x^2} + \frac{\partial^2 A}{\partial y^2} + \frac{\partial^2 A}{\partial z^2}$

   b. Cylindrical coordinates

Coordinates	(ρ,ϕ,z)
Vector	$\mathbf{A} = A_\rho a_\rho + A_\phi a_\phi + A_z a_z$
Gradient	$\nabla \mathbf{A} = \frac{\partial A}{\partial \rho} a_\rho + \frac{1}{\rho}\frac{\partial A}{\partial \phi} a_\phi + \frac{\partial A}{\partial z} a_z$
Divergence	$\nabla \cdot \mathbf{A} = \frac{1}{\rho}\frac{\partial}{\partial \rho}(\rho A_\rho) + \frac{\partial A_\phi}{\partial \phi} + \frac{\partial A_z}{\partial z}$

Curl	$\nabla \times A = \frac{1}{\rho}\begin{vmatrix} a_\rho & \rho a_\phi & a_z \\ \frac{\partial}{\partial \rho} & \frac{\partial}{\partial \phi} & \frac{\partial}{\partial z} \\ A_\rho & \rho A_\phi & A_z \end{vmatrix}$
	$= \left(\frac{1}{\rho}\frac{\partial A_z}{\partial \phi} - \frac{\partial A_\phi}{\partial z}\right) a_\rho + \left(\frac{\partial A_\rho}{\partial z} - \frac{\partial A_z}{\partial \rho}\right) a_\phi + \frac{1}{\rho}\left(\frac{\partial}{\partial x}(\rho A_\phi) - \frac{\partial A_\rho}{\partial \rho}\right) a_z$
Laplacian	$\nabla^2 \mathbf{A} = \frac{1}{\rho}\frac{\partial}{\partial \rho}\left(\rho \frac{\partial A}{\partial \rho}\right) + \frac{1}{\rho^2}\frac{\partial^2 A}{\partial \phi^2} + \frac{\partial^2 A}{\partial z^2}$

c. Spherical coordinates

Coordinates	(r, θ, ϕ)
Vector	$\mathbf{A} = A_r a_r + A_\theta a_\theta + A_\phi a_\phi$
Gradient	$\nabla \mathbf{A} = \frac{\partial A}{\partial r} a_r + \frac{1}{r}\frac{\partial A}{\partial \theta} a_\theta + \frac{1}{r \sin\theta}\frac{\partial A}{\partial \phi} a_\phi$
Divergence	$\nabla \cdot \mathbf{A} = \frac{1}{r^2}\frac{\partial}{\partial r}(r^2 A_r) + \frac{1}{r \sin\theta}\frac{\partial}{\partial \theta}(A_\theta \sin\theta) + \frac{1}{r \sin\theta}\frac{\partial A_\phi}{\partial \phi}$
Curl	$\nabla \times A = \frac{1}{r^2 \sin\theta}\begin{vmatrix} a_r & r a_\theta & (r \sin\theta) a_\phi \\ \frac{\partial}{\partial r} & \frac{\partial}{\partial \theta} & \frac{\partial}{\partial \phi} \\ A_r & r A_\theta & (r \sin\theta) A_\phi \end{vmatrix}$
	$= \frac{1}{r \sin\theta}\left(\frac{\partial}{\partial \theta}(A_\phi \sin\theta) - \frac{\partial A_\theta}{\partial \phi}\right) a_r + \frac{1}{r}\left(\frac{1}{\sin\theta}\frac{\partial A_r}{\partial \phi} - \frac{\partial}{\partial r}(r A_\phi)\right) a_\theta$
	$+ \frac{1}{r}\left(\frac{\partial}{\partial r}(r A_\theta) - \frac{\partial A_r}{\partial \theta}\right) a_\phi$
Laplacian	$\nabla^2 A = \frac{1}{r^2}\frac{\partial}{\partial r}\left(r^2 \frac{\partial A}{\partial r}\right) + \frac{1}{r^2 \sin\theta}\frac{\partial}{\partial \theta}\left(\sin\theta \frac{\partial A}{\partial \theta}\right) + \frac{1}{r^2 \sin\theta}\frac{\partial^2 A}{\partial \phi^2}$

2. Vector identity

a. Triple products

$$\mathbf{A}(\mathbf{B} \times \mathbf{C}) = \mathbf{B}(\mathbf{C} \times \mathbf{A}) = \mathbf{C} \cdot (\mathbf{A} \times \mathbf{B})$$

$$\mathbf{A}(\mathbf{B} \times \mathbf{C}) = \mathbf{B}(\mathbf{A} \cdot \mathbf{C}) - \mathbf{C}(\mathbf{A} \cdot \mathbf{B})$$

b. Product rules

$$\nabla(fg) = f(\nabla g) + g(\nabla f)$$

$$\nabla(\mathbf{A} \cdot \mathbf{B}) = \mathbf{A} \times (\nabla \times \mathbf{B}) + \mathbf{B} \times (\nabla \times \mathbf{A}) + (\mathbf{A} \cdot \nabla)\mathbf{B} + (\mathbf{B} \times \nabla)\mathbf{A}$$

$$\nabla \cdot (f\mathbf{A}) = f(\nabla \cdot \mathbf{A}) + \mathbf{A} \cdot (\nabla f)$$

$$\nabla(\mathbf{A} \times \mathbf{B}) = \mathbf{B} \cdot (\nabla \times \mathbf{A}) - \mathbf{A} \cdot (\nabla \times \mathbf{B})$$

$$\nabla \times (f\mathbf{A}) = f(\nabla \times \mathbf{A}) - \mathbf{A} \times (\nabla f) = \nabla \times (f\mathbf{A}) = f(\nabla \times \mathbf{A}) + (\nabla f) \times \mathbf{A}$$

$$\nabla \times (\mathbf{A} \times \mathbf{B}) = (\mathbf{B} \cdot \nabla)\mathbf{A} - (\mathbf{A} \cdot \nabla)\mathbf{B} + \mathbf{A}(\nabla \cdot \mathbf{B}) - (\nabla \cdot \mathbf{A})$$

c. Second derivative

$$\nabla \cdot (\Delta \times \mathbf{A}) = 0$$

$$\nabla \times (\nabla f) = 0$$

$$\nabla \cdot (\nabla f) = \nabla^2 f$$

$$\nabla \times (\nabla \times \mathbf{A}) = \nabla(\nabla \cdot \mathbf{A}) - \nabla^2 \mathbf{A}$$

d. Addition, division, and power rules

$$\nabla(f + g) = \nabla f + \nabla g$$

$$\nabla \cdot (\mathbf{A} + \mathbf{B}) = \nabla \cdot \mathbf{A} + \nabla \cdot \mathbf{B}$$

$$\nabla \times (\mathbf{A} \times \mathbf{B}) = \nabla \times \mathbf{A} + \nabla \times \mathbf{B}$$

$$\nabla\left(\frac{f}{g}\right) = \frac{g(\nabla f) - f(\nabla g)}{g^2}$$

$$\nabla f^n = nf^{n-1}\nabla f\left(n = \text{integer}\right)$$

3. Fundamental theorems

a. Gradient theorem

$$\int_a^b (\nabla f) \cdot dl = f(b) - f(a)$$

b. Divergence theorem

$$\int_{volume} (\nabla \cdot \mathbf{A}) dv = \oint_{surface} \mathbf{A} \cdot ds$$

c. Curl (Stokes) theorem

$$\int_{surface} (\nabla \times \mathbf{A}) \cdot ds = \oint_{line} \mathbf{A} \cdot dl$$

d. $$\oint_{line} fdl = -\oint_{surface} \nabla f \times ds$$

e. $$\oint_{surface} fds = -\oint_{volume} \nabla f dv$$

f. $$\oint_{surface} \mathbf{A} \times ds = -\oint_{volume} \nabla \times \mathbf{A} dv$$

Index

J

K

L

M

O

P

Q

R

For Product Safety Concerns and Information please contact our EU representative GPSR@taylorandfrancis.com Taylor & Francis Verlag GmbH, Kaufingerstraße 24, 80331 München, Germany

Printed and bound by CPI Group (UK) Ltd, Croydon, CR0 4YY
01/05/2025
01858618-0001